Recent Advances in Structural Integrity Analysis: Proceedings of the International Congress (APCF/SIF-2014)

Recent Advances in Structural Integrity Analysis: Proceedings of the International Congress (APCF/SIF-2014)

Darlington Campus, University of Sydney, Australia

December 9 - 12, 2014

The APCF/SIF-2014 Congress uniting:
Asian-Pacific Conference on Fracture and Strength (APCFS-2014)
International Conference on Structural Integrity and Failure (SIF-2014)

Hosted by:
The University of Sydney

Co-organised by:
Australia Fracture Group (AFG)
Chinese Mechanical Engineering Society Materials Institution (CMES-MI)
Korean Society of Mechanical Engineers, Materials and Fracture Division (KSME-MFD)
The Japanese Society of Mechanical Engineers, Materials and Mechanics Division (JSME-MMD)

ELSEVIER

AMSTERDAM • BOSTON • CAMBRIDGE • HEIDELBERG • LONDON
NEW YORK • OXFORD • PARIS • SAN DIEGO
SAN FRANCISCO • SINGAPORE • SYDNEY • TOKYO
Woodhead Publishing is an imprint of Elsevier

WP WOODHEAD PUBLISHING

Woodhead Publishing is an imprint of Elsevier
80 High Street, Sawston, Cambridge CB22 3HJ, UK
225 Wyman Street, Waltham, MA 02451, USA
Langford Lane, Kidlington, OX5 1GB, UK

First published 2014, Woodhead Publishing

British Library Cataloguing in Publication Data
A catalogue record for this book is available from the British Library.

Library of Congress Control Number: 2014955010

ISBN 978-0-08-100203-2 (print)
ISBN 978-0-08-100225-4 (online)

For information on all Woodhead Publishing publications visit our website at http://store.elsevier.com

Produced from electronic copy supplied by authors.

Transferred to Digital Printing in 2015

APCFS/SIF-2014 CONGRESS CHAIRS AND COMMITTEES

** As this book was going to press, the organizers received the sad news of the death of Professor Paolo Lazzarin*

Okazaki M	Nagaoka University of Technology
Rose F	DSTO, Australia
Sakagami T	Kobe University
Shi H	Tsinghua University
Sugeta A	Hiroshima University
Sun J	Xi'an Jiaotong University
Takahashi J	The University of Tokyo
Wang C	RMIT University
Wang TJ	Xi'an Jiaotong University
Williams G	Imperial College of London
Wu JS	Hong Kong University of Science & Technology
Yoshikawa N	The University of Tokyo
Yosibash Z	Ben-Gurion University of Negev
Zhen L	Harbin Institute of Technology
Zhou, L	Hong Kong Polytechnic University

Technical Committee

Arai M	Nagoya University
Baba H	IHI Corporation
Biwa S	Kyoto University
Chen X	Tianjin University
Das R	CSIRO, Australia
Ding H	Wuhan University
Dyskin A	The University of Western Australia
He Y	Air Force Engineering University, P.R. China
Ho SY	DSTO, Australia
Hirakata H	Osaka University
Hu XZ	The University of Western Australia
Inaba K	Tokyo Institute of Technology
Iwasaki T	Hitachi, Ltd
Izumi S	The University of Tokyo
Kim HS	The University of Newcastle
Li JC	The University of Technology, Sydney
Li L	Hunan University
Ma J	The University of South Australia
Mizutani Y	Tokyo Institute of Technology
Notomi M	Meiji University
Ogawa T	Aoyama Gakuin University
Pasternak E	The University of Western Australia
Shibutani T	Yokohama National University
Sumigawa T	Kyoto University
Takano N	Keio University
Tran B	ASC, Australia
Uematsu Y	Gifu University
Umeno Y	The University of Tokyo
Wang L	Northeastern University, China
Wang ZD	East China University of Science and Technology
Wu Y	Hefei University of Technology
Xiong W	Huazhong University of Science and Technology
Yan C	Queensland University of Technology
Yan W	Monash University
Yang C	The University of Western Sydney
Yang J	Inner Mongolia University of Science & Technology
Yang X	Beihang University
Yoneyama S	Aoyama Gakuin University
Zhang S	ADFA, The University of New South Wales
Zhang X	East China University of Science & Technology
Zhao J	Dalian University of Technology

CONTENTS

IMPACT AND INTEGRITY

BIO-MATERIALS AND COMPOSITES

ENVIRONMENTAL EFFECTS

SMART MATERIALS/STRUCTURES

STRUCTURE-PROPERTY RELATIONSHIP

TESTING METHODS

FOREWORD

This book is a compendium of papers submitted and presented at the APCF/SIF-2014 Congress held in Sydney, Australia from 9 to 12 December 2014. All papers have been peer-reviewed by the experts in the relevant areas. The APCF/SIF-2014 Congress united the Asian-Pacific Conference on Fracture and Strength 2014 (APCFS-2014) and the International Conference on Structural Integrity and Failure (SIF-2014). The Congress was aimed to provide a unique opportunity for academics, engineers and postgraduate students to meet, present and discuss the latest research developments, challenges and trends in structural integrity.

Structural integrity is a key issue in the aerospace, power generation, transport, marine and many other industries. Structural integrity evaluation is based on fundamental understanding of failure mechanisms such as fracture, fatigue, creep, buckling, corrosion, etc. It largely relies on advances in damage growth modelling, strength prediction, defect detection and structural health monitoring techniques. The book contains 117 papers, covering key aspects of structural integrity problems with a particular emphasis on the characterisation of complex mechanisms of fracture, fatigue and creep, structure-property relationship, multi-scale modelling and the development of more accurate technologies for structural damage evaluation.

The Congress was hosted by The University of Sydney and co-organized by Australia Fracture Group (AFG), the Chinese Mechanical Engineering Society, Materials Institution (CMES-MI), the Korean Society of Mechanical Engineers, Materials and Fracture Division (KSME-MFD) and The Japanese Society of Mechanical Engineers, Materials and Mechanics Division (JSME-MMD). The Congress followed the series of the previous very successful APCF and SIF international forums, in particular, APCFS 2012, Busan and the 8th SIF, Melbourne, 2013.

The book is the result of contributions from many researchers from different laboratories, universities and research institutions. The Editors wish to express their most grateful thanks to all authors of papers included in the book. Further, the Editors wish to thank many of their colleagues for their kind support and help, including Hayashi M, Johnston A, Kim YJ, Kishimoto K, Mai YW, Rose F, Tu ST, and Wang C.

Finally, the Editors wish to thank Deng S (The University of Sydney) and Sue Gruzelier who provided great help in receiving, compiling and editing of the papers and the associated publication agreements; without their assistance, this publication of the Book would not be possible.

Ye L (The University of Sydney)
Kotousov A (The University of Adelaide)
Chang L (The University of Sydney)

Optimization and fracture mechanism analysis of TC17 titanium alloy simulated-blade with two-sided laser shock processing

X. Nie, Y. Li, W. He, L. Zhou
Science and Technology on Plasma Dynamics Laboratory,
Air Force Engineering University, P.R. China

ABSTRACT

Laser shock processing (LSP) is an innovative surface treatment technology, which can effectively improve the fatigue performance of metals. In order to apply this technology on aero-engine compressor blade to improve its fatigue resistance, a TC17 titanium alloy simulated-blade was designed and trested by LSP. According to the finite element analysis and fatigue test results, the LSP procedure was optimized. And the fatigue strength was effectively improved by the optimized LSP procedure, compared to the first LSP procedure. The fracture mechanisms of fatigue crack initiation and growth with different LSP procedures were incestigated and compared.

1. INTRODUCTION

Laser shock processing (LSP) is an innovative surface treatment technology, which can improve the fatigue resistance of metals and alloys (1). Compared with conventional shot peening (SP), LSP has some special advantages, such as deeper compressive residual stress, small cold work rate, lower surface roughness and better controllability. Beacause of the above advantages, it becomes more and more popular in the surface treatment field. What's more, LSP has been successfully applied on fan/compressor blades of military aero-engines to improve HCF performance and foreign object damage resistance. Many studies have been carried out to discuss the effects of LSP on fatigue property for different metals and alloys (2,3). And some reseaches focused on the effects on the fatigue crack initiation and growth (4-6). In our previous work, LSP was successfully applied on some titanium alloys to improve the fatigue strength, and the strengthening mechanism was also discussed(7,8).

However, the above experimental studies just investigated the effects of a specified LSP procedure, always with a simple laser-peened path, on the fatigue performance. But there were few experimental studies about the effects of different LSP procedures including laser parameters and laser shocked-path. In this paper, LSP procedure for TC17 simulated-blade was optimized according to the work stress state. And the simulated-blades were treated with two LSP procedures. The fatigue limits with different LSP procedures were compared by the up-and-down method during fatigue tests. The fracture mechanisms of fatigue crack initiation and growth with different LSP procedures were compared.

2. EXPERIMENTAL PROCEDURE

2.1 Laser shock processing

The LSP process utilizes laser pulse irradiated at the target surface covered by opaque ablating layer and transparent confining layer. When the laser beam passes through the transparent layer and strikes the surface, the ablating layer absorbs the laser and immediately vaporized into the plasma. The rapidly expanding plasma leads to the formation of shock wave which strikes the material and propagates into the material with an intensity of several GPa. If the shock pressure is greater than the dynamic field strength, plastic deformation will be generated with compressive residual stress and microstructure changes in the surface. Because of the symmetrical structure, the simulated-blades were treated by two-sided LSP shown in reference (7).

2.2 Material and specimen

TC17 titanium alloy is widely used in Chinese aviation field, such as aero-engine fan and compressor blade. In order to simulate the work stress state of aero-engine compressor blade, the TC17 titanium alloy specimen was designed and machined into the special structure, namely simulated-blade shown in Fig.1(a). The simulated-blades were machined by Gas-turbine research institute of China. Fig.1(b),(c) is the first-order modal displacement contour and equivalent effective stress contour by finite element analysis (FEM). The first-order modal is a cantilever vibration modal. The greatest vibration amplitude locates at the blade-tip. The calculated first-order frequence is 338 HZ, which is well consistent with the factual resonance frequence in vibratory fatigue tests, 332HZ~345HZ. In addition, the maxiam vibration stress locates at the blade-root rounding near the R transition line, especially the center region. In response to the stress analysis results, the LSP region was confirmed as the blade-root rounding area, the dashed region in Fig.2(a), which is 30 mm×14.4 mm.

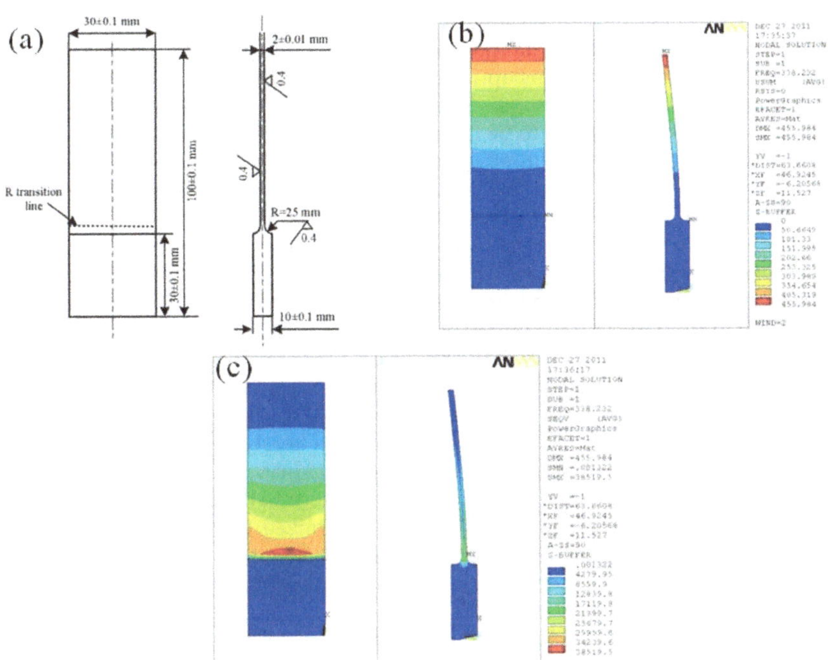

Fig.1 Schematic diagram and first-order modal of simulated-blade

3. RESULTS AND DISCUSSION

3.1 LSP procedure optimization and fatigue performance

The simple snaky laser shocked-path was applied in first LSP procedure (Fig.2(a)) with detailed laser parameters as following: laser energy/6J, laser duration/20ns, laser spot diameter/3mm, laser fluence/4.24GW/cm^2, overlapping rate/50%, 1 impact. Fifteen simulated-blades were treated by LSP. After LSP, thirty simulated-blades without and with LSP were used to conduct vibration fatigue tests on D-300-3 electric vibration system shown in reference (7). In the fatigue test, the test stress was measured by a strain gage which is sticked on the simulated-blade root with highest work stress as shown in Fig.1(c). At the same time, the amplitude of simulated-blade tip in the fatigue test was measured by a laser displacement sensor. Then, the relataionship between the highest work stress and amplitude of simulated-blade tip was established. And we just monitored the amplitude of simulated-blade tip in test. The fatigue test results were processed by the fatigue up-and-down method.

The original fatigue limit of TC17 titanium alloy simulated-blade is 405.7MPa. However, the fatigue limit with first procedure is only 360MPa, decreased by 11.3%. According to the fluorescent test results, it is found that the fatigue crack initiates at the blade edge within the LSP region, far from the blade-root rounding. In order to analyze the the cause of the fatigue crack initiation, a numerical simulation work of the first LSP procedure was conducted. The simulation results indicate that the maximal equivalent effective plastic strain locates at blade edge, where there is a material protrusion resulted from the accumulated plastic deformation. The accumulated plastic deformation may result in the complicated residual stress field generated. Even more, tensile residual stresses may be generated in the material protrusion area, which leads to fatigue crack initiation. Thus, the main cause of fatigue crack initiation is not the vibration stress, and it may be induced by the unbenefited residual stresses.

According to the cause of the fatigue performance deterioration with the first LSP procedure, the LSP procedure was optimized as shown in Fig.2(b). In order to advoid the accumulated plastic deformation generated at blade edge, the center region undergone great work stress was designed to be laser peened with a great intensity, but with a low intensity for the simulated-blade edge region. The design target is to induce greater residual compressive stresses in the center region and reduce the accumulated plastic strain at the blade edge. In the previous research work, it is found that there is direct relationship between the laser fluence and laser induced plastic deformation. The greater laser fluence is, the greater plastic deformation induced by LSP is. Therefore, an optimized distinctive LSP procedure was confirmed with four LSP regions, region 1 (3J/20ns/Φ2.4mm/50%/1 impact) with high laser fluence (3.32GW/cm^2) and region 2/3/4 (2J/20ns/Φ2.4mm/50%/1 impact) with low laser fluence (2.21GW/cm^2). And the treatment sequence is region 1 first, then region 2, region 3 and region 4. The great compressive residual stresses are produced in region 1 for the resistance to the greatest work stress. And lower compressive stresses are generated in region 2/3/4 for the transition of great compressive stresses in region 1, preventing stress concentration. Fifteen simulated-blades treated with the optimized LSP procedure were used to conduct fatigue tests. The fatigue limit with the optimized LSP procedure is 462.9MPa, 14.1% incresed by compared with the orginal fatigue limit. And the fatigue crack initiates at the center region, not the blade edge. In summary, the first LSP procedure can induce great compressive stresses in the LSP region, but with great accumulated plastic deformation, even tensile residual stress at blade edge resulting in fatigue cracking. In contrast, the optimized LSP procedure can induce great compressive stresses in region 1 and realize the gradual transition of the compressive stresses in region2/3/4, without great accumulated plastic deformation generated at blade edge.

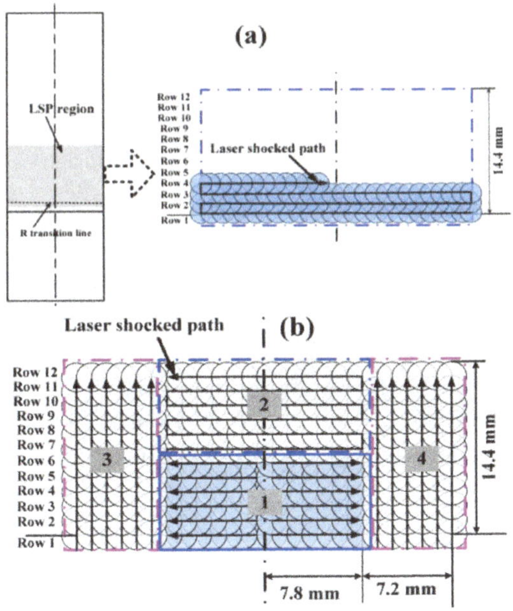

Fig.2 Two LSP procedures and laser shocked path of simulated-blade

3.2 Fracture mechanism analysis

The fractographys of simulated-blades treated by two LSP procedures were observed by a JEOL/JSM-6360LV scanning electron microscope (SEM). Before the observation, the raptured simulated-blades were cleaned with ethanol in the ultrasound cleaning machine. Fig.3(a) is the typical fractography with first procedure. It shows that the fatigue crack initiates at the blade edge and propogates into the center/depth. In the period of fatigue crack initiation and transition, there are many cleavage planes, subtle fatigue striations. There are many ridges in the fatigue crack growth region, but not very orderly in Fig.3(b), which may be ascribed to the irreguar residual stress distribution. Dense fatigue striations with secondary cracks are generated during fatigue crack growth in Fig.3(c), which results from complicated residual stress state.

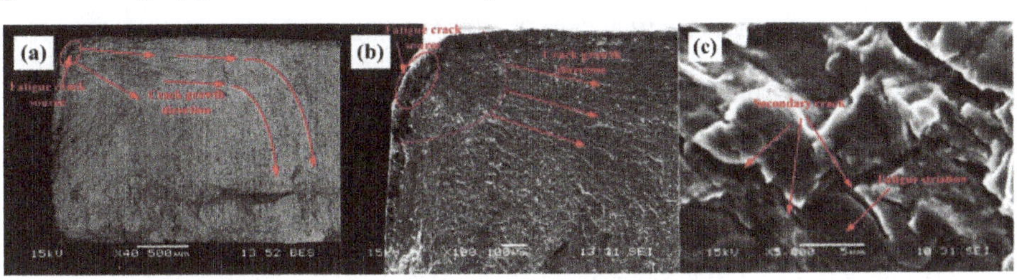

Fig.3 Typical fractography with the first LSP procedure

Differently, the fatigue crack initiates at the surface of simulated-blade center part near the blade root rounding with optimized procedure (Fig.4(a)). The ridges are clearer with layered distribution, and denser compared to that with first procedure (Fig.3(b)). It can be concluded that more time of the fatigue crack initiation is needed. The fatigue crack growth presents the sectorial radial shape with a larger crack extension region in Fig.4(b). The magnified micrograph of stable growth region indicates that there are lots of cleavage steps and sublte fatigue striations in the cleavage plane in Fig.4(c). Besides, many secondary fatigue cracks were generated between cleavage steps. The fatigue striations get denser and the distance among them becomes smaller. It means that the stress intensity factor ΔK at the crack front is small relatively, and the fatigue crack growth rate decreases, which is results from the positive action of residual compressive stress. Some fatigue cracks are compelled to change growth direction and even arrested under residual compressive stress, forming the secondary cracks.

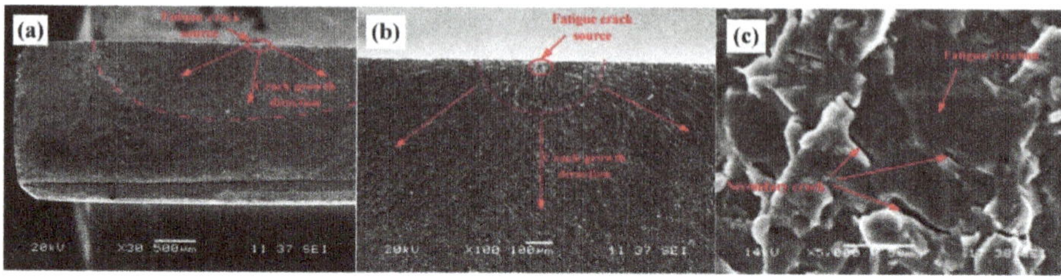

Fig.4 Typical fractography with the optimzed LSP procedure

4. CONCLUSIONS

The TC17 titanium alloy simulated-blade was treated by two LSP procedures and the second procedure was optimized from the first LSP procedure. The differences on the fracture mechanism between two LSP prodedures were discussed. The fatigue limit of TC17 titanium alloy simulated-blade can be improved from 405.7 MPa to 462.9 MPa, increased by 14.1%, by the optimized LSP procedure. In contrast, the fatigue limit decreased by 11.3% with the first LSP procedure. The fatigue crack initiates at the simulated-blade edge when the first LSP procedure and at the center when the optimized LSP procedure. With respect to the first LSP procedure, there are more orderly stripes, denser fatigue striations with the second LSP procedure.

REFERENCES

[1] Charles S. Montross, Tao Wei, L. Ye, G. Clark, Y. W. Mai. (2002) Int. J. Fatigue, **24**, 1021-1036.

[2] B. P. Fairand, B. A. Wilcox, W. J. Gallagher, D. N. Williams. (1972) J. Appl. Phys. **431**, 1497-1502.

[3] P. Peyre, R. Farrbo, P. Merrien, H. P. Lieurade. (1996) Mater. Sci. Eng. A. **210**, 102-113.

[4] J. J. Ruschau, R. John, S. R. Thompson, T. Nicholas. (1999) Int. J. Fatigue. **21**, S199-209.

[5] R. K. Nalla, I. Altenberger, U. Noster, G. Y. Liu, B. Scholtes, R. O. Ritchie. (2003) Mater. Sci. Eng. A. **355**, 216-230.

[6] X.D. Ren, Y.K. Zhang, H.F. Yongzhuo, L. Ruan, D.W. Jiang, T. Zhang, K.M. Chen. (2011) Mater. Sci. Eng. A. **528**, 2899-2903.

[7] Xiangfan Nie, Weifeng He, Liucheng Zhou, Qipeng Li, Xuede wang. (2014) Mater. Sci. Eng. A. **594**, 161-167.

[8] Xiangfan Nie, Weifeng He, Shunlai Zang, Xuede wang, Jie Zhao. (2014) Surf. Coating Technol. **253**, 68-75.

Investigation of fracture mechanism of electroplated coating subjected to contact loading

M. Niwa, H. Toyama, A. Yonezu
Department of Precision Mechanics, Chuo University, Japan

ABSTRACT

This study investigated the contact fracture mechanism of electroplated coating on stainless steel substrate subjected to contact loading. Ball indentation test with large contact force was performed, such that the brittle produced complicate coating fracture. The fracture nucleation process was investigated using the acoustic emission technique. In addition, finite element method with cohesive zone model was carried out to compute stress field, and then simulated the crack initiation and progression during the test. This study clarified the mechanism of coating crack and evaluated the fracture strength.

1. INTRODUCTION

Hard thin films or surface coatings on ductile metallic substrates are often used for contact and slide wear protection. Therefore, the mechanism of their contact fracture is critical for ensuring their mechanical performances. Among the electroplated hard coatings, Ni–P material can achieve high hardness, high strength and other superior mechanical properties, providing excellent performance (such as wear and corrosion resistance) for metallic ductile substrate.

In this study, ball indentation test was carried out to simulate contact fracture of electroplated Ni–P coating on stainless steel substrate. Acoustic emission technique was utilized to monitor the timing of coating cracks. In addition, stress field upon indentation was computed by finite element method (FEM), where the cohesive zone model (CZM) was used to simulate crack nucleation and propagation. This study investigated the contact fracture mechanism and evaluated fracture strength of the Ni-P coating.

2. MATERIALS AND EXPERIMENTAL METHOD

The materials used in this study is electroplated Ni-P deposited onto SUS304. After electroplating, heat treatment of 350°C was performed for 1 hour in a vacuum. According to Refs. (1, 2), post-heat treatment is crucial for the mechanical properties of the coating. The hardness coating (*HV*=800 - 1000) is achieved (1, 2) for wear protection, whereas the fracture toughness exhibits the lowest value (2). The coating thickness is about 180 μm. The coating is well bonded with the steel substrate. The mechanical properties (elastoplastic properties) of both Ni-P coating and steel substrate were evaluated by microindentation tests (3). Our previous study evaluated the mechanical properties of Ni-P coating and SUS304 i.e., the Young's modulus E = 217 and 175 GPa, and the yield stress σ_y = 2.4 and 0.22 GPa, respectively. For the experiment, ball indentation test with the diameter of 10 mm was carried out in order to produce the coating crack. The maximum force is set to 2000 N. The number of test was two times. Their detail information on the experimental equipment was described in Ref (3).

3. EXPERIMENTAL RESULTS

Figure 1 show the micrographs of specimen surface after the test. It shows complicated coating crack morphology: one type is circumferential crack, namely "ring crack", and the other type initiates from the ring crack and propagate radially, namely "radial crack". However, there is no delamination (coating spalling) thanks to the strong adhesive strength of the coating/substrate system. Figure 2 shows the indentation *F–h* curve with the generation timing of AE. It is found that the first AE was detected at about $F = 1200$ N and several AEs were subsequently monitored up to about $F = 1500$ N. Furthermore, the unloading process was found to start AE generation from $F = 1600$ N. Therefore, it is expected that the coating cracks occur during both loading and unloading. The stress field at crack nucleation is discussed in Section 4.1.

As shown in Fig.1, the radial crack propagates along the radial direction. It is expected that the radial crack length reaches the maximum value at the end of the unloading. The stop of crack propagation is due to the balance between stress intensity factor around the crack tip and the fracture toughness (K_C). Thus, this radial crack length is related to the fracture toughness of the material. This discussion will be described in Section 4.2.

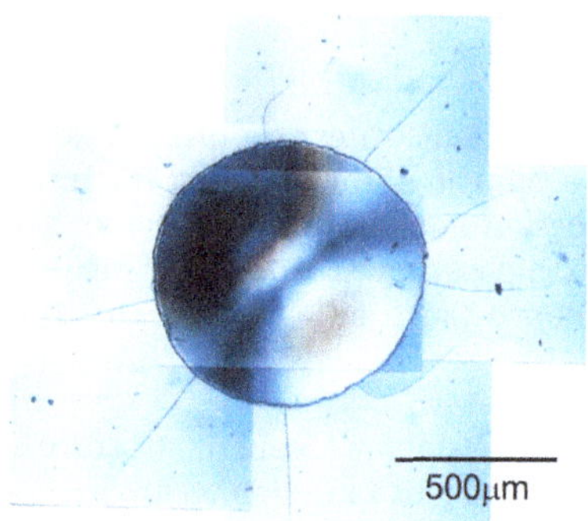

Fig. 1. Micrograph of impressions tested with $F_{max} = 2000$ N.

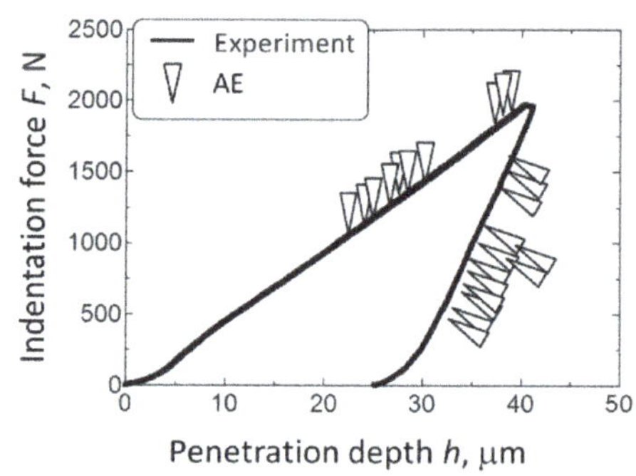

Fig. 2. Indentation curve with AE generation timing.

4. DISCUSSION

4.1 Crack initiation

By taking advantage of symmetry, an axisymmetric model is established for FEM analysis with the commercial code MARC and MENTAT. Figure 3 represents a contour map of the indentation stress field computed by FEM simulation when the first AE was detected (as discussed in Fig. 2). Figure 3 (a) shows the map for the radial stress component (σ_{rr}), which is responsible for ring crack, and Figure 3 (b) is the circumferential component ($\sigma_{\theta\theta}$) for radial crack. For σ_{rr}, a large tensile stress (up to about 1.8 GPa) occurs outside the contact region, where the ring crack may produce if such a tensile stress is sufficiently high. Indeed, such a prominent tensile stress is contributed by the large local bending curvature of the film, assisted by the extensive plastic deformation of the substrate. Therefore, the first AE is from the ring crack, suggesting that the ring crack initiated first.

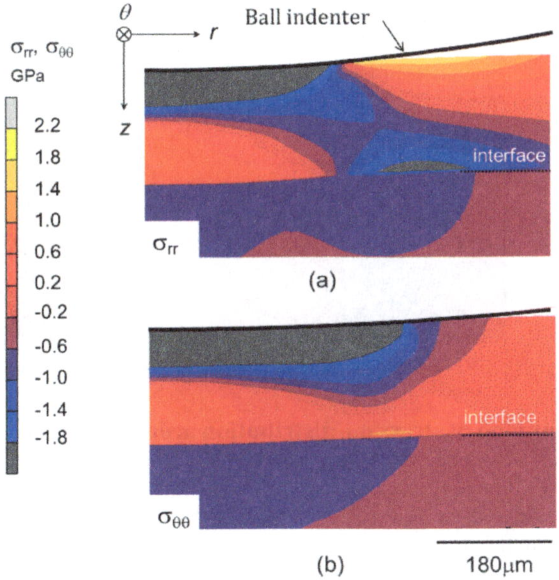

Fig. 3. Counter map of stress distribution when the first AE was detected. (a) normal stress along radial direction, σ_{rr}, and (b) normal stress along circumferential direction, $\sigma_{\theta\theta}$.

We next investigated the stress field of σ_{rr} at the first AE generation (F = 1200 N). The maximum σ_{rr} stress is estimated to be 1.8 GPa. This corresponds to the critical value of coating strength. Thus, the next question is how the radial crack initiates after the ring crack formation. In order to investigate how the radial crack forms, the existence of ring crack must be incorporated with the stress analysis during indentation. Thus, we employed the cohesive zone model (CZM) in the FEM to compute the stress field in conjunction with ring crack formation. The present CZM employed an exponential law (called the Smith-Ferrante type), since it can readily apply to the simulation of crack propagation in a typical brittle solid. This exponential law requires two independent materials parameters, i.e., the maximum stress σ_{max} and the crack growth resistance *Kc*. Since the σ_{max} roughly corresponds to the critical stress (fracture strength), σ_{max} is set to be 1.8 GPa. However, the other parameter *Kc* is unknown.

As an example case, *Kc* is set to 5.5 MPa $m^{1/2}$, such that the ring crack (due to σ_{rr} component) propagates from the surface to the interface. Figure 4 shows the snapshot of the normal stress $\sigma_{\theta\theta}$ distribution after the ring crack formation by CZM. In both Fig. 4 (a) and (b) "crack position" indicates the ring crack formation (i.e. the crack propagates through the interface). Figure 4 (a) shows the stress field during the loading process at F = 1500 N, while Fig. 4 (b) shows the result during the unloading process at 0 N (full unloading). Near the ring crack tip (at the intersection of the interface), large $\sigma_{\theta\theta}$ is found to develop. This magnitude of $\sigma_{\theta\theta}$ increases during the unloading process and reaches the maximum value at full unloading. Therefore, the ring crack formation encourages larger $\sigma_{\theta\theta}$ component, which may produce the radial crack. In other words, the large $\sigma_{\theta\theta}$ develops upon unloading, and the radial crack initiates at the right side of the ring crack tip (see Fig.4).

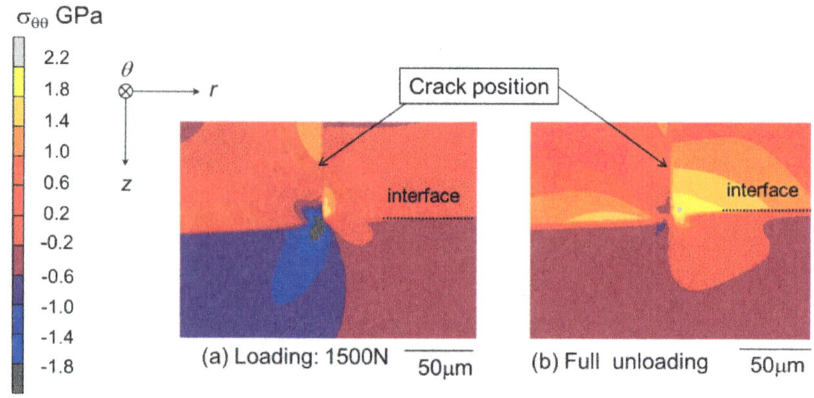

Fig. 4. Contour map of stress $\sigma_{\theta\theta}$ distribution around the impression during indentation loading and unloading.

To investigate timing of the radial crack initiation, the change in the maximum $\sigma_{\theta\theta}$ value is investigated as a function of the indentation force (during unloading process) as shown in Fig. 5. As expected, the maximum $\sigma_{\theta\theta}$ increases with decreasing indentation force. Referring to Fig. 2, the AE occurrence was observed at about F = 1600 N under unloading. When the force F is 1600 N during the unloading, the maximum $\sigma_{\theta\theta}$ value corresponds 1.82 GPa. This $\sigma_{\theta\theta}$ value is reasonably agreement with the critical stress σ_{rr} for ring crack (1.8 GPa). Therefore, it is concluded that radial crack is produced by $\sigma_{\theta\theta}$, which develops near the interface and the ring crack tip. This suggested that the ring crack formation (during loading) is crucial for subsequent radial crack nucleation during unloading.

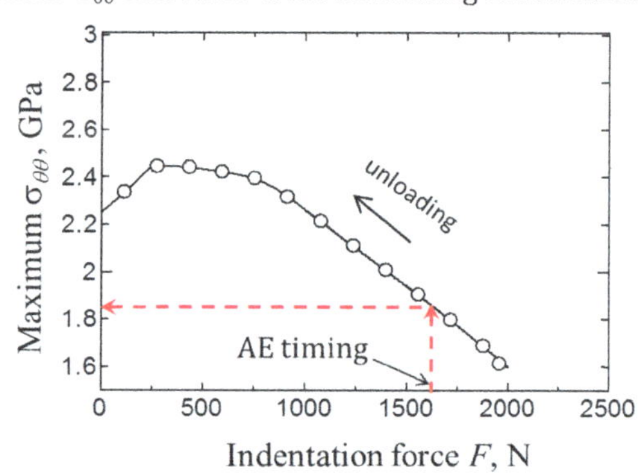

Fig. 5. Changes in maximum stress $\sigma_{\theta\theta}$ as a function of indentation force during unloading.

4.2 Radial crack propagation

To simulate radial crack propagation, 3D-FEM model (one - quarter model) was created. For ring crack formation, the slit with node-separating is artificially introduced at the location of ring crack. Note that the computed stress $\sigma_{\theta\theta}$ field does not change in between ring crack formation by CZM (Fig. 4) and the present model with artificial crack (slit).

For the radial crack propagation, CZM element is introduced along radial direction from the impression rim (i.e. the slit of ring crack formation). For CZM element, the constitutive equation of traction - separation law is exponential type. The maximum stress σ_{max} is set to 1.8 GPa, which was determined in the section 4.1. Figure 6 shows the result of Kc = 2.3 MPa $m^{1/2}$, showing the contour map of damage parameter D. Here, the parameter of D indicates a crack propagation area. This figure shows the series of crack propagation during unloading process. The radial crack initiates at the ring crack tip and propagates up to the coating surface. The radial crack propagates radially. It is also found that the radial crack length is strongly influenced by fracture toughness (Kc). Thus, this study conducted various computations to explore actual fracture toughness of the coating.

In the present computations, the crack length of Kc = 2.0 MPa $m^{1/2}$ is 927 μm, while the Kc = 2.3 MPa $m^{1/2}$ is length of 788 μm. On the contrary, the experimental data indicated the averaged crack length was 395 μm (the range of 302 - 517 μm). Thus, the actual fracture toughness may be 3-4 MPa $m^{1/2}$, which is in the range of previous data obtained from the conventional facture mechanics test. They reported that Kc is 3.7 MPa $m^{1/2}$ (4). In the future, we will clarify the actual fracture toughness of the coating.

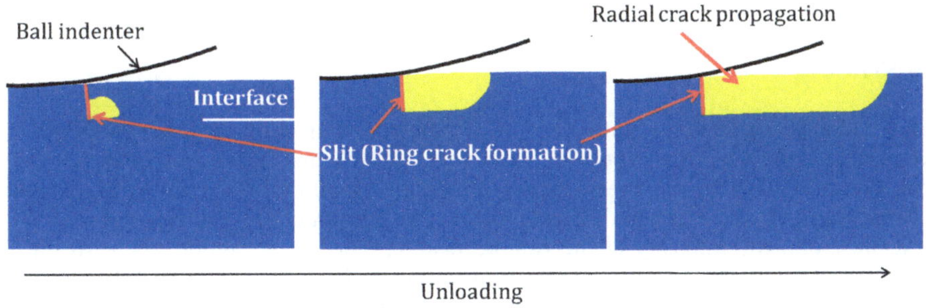

Fig. 6. The contour map of damage parameter *D* (crack propagation) during the unloading process.

5. CONCLUSION

This study investigated the contact fracture mechanism of electroplated Ni-P coating on stainless steel substrate, which is important for its application as contact/sliding member for wear resistance. By using the AE and FEM, the mechanism of the complicate coating crack system is clarified, the stress criterion for each cracking system is quantified. The present method based on combined framework of experiment and computation is possible to evaluate fracture strength and toughness of the brittle Ni-P coating. Based on these findings, further systematic study may provide the suggestion of controlling or preventing the coating cracks.

This work is supported by JSPS KAKENHI (Grant No. 26420025) and JGC-S Scholarship Foundation.

REFERENCES

(1) Yashiki T, Nakayama T, Kato J, The Iron and Steel Institute of Japan. 81 (1995) 48–53 (in Japanese).

(2) Morikawa Y, Cho H, Nakayama T, Takemoto M, The Iron and Steel Institute of Japan. 82 (1996)935-940 (in Japanese).

(3) Yonezu A, Niwa M, Ye J, Chen X, Materials Science & Engineering A 563 (2013) 184-192, DOI: 10.1016/j.msea.2012.11.054

(4) Zhou QJ, He JY, Li. JX, Chu. WY, Qiao. LJ, Materials Letters 60 (2006) 349-351, DOI: 10.1016/j.matlet.2005.08.049

Effect of vertex singularities on the displacement and strain fields near a crack front

Z. He[a], A. Kotousov[a], L.R.F. Rose[b]
[a] School of Mechanical Engineering, The University of Adelaide, Australia
[b] Aerospace Division, DSTO, Australia

ABSTRACT

The current paper investigates the linear-elastic strain and displacement fields in vicinity to a crack front. It is demonstrated that a finite transverse strain exists along the front of a through-the-thickness crack. It is demonstrated that this finite strain is linked to the vertex singularities; and its magnitude is a function of the elastic properties, plate thickness and the intensity of the remote loading. Therefore, neither plane strain nor plane stress is an accurate assumption in the analysis of the strain and displacement field near the crack tip except in the case when the plate thickness is sufficiently large.

1 INTRODUCTION

Analysis of plate components weakened by through-the-thickness cracks subjected to in plane loading is often based on the plane stress or plane strain assumptions. These simplifications essentially reduce the dimensionality of the actual three-dimensional problem and facilitate obtaining many fundamental results of the classical Fracture Mechanics as well as exact solutions to practically important problems (1). However, the experimental evidences collected over the past few decades have indicated that the three-dimensional stress/strain states associated with the plate thickness effect, vertex singularities and coupling of fracture modes can play an important role in fracture phenomena (2-5). Subsequently, this paper focuses on the investigation of the three-dimensional displacement and strain fields in the close vicinity of a crack tip.

The problem under consideration has been previously investigated with analytical, experimental and numerical approaches. These investigations include a pioneering study of Hartranft and Sih on the application of the eigenfunction expansion approach to three-dimensional crack problems (6), which was further extended by Omer and Yosibash (7), an analytical solution for a semi-infinite crack (8) as well as a semi-analytical approach for the analysis of three-dimensional crack problems with the distributed dislocation technique and the three-dimensional solution for an edge dislocation (9). A number of very accurate experimental studies have also been conducted in the past using advanced tools, such as interferometry (10). The first comprehensive numerical study of the displacement and stress fields near the crack front was conducted by Nakamura and Parks (11). The approach and outcomes of this study have been verified and extended by many other researchers, e.g. Berto et al. (12). In particular, all these studies agree that the region of the three-dimensional effects is confined to approximately one-half of the plate thickness around the crack front, before converging into a two-dimensional plane stress field at a radial distance equal, roughly, to the plate thickness (13-14).

Previous numerical attempts have also identified that near to the intersection of the crack front and a free surface the stress state is significantly affected by, so-called, vertex (or 3D corner) singularity. This effect was first described in the late 1970s and early 1980s by Benthem and a number of other researchers, who employed a finite difference scheme and the eigenfunction expansion method to demonstrate that at the intersection of the crack front and a free surface, the square root singularity disappears, and at such a point, one has to deal with a three-dimensional vertex singularity (15). As opposed to the edge singularities, which have a universal square root power, the strength of vertex singularities is affected by Poisson's ratio of the material. The main results for the behaviour of the vertex singularity have been obtained with semi-analytical techniques or direct numerical methods, such as the Finite Element Method (16).

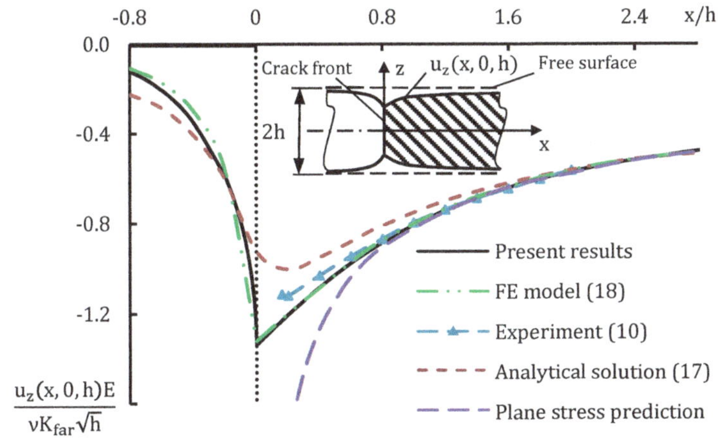

Figure 1 Normalised out-of-plane displacement of the free surface along the crack direction as a function of the normalised distance x/h

As mentioned above, the focus of the current paper is on the effect of vertex singularities on the transverse (out-of-plane) displacement and strain fields near the crack front. Figure 1 presents outcomes of experimental (10), analytical (17) and numerical (18) studies on the transverse displacement $u_z(x, 0, h)$, at the free surface ($z = h$) in the vicinity of the crack front located at $x = 0$. The through-the-thickness crack in these studies was subjected to mode I loading and the material in all these studies is considered to be linear-elastic (17, 18) or sufficiently brittle (10). As it can be seen from this figure, far from the crack tip the transverse displacements follow the plane stress solution, whilst in a close vicinity of the crack tip the displacement field is different to what is predicted by either the classical plane stress, $u_z \sim r^{-1/2}$, or plane strain solutions, $u_z \equiv 0$. The aim of this paper is to demonstrate that these non-zero transverse displacements at the tip of the crack are associated with the vertex singularities. The finite element modelling approach, adopted in the current work, will be presented in the next section.

2 FINITE ELEMENT MODEL

The numerical study considers an arbitrary semi-infinite plate containing a through-the-thickness crack loaded in mode I. The finite element analysis was conducted using the ANSYS 14 package utilising a mesh consisting of an initial arrangement of 15-node trapezoidal elements at the notch tip, surrounded by a radial array of 20-node brick elements, where each element spanned an angular sweep of 7.5°. The half thickness of the model consists of 30 layers of elements, for which the thickness is gradually decreased from the mid-plane to the free surface to better represent the variations of the stresses and displacements near the free surface, which is affected by the vertex singularities. To simulate remotely applied mode I loading, the displacement boundary conditions, in accordance with classical 2D Williams' solution (19) for plane stress, were applied far from the crack tip. The developed FE model was validated against the previously published results from various studies (as shown in Figure 1).

3 EFFECT OF VERTEX SINGULARITIES

In the beginning, we investigated the radial distribution of the transverse displacement u_z near the crack front. One interesting finding is that the displacement field very close to the crack tip does not depend on the angular position and constant along the radial direction confined by one tenth of the plate thickness ($r/h < 0.1$). This transverse displacement is a function of the coordinate z along the thickness only. Typical simulation results of the dimensionless transverse displacement for Poisson's ratio $\nu = 0.3$ are shown in Figure 2.

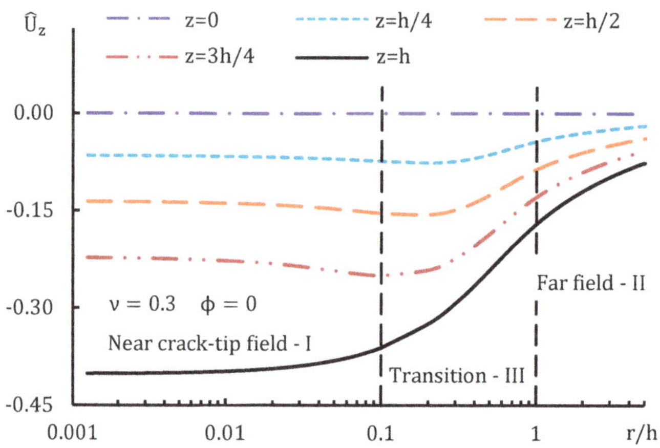

Figure 2 Distribution of the dimensionless transverse displacement $\hat{U}_z$ along the bisector line $\phi = 0$, at different positions z along the crack front

The numerical results are presented in Figure 2 utilise a dimensionless displacement function, which is defined as:

$$\hat{U}_z = \frac{u_z(r, \phi, z)E}{K_{far}\sqrt{h}} \tag{1}$$

where, K_{far} is the remotely applied stress intensity factor, $E = 2\mu/(1+\nu)$ is Young's modulus, h is the half plate thickness, r, ϕ and z are the cylindrical polar coordinates with the origin located at the middle of the crack front.

From Figure 2 one can see three characteristic regions: I - near the crack-tip region encapsulated by a radial distance of approximately 0.1h from the crack front; II - far field region, where the displacement and stress fields correspond to plane stress solution.

This region is located outside the cylindrical domain $r > 2h$ encapsulating the crack front. The third characteristic region (III) is the transition zone, in which the displacement converges from plane stress to the near crack-tip transverse displacement field.

Next, we provide evidences to demonstrate that this near crack-tip field is related to the vertex singularities. Figure 3 shows the normalised transverse displacement as a function of the position along the crack front z for different Poisson's ratios ($\nu = 0.1 \sim 0.5$). $z = 0$ ($\xi = 1$) corresponds to the mid-plane and $z = h$ ($\xi = 0$) represents the free surface. It is found that the transverse displacements near the vertex point follow the vertex singular solutions, or $u_z \sim \xi^{\lambda_c}$ (12), where ξ is the through-the-thickness coordinate with the origin at the free surface, or $\xi = 1 - z/h$ and λ_c is the strength of the vertex singularity. The values of the strength of the vertex singularity λ_c for the selected Poisson's ratios (listed in Figure 3) are compared with the previously published results. It is found that the differences between the present results and those from references (20-22) are less than 1%, which can be considered as an excellent agreement.

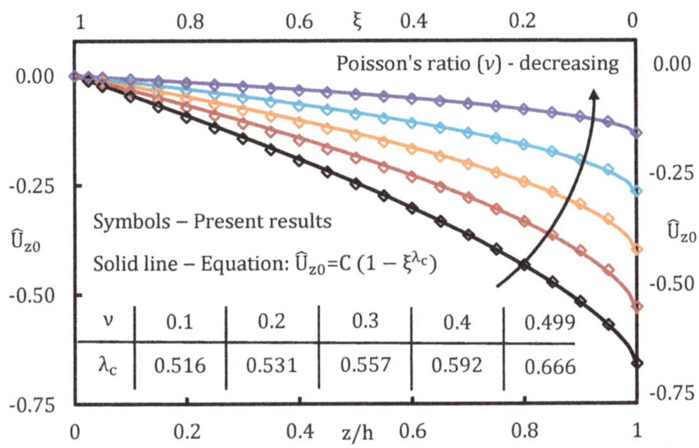

Figure 3 Dimensionless transverse displacement, $\hat{U}_{z0}$, along the crack front in region I for various Poisson's ratio (0.1; 0.2; 0.3; 0.4 and 0.499)

For all Poisson's ratios, the dimensionless function $\hat{U}_z$ very near the crack front (region I) can be rather accurately represented by the following equation:

$$u_z(r, \phi, z) = u_z(z) \approx -\frac{1.34 \cdot \nu}{E} K_{far} \sqrt{h} \cdot \left(1 - \xi^{\lambda_c}\right) \quad (2)$$

The strain, associated with this out-of-plane displacement, at the middle of the crack front ($x = 0, y = 0, z = 0$) can be approximated as:

$$\varepsilon_z^{mid} \approx -\frac{1.34 \cdot \nu \cdot \lambda_c}{E\sqrt{h}} K_{far} \quad (3)$$

It is clear that the plane strain solution assuming that $\varepsilon_z = 0$, can only be recovered for sufficiently thick plates, when the effect of the vertex singularities is negligible; in all other cases non-zero displacements and strains exist at the crack tip. Plane stress solution cannot be recovered at the crack tip even for very thin plates. Moreover, the strain and displacement fields will be controlled by the vertex singularities.

4 CONCLUSION

In this work we demonstrated that at the crack front the transverse displacements and strains are finite and are functions of the position along the crack front. It is also demonstrated that the existence of these finite displacements and strains is associated with the vertex singularities. Near the free surface the transverse displacements and strains follow the three-dimensional vertex singular solutions and in the mid-section of the plate the transverse strain is relatively constant. The latter further justifies the analytical solutions obtained with the first order plate theory (Kane and Mindlin theory), which assumes a constant strain across the plate thickness. In many studies it was proofed that these analytical solutions can describe the three-dimensional stress and strain fields adequately.

REFERENCES

[1] Kotousov, A. (2007) *Int. J. Solids Struct.* **44**, 8259-8273.

[2] Kotousov, A. (2010) *Int. J. Solids Struct.* **47**, 1916-1923.

[3] Kotousov, A., Lazzarin, P., Berto, F. & Harding, S. (2010) *Eng. Fract. Mech.* **77**, 1665-1681.

[4] Pook, L. P. (2013) *Fatigue Fract. Eng. Mater. Struct.* **36**, 699-723.

[5] Pook, L. P. (2013) *Eng. Fract. Mech.* **77**, 1619-1630.

[6] Hartranft, R. J. & Sih, G. C. (1969) *J. Math. Mech.* **19**, 123-138.

[7] Omer, N. & Yosibash, Z. (2005) *Int. J. Fract.* **136**, 1-36.

[8] Yang, W. & Freund, L. B. (1985) *Int. J. Solids Struct.* **9**, 977-994.

[9] Kotousov, A. & Wang, C. H. (2002) *Int. J. Eng. Sci.* **40**, 1775-1790.

[10] Humbert, L., Valle, V., & Cottron, M. (2000) *Int. J. Solids Struct.* **37**, 5493-5504.

[11] Nakamura, T. & Parks, D. M. (1998) *J. Appl. Mech.* **55**, 805-813.

[12] Berto, F., Lazzarin, P. & Kotousov, A. (2011) *Mech. Mater.* **43**, 332-341.

[13] Berto, F., Lazzarin, P. & Kotousov, A. (2011) *Int. J. Fract.* **167**, 119-126.

[14] Harding, S., Kotousov, A., Lazzarin, P. & Berto, F. (2010) *Int. J. Fract.* **164**, 1-14.

[15] Benthem, J. P. (1977) *Int. J. Solids Struct.* **13**, 479-492.

[16] Pook, L. P. (2000) *Fatigue Fract. Eng. Mater. Struct.* **23**, 141-150.

[17] Kotousov, A. & Tan, P. J. (2004) *Int. J. Fract.* **127**, L97-L103.

[18] Pfaff, R. D., Washabaugh, P. D. & Knauss, W. G. (1995) *Int. J. Solids Struct.* **32**, 939-955.

[19] Williams, M. L. (1957) *J. Appl. Mech.* **24**, 109-114.

[20] Shivakumar, K. N. & Raju, I. S. (1990) *Int. J. Fract.* **45**, 159-178.

[21] Benthem, J. P. (1980) *Int. J. Solids Struct.* **16**, 119-130.

[22] de Matos, P. F. P. & Nowell, D. (2008) *Int. J. Fract.* **30**, 1930-1943.

Calculating the essential work of fracture of proton exchange membranes using a finite element analysis

T. Vermot des Roches, M. Omiya
Keio University, Department of Mechanical Engineering, Japan

ABSTRACT

The membrane is one of the critical components of a Proton Exchange Membrane Fuel Cell, as its reliability is necessary to ensure a proper functioning of the fuel cell. The objective of the present study is to calculate the essential work of fracture of the proton exchange membrane by way of a numerical simulation. This should help providing a better characterization of its fracture behavior. The originality of this study lies in using a finite element model to this purpose, with the advantage of necessitating less experimental inputs for the calculation of the work of fracture.

1 INTRODUCTION

The ever increasing pressure to decrease the dependency of power generation and storage on fossil fuels gives incentive to find alternative power sources. Thanks to their high efficiency and environmentally friendly characteristics, proton exchange membrane fuel cells (PEMFCs) are considered a promising technology and have received a lot of attention. This is particularly true in the sector of transportation, as the high energy concentration of dihydrogen allows vehicles to have a high autonomy with little fuel weight. However, a higher entry price and a limited durability when compared to more conventional power units make it difficult for the PEMFC technology to be adopted globally. In order to achieve a greater lifetime for PEMFCs, it is necessary that the failure mechanisms of its components be clearly understood.

The membrane electrode assembly (MEA), composed of the two electrodes and of the proton exchange membrane, Nafion in this study, is the central component of a PEMFC: being the siege of the chemical reaction, it helps conducting protons between the electrodes while acting as an electronic insulator and as a barrier for the reactants. As such, any failure or degradation of the MEA will compromise the function of the PEMFC. Incidentally, the MEA is also the most expensive component of a fuel cell, and its most common failure point(1). Despite its importance, the properties of the MEA are not well documented. Since the MEA is the limiting factor for the durability of the fuel cell, there is a great incentive to develop more accurate models that would facilitate our understanding of the MEA.

This study investigates the essential work of fracture (EWF) of the proton exchange membrane in environmental conditions relevant to the actual function of the fuel cell.

The EWF is commonly used to determine the toughness of ductile polymers, which is a key property when assessing the durability and reliability of such meterials. The Double Edge Notched Tensile (DENT) test is used to this end. Based on experimental results, a finite element model is developed to provide an alternative method of calculating the work of fracture.

2 ESSENTIAL WORK OF FRACTURE

The essential work of Fracture is an important material property to calculate when studying the fracture of polymers. The interest of the EWF approach over other methods such as the J-integral lies in the relative simplicity of the experimental protocol. A double-edge notch tensile test has to be performed on the notched membrane, as described in Fig. 1(a). The total fracture work is then calculated. In order to account for plastic dissipation and other dissipated energies outside of the EWF, several experiments have to be conducted with different ligament lengths. By extrapolating the results to a theoretical ligament length of 0, where no plastic deformation occurs, the EWF can be found. In order to obtain accurate measurement of the EWF with this method, it is necessary to obtain plane stress conditions in the specimen and to avoid edge effects(2). Because of these restrictions, the valid range of ligament lengths can be fairly limited in the experiments.

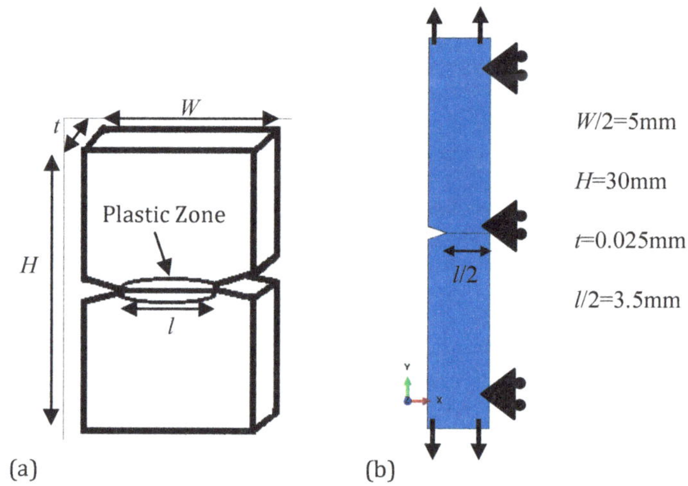

Fig. 1 DENT test specimen: (a) Schematic;
(b) Calculation model with boundary conditions

3 FINITE ELEMENT MODEL

3.1 Modeling the behavior of Nafion

Nafion is a perfluorosulfonic acid polymer with a good chemical stability. Its characteristics make it a material of choice as a proton exchange membrane(3). However, relatively little research has been done on its mechanical properties, especially on its failure behavior. Nafion exhibits a non-linear viscoplastic behavior, with a strong dependency on both humidity and temperature, making it important to account for these parameters during experiments(4).

The most accurate numerical model for Nafion found in the literature was proposed by Prof. M.C. Boyce in 2010(3) and uses a variety of components to account for both intermolecular and network deformation of Nafion. In the present study, this model was used for the finite element modeling of Nafion, though some of the parameters were modified in order for the simulated behavior to match as closely as possible the experimentally observed stress-strain behavior of the membrane. The model was found to reproduce the behavior of Nafion accurately for the purpose of this study, as shown in Fig. 2.

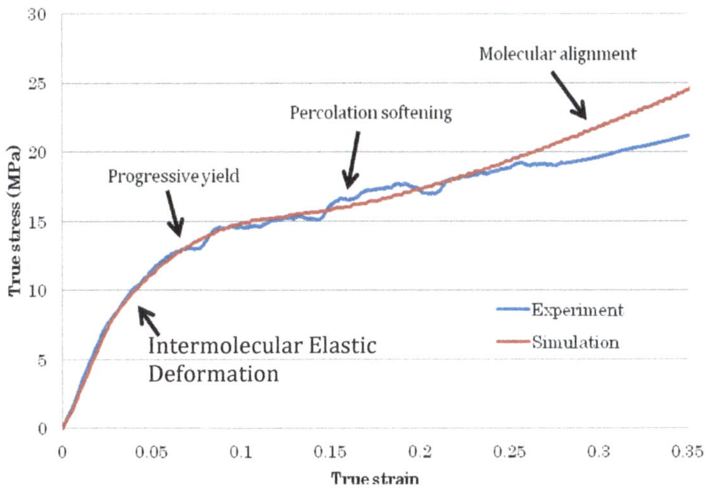

Fig. 2 Result of the simulation of Nafion behavior at 25ºC and 20%RH

3.2 Cohesive zone model

The characterization of the cohesive zone in the model is of particular importance here as this is what will define the fracture behavior in the simulation. In addition to that, analyzing the results for a given cohesive zone model can provide insights for an experimental protocol that diminishes unwanted effects such as plastic deformation or friction.

The model employed here uses a maximum stress criterion, where cohesive elements become separated after the maximum stress has been reached. As there does not exist as of now enough data on the fracture of Nafion, this criterion needs to be first estimated through experiment.

In the case of the DENT experiment, friction was found to not have any appreciable effect on the fracture energy being dissipated.

3.3 Method of calculating the EWF

In this study, a numerical scheme is proposed in order to calculate the EWF. This method has the advantage of only needing a single experimental input: the applied strain at the time of fracture for a given ligament length. Figure 1(b) shows the model that was used in this study. Since the numerical model allows to accurately monitoring the distribution of dissipated energy in the material, the EWF can be calculated. The theory assumes that the fracture energy can be separated into essential and non-essential (geometry dependant) work:

$$W_F = W_E + W_P \tag{1}$$

where W_F is the total dissipated energy, W_E is the essential work of fracture and W_P is the non-essential work comprised of the energy dissipated in the plastic zone. Since the size of the plastic zone is proportional to the ligament length(5), Eq.1 can be rewritten as

$$w_F lt = w_E lt + \beta w_P l^2 t \Rightarrow w_F = w_E + \beta w_P l \tag{2}$$

where w_F, w_E and w_P are the specific works corresponding to those in Eq.1, t is the specimen thickness, l is the ligament length and β is the shape factor of the plastic zone. β is unknown, however by plotting w_F against the ligament length and extrapolating to a length of 0, it is possible to isolate w_E, which is the specific essential work of fracture.

4 RESULTS AND DISCUSSION

Figure 3 shows the result of a DENT test that was conducted on a Nafion specimen at ambient conditions, together with the result of the equivalent numerical simulation. The critical strain, which is the strain that needs to be applied globally to initiate the fracture, is measured at ε_c=0.0646. The global stress is then 15.4MPa; however because of stress concentration at the notch tip, the actual fracture stress, which is used as a criterion for the crack formation, is higher.

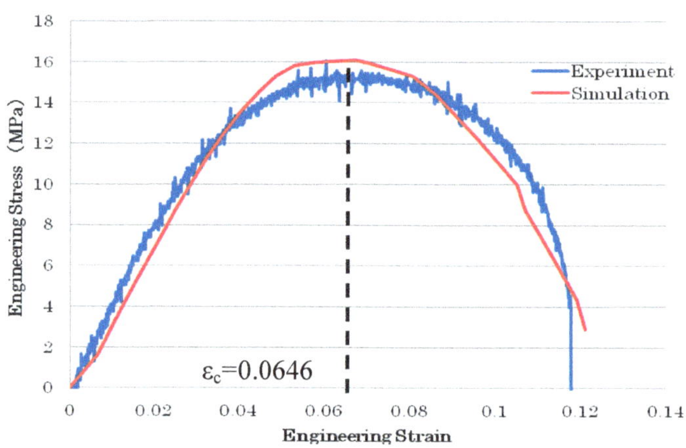

Fig. 3 Stress-Strain curve for Nafion during a DENT test

The true stress calculated at the notch tip at the time of failure is found to be between 21MPa and 23MPa on all simulations, which is coherent with the maximum stress of 23MPa for Nafion indicated on the material data sheet. These parameters were used to conduct the finite element analysis and w_F was calculated, yielding the results depicted in Fig. 4. Extrapolating the curve to 0 leads to an EWF of approximately 19.9 kJ/m^2. This value is to be compared to the value of 20.5 kJ/m^2 found in the literature for a Nafion membrane with the DENT test(5). While the methods used in Ref.5 involved more complex experimental protocols, this study requires comparatively fewer experimental inputs, making it worthwhile for calculating the EWF of polymer membranes.

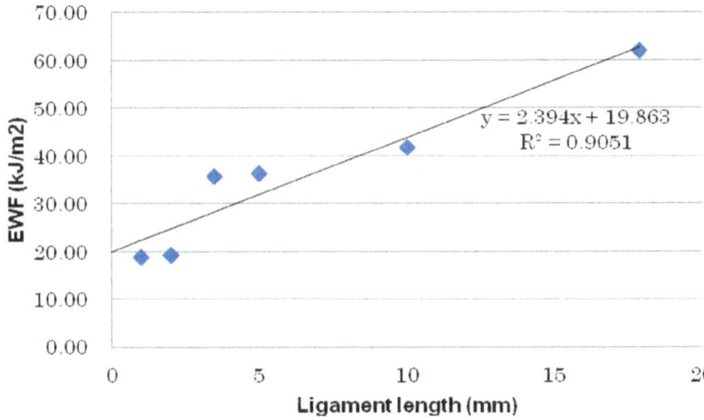

Fig. 4 Total Work of Fracture calculated for different simulations using different ligament lengths and mesh sizes. The EWF is calculated at 19.9 kJ/m²

5 CONCLUSION

In this study, a numerical method was proposed in order to calculate the EWF of a Nafion membrane. The method relies on simulating a DENT test using a finite element model that can reproduce the behavior of Nafion. The only parameter that is needed in order to simulate the fracture is the critical stress of the cohesive zone, which can be simply obtained through experiment. Because of this, the proposed method is significantly quicker in calculating the EWF than the alternatives found in the literature. The EWF for Nafion was found to be 19.9 kJ/m^2, which is coherent with previously reported results. This suggests that the method is accurate at calculating the EWF of polymers membranes.

REFERENCES

[1] DOE progress report to Congress, Jan. 2009

[2] W.K.Y. Poon, E.C.Y. Ching, C.Y. Cheng, R.K.Y. Li, Measurement of plane stress essential work of fracture (EWF) for polymer films: effects of gripping and notching methodology, *Polymer Testing 20 (2001) 395–401*

[3] M. N. Silberstein, M. C. Boyce, Constitutive modeling of the rate, temperature, and hydration dependent deformation response of Nafion to monotonic and cyclic loading, *Journal of Power Sources 195 (2010) 5692–5706*

[4] T. Vermot des Roches, M. Omiya, Failure Prediction for Membrane Electrode Assembly, *Journal of Computational Science and Technology; ISSN:1881-6894; VOL.7; NO.2; PAGE.221-230; (2013)*

[5] Yongqiang Li et al., Characterizing the fracture resistance of proton exchange membranes, *Journal of Power Sources 185(2008)374-380*

Stress singularity analysis and experimental determination of bonding strength of viscoelastic/ viscoelastic bi-material interface

Z. Xia, M.A.K. Chowdhuri
Department of Mechanical Engineering, University of Alberta, Canada

ABSTRACT

This paper presents an analytical solution for determination of stress singularity of a viscoelastic/viscoelastic bonded joint with a spherical interface. The elastic-viscoelastic corresponding principle is used to find the solution of the viscoelastic/viscoelastic bonded joint from the available solution of an elastic/elastic bonded joint. It is showed that the stress singularity can be eliminated if the bonding angle at the corner of the interface is less than a critical bonding angle value. Then an innovative experimental method is applied to specimens with the spherical interface with a bonding angle less than the critical bonding angle. And the bi-axial tension-shear bonding strength for the PVC/Epoxy interface is determined in a form of quadratic equation.

1. INTRODUCTION

Bi-material interfaces are found in many advanced materials and structures. The interface bonding strength plays a decisive role in the global and functional properties of these materials and structures, since damages or failures are most likely to occur at the bi-material interfaces.

The measurements of the interface bonding strength of bi-materials is also more complicated than measurement of tensile or shear strength of a homogeneous material. In various suggested bonding strength measuring methods, such as ASTM D3165 [1] for shear strength test, ASTM D897 [2} for tensile strength test, in fact, non-uniform multi-axial stress status at the bonding areas and stress singularity at the interface corners can be generated. As a result, the strengths values obtained as ratios of de-bonding load to interface area represent an averaged characteristic, and do not represent the actual bonding strengths values of the interface. Therefore, to more accurately determine bi-material interface bonding strength, a prerequisite for the test specimen is to eliminate the stress singularity at the interface.

In this paper the elastic-viscoelastic analogy method is used to find the solution of the viscoelastic/viscoelastic bonded joint from the solution of corresponding elastic/elastic bonded joint. It is showed that the stress singularity can be eliminated if the bonding angle at the corner of the interface is less than a critical bonding angle value. Then an innovative experimental method is applied to specimens with the spherical interface with a bonding angle less than the critical bonding angle. As an example, the bi-axial tension-shear bonding strength for the polyvinylchloride and epoxy (PVC/Epoxy) interface is determined in a form of quadratic equation.

2. ANALYTICAL SOLUTION FOR ELASTIC/ELASTIC BONDED JOINT

Consider an axisymmetric cylindrical model of two elastic isotropic and homogeneous materials bonded together with a spherical interface as shown in Fig 1. The stress field near the interface corner O can be expressed into the following form [3]:

$$\sigma_{ij}(r,\theta) = Kr^{\lambda-1} f_{ij}(\theta,\gamma,\alpha,\beta,\lambda) \tag{1}$$

In the above expression, r,θ are the polar coordinate as shown in Fig. 1, K is the generalised stress intensity factor, f_{ij} is a function of angular coordinate θ, λ-1 is the order of the stress singularity, α,β are Dundurs' parameters [4] which depend on combinations of the elastic constants of the two materials. The λ is the eigen value and can be determined from the following eigenvalue equation [3].

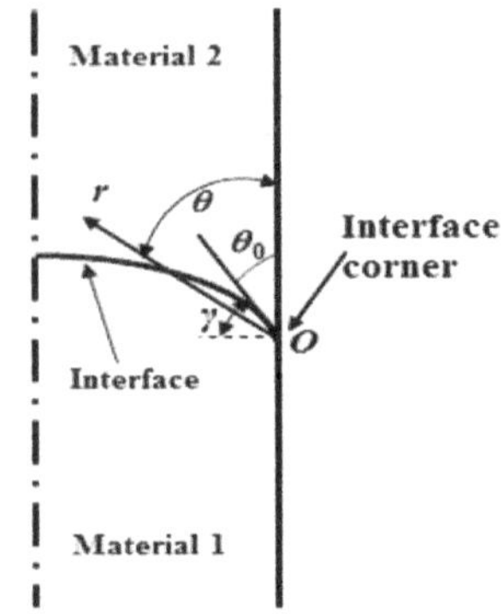

Fig. 1: Axi-symmetric model of bi-material interface

$$\begin{aligned}&\sin^2\lambda\pi - 2(2\lambda^2\cos^2\gamma - 1)\alpha\sin\lambda\pi\sin 2\lambda\gamma + 4\lambda^2\beta\sin\lambda\pi\sin 2\lambda\gamma\cos^2\gamma\\ &+4\lambda^2\cos^2\gamma(1-\cos\lambda\pi\cos 2\lambda\gamma - 2\lambda^2\cos^2\gamma).\alpha\beta + \{\sin^2 2\lambda\gamma + 4\lambda^2(\lambda^2-1)\cos^2\gamma - \lambda^4\sin^2 2\gamma\}.\alpha^2\\ &+\{4\lambda^2\cos^2\gamma(\lambda^2\cos^2\gamma - 1 + \cos\lambda\pi\cos 2\lambda\gamma) + (\cos\lambda\pi - \cos 2\lambda\gamma)^2\}.\beta^2 = 0\end{aligned} \tag{2}$$

If $0<\lambda<1$, the stress component goes to infinity when $r\to 0$, i.e. there exists singular stress state. Thus the condition for elimination of stress singularity from the interface corner is, $\lambda>1$. The angle, $\theta_0 = 90° - \gamma$ is called bonding angle defined as the angle between the tangent of the sphere interface at the free edge to the generator of the cylindrical surface as shown in Fig. 1 For a given pair of materials, there exists a critical value of the bonding angle, θ_c, which delineates the singular and non-singular stress field near the free edge. Therefore, the stress singularity can be avoided if the bonding angle is less than the critical value and a finite interface stress field along the spherical bi-material interface can be determined.

3. ANALYTICAL SOLUTION FOR VISCOELASTIC/VISCOELASTIC (PVC/EPOXY) BONDED JOINT

In this study, Prony series with three terms are used to model the viscoelastic shear modulus and Poisson's ratio of both PVC and epoxy materials:

$$\mu(t) = A_1 + A_2 e^{-\phi_1 t} + A_3 e^{-\phi_2 t} \tag{3}$$

$$\upsilon(t) = B_1 - B_2 e^{-\phi_3 t} - B_3 e^{-\phi_4 t} \tag{4}$$

where A_i, B_i are coefficients and $\phi_i = 1/t_i$; t is the relaxation time.

The elastic-viscoelastic analogy [5] states that elastic solutions can be converted to Laplace transformed viscoelastic solutions through the replacement of elastic moduli and elastic Poisson's ratio by the *transformed parameter*, s multiplied transforms of the viscoelastic moduli and Poisson's ratio. Using Laplace transform, the transformed shear modulus $\mu^*(s)$, transformed Poisson's ratio $\upsilon^*(s)$, transformed Dundurs Parameters, $\alpha^*(s)$, $\beta^*(s)$ and the equation (2) in the transformed domain is given by

$$\begin{aligned}&\sin^2\lambda\pi+2(2\lambda^2\cos^2\gamma-1)\,s\alpha^*(s)\sin\lambda\pi\sin 2\lambda\gamma-4\lambda^2\,s\beta^*(s)\sin\lambda\pi\sin 2\lambda\gamma\cos^2\gamma\\&+4\lambda^2\cos^2\gamma(1-\cos\lambda\pi\cos 2\lambda\gamma-2\lambda^2\cos^2\gamma).s\alpha^*(s).s\beta^*(s)\\&+\{\sin^2 2\lambda\gamma+4\lambda^2(\lambda^2-1)\cos^2\gamma-\lambda^4\sin^2 2\gamma\}.(s\alpha^*(s))^2\\&+\{4\lambda^2\cos^2\gamma(\lambda^2\cos^2\gamma-1+\cos\lambda\pi\cos 2\lambda\gamma)+(\cos\lambda\pi-\cos 2\lambda\gamma)^2\}.(s\beta^*(s))^2=0\end{aligned}\tag{5}$$

Doing inverse Laplace transform and carrying out tedious algebraic manipulations, for the viscoelastic/viscoelastic interface, the above equation in time domain becomes

$$\begin{aligned}&\tau_1\sin^2\lambda\pi+\tau_2.2(2\lambda^2\cos^2\gamma-1)\sin\lambda\pi\sin 2\lambda\gamma-\tau_3.4\lambda^2\sin\lambda\pi\sin 2\lambda\gamma\cos^2\gamma\\&+\tau_4.4\lambda^2\cos^2\gamma(1-\cos\lambda\pi\cos 2\lambda\gamma-2\lambda^2\cos^2\gamma)+\tau_5.\{\sin^2 2\lambda\gamma+4\lambda^2(\lambda^2-1)\cos^2\gamma-\lambda^4\sin^2 2\gamma\}.\alpha^2\\&+\tau_6.\{4\lambda^2\cos^2\gamma(\lambda^2\cos^2\gamma-1+\cos\lambda\pi\cos 2\lambda\gamma)+(\cos\lambda\pi-\cos 2\lambda\gamma)^2\}=0\end{aligned}\tag{6}$$

where $\tau_1\cdots,\tau_6$ are long expressions of function of time t. Due to space limitation, their full expressions are omitted but can be requested from the first author.

Given a bonding angle $\theta_0=90°-\gamma$, the corresponding eigen value function $\lambda(t)$ can be solved from Eq.(6). Figure 2 shows the results for different bonding angles. It can be seen that for the PVT/Epoxy interface, when the bonding angle in Fig. 1 is designed to be less than 47°, $\lambda(t)>1$ and the stress singularity can be eliminated.

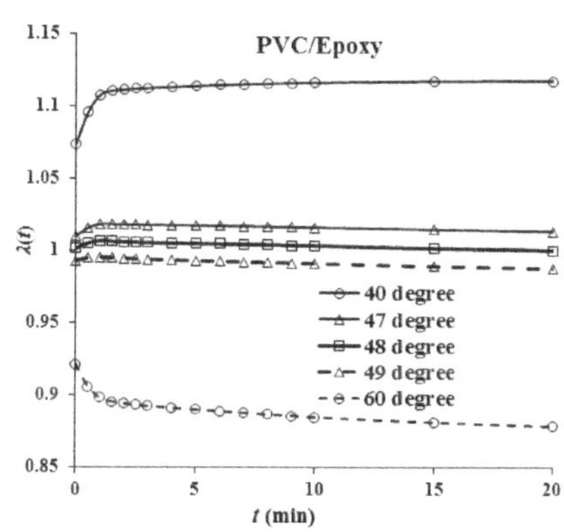

Fig. 2: Variation of λ(t) with time for different bonding angles of 40°, 47°, 48°, 49° and 60°

4. DETERMINATION OF BONDING STRENGTH OF PVT/EPOXY INTERFACE

Special cylindrical specimens with spherical bi-material interface of PVT and Epoxy are manufactured. The bonding angle is chosen to be 40°, which is less than the critical bonding angle of 47°. By this design the stress singularity can be eliminated under any combination of global tensile and torsional loads. Both tensile and shear stresses are not uniformly distributed along the spherical interface. However, the detailed stress distribution can be obtained through finite element analysis (FEA) on the specimen. Tests with different ratios of tensile and torsional loads are conducted on the specimens until failure of each specimen. Experimental results show that the failures always happened or initiated at the interfaces of the specimens. For each loading case, the distribution of interface tensile and shear stress components can be obtained from the FEA. Since the distributions are not uniform, the debonding should start at certain critical point on the interface. The critical point should be determined based on a biaxial tension-shear bonding strength criterion. However, the criterion is unknown before processing the test data. Therefore, the following iteration steps are adopted to determine the critical points and the biaxial tensile-shear bonding strength criterion (strength envelope in tension-shear stress plane):

1) Assume a biaxial tension-shear debonding criterion, for example, based on the von-Mises equivalent stress criterion. For each ratio of tension-shear loading case, the interface tensile and shear stress distribution along the interface can be found from the FEA. Then based on the assumed von-Mises equivalent stress criterion the critical point's location on the interface and this point's tensile and shear stresses values can be found from the FEA results;
2) Plot these critical points on the σ-τ plane. From the coordinates of these critical points, an updated criterion can be determined by using the best fitting technique. Repeating above steps until a stable criterion is obtained. In practice, through only 2 or 3 iterations a convergent result can be obtained.

Table 1: Failure loads for each combined tensile-torsional loadings

No	Tensile load (N)	Torsional load (N-m)
1	0	33.98
2	766.80	34.06
3	1620.21	28.10
4	2552.72	18.82
5	3112.75	0

Five different combination of tensile-torsional loads were applied on the PVT/Epoxy specimens and the failure loads for each load combination is listed in Table 1.

Then, viscoelastic finite element analysis is carried out for each of above loading case and the interface tensile and shear stress distribution at each corresponding failure load level were obtained. For example, Fig. 3 shows the interface stress components distributions for the pure tensile loading case.

To obtain the interface bonding strength envelope through iteration, the following criterion based on von Mises equivalent stress (in the unit of MPa) is first assumed:

$$(\frac{\sigma_n}{26.22})^2 + (\frac{\tau}{15.14})^2 = 1 \qquad (7)$$

Figure 4 shows the points obtained considering the von Mises equivalent stress criterion and the von Mises equivalent stress curve expressed by Equation (7). It is noted that the stress data points are not fitted well by the von-Mises equivalent stress criterion. Next, the second trial criterion is found by applying the best fit curve technique on the 1st set of stress data points, which provide the following equation:

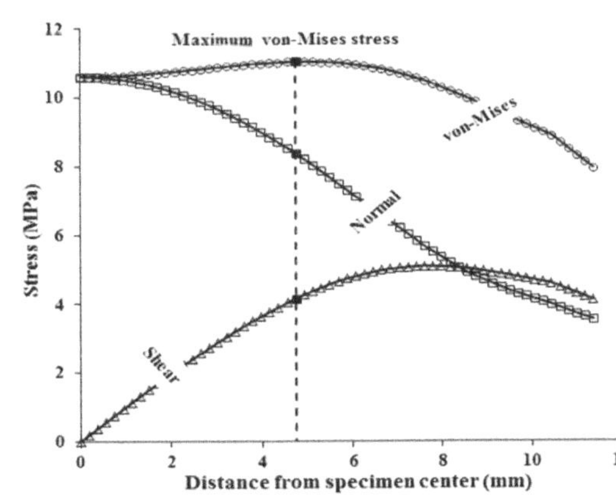

Fig. 3: Stress distribution along the interface for pure tensile loading

$$(\frac{\sigma_n}{7.95})^2 + (\frac{\tau}{15.14})^2 = 1 \qquad (8)$$

The third criterion is obtained by fitting the 2nd set of stress data points and it is the converged one:

$$(\frac{\sigma_n}{9.63})^2 + (\frac{\tau}{15.14})^2 = 1 \quad (9)$$

Thus, the final biaxial tensile-shear bonding strength of the PVT/Epoxy interface can be expressed in the form of equation (9). The strength envelop is shown in Fig. 5.

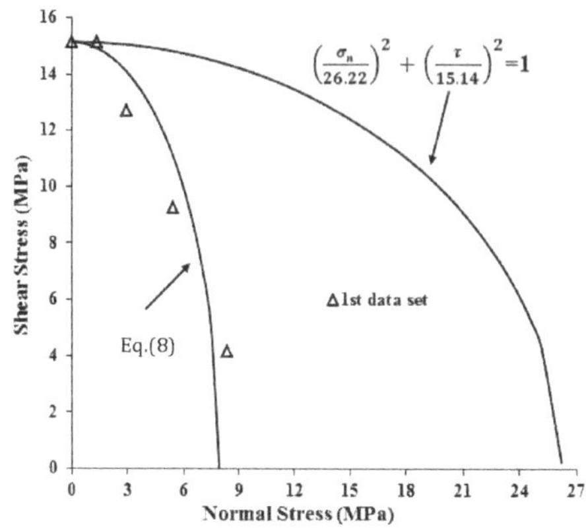

Fig. 4: The stress points obtained considering the von-Mises equivalent stress criterion and the von-Mises equivalent stress curve

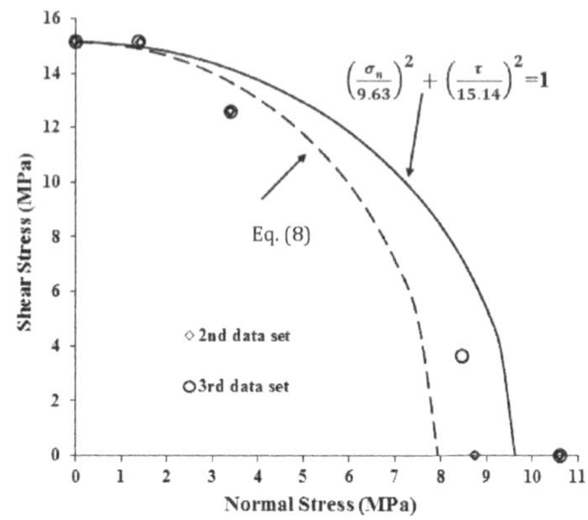

Fig. 5: The iterated second and third (final) criterion (bonding strength envelope)

5. CONCLUSIONS

In this paper, the analytical solution for a spherical viscoelastic/viscoelastic interface is first obtained by using elastic-viscoelastic analogy method. It is concluded that the singular stress state at the interface can be avoided if the bounding angle at the corner of the interface is less than a critical value. With such designed specimens without the stress singularity, the interface tension-shear bonding strength between two viscoelastic materials can be more accurately determined. In practical applications, the shape of bi-material interface should be designed to avoid singular stress state in order to obtain higher resistance to failure.

REFERENCES

[1] ASTM D3165 (2007). ASTM International, West Conshohocken, PA.
[2] ASTM D897 (2008). ASTM International, West Conshohocken, PA.
[3] Chowdhuri, M. A. K., Xia, Z. (2013) *Composites Part B: Engineering*, **44**, 253-259.
[4] Dundurs J. (1969) Mathematical theory of dislocations. ASME, NY.
[5] Christensen, R.M. (1982) Theory of Viscoelasticity - An Introduction, second edition, Academic Press, New York.

Ductile fracture simulation for toughness specimens from low alloy and stainless steel pipes

H.W. Ryu, J.J. Han, K.D. Bae, Y.J. Kim
Mechanical Engineering, Korea University, Korea

ABSTRACT

This paper describes ductile fracture simulation for Battelle fracture toughness and full-scale pipe tests (1). Materials in this research are low carbon steel and TP304 stainless steel. Tensile and Compact tension (CT) specimens are simulated to determine the fracture criteria with finite element (FE) method. Then, pipes containing through wall crack and surface crack are predicted by proposed damage method. The results from simulations are compared with test data for verification of proposed method and to check the applicability on industrial field. It is shown that predicted simulation results are compared with experimentally measured values, showing overall good agreements.

1 INTRODUCTION

Pipe test with same scale of piping line is an accurate and reliable way for assessment of structural integrity. However, it takes high cost and efforts to evaluate the integrity of pipes containing crack-like defects. Fracture toughness tests are used to substitute for full-size pipe test to predict ductile fracture behaviour. For example, CT test is extensively used to estimate the resistance of crack growth for the material. One more economical way to predict the characteristics of fracture is to use finite element (FE) damage analyses. In recent years, the FE method (2) based on stress-modified fracture strain model (3, 4) has been developed to simulate ductile fracture behaviour.

In this study, by using proposed damage analysis, the fracture criteria are determined for each low carbon steel (SA-333 Gr. 6) and TP304 stainless steel (SA-376). Then, two types of pipes with different crack shape are simulated with 3-D FE method. The results from ductile fracture simulation are compared with test data for verification of proposed method. It is shown that predicted simulation results are compared with experimentally measured values, showing overall good agreements.

2 DUCTILE FRACTURE SIMULATION

The damage analysis conducted in this research is based on the empirical ductile fracture model. It is one of stress-modified fracture strain model and published by Rice and Tracey (5). In this model, fracture strain ε_f related with the void enlargement is assumed to be depend exponentially on the triaxial stress state σ_m/σ_e:

$$\varepsilon_f = A\exp\left(-C\frac{\sigma_m}{\sigma_e}\right)+B \quad ; \quad \frac{\sigma_m}{\sigma_e}=\frac{\sigma_1+\sigma_2+\sigma_3}{3\sigma_e} \tag{1}$$

where σ_i (i=1-3) are principle stress components and A, B and C are material constants. According to the analytical research from Rice and Tracey, the material constant C is approximately -1.5. Damage ω can be calculated by summing incremental damage $\Delta\omega$, given by:

$$\Delta\omega = \frac{\Delta\varepsilon_e^p}{\varepsilon_f} \tag{2}$$

where $\Delta\varepsilon_e^p$ is the equivalent plastic strain increment, calculated from FE analysis in ABAQUS user subroutines (6). Fig. 1 shows a simple schematic illumination of stress reduction procedure on integration point of FE element. When the element in ductile fracture simulation is determined to be failed by fracture criteria, the load carrying capacity of element is forced to reduce to cut-off value. Decreasing slope and cut-off value are properly determined to avoid the instability of FE simulation.

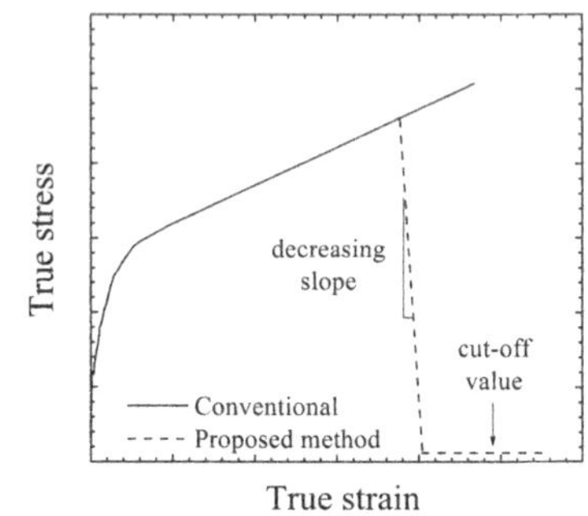

Fig. 1 Schematic description of stress reduction in ductile fracture simulation.

2.1 Fracture criteria

As described in Fig. 2, FE analyses are conducted for tensile test specimen at 288ºC. From the results of simulation, fracture strain and stress triaxiality are obtained at the local point where the initiation of failure is expected. The point of smooth bar in Fig. 4 is determined at low stress triaxiality. Then, the point of high stress triaxiality can be chosen by the results of simulation for CT specimen. A simple calibration was conducted to determine fracture strain ε_f and critical accumulated damage ω_c in fracture criteria. Fig. 3 describes the representative results of damage simulation on CT specimen with a simple calibration. The results show overall agreements with test data. Then, determined criteria are used to predict load - load line displacement (*P-LLD*) curves for circumferentially cracked pipes.

Three criteria are determined for low alloy steel (SA333 Gr. 6). Because this calibration is under the system of indeterminate, a number of criteria can be used for ductile fracture simulation and eq. (3)-(5) is representatively selected. Opt. 1 means the simplest way to use by fixing the constant B equal to zero. Only a critical damage ω_c need to be determined by CT simulation. Opt. 2 is more precise, but, complicated calibration is needed to get constants. Finally, calibrated fracture criteria by comparing with test data are depicted in Fig. 4.

$$\text{Opt. 1: } \varepsilon_f = 1.97182 \times \exp\left(-1.5 \times \frac{\sigma_m}{\sigma_e}\right) \tag{3}$$

$$\text{Opt. 2-1: } \varepsilon_f = 1.82417 \times \exp\left(-1.5 \times \frac{\sigma_m}{\sigma_e}\right) + 0.0671 \tag{4}$$

$$\text{Opt. 2-2: } \varepsilon_f = 1.70814 \times \exp\left(-1.5 \times \frac{\sigma_m}{\sigma_e}\right) + 0.11983 \tag{5}$$

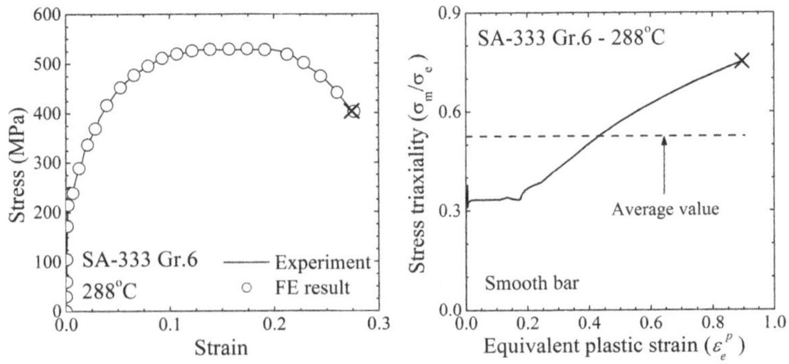

Fig. 2 Tensile test data and FE result to determine uniaxial ductility.

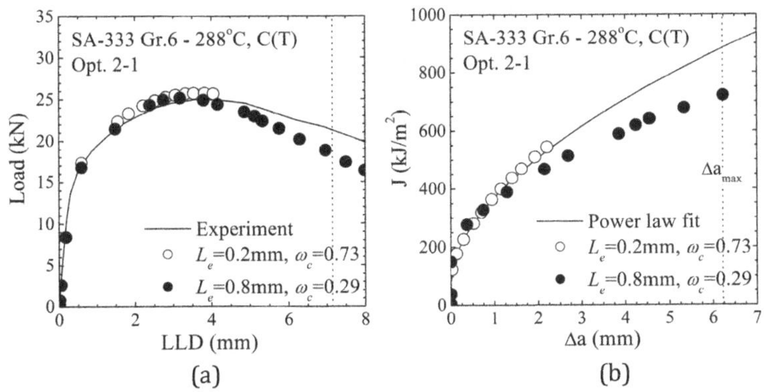

Fig. 3 Compact tension test data and FE results;
(a) load – load line displacement curve and (b) J resistance curve.

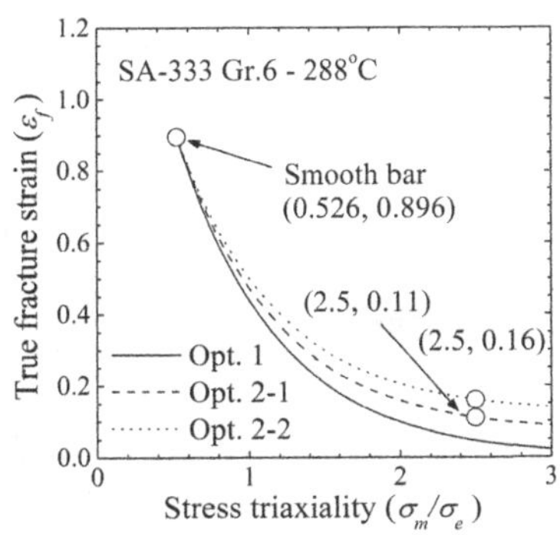

Fig. 4 Criteria for ductile fracture simulation; stress-modified fracture models for SA333 Gr. 6 steel at 288°C.

2.2 Change of Element size

A proper element size for the specific material is assumed in conventional damage analysis such as GTN model and cohesive zone model. The element size corresponds the case of critical accumulated damage ω_c equal to unity value. But, in proposed model, element size can vary with modifying the critical value of accumulated damage. Then, a big size of element size such as L_e=0.8 mm can be applicable to full-scale pipe simulation.

From the simulation of CT specimen, as shown in Fig. 3, a critical damage value that reproduce ductile fracture behavior can be determined for each criterion (Opt. 1 and Opt. 2). This element-size-dependent critical damage model provides a solution to overcome the problem due to numerical instability.

3 FE ANALYSIS AND RESULTS

A quarter model for tensile smooth bar and CT specimens was used considering symmetry conditions. Eight-node brick elements (C3D8) in ABAQUS were uniformly spaced in the cracked section. And, damage analysis was performed with the non-linear geometry change option. For CT specimen, *P-LLD* and *J* resistance curves are plotted in Fig. 3. Predicted curves with damage analysis of SA333 Gr. 6 are compared with experimentally measured values, showing overall good agreements. Finally, with combining simulation results of tensile and this CT specimens, the fracture criterion is determined on Opt. 1 and Opt. 2 in Fig. 4. The test data of circumferentially cracked pipes shown in Table 1 can be predicted by FE analysis with determined fracture criterion. The *P-LLD* and *Δa-LLD* curves of test data and FE result are depicted in Fig. 5 with good agreement. In Fig. 6, two types of crack are under the good prediction on both of crack initiation load and maximum load by using proposed damage method. The damage analysis for different material such as stainless steel (SA376 TP304) were also conducted to make fracture criteria and validate this method and show good agreement in general.

Table 1 Summary of dimensions for circumferential cracked SA333 Gr. 6 pipes at 288°C.

Specimen	D_o (mm)	t (mm)	r/t	a/t	θ/π
Through-wall crack (4131-7)	273.1	18.3	6.96	-	0.346
Surface crack (4115-1)	256.2	17.3	6.90	0.700	0.42
Surface crack (4115-2)	272.0	17.1	7.45	0.710	0.43
Surface crack (4131-8)	270.6	15.1	8.46	0.678	0.48

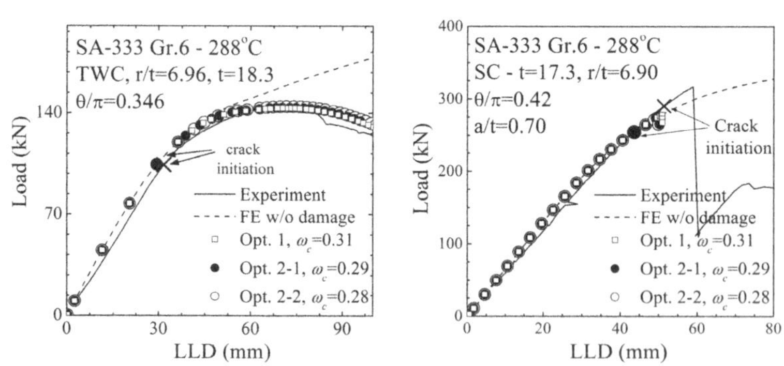

Fig. 5 Results of (a) through-wall cracked and (b) surface cracked pipe test data and FE results.

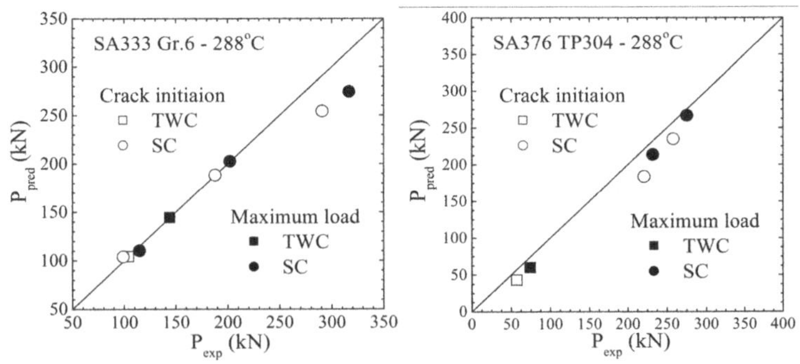

(a) (b)

Fig. 6 Comparison of test data with predictions for low alloy and stainless steel; (a) low alloy steel (SA333 Gr. 6) and (b) stainless steel (SA376 TP304).

4 CONCLUSION

FE damage analyses for two types of steels were conducted for fracture toughness specimen and circumferentially cracked pipes. The criteria for ductile fracture simulation can be determined by tensile and fracture toughness test data only. The proposed method can reliably predict different types of circumferentially cracked pipes such as through-wall crack and surface crack and different materials such as low alloy and stainless steel. This method offers significant advantages in estimates for ductile fracture behaviour.

ACKNOWLEDGEMENT

This research was financially supported by the Ministry of Trade, Industry & Energy (MOTIV), Korea Institute for Advancement of Technology (KIAT) and Dongnam Institute for Regional Program Evaluation through the Leading Industry Development for Economic Region (No. 20131520202170).

This research was supported by National Research Foundation of Korea(NRF) funded by the Ministry of Science, ICT and Future Planning. (NRF-2013M2B2B1075733)

REFERENCES

(1) Pipe Fracture Encyclopaedia. 1997. Volume 3: Pipe Fracture Test Data. Battelle.
(2) Oh, C.-S., Kim, N.-H., Kim, Y.-J., Baek, J.-H., Kim, Y.-P., and Kim, W.-S., 2011. A finite element ductile failure simulation method using stress-modified fracture strain model. Engineering Fracture Mechanics 78:124-137.
(3) McClintock, F. A., 1968. A Criterion for Ductile Fracture by the Growth of Holes. Journal of Applied Mechanics 35:363-371.
(4) Hancock, J. W. and Mackenzie A. C., 1976. On the mechanisms of ductile failure in high-strength steels subjected to multi-axial stress-states. Journal of the Mechanics and Physics of Solids 24:147-160.
(5) Rice, J. R. and Tracey, D. M., 1969. On the ductile enlargement of voids in triaxial stress fields. Journal of the Mechanics and Physics of Solids 17:201-217.
(6) ABAQUS Version 6.11, 2011. Analysis User's Manual. Dassault Systemes Simulia Corp., Providence, RI.

Cracks of classical modes in small-scale Cosserat continuum

A. Dyskin [1], E. Pasternak [2], M. Esin [2]
[1] School of Civil, Environmental and Mining Engineering, The University of Western Australia, Australia
[2] School of Mechanical and Chemical Engineering, The University of Western Australia, Australia

ABSTRACT

We propose a fracture growth mechanism based on mutual rotations of the particles leading to breakage of inter-particle bonds. This criterion can be expressed in terms of the moment stresses, which at the crack tip show a singularity of 3/2 power for Modes I, II and III – stronger than the conventional one. The moment stress singularity leads to relative particle rotations and bending of inter-particle bonds.

1 INTRODUCTION

In Linear Elastic Fracture Mechanics the criteria of crack growth are based on the stress intensity factors, which control the stress singularities at the crack tip. The stress singularities reflect the case of extremely small process zone when it size can be neglected. The absence of the characteristic sizes near the crack tip leads to the power law stress singularities [1]. In heterogeneous materials where the microstructural size l is not small and cannot be neglected a process zone is introduced as a model of microscopic fracture processes at the crack tip [2]. Generally the process zone concept presumes that information of its behaviour is known. Furthermore the loading curves are required in the form of stress vs. displacement discontinuity. Since the microscopic testing or modelling of the elements of the material in the process zone can only deliver stress-strain dependence, an additional length parameter (e.g. the width of the process zone) is needed to obtain stress-displacement relationship. To overcome this complication an intermediate model was proposed which puts forward the crack growth criterion based on the local material failure caused by the stress at a distance l from the crack tip [3]. This is illustrated at Figure 1a.

Approximation of the stress distribution at the crack tip by a power law asymptotics for instance such as $K_I/\sqrt{2\pi r}$ is strictly speaking possible only for $r>>l$, where r is the distance from the crack tip. It is essential that the asymptotic approximation is usually possible up to the limit $r\sim l$, which we use here following [4-6].

Typically the local failure is presumed to occur in brittle materials under tensile stress (Figure 1b) or, in ductile materials, under shear stress. In particulate materials such as rock and concrete, another type of local failure is conceivable – failure due to bending breakage of inter-particle (inter-grain) bonds (Figure 1c). The bending is caused by mutual rotation of the particles under stress non-uniformity. The simplest case of stress non-uniformity – linearly changing stress – can be modelled by introducing moment stress (μ in Figure 1c) reciprocal to the rotation gradient. This brings us to the Cosserat theory. In what follows we consider cracks in Cosserat continuum, but in light of the abovementioned concept consider bending failure caused by stresses at distance l from the crack tip.

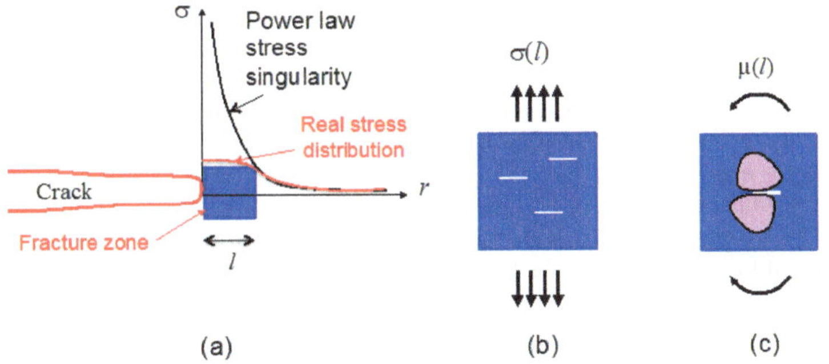

Figure 1. Crack growth due to local failure caused by stress at distance *l* from the crack tip: (a) replacement of stress distribution with the power law stress singularity; (b) local tensile failure; (c) local failure in particulate materials caused by mutual grain rotation.

2 COSSERAT MODEL OF PARTICULATE MATERIAL AND THE ASSOCIATED SCALES

Isotropic Cosserat continuum is defined by the following equilibrium and constitutive equations in co-ordinate frame (x_1, x_2, x_3), e.g. [7]:

$$\begin{aligned} &\sigma_{ji,j} = 0, \quad \mu_{ji,j} + \varepsilon_{ijk}\sigma_{jk} = 0, \quad i,j = 1,2,3 \\ &\sigma_{ji} = (\mu+\alpha)\gamma_{ji} + (\mu-\alpha)\gamma_{ij} + \lambda\gamma_{kk} \\ &\mu_{ji} = (\gamma+\varepsilon)\kappa_{ji} + (\gamma-\varepsilon)\kappa_{ij} + \beta\kappa_{kk} \end{aligned} \tag{1}$$

where σ_{ij} and μ_{ij} are stress and moment stress, ε_{ijk} is the alternating tensor. The deformation measures - the strain and curvature-twist tensors are

$$\gamma_{ji} = u_{i,j} - \varepsilon_{kji}\varphi_k, \quad \kappa_{ji} = \varphi_{i,j} \tag{2}$$

where u_i and φ_i are independent displacement and rotation vectors respectively, subscript $,j$ denotes differentiation with respect to x_j, and μ, α, γ, ε, λ, β are the Cosserat elastic moduli.

Substituting the constitutive equations into the equations of equilibrium and using the representation of deformation measures (2) one obtains the equilibrium equations in displacements and rotations in the following vector form:

$$\begin{aligned} &(\lambda+2\mu)\,\mathrm{grad\,div}\,\mathbf{u} - (\mu+\alpha)\,\mathrm{rot\,rot}\,\mathbf{u} + 2\alpha\,\mathrm{rot}\,\boldsymbol{\varphi} = 0 \\ &(\beta+2\gamma)\,\mathrm{grad\,div}\,\boldsymbol{\varphi} - (\gamma+\varepsilon)\,\mathrm{rot\,rot}\,\boldsymbol{\varphi} + 2\alpha\,\mathrm{rot}\,\mathbf{u} - 4\alpha\,\boldsymbol{\varphi} = 0 \end{aligned} \tag{3}$$

The presence of moduli γ, ε, β which have dimensions of Pa/m^2 leads to the appearance of characteristic lengths such as:

$$l^2 = \frac{(\mu+\alpha)(\gamma+\varepsilon)}{4\mu\alpha}, \quad l_2^2 = \frac{\gamma+\varepsilon}{4\alpha}, \quad \tilde{l}_2^2 = \frac{\beta+2\gamma}{4\alpha}, \quad l_1^2 = \sqrt{l^2 - l_2^2} \tag{4}$$

This distinguishes the Cosserat continuum from the classical one. Therefore cracks in the Cosserat continuum can have power law asymptotics at the crack tip only in two cases, for $r \ll l_{min}$ or $r \gg l_{max}$, where l_{min}, l_{max} are the smallest and largest Cosserat characteristic length.

In order to estimate the Cosserat lengths we model, following [8, 9], the isotropic particulate material as an assembly of spherical particles of diameter D connected to each other by elastic bonds, randomly distributed over the particle surface with the coordination number k. We assume that each bond is characterised by normal stiffness k_n, the shear (tangential) stiffnesses k_s, the same in both directions, a twist stiffness $k\varphi_n$ and two equal bending stiffnesses $k\varphi_s$. The stiffnesses are determined by modelling the links as elastic cylinders and developing the stress distributions over the cylinder cross-section into the Taylor series keeping only linear terms. Taking into account symmetry the classical stiffnesses, k_n, k_s, are found using uniform stress distributions while the twist $k\varphi_n$ and bending stiffnesses $k\varphi_s$ are found from linear stress distributions. Assuming the bond to be a cylinder of height h and radius b one obtains the following expressions for the stiffnesses:

$$k_n = \pi b^2 E_b/h, \quad k_s = \pi b^2 G_b/h, \quad k_{\varphi_s} = \pi b^4 E_b/4h, \quad k_{\varphi_n} = \pi b^4 G_b/2h \tag{5}$$

Here E_b and G_b are the Young's modulus and shear modulus of the material of the bond.

Using the homogenisation by differential expansions [8, 9], we obtain the following relationship between the Cosserat moduli and the bond stiffnesses:

$$\begin{aligned} \mu &= \frac{kV_s}{5\pi D}\left(k_n + \frac{3}{2}k_s\right), \quad \alpha = \frac{kV_s}{2\pi D}k_s, \quad \lambda = \frac{kV_s}{5\pi D}(k_n - k_s) \\ \gamma &= \frac{kV_s}{5\pi D}\left(k_{\varphi_n} + \frac{3}{2}k_{\varphi_s}\right), \quad \varepsilon = \frac{kV_s}{2\pi D}k_{\varphi_s}, \quad \beta = \frac{kV_s}{5\pi D}\left(k_{\varphi_n} - k_{\varphi_s}\right) \end{aligned} \tag{6}$$

Substituting Eq. (5) into Eq. (6) and the result into Eq. (4) leads to the following estimate

$$l^2 = \frac{1}{10}b^2\frac{(E_b + 4G_b)(2E_b + G_b)}{(2E_b + 3G_b)G_b} \sim b^2 \tag{7}$$

with other Cosserat lengths being of the same order. Given that the bond diameter cannot be greater than the particle we have $2b<D$. Therefore $l \sim D$.

The above analysis shows that the Cosserat lengths are of the order of the grain size. As the process zone size should be of the same order we deal with asymptotics of distances from the crack tip $r \gg l$. (This is opposite to existing studies of cracks in the Cosserat continuum, e.g. [10] where $r \ll l$ was presumed.) In other words, if r is fixed we deal with asymptotics $l \to 0$. This brings us to what we call the *small-scale Cosserat continuum* [6].

3 SMALL-SCALE COSSERAT CONTINUUM

The limiting transition of Cosserat characteristic lengths tending to zero corresponds to tending to zero the Cosserat moduli γ, ε, β. This limiting transition being applied to Eqs. (3) leads to

$$
\begin{aligned}
&(\lambda + 2\mu)\operatorname{grad}\operatorname{div}\mathbf{u} - \mu\operatorname{rot}\operatorname{rot}\mathbf{u} = 0 \\
&\operatorname{rot}\mathbf{u} - 2\boldsymbol{\varphi} = 0
\end{aligned}
\qquad (8)
$$

The first equation in (8) is the Lame equation of the classical elasticity, while the second equation gives the expression of the rotation field through the displacement field. These two equations give the main term of the asymptotics of small Cosserat lengths. In order to obtain this term one needs to solve the classical elastic problem, find the displacement field, determine using the second equation of (8) the rotation field and then using relations (2) and the constitutive equations from (1) determine the moment stress tensor μ_{ij}.

4 CRACKS OF CLASSICAL MODES. MOMENT STRESS SINGULARITY

Applying the described procedure to the crack problems one can find the distribution of the moment stresses at the tip of cracks of three classical modes (the conventional stress singularity remains the same).

Mode I crack:

$$
\mu_{23} = -\frac{2K_I l_2^2 \alpha}{\sqrt{2\pi} r^{3/2}} \frac{1-\nu}{\mu}, \quad \mu_{ij} = 0, \quad i \neq 2, \quad j \neq 3 \qquad (9)
$$

Mode II crack:

$$
\mu_{13} = \frac{2K_{II} l_2^2 \alpha}{\sqrt{2\pi} r^{3/2}} \frac{1-\nu}{\mu} \quad \mu_{ij} = 0, \quad i \neq 1, \quad j \neq 3 \qquad (10)
$$

Mode III crack

$$
\mu_{11} = -\frac{K_{III}\gamma}{2\sqrt{2\pi} r^{3/2}\mu} \quad \mu_{22} = -\mu_{11} \quad \mu_{ij} = 0, \quad i,j \neq 1, \quad i,j \neq 2 \qquad (11)
$$

Here ν is the Poisson's ratio.

It is seen that different crack models produce different non-vanishing components of the moment stress. Figure 2 shows the types of failure produced. Mode I produces failure that effectively continues the crack growth. Mode II produces microcracks that attempt to kink. Mode III produces multitude of microcracks that are inclined at 45^0 to the crack plane.

Another interesting feature is that the moment stress has singularity of order of 3/2 which is higher than the conventional stress singularity. Since the energy release rate is still the same as for a conventional crack the moment stresses do not contribute to the energy. Therefore the non-integrable stress singularities do not lead to infinite energy release rate.

5 CONCLUSIONS

We propose a fracture growth mechanism based on mutual rotations of the particles leading to breakage of inter-particle bonds. This criterion can be expressed in terms of the moment stresses, which at the crack tip show a power law singularity with exponent -3/2 for Modes I, II and III – stronger than the conventional one. This strong non-integrable singularity does not lead to infinite energy since in this approximation the moment stresses at the crack tip do not contribute to the energy release rate.

The moment stress singularity leads to relative particle rotations and bending of inter-particle bonds. This produces microcracks that either work to extend the crack (Mode I case) or produce experimentally observed kinks (Modes II) and multitudes of microcracks (Mode III).

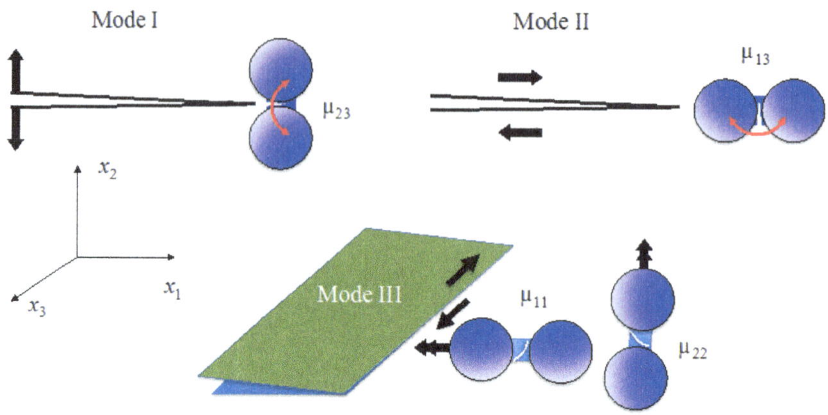

Figure 2. Schematics of bending (Modes I and II) and twisting (Mode III) local failures produced by cracks of conventional modes. The double arrow in Mode III picture signifies positive direction of rotation.

6 REFERENCES

[1] Caballero, A. & Dyskin, A. V. (2008) Mesh scalability in direct finite element simulation of brittle fracture, *Engineering Fracture Mech.* **75**, 4066-4084.

[2] Hillerborg, A. (1983) Analysis of one single crack. In: *Fracture Mechanics of Concrete*, F.H. Wittmann (ed.), Elsevier, Amsterdam, 223-249.

[3] Dyskin, A. V. (1997) Crack growth criteria incorporating non-singular stresses: Size effect in apparent fracture toughness. *Intern. J. Fracture*, **83**, 191-206.

[4] Dyskin, A. V. & Pasternak, E. (2008) Rotational mechanism of in-plane shear crack growth in rocks under compression. *Proc of the 1st Southern Hemisphere International Rock Mechanics Symposium* SHIRMS 2008 (Y. Potvin, J. Carter, A. Dyskin and R. Jeffrey eds.), Australian Centre for Geomechanics, **2**, 99-110.

[5] Pasternak, E. & Dyskin, A. V. (2009) Intermediate asymptotics for scaling of stresses at the tip of crack in Cosserat continuum. *Proc. 12th Intern. Conf. Fracture ICF12*, Ottawa, 12-17 July 2009. Paper T40.014, 10 pp.

[6] Dyskin, A.V. & Pasternak, E. (2013) Mechanism of in-plane fracture growth in particulate materials based on relative particle rotations. *Proc. 13th International Conference on Fracture*, June 16–21, 2013, Beijing, China. Paper S09-003.

[7] Nowacki W. Theory of Micropolar Elasticity. Springer. Wien; 1970.

[8] Pasternak, E., Muhlhaus, H-B. & Dyskin, A. V. (2004) On the possibility of elastic strain localisation in a fault. *Pure and Appl. Geophys.* **161**, 2309-2326.

[9] Pasternak E. & Muhlhaus, H-B. (2005) Generalised homogenisation procedures for granular materials, *Engineering Mathematics* **52**, 199-229.

[10] Sternberg, E. & Muki, R. (1967) The effect of couple stress on the stress concentration of a crack. *Int. J. Solids Struct.* **3**, 69-95.

Ductile fracture simulation for A106 Gr.B carbon steel under high strain rate loading condition

H.S. Nam [1], J.S. Kim [1], J.J. Han [1], J.W. Kim [2], Y.J. Kim [1]
[1] Mechanical Engineering, Korea University, Korea
[2] Nuclear Engineering, Chosun University, Korea

ABSTRACT

This paper provides simulation of ductile crack growth under high strain rate loading condition using a stress-modified fracture strain models. The stress-modified fracture strain model is determined to be incremental damage in terms of stress triaxiality (σ_m/σ_e) and fracture strain (ε_f) for dimple fracture from tensile test result with FE analyses technique. In order to validate stress-modified fracture strain model in dynamic loading conditions, this paper compares FE results with test results fitted by Johnson-Cook model. The calibrated damage model predicts CT test result under high strain rate. Simulated results agree well with experimental data.

1 INTRODUCTION

In order to avoid sudden dynamic accident in nuclear industries, it is essential to understand mechanical behavior under high strain rates. In many engineering problems, failures are subjected to dynamic loading that may occur with high strain rate, such as the earthquake loading. In the range of high strain rates, the mechanical behavior of materials is characterized by an increased in strain rate sensitivity. Due to its importance, a variety of institutions such as Battelle and CRIEPI performed tensile test, fracture toughness test (1-3) under dynamic loading condition.

Recently a simple finite element method in a quasi-static test has been proposed to implement fracture simulation based on the well-known stress modified fracture strain model (4,5). The stress- modified fracture strain model is determined to be incremental damage in terms of stress triaxiality (σ_m/σ_e) and fracture strain (ε_f) for dimple fracture from tensile test result with FE analyses technique. Since dynamic loading effect is especially important to assess pipe with crack-like defect, this work propose the integrated model which combines quasi-static with dynamic loading effect.

2 DYNAMIC MATERIAL PROPERTIES

In order to analyze the crack tip stress and strain field under dynamic loading condition, this paper consider A106 Gr. B test result in Pipe Fracture Encyclopedia (1). And these test results were characterized using the Johnson-Cook model (6).

2.1 Test result

The tensile properties for A106 Gr. B were examined over a wide range of strain rates in an attempt to characterize the strain rate effect. Tensile tests were conducted on flat, pin-loaded specimens having a width of 6.35mm, thickness of 3.18mm, and length of 25.4 in the gage section. The tensile axis of the specimen was parallel with the pipe axis. Specific details of a tensile specimen is depicted in Fig. 1 (a). The tensile tests were conducted in a servo-hydraulic machine at strain rates of 4 x 10^{-4}/s, 3.4/s and 11.6/s. Strain was measured using an optical extensometer which follows flags located at the opposite ends of the gage section. Fig. 1 (b) shows engineering stress-strain curves for three tensile specimens tested at 288°C at three different strain rate. At high strain rate, both the ultimate tensile strength and the fracture elongation were significantly lower than for quasi-static testing rates.

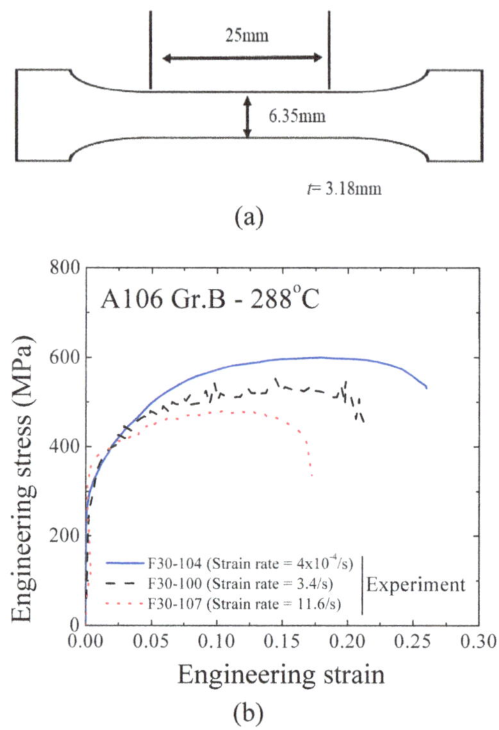

Fig. 1 (a) Specific details of a tensile specimen
(b) Engineering stress-stain curve for A106 Gr. B

For fracture toughness test, four tests were performed, as summarized in Pipe Fracture Encyclopedia. In this paper, one test was considered to analyze high strain rate effect. Fracture toughness test was performed at 288°C on CT specimens machined from full scale A106 Gr. B pipe. CT specimen having a full-thickness of 0.5T. The specimens were oriented such that crack growth was in the circumferential direction (L-C orientation). In dynamic fracture toughness test, the displacement rate was selected to cause initiation in approximately 0.2 seconds.

2.2 Johnson-Cook model

The Johnson-Cook model (6) is a function of von mises tensile flow stress, in accordance with strain hardening, strain rate hardening, and thermal softening. Eq (1) represents a Johnson-Cook model,

$$\sigma_{eq} = (A + B(\varepsilon_{eq}^{pl})^n)\left(1 + C\ln(\frac{\dot{\varepsilon}}{\dot{\varepsilon}_o})\right)\left(1 - \left(\frac{T - T_o}{T_{melt} - T_o}\right)^m\right) \quad (1)$$

where ε is the equivalent plastic strain, $\dot{\varepsilon}$ is the plastic strain rate and $\dot{\varepsilon}_o$ is a reference strain rate. The parameter A is the initial yield strength of the material at quasi-static strain rate. The parameter B and n represents the flow stress on strain hardening behavior at quasi-static strain rate. And the parameter C represents strain rate effect, and m represents thermal softening effect. T_{melt}, T_o each represents melting temperature and references temperature. Because this paper analyzes experimental results under constant temperature condition, coefficients such as m, T_{melt}, T_o were not considered. In this paper, three different strain rate tensile tests are fitted using Johnson-Cook model. F30-104 test was performed under quasi-static condition. Based on the test results, the coefficient values can be determined (A, B, n). Besides that, this paper also includes coefficients (C, $\dot{\varepsilon}_o$) computed by using tensile strength of three different strain rates. Fitting equation shows in Eq (2).

$$\sigma_{eq} = (294 + 818.72(\varepsilon_{eq}^{pl})^n)\left(1 - 0.0644\ln(\frac{\dot{\varepsilon}}{0.68})\right) \quad (2)$$

3 SIMULATION OF DUCTILE FRACTURE UNDER HIGH STRAIN RATE

The damage model used in this paper is based on the phenomenological ductile fracture model which is also known as the stress-modified fracture strain model. Through analytical approximate study, Johnson and Cook suggested following equation. In this model, fracture strain ε_f for dimple fracture is assumed to depend exponentially on the stress triaxiality σ_m/σ_e, strain rate:

$$\varepsilon_f = \left[A\exp\left(-C\frac{\sigma_m}{\sigma_e}\right) + B\right]\cdot\left[1 + D\ln\left(\frac{\dot{\varepsilon}}{\dot{\varepsilon}_o}\right)\right] \quad (3)$$

where A, B, C and D are material constants. Damage ω can be calculated by summing incremental damage $\Delta\omega$, given by

$$\Delta\omega = \frac{\Delta\varepsilon_e^p}{\varepsilon_f} \quad (4)$$

Where $\Delta\varepsilon_{ep}$ is the equivalent plastic strain increment, calculated from FE analysis implemented in ABAQUS using user subroutines. To calibrate the stress-modified fracture strain model under dynamic loading condition, calibrations were made by comparing FE results with experimental results of tensile and the fracture toughness tests. Conventional 3-D elastic-plastic FE analysis was performed to simulate flat plate tensile test according to test speed. The analyses were conducted by using ABAQUS Standard 6.13 (7).

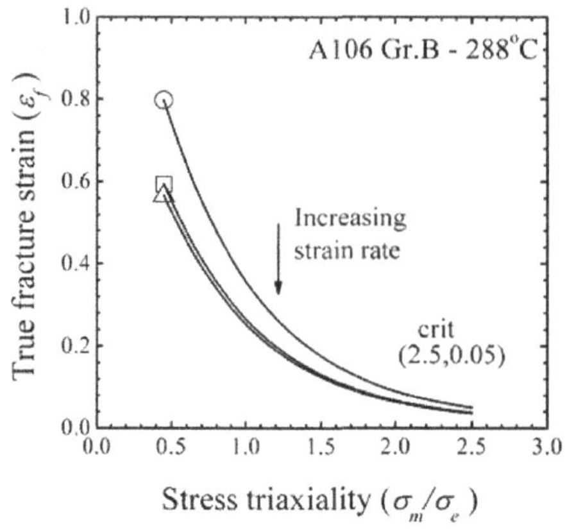

Fig. 2 Criteria for ductile fracture simulation

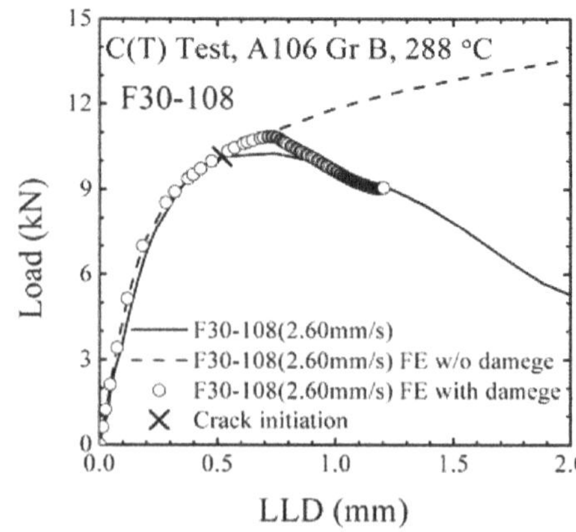

Fig. 3 Comparison of experimental data with simulated results

To analyze the failure behavior under dynamic loading condition, dynamic option were selected. To determine the stress-modified fracture strain, more data points are needed, for instance, using notched bar tensile tests with different notch radii. When notched bar tensile tests are not available (as is the case in the present problem), we need to simplify the form of Eq. (3). Through analytical approximate study, Rice and Tracey (8,9) suggested that the constant *C* in Eq. (3) is approximately -1.5. And By the theoretical value of the Prandtl fiel, a value at high stress triaxciality of σ_m/σ_e =2.5 is assumed. Determined fracture strain model is shown in Fig. 2 based on tensile test & FE data. The above failure simulation technique is implemented in the commercial FE program, ABAQUS using the UHARD user subroutines. Simulated CT specimen results are compared with experimental data in Fig. 3. For Load-LLD data, conventional elastic-plastic FE results without damage analysis are also compared. Simulated FE results with damage analysis agree with experimental data.

4 CONCLUSIONS

In this paper, a numerical method to simulate failure behavior of A106 Gr. B carbon steel under high strain rate loading condition fracture toughness tests is proposed to predict ductile crack growth under high strain rate condition. Simulated results agree well with experimental data.

ACKNOWLEDGMENTS

This research was supported by Basic Science Research Program through the National Research Foundation of Korea (NRF) funded by the Ministry of Science, ICT & Future Planning (MSIP) (No. 2007-0056094, 2013M2A8A1040924, 2013M2A7A1076396).

REFERENCES

[1] Pipe Fracture Encyclopedia 1997. Volume 3 : Pipe Fracture Test data. Battelle

[2] U. Andrade, M. A Meyers, K. S. Vecchio and A. H. Chokshi, "Dynamic recrystallization in high-strain, high strain-rate plastic deformation of copper", Acta metal mater, Vol. 42, pp 3183-3195, 1994

[3] F. Rivalin, A. Pineau, M. Di Fant, J. Besson, "Ductile tearing of pipeline-steel wide plates I. Dynamic and quasi-static experiments", Engineering Fracture Mechanics., Vol. 68, pp 329-345, 2001

[4] Oh, C.S, Kim, N.H, Kim, Y.J, Baek, J.H., Kim, Y.P., Kim, W.S., "A finite element ductile failure simulation method using stress-modified fracture strain model." Engineering Fracture Mechanics., Vol. 78, pp. 124-137, 2012

[5] Kim, N. H., Oh, C.S., Kim, Y.J., Yoon, K. B., Ma, Y.H., "Comparison of fracture strain based ductile failure simulation with experimental results." International Journal of Pressure Vessels and Piping, Vol. 88, pp. 434-447, 2011

[6] G R. Johnson and W H. Cook, "Fracture characteristics of three metals subjected to various strains, strain rates, temperature and pressures", Engineering Fracture Mechanics., Vol. 21, pp 31-48, 1985

[7] ABAQUS version 6.13. User's manual, Dassault Systems, 2013

[8] Hancock, J. W., Cowling, M. J., "Role of state of stress in crack-tip failure processes." Metal Science, Vol. 14, pp. 293-304, 1980

[9] Rice, J. R., Tracey, D. M., "On the ductile enlargement of voids in triaxial stress fields." Journal of Mechanics and Physics of Solids, Vol. 17, pp. 201-217, 1969

Effect of plasticity hardening exponent and creep exponent on crack tip stress under elastic-plastic-creep conditions for SE(B) specimen

H.S. Lee [1], J.H. Je [1], Y.J. Kim [1], R.A. Ainsworth [2], P.J. Budden [3]
[1] Mechanical Engineering, Korea University, Korea
[2] The University of Manchester, UK
[3] Assessment Technology Group, EDF Energy, UK

ABSTRACT

This paper investigates an effect of plasticity hardening exponent (m) and creep exponent (n) on crack tip stress under elastic-plastic-creep conditions for Single-Edge-Notched (SE(B)) specimen, via two-dimensional (2-D) finite element (FE) analyses. For analysis, finite element results compare with Riedel and Rice equation. In case of creep exponent (n) equal or bigger than plasticity hardening exponent (m), crack tip stress field under elastic-plastic-creep conditions is predicted by Riedel and Rice equation using creep exponent since crack tip stress field is affected by creep exponent than plasticity hardening exponent. On the other hand, in case of n less than m, crack tip stress field under elastic-plastic-creep conditions is predicted by Riedel and Rice equation using m since crack tip stress field is affected by m than n.

1 INTRODUCTION

To assessment cracked components at elevated temperatures, creep crack growth needs to be analysed. For materials under steady state creep conditions, creep crack growth can be characterized by C*-integral (1-2). Under small scale transient creep conditions, creep crack growth can be characterized by the transient creep C(t)-integral. C(t) parameter describes the near crack tip stress fields (3-5). Riedel and Rice proposed equation that describes crack tip stress fields using C* and C(t) parameters (6). Until recently, a number of works have been reported in the literature up to present on crack tip stress field under elastic-creep conditions. However, a little investigation on crack tip stress field under elastic-plastic-creep conditions are performed. This paper investigate crack tip stress field under elastic-plastic-creep conditions to assessment cracked components at elevated temperatures. Furthermore, this paper investigates effect of plasticity hardening exponent and creep exponent on crack tip stress field.

2 FINITE ANALYSIS METHOD

2.1 Geometry and FE Model

Finite element analysis was simulated by using ABAQUS v6.13 which is commercial software. Fig. 1 (a) shows finite element model of single edge notched specimen. 1/4 model corresponding to symmetry condition is used and CPE8R (plane strain) element is used.

Schematic illustration of SE(B) specimen and loading condition are shown Fig. 1 (b). W which is the width of specimen was assumed as 50mm. a which is the crack length of specimen was assumed as 25mm. *M* is pure bending moment applied to specimen.

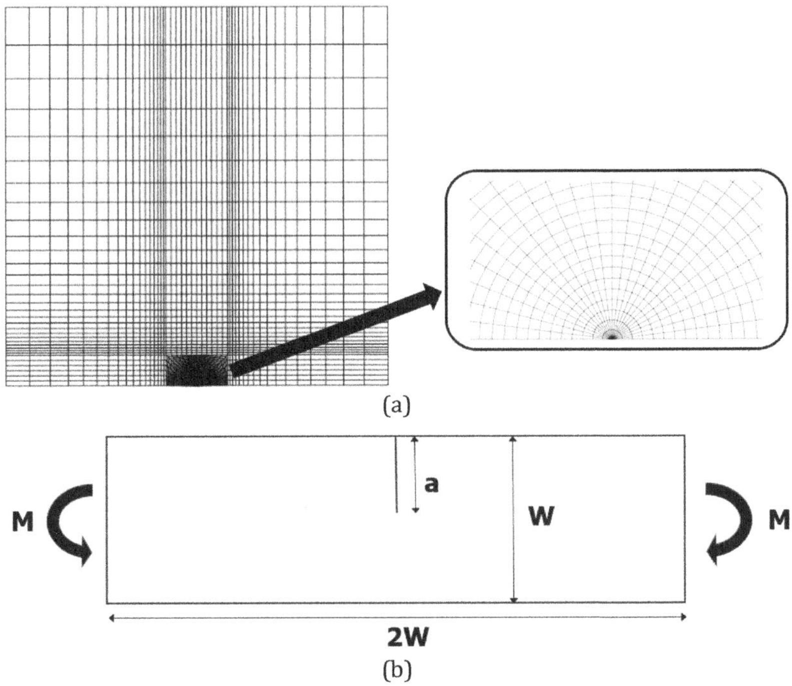

Figure 1. Mesh, geometry of SE(B) specimen

2.2 Material properties

Material properties used for analysis were assumed to follow Ramberg-Osgood relation which is power-law function form. Ramber-Osgood relation is described in Eq. (1).

$$\varepsilon = \varepsilon_e + \varepsilon_p + \varepsilon_c$$
$$\varepsilon_e = \frac{\sigma}{E} \qquad (1)$$
$$\varepsilon_p = \alpha \left(\frac{\sigma_O}{E} \right) \left(\frac{\sigma}{\sigma_O} \right)^m$$

In here, ε_e, ε_p and ε_c are elastic strain, plastic strain and creep strain each. *m* and α is material constant. Elastic modulus, *E* and yield stress, σ_O was 200 GPa and 300 MPa each. Plasticity hardening exponent, *m* was assumed for 5, 10. $0.002E/\sigma_O$ was used for α.

The creep deformation is assumed to follow a simple power law which is described below in Eq. (2).

$$\dot{\varepsilon}_c = B\sigma^n \qquad (2)$$

Creep exponent, *n* was assumed for 5, 10. To investigate the effect of *n* and *m*, the value of *n* and *m* may be systematically varied.

Table 1. Material properties

E	σ_0	α	ν	m	n	B
200 GPa	300 MPa	4/3	0.3	5,10	5	3.2E-15
					10	3.2E-25

3 RESULTS

For investigation effect of plasticity hardening exponent and creep exponent on crack tip stress field, finite element simulated results are compare with equation which proposed by Riedel and Rice (6).

$$\frac{\sigma_{ij}}{\sigma_{ref}} = \left[\frac{C(t)}{I_n \sigma_{ref} \dot{\varepsilon}_{ref} r}\right]^{\frac{1}{n+1}} \tilde{\sigma}_{ij}(n,\theta)$$
$$\dot{\varepsilon}_{ref} = B\sigma_{ref}^n$$
$$\sigma_{ref} = \frac{M}{M_L}\sigma_O \qquad (3)$$

M_L is limit moment of SE(B) specimen.

Fig. 2 - Fig. 5 show finite element results and Riedel and Rice equation about crack tip stress field.

As expected, in case of m=n and m<n, FE results agree well with RR equation which relate to creep exponent n, as shown in Fig. 2, Fig. 3 and Fig. 4. Also, in case of n>m, FE results agree well with RR equation which relate to creep exponent. These results imply that crack tip stress field is affected by creep exponent than plasticity hardening exponent.

On the other hand, in case of m>n, FE results agree well with RR equation which relate to plasticity hardening exponent m, as shown in Fig. 5. These results imply that crack tip stress field is affected by plasticity hardening exponent than creep exponent.

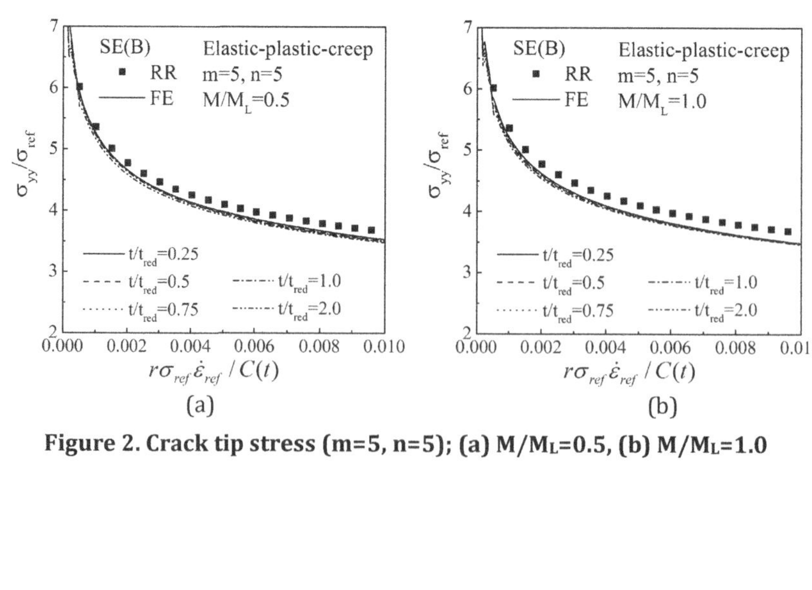

Figure 2. Crack tip stress (m=5, n=5); (a) M/M_L=0.5, (b) M/M_L=1.0

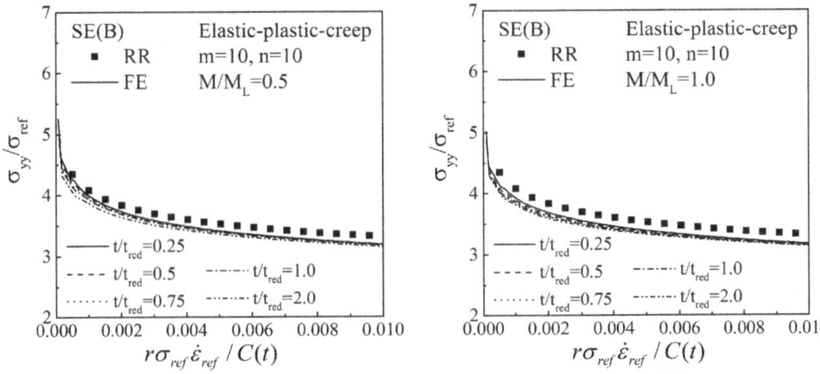

(a) (b)

Figure 3. Crack tip stress (m=10, n=10); (a) M/ML=0.5, (b) M/ML=1.0

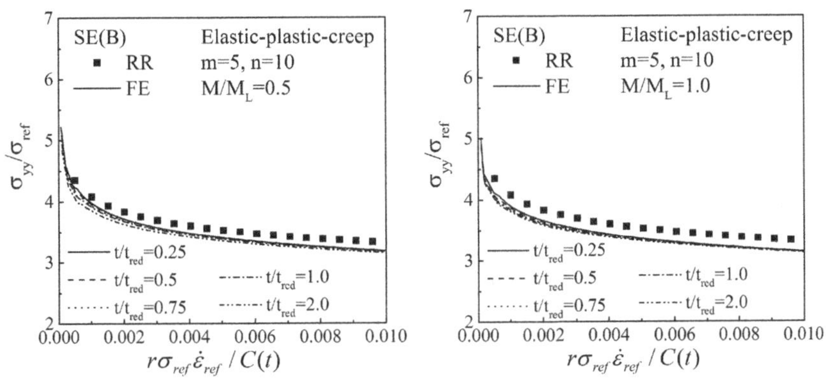

(a) (b)

Figure 4. Crack tip stress (m=5, n=10); (a) M/ML=0.5, (b) M/ML=1.0

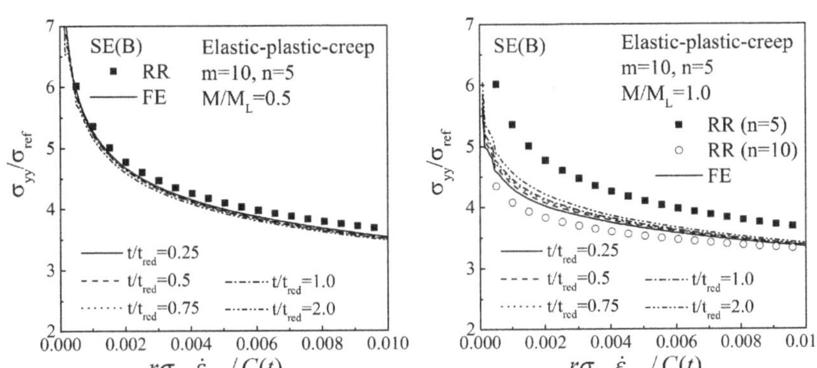

(a) (b)

Figure 5. Crack tip stress (m=10, n=5); (a) M/ML=0.5, (b) M/ML=1.0

4 CONCLUSIONS

In this paper, effect of plasticity hardening exponent and creep exponent on crack tip stress under elastic-plastic-creep conditions is investigated using FE analysis. Finite element results compared with Riedel and Rice equation. For $m=n$, $m<n$ case, FE results agree well with RR equation which relate to n. For $m>n$ case, FE results agree well with RR equation which relate to m. Also, this study confirms that crack tip stress did not nearly change with increasing time in transient creep.

ACKNOWLEDGMENTS

This research was supported by National Research Foundation of Korea (NRF) funded by the Ministry of Science, ICT and Future Planning. (NRF-2013M2B2A9A03051295, NRF-2013M2B2B1075733)

REFERENCES

(1) Miller, A. G. (1998) *Int. J. Pressure Vessels Piping.* **32,** 191–327. doi: 10.1016/0308-0161(88)90073-7
(2) Evans, R. W. & Wilshire, B. (1985) *Institute of Metals.* Creep of metals and alloys.
(3) Kim, Y. J. (2001) *Int. J. Pressure Vessels Piping.* **78**, 661–666. doi: 10.1016/S0308-0161(01)00080-1
(4) Tada, H., Paris, P. & Irwin, G. (1985) *Paris Production Inc.* The stress analysis of cracks handbook.
(5) Miller, A. G. & Ainsworth, R. A. (1989) *J. Engng. Fract. Mech.* **32**, 233–247. doi: 10.1016/0013-7944(89)90296-8
(6) Ridel, H. & Rice, J. R. (1980) *American Society for Testing and Materials.* 112-130. ASTM STP 700.

A new approach for evaluating stress intensity factor based on thermoelastic stress analysis

Y. Izumi [1], T. Sakagami [2], K. Yasumura [2], D. Shiozawa [2]
[1] Department of Mechanical Systems Engineering, The University of Shiga Prefecture, Japan
[2] Department of Mechanical Engineering, Kobe University, Japan

ABSTRACT

Thermoelastic stress analysis (TSA) using by infrared thermography has been widely used as an effective full-field stress measurement technique. The present authors applied TSA technique to nondestructive evaluation of fatigue cracks in steel bridges, in which fatigue cracks were detected based on singular stress fields observed around crack tips and structural integrity was evaluated based on stress intensity factor calculated from observed near tip stress field. In this paper, a new approach for evaluating stress intensity factor based on TSA technique is proposed. The coefficients including stress intensity factor in the near tip stress field function expressed in higher order terms are determined by the least square fitting using experimentally obtained data by TSA. It was found that values of stress intensity factors K_I and K_{II} were obtained in good accuracy.

1 INTRODUCTION

Recently, fatigue crack propagation in aged structures has become a serious problem that can lead to their catastrophic failure. For large-scale steel structures such as highway bridges, nondestructive evaluation of fatigue damage is necessary to ensure safety and to estimate the remaining life of these structures. As conventional nondestructive testing (NDT) techniques for steel bridges, visual testing, magnetic particle testing and ultrasonic testing have been commonly employed. However, these techniques are time- and labor- consuming techniques, furthermore they can only be employed for crack detection, and can not be used to directly evaluate the remaining strength based on fracture mechanics. Thermoelastic stress analysis (TSA) by infrared thermography has been widely used as an effective full-field experimental stress measurement technique. The present authors applied TSA technique to nondestructive evaluation of fatigue cracks in steel bridges, in which fatigue cracks were detected based on singular stress fields observed around crack tips and structural integrity was evaluated based on stress intensity factor calculated from observed near tip stress field [1,2]. In this paper, a new approach for evaluating stress intensity factor based on TSA technique is proposed. The coefficients including stress intensity factor in the near tip stress field function expressed in higher order terms are determined by the least square fitting using experimentally obtained data by TSA. The proposed technique is employed for determining stress intensity factors K_I and K_{II} for mixed-mode crack from stress distributions measured by TSA technique.

2 THEMOELASTIC STRESS ANALYSIS

Dynamic stress change causes very small temperature change under the adiabatic condition in solid. This phenomenon is known as the thermoelastic effect and is described by Lord Kelvin's equation relating the temperature change (ΔT) to a change in sum of principal stresses ($\Delta\sigma$) under cyclic variable loading as follows.

$$\Delta T = -\frac{\alpha}{\rho C_p} T \Delta\sigma \tag{1}$$

α : Coefficient of thermal expansion
ρ : Mass density
C_p : Specific heat at constant pressure
T : Absolute temperature

A change in the sum of the principal stresses ($\Delta\sigma$) is obtained by measuring the temperature change (ΔT) using infrared thermography.

3 STRESS INTENSITY FACTOR EVALUATION BASED ON THERMOELASTIC ANALYSIS

Stress distribution around crack tip under mixed mode loading is given by Airy's stress function as follows.

$$\begin{aligned}\Phi(r,\theta) = &\sum_{n=1,3\ldots} r^{1+n/2}\left(C_{1n}\left(\cos\frac{n-2}{2}\theta - \frac{n-2}{n+2}\cos\frac{n+2}{2}\theta\right) + C_{2n}\left(\sin\frac{n-2}{2}\theta - \sin\frac{n+2}{2}\theta\right)\right)\\ &+ \sum_{n=2,4\ldots} r^{1+n/2}\left(C_{1n}\left(\cos\frac{n-2}{2}\theta - \cos\frac{n+2}{2}\theta\right) + C_{2n}\left(\sin\frac{n-2}{2}\theta - \frac{n-2}{n+2}\sin\frac{n+2}{2}\theta\right)\right)\end{aligned} \tag{2}$$

where r is a distance from the crack tip and θ is an angle from crack existing line. Taking account of

$$K_{\mathrm{I}} = C_{11}\sqrt{2\pi}, \quad K_{\mathrm{II}} = C_{21}\sqrt{2\pi} \tag{3}$$

equation of the sum of principal stresses around crack tip under mixed-mode loading is derived from Eq.(2) for the 1st to 4th order terms as follows.

$$\begin{aligned}\sigma_r + \sigma_\theta = &\left(\sqrt{\frac{2}{\pi r}}\cos\frac{1}{2}\theta\right)K_{\mathrm{I}} + C_{12} + \left(\sqrt{r}\cos\frac{1}{2}\theta\right)C_{13} + (r\cos\theta)C_{14}\\ &-\left(\sqrt{\frac{2}{\pi r}}\sin\frac{1}{2}\theta\right)K_{\mathrm{II}} + \left(\sqrt{r}\sin\frac{1}{2}\theta\right)C_{23} + (r\sin\theta)C_{24}\end{aligned} \tag{4}$$

Coefficients K_{I}, C_{12}, C_{13}, C_{14}, K_{II}, C_{23} and C_{24} in Eq.(4) can be calculated by the least square fitting using numerous data of the sum of principal stresses obtained by TSA technique, thus stress intensity factors K_{I} and K_{II} can be determined.

4 NUMERICAL SIMULATION OF STRESS INTENSITY FACTOR EVALUATION

Numerical simulations of stress intensity factor evaluation were carried out using boundary element method (BEM) analysis model as shown in Fig. 1. Stress distribution of the steel plate with an inclined crack was obtained by BEM. Inclined angles of the cracks φ were set to be 31°, 45°, 59° and 75°. The values of stress intensity factors K_{I} and

K_{II} were determined by the proposed least square fitting method using BEM analysis data on stress distribution in the region where r=5mm~15mm and θ=-120°~120° shown in Fig. 2.

Considering the experimental errors, stress intensity factor determination was conducted when stress values contained random errors in the range of 20MPa, as well as the calculation without errors. Obtained results are shown in Table 1. It is found that stress intensity factors K_I and K_{II} values were obtained in good accuracy compared with handbook values [3].

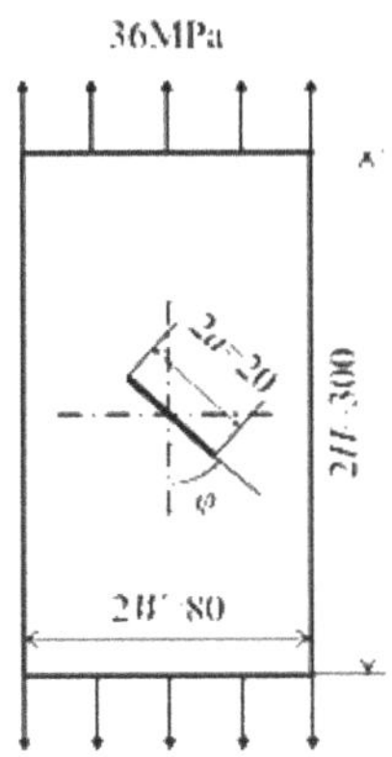

Fig. 1 BEM analysis model (mm).

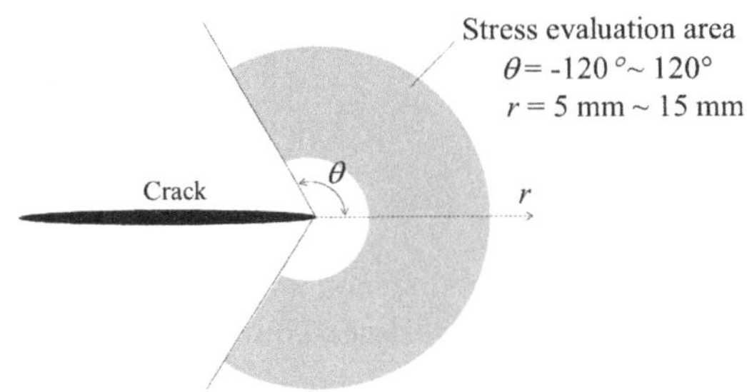

Fig. 2 Stress evaluation area.

Table 1 Stress intensity factor K_I and K_{II} determined from BEM results. (K_I, K_{II} : in MPa $m^{1/2}$).

Angle φ [°]	BEM 20MPa error		BEM No error		Handbook value	
	K_I	K_{II}	K_I	K_{II}	K_I	K_{II}
31	1.74	3.25	1.77	2.95	1.70	2.84
45	2.99	3.07	3.17	3.27	3.36	3.25
59	4.64	2.89	4.66	2.85	5.01	2.79
75	5.94	1.53	6.01	1.62	6.19	1.60

5 STRESS INTENSITY FACTOR EVALUATION BASED ON TSA MEASURED RESULTS

Experimental studies were carried out for steel plate with an inclined slit introduced by electro discharge machining. Configurations of the specimen and measurement area of TSA are shown in Fig. 3, as well as the loading conditions. Stress distributions measured by TSA system are shown in Fig. 4. In an IR-measurement, SC7500 (FLIR company, detector: InSb, NETD: 20mK) was employed and flat-black paint was employed to avoid reflection and to increase the emissivity. Stress intensity factors K_I and K_{II} were calculated from stress data in the region where r=5mm~15mm and θ=-120°~120°. Obtained results are shown in Table 2. It is found that values of stress intensity factors K_I and K_{II} were obtained in good accuracy compared with handbook values [3] demonstrating the feasibility of the proposed technique.

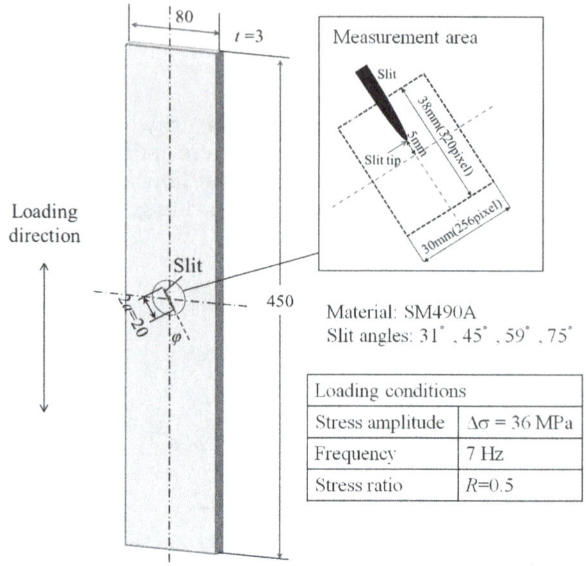

Fig. 3 Specimen configurations and test conditions.

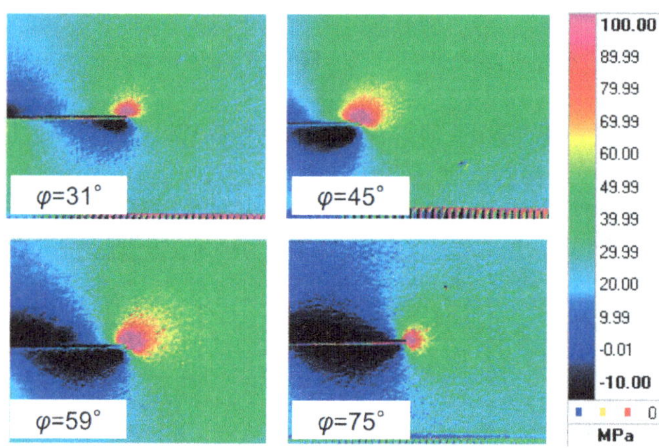

Fig. 4 Stress distributions measured by TSA system.

Table 2 Stress intensity factors K_I and K_{II} obtained by TSA measurement.

Angle φ 【°】	Test 1		Test 2		Test 3		Handbook value	
	K_I	K_{II}	K_I	K_{II}	K_I	K_{II}	K_I	K_{II}
31	2.41	3.25	1.70	2.82	2.42	3.12	1.70	2.84
45	2.98	3.20	2.98	2.88	3.68	3.32	3.36	3.25
59	4.22	2.87	5.47	3.35	5.17	3.26	5.01	2.79
75	4.77	1.40	5.06	1.20	5.70	1.48	6.19	1.60

6 CONCLUSIONS

In this paper, a new TSA based approach for evaluating stress intensity factor by the least square fitting of the coefficients in higher order near tip stress distribution function was proposed. It was found that values of stress intensity factors K_I and K_{II} were obtained in good accuracy demonstrating the feasibility of the proposed technique.

REFERENCES

[1] Sakagami, T., Izumi, Y., Kubo, S., Application of infrared thermography to structural integrity evaluation of steel bridges, Journal of Modern Optics, Volume 57, Issue 18, pp. 1738-1746, 2010.

[2] Sakagami, T., Izumi, Y., Kubo, S., Successful Application of Thermoelasticity to Remote Inspection of Fatigue Cracks, Tom Proulx (Ed.), Thermomechanics and Infra-Red Imaging, Volume 7, Proceedings of the 2011 Annual Conference on Experimental and Applied Mechanics, pp.99-106, 2011, Springer.

[3] Murakami, Y. (Ed. in chief), Stress Intensity Factors Handbook, Vol.1 and 2, 1987, The Society of Materials Science, Japan.

Environmental stress cracking behavior of high crystalline polypropylene in different surface active agents using modified notched constant load test

J.W. Wee, Y.J. Zhao, B.H. Choi
Department of Mechanical Engineering, Korea University, Korea

ABSTRACT

Environmental stress cracking (ESC) of polymeric materials is induced by the penetration and diffusion of surface agents in the stressed polymers, which lubricates the mobility of polymer chains. In this paper, ESCR tests of high crystalline polypropylene (HCPP) were conducted by use of modified notched constant load test in contact with two different scent oils. The relationship between the lifetime and the loading condition were observed for two types scent oil. Swelling tests were also conducted to compare the diffusivity of two scent oils. The higher diffusivity of limonene oil results in shorter lifetime than in case of lavender oil.

1 INTRODUCTION

Environmental stress cracking (ESC) phenomena is one of the common causes of long-term failure of polymeric materials (1, 2). ESC failure mechanism of polymer is different from stress corrosion cracking (SCC), in terms of no existence of the chemical degradation of polymers by contacting with adverse environments. ESC is induced by sorption and diffusion of the particular liquid or gas state non-reactive agents into the mechanically-loaded polymer medium, particularly into the amorphous region. Generally this phenomena makes molecular chains in amorphous region become soften and plasticized. In case of high density polyethylene (HDPE), diffused detergents play a role of a kind of lubricant to the tie chains in amorphous regions so that inter-lamella cohesive strength deteriorates due to the loosening of tie molecules (3). It may lead the unexpected failure of polymers in lower loading states than their expected strength. The examples of agents inducing ESC include detergents, soaps, oils, and so on. Since these environments are very common in considering the wide applications of polymers, ESC has attracted many attentions in the polymer industry, and intensive studies have been conducted by many researchers.

However, there remains still no full understanding on the ESC phenomena because of the complicated relations between involved parameters such as agent diffusivity, viscoelastic behaviour, void formation and growth, and so on (4). Moreover, in spite of the importance of ESC as a long term property of polymers, a lot of studies have been concentrated on PE and limited number of glassy polymers. Considering the use of variety of polymers in today, in-depth researches on the nature of ESC of the diverse polymers are necessary.

In this study, ESCR of high crystalline polypropylene (HCPP) in two different scent oils was measured by modified ASTM D 5397 methodology (5). Different diffusivity in each oil leads to the discrepancy in lifetime in same loading condition. To compare the swelling effect of oils of diffusion, volume changes of HCPP in each oil were also measured.

2 MATERIALS AND TEST METHODS

2.1 Materials

ESCR tests were performed with injection moulded HCPP samples following ASTM D 1822 type-L (6). Samples were prepared in thickness of 1.9 mm and initial notch was introduced for accelerated test condition and containment of almost same imperfection with size of roughly 0.2 mm. Two different scent oils were used as non-reactive surface diffusive agents, i.e. limonene and lavender oils. For the swelling test, block shape HCPP with initial volume of roughly 40 mm^3 was prepared.

2.2 Test Methods

Each sample was subjected to constant stress to failure in two different oils at elevated temperature of 50°C. The range of 7~14 MPa was chosen for this constant stresses. Until the final failure, the variation of displacement of each loaded samples was recorded to obtain the creep strain - time plots. From these plots, crack initiation time (t_i) can be obtained as well as the time to failure (t_f) as shown in Fig. 1. To compensate the slight variation of the initial notch size, initial stress intensity factor (SIF) was calculated and it was used to represent loading condition. Consequently, initial SIF - t_i and t_f plots were obtained as plotted in Fig. 2. Post-fracture analysis was also conducted by scanning electron microscope (SEM). To understand the diffusion behaviour of the two oils into HCPP medium, swelling test was performed for 200 hours at 50°C.

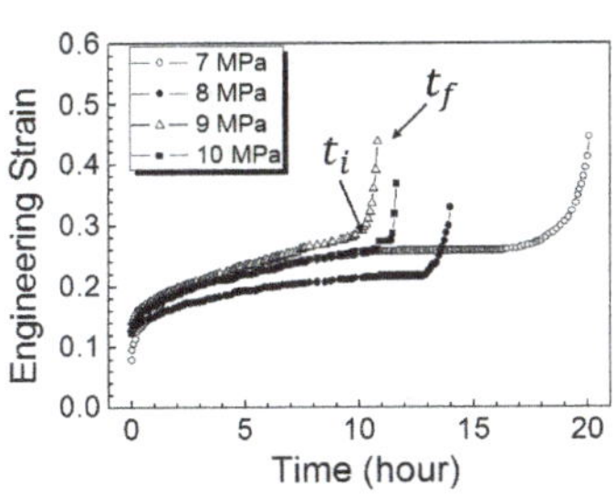

Fig. 1. Creep strain – time plots. Creep strain is dramatically increased with crack initiation, t_i.

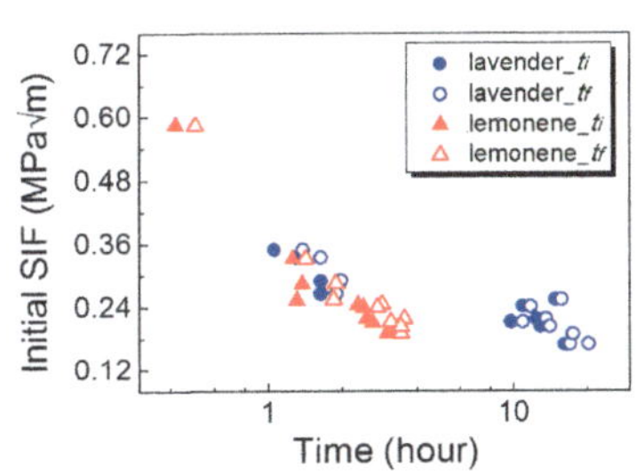

Fig. 2. Measured t_i and t_f for HCPP in lavender and limonene oil.

3 RESULTS AND DISCUSSION

3.1 ESCR result – Relationship between initial SIF and t_i, t_f

Fig. 2 shows the time to crack initiation and failure in different oils. At the low loading condition, HCPP in lavender oil shows longer lifetime than that in limonene oil. As severe loading condition is subjected, t_i and t_f values are quite overlapped which means no discrepancy between two oils. In the lavender case, the non-data space between two separation groups should be tested in the future work. Also, there are small difference between crack initiation and final fracture. That is, the total lifetime is mainly consisted of crack initiation time and slow crack growth portion is relatively small.

3.2 Post-fracture analysis

After the ESCR test, fracture surfaces were observed by scanning electron microscope (SEM). Fig. 3 shows the representative fracture surface in active agent contacting condition. There is no significant difference between lavender and limonene oil cases. At the early stage of crack propagation, highly rough fracture surfaces were observed, and after the each micro cracks were combined, crack propagates radial direction due to the stress-triaxiality at the middle region of the sample. These initial multiple crack propagations in early stage, which is a typical phenomenon of ESC, may be resulted from the heterogeneity of HCPP, i.e. separated amorphous and crystalline phase as illustrated in Fig. 4a. In the early loaded state, each micro crack is generated at different position, even different height due to the material heterogeneity (Fig. 4b). Generally these micro cracks are coalesced gradually as crack propagation. Some samples with highly irregular initial notch showed separated cracks until the final fracture (Fig. 4c).

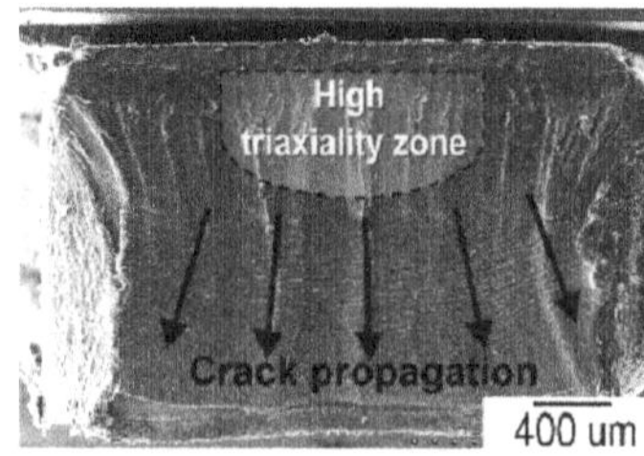

Fig. 3. Fracture surface of HCPP and radial crack propagation.

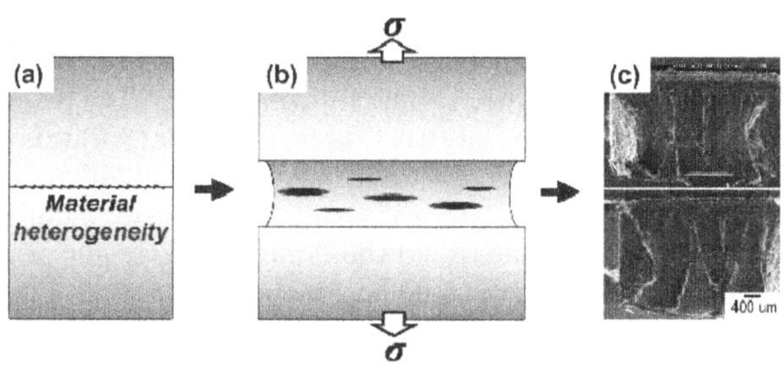

Fig. 4. Effect of uneven initial notch; (a) initial unloaded state, (b) micro voids generated at the different positions in early stage of loading, and (c) uneven fracture surfaces.

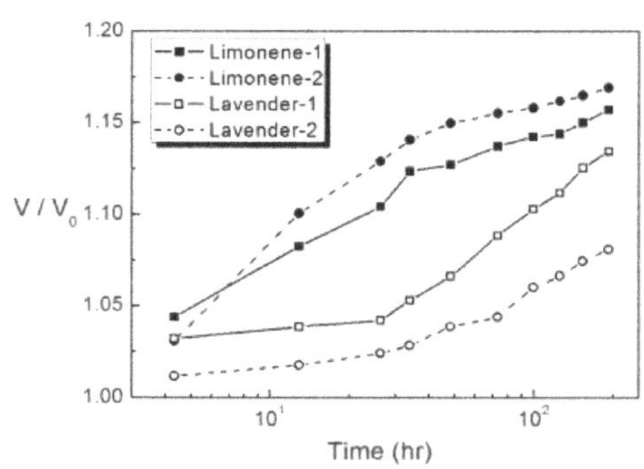

Fig. 5. Volume increase of HCPP blocks by swelling in different oils.

3.3 Swelling test data of HCPP block for limonene and lavender oil

To compare the diffusion rate of scent oils, swelling tests were performed. Fig. 5 plots the volume change of block shape of HCPP with initial volume of roughly 40 mm^3, in lavender and limonene oils for 200 hours at 50°C. It is obvious that the diffusion rate is higher in case of HCPP in limonene, i.e. swelling phenomenon is much enhanced in limonene than lavender oil. Limonene oil consists of D-limonene, beta-pinene, and alpha-pinene with molecular weight (MW) of the same as 136.23 g/mol. However, lavender oil consists of beta-pinene and 1-octene-3yl-acetate which has higher MW of 170.25 g/mol. Due to this cumbersome molecule in the lavender oil, diffusion is much slower than limonene oil.

3.4 Discussions

ESC phenomenon is intensively accelerated when some degree of loading is applied. The limiting case of small load is swelling test condition, which requires the significant time for sorption. Diffusion of oils inducing ESC effect is concentrated on the amorphous region in case of semi-crystalline polymer such as HCPP. As mentioned above, there are threshold loading conditions to introduce the overlapped lifetime in different scent oils. Above the threshold load (threshold SIF) condition, amorphous regions in HCPP get to have enough space for the diffusion of even large molecules in oils. Therefore, this threshold SIF depends on the molecular size of active agents as illustrated schematically in Fig. 6. In case of limonene oil, consisting of relatively smaller molecules than lavender oil, threshold SIF smaller than lavender case. More tests may be required to find the threshold SIF of limonene oil. On the contrary, the threshold SIF of lavender oil case is higher than limonene case, due to the additionally included large molecule, i.e. 1-octene-3yl-acetate. Reference line is also indicated as air case without ESC effect.

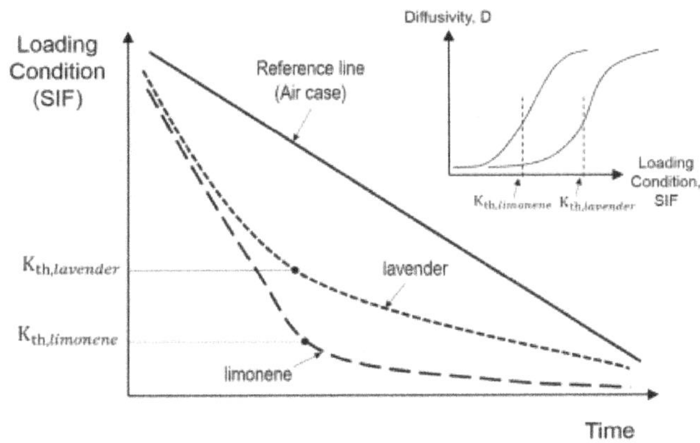

Fig. 6. Schematic diagram of the SIF-life time curves.
Additional plot shows diffusivity dependence on loading condition.

The fracture surface analysis shows the initially generated multiple small cracks and gradually merged main crack. This multiple cracks are typical feature of the environmentally induced cracking such as ESC. It mainly induced by the material heterogeneity, i.e. separated amorphous regions in crystalline regions in HCPP medium. Therefore, the micro cracks may initiate in the amorphous phases, which is the weaker point to ESC at the notch tip region. Individual micro cracks get to gradually merge into the main crack, and finally the main crack propagates in radical direction as displayed in Fig. 3. It is worth noting that micro cracks are generated even at the different positions in the axial loading direction. When this separation distance in axial direction between micro cracks are relatively long, no merged main crack may be made until final fracture as shown in Fig. 4c.

Finally, the observed small portion of slow crack growth (SCG) can be an evidence of the ductile failure of ESC fracture. In general, in case of quasi-brittle fractures, there are relatively large portion of time for slow crack growth in total lifetime after the crack initiation. However, in our case, total lifetime t_f is slightly longer than crack initiation time t_i. It means that that the ductile failure can be a dominant fracture mechanism in ESC-induced fracture.

4 CONCLUSION

ESC is induced by diffusion of particular surface agents with no chemical reaction, leading to weakening the van der Waals forces between molecular chains, especially in amorphous regions. For the further understanding on the ESC phenomena, creep and swelling tests of HCPP in two different liquid oils at elevated temperature were performed. The main results and discussions in this study can be summarized as follows.

First, the ESC phenomenon is also accelerated with loading conditions, i.e. initial SIF. As severe loading condition is applied, free volume between molecular chains in amorphous region increases, and it induces the high diffusivity of oils. Particularly, there is a threshold SIF depending on the agent molecular size. As the average molecular size of the liquid agents grows, threshold SIF to diffuse well is increased. This threshold SIF concept can be a useful design parameter as a critical load on the application of polymers in the ESC-inducing environments.

Second, multiple small cracks are made in crack initiation stage of ESC failure. It arose from the material heterogeneity of HCPP, such as discrete amorphous regions surrounded by crystalline region. In the weak regions to the ESC, micro cracks are generated and finally merged together as a main crack.

Finally, relatively small portion of SCG in t_f indicates that the ESC induced failure is ductile fracture, not a quasi-brittle fracture. This result is in accordance with the ESC mechanism. The diffused oil molecules reduce the van der Waals force between the polymer molecular chains and it plasticize the motion of the chains mainly in the amorphous region, resulting in large deformation before fracture.

ACKNOWLEDGEMENT

The author would like to thank LG electronics Inc. for providing test samples.

REFERENCES

[1] Robeson, L. M. (2013) *Polym. Eng. Sci.* **53**, 453-467. doi:10.1002/pen.23284
[2] Choi, B. H., Weinhold, J., Reuschle, D. & Kapur, M. (2009) *Polym. Eng. Sci.* **49**, 2085-2091. doi:10.1002/pen.21458
[3] Lustiger, A. & Markham, R. (1983) *Polymer.* **24**, 1647-1654. doi:10.1016/0032-3861(83)90187-8
[4] Arnold, J. C. (1995) *Mat. Sci. Eng. A.* **197**, 119-124. doi:10.1016/0921-5093(94)09759-3
[5] ASTM Standard D 5397, *ASTM Int.* **04.13** doi:10.1520/D5397-07R12 (2012)
[6] ASTM Standard D 1822, *ASTM Int.* **08.01** doi:10.1520/D1822 (2013)

Fracture and wear in shredder hammer tungsten carbide tips in the sugar cane shredding process

L. Yin[1], D. Authurs[1], S. Wright[1], L. Santarossa[2]
[1] Matter & Materials, College of Science, Technology & Engineering, James Cook University, Australia
[2] Wilmar Sugar Australia, Australia

ABSTRACT

This paper reports on facture and wear of damaged shredder hammer tips manufactured from cemented tungsten carbide with an ultrafine grain structure and Ni-Ti-Cr as a binder instead of Co, for sugar cane shredding. Optical imaging and scanning electron microscopy were applied for measurement of fracture and wear regions of damaged shredder hammer tips at macro and microstructural scales. Our results reveal that edge and large scale fractures in the form of conchoidal fracture were caused by impacts from hard objects such as rocks or tramp metal materials in the shredding process. Minor wear occurred due to abrasive wear of tungsten carbide grains and fatigue-induced cracks in the metal binders. This study shows that fracture was the main failure mode of the trial cemented tungsten carbide used that caused the instant failure of the shredder hammer tips, while minor wear also occurred.

1 INTRODUCTION

Sugar cane is a tall perennial grass grown in warm temperate to tropical regions in Australia and worldwide. Generally, mature sugar cane stalks are 2–4 m high and have diameters of 20–50 mm. They contain multiple nodes and internodes, and are generally cut into sugarcane billets of around 0.2 m long in the field to facilitate transportation to the factory. Cane shredders are used in the Australian sugar industry to shred sugar cane billets against shredder grid bars to pulverize the cane billets into a fibrous mat. The aim is to rupture parenchyma cells in the cane as much as possible to allow for maximum extraction of sugar in the crushing station. A shredder generally comprises of 90–250 swing hammers that rotate on a rotor at around 1000 rpm and have swing diameters ranging from 1.52 m to 1.83 m [1]. The velocity of the shredder hammer tips ranges from 90–100 m/s and the shredder is driven by a prime mover of several megawatts [1]. Shredders in the Australian sugar industry can process sugar cane at 7,000–15,000 tonnes/day [1]. Sugar cane shredding accounts for approximately 20% of the total energy requirements in a typical sugar mill. It is the most energy consuming process in the sugar industry and has the largest effect on sugar extraction and costs [2].

Shredder hammers tips undergo significant deterioration during processing of sugar cane due to the energy absorbed in the shredding process, the abrasive nature of the soil particles mixed with the sugar cane and the acidic pH of the cane juice. A typical shredder swing hammer has a nominal mass of 20 kg and a replaceable hammer tip of approximately 3 kg mass. A hammer tip is typically manufactured from a carbon steel base mechanically fastened to the hammer, a white iron protective block and tungsten carbide tiles. Shredding forces are mainly applied to shredder hammer tips, resulting in high wear and failure rates in the tungsten carbide tiles. The cost for tungsten carbide tiles accounts for 70–80% of the total costs for shredder hammer tips. The failure of the tungsten carbide tiles costs the sugar industry millions of dollars annually; yet, the quantitative cause of failures is not well understood. This paper presents the quantitative evaluation of the failure mechanisms of damaged shredder hammer tips manufactured from cemented tungsten carbide with Ni-Ti-Cr bound ultrafine microstructures in sugar cane shredding.

2 EXPERIMENTAL PROCEDURE

The shredding process was carried out at Invicta Mill, Giru, Queensland, Australia, using an industrial shredder containing 144 shredding hammer tips rotating at 96 m/s with a rotating diameter of 1.52 m. For the duration of the trial, the 4 MW shredder was fed at a cane feed rate of around 400 tonnes/hour at room temperature. Accordingly, approximately 290 N force was applied by each shredding hammer tip to the cane billets. All replaceable hammer tips were manufactured from cemented tungsten carbide with three 30 mm × 30 mm × 12 mm tiles (Sandvik) that were silver soldered to a white iron, manufactured to the requirements of ASTM A532 17-2 Cr-Mo. The white iron was vacuum brazed to a threaded carbon steel forging that was bolted to the shredder hammers [1]. Among the batch of 144 tips, there were six trial tips manufactured from cemented tungsten carbide with ultrafine microstructures with Ni-Ti-Cr as a binder (Sandvik), instead of Co as a binder being typically used. The carbide material of the trial tips had a Vickers hardness of 1600 (15.69 GPa) and transverse rupture strength of 3800 MPa. After processing of 21,200 tonnes of sugar cane, the tips were inspected and it was found that the trial tips had failed prematurely. The typical life expectancy for shredder hammer tips in this machine is 300,000–400, 000 tonnes depending of cane feed quality. The six trial tips were then removed from the machine and subjected to fracture and wear studies.

Digital optical images of all trial tungsten carbide tips were taken from top and side views. These images were then imported to Image J (NIH, USA), a java-based image processing program for imaging analysis. Areas of fracture or wear were calculated over the top faces of the tips. The fracture or wear fraction was calculated as the ratio of the fracture/wear area and the total face area of the tip. Selected carbide tiles were also removed from the white iron bases and then cleaned using acetone in an ultrasonic bath. The cleaned tungsten carbide samples were examined using scanning electron microscopy (Jeol JSM5410LV, USA). The chemical composition of the carbide tips were also analysed using x-ray energy dispersive spectrometry via an electron prober micro-analyser (Jeol JXA 8200, EPMA, USA).

3 RESULTS AND DISCUSSION

The x-ray energy dispersive spectrometry analytical results show that the tungsten carbide tips contained 81.5 mass% tungsten carbide, 13.5 mass% Ni, 2.8 mass% Ti and 2.2 mass% Cr.

Figure 1 shows the damage features of the used tungsten carbide tips, including edge fracture, double edge fracture, large scale fracture, and minor wear with cracks. Among the six trial tips, fracture was prevalent. Three tips had a single edge fracture, one tip had a double edge fracture, and one tip had a large scale fracture. All fractured morphologies reveal brittle characteristics in which conchoidal fracture occurred. On these fractured tips, minor wear was also observed. Minor wear with cracks also occurred on a fracture-free tip.

Figure 1. Damage features of tungsten carbide tips: edge fracture, double edge fracture, large scale fracture, and minor wear with cracks.

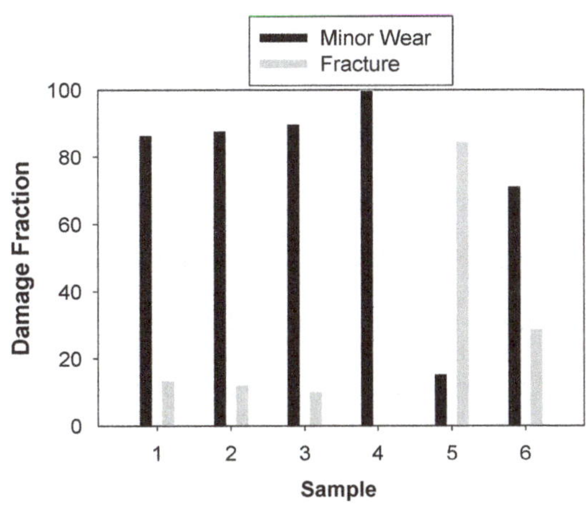

Figure 2. Fracture and minor wear fractions.

Figure 2 shows the fracture/minor wear fractions of these tips, in which the edge fracture damage occupied over 10% of the work areas and the large scale damage covered 85% of the work area. These fractured areas significantly reduced the functionality of the shredder tips.

Figure 3 shows scanning electron micrographs of the fractured and worn tungsten carbide areas. Figure 3a reveals the large scale fracture morphology in which deep fractures across the large tungsten carbide bulk solid and fractured clusters were observed. This indicates that the tungsten carbide tips were exposed to extreme impact by hard rock or tramp material at very high speeds. Figure 3b shows that the damaged microstructures in ultrafine WC grains with grain sizes of < 1 μm were observed and fractures occurred at the WC-metal interface. It also shows the breakaway of areas of material. WC grain fragmentation was hardly observed due to the ultrafine microstructure. Figure 3c shows the wear morphology of the cemented carbide, in which microcracks were also observed. Figure 3d reveals that abrasive wear occurred accompanied by fatigue-induced cracks at the WC-metal interface and breakaway of material.

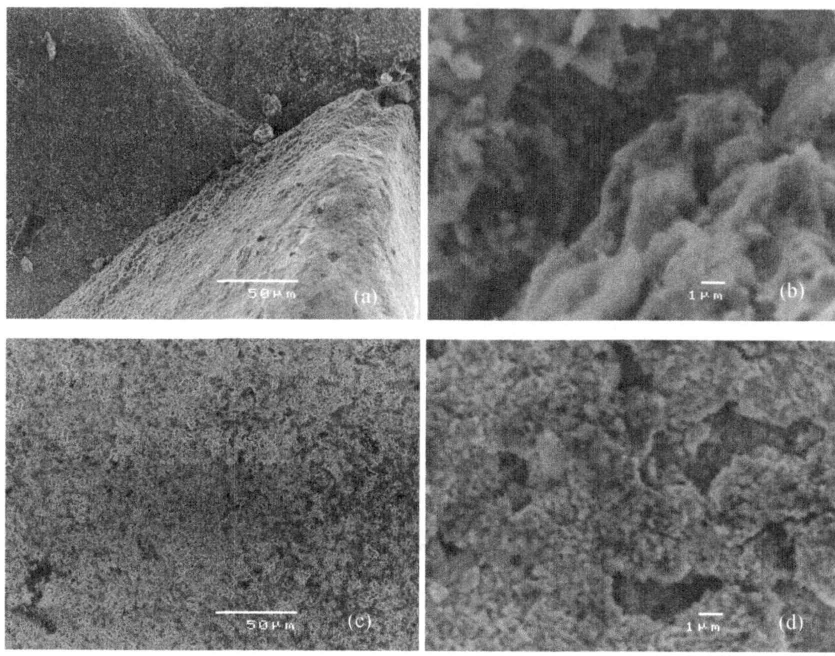

Figure 3. Scanning electron micrographs of (a) a large scale fracture of tungsten carbide tip, (b) fracture at the microstructural level, (c) wear morphology of tungsten carbide tip, and (d) wear at the microstructural level.

The vast majority of applications for cemented carbides exploit their high resistance to wear when subjected to abrasion and erosion [3–5]. Typically shredder hammer tips used in the sugar industry are manufactured from cemented carbides with ultrafine microstructures with Co as a binder. This material is tolerant to a degree of impact during operations. This study reveals that fracture in the trial shredder tips with Ni-Ti-Cr as a binder for ultrafine carbide can be more severely affected by impact in heavy production processes such as shredding of sugar cane. The selection of materials for the shredding of sugar cane must consider the heavy applied loads, and impact from rock and tramp materials at high speeds. An understanding of the cemented carbide responses to complex tribological environments has major commercial implications to the sugar industry. However, the development of such an understanding is complex due to cemented carbides having a wide spectrum of microstructures which are strongly dependent on the parameters of tungsten carbide grains as well as the variables of the binders. In particular, ultrafine grades with Ni-Ti-Cr as a binder have the exhibited property combination, which represents a significant departure from those of the more conventional of WC-Co hard metals [6–9]. This work provides some of the first quantitative data on fracture and wear fractions of cemented carbide tips with Ni-Ti-Cr as binders for the cane shredding process using materials characterization techniques. In the future more comprehensive comparison studies of a wide range of cemented carbide tips will be conducted for optimal materials selection in the sugar industry.

4 CONCLUSIONS

This paper presents a quantitative study of fracture and wear in damaged shredder hammer tips manufactured from cemented ultrafine grain carbide with a specific proportion of Ni-Ti-Cr as a binder. In comparison to the conventional WC-Co tips, the Co-free cemented carbide tips have a much higher hardness resulting in minor abrasive wear during the shredding process. However, severe brittle fracture occurred in these Co-free tips during shredding, which caused instant tip failure. This indicates that the specific proportion of Ni-Ti-Cr-cemented tungsten carbides with ultrafine structures that were tested, are potentially suitable for lower impact shredding processes in which sugar cane billets are not contaminated by rock or tramp materials. Further research will be conducted to examine other proportions of Ni-Ti-Cr in cemented tungsten carbide that may perform more effectively as binders for the specific application of the sugar cane shredding.

ACKNOWLEDGEMENTS

The authors thank Dr. Kevin Blake of the Advanced Center at James Cook University for experimental assistance. This work was supported by Wilmar Sugar Australia.

REFERENCES

1. Santarossa, L. G. & Anderson, S. I. (2014) *Int. Sugar J.* **28**, 280–289
2. Siddhartha Bhatt, M., & Rajkumar, N. (2001) *Appl. Therm. Eng.* **21**, 1707–1719. doi: 10.1016/S1359-4311(01)00027-8
3. Yin, L., Pickering, J. P., Ramesh, K., Huang, H., Spowage, A. C., & Vancoille, E. Y. J. (2005) *Int. J. Adv. Manuf. Technol.* **26**, 766–773. doi: 10.1007/s00170-004-2056-0
4. Yin, L., Spowage, A. C., Ramesh, K., Huang, H., Pickering, J. P., & Vancoille, E. Y. J. (2004) *Int. J. Machine Tool Manuf.* **44**, 533–543. doi: 10.1016/j.ijmachtools.2003.10.022
5. Yin, L., Vancoille, E. Y. J., Ramesh, K., Huang, H., Pickering, J. P., & Spowage, A. C. (2004) *Proc. Inst. Mech. Eng., B J. Eng. Manuf.* **218**, 419–429. doi: 10.1243/095440504323055533
6. Allen, C., Sheen, M., Williams, J., & Pugsley, V. A. (2001) *Wear* **259**, 604–610. doi: 10.1016/S0043-1648(01)00667-6
7. Hegeman, J. B. J. W., De Hosson, J. Th. M., & de With, G. (2001) *Wear* **248**, 187–196. doi: 10.1016/S0043-1648(00)00561-5
8. Gee, M. G., Gant, A., & Roebuck, B. (2007) *Wear* **263**, 137–148. doi: 10.1016/j.wear.2006.12.046
9. Murthy, J. K. N., Rao, D. S., & Venkataraman, B. (2001) *Wear* **249**, 592–600. doi: 10.1016/S0043-1648(01)00682-2

An investigation into variations in roughness-induced crack closure in high strength aircraft alloys under fatigue loading

K.F. Walker[1,2], C.H. Wang[2], J.C. Newman, Jr.[3]
[1] DSTO, Australia
[2] RMIT University, Australia
[3] Mississippi State University, USA

ABSTRACT

Crack closure, driven predominantly by plasticity and roughness, has been identified as a critically important issue affecting fatigue crack growth in high strength alloys. This paper details an investigation into the closure behaviour four materials which exhibit a broad range of crack surface roughness. The work includes quantitative evaluation of the crack profiles using optical interferometry, compliance measurement of crack closure levels under constant amplitude loading, and quantitative fractography interpretation of closure effects during spectrum loading. Roughness was found to account for a major rise in closure under constant amplitude conditions, but was found to be much less significant under spectrum loading.

1 INTRODUCTION

The discovery of the fatigue crack closure phenomenon by Elber [1] has been extremely important in understanding fatigue crack growth in aircraft structures manufactured from high strength metallic alloys. Several mechanisms can contribute to closure including plasticity, roughness and debris formation. Newman developed the FASTRAN crack growth analysis computer code [2] based on the most significant of these; plasticity. FASTRAN is a life prediction code based on the Dugdale yield zone model [3] but modified to leave plastic deformations in the wake of the advancing crack. A key feature of the code is the ability to model three-dimensional constraint effects. A constraint factor, α, is used to elevate the flow stress σ_o at the crack tip to account for the influence of stress state ($\alpha\sigma_o$) on plastic-zone size and crack surface displacements. The plasticity only approach has been found to correlate experimental observations with analysis very well for the materials which exhibit smooth crack surfaces, such as 7249-T76511 [4], and 7075-T6 [5]. Consistent constraint assumptions for both constant amplitude (CA) and spectrum loading are used. But limitations of the plasticity only approach as used in FASTRAN have become evident when applied to materials which exhibit very rough crack surfaces [6-8]. Newman et al. [6] successfully collapsed the CA crack growth rate test data to a unique ΔK_{eff} relation for 7050-T7451 material, but the correlation required a very low constraint factor α=1.3. The low value of α was proposed to account for roughness closure effects not being explicitly modelled in FASTRAN. Walker et al. [8] found a similar result for the coarse-grain β-annealed Ti-6Al-4V ELI alloy which exhibited extremely rough crack surfaces.

The low α value also worked well when carried through and applied to FASTRAN analysis for spectrum loading [8, 9]. Previous research efforts have focused on developing modelling for the roughness contribution such that a more reasonable and justifiable value for α can be used for the plasticity component. That approach was moderately successful for β-Ti-6Al-4V material [8]. But recent improvements in FASTRAN [2] now suggest that the approach may not be as effective as previously thought. Additionally, recent research at DSTO has identified significant, fundamental differences in fatigue crack path and rate at the smallest possible length scales on a cycle-by-cycle basis for CA compared with spectrum loading [10]. This paper details an investigation into closure effects for materials with a broad range of crack surface roughness characteristics under conditions of both CA and spectrum loading.

2 MATERIALS, TEST SPECIMENS AND CRACK GEOMETRIES CONSIDERED

Four high strength aircraft alloys with a range of crack surface roughness characteristics were considered (see Figure 1). Details are as follows:

a. 7249-T76511, very smooth, typical crack surface roughness R_a = 4 μm. Compact tension, C(T) specimens, through thickness crack, both CA and spectrum loading.
b. 7075-T6, smooth, typical crack surface roughness R_a = 8 μm. C(T) specimens, through thickness crack, both CA and spectrum loading.
c. 7050-T7451, rough, typical crack surface roughness R_a = 20 μm. C(T) specimens, through thickness crack for CA, low K_T dog-bone tension specimen with surface crack for spectrum loading.
d. B-annealed Ti-6Al-4V ELI, extremely rough, typical crack surface roughness R_a = 50 μm. Eccentrically loaded single edge notch tension (ESE(T)) specimens, through thickness crack for both CA and spectrum loading.

a. 7249-T76511, very smooth, B=3.2mm C(T) Typical R_a = 4 μm   b. β-Ti-6Al-4V, ext. rough, B=6.35 mm ESE(T) Typ. R_a = 50 μm

Figure 1 : Fatigue crack surfaces for extreme roughness cases

EXPERIMENTAL METHODS

2.1 Compliance based closure measurement

Compliance based methods were used to measure crack closure under CA conditions for the through thickness cracks in the bend specimens, i.e. C(T) and ESE(T) configurations.

2.2 Marker bands and quantitative fractography

Almost all of the data available in the literature relating to measured crack closure and crack closure modelling pertains to CA loading. This is especially true for small cracks which initiate at natural features. However a novel test and analysis program was recently undertaken at DSTO to investigate crack closure for small cracks from natural features in 7050-T7451 under spectrum loading [11]. A specially engineered load sequence with bands of CA at various mean stress levels (local R=0.5, 0.4, 0.3, 0.2, 0.1 and 0.0) was conducted using low-K_T coupons from 7050-T7451 material. Quantitative fractography was used to determine crack growth rates during the CA bands and this was related to the appropriate ΔK_{eff} value and closure was therefore inferred.

2.3 Optical interferometry

Optical interferometry was used to determine the crack surface profile and roughness characteristics. This was performed using a Bruker NPFlex Optical Profiling Microscope using 3-D white light vertical scanning interferometry. A complete scan of the crack surface can be obtained, and data on a nominated path broken down into profile, waviness and roughness can also be obtained.

ANALYTICAL MODELLING

Crack growth modelling was performed with FASTRAN [2] which explicitly accounts for plasticity induced closure only. Because the user can effectively control the crack closure level in FASTRAN through empirical adjustments of the constraint factor α, it is possible to approximate the effect of other contributions like roughness. Two models which explicitly account for the roughness contribution to closure (1. Crapps/Daniewicz, and 2. Zhang) were developed and evaluated for constant amplitude conditions (for 7050-T7451 and β-Ti-6Al-4V respectively) and the results compared with FASTRAN analysis and experimental data.

FASTRAN [2] is a fatigue crack growth analysis code based on the crack closure concept and a Dugdale or strip yield model. Experience with a significant number of materials and geometries have shown that α values of around 1.8-2.0 which are consistent with 3-D elastic-plastic Finite Element Analysis (FEA) [12, 13] correlate test measurements with analyses very well.

Crapps and Daniewicz [14] developed a model to account for roughness-induced closure. Their model is a strip-yield model like FASTRAN, but modified to account for roughness by lengthening and shortening the elements along the crack due to crack face sliding. Asperity angle and frequency were the significant parameters. Asperity angle was found to be the most significant factor. A 15 degree asperity at R=0.1 constant amplitude loading increased the ratio of opening load P_{op} to maximum load P_{max} from 0.4 to 0.6. The significance of this result is discussed later.

Zhang etal [15] developed a combined roughness and plasticity model based on the Dugdale model for the plasticity component and the Suresh/Ritchie geometric model for the roughness component. The model was further developed and applied to the β-Ti-6Al-4V material by Walker etal [8]. The significance of that result in the context of the present work is also discussed later.

3 RESULTS

3.1 Constant amplitude (CA) loading

Measured crack opening values for the four materials tested under constant amplitude loading at R=0.1 are shown in Figure 2. FASTRAN analysis results at two values of α (1.0 and 2.0) are also shown. Three-dimensional (3D) elastic-plastic FEA studies have shown that the value of constraint for perfectly flat crack surfaces (no roughness) is around 1.8-2.0 [12, 13]. The materials which exhibit smoother crack surfaces also exhibit lower crack opening load, and FASTRAN analysis using 3-D constraint α of 2.0 correlated very well against the experimental observations (see Figure 2). The materials which exhibit rough crack surfaces exhibit elevated crack opening, and an artificially low value of α (around 1.0) is proposed (see Figure 2). These results suggest that significant roughness-induced closure effects are present during CA loading for the materials which exhibit rough crack surfaces (7050 and β-Ti), but not for the materials which exhibit smoother crack surfaces (7249 and 7075). Optical interferometry results for a 7050-T7451 coupon subjected to R=0.1 constant amplitude loading revealed an average and median asperity angle of 15 degrees, which when applied in the Crapps/Daniewicz model raised the crack opening P_{op}/P_{max} from about 0.4 to 0.6, which is close to the observed difference between the materials which developed rough compared with smooth crack surfaces as shown in Figure 2. This is an independent validation of the Crapps/Daniewicz approach. A development of the Zhang model (by the authors [8]) identified a similar elevation in crack closure due to roughness under constant amplitude conditions for the β-Ti-6Al-4V ELI material. The Zhang model is based on a combination of plasticity-induced closure with a geometric roughness model based on the Suresh and Ritchie approach [16].

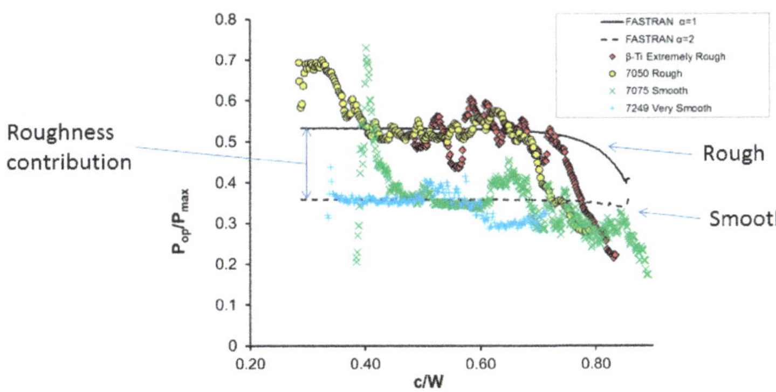

Figure 2 : Crack opening load at constant amplitude R=0.1

3.2 Spectrum loading

As described earlier, the PICC only approach has been found to work very well for the materials which exhibit smooth crack surfaces, i.e. 7249-T76511 [4], and 7075-T6 [5]. FASTRAN modelling correlated the experimental results well using the same constraint assumptions for both CA and spectrum loading.

But an important question remains; what about materials with rough fatigue-crack surfaces? FASTRAN analyses were conducted for various spectrum loading scenarios. If additional closure caused by roughness were to be present for the spectrum loading case as for the constant amplitude loading, then a low value of α of about 1.0 would be expected for the correlation. Instead, it was found that high constraint, similar to that expected for materials which exhibit very smooth crack surfaces, α=2.0 was required for 7050-T7451 and β-annealed Ti-6Al-4V ELI. Additionally, FASTRAN analyses were conducted on a cycle-by-cycle basis for the marker band case described in Section 2.2. Again, it was found that α=2.0 correlated the analysis to test observations well, which was contrary to the constant amplitude case for the 7050-T7451 material. The results suggest that the roughness-induced closure effects which were evident under CA loading were no longer present under spectrum loading.

4 DISCUSSION AND CONCLUSION

Crack surface roughness was found to cause elevated crack opening loads for 7050-T7451 and β-Ti-6Al-4V ELI materials under CA loading conditions. The increase could be accounted for analytically reasonably well using either the Crapps/Daniewicz approach in the case of 7050-T7451, or a development of the Zhang model in the case of β-Ti-6Al-4V ELI. The plasticity only FASTRAN model could also correlate the closure levels for those materials, but only if an artificially low value of 3-D constraint α of about 1.0 (plane stress) was applied. For the materials considered here which exhibited smooth crack surfaces (7249-T7651 and 7075-T6), FASTRAN modelling under CA conditions has separately been found to match the results well using values of constraint α of around 1.8-2.0, suggesting that roughness does not contribute to fatigue-crack closure for those materials. In the case of spectrum loading, FASTRAN analyses correlated the crack growth behaviour very well using values of constraint α of 1.8-2.0 for <u>all four</u> materials. The results presented here suggest that roughness closure effects which may be apparent under CA loading are much less significant under spectrum loading conditions, regardless of the fact that the material may exhibit significant crack surface roughness. These are very significant findings because CA data are often used for analyses of structures under spectrum loading, and non-conservative results are therefore possible if beneficial roughness closure effects are assumed but do not occur for the spectrum loading case. Further work is planned including elastic-plastic FEA for non-straight (rough) cracks, and experiments to measure the crack opening loads during spectrum loading.

5 REFERENCES

[1] Elber, W., ASTM STP 486 American Society for Testing and Materials, West Conshohocken, Pennsylvania, USA, 1971: p. 230-242.

[2] Newman, J.C., Jr., 2013, Fatigue & Fracture Associates LLC.

[3] D.S.Dugdale, Journal of Mechanics and Physics of Solids, 1960. **8**(2): p. 100-104.

[4] Newman, J.C.J., Walker, K. F., and Liao, M., in *Aircraft Structural Integrity Program*. 2013: Bonita Springs, FL, USA.

[5] Walker, K., and Newman, J.C., Jr., in *The Aircraft Structural Integrity Program Conference*. 2012: San Antonio Texas USA.

[6] Newman, J.C., Jr., Yamada, Y., and Newman, J. A., Journal of ASTM International, 2010. **7**(4).

[7] Walker, K.F., and Barter, S.A., in *International Committee on Aeronautical Fatigue*. 2011: Montreal, Canada.

[8] Walker, K.F., Wang, C.H., and Newman,J.C., Jr., Advanced Materials Research, 2014. **891-892**: p. 319-324.

[9] Newman, J.C., Jr., Shaw, J.W., Annigeri, B.A., and Zeigler, B.M. in *Turbine Technical Conference and Exposition ASME Turbo Expo*. 2012. Copenhagen, Denmark.

[10] White, P.A., Barter, S.A., and Molent, L., International Journal of Fatigue, 2007. **30**: p. 1267-1278.

[11] Barter, S., Burchill, M., and Jones, M., Advanced Materials Research, 2014. **891-892**: p. 93-99.

[12] Newman, J.C., Jr., Bigelow, C.A., and Shivakumar, K.N., Engineering Fracture Mechanics, 1993. **46**(1): p. 1-13.

[13] Newman , J.C., Jr., Crews, J.H., Bigelow, C.A., and Dawicke, D.S., in *Constrain effects in fracture: Theory and applications, ASTM STP 1244*, M. Kirk and A. Bakker, Editors. 1995, ASTM. p. 21-42.

[14] Crapps, J., and Daniewicz, S.R., International Journal of Fatigue, 2012. **45**: p. 15-30.

[15] Zhang, X.P., Li, J. C., Wang, C. H., Ye, L., and Mai, Y. W., International Journal of Fatigue, 2002. **24**(5): p. 529-536.

[16] Suresh, S., and Ritchie, R.O., Metallurgical Transactions A, 1982. **13A**(9).

An analysis of elasto-plastic fracture criteria

A. Kotousov, A. Khanna, S. Bun
School of Mechanical Engineering, The University of Adelaide, Australia

ABSTRACT

The aim of this paper is to critically evaluate, how well the most common non-linear fracture criteria actually describe the general experimental trends, which were confirmed in a large number of fracture tests, conducted over the past fifty years. In particular, we examine the agreement between these general experimental trends and the theoretical predictions based on the critical value of the crack-tip stress at some certain distance, J-integral, crack tip opening displacement and crack-tip-opening angle. All theoretical predictions are derived from the classical strip yield model, which is considered to be adequate for the purpose of the current study.

1 INTRODUCTION

When the characteristic size of the crack-tip process zone is sufficiently small compared with other dimensions of a structural component, the stress intensity factor is the only parameter controlling the stress state in the vicinity of the crack tip and therefore, fracture initiation from this region. In this case, Linear Elastic Fracture Mechanics (LEFM) is an established method of fracture evaluation of structural components weakened by cracks. However, in the presence of a large yielding region or process zone ahead of the crack tip, the crack initiation and extension conditions are considerably more difficult to investigate and formulate in terms of a single critical value characterizing the fracture resistance of the material (1). One of the reasons behind these difficulties is associated with the selection of an appropriate parameter characterizing the elasto-plastic crack tip behavior, which also has a direct link to fracture initiation conditions, and, by definition, is geometry independent (2-4).

A number of non-linear or elasto-plastic fracture controlling parameters were proposed in the past; and these include but are not limited to: crack-tip stress or strain, tearing modulus, finite element crack tip node force, crack-tip-opening displacement or angle, crack-tip force, energy-release rates, elasto-plastic energy, computational process zone energy release rate and J-integral. In this work, we will examine the compliance of the most common non-linear fracture criteria with the general empirical trends in fracture of structural materials, normally observed in experimental studies and standard mechanical tests. Such examination could help, for example, in the selection of an appropriate fracture criterion to describe particular experimental results, or the effect of certain test parameters (geometry or loading conditions), or to predict fracture behavior of the component beyond the tested bounds. The latter is a typical situation encountered in many Fracture Mechanics applications when fracture properties are normally obtained with a laboratory scale specimen but the actual application has a very different scale, geometry or loading conditions.

2 GENERAL EMPIRICAL TRENDS

2.1 Recovery of LEFM

The classical LEFM criterion can be written in terms of the stress intensity factor as

$$K = K_C , \quad (1)$$

where K is the stress intensity factor at a crack tip and K_C is the critical value of the stress intensity factor or apparent fracture toughness, which can be related to the plate thickness of the tested specimen or structure. At sufficiently low applied stress levels, σ_{ap}, the size of the plasticity region ahead of the crack tip is relatively small in comparison with all other characteristic dimensions of the problem geometry (including the crack length and except the plate thickness). Therefore, at this low failure stress, the non-linear elasto-plastic criteria are expected to recover the classical LEFM criterion (1).

2.2 Reduction in apparent fracture toughness at high level of applied stress

Many experimental studies conducted in the past have recognized that the use of the stress intensity factor, K, for fracture assessment at high applied to yield stress ratios, σ_{ap}/σ_Y, underestimates the crack tip loading and, therefore, some plasticity correction is normally required. In general, this correction is a function of the material properties of the component, crack geometry and type of loading; and it is explicitly incorporated within the so called Failure Assessment Diagram (FAD), used in many industrial standards and failure assessment codes such as British BS 7910 or US and Canadian API 1104 and CSA Z662 that pursue the FAD philosophy (5). Different standards apply different approaches and adopt different master curves or FAD to describe this tendency. However, the monotonic reduction of the critical stress intensity factor or apparent fracture toughness with an increase of the applied stress level is normally undisputed.

2.3 Decrease of fracture toughness with increasing section thickness

An increase in the thickness of a test specimen generally promotes a lower energy absorbing fracture process by promoting void nucleation and growth or inducing a cleavage mechanism of crack propagation. Another common explanation of the effect of plate thickness on fracture resistance evokes the statistical weakest link size concept. Based on this concept, increasing specimen thickness also increases the possibility of the presence of embrittled microstructural zones along the crack front, which significantly reduces the material resistance to crack propagation. Both interpretations lead to a similar conclusion, namely, that the apparent fracture toughness decreases with increasing the section thickness of fracture specimen for both brittle cleavage as well as ductile fractures (6).

2.4 Size effect

The size effect considered below is primary related to the alteration of the strength with the change of the in-plane sizes of the specimen or structure rather than the section thickness considered previously in 2.3. It is not surprising that various analytical models have been developed to explain these experimental findings and to provide a theoretical basis for many design applications of fracture mechanics. Among these theoretical models, one of the most commonly known is the Size Effect Law proposed by Bazant (9) and Bazant and Pfeiffer (10). Other size effect models include the Multi-Fractal Scaling Law proposed by Carpinteri and Chiaia (11, 12) and the size strength scaling associated with the three-dimensional stress states at crack tip (7, 8). All these experimental and analytical studies for brittle or ductile materials predict a decrease in the critical stress (or stress to cause initiate fracture) of a specimen of the same proportions but having lager (or scaled) dimensions. Therefore, one would expect that the elasto-plastic criteria should also reflect or describe this empirical tendency.

3 COMMON ELASTO-PLASTIC FRACTURE CRITERIA

3.1 Crack tip stress criterion

There were many fracture criteria developed in the past based on reaching a critical stress level, σ_{cr}, at some distance, δ, ahead of the crack tip. For example, in (13), the initiation of unstable fracture is defined to occur when the tensile stress reaches a critical value at $\delta = 4 \times \text{CTOD}$ ahead of the crack tip. In this paper, it is also argued that the particular choice of 4, 5, or 6 CTOD, is unimportant as the stress field scales identically with macroscopic loading parameters. This stress value is normally determined by the three-dimensional elasto-plastic finite element method (14), which makes the process of the evaluation of fracture initiation quite tedious, computationally extensive, and sometimes, quite ambiguous The latter is attributed to the influence of the boundary conditions and mesh refinement on the calculated stress field ahead of the crack front as well as to the selection of yield criterion and constitutive equation to describe the evolution of plastic deformations.

3.2 CTOD - Crack Tip Opening Displacement

This parameter has been independently proposed by Cottrell and Wells in 1961. It is based on the crack-face opening that occurs upon loading. This criterion suggests that a critical crack-tip opening displacement, CTOD_c, which is both geometry-independent and a property of the material, is required for unstable crack extension. Such an approach has the attraction of invoking a physical representation of fracture process occurring at the crack tip and only requires a small scale testing for determination of CTOD_c. Various methods have been developed to determine the crack-tip opening displacement, typically by applying appropriate geometrical correction factors to displacement measurements remote from the crack tip (which is virtually inaccessible). The CTOD criterion has been successfully applied to numerous structural applications, such as aircrafts, welded structures and pipelines.

3.3 CTOA - Crack Tip Opening Angle

Numerous investigators have experimentally measured CTOD or CTOA during the fracture process. Luxmoore et al. (15) have experimentally demonstrated that CTOA was constant from the onset of stable crack growth in two aluminum alloys, but found different values for different crack geometries. The CTOA concept is very appealing because of its readily-grasped physical significance and the opportunity that it offers for direct measurement (1). It should be recognized that there are two different definitions of CTOA: a crack tip angle that reflects the actual slope. It can be linked directly to the fracture process but the measurement of this parameter represents a formidable measurement task. Another definition for CTOD is an averaged value of the slope, which can be measured but it has less physical justification.

3.4 J-integral

The J-integral, as originally proposed by Rice, is a path-independent contour integral which may be used to characterise near-crack-tip deformation filed in linear and non-linear elastic materials. The basis for using J to characterise fracture stems from the premise that a critical value of the J-integral, J_c, is required for crack extension. The J-integral originally emerged as a fracture criterion for small scale plasticity conditions at crack tip where it served more or less as an extension of LEFM. In this case it can be interpreted as an energy release rate or as an independent crack tip parameter. Later, J-integral was proposed as a fracture criterion in the presence of large scale plasticity for characterisation of fracture initiation as well as for stable crack growth utilising geometry-independent J-resistance curves. In this paper we are interested in the capabilities of J-integral to evaluate the fracture initiation only.

4 METHOD OF ANALYSIS

The examination of the above criteria is carried out on the basis of the classical Dugdale model. An exact analytical solution exists for a crack in a wide plate loaded in mode I in the case of plane stress (very thin plates) and plane strain (very thick plates) conditions prevailing at the crack tip. This model provides an approximate plasticity correction, which appears to offer a good prediction of the size of the plastic zone, stress field and crack opening displacement [16]. The use of the analytical model has many advantages in comparison with 3D numerical simulations; these advantages include the simplicity and clarity, as well as the possibility to independently reproduce the outcomes of such modelling. As mentioned above, the outcomes of 3D numerical simulations can be significantly affected by the boundary conditions, mesh refinement as well as the selected constitutive equation describing plasticity effects at the tip of the crack. In addition, there is some ambiguity in the definition of fracture controlling parameters, which can vary significantly along the crack front. Therefore, there is a strong reasoning behind the selection of an analytical approach for evaluating of the general trends, which is the focus of the current work.

5 DISCUSSION

Table 1 summarizes the outcomes of the undertaken evaluation of the various elasto-plastic fracture criteria with the generalized Dugdale model as described above. It can be seen from this table that none of the considered criteria complies with all experimental tendencies. The best agreement with experimental trends is found for J-integral based criterion. The second best is for CTOD-based fracture criterion, which does recover the LEFM criterion, but for plane stress conditions only. These results are in-line with the popularity of these criteria for the assessment of elasto-plastic fracture. If we compare these two criteria, CTOD and J-integral, applications using J-integral do currently dominate. This fracture controlling parameter has significant advantages. It offers computational ease (its virtual independence of finite element type and element size), and it can provide a lower bound estimate of fracture conditions for many practical situations (1). However, both of these criteria demonstrate an opposite tendency to the experimental results (trends) with respect to the effect of the section thickness on the apparent fracture toughness for the problem under consideration.

Table 1: Compliance of Non-linear fracture Mechanics criteria with the established experimental trends

	Elasto-Plastic	Fracture criterion		
Empirical Tendency	Tip Stress	CTOD	CTODA	J
I. Recovery of LEFM criterion	No	Yes*	No	Yes
II. Effect of Applied Stress	No	Yes	No	Yes
III. Section Thickness Effect	Yes	No	Yes	No
IV. Scale Effect	No	Yes	No	Yes

*Only for plane stress conditions (very thin plates)

On the other hand, the two other criteria (crack tip stress and CTOA) do not comply with any of the experimental tendencies, except for Tendency III. These criteria are capable of describing the effect of the thickness on the apparent fracture toughness. In many experimental studies it was admitted that CTOA-based criterion can also be valid for extensive stable crack growth (where the applications J-integral are limited). This, probably, explains the limited area of the applicability or popularity of these two criteria in the assessment of crack initiation. But again, their limited use can be well justified by the fact that the most popular criteria (J-integral and CTOD) fail to describe the section thickness effect.

6 CONCLUSION

The main outcome of this analysis (experimental trends vs. theoretical predictions) is that none of the considered elasto-plastic criteria describe all common experimental trends. The J-integral based criterion provides the best predicting capabilities within the utilised assumptions, followed by the CTOD criterion. Both of these criteria fail to describe the effect of the thickness on the apparent fracture toughness. The critical crack tip stress and critical CTOA criteria generate this theoretical tendency, which is in-line with the experimental observations. However, the latter criteria fail to describe the decrease in strength with increasing the applied load or specimen size, and should not be used to analyse or predict fracture when a significant change in the applied stress or dimension of the specimen is involved in the testing or fracture evaluation.

Despite that the conducted analysis is based on many assumptions and simplifications associated with analytical modelling; the outcomes of this study can serve as a guideline to the analysis and prediction of elasto-plastic fracture phenomena and test results with various fracture criteria. Obviously, other criteria can be critically examined by utilising the current framework or other common experimental evidences can be incorporated into the examination. In addition, one interesting development of this work could be a development of a new elasto-plastic criterion which complies with all experimental tendencies and has a physical justification similar to J-integral or CTOA.

REFERENCES

[1] Kanninen, M.F., Popelar, C.H. and Broek, D. (1981) *Nucl. Eng. Des.* **67**, 27 - 55.
[2] Anderson, T.L. (1995) CRC Press, Boca Ration.
[3] Broberg, B.K. (1999) Academic Press, New York.
[4] Broek, D. (1982) Martinus Nijhoff, The Hague.
[5] Zerbst, U., Ainsworth, R.A. and Schwalbe, K.-H. (2000) *Int. J. Press. Vessels Pip.* **77**, 855 - 867.
[6] Wallin, K. (1985) *Eng. Fract. Mech.* **22(1)**, 149-163.
[7] Kotousov, A. (2010) *Int. J. Solids Struct.* **47**, 1916-1923.
[8] Kotousov, A., Lazzarin, P., Berto, F. and Harding, S. (2010) *Eng. Fract. Mech.* **77**, 1665-1681.
[9] Bazant, Z.P. (1984) *J. Eng. Mech.* **110(4)**, 518-535.
[10] Bazant, Z. P. and Pfeiffer, P. A. (1987) *ACI Mater. J.* **84**, 463-480.
[11] Carpinteri, A. and Chiaia, B. (1995) *Mater. Struct.* **28**, 435-443.
[12] Carpinteri, A. and Chiaia, B. (1996) *Mater. Struct.* **29**, 259-266.
[13] Dodds Jr., R.H., Anderson, T.L., Kirk, M.T. (1991) *Int. J. Fract.* **48**, 1-22.
[14] Meshii, T., Lu, K. and Takamura, R. (2013) *Eng. Fract. Mech.* **104**, 184–197.
[15] Luxmoore A., Light M.F. and Evans W.T. (1977) *Int. J. Fract.* **13**, 257–9.
[16] Hutchinson, J. W. (1979) Depart. Solid. Mech, Tech. University of Denmark.

A high-frequency fatigue accelerated measuring method for P-S-N curve and fatigue limit

G.Y. Wang, M.T. Ma, Q.S. Jin, Z.G. Li
China Automotive Engineering Research Institute Co. Ltd., P.R. China

ABSTRACT

An accelerated measuring method for estimating P-S-N curve and fatigue limit was introduced amply in paper. P-S-N curve and fatigue limit of a steel plate applied for automotive crossbeam were separately estimated with standard grouped measuring method, up-down measuring method and the accelerated measuring method. The result shows that the errors are within 10%.

1 INTRODUCTION

P-S-N curve is achieved generally with Standard Grouped Measuring Method,(SGMM) and the fatigue limit is achieved generally with Up-Down Measuring Method (UDMM)[1-3]. Though these methods have been approbated and adopted for measuring fatigue property of a metal material, long testing period and high cost are inevitable drawbacks. The paper introduces the principle of the test method, statistical analysis of the data.

2 MEASURING PRINCIPLE

It is assumed that the *S-N* curve consists of an inclined straight line in the finite fatigue life range and a horizontal straight line in the infinite fatigue life regime. This is often realistic for many engineering materials, when the data are represented using appropriate coordinates, generally on semi-log or log-log paper. Since a straight line can be determined by connecting two different points, In order to determine S-N curve, the method introduced in paper employs a Quarter Cycle Intercept (QCI) point, which means that failure occurs after one-fourth cycle.

As shown in Fig.1, point B denotes a figure life N_i corresponding to a stress level S_i. Point A denotes QCI. Joint point A and point B, forming a line AB. Line AB intersects other n-1 stress levels, forming n-1 points, i.e. point C, named suppositional data points. And these suppositional data points are used to fit P-S-N curve as experiment data points. The method for measuring P-S-N curve is named Suppositional Sample Measuring Method (SSMM). Similarly, line AB and vertical line to specify circulation base number N_0, e.g. 10^7 cycles, intersect at point D. The ordinate of point D represents the fatigue limit of the "i"th specimen. The method for measuring fatigue limit is named Fixed Emitting Point Method (FEPM).

The coordinate of QCI is determined with experiment points, the method as follows:

First, the mean of fatigue life is calculated from the following equation:

$$\overline{X_i} = \frac{1}{m_i} \sum_{j=1}^{m_i} \lg N_{ij} \tag{1}$$

where m_i is the sub-total number of testing data points at the "i"th stress level and N_{ij} is the fatigue life of the specimen no. j in this group.

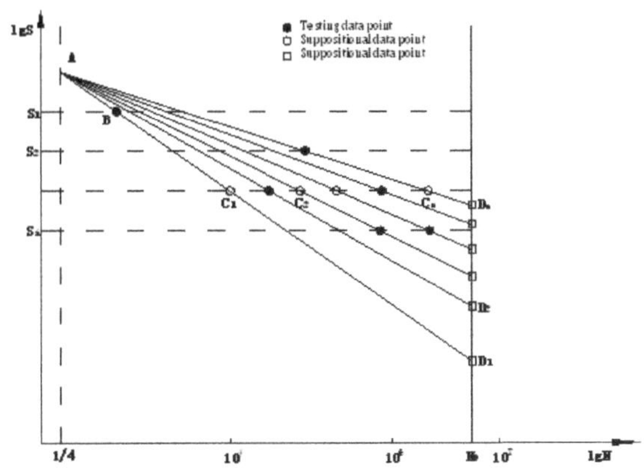

Fig.1 Schematic of the SSMM and the FEPM

Then, in log-log coordinate system data pairs $(\overline{X_i}, \lg S_i)$ are taken to fit the median S-N line of the sample with least square method.

$$\lg S_i = \alpha + \beta \overline{X_i} \tag{2}$$

where $\overline{X_i}$ is the mean of fatigue life of sample loaded at "i"th stress level. S_i is the test load expressed in force, moment or stress; α and β are intercept and slope of the line respectively.

The correlation coefficient of Eq.2 should be checked, in order to decide if more test data are asked and the required minimum value is reached.

The ordinate of point A is obtained from Eq.2, which can be written as:

$$\lg S_A = \alpha + \beta \lg \frac{1}{4} \tag{3}$$

According to geometric relationship, the fatigue life of suppositional data point C is obtained from the following equation:

$$\lg N_c = \lg \frac{1}{4} + \frac{(\lg S_c - \lg S_A)\left(\lg \frac{1}{4} - \lg N_B\right)}{\lg S_A - \lg S_B} \tag{4}$$

According to geometric relationship, the fatigue limit of suppositional data point D is obtained from the following equation:

$$\lg S_D = \lg S_A + \frac{(\lg N_0 - \lg 1/4)(\lg S_A - \lg S_c)}{\lg 1/4 - \lg N_c} \tag{5}$$

3 DATA ANALYSIS

It is necessary to analyze the normal distribution of data points. The mapping method and analytic method are used in the method introduced in paper analyses as the same as SGMM [2]. The analysis method of the lower limit of fatigue life and the lower limit of fatigue strength at different survival probability as follows:

First, the mean fatigue life or the mean fatigue limit of data points and the standard deviation are estimated from the following equation.

$$\bar{x} = \frac{1}{n}\sum_{i=1}^{n} x_i \tag{6}$$

$$\sigma_x = \sqrt{\frac{1}{n-1}\sum_{i=1}^{n}\left(x_i - \bar{x}\right)^2} \tag{7}$$

where n is the sub-total number of data points, σ_x is the standard deviation, $\bar{x}$ is the mean and x_i is the fatigue life values of data points or fatigue limit values of data points.

Then, the coefficient of variation of fatigue life or fatigue limit is estimated from the follow equation:

$$\lambda = \frac{\sigma_x}{\bar{x}} \tag{8}$$

where λ is the coefficient of variation.

And the sub-total number of data points must meet with the following equation:

$$\delta_x = \frac{\lambda \cdot t_{\alpha,v}}{\sqrt{n}} \tag{9}$$

where δ_x is error limit, $t_{\alpha,v}$ is the value of t distribution.

Finally, the lower limit of fatigue life or the lower limit of fatigue strength is estimated at a given survival probability from the following equation:

$$x_p = \bar{x} - k_{(p,1-\alpha,v)}\sigma_x \tag{10}$$

where x_p is the lower limit of fatigue life or the lower limit of fatigue strength, and the coefficient $k(p, 1 - \alpha, v)$ is the one-sided tolerance limit for a normal distribution.

4 MEASURING VALIDATION

In order to validate the validity of the method introduced in paper, a fatigue test on a steel plate applied for automotive crossbeam has been achieved. P-S-N curve has been estimated with SGMM, and fatigue limit has been estimated with UDMM. During the test, the stress ration R=-1, the appointed number of cycle is 10^7 cycles. The results are shown in Tab.1 and Tab.2. And the first four data points of each stress level in Tab.1 were employed for estimating fatigue life and fatigue limit with the accelerated measuring method. The mean of fatigue life, the lower limit of fatigue life at survival probability 95% and 99%, the mean of fatigue limit and the lower limit of fatigue limit at survival probability 95% and 99% were estimated separately. The results are given in Tab.3-Tab.6.

The results given in Tab.3 and Tab.4 show that comparing with SGMM, the relative error of mean fatigue life estimated with SSMM is less than 5%. The relative error of the lower limit of fatigue life estimate with the SSMM at given survival probability is less than 10%.

The results given in Tab.5 and Tab.6 show that comparing with UDMM, the relative error of mean fatigue limit estimated with FEPM is less than 5%. The standard deviation estimated with FEPM is smaller than that estimated by UDMM, which indicates the dispersibility of data points estimated with FEPM is better. The relative error of lower limit of fatigue limit estimated with FEPM at given survival probability is less than 10%.

Tab.1 Testing results on a steel plate for automotive crossbeam with SGMM

Sequence number	Stress level/MPa	Cycle number	Sequence number	Stress level/MPa	Cycle number
1	417	161103	18	367	921517
2	417	159066	19	342	1352560
3	417	123500	20	342	1562320
4	417	132198	21	342	1289900
5	417	116421	22	342	950365
6	417	150053	23	342	1014550
7	392	235200	24	342	1306330
8	392	392102	25	317	2523650
9	392	305233	26	317	2089470
10	392	288167	27	317	1957710
11	392	257656	28	317	1858150
12	392	315620	29	317	2364400
13	367	693509	30	317	2039030
14	367	508823	31	300	4662980
15	367	623365	32	300	4022380
16	367	582270	33	300	3320200
17	367	703340	34	300	4818820

Tab.2 Testing results on a steel plate for automotive crossbeam with UDMM

Sequence number	Stress level/MPa	Cycle number	Sequence number	Stress level/MPa	Cycle number
1	292	4243851	9	292	>10000000
2	275	4874257	10	309	3047291
3	258	>10000000	11	292	4268925
4	275	>10000000	12	275	4901352
5	292	4082652	13	258	>10000000
6	275	>10000000	14	275	>10000000
7	292	4637874	15	292	3988253
8	275	>10000000	16	275	>10000000

Tab.3 Results of mean fatigue life estimated separately with SGMM and SSMM

Survival probability	Measuring method	Stress level/MPa					
		417	392	367	342	317	300
50%	SSMM	5.4867	5.6412	5.8059	5.9822	6.1718	6.3096
	SGMM	5.144	5.4699	5.8199	6.0893	6.3277	6.6193
	Relative error	6.66%	3.13%	-0.24%	-1.76%	-2.46%	-4.68%

Tab.4 Results of lower limit of fatigue life at given survival probability estimated separately with SGMM and SSMM (confidence level is 95%)

Survival probability	Measuring method	Stress level/MPa					
		417	392	367	342	317	300
95%	SSMM	4.9937	5.1358	5.2871	5.4491	5.6234	5.7499
	SGMM	4.8415	5.0725	5.5085	5.6671	6.0664	6.2396
	Relative error	3.14%	1.25%	-4.02%	-3.85%	-7.30%	-7.85%
99%	SSMM	4.8076	4.9449	5.0911	5.2477	5.4163	5.5385
	SGMM	4.7309	4.9273	5.3947	5.5127	5.9708	6.0996
	Relative error	1.62%	0.36%	-5.63%	-4.81%	-9.29%	-9.20%

Tab.5 Results of mean fatigue limit estimated separately with UDMM and FEPM

Survival probability	Measuring method	Fatigue limit/MPa	Standard deviation /MPa
50%	UDMM	281.38	10.70
	FEPM	274.91	5.61
	Relative error	-2.30%	—

Tab.6 Results lower limit of fatigue limit at given survival probability estimated separately with UDMM and FEPM (confidence level is 95%)

Survival probability	Measuring method	Lower limit of fatigue limit/MPa
95%	UDMM	247.27
	FEPM	261.96
	Relative error	5.94%
99%	UDMM	234.80
	FEPM	257.07
	Relative error	9.48%

5 CONCLUSION

It is feasible to estimate the fatigue life with SSMM and the fatigue limit with the FEPM introduced in paper. The accelerated measuring method introduced in paper costs fewer time.

6 ACKNOWLEDGEMENT

This work was carried out with the financial support of Chongqing Basic and Frontier Research Program (cstc2013jcyjC60001). This support is greatly acknowledged.

REFERENCES

[1] Gao, Z. Z. & Xiong, J. J. The present situation and prospects for the fatigue/fracture reliability study, Journal of mechanical strength, 1995, 17(3), 61-79.

[2] ISO 12107: 2003 Metallic materials—Fatigue testing—Statistical planning and analysis of data.

[3] Feng, M.B. & Chen, Y. L., Fixed Emitting Point Method-An Accelerated Test Program to Determine Fatigue Limit of Automotive Components, SAE paper 2004-01-1646.

Study on fatigue damage of the Cu32W68 pseudo alloy

Y.T. Huang[1], W.Z. Chen[1,2], W. Li[1], Q.H. Guo[1], Q. Yan[1]
[1] College of Materials Science & Engineering, Fujian University of Technology, P.R. China
[2] College of Materials Science & Engineering, Xiamen University of Technology, P.R. China

ABSTRACT

The three-point bending fatigue test of Cu32W68 alloy, a kind of pseudo alloy composed of tungsten and copper prepared by sintering and infiltration method, was carried out at ambient temperature. The characteristic curve of fatigue damage and quantitative tracking of Cu32W68 alloy under cyclic loading were measured. The results show that the fatigue damage process of Cu32W68 alloy under three-point bending test can be divided into three stages of linear variation EF, FG, GH based on the changing of materials' damage parameter D, which occupy the 49%, 32% and 19% of the fatigue life, respectively.

Keywords: Cu32W68; electric contact material; fatigue damage; materials' damage parameter D; fatigue crack evolution

1. INTRODUCTION

Copper-tungsten alloy (CuW alloy) is a series of two-phase hybrid organization for tungsten and copper phase, which is a kind of pseudo alloy as copper and tungsten present mutual insolubility characteristic, poor wettability, so that could not form intermetallic compounds. CuW alloys are widely used in high voltage electrical contacts, which are the key elements to determine the reliability and service life of the switch's operation, because of their high hardness, high strength, good arc erosion resistance and welding resistance performance(1,2). The integral contact material can provide prolong lifetime and work reliability in their service(3,4).

Researchers have shown their huge interest in the copper-tungsten alloy and have made much works in the wear and impact properties of the contact in recent years(5,6). However, high voltage electrical apparatus in operation have to endure enough self-force to switch on-off the moving contact and the fixed contact pressure. So the contact afforded the fatigue damage as well as the wear and impact during the performance. But to the best of our knowledge, there are little investigation on the fatigue damage of the CuW pseudo alloys and the relationship with the microstructure and fatigue damage in the dynamic load for the CuW alloy.

In this paper, Cu32W68 pseudo alloys were prepared by sintering-infiltration method. The rules of fatigue damage and the microstructure evolution during fatigue process were detail investigated. Significantly, the fatigue damage process was by the damage mechanics based on the macro-phenomenological theory.

2. EXPERIMENTAL

Cu32W68 alloy was prepared by sintering and infiltration method. Cathode copper powder and tungsten powder with the diameter of 75 μm and 7.5 μm, respectively, were used as the original materials. Copper powders with three mass percentage used as induced copper powder were first milled with tungsten powders for 4h, while stirring in the blended. The mixture was reduced and sintered at 1500℃under the $1.6m^3/h$ H_2 atmosphere. The size of samples was 13 mm×13 mm×70 mm after sintering.

The three-point bending fatigue test was measured at ambient temperature by using the INSTRON1342 fatigue testing machine. The changing loaded cyclic loads were controlled by sinusoidal pulsating load with the 5Hz test frequency. The two types of cyclic loads were -1.0~-8.5 kN. The sign of stress and displacement under the cyclic load were dynamic measured by the displacement sensors and load sensors from the fatigue testing machine.

Here, materials' damage parameter D is used as an important parameter in the characterization for fatigue damage process, which is defined as, $D=A/A_0$, the ratio between the initial value A_0 of dimensionless parameter in cycle loads' working and the measured value A in the testing process. The materials' damage parameter D could sensitively react the specimens' fatigue damage process through large of macro and micro test observation. Based on the Dobson model, the authors fundamentally improved the accuracy by using the signal feedback and compensation technique in systems analysis theory. According to our previous research work, a mathematical model on dynamic stress of Cu32W68 alloy materials and the resulting displacement data has been built. In order to ensure the minimum error between the establishment of mathematical models and real situation, taking the technique of continuously makes the measured data modify the built mathematical models, that is, the real-time feedback and compensation technique in the error of systems analysis, so they can ensure the accuracy of the built mathematical models.

The microstructure of fatigue cycle process was evaluated by a Philips XL30ESETMP environment scanning electron microscope.

3. RESULTS AND DISCUSSION

3.1 Fatigue damage process of the Cu32W68 alloy

Fig.1 shows the fatigue damage parameter with different cycle index of the load under the cyclic loads of -1.0~-8.5 kN for the Cu32W68 alloy. In detail, the fatigue damage parameter D sharply goes down with the increasing of the cycle index when the cycles from 0 to 560, and then, the damage parameter D shows slight decline or almost remains unchanged until the cycles reach 930. However, the damage parameter D sharply goes up with the increasing of cycles, which means that fatigue crack rapid expansion and then fracture of the alloy occurs at this time. According to the relationship between the fatigue damage parameter D with loading cycle index, the fatigue damage process of the Cu32W68 alloy can be divided into three processes of linear variation EF, FG, GH based on the changing of D. In the first stage (EF), within 0 to about 560 cycles, it is the stage for fatigue crack initiation. By quantitative analysis, this stage accountes for approximately 49 percent of the overall fatigue service life of the alloy. In the second stage (FH), the loads cycles are about 370 cycles and account for 32 percent of the specimens' overall fatigue service life, which means the stage for fatigue crack development. The final stage (HG), the total cycles under the loads are about 220 cycles, and this damage process accounts for about 19 percent of the Cu32W68's overall fatigue service life, which means that the fatigue crack rapid expands to fracture.

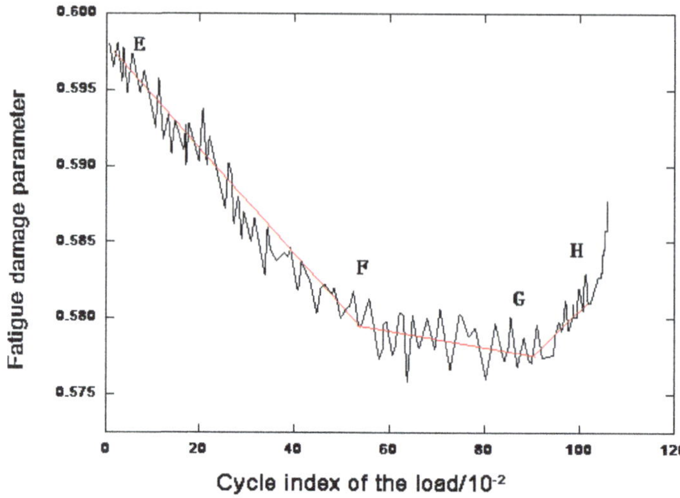

Fig.1 Fatigue damage parameters with different cycle index of the load

3.2 Morphology of the fatigue damage process

The surface morphology of Cu32W68 alloys are significantly affected by the fatigue cycles. The images of the morphology of Cu32W68 pseudo alloys with original state (a), after 200 cycles(b), slip bands appeared(c), fatigue crack initiation(d) and fatigue crack confluence(e) are shown in Fig. 2, respectively. It can be found the original microstructure before the fatigue cycle from the Fig. 2(a). The Cu32W68 pseudo alloy is composed of two-phase copper and tungsten. By infiltration and solidification, Cu coated around uniform structure of white tungsten-matrix particles with clear interface and intact bonding without inclusion of holes. The range of particle size is from 3μm to 12μm and the average particle size is 8 μm.

Under the cyclic loading, copper phase deformed first. As shown in the Fig. 2(b), copper phase displaced after 200 fatigue cycles while the tungsten particles are not deformed.. At this stage, the copper-tungsten interface is still straight and complete. There were not defects such as cracks and holes present after going through the fatigue. The metallurgical structure is still compaction and the copper-tungsten interface is well-combined. According to our other mechanical properties tests, the hardness of the Cu32W68 pseudo alloys increases with the fatigue cycles.

With the increasing of cyclic loading, the fatigue crack of Cu32W68 alloy is first initiated in the Cu phase and then developed. When the fatigue cycles till 560, corresponding to the first fatigue cyclic stage (EF), a large amount of glide band appears in the copper phase during the cycle process, as shown in Fig. 2(c). Cu32W68 pseudo alloys are made up of two- immiscible phase of copper and tungsten. While the tungsten phase has the high-intensity and poor plasticity, the copper phase has the poor-intensity and high plasticity. Therefore, it is not strange that the deformation and glide bands appeared first in the copper phase after the cyclic loads for the different properties of two-phases. It is found that the fatigue crack of Cu32W68 alloy is always deformed in copper particle and there are no sign of fatigue crack initiation on the copper-tungsten interface or inside the tungsten.

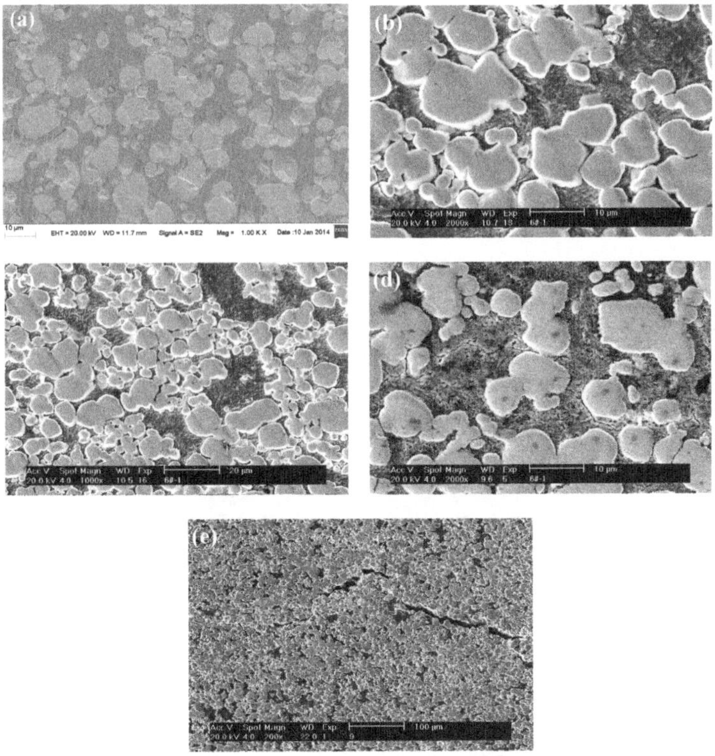

Fig.2 SEM images of Cu32W68 pseudo alloys with none(a), 200 cycles(b), slip bands appeared(c), fatigue crack initiation(d) and fatigue crack confluence(e)

The deformation of the copper phase in cyclic loads closes to saturation after the first stage. In the second stage (FH) of the fatigue cycles, as shown in Fig. 2(d), the fatigue cracks initiate in the copper phase and Cu32W68 alloy penetrates into FH stage of fatigue damage under cyclic loads.

After multiple cracks initiate inside the copper crystal in FH stage with the constant increasing of fatigue cycles, the third stage (HG) appears, as shown in the Fig. 2(e). The micro cracks of the copper phase gradually merge, cut through with the further increasing of cycles, then the macro cracks appear on the surface of the specimens. It is the stage of accelerated development for the fatigue damage and finally rapid fracture, as shown in the Fig. 2(e).

4. CONCLUSIONS

The Cu32W68 pseudo alloy prepared by sintering-infiltration method is composed of two-phase copper and tungsten. Cu coated around uniform structure of white tungsten-matrix particles, the average tungsten particle size is 8 μm. The fatigue damage process of Cu32W68 alloy under three-point bending test can be divided into three stages of linear variation EF, FG, GH, according to the relationship between the fatigue damage parameter D with loading cycle index. They occupy the 49%, 32% and 19% of the whole fatigue life, which are corresponding to the fatigue crack initiation, development and fracture during the fatigue damage process, respectively. The microstructure evolution of the fatigue damage process shows that the fatigue crack of Cu32W68 alloy is first deformed and initiated in the Cu phase and then developed with the increasing of cyclic loading. It is found that both the early slipping and fatigue crack initiation always occur in copper phase, neither in the interface of the copper-tungsten nor inside the tungsten. The fatigue crack initiation process (FH stage) and the fatigue crack expansion process (FH stage) accounted for 81 percent of whole life. Therefore, the fatigue life of Cu32W68 alloy was mainly decided by copper phase.

ACKNOWLEDGEMENTS

This work is supported by the National Natural Science Foundation of China (No. 51271058) and of the Start-up Foundation of Fujian University of Technology (GY-Z13106).

REFERENCES

(1) Yanlong W, Shuhua L, Peng X, et al. FEM simulations of tensile deformation and fracture analysis for CuW alloys at mesoscopic level[J]. Computational Materials Science, 2011, 50(12): 3450-3454.
(2) Weichan C, Shuhua L, Zhuangfeng G, et al. Effect of Fe on vacuum breakdown properties of CuW alloys[J]. International Journal of Refractory Metals and Hard Materials, 2011, 29: 656-661.
(3) Kun Q, Shuhua L, Peng X, et al. In situ synthesis and electrical properties of CuW-La_2O_3 composites[J]. International Journal of Refractory Metals and Hard Materials, 2012, 31: 147-151.
(4) Liang S, Wang X, Wang L, et al. Fabrication of CuW pseudo alloy by W–CuO nanopowders[J]. Journal of Alloys and Compounds, 2012, 516: 161-166.
(5) Yang X, Zou J, Liang S, et al. Effect of Alloying Interlayer on Interfacial Bond Strength of CuW/CuCr Integral Materials[J]. Materials Science and Engineering: A, 2010, 527(21): 5631-5636.
(6) Wang Y, Liang S, Ren J. Analysis of meso-scale damage and crack for CuW alloys induced by thermal shock[J]. Materials Science and Engineering: A, 2012, 534: 542-546.
(7) Dobson S, Noori M, Hou Z, et al. Modeling and random vibration analysis of SDOF systems with asymmetric hysteresis[J]. International Journal of Non-Linear Mechanics, 1997, 32(4): 669-680.

Small fatigue crack propagation in Al-Cu alloy laminated structure via ultrasonic consolidation

X.H. He[a], H.J. Shi[a], Y.D. Zhang[b], W.X. Fu[b], Z.G. Yang[b], C.E. Wilkinson[c]
[a] AML, School of Aerospace, Tsinghua University, P.R. China
[b] Key Laboratory for Advanced Materials of Ministry of Education, School of Materials Science and Engineering, Tsinghua University, P.R. China
[c] The Boeing Co., USA

ABSTRACT

In this work, the small fatigue crack behavior of the Al-Cu alloy laminated structure, produced by ultrasonic consolidation, was investigated by in-situ scanning electron microscopy method. The influence of local microstructure on the crack propagation of the laminates after different heat treatments has been compared and discussed. It was found that both the precipitations and the interfaces of the layers could retard crack growth and alter the growth direction, and the effect of interfaces was much stronger. The barrier effect levels of the interfaces on impeding the fatigue crack growth depended on the ratio of the toughness of the materials to interlaminar bond strength, which was affected by the heat treatment.

1 INTRODUCTION

Laminated metal structures have been widely utilized in aerospace, navigation and defense industries due to their unique properties, e.g. impact behavior, fracture toughness (1). Ultrasonic consolidation (UC) process is an innovative solid-state manufacturing technology, with a number of advantages over the traditional techniques, e.g. internal geometry capability, less consumed energy and residual stresses. Particularly, UC allows embedding fibers, wirings and electronics sensors into the laminates to make smart materials (2, 3). However, most of the investigations on laminated structures were focused on the fracture and impact toughness (4), but the fatigue properties were limited studied (5, 6). Besides, the researches so far on the UC produced laminates were concentrated on the optimization of parameters by peel test and microstructure observation (7-9), but the studies on the mechanical properties were scarce and not in-depth (2, 10), let alone the fatigue property.

This work uses the in-situ SEM technique to investigate the small fatigue crack growth (SFCG) behavior in the UC produced Al-Cu laminated structure after different heat treatments (HT). Particular emphasis focuses on examining the fatigue cracking mechanism and the effect of local microstructure on crack growth.

2 MATERIALS AND EXPERIMENTAL DETAILS

The nominal composition of the Al-Cu alloy foil used in the current work is (in wt.%): 0.5Si, 0.5Fe, 3.8-4.9Cu, 0.3-0.9Mn, 1.2-1.8Mg, 0.1Cr, 0.25Zn, 0.15Ti and Al balance. The laminated structure is successfully produced layer by layer via UC process. The dog-bone shape slab specimens are cut along the longitudinal orientation of the metal foils, with a 1.4mm by 1.5mm gauge cross section. A U-shaped notch is prepared by WEDM on one side of the specimens. The specimens are divided into two groups and heat-treated to a T4 and a T62 temper, respectively. The surfaces of all the specimens are polished to a mirror surface and then etched in a solution of 1%HF + 1.5%HCl + 2.5%HNO_3 + 95%H_2O to reveal the microstructure.

Optical microscopy (OM) and scanning electron microscopy (SEM) are first performed to observe the microstructure of the laminated structure after etching. The in-situ fatigue tests are performed at room temperature (RT) in the vacuum chamber of the SEM using a specially designed servo-hydraulic testing system. A maximal stress of 300MPa is applied throughout the tests for all specimens. The load waveform utilized is sinusoidal with stress ratio R=0.1 and frequency f=5Hz.

3 RESULTS

3.1 Microstructural characteristics

The OM observations of the microstructures on a cross-section of the laminates are shown in Fig.1. The whole bonding quality of the interfaces after both heat treatments is well. Although there are a few small defects at the interfaces, no initial delaminations occur in the laminated structure. The grain sizes of the specimens are not uniform, but there was no obvious difference of the grain sizes between the specimens after T4 and T62 HT. Generally, the grains near the interfaces were much smaller, as shown in Fig. 1(a) and (b). Moreover, precipitation compounds were widely distributed in the grains, as indicated in Fig. 1(c).

Figure 1. OM of laminated structures after (a) T4 HT and (b) T62 HT; (c) SEM observation of precipitations in the structure. (The black arrows indicate the locations of interfaces.)

3.2 SFCG of specimens after T4 HT

The typical microstructure of the laminated specimen after T4 HT is shown in Fig. 2a. There are a few of defects at interfaces and precipitation compounds distributing in the whole specimens. The stress concentration at the prepared U-shaped notch compels the crack to initiate from the notch root, as shown in Fig. 2b. Although there are also small fatigue cracks at the precipitations after over 23,000 cycles, they cannot become the main crack. It is worth noticing that the crack don't break the precipitations but tend to change its path and bypass the small blocky precipitation. This is because that the stress in the initial stage is not enough for the strength of precipitations. Along with the crack growth, some obvious slip traces are observed near and in front of the crack. As the crack propagated along these slip traces, a subsurface precipitation is revealed, which explained why the slip trace appeared preferentially at that site.

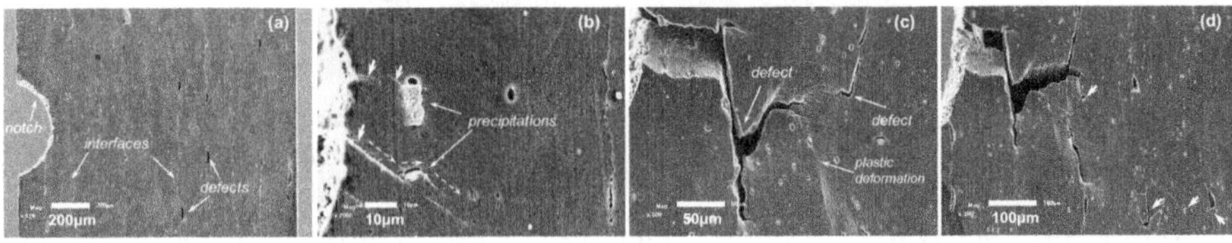

Figure 2. Fatigue short crack growth in laminated specimen after T4 HT. (a) 0 cycles; (b) 22954 cycles; (c) 28241 cycles; (d) 28837 cycles.

The interface of foils is a unique part in the laminated structure, thus and its influence on the crack growth of the laminated specimens is of great concern. The main crack is strongly blocked after approaching the first interface. During the subsequent thousands of cycles, the transvers propagation of the fatigue crack stops, instead, the crack begins to propagate along the interfaces and generate evident delamination. Meanwhile, the main crack continued to impinge the interface and its crack tip opening displacement (CTOD) visibly increased as illustrated in Fig. 2c, indicating evident effect on retarding the growth of the transverse crack. At this stage, the longitudinal crack along interface was dominated and propagated faster. This retardation by local interface delamination has also been reported in the literature (6). However, the interface crack doesn't propagate any further after a number of cycles. Some short cracks initiate from the defect tips along the interface and start to grow transversely, one of which cracks become the main crack and rapidly propagated (Fig. 2c). At the same time, obvious plastic deformation with evident slip traces appears in the front region of the crack, because of the high stress near the crack tip. In addition, defects at the second interface and fractured precipitations due to deformation incompatibility also emerge, as indicated by the short arrows in Fig. 2c. It is interesting to find that, the slip traces occur near the defects or point at the defects. Therefore, it is suggested that the slip traces indicate the crack growth path, i.e., the crack tend to connect the defects together. The inference is confirmed by the crack propagation behavior during the following cycles. And the rugged growth path of crack is attributed to the large plastic deformation and deformation incompatibility. When arriving at the second interface, the crack behavior was similar to that at the first interface. Due to the high stress level, lots of defects at the other interfaces and small cracks caused by the fracture of precipitations far from the main crack also emerged obviously (Fig. 2d). These short cracks coalesced with the main crack to induce the final rupture.

3.3 SFCG of specimens after T62 HT

The typical microstructure of the laminated specimen after T62 HT is shown in Fig. 3a, and the charateristics are similar to those after T4 HT. Since the first interface is close to the notch and the precipitation at the interface also generates stress concentration, the initial short cracks not only from the notch root, but also from the precipitations at the first interface, as shown in Fig. 3b. These initial short cracks coalesce, bypass some subsurface precipitations (small arrows in Fig. 3b) and improved the main crack. Although the crack also tends to propagate along the interface as same as that in specimen after T4 HT, the propagation distance is very short and the transverse crack is still dominated. The main crack crosses the first three interfaces and propagates forward stably and transgranularly, without obvious increasing of CTOD. That is, the interface shows a weak effect on the SFCG in the initial stage.

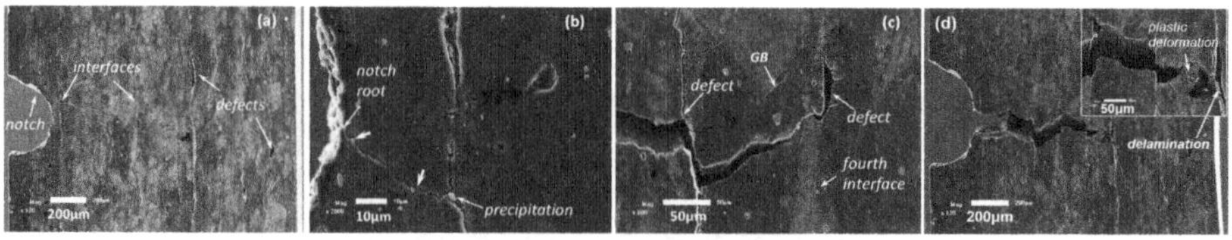

Figure 3. Fatigue short crack growth in laminated specimen after T62 HT. (a) 0 cycles; (b) 13763 cycles; (c) 18254 cycles; (d) 18903 cycles.

The crack growth behaviors are similar when the crack goes through the fourth and fifth layers, as shown in Fig. 3c. First, a short crack initiates from the defects at the interfaces, as indicated by small arrows in Fig. 3c. Then, the crack propagates forward and tends to coalesce with the defects at the next interface. Although some defects at the GBs emerged apparently during the propagation as indicated in Fig. 3c, the cracking is still transgranular rather than intergranular. After reaching the fifth interface, the CTOD increases evidently instead of the crack growth. Severe plastic deformation appears at the crack tip, as shown in Fig. 3d. Moreover, the delamination at the sixth interface was obvious before the final fracture.

4 DISCUSSIONS

According to the observations in Section 3, it can be easily concluded that the interfaces effectively act as barriers to the crack growth. Due to the bigger CTOD and larger range of delaminations, the blockage effect of the interfaces of specimens after T4 HT is much stronger than that after T62 HT. In order to explain the underlying mechanism, the basic mechanical properties of Al-Cu alloy used in this work are tested according to ASTM standards and the bond strength of the laminated specimens is tested by the method in the literature (7). It should be noted that the CTOD at the maximum load is used to characterize the fracture toughness in the ASTM standard. The results in Table 1 show that: the materials after T4 HT have much larger elongation and fracture toughness than those after T62 HT; however, the tensile strengths of materials and bond strengths of the laminated structures after two different HTs are very approximate. The higher fracture toughness means the larger energy required for crack propagation in the materials, while the higher bond strength indicates the more energy needed for the interface delamination. Therefore, the specimens after T4 HT, with higher toughness, require more energy for the transverse crack growth in the next layer than that for the interface delamination. The energy is first consumed in the earlier and larger range of delaminations rather than the new crack initiation in the next layer, and much more extra time is needed for the accumulation of enough energy for the transverse crack propagation. The specimens after T62 HT exhibit the opposite behaviors. Hence, in our work, it was thought that the ratio of toughness to bond strength (i.e. $n=\delta/P$) plays a critical role in the crack propagation near the interfaces. The higher n means the delaminations will occur earlier and more obviously, while the lower n indicates the crack is more inclined to cross the interface with no or slight deflection and propagate in the next layer.

5 CONCLUSIONS

The small crack growth behaviors in the UC made Al-Cu laminated structure after a T4 and a T62 heat treatment are investigated by in-situ SEM technique at RT. The SFCG in this laminated structure is discontinuous, transgranular and shows great sensitivity on the microstructures. Particularly, the interfaces obviously act as barriers to prevent the crack propagation and also alter the crack path. The transverse crack is impeded and turns to propagate along the interfaces for a finite length, when arriving at the interfaces. Some short cracks caused by the fracture of precipitations and defects at the interfaces coalesce with the main crack and lead to the final rupture. Moreover, the ratio of the materials toughness to the interfaces bond strength of specimens after different HTs determine the crack behavior at the interfaces, i.e. whether delaminate firstly or just propagate forward with no or slight deflection. Concerning the similar bond strength, the impeditive influence of the interfaces on the cracking in specimens after T4 HT is stronger, due to the higher toughness than the specimens after T62 HT.

Table 1 Mechanical properties of Al-Cu monolithic specimens and bond strength of laminated specimens

HT	Al-Cu alloy monolithic			Laminated
	Tensile strength σ_{UTS} (MPa)	Elongation (%)	Fracture toughness δ^a (μm)	Bond strength P (N)
T4	457	21	36	10290
T62	449	6	18	10560

[a] δ represents the CTOD at the maximum load applied on the specimen in fracture toughness test.

ACKNOWLEDGEMENTS

The financial supports from Tsinghua-Boeing Joint Research Center (TBRC) and the National Natural Science Foundation of China (Nos. 11372151 and 51071094) are highly acknowledged.

REFERENCES

Liu, C. Y., Wang, Q., Jia, Y. Z., Jing, R., Zhang, B., Ma, M. Z. & Liu, R. P. (2012) *Mater. Sci. Eng. A* **556**, 1-8. doi: 10.1016/j.msea.2012.06.046.

Yang, Y., Ram, G. D. J. & Stucker, B. E. (2009) *J. Mater. Process. Technol.* **209**, 4915-4924. doi: 10.1016/j.jmatprotec.2009.01.014.

Cheng, X., Datta, A., Choi, H., Zhang, X. & Li, X. (2007) *J. Manuf. Sci. Eng.* **129**, 416-424. doi: 10.1115/1.2515456.

Cepeda-Jimenez, C. M., Carreno, F., Ruano, O. A., Sarkeeva, A. A., Kruglov, A. A. & Lutfullin, R. Y. (2013) *Mater. Sci. Eng. A* **563**, 28-35. doi: 10.1016/j.msea.2012.11.052.

Hassan, H. A., Lewandowski, J. J. & Abd El-Latif, M. H. (2004) *Metall. Mater. Trans. A* **35A**, 45-52. doi: 10.1007/s11661-004-0107-7.

Liu, H. S., Zhang, B. & Zhang, G. P. (2011) *Scripta Mater.* **65**, 891-894. doi: 10.1016/j.scriptamat.2011.08.001.

Zhang, C., Deceuster, A. & Li, L. (2009) *J. Mater. Eng. Perform.* **18**, 1124-1132. doi: 10.1007/s11665-008-9342-1.

Li, D. & Soar, R. C. (2009) *J. Eng. Mater. Technol.-Asme.* **131**. doi: 10.1115/1.3030946.

Hopkins, C. D., Dapino, M. J. & Fernandez, S. A. (2010) *J. Eng. Mater. Technol.-Asme.* **132**. doi: 10.1115/1.4002073.

Mariani, E. & Ghassemieh, E. (2010) *Acta Mater.* **58**, 2492-2503. doi: 10.1016/j.actamat.2009.12.035.

Fatigue life of lead free solder material with very sharp notch

T. Kawakami, T. Kinoshita, H. Oriyama
Mechanical Systems Enginerring, Toyama Prefectural University, Japan

ABSTRACT

Lead free solder materials have been wildly used on printed circuit boards in electronic apparatus. The solder joints sometimes are opened under thermal cyclic loads. The fatigue crack is initiated usually around the edge of the interface where the strain concentrates severely. In this study, Sn-3.0Ag-0.5Cu test pieces with V shape notch were supplied to fatigue tests, and inelastic simulation was performed to obtain the strain distribution around the notch. The low cycle fatigue test result with or without a notch and the simulated result were compared, and discussed the prediction method for fatigue life of lead free solder joint.

1 INTRODUCTION

Lead free solder materials have been wildly used in the electronic industry. The solder joints sometimes are opened under thermal cyclic loads as low cycle fatigue phenomena. The fatigue crack is initiated usually around the edge of the interface between the solder joint and printed circuit board (PCB) or device. Because of very severe strain concentration is occurred around the edge, where is sometimes the stress strain singular point. Many thermal cyclic tests with actual mounted PCB and numerical simulations were performed, it was proposed that the fatigue life of such a solder joint can be predicted by the inelastic strain range at the inside point 50μm in depth from the edge of the interface of solder joint[1,2]. However, there are many uncertainty factors, joint shapes, thermal load conditions and thermal properties. In this study, test pieces of Sn-3.0Ag-0.5Cu material with V shape notch were supplied to low cycle fatigue tests and the fatigue curve was obtained at room temperature. And inelastic stress strain simulation was carried out on test conditions to verify the 50μm rule for low cycle fatigue life of solder materials with severe strain concentrated portion excluding uncertainty factors.

2 LOW CYCLE FATIGUE TEST

Low cycle fatigue tests had been carried out and fatigue curve was obtained for Sn-Ag-Cu solder materials at room temperature and under total strain rate 10^{-4}/s with smooth test specimens by electro-hydraulic servo fatigue testing machines. The diameter of the test specimen was 10mm or 6mm.

The shape of the test specimen with V notch was illustrated in Fig.1. V notch was machined with threading tool. The radius at the bottom of notch could be made keener and it was controlled as 50μm. The test equipment (Load Capacity; 10kN) and test specimen was shown in Fig.2. The gage length of the extensometer was 10mm. Low cycle fatigue tests were achieved at room temperature and under total nominal strain rate 10^{-4}/s. Test results were shown in Fig.3 together with the result of smooth test specimens according to Coffin-Manson rule. The equations for the fatigue curve without and with V notch were derived as follows,

$\Delta\varepsilon_{in} \cdot N_f^{m} = C$ where m =0.572, C =0.790 for without notch...........(1),
$\Delta\varepsilon_{in} \cdot N_f^{m'} = C'$ where m'=0.221, C'=0.443 for with notch................(2).

The Inclination of fatigue curve with V notch was gentler than that of smooth specimen on double-logarithmic graph, and the effect of V notch decreased in long life conditions. The value of the reduction factor β' was 1.45 at N_f=1,000, which is a typical number in thermal cycle test (TCT) for electronic apparatus.

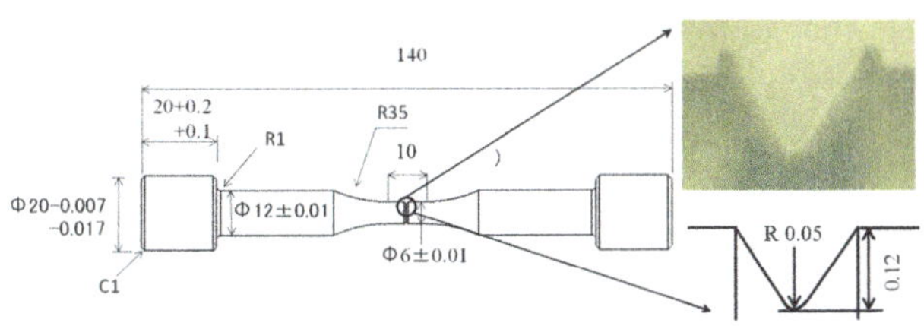

Fig.1 Test Specimen with V Notch Configuration

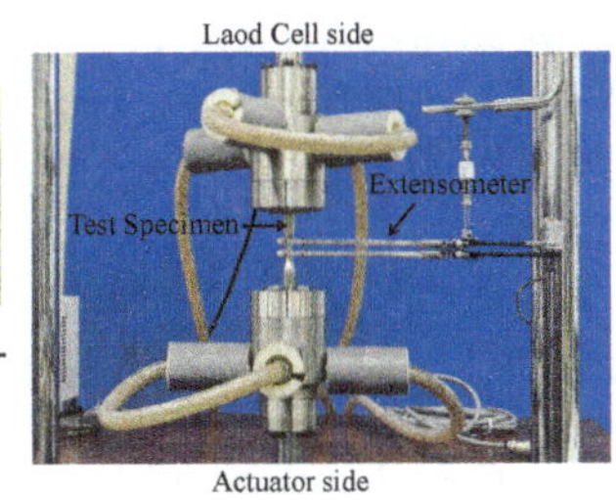

Fig.2 Test Equipment

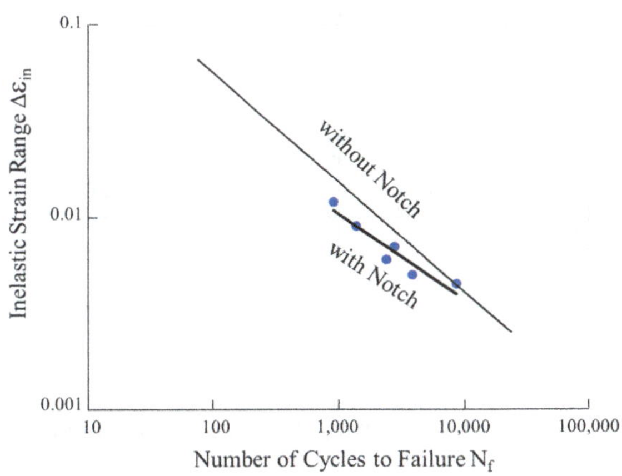

Fig.3 Low Cycle Fatigue Lives for Sn-3.0Ag-0.5Cu Solder Material with V Notch

3 NUMERICAL STRESS STRAIN SIMULATIONS AROUND THE V NOTCH

In order to obtain the inelastic strain range in the vicinity of V notch, inelastic numerical simulation based on the finite element method was carried out. The simulation code was ADVENTUREClaster® that is a large scale parallel computing general purpose code[3)].

3.1 FEM model

The 1/8 area of the inside of the gage length of the extensometer was modelled as shown in Fig.3. Symmetric surfaces were kept the each plane as boundary conditions.

And right side surface was shifted by ±0.025mm (total strain range; 0.01, inelastic strain range; 0.009, 200s/cycle) in the axial direction 2.5 cycles at room temperature as shown in Fig.4. FEM model was shown in Fig.5. Isoparametric element was adopted.

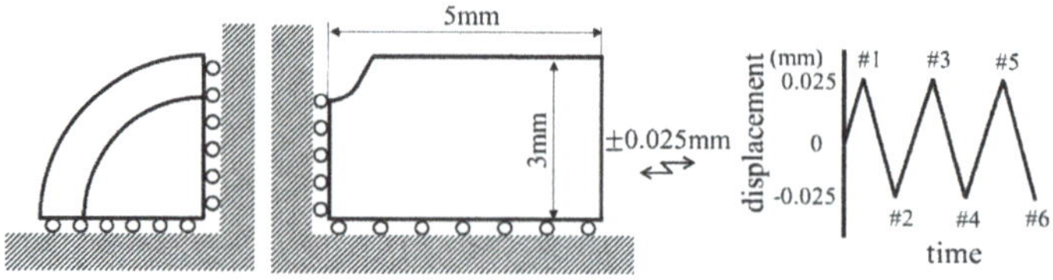

Fig.4 Analyzed Area and Load Conditions

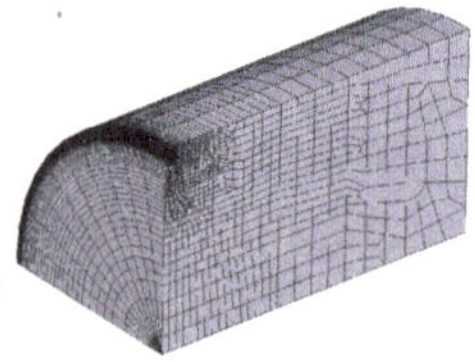

Fig.5 FEM Model

3.2 Material properties

Material properties of Sn-3.0Ag-0.5Cu solder material are plugged in ADVNTURECluster. The material model is non-unified constitutive model, and it can simulate the behaviour depending on time for transient and steady strain rates of Sn-3.0Ag-0.5Cu solder material accurately[4)].

3.3 FEA results

The axial stress and strain distributions were shown in Fig.6 and Fig.7. Distributions at tensile peak time #5 were Fig.6 and distributions at compressive peak time #6 were in Fig.7. Cumulative equivalent strain distributions were also shown in Fig.8. Stress and strain severely concentrated at the bottom of the notch. Inelastic strain range was calculated as the difference between peak times in Fig.8 on proportional cyclic loading conditions. Inelastic strain range distribution from the bottom in radius direction was shown in Fig.9.

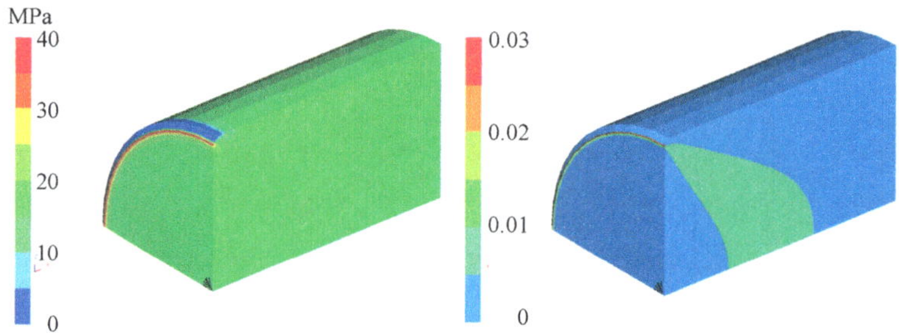

Fig.6 Axial Stress and Strain Distributions at Tensile Peak Time #5

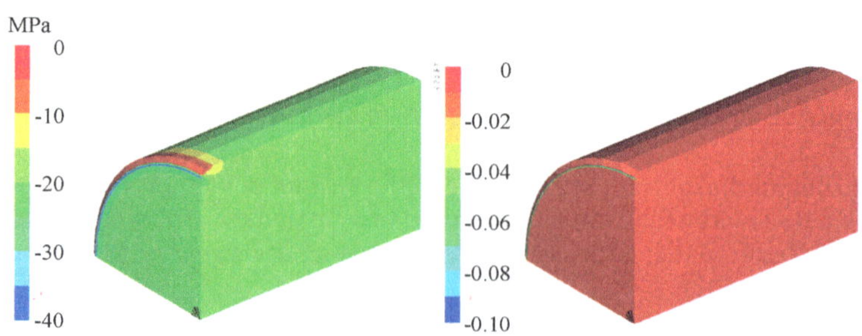

Fig.7 Axial Stress and Strain Distributions at Compressive Peak Time #6

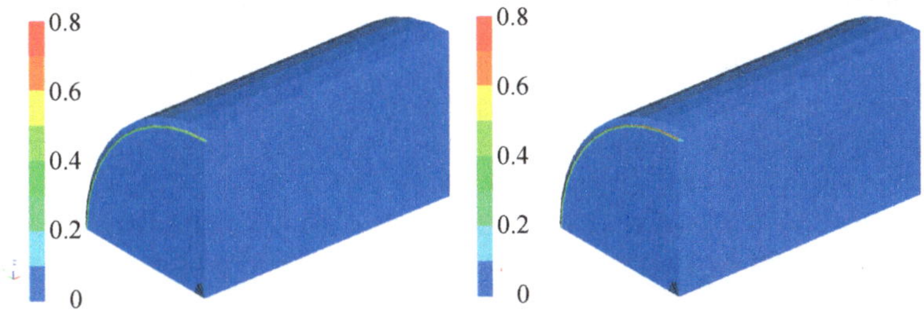

Fig.8 Cumulative equivalent strain distributions at Tensile and Compressive Peak Times #5/6

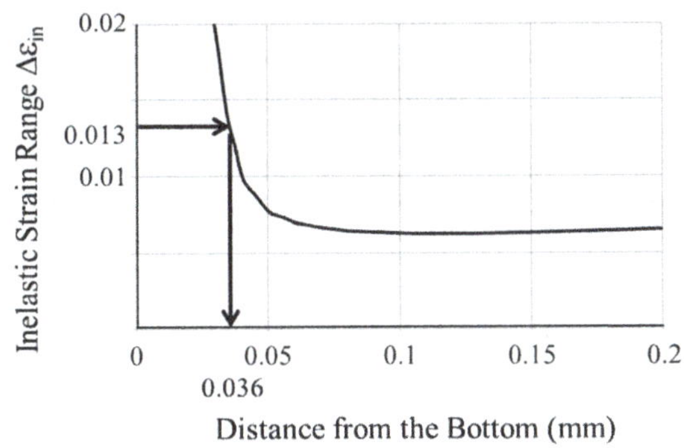

Fig.9 Inelastic Strain Range in the vicinity of the V Notch Bottom

4 VERIFICATION OF 50μm RULE

The value of concentrated inelastic strain range at the bottom of the notch was 0.12 that was 13 times of nominal inelastic strain range 0.09. If this inelastic strain range value was applied to eq.1 for the fatigue curve of smooth test specimen shown in Fig.1, the number of life became only 120 cycles. And the predicted the number of life was too shorter than the number of life for the test specimen with V notch 1,370 cycles by eq.2 derived from experiments. By calculating backward from the actual life 1,370 cycles for the fatigue curve of smooth test specimen, inelastic strain range should be estimated as 0.013. And calculating backward from the inelastic strain range 0.013 with Fig.9, the value at 40μm in the depth from the bottom of the notch corresponded to the inelastic strain range 0.012. Then, 50μm rule can be considered as a valid estimation method for low cycle fatigue life estimation of solder joint.

5 CONCLUSIONS

- Low cycle fatigue tests were carried out with Sn-3.0Ag-0.5Cu solder material test specimens having V notch at room temperature.
- Inelastic stress strain analysis was performed based on large scale parallel computing FEM, and inelastic strain range distribution around the bottom of the V notch was obtained.

- Life estimation based on the inelastic strain range at the bottom of V notch was much too conservative.
- Comparing the number of life for with V notch with low cycle fatigue life of Sn-3.0Ag-0.5Cu solder material, life estimation based on the inelastic strain range at inside point by 40μm in the depth from bottom agreed with experimental result for V notch.
- The validity of 50μm rule for solder joints could be shown.

REFERENCES

1) Mukai, M., Kawakami, T., Takahashi, K., Kishimoto, K. & Shibuya, T. (1998) JSME International J., Vol.41, No.2(A), 260-266.
2) Mukai, M., Kawakami, T., Hiruta, Y., Takahashi, K., Kishimoto, K. & Shibuya, T. (1998) Trans. ASME J. of Electronic Packaging, Vol.120 No.2, 207-212.
3) Akiba, H., Ohyama, T., Shibata, Y., Yuyama, K., Katai, Y., Takeuchi, R., Hoshino, T., Yoshimura, s., Noguchi, H., Gupta, M., Gunnels, J. A., Austel, V., Sabharwal, Y., Garg, R., Kato, S., Kawakami, T., Todokoro, S. & Ikeda, J. (2006) Pro. of SC06.
4) Kobayashi, M., Mukai, M., Takahashi, H., Ohono, N., Kawakami, T. & Ishikawa, T. (2003) International J. for Numerical in Engineering, Vol.58 No.10, 1523-1543.

Evaluation of the effects of low temperature nitriding on 4-points bending fatigue properties of Ti-6Al-4V alloy

Y. Nakamura [1], S. Yoshida [1], S. Kolicjo [2], A. Ueno [3]
[1] Graduate School of Science and Engineering, Ritsumeikan University, Japan
[2] Faculty of Engineering, Kobe University, Japan
[3] College of Science and Engineering, Ritsumeikan University, Japan

ABSTRACT

In order to improve both of the fatigue and wear resistance of titanium alloy (Ti-6Al-4V), a low temperature nitriding process was developed. This work will evaluate the effects of low temperature nitriding on the microstructure and fatigue properties of titanium alloy. Low temperature nitrided (600 °C) specimen showed higher fatigue strength compared to the conventional nitrided (850 °C) one. This was because low temperature nitriding can suppress the grain-coarsening of Ti-6Al-4V alloy during nitriding. Then, 4-points bending fatigue fracture mechanism of Ti-6Al-4V was discussed based on the microstructures obtained and from viewpoints of fractography.

1 INTRODUCTION

Titanium alloy; highlighted in this study, has been used as various components, such as turbine blades and bio-materials due to its high specific strength and excellent corrosion resistance. However, titanium shows high friction coefficient and poor wear resistance. It is necessary to improve tribological properties of titanium alloy by surface modification. Mainly, nitriding has been widely applied in various fields of engineering to improve wear resistance of material. Nitriding can increase surface hardness and form two types of surface layers; a nitrogen compound layer and a nitrogen diffusion layer. The nitrogen compound layer especially improves the tribological properties of a material due to its high hardness.

However, nitriding decreases the fatigue strength of titanium because grain-coarsening occurs during the process [1], which should be performed at high temperature. Therefore, nitriding at lower temperature is expected to suppress grain-coarsening of titanium alloy, resulting in improving both of the fatigue and tribological properties of titanium alloy. The authors [2] have previously developed the low temperature nitriding to suppress the grain-coarsening of commercially pure titanium (CP titanium) during the process. It was clarified that fine grains remained in the low temperature nitrided CP titanium after nitriding and nitrided layer was formed on CP titanium nitrided at the temperature higher than 600 °C.

The aims of this study are to characterize the surface layer formed on the low temperature nitrided titanium alloy (Ti-6Al-4V) and to clarify the effects of low the temperature nitriding on the 4-points bending fatigue properties of titanium alloy.

2 EXPERIMENTAL PROCEDURES

2.1 Material and specimen

The material used in this study was the α + β titanium alloy (Ti-6Al-4V) with the chemical composition shown in Table 1. Mechanical properties of this alloy are shown in Table 2. Titanium plates were machined into the specimens with 3 mm in width and 20 mm in length using a wire-electrical discharge machine. These specimens were polished with emery papers (#320 to #4000) to 1 mm in thickness and mirror-finished using SiO_2 suspension.

Table 1 Chemical composition

(mass%)

Al	V	Fe	H	N	O	C	Ti
6.31	4.13	0.12	0.002	0.006	0.11	0.024	Bal.

Table 2 Mechanical properties

Tensile strength, MPa	0.2%Proof stress, MPa	Elongation, %	Reduction of area, %
961	895	16.0	30.7

2.2 Plasma nitriding

The plasma nitriding was performed at relatively low temperature (550 and 600 °C) for 5 h in H_2 and N_2 atmosphere (N550 and N600 series). Moreover, the specimens treated with conventional nitriding at 850 °C for 5 h were also prepared (N850 series).

2.3 Characterization of nitrided layer

Surface hardness of the low temperature nitrided specimens was measured using a micro-Vickers hardness tester at a load of 0.098 N. The surface microstructures of the specimens were characterized using an optical microscope, scanning electron microscopy (SEM) and X-ray photoelectron spectroscopy (XPS). The crystal structures of the specimens were identified using X-ray diffraction (XRD) with CuKα radiation (wavelength: 0.154 nm). Moreover, electron backscatter diffraction technique (EBSD) analysis was conducted to investigate the microstructural changes of titanium alloy during the nitriding process.

2.4 Fatigue test

4-points bending fatigue tests were performed in ambient air without any controls of the temperature and the moisture at a frequency of 10 Hz under the stress ratio R = 0.1. After testing, the fracture surfaces were observed using SEM. In this study, the fatigue limit was determined following the JSMS standard [3].

3 RESULTS AND DISCUSSION

3.1 Characterization of nitrided layer

Fig. 1 shows the cross-sectional SEM micrographs of the (a) N600 and (b) N850 series etched with Kroll's solution, respectively. Nitrogen compound layer was evident at the nitrided surfaces of the both series, as indicated with arrow marks in Fig. 1. The N850 series possessed a thicker compound layer (3.2 μm) in comparison to the N600 series (0.5 μm). This was because the amount of diffused nitrogen was increased with nitriding temperature. Such a compound layer was not observed at the surface of the N550 series.

XRD analysis was performed to examine the characteristics of the microstructure of the nitrided Ti-6Al-4V alloy in more detail. Fig. 2 shows the X-ray patterns for the surfaces of the un-nitrided and nitrided specimens. The N600 series exhibited the diffraction peaks of substrate and Ti_2N, and the diffraction peaks of TiN and Ti_2N were evident in the N850 series. In contrast, the specimen nitrided at 550 °C exhibited only diffraction peaks due to α-titanium and β-titanium. It was clarified that nitrogen compound layer was formed on Ti-6Al-4V alloy nitrided at the temperature higher than 600 °C. Same tendency was observed in the previous reports about the CP titanium [2].

Fig. 3 shows the results of hardness tests. Vickers hardness *HV* measured on the top surface and indentation depth *d* estimated based on the size of indentation mark are shown in this figure. Surface hardness of titanium alloy was increased and indentation depth was decreased with nitriding temperature. This was because thickness of compound layer was increased with nitriding temperature, as shown in Fig. 1. However, the N550 series without compound layer had higher hardness compared to the un-nitrided specimen. As results of XPS analysis, nitrogen element was detected in the N550 series. XRD and XPS analyses indicate that a nitrogen diffusion layer without nitrogen compounds is formed on the N550 series.

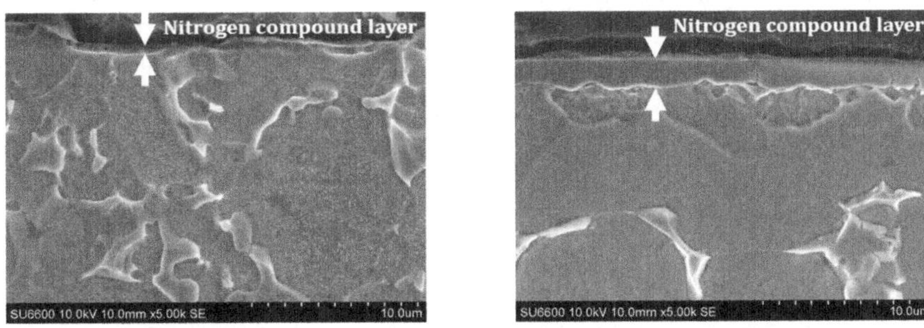

(a) N600 **(b) N850**

Fig.1 Cross-sectional SEM micrographs of the nitrided specimens

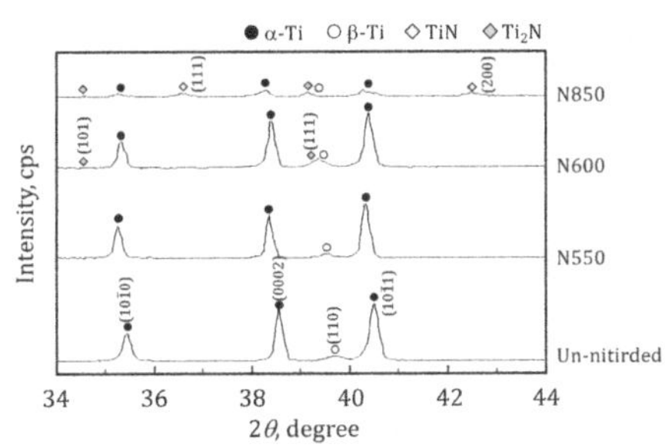

Fig.2 XRD patterns

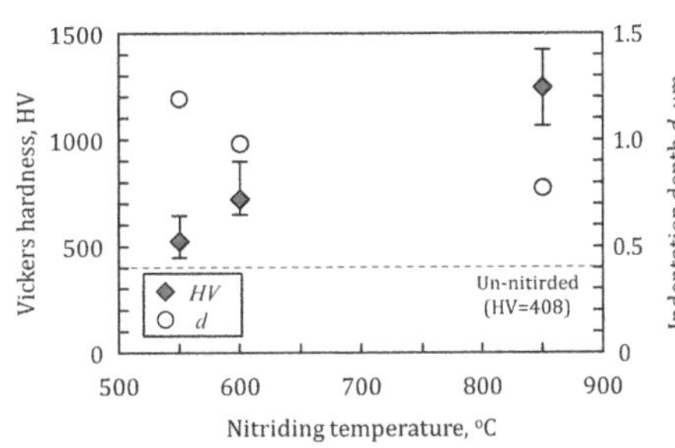

Fig.3 Results of hardness tests

3.2 Evaluation of microstructural changes during nitriding

In this section, the behaviour of microstructural changes by low temperature nitriding was investigated. Fig. 4 shows the inversed pole figure (IPF) and image quality (IQ) maps of the un-nitrided, N550, N600 and N850 series beneath the surface layer. Nitriding at 850 °C coarsened grains of titanium alloy and average grain size of the N850 series was 5.7 μm; larger than that of the un-nitrided series (2.2 μm). On the other hand, the crystal grain size of the N550 (2.4 μm) and N600 series (2.5μm) were almost the same as the un-nitrided one. These results indicate that low temperature nitriding can suppress the grain-coarsening of Ti-6Al-4V alloy. Consequently, the low temperature nitriding at 600 °C achieved both of the suppression of the grain-coarsening and formation of the nitrogen compound layer on Ti-6Al-4V alloy.

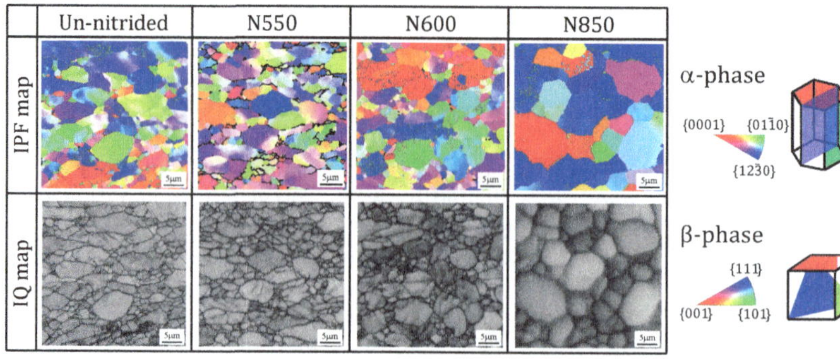

Fig.4 Results of EBSD analysis

3.3 Results of 4-point bending fatigue test

The 4-point fatigue tests were conducted to examine the fatigue properties of titanium alloy treated with the low temperature nitriding, which could suppress grain-coarsening. Fig. 5 shows the results of high-cycle fatigue tests. In Fig. 5 (a), each *S-N* curve was determined by accepting the *S-N* model with fatigue limit in the JSMS standard regression models [3]. The fatigue limit σ_W of the N600 series (711 MPa) was higher than that of the N850 series (474 MPa). This was because the fine grains remained in the N600 series, as shown in Fig. 4. Therefore, the N600 series was almost the same fatigue limit as the un-nitrided series (710 MPa). This result means that low temperature nitriding prevents the reduction of the fatigue limit of Ti-6Al-4V alloy.

However, there were remarkable differences in fatigue life N_f at the finite life region between the N600 and un-nitrided series. Critical number of stress cycles giving fatigue limit N_W of the N600 series (1.2 x 10^5) was much lower than that of un-nitrided one (8.5 x 10^6). In order to investigate the effects of compound layer on the fatigue properties of Ti-6Al-4V alloy, fatigue tests were also performed for the N550 series without compound layer. In Fig. 5 (b), the number of fatigue test data of the N550 series was not so enough to analyse *S-N* curve and fatigue limit. The highest stress value without fatigue failure in the N550 series (858 MPa) was higher than the fatigue limit of the N600 series (711 MPa). These results imply that nitrogen diffusion layer without nitrogen compounds improves the fatigue properties of Ti-6Al-4V alloy and that nitrogen compound layer reduces the fatigue life.

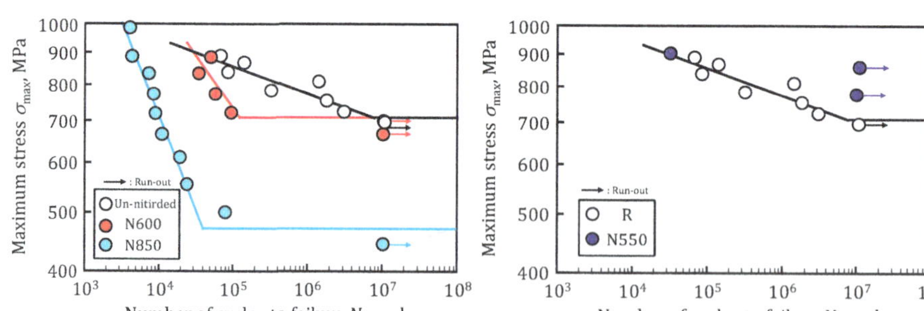

(a) (b)

Fig.5 Results of 4-points bending fatigue tests

3.4 Effects of nitrogen compound layer on the fatigue properties

Fig. 6 shows the SEM micrographs of fracture surface of low temperature nitrided and conventional nitrided specimens. In this study, every specimen exhibited the surface fracture modes. In Fig. 6 (b), characteristic flat area; corresponding to the thickness of compound layer, was clearly observed at the N850 series. Tokaji et. al [4] reported that fatigue cracks were initiated suddenly with a relatively large size at the nitrided surface, resulting in reducing the fatigue strength of titanium alloy. Based on these results and report, nitrogen compound layer shows brittle fracture during fatigue tests and adversely affects the fatigue properties of titanium alloy.

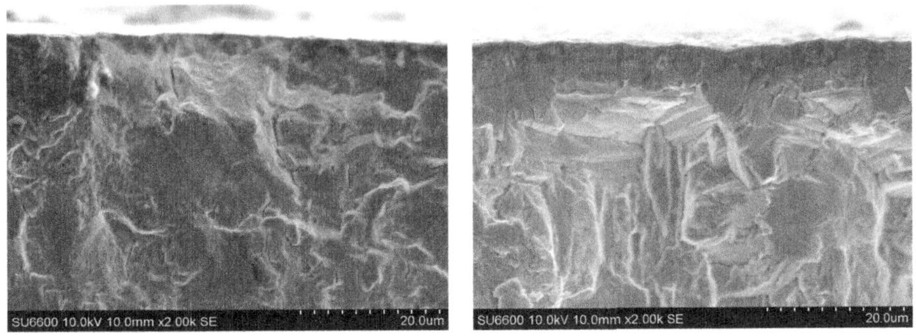

(a) N600 (σ_{max} = 887 MPa, N_f = 5.0 x 10^4) **(b) N850 (σ_{max} = 888 MPa, N_f = 4.3 x 10^3)**

Fig.6 Typical feature of fracture surfaces

4 SUMMARY

The microstructure of the low temperature nitrided Ti-6Al-4V alloy was characterized, and its effects on the 4-points bending fatigue properties of Ti-6Al-4V alloy was examined. The following conclusions were reached:

1. Nitrogen compound layer is formed on the surface nitrided at the temperature higher than 600 °C, resulting in increasing surface hardness of titanium alloy.
2. Low temperature nitrided (600 °C) specimen showed higher fatigue limit compared to the conventional nitrided (850 °C) one. This is because low temperature nitriding can suppress the grain-coarsening of Ti-6Al-4V alloy.
3. The low temperature nitriding enables to suppress the reduction of the fatigue limit of Ti-6Al-4V alloy. However, the fatigue life of the low temperature nitrided titanium alloy at a finite life region is decreased compared to un-nitrided one due to the existence of the nitrogen compound layer.
4. Low temperature nitriding (550 °C) increases the fatigue strength of a titanium alloy due to the formation of the nitrogen diffusion layer without nitrogen compounds.

5 REFERENCES

[1] Morita, T., Takahashi, H., Shimizu, M., & Kawasaki, K. (1997) *Fatigue & Fracture of Engineering Materials & Structures.* **20**, 85-92. doi:10.1111/j.1460-2695.1997.tb00404.x

[2] Kikuchi, S., Nakamura, Y., Ueno, A., & Ameyama, K. (2014) *Advanced Materials Research*. **891-892**, 656-661.
doi:10.4028/www.scientific.net/AMR.891-892.656

[3] JSMS Committee on Fatigue of Materials and JSMS Committee on Reliability Engineering, Standard evaluation method of fatigue reliability for metallic materials -Standard Regression Method of *S-N* Curves- [JSMS-SD-6-08], The Society of Materials Science Japan, 2008.

[4] Shibata, H., Tokaji, K., Ogawa, T., & Hori, C. (1994) *International Journal of Fatigue*. **16**, 370-376. doi:10.1016/0142-1123(94)90448-0

ΔK_{th} estimation of aluminum die-casting alloy by means of $\sqrt{area}$ method

A. Ueno[1], M. Nishida[2], S. Miyakawa[3], K. Yamada[3], S. Kikuchi[4]

[1] College of Science and Engineering, Ritsumeikan University, Japan

[2] Nippon Steel & Sumitomo Metal Corp., Japan

[3] Denso Co., Ltd, Japan

[4] Faculty of Engineering, Kobe University, Japan

ABSTRACT

In order to evaluate quantitatively the effects of casting defects on the fatigue properties of aluminum die-casting alloy, a ΔK_{th} for aluminum die-casting alloy containing large casting defects was proposed by means of $\sqrt{area}$ method. Fatigue tests were carried out by using a rotating bending fatigue testing machine for the specimen having four artificial drilled holes whose $\sqrt{area}$ were ranged from 300 to 2000 μm. According to the fatigue limits of the hourglass type specimen and of the smooth specimens having four artificial drilled holes, equation for ΔK_{th} estimation for aluminum die-casting alloy was proposed by means of $\sqrt{area}$ method.

1 INTRODUCTION

Recently, global warming is caused by increasing the amount of greenhouse gas. In order to improve energy-efficient for automobile, aluminum die-casting alloy with low cost and lightweight components has been widely used as automobile components. Moreover, it is required that the structures use for a long period to reduce consumption in resources and energy. However, it is well-known that some large casting defects which are created by process of manufacture decrease fatigue strength of aluminum die-casting alloy. Therefore, it is important to evaluate the reliability of strength of aluminum die-casting alloy.

At present, studies of ΔK_{th} estimation and quantitative evaluation based on fracture mechanics method in aluminum casting alloy were reported. In a general way, fatigue limit estimation of metallic materials was obtained by experimental rule. Murakami (1) reported that fatigue limit was able to evaluate using the $\sqrt{area}$ for ferrous materials. Moreover, Noguchi (2) proposed fatigue limit estimation method for non-ferrous materials by modifying the Murakami's method. Although it is well known that the $\sqrt{area}$ method is not applicable to defects whose size is larger than 1000 μm. The purpose of this study is to propose a fatigue limit (3) and ΔK_{th} estimation of aluminum die-casting alloy having large defects by means of the $\sqrt{area}$ method.

2 EXPERIMENTAL PROCEDURES

2.1 Materials and specimen

The materials used in this study was aluminum die-casting alloy (JIS ADC12). Chemical compositions of this material are shown in Table 1. The tensile strength and Vickers hardness were 296 MPa and 90, respectively. In order to investigate relationship between fatigue strength and defect size, specimen having four artificial drilled holes, whose $\sqrt{area}$ were ranged from 300 μm to 2000 μm, were used (Fig. 1, Fig. 2). Four different size holes were drilled on spiral line with 90° interval. Each hole was subjected to different bending moment. Specimens were buff-finished with alumina abrasive grain size of 1 μm after the grinding process. Finally, specimen were heat-treated in vacuum at 473 K for two hours to eliminate residual stress.

Table 1 Chemical compositions (mass%)

Si	Cu	Fe	Zn	Mn	Mg	Al
9.4~10.4	1.9~2.5	0.72~0.93	0.42~0.55	0.16~0.38	0.21~0.28	Bal.

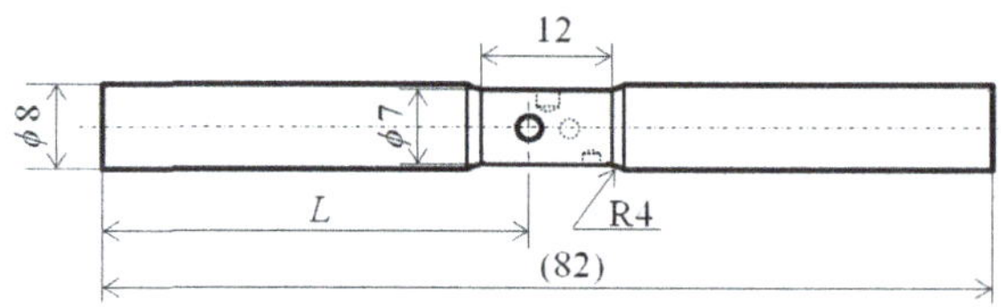

Fig. 1 Shape and dimensions of specimen

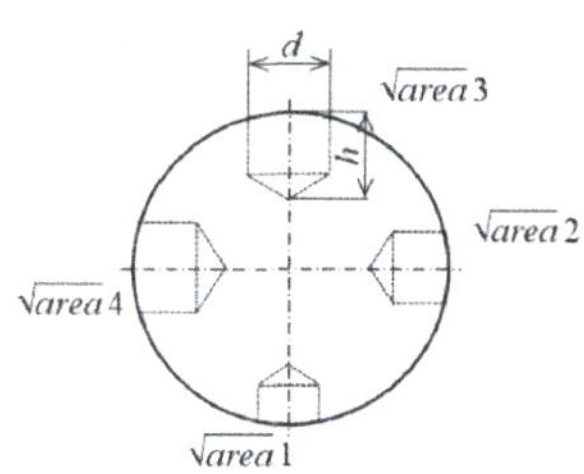

$\sqrt{area}$, μm	d, μm	h, μm
1900~2000	2.2	2.1
1700~1800	2.0	2.0
1500~1600	1.8	1.7
1300~1400	1.5	1.4
1100~1200	1.3	1.2
900~1000	1.1	1.0
700~800	0.9	0.8
500~600	0.7	0.7
300~400	0.4	0.4

Fig. 2 Shape and dimensions of artificial defects

2.2 Fatigue tests

Dual-spindle cantilever-beam type rotating bending fatigue testing machine (Fig.3), rotating speed of the spindle is 52.5 Hz, was used in this study. Eccentricity at the loading point of the specimens was kept within ±20 μm. Fatigue tests up to 10^9 cycles were carried out in ambient air and temperature. After fatigue tests, fracture surfaces and each artificial defects were observed by means of SEM.

Fig. 3 Dual-spindle cantilever-beam type rotating bending fatigue testing machine

2.3 $\sqrt{area}$ method

Murakami *et al.* (1) has been proposed the $\sqrt{area}$ method for estimating a fatigue limit of steel. (Eq. (1)).

For surface defect: $$\sigma_w = \frac{1.43\,(120 + HV)}{(\sqrt{area})^{1/6}} \qquad (1)$$

Also they proposed an equation for calculating the stress intensity factor K by means of $\sqrt{area}$;

For surface defect: $$K = 0.65\sigma\sqrt{\pi\sqrt{area}} \qquad (2)$$

By substituting Eq.(1) to Eq.(2), they proposed the ΔK_{th} estimation equation by means of $\sqrt{area}$;

For surface defect: $$\Delta K_{th} = 3.3 \times 10^{-3}(HV + 120)(\sqrt{area})^{1/3} \qquad (3)$$

where *area* (μm^2) is a projected area of defect onto a plane perpendicular to the applied stress and HV is Vickers hardness number. It is well known that these equations are very useful to estimate a fatigue limit σ_w and/or threshold stress intensity of small fatigue crack ΔK_{th} of ferrous materials having some defects that size less than 1000 μm.

On the other hand, it is known that equations (1) and (2) show overestimation for applying to non-ferrous materials, such as aluminum alloy (2). Also, applicability of equations (1) and (2) to large defects whose area is larger than √1000 μm is unknown.

3 EXPERIMENTAL RESULTS AND DISCUSSION

3.1 ΔK-N characteristics

Figs 4(a) and 4(b) show example of test results using specimens with artificial defects.

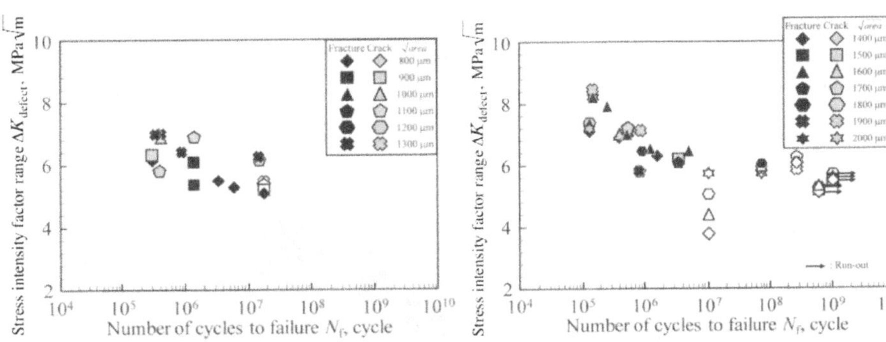

(a) For cases of $\sqrt{area}$ ranged from 800 μm to 1300 μm

(b) For cases of $\sqrt{area}$ ranged from 1400 μm to 2000 μm

Fig. 4 ΔK-N characteristics

Four polygonal marks arranged in one line indicate bending stress of each artificial defect subjected different bending moment. Black marks indicate the defect of fracture origin (Fig. 5(a)). White marks indicate the defect without fatigue crack after fatigue test. And, gray marks indicate the defect some fatigue crack initiated (Fig.5(b)). Also, four polygonal marks having arrow at $N = 10^9$ cycles indicate run-out specimen.

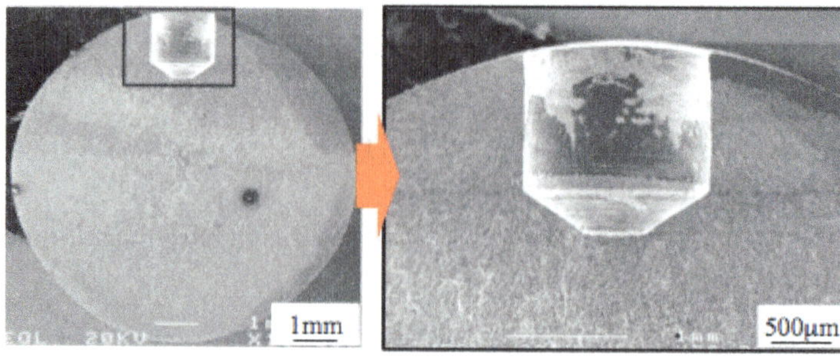

(a) Fracture surface (σ_a = 84 MPa, $\sqrt{area}$ =1400 μm)

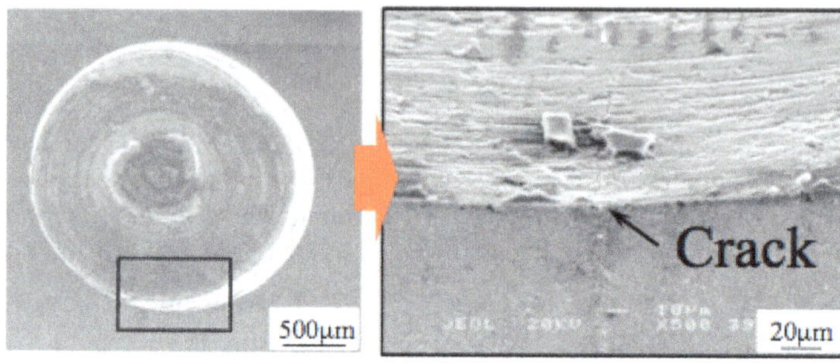

(b) Artificial defect (σ_a = 71 MPa, $\sqrt{area}$ = 1800 μm)

Fig. 5 Examples of SEM micrographs

3.2 ΔK_{th} estimation

In ΔK-$\sqrt{area}$ diagram obtained using specimens having four artificial drilled holes (Fig. 6), we assumed that the ΔK_{th} of each size artificial defects were able to define as an average ΔK_{defect} value between fractured(black circle) and without crack(white circle) specimen or value between cracked (gray circle) and without crack specimen. After extracting only average ΔK_{defect} value, redrawn diagram is shown in Fig. 7. It is found that the inclination of a regression line changed around $\sqrt{area}$ of about 1400 μm.

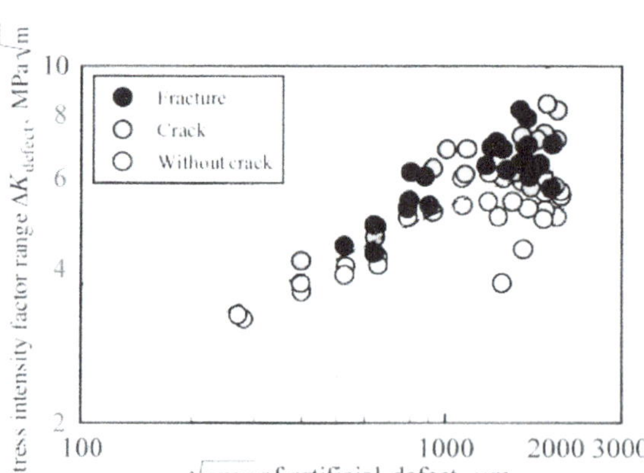

Fig. 6 Relationship between ΔK_{defect} and $\sqrt{area}$

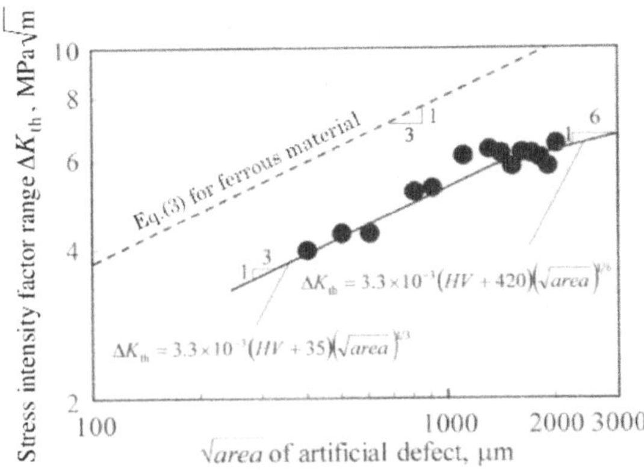

Fig. 7 Relationship between ΔK_{th} and $\sqrt{area}$

Therefore, following two regression lines were obtained:

For $\sqrt{area} \le 1400\ \mu m$:

$$\Delta K_{th} = 3.3 \times 10^{-3}(HV + 35)(\sqrt{area})^{1/3}$$

For $\sqrt{area} > 1400\ \mu m$:

$$\Delta K_{th} = 3.3 \times 10^{-3}(HV + 420)(\sqrt{area})^{1/6}$$

4 SUMMARY

For estimating the ΔK_{th} of aluminum die-casting alloy having large size defects, fatigue tests were carried out using specimen having four artificial small drilled holes, whose $\sqrt{area}$ were ranged from 300 μm to 2000 μm. New modified equations for estimating ΔK_{th} of an aluminum die-casting alloy were proposed according to approximate from relationship between ΔK_{th} and $\sqrt{area}$ of artificial defect.

REFERENCES

[1] Murakami, Y. and Endo, M. (1994) International Journal of Fatigue, **16**, 163-182.

[2] Noguchi, H. and Morishige, K. (2007) Proc. of the Japan Society of Mechanical Engineers, **56**, 137-138.

[3] Ueno, A., Itami, N., Onishi, T., Nishida, M. and Kikuchi, S. (2012) Proc. of the 8th International Forum on Advanced Materials Science and Technology, CD-ROM.

Effect of ultrasonic shot peening treatment on fatigue behavior of AZ61 magnesium alloy

K. Sanada [1], T. Kakiuchi [2], Y. Uematsu [2], K. Hattori [3], Y. Watanabe [3]
[1] Mechanical and Systems Engineering Division, Graduate School of Gifu University, Japan
[2] Department of Mechanical Engineering, Gifu University, Japan
[3] Toyo Seiko Co., Ltd, Japan

ABSTRACT

Ultrasonic shot peening (USP) treatment was performed on the wrought magnesium (Mg) alloy AZ61 with the different conditions and rotary bending fatigue tests were performed to investigate the effect of USP treatment on the fatigue behavior. The experimental results revealed that the fatigue strength of Mg alloy was strongly dependent on the USP condition. The USP with the high arc height using the large shots resulted in decreasing the fatigue strength. However the USP with the appropriate condition, in which the arc height is not so high but sufficient, improved the fatigue strength appreciably.

1 INTRODUCTION

The magnesium (Mg) is the lightest among the metals which are utilized practically. The Mg alloy has the excellent specific strength and rigidity compared with the other alloys and they are expected to be applied in the structural components in the automotives and the airplanes. The high fatigue strength is required in the structural components. Thus the improvement of fatigue strength of Mg alloy is important in engineering. The shot peening (SP) treatment is one of the well-used methods to improve the fatigue strength of metallic alloy. The SP produced the work hardening and the compressive residual stress near the surface, which resulted in improving the fatigue strength. However it also deteriorates the surface as such that the surface is roughened, which could result in decreasing the fatigue strength. The effect of shot peening on the fatigue strength of Mg alloy has been studied (1) - (4). In these previous studies, it has been shown that the optimum shot peening condition to increase the fatigue strength of Mg alloy is narrow. It was indicated that surface damages were aggravated especially in the Mg alloy due to the limited deformability of HCP crystal structure (2). Thus a shot peening could sometimes make a detrimental effect on the fatigue strength of Mg alloy. The ultrasonic shot peening (USP) treatment is one of the SP treatments in which the ultrasonic vibrator is utilized to accelerate the shots. It is regarded that the USP could suppress the deterioration in the surface roughness while it produces the comparable compressive residual stress compared with the conventional SP.

In this study, the USP treatment was performed on the wrought Mg alloy AZ61 with the different conditions and rotary bending fatigue tests were performed to investigate the effect of USP treatment on the fatigue behavior.

2 EXPERIMENTAL MATERIAL AND PROCEDURE

2.1 Experimental material

The test material used in this study was a wrought Mg alloy AZ61. The material was received as a round bar of φ15 mm. Tables 1 and 2 show the chemical compositions and the mechanical properties, respectively.

Table 1 Chemical compositions of material (wt.%).

Al	Zn	Mn	Si	Fe	Cu	Ni	Mg
6.0	0.57	0.38	0.008	<0.002	<0.002	<0.002	Bal.

Table 2 Mechanical properties of material.

0.2%proof stress $\sigma_{0.2}$ (MPa)	Tensile strength σ_B (MPa)	Elongation δ (%)	Elastic modulus E (GPa)	Vickers hardness HV
230	336	22	39	65

2.2 Experimental specimen

The test specimens were machined by cutting as that the longitudinal axis was parallel to the extrusion direction. The diameter of gauge section was φ5 mm.

2.3 SP treatment conditions

Table 3 shows the conditions of USP treatment. For example, USP-1.0-300 denotes the specimen which was USPed using the shots of 1.0 mm and the coverage of 300 %. Some specimens of USP-1.0-300 and USP-2.0-300 were polished to remove the surface layer and to smooth the surface roughness, which is denoted as USP-1.0-300-P and USP-2.0-300-P, respectively. The depth of polishing was set to be 20 and 35 μm in USP-1.0-300-P and USP-2.0-300-P, respectively, which was determined by the preliminary test as it was enough to smooth the surface of specimen to be mirror-finished. The DSP denotes the specimen which was treated by double USP, in which the first USP was the same with USP-1.0-300 following by the second USP as the same with USP-0.6-300. The specimens which were treated by the conventional SP, in which the air was utilized to accelerate the shots, were also prepared for comparison, which is denoted as "ASP". Table 4 shows the conditions of ASP treatment. The specimen of base material was polished over the depth of 50 μm and mirror-finished before the fatigue test to remove the surface layer which was affected by machining and to smooth the surface and the SP and USP specimens were not polished except for USP-1.0-300-P and USP-2.0-300-P.

2.4 Experimental procedure

Fatigue tests were performed using a rotary cantilever bending fatigue testing machine under the conditions with the stress ratio R = -1 and the frequency f = 53 Hz in the laboratory air. After the fatigue tests, all fracture surfaces were observed using a scanning electron microscope (SEM).

Table 3 Ultrasonic shot-peening conditions.

Specimen	Diameter of shot (mm)	Coverage (%)	Arc height (mmA)	Time (sec)	Polish depth (μm)
USP-0.6-300	0.6	300	0.080	120	–
USP-1.0-300	1.0	300	0.150	120	–
USP-2.0-300	2.0	300	0.277	150	–
USP-1.0-80	1.0	80~90	0.076	20	–
USP-2.0-200	2.0	200	0.227	100	–
USP-1.0-300-P	1.0	300	0.080	120	20
USP-2.0-300-P	2.0	300	0.150	150	35
DSP	1.0→0.6	300→300	0.148 →0.087	120 →120	–
Base material	–	–	–	–	50

Table 4 Air blast shot-peening conditions.

Specimen	Diameter of shot (mm)	Air pressure (MPa)	Coverage (%)	Arc height (mmN)
ASP	0.120	0.16	300	0.09

3 EXPERIMENTAL RESULTS AND DISCUSSION

Fig. 1 shows the *S-N* diagram. The fatigue limit at 10^7 cycles of base material was 134 MPa. The fatigue strengths of SPed and USPed specimens were strongly dependent on the peening condition.

The fatigue limit of ASP was 144 MPa, which was 10 MPa higher than that of base material.

The fatigue limits of USP-0.6-300, USP-1.0-300 and USP-2.0-300 were 144, 154 and 124 MPa, respectively. The USP using the shots of 0.6 mm and 1.0 mm could improve the fatigue strength and the improvement using the larger shots of 1.0 mm was larger than that using the smaller shots of 0.6 mm and than that of ASP. On the other hand, the USP using the larger shots of 2.0 mm decreased the fatigue strength. Comparing these three specimens, the USP treatment using the shots of middle size was the most effective to improve the fatigue strength. From the Vickers hardness tests, the maximum value of hardness in USP-2.0-300 was approximately 100 HV, which was higher than that in base material, 65 HV, and it was as high as that in USP-1.0-300.

The fatigue limits of USP-1.0-300-P and USP-2.0-300-P, which are the specimens whose surface layers were removed by polishing, were both 172 MPa and the fatigue strengths were improved largely. It indicates that the surface layer produced by the USP made a detrimental effect on the fatigue strength, particularly in USP-2.0-300.

The fatigue limit of USP-2.0-200 was 144 MPa, which was higher than not only that of base material but also that of USP-2.0-300. Thus, in this case, the lower coverage was more effective to improve the fatigue strength. On the other hand, the fatigue limit of USP-1.0-80 was 124 MPa, which was lower than that of USP-1.0-300 and moreover it was even lower than that of base material. In this case, the lower coverage made a detrimental effect on the fatigue strength. In this condition, the peening intensity was not sufficient because of the too low coverage.

The fatigue limit of DSP was 172 MPa, which was as high as those of USP-1.0-300-P and USP-2.0-300-P. It is considered that the sufficient work hardening and the compressive residual stresses were obtained by the first USP and then the surface layer, which made a detrimental effect, was modified by the following second USP.

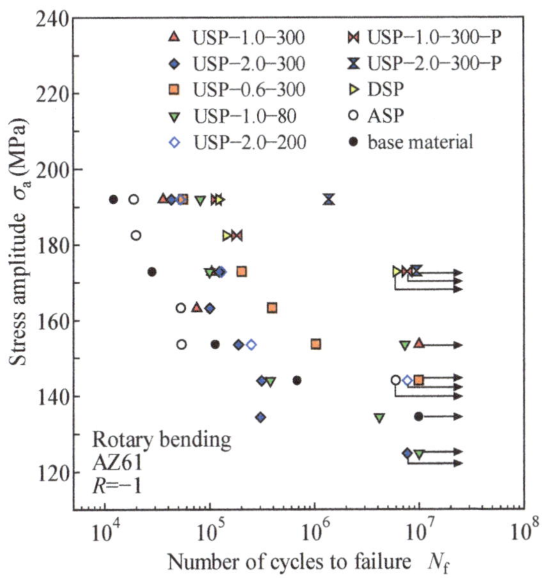

Fig. 1 *S-N* diagram.

Fig. 2 shows the SEM micrographs showing an example of fracture surfaces of USP-1.0-300 specimen. Fig. 2(a) shows the overview and Fig. 2(b) shows the magnified view of one of the crack initiation sites. The fatigue cracks initiated from more than one sites and all of the cracks initiated from the surface in this specimen. The cracks initiated from the surface not only in USP-1.0-300 specimens as shown in Fig. 2 but in all of the specimens in this study.

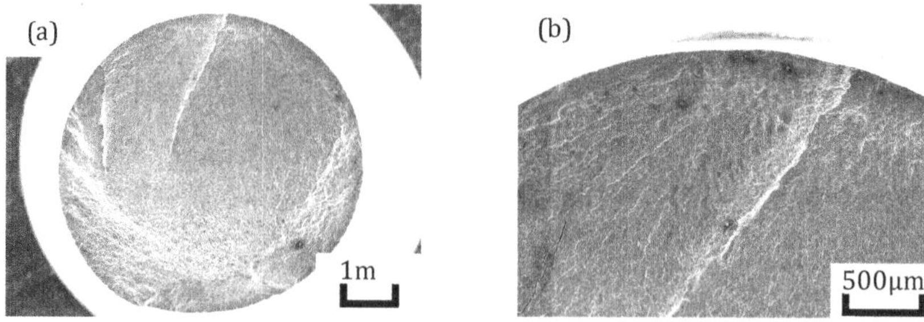

Fig. 2 SEM micrograph showing fracture surface of USP-1.0-300 specimen (σ_a = 163 MPa, N_f = 7.5×10^4): (a) overview, (b) magnified view of crack initiation site.

4 CONCLUSION

In this study, the effect of USP on the fatigue behavior of Mg alloy AZ61 was investigated. From the experimental results, it was revealed that the fatigue strength was strongly affected by the condition. While the USP with the excessive intensity could deteriorate the fatigue strength, the appropriate condition could improve the fatigue strength largely. Particularly the surface layer produced by the USP could make a detrimental effect and thus the additional surface modification could enhance the fatigue strength.

REFERENCES

[1] Wagner, L. (1999) *Materials Science and Engineering A*, **263**, 210-216.
[2] Zhang, P. & Lindemann, J. (2005) *Scripta Materialia*, **52**, 485-490.
[3] Fouad, Y., Mhaede, M. & Wagner, L. (2010) *Fatigue & Fracture of Engineering Materials & Structures*, **34**, 403-407.
[4] Bhuiyana Md. S., Mutoh, Y. & McEvily, A.J. (2012) *Materials Science and Engineering A*, **549**, 69-75.

The effect of specimen thickness on fatigue crack growth under variable amplitude loading in 7075-T7351 Aluminium

C. Wallbrink
Defence Science and Technology Organisation, Australia

ABSTRACT

An investigation has been conducted to examine the effect of material thickness on fatigue crack growth in 7075-T7351 Aluminium under variable amplitude loading. The results of the investigation have demonstrated that a significant effect on fatigue crack growth is observed under variable amplitude loading for specimen thicknesses ranging from 2 mm to 6.4 mm. The observed behaviour was modelled by modifying the constraint factors used in the FASTRAN crack closure model. The present observation that specimen thickness has a significant effect on fatigue crack growth highlights the importance of selecting appropriate material data and a suitable fatigue crack growth model to use in subsequent fatigue crack growth analysis.

1 INTRODUCTION

This work supports a collaborative programme between the Royal Australian Air Force and the Royal Air Force of the UK in which a full-scale wing fatigue test of a C-130 transport aircraft is being conducted. Specifically the aim of the testing is to aid in validation of the RAAF C-130J full scale wing fatigue test interpretation process. Limited information is presented in the literature that investigates the thickness effect or its implications. The ASTM standard on crack growth rate measurement [1] does not consider specimen thickness except to mitigate against specimen buckling. The standard does however mention that mixed results have been published due to the effects of thickness and as such thickness should be considered in research and design. A previous investigation using constant amplitude loading for 7075-T7351 Aluminium concluded that there was an insignificant effect on fatigue crack growth due to specimen thickness [2]. However, it was not known what behaviour should be expected under variable amplitude loading. Given the importance of delivering a robust methodology for the C-130 test interpretation, the present investigation was undertaken to assess the effect of the thickness on fatigue crack growth under variable amplitude loading.

2 EXPERIMENTAL TEST PROGRAM

The centre-crack specimens were manufactured from 0.25" AA7075-T7351 plate and cut to size. The dimensions of the test specimens were 380 mm long and 100 mm wide milled to the desired thickness by removing material from both sides of the plate. A hole of 3 mm in diameter was drilled into the centre with wire cut slits 0.5 mm in depth cut to either side of the hole. Three specimen thicknesses were considered in this investigation, 6.4, 4.1 and 2.0mm, chosen to be representative of the thicknesses of fatigue critical areas on the C-130J.

Variable amplitude loading in the form of two RAAF C-130 usage spectra were chosen. The spectra are representative of loading in the centre wing lower surface and outer wing lower surface panels of the C-130J (with peak stresses of 114.5 MPa and 106.1 MPa respectively). The specimens were pre-cracked under constant amplitude (R=0.1) loading at 70% of the maximum spectrum load including load shedding down to 50% to grow a pre-crack of 2mm. A direct current potential drop (DCPD) method was used to measure the total crack growth during the test via the use of the Johnson equation [3]. Results presented here are reported as half-length measurements approximating crack growth from the centre of the test specimen.

As the test spectrum contains significant compression loads and small specimen thicknesses, anti-buckling guides were required to prevent buckling of the specimen under compressive loads. Teflon tape was used between the buckling guides and the specimen to preserve electrical insulation and allow the DCPD system to function. Bolts on the buckling guide were lightly tightened so as to just support the weight of the buckling guide on the specimen and to prevent slippage during testing.

3 RESULTS

The results shown in Figure 1 and 2 show a distinct effect of thickness on fatigue life, contrary to the observations in [2]. The effect appears to be very pronounced, with a factor of two to three between the final fatigue lives for the various thicknesses tested.

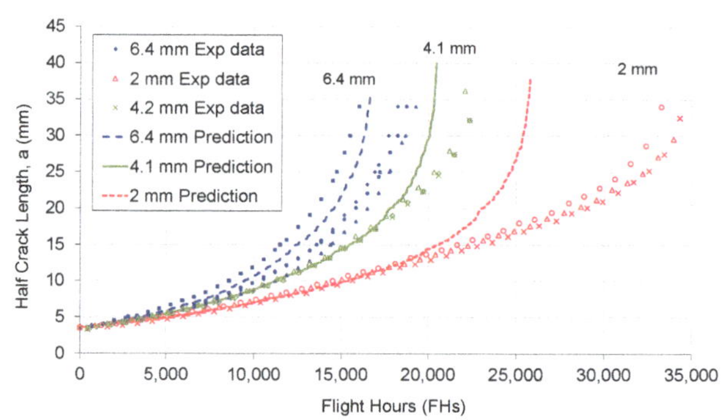

Figure 1: Half length crack growth under the centre wing lower surface spectrum compared to FASTRAN predictions for various specimen thicknesses

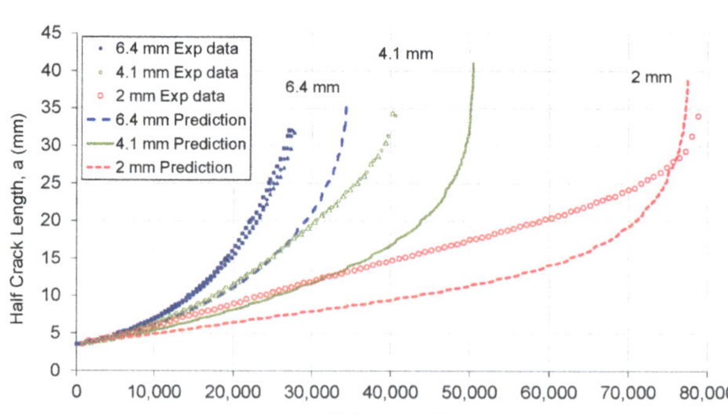

Figure 2: Half length crack growth under the outer wing lower surface spectrum compared to FASTRAN predictions

4 CRACK GROWTH ANALYSIS USING FASTRAN

The input parameters used in FASTRAN 3.8a are presented in Table 1. These parameters describe the material properties obtained from [4] and coupon geometry. The crack growth rate curve is presented in Figure 3. FASTRAN 3.8a uses a strip yield model to estimate the level of plastically induced crack closure which enables estimation of the load sequence on fatigue crack growth predictions. The crack tip behaviour (plane-stress or plane-stain) is controlled by the constraint factor α.

Table 1: FASTRAN 3.8a Input Parameters for the aluminium 7075-T7351 coupons

Input Parameter	Value
Yield Stress (MPa)	393
Ultimate Stress (MPa)	476
Young's Modulus (MPa)	71016
Poisson's Ratio	0.33
Fracture Toughness K_c (MPa√m)	65.9
ΔK Threshold (C_3) (MPa√m)	0.74
Width (mm)	50
Thicknesses (mm)	2, 4.1, 6.4
Initial Crack Length (mm)	1.27
Notch Length (mm)	1.27
Notch Height (mm)	0.0001

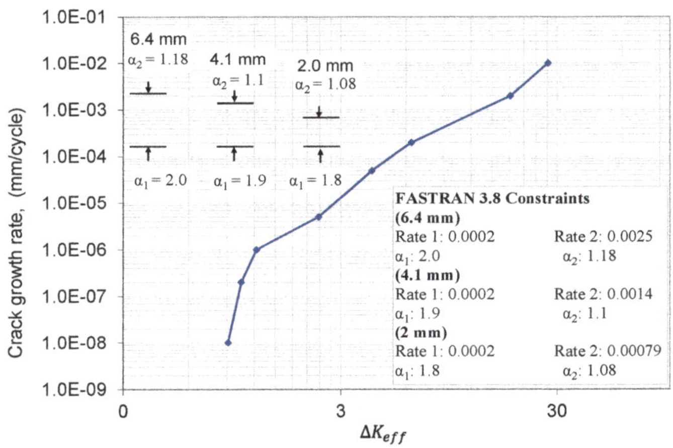

Figure 3: Crack growth rate versus ΔK_{eff} curve for 7075-T7351

4.1 Background to FASTRAN constraint adjustment

Adjusting the constraint factors used in FASTRAN 3.8a is commonly used to calibrate the model as has been done in [3] for the C-130 wing fatigue test interpretation. FASTRAN does not provide an internal method to adjust these parameters to account for thickness, but must be manually altered. FASTRAN 3.8a uses two constraint factors designated α_1 and α_2 to control the transition from plane-strain to plane-stress which is described by Newman [5]. Newman defines a minimum crack growth rate below which α_1 is used in the evaluation of the plastic zone. A maximum crack growth rate is also defined above which α_2 is used. Between the minimum crack growth rate and the maximum crack growth rate the level of constraint is determine through a linear interpolation between α_1 and α_2. Newman has proposed an equation to help modify the constraint factors used in prediction to account for the effect of thickness on fatigue crack growth [5]. The equation is as follows:

$$\left(\Delta K_{eff}\right)_T = 0.5\sigma_o\sqrt{B} \quad (1)$$

where ΔK_{eff} is the effective stress intensity range, σ_o is the flow stress defined as the average of the yield stress and ultimate tensile stress and B is the section thickness. $\left(\Delta K_{eff}\right)_T$ gives an indication of the effective stress intensity factor around which the crack tip deformation transitions from one of plane-strain to plane-stress. Cracks with low stress intensity factors and small plastic zone sizes tend to have higher constraint whereas higher stress intensity factors or a decrease in section thickness can lead to lower constraint. Although $\left(\Delta K_{eff}\right)_T$ indicates the region of constraint loss, it does not give an indication of the size of the region. Methods to estimate the size of this region have yet to be developed. A method of trial-and-error is still the most common approach [5].

Wilhem [6] indicates that shear lip development appeared to occurred at a consistent ΔK_{eff} value for a given material. Thus the beginning of the constraint loss regime can be considered to be independent of the thickness of the material. This tells us that the minimum rate used in FASTRAN to define the beginning of the constraint loss regime accompanied by the appropriate α_1 will likely remain constant.

4.2 FASTRAN thickness constraint factor development

Constraint factors were developed through a process of trial-and-error and are presented on the crack growth rate curve presented in Figure 3. Indeed, a constant crack growth rate corresponding to α_1 was found to work well in the analysis. Crack growth analysis for each of the spectra and thicknesses considered are presented in Figure 1 and 2. Good correlation between prediction and experimental result is achieved in Figure 1. It should be noted that the crack growth predictions in Figure 2 consistently give non-conservative results. This is mainly due to the choice of constraint factors which give emphasis to the centre wing lower surface spectra, which is expected to be more critical for fleet management.

5 DISCUSSION

Several explanations for the observed thickness effect on crack growth rate exist in the literature. By far the most prevalent explanation uses plastically induced crack closure. As the thickness of the specimen reduces, the stresses move from a state of plane-strain to a state plane-stress resulting in an increase in plastically induced crack closure and a subsequent reduction in crack growth rate. Other explanations involve residual stresses in the plastic zone, crack path deviation and crack path bifurcation. The results here are in contrast to those in [2] which did not show a pronounced thickness effect. It is postulated that the geometry of the specimen in [2] (being a compact test specimen) has contributed to the observed behaviour along with the chosen stress ratios and the thickness range investigated. To investigate this further, testing should be conducted with the present test specimens under constant amplitude loading. As many of the crack growth algorithms used today do not account directly for the effect of specimen thickness on crack growth rates, the present results highlight a challenge in predicting crack growth of a specimen under variable amplitude loading using material data obtained from a different thickness. In an attempt to model the observed behaviour modifications were made to the constraint factors used in the well-known closure model

FASTRAN. Preliminary results were promising and open the door for further developments to improve the robustness of fatigue crack growth predictions.

6 CONCLUSION

A thickness effect in 7075-T3751 bare plate ranging in thickness from 2 mm to 6.4 mm has been observed on fatigue crack growth under representative C-130 load spectra at centre and outer wing locations. For all the coupons tested the thinner coupons display longer fatigue lives. The present results highlight the importance of appropriate material data selection in fatigue crack growth predictions. Further, many fatigue crack growth models ignore the effect of thickness, but the present results indicate that careful consideration should be given to the effect of thickness and the subsequent appropriate application of a fatigue model.

7 ACKNOWLEDGEMENTS

The author would like to acknowledge the contribution of Matthew McCoy in developing the test plan and conducting FASTRAN analysis to determine appropriate constraint factors and the support and helpful discussions with Robert Ogden and Chris Loader. The author would also the like to acknowledge the help and support of David Taylor and Matthew Pelosi in the DSTO Fatigue and Fracture Laboratory.

8 REFERENCES

1. ASTM (2001) *Standard Test Method for Measurement of Fatigue Crack Growth Rates.* DOI: 10.1520/E0647-00, West Conshohocken, ASTM International.
2. Zhuang, W., Liu, Q. and Hu, W. (2010) The Effect of Specimen Thickness on Fatigue Crack Growth Rate and Threshold Behaviour in Aluminium Alloy 7075-T7351. In: *6th Australasian Congress on Applied Mechanics,* Perth, Australia.
3. McCoy, M. and Ogden, R. (2011) *Coupon Test Program in Support of the Royal Australian Air Force C-130J-30 Hercules Full Scale Fatigue Test Interpretation.* DSTO-TR-2557.
4. *Metallic Materials Properties Development and Standardization.* (2003) MMPDS-01, US Department of Defence.
5. Newman Jr, J. C. (1992) *Effects of Constraint on Crack Growth under Aircraft Spectrum Loading.* NASA Technical Memorandum 107677, NASA.
6. Wilhem, D. P. (1967) Investigation of Cyclic Crack Growth Transitional Behavior. In: *Fatigue Crack Propagation ASTM, STP 415.* Fatigue Crack Propagation ed. 365-383.

Fatigue behaviour of web penetration details with a slit in steel girder

M. Sakano[1], N. Yoshida[1], H. Konishi[2], T. Fujii[3]
[1] Department of Civil and Environmental Engineering, Kansai University, Japan
[2] Japan Bridge Association, Japan
[3] Ministry of Land, Infrastructure and Transport, Kinki Regional Development Bureau, Japan

ABSTRACT

Fatigue cracking in steel girder web penetration details is so dangerous that it can break steel girders. Since a number of highway bridges with such web penetration details may exist in Japan, it is of urgent importance to grasp these fatigue strength properties. The purpose of this study is to clarify fatigue behaviour of steel girder web penetration details with a slit through fatigue tests of specimens which have steel girder web penetration details with a slit.

1 INTRODUCTION

Fatigue cracking in steel girder web penetration details is so dangerous that it can break steel girders. One-meter-long crack was detected in Yamazoe Bridge in 2006 (1). Since a number of highway bridges with such web penetration details may exist in Japan, it is of urgent importance to grasp these fatigue strength properties. However, few fatigue tests have been reported on steel girder web penetration details (2), (3). The purpose of this study is to clarify fatigue behaviour of steel girder web penetration details with a slit through fatigue tests of specimens which have steel girder web penetration details with a slit.

2 TEST METHOD

2.1 Specimen

We designed and fabricated 2-type steel girder specimens with web penetration details where a cross beam bottom flange was connected to each top or bottom surface of a slit by fillet welding. Figure 1 shows configurations and dimensions of a specimen with a cross beam bottom flange connected to the top surface of the slit, and the location of strain gauges. The specimen has a slit in the lower center of steel girder web. The material is JIS 490YA steel. Submerged arc welding was used to connect web and flanges, while CO_2 arc welding was used to connect the other members. Table 1 shows the mill sheet of steel plates used in the specimen.

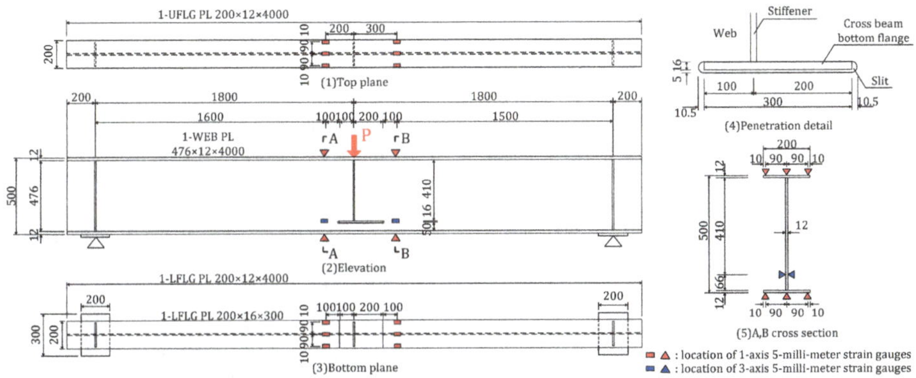

Figure 1. Configurations and dimensions of specimen and location of strain gauges.

Table 1. The mill sheet of specimen.

Member	Plate thickness (mm)	Mechanical properties			Chemical composition(%)					
		Y.P. (MPa)	T.S. (MPa)	EL (%)	C ×100	Si ×100	Mn ×100	P ×1000	S ×1000	Ceq.1 ×1000 Ceq.2 ×1000
Cross beam bottom flange	16	436	543	26	16	41	140	15	6	41 —
Upper flange	12	451	554	22	15	20	117	13	2	22 36
Lower flange Web Sole plate Stiffener	12	407	523	24	15	19	115	18	5	22 35
Spec	MIN MAX	365 —	490 610	15 —	— 20	— 55	— 165	— 35	— 35	— 26

2.2 Static loading test

First, we conducted Static Loading Test in 3-point bending condition to grasp the stress distributions around web penetration details. 3-axis strain gauges were pasted on both surfaces of the web plate horizontally 100mm away from cross beam bottom flange edges to avoid the influence of the stress concentration near the welded joint. Three 1-axis strain gauges were pasted on the top surface of top flange and the bottom surface of bottom flange in the cross section where 3-axis strain gauges were pasted. The test load was set to 100kN so that the maximum tensile stress of about 50MPa might be generated in the bottom flange.

2.3 Fatigue test

Fatigue test was conducted in 3-point bending condition as static loading test to clarify fatigue cracking behaviour and fatigue strength. The loading frequency was 6Hz. The load range was set to be 100kN, with the maximum load of 300kN and the minimum load of 200kN. In the case of a fatigue crack was propagated in one of two tested area, a stop hole was drilled and a high tension bolt was tightened to continue fatigue tests. Eddy Current Test (ET) and Magnetic Particle Test (MT) were applied to detect fatigue cracks at weld toe along cross beam bottom flange edges.

3 TEST RESULTS

3.1 Static loading test results

Photo 1 shows loading test set up and Figure 2 shows principal stress distributions around web penetration details. Nominal stresses were obtained from bending moment and shearing force calculated according to the beam theory neglecting the cross beam bottom flange and stiffeners. It was confirmed that the measured maximum principal stresses nearly equal to the calculated values, although the measured value was about 2 - 6% larger than the calculated one.

Photo 1. Loading test set up.

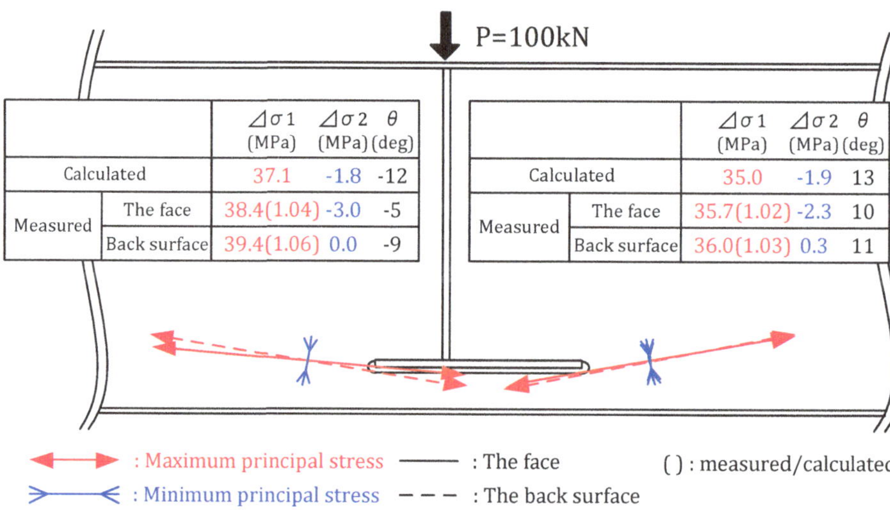

		$\Delta\sigma1$ (MPa)	$\Delta\sigma2$ (MPa)	θ (deg)
Calculated		37.1	-1.8	-12
Measured	The face	38.4(1.04)	-3.0	-5
	Back surface	39.4(1.06)	0.0	-9

		$\Delta\sigma1$ (MPa)	$\Delta\sigma2$ (MPa)	θ (deg)
Calculated		35.0	-1.9	13
Measured	The face	35.7(1.02)	-2.3	10
	Back surface	36.0(1.03)	0.3	11

Figure 2. Principal stress distributions around web penetration details.

3.2 Fatigue test result

Figure 3 shows relationship between the fatigue crack length and the number of loading cycles. Cracks with the length of 8mm were detected at four web-side toes of the fillet welds along cross beam bottom flange edges at 0.2 million cycles. Fatigue cracks were propagated into the web plate on both surfaces in the section A at 0.8 million cycles, on the face in the section B at 1.2 million cycles and on the back surface in the section B at 2.3 million cycles. As the crack in the section A was growing, the crack in the section B was growing slowly because the load could not be carried to the section B.

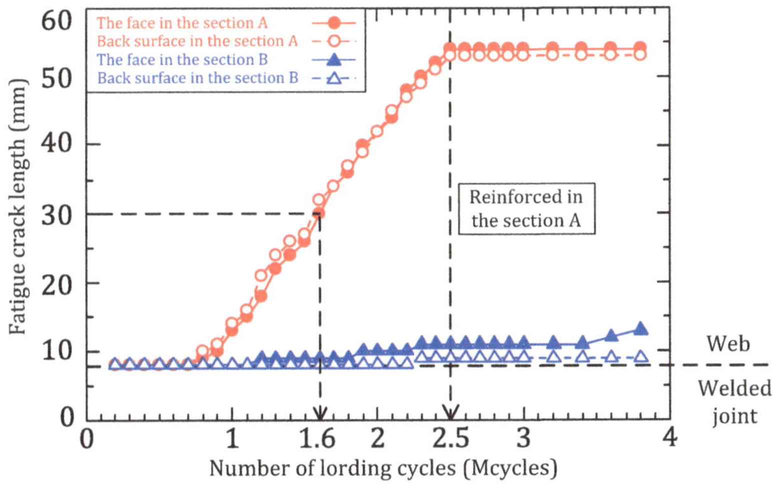

Figure 3. Relationship between fatigue crack length and number of loading cycles.

Figure 4a shows a fatigue crack propagated to the length of 54mm at 2.5 million cycles. The cracks were propagated into the web plate almost perpendicularly to the maximum principal stress direction. Figure 4b shows a stop hole and a high tension bolt. A stop hole was drilled and a high tension bolt was tightened in order to stop fatigue crack propagation in the section A at 2.5 million cycles. Figure 5 shows the relationship between bending stress range and fatigue life of the specimen. Fatigue crack detection life (Nd) is 1/6 - 1/4 of JRA class H', fatigue life when the crack has been propagated into the web plate (Nw) is just on class H' line, and fatigue life when the crack length reaches 30mm (N_{30}) satisfies class H'.

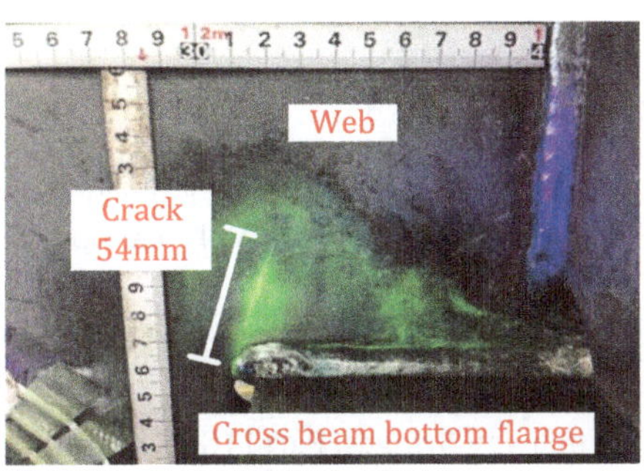

Figure 4a. Fatigue crack at 2.5 million cycles.

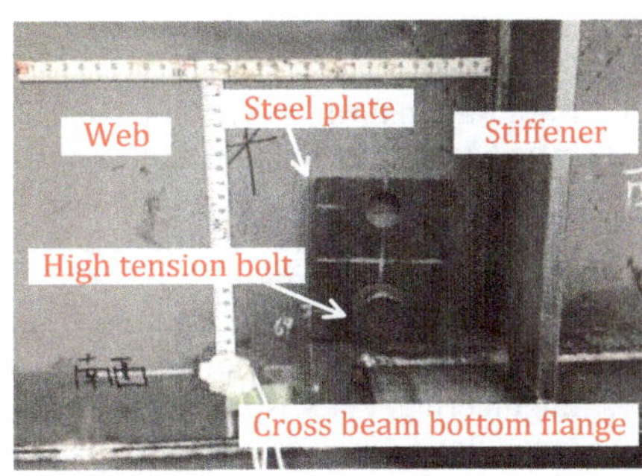

Figure 4b. A stop hole and a high tension bolt.

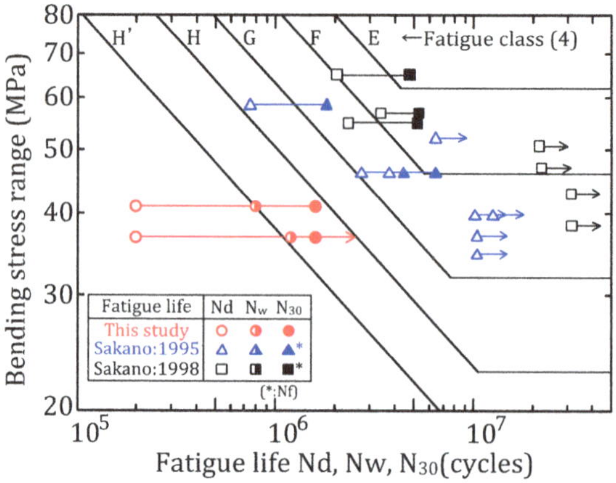

Figure 5. Fatigue life of the specimen (Bending stress range).

4 CONCLUSIONS

The principal results obtained through this study are as follows;

1) Fatigue cracks have been initiated at the web-side toe of the fillet weld along cross beam bottom flange edges, and propagated into the web plate almost perpendicularly to the maximum principal stress direction.
2) Fatigue crack detection life (Nd) is 1/6 - 1/4 of JRA class H', fatigue life when the crack has been propagated into the web plate (Nw) is just on class H' line, and fatigue life when the crack length reaches 30mm (N_{30}) satisfies class H'.

REFERENCES

[1] Nara National Highway Homepage, 2014. (in Japanese) http://www.kkr.mlit.go.jp/nara/

[2] M. Sakano, I. Mikami, H. Murayama, K. Moriwaki : Fatigue Strength of Railway Plate Girder with Web Penetration, Journal of Structural Engineering, Vol.41A, pp.965-973, 1995. (in Japanese)

[3] M. Sakano, M. Hozumi, T. Shimori, I. Mikami : Long Term Fatigue Strength of Web Gusset Joint in Floor-Beam-to-Main-Girder Connection, Steel Construction Engineering, Vol.5, No.18, pp.31-40, 1998. (in Japanese)

[4] Japan Road Association : Specifications for Highway Bridges, Part I (General) and II (Steel bridges), 2012. (in Japanese)

Accelerated method of propagation threshold of intergranular stress corrosion cracking by using fretting fatigue

Y. Otsuka [1], M. Hiruta [2], H. Suematu [3], Y. Mutoh [1]
[1] Department of System Safety, Nagaoka University of Technology, Japan
[2] Graduate School of Nuclear System Safety Engineering, Nagaoka University of Technology, Japan
[3] Department of Nuclear Safety, Nagaoka University of Technology, Japan

ABSTRACT

Intergranular stress corrosion cracking tests by using fretting contacts with high stress ratio were conducted. All of the initiated cracks were intergranular manner due to crevice corrosion and removal of surface oxide layer by frictions. The crack growth rate revealed higher value than that of large cracks, which demonstrated small crack effect on intergranular cracking.

1 INTRODUCTION

Intergranular stress corrosion cracking (IGSCC) highly relates to the nucleation of cracks in the heat affected zone(HAZ) or cold-worked zone of austenitic stainless steels, especially in the cases of aged nuclear pressure reactor components[1]. In order to conduct an integrity assessment on existing cracks, threshold value of IGSCC is indispensable to determine "safe" crack length. However, the difficulty in measuring such the threshold values of IGSCC have prevented the integrity assessment from quantitative SCC life evaluation by using fracture mechanics approach. Our group has proposed a new accelerated method of propagation threshold of intergranular stress corrosion cracking by using fretting fatigue[2,3]. Our proposed method can generate only IGSCC in shorter time. The purpose of this study is to elucidate the nucleation conditions of IGSCC of sensitized austenitic stainless steels by fretting fatigue and also show the small crack effect of IGSCC in crack initiation periods.

2 EXPERIMENTAL METHOD

Sensitized JIS SUS304 is used. Figure 1 shows the dimensions of testing specimen and contact pad used in fretting fatigue. In subsequent SCC test, only the specimen is used. The specimen is machined by mechanical milling and electrical discharge machining from a plate of 3mm thickness. The surface of the specimens was polished by using emery papers from #240 to #1200 and subsequent buff polishing by diamond paste of 6,3,1 μm, respectively. Sensitization heat treatment applied on the specimen by 923K, 3.6ks in vacuum furnace.

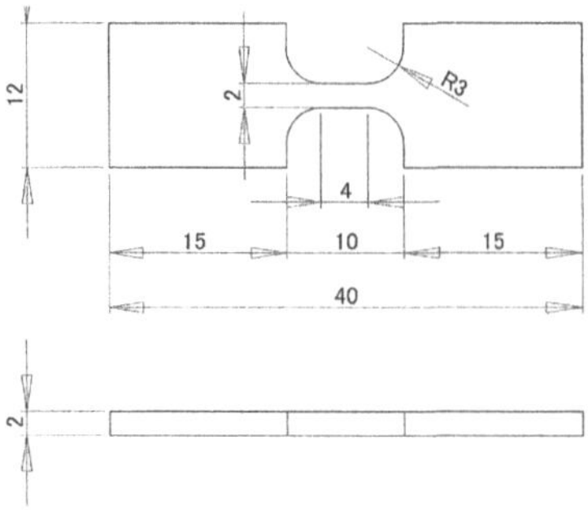

(a) Specimen

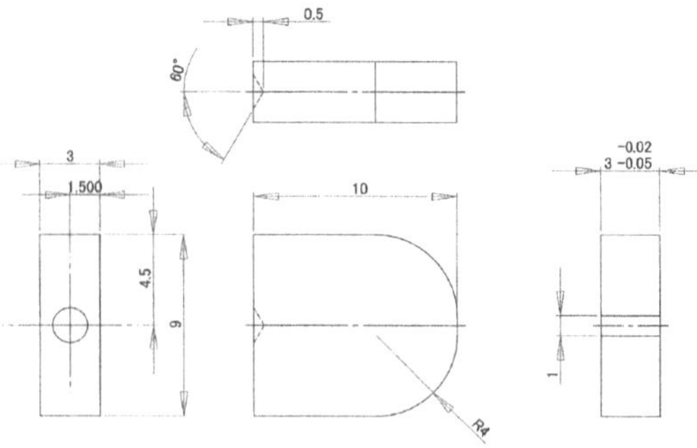

(b) Contact pad

Figure 1 Specimen and contact pad size (Unit; mm)

A servo-hydraulic fatigue testing machine (load capacity; 50kN, Shimadzu Cooperation, Japan) with autoclave (TOSHIN KOGYO, Japan) is used for fretting fatigue test. Testing environment was pressurized high temperature water of 561K, 7.3 MPa, which simulated the environment in a nuclear reactor of boiled water reactor (BWR). Loads were set to apply maximum stresses of 177,265,377MPa. Numbers of loading were 5×10^4,5×10^5,6.5×10^6 by stress ratio R =0.9 and loading frequency f= 20 Hz. Contact pressure was 100 MPa, which was set in advance of heating because strain measurement by using conventional strain gage in pressurized high temperature water was impossible.

IGSCCs on the specimens were measure by SEM images (HITACHI Miniscope TM3000) Stress intensity factors were calculated by using Ishida-Noguchi's equation on 3 dimensional semi-elliptical surface cracks subjected to normal loading. Stresses in the calculations were determined by elastic-plastic finite element analyses. Aspect ratio of the IGSCCs was assumed to be 1. In addition, interactions between IGSCCs were ignored.

Stress corrosion cracking tests by using the pre-cracked specimen by fretting fatigue were subsequently conducted. The testing machine was the same as the one in the case of fretting fatigue tests. Loads in SCC test were set in order to apply a maximum nominal stress same amd higher than the one in the fretting fatigue tests. The SCC test was interrupted after 500 hours and then crack propagation behavior was observed by SEM.

3 RESULT AND DISCUSSIONS

Figure 2 shows the crack shape on the surface of specimen after fretting fatigue test. Jig-zag like shape of the cracks go along with the pass of grain boundary, which means that all the cracks were IGSCCs. No transgranular SCC(TGSCC), which go straightly, was observed. The result in figure 3 clearly showed that the proposed method by using fretting fatigue can successfully nucleate only IGSCC in optimized loading conditions.

Figure 3 shows Gumbel cumulative distributions of IGSCCs. All the three regression curves are statistically significant by single regression analyses (Adjusted R^2=0.86,0.98,0.96 for N=5×10^4, =5×10^5, 6.5×10^6. F-statistics were 870(degree of freedom dF=50), 2039(dF=50), 83(dF=12), respectively. All three curves were significant at significance level p = 0.05. The curve of N= 6.5×10^6 has a different gradient from those in left two curves probably due to the coalescence of IGSCCs during longer loading period.

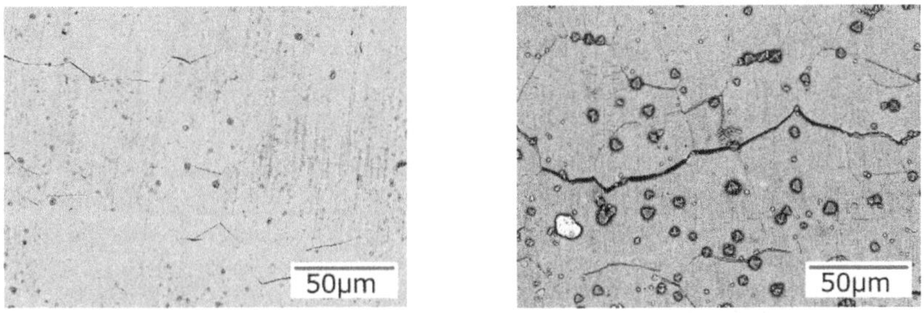

(a) N=5×10[4] (b) N=6.5×10[6]

Figure 2 SEM images of intergranular cracking by optimized loading conditions

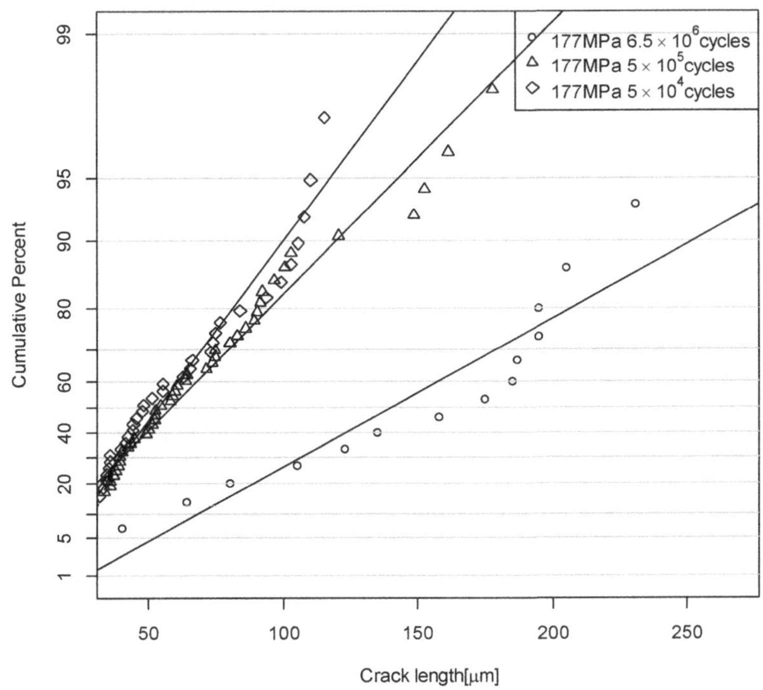

Figure 3 Gumbel distributions of intergranular pre-cracks.
Fretting fatigue conditions; 177MPa, R=0.9 f=20Hz

The curves in figure 3 can predict maximum crack length by 99% cumulative length.

Figure 4 shows the relationship between crack growth rate and stress intensity factors. Propagation behavior of pre-cracks by fretting fatigue clearly shows the small crack growth behavior in stress concentration area. The small crack behavior is caused by combined effect of removal of surface oxide films by fretting contact, localized maximum principal stress and crevice corrosion. We then conduct SCC test without fretting. The dashed lines in the figure show the experimentally extrapolating line. If the loading condition exceeds the dashed area, IGSCCs propagated and finally leaded to entire fracture. In the cases of predicted loading condition, IGSCCs only slowly propagate at the rate of approximately 10^{-12} m/s, which can be regarded as threshold value.

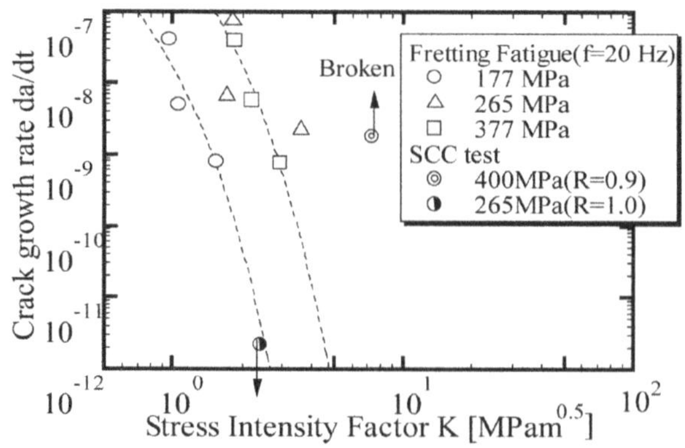

Figure 4 Crack growth rates of pre-cracks by Fretting fatigue and IGSCC by normal loading

4 SUMMARY

Proposed accelerated method of IGSCC by fretting fatigue can observe threshold valued of only IGSCC. In addition the result revealed that threshold value of IGSCC is much lower than the value of IG+TGSCC in conventional testing. Longer SCC test is now conducted in order to solidity the validity of testing method we proposed.

ACKNOWLEDGEMENT

The authors deeply appreciate the partial financially support by JST A-STEP FS stage (Contract No.; AS242Z02617K) and UCHIDA energy science research foundation.

REFERENCES

[1] Horn, R.M., *et.al.,* (1997) Nuclear Engineering and Design. **174**, 313–325.

[2] Saengsai, A., Otsuka, Y., & Mutoh, Y. (2011) Advanced Materials Research. **399-401**, 2276–2282.

[3] Saengsai, A., Otsuka, Y., & Mutoh, Y. (2011) Journal of Solid Mechanics and Materials Engineering. **5**(12), 898–907.

Effects of environment on fatigue crack growth behavior of 2000 and 7000 series aluminum alloys

R. Yamada[1], S. Ishizawa[2], G. Itoh[2], A. Kurumada[2], M. Nakai[1,3]
[1] Graduate School of Science and Engineering, Ibaraki University, Japan
[2] Department of Mechanical Engineering, Ibaraki University, Japan
[3] Kobe Steel, Ltd, Japan

ABSTRACT

Fatigue crack growth behaviour of high strength aluminum alloys for aircraft components was studied experimentally in relation to testing environment. Crack path was mainly transgranular. Crack growth rate was slightly larger in 7075-type alloy with higher fraction of intergranular path than 2024-type alloy, when tested in moist air. The rate of the 7075-type alloy was decreased by changing the environment to the flow of nitrogen gas. In moist air, hydrogen embrittlement based on intragranular cracking has been known to occur in 7000 series alloys. Thus, the large fatigue crack growth rate of the 7075-type alloy as attributable to hydrogen embrittlement.

1 INTRODUCTION

Since the high-strength aluminum alloys have good workability and high specific strength, they have been the primary structural material for aircrafts. Among them, 2024 of 2000 (Al-Cu-Mg) series and 7075 of 7000 (Al-Zn-Mg) series have been frequently used. The 7000 series alloys have the highest strength in the aluminum alloys, but lower fatigue properties than 2000 series alloys, as shown in Fig 1. (1) Thus, 2000 series alloys accounts for a large proportion of the aircraft components for their better fatigue properties than 7000 series alloys have not been elucidated yet. Besides, the 7000 series alloys are far more sensitive to hydrogen embrittlement than 2000 series alloys, as shown in Fig.2.(2) Both in the fatigue and hydrogen embrittlement the crack growth is time-dependent. From this, it can be considered that the two phenomena are closely related to each other. In order to explain the difference in the fatigue crack growth behavior in the two series alloys, the effect of the test environment on the fatigue crack growth of the two series alloys will be investigated in this study, considering the relationship to hydrogen embrittlement.

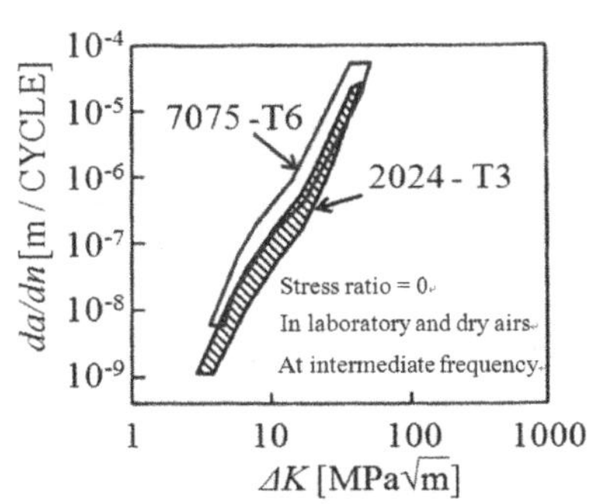

Fig. 1 Fatigue crack growth rate, da/dn, vs. stress intensity factor range, ΔK, curves for 7075-T6 and 2024-T3 (1).

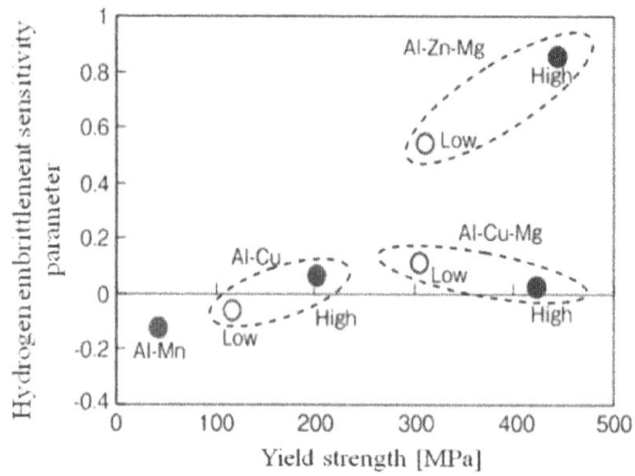

Fig. 2 Hydrogen embrittlement sensitivity parameters of the alloys plotted as a function of yield strength(2).

2 SPECIMENS

Shown in Table 1 are the composition of the alloys used in this study and the standard composition range of 7075 alloy. The alloy specimens No.1 and No.2 are based on the composition of the 7075 and 2024 alloys, respectively. In order to investigate basic principle, *i,e.*, the effect of the strengthening phases, the amount of impurity elements (Fe and Si) and the additive (Cr for 7075 or Mn for 2024) have been minimized. In addition to the two alloy specimens, a 2024 alloy was melted and DC – cast in the laboratory. The three alloy ingots were hot-rolled to a thickness of 7mm. After scalping to 6mm thickness, CT test pieces were cut by electric discharge machining in L-T direction. The CT test pieces were then subjected to T6 or T4 temper. The T6 temper for No.1 consisted of solution treatment for 1h at 470°C, water quenching and subsequent artificial aging for 24h at 120°C, while the T4 for No.2 and 2024 of solution treatment for 1h at 495°C, water quenching and subsequent natural aging for 96h.

Table 1 Composition of the specimens in mass%.

	Si	Fe	Cu	Mn	Mg	Cr	Zn	Ti	Zr	Al
No.1	0.005	0.005	1.391	—	2.425	0.008	5.693	0.018	—	Bal.
7075*	≦0.4	≦0.5	1.2-2.0	≦0.3	2.1-2.9	0.18-0.28	5.1-6.1	≦0.20	—	Bal.
No.2	0.001	0.038	4.323	0.001	1.624	0.004	0.005	0.021	0.004	Bal.
2024	0.189	0.230	4.401	0.58	1.6	0.01	0.01	0.016	0.004	Bal.

* Standard alloy composition ranges.

3 EXPERIMENTAL PROCEDURE

Both sides of the CT specimen of 6mm thick were mirror-finished. After that, one side was etched in Keller's reagent (concentrated hydrochloric acid 0.5ml, concentrated nitric acid 2.5ml, hydrofluoric acid 1.5ml, distilled water 95ml) to reveal the microstructure. Fatigue crack growth test, was carried out in a sinusoidal stress wave of a frequency of 20Hz with a stress ratio of R (= Kmin / Kmax) =0.1 in air with MA(relative humidity of 90% at 30°C) and DNG(relative humidity of 40% at 25°C). In the MA environment, hydrogen embrittlement has been reported to occur in 7000 series alloys.

(2) In addition to the assessment of the crack growth rate (da/dn) against stress intensity factor range, the crack path wasw observed with an optical microscope and fractography was made with a scanning electron microscope.

4 RESULTS AND DISCUSSION

Figure 3 shows the plots of da/dn vs. ΔK for the three specimens together with the previously reported data for 7075 and 2024 alloys. The effect of test environment in No.1 is also shown. Crack growth rate of No.1-MA is the largest, the crack growth rate becomes smaller in the order of No.1-MA, No.1-DNG, No.2-MA and 2024-MA. Crack path was mainly transgranular, as shown in Fig. 4. However, the ratio of intergranular to transgranular, is highest in No.1-MA, as shown in Fig. 5. This is also confirmed in the fractognaphic observation, where transgranular fracture surface is found to consist of quasi-cleavage and dimple fracture surfaces, as shown in Fig. 6. The ratio of quasi-cleavage to dimple is larger in accordance with the order of crack growth rate.

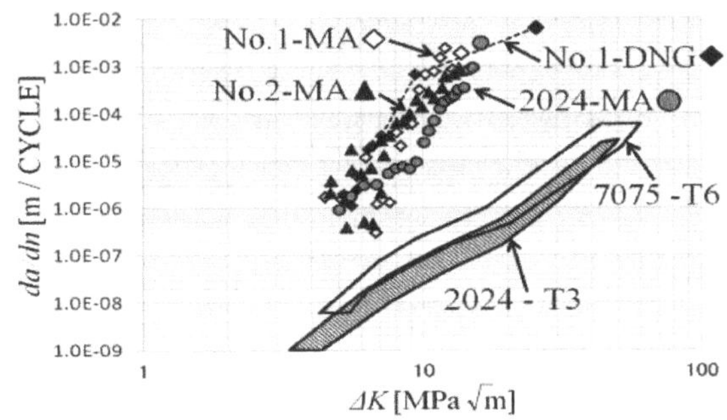

Fig. 3 Fatigue crack growth rate, da/dN, vs. stress intensity factor range, ΔK, curves.

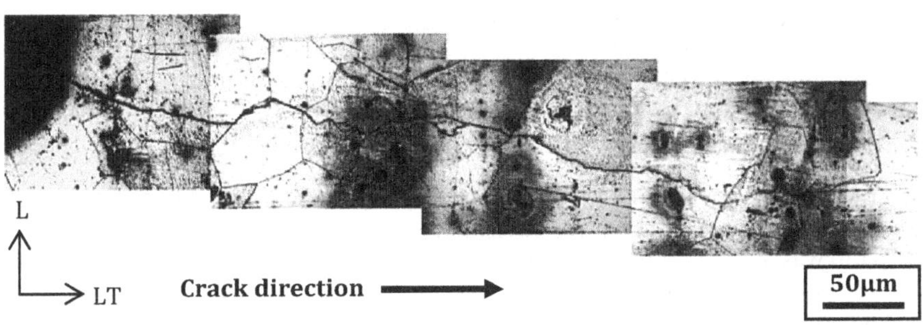

Fig. 4 Optical microscopic image of fatigue crack growth path in No.1-MA.

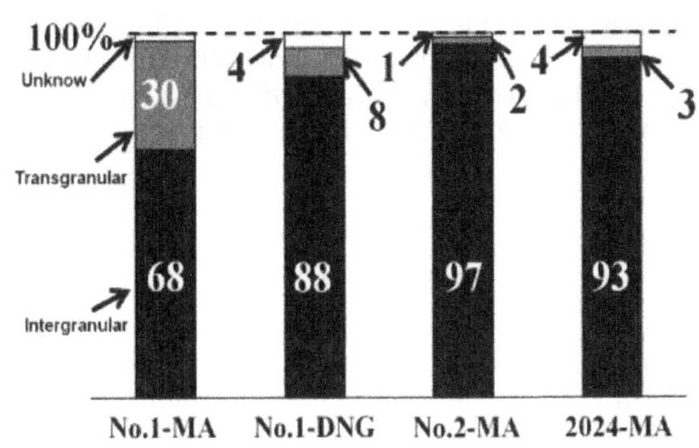

Fig. 5 Ratio of crack path type : intergranular to transgranular.

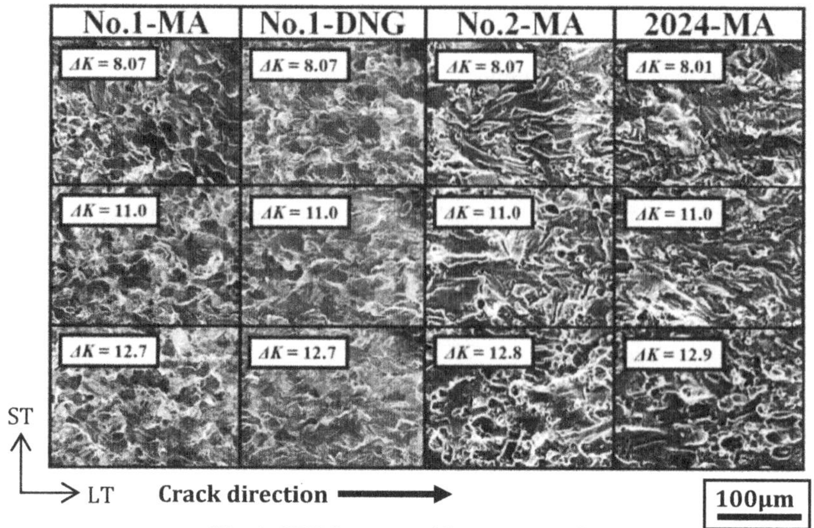

Fig. 6 SEM images of fracture surfaces.

5 SUMMARY

The 7000 series alloys have the highest strength in the aluminum alloys, but lower fatigue properties than 2000 series alloys. Thus, 2000 series alloys accounts for a large proportion of the aircraft components for their better fatigue properties than 7000 series alloys have not been elucidated yet. Besides, the 7000 series alloys are far more sensitive to hydrogen embrittlement than 2000 series alloys. In order to explain the difference in the fatigue crack growth behavior in the two series alloys, the effect of the test environment on the fatigue crack growth of the two series alloys will be investigated in this study, considering the relationship to hydrogen embrittlement.

Crack growth rate was slightly larger in No.1 than No.2, when tested in MA. The rate of No.1 was decreased by changing the environment to the flow of nitrogen gas. Although crack path was mainly transgranular, the fraction of intergranular crack path was largest in No.1 tested in MA. From this fact and the already-known fact that hydrogen embrittlement based on intragranular cracking occurs in 7000 series alloys tested in MA, the larger fatigue crack growth rate of No.1 was presumed to be correlated with hydrogen embrittlement. Therefore, if the hydrogen embrittlement in the 7000 series alloys in able to be checked up to the level of 2000 series alloys, the fatigue crack growth property will also be improver. Then the 7000 series alloys will be used in most parts of an aircraft instead of 2000 series alloys became of their higher strength, leading to fuel cost saving as well as alleviation of the load against global warming through weight reduction.

REFERENCES

[1] T. Kishi and T. Ohhira : J.JILM, 31(1981), 628-634.
[2] K. Ichitani and K. Koyama : Furukawa-Sky Review, 5(2009), 20-28.
[3] Registration Record of International Alloy Designations and Chemical Composition Limits for Wrought Aluminum and Wrought Aluminum Alloys, the Aluminum Association, 1976 (Revised 1980).

Thermoelastic monitoring of fatigue degradation in aluminium alloy supersonic particle deposition coatings

J. Choi, W. Zhuang, N. Rajic
Aerospace Division, Defence Science and Technology Organisation, Australia

ABSTRACT

Supersonic particle deposition is an established metallic repair technology. However, relatively little is known about the durability of supersonic particle deposited coatings under fatigue. The present paper describes an investigation into the fatigue performance of aerospace-grade aluminium alloy supersonic particle deposition coatings using thermoelastic response measurements as the diagnostic methodology. The measurements were obtained using a novel microbolometer based system developed by the Australian Defence Science and Technology Organisation. It is shown that this approach can furnish reliable indications of fatigue cracking in the coating and the evidence presented suggests that it can also discriminate the position of the crack tip in the coating from that in the substrate.

1 INTRODUCTION

Supersonic Particle Deposition (SPD), otherwise known as cold spraying, is an advanced repair technology for metallic structures (1, 2). Although SPD is an established aircraft repair technology, its use is largely confined to dimensional restoration and environmental protection for tertiary aircraft structures (3, 4). In recent years, there has been a move to develop and certify the SPD process as a structural repair technology for load-bearing components (5-7). This more ambitious application requires additional knowledge with respect to the static strength and fatigue durability of SPD coatings.

Previous research has shown that thermoelastic stress analysis (TSA), a full-field stress analysis method based on the thermoelastic effect, can offer useful insight into the structural performance of such coatings (5). Although the use of TSA to investigate fatigue is well established (8), prior work in this area has relied exclusively on a commercial form of TSA technology that relies on cryogenically cooled infrared imagers. These are relatively expensive and cumbersome devices that are ill-suited to long-term continuous monitoring. For example, the use of a liquid Nitrogen cooled system would require a periodic top-up of the dewar, whilst the use of a closed-cycle cooled system would need to take account of the fact that such systems have a finite life. The present work utilises an uncooled microbolometer based system (9). This system is far more compact and lower in cost than a photon-imager system, which permits one to apply TSA in a long-term monitoring role without the expense and inconvenience associated with cryogenically cooled imagers.

In the present paper, SPD coatings were applied to aluminium alloy 7050-T7451 coupons, which were tested under fighter spectrum loading. Initiation and propagation of fatigue cracks in the SPD coatings and substrate were assessed in quasi-real-time by means of periodic thermoelastic response measurement. These measurements were compared to crack growth estimates obtained from post-test quantitative fractography (QF).

2 METHOD

2.1 Coupon dimensions

A single-edge notch rectangular coupon was used as a substrate to test the SPD coatings. The notch depth (D) was 4.0 mm with the notch radius (r) of 2.0 mm.

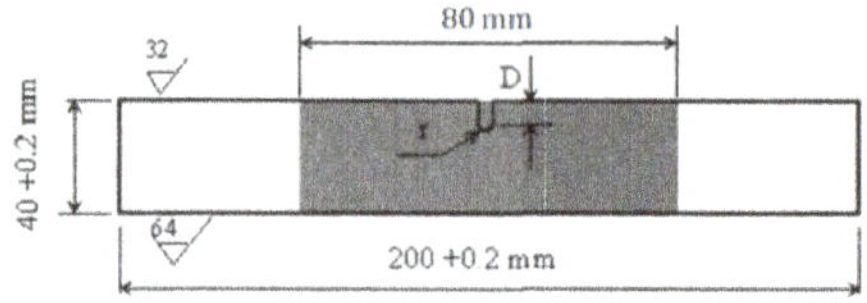

(a)

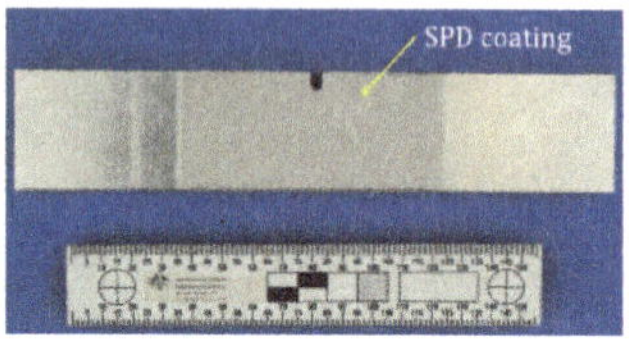

(b)

Figure 1: (a) Dimensions of substrate coupon, (b) Photograph of the coated coupon

2.2 Supersonic particle deposition

The SPD coatings were deposited using a stationary SPD system with a polycarbonate nozzle. The process temperature was 400 °C and the main gas pressure was 40 Bar. The SPD coatings were produced with an aluminium alloy 7075 powder with a particle diameter of between 30 to 40 μm. The coatings were approximately 1 mm thickness on each side of the substrate surface.

2.3 Test apparatus

A 100 kN MTS servo-hydraulic testing machine was used to load the coupons. The thermoelastic response measurements were made using a MiTE system (10) equipped with a FLIR A315 microbolometer. A thin coating of black paint was applied to each coupon to ensure a uniformly high surface infrared emissivity.

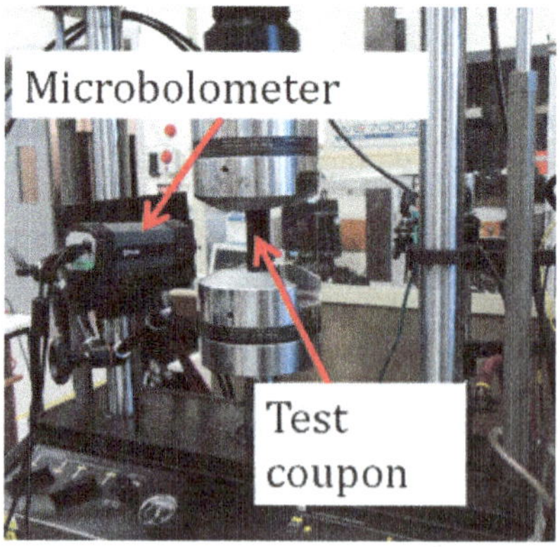

Figure 2: The microbolometer setup on MTS testing machine

A variable amplitude spectrum representative of fighter aircraft wing root loads was applied at an average frequency of approximately 5 Hz. Marker band underloads were included in the spectrum to facilitate post-test analysis of the crack growth rate using QF (11).

3 RESULTS

3.1 Fractography results

Figure 3 shows a typical failure surface of the coupons while the microscope images of the failure surface are shown in Figure 4. The fatigue progression markings caused by the constant amplitude marker band underloads are not easily identifiable in the fracture surface of the SPD coating as shown in Figure 4a. However, these marker bands are clearly visible in the substrate material, shown in Figure 4b, which facilitated post-test QF crack growth analysis.

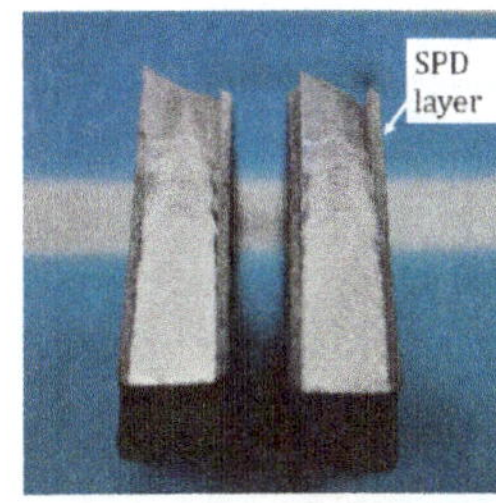

Figure 3: Typical failure surface of coupons

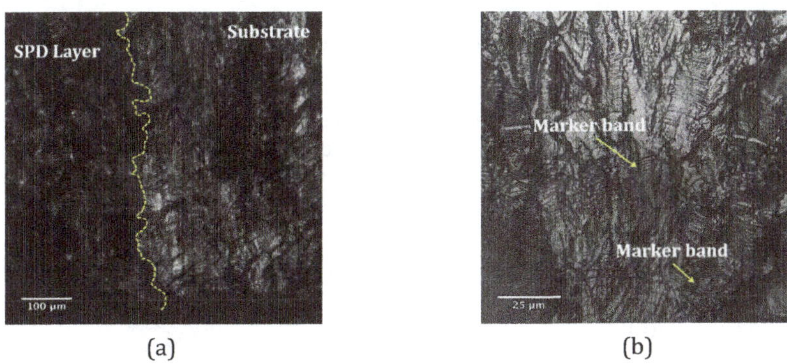

Figure 4: Microscope images of the crack surface: (a) on the boundary of SPD coating and (b) zoomed-in image of the marker bands in substrate material

3.2 Thermoelastic results

Figure 5 shows a representative thermoelastic response for a relatively short crack, where Figure 5a and Figure 5b correspond respectively to the signals in-phase and in quadrature to the applied loading. A previous study has shown that the quadrature signal provides better localisation of the crack tip (12) particularly under severe heat conduction, so for the present work the crack tip position was deduced from the position of the quadrature peak. Figure 6 compares the trace of the quadrature peak with estimates of crack growth obtained from post-test QF examination of the fracture surface. Whilst it is clear from the comparison that QF has a distinct advantage in resolution, which one should expect given the circumstances under which the loading was applied (tailored for post-test examination), Figure 7 illustrates the capacity of a thermoelastic response measurement to furnish other unique and useful insights. Figure 7c is an overlay of two thermoelastic response measurements – one taken in-phase with the load (Figure 7a) and therefore (approximately) proportional to the bulk stress distribution and the other in quadrature but processed in a manner that accentuates the crack trace on the surface of the coupon (Figure 7b). The tip of the surface crack, as shown in Figure 7b, is evidently well advanced of the in-phase response peak shown in Figure 7a, suggesting that the crack in the SPD coating has grown ahead of the crack in the substrate. This is plausible from a thermoelastic perspective as the SPD coating is relatively thin and highly conductive which would permit the thermoelastic response of the substrate to rapidly diffuse through to the surface. An extensive survey of the full set of data is yet to be completed so it is unknown if more evidence of such a phenomenon exists. It is certainly worth further exploration and supplementary testing. Indeed, one would expect that if rates of crack growth in the coating and substrate do differ, one should be able to determine this by testing the coupons at two different loading frequencies in rapid succession - once at a relatively high frequency to characterise the thermoelastic response of the SPD coating and once at a relatively low frequency to permit measurement of the substrate thermoelastic response.

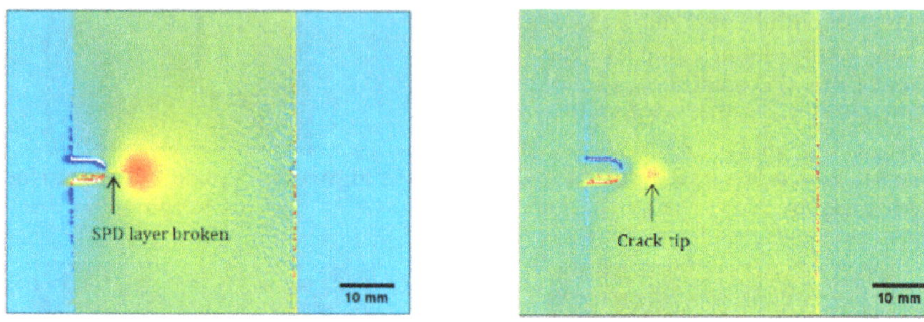

(a) (b)

Figure 5: Thermoelastic response at approximately 16.2 blocks of loading (a) in-phase, (b) quadrature

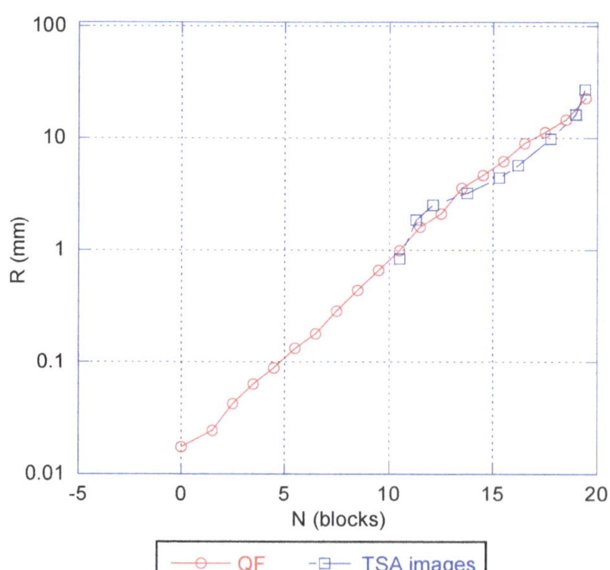

Figure 6: Crack growth measurement from post-test QF and thermoelastic monitoring (shown in log-linear scale and R is crack size from the notch root)

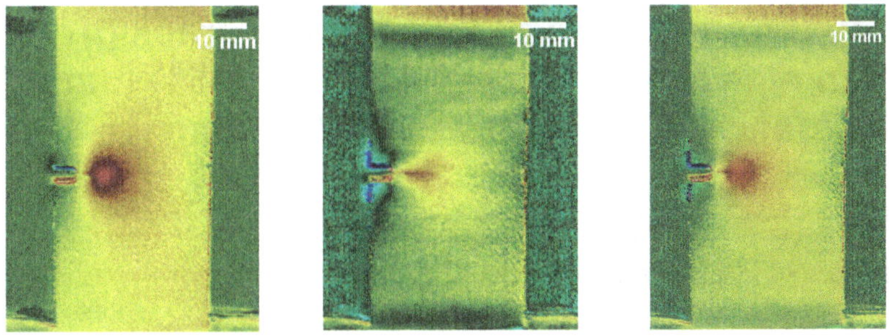

(a) (b) (c)

Figure 7: Thermoelastic response at approximately 16.5 blocks of loading: (a) in-phase, (b) processed thermograph and (c) overlay of (a) and (b)

4 CONCLUSIONS

Measurements obtained with a novel microbolometer-based thermoelastic monitoring system have been shown to yield useful insights into fatigue cracking in aluminium coupons reinforced with an SPD coating. Evidence was presented that suggests that cracks in an SPD coating can grow in advance of a crack in the substrate. This finding is preliminary and needs to be verified by independent means. However, if confirmed, it warrants further investigation of the proposed use of SPD as a structural reinforcement. This should also include consideration of the implications of differential rates of crack growth for in-service non-destructive inspection of SPD coated structural components.

REFERENCES

[1] Papyrin, A., et al. (2007) *Cold spray technology* Papyrin, A. ed., Amsterdam; London: Elsevier

[2] Sing, H., Sidhu, T. S. and Kalsi, S. B. S. (2012) *Frattura ed Integrità Strutturale* **22** 69-84, doi: 10.3221/IGF-ESIS.22.08

[3] DOD-MFFP (2011) *Materials Deposition, Cold Spray*. In *Department of Defense Manufacturing Process Standard*. USA. Department of Defense

[4] Gerrard, D., et al. (2007) Cold Spray Coating Technology for the Corrosion Protection and Repair of JSF/Aerospace Components. In: *ASM TSS*

[5] Jones, R., et al. (2011) *International Journal of Fatigue* (9) 1257, doi: 10.1016/j.ijfatigue.2011.03.013

[6] Matthews, N., Jones, R. and Sih, G. C. (2013) In: *Internation Conference on Airworthiness and Fatigue*, Beijing, China

[7] Zhuang, W., et al. (2013) In: *Aircraft Airworthiness and Sustainment*, Brisbane, Australia: 23-25 July, 2013

[8] Díaz, F. A., Yates, J. R. and Patterson, E. A. (2004) *International Journal of Fatigue* **26** (4) 4// 365-376, doi: 10.1016/j.ijfatigue.2003.08.018

[9] Rajic, N. and Rowlands, D. (2013) *Quantitative InfraRed Thermography*, doi: 10.1080/17686733.2013.800688

[10] DSTO. *Microbolometer Thermoelastic Evaluation (MiTE)*. (2013) Updated 11/12/2013 [Accessed 01/08/2014]; Available from: http://www.dsto.defence.gov.au/mite/

[11] Barter, S., Molent, L. and Wanhill, R. J. H. (2009) In: *25th ICAF Symposium*, Rotterdam

[12] Rajic, N. and Galea, S. (2014) *in-press Structural Health Monitoring*

Improvements to predicting fatigue crack growth rates in aluminium alloy (AA7050-T7451) loaded with a standard transport aircraft spectrum

M. Burchill [1], K. Walker [1], S. Barter [1], C. Wang [2], A. Khadka [2]
[1] Defence Science and Technology Organisation, Australia
[2] Royal Melbourne Institute of Technology, Australia

ABSTRACT

The accurate prediction of fatigue crack growth is vital to design, structural airworthiness and safety; hence of great interest to fleet manufacturers, operators and maintainers. The Defence Science and Technology Organisation (DSTO) has been developing a novel testing method to obtain data to improve fatigue life prediction capabilities. This paper presents research undertaken at DSTO and Royal Melbourne Institute of Technology, into a new material model for predicting crack growth rates for a typical flight spectrum. A small coupon fatigue test study was undertaken, and post-test crack growth measurements were found to compare favourably with the proposed updated model's results.

1 INTRODUCTION

Previous work by many researchers, including [1,2] have identified deficiencies in standard long-crack material data when used with fatigue life prediction codes based on linear elastic fracture mechanics (LEFM) principles. Typically, standard material data under predicts the growth of short natural cracks (near the fatigue threshold) when compared to crack measurements from coupon or component fatigue testing. The importance of correctly predicting short crack growth lives, particularly under variable amplitude (VA) spectra has long been recognised as important to accurate fatigue life predictions of all aircraft, and of specific interest to Defence Science and Technology Organisation's (DSTO) work on Royal Australian Air Force (RAAF) aircraft, [3-5].

The most fundamental and important input used in fatigue crack growth analysis models, such as AFGROW [6] and FASTRAN [7], is the baseline, constant amplitude (CA) crack growth rate per cycle (da/dN) data. Crack growth rate is typically defined as a function of the applied cyclic stress intensity factor range (ΔK) with corrections for: (i) mean stress or stress ratio ($R = \sigma_{min}/\sigma_{max}$); (ii) ΔK values at the *threshold* of crack growth, ΔK_{th}; and (iii) maximum stress intensity of the load cycle, K_{max} approaching the fracture toughness of the material, K_C.

Since cracks from typical material discontinuities spend the majority of their lives growing at rates just above ΔK_{th} [5], this region of the crack growth data curves is of great importance to the successful prediction of the total lives of cracks (up to 10mm depth). To address this for a common aluminium alloy that is used for fatigue critical

RAAF airframe components: AA7050-T7451 (AA7050), researchers at DSTO and RMIT have proposed enhancements to fatigue life prediction through the development of new fatigue material data, and prediction code refinements.

This new material data has, when used in the above noted codes, been found to improve the accuracy of fatigue crack growth predictions, for coupons tested with a combat aircraft spectrum [8, 9]. Such spectra typically contain many large peak loads. Here, to increase confidence in the robustness of this data, a further coupon test and analysis program was undertaken, where the standard transport spectrum: mini-TWIST (The Transport Wing Standard Load Programme) [10] was used as the loading.

2 THRESHOLD AND SHORT CRACK (TASC) PROGRAM AT DSTO

From extensive DSTO experience in testing, inspecting and analysing fatigue cracks in common aircraft alloys [11-13], a short crack growth rate measuring method was developed that is an alternative to those suggested in standards, such as ASTM-E647 (American Society for Testing and Materials [14]). This new method uses test load sequences selected to produce local crack path and topological changes, to make *features* on the fracture surface that can be related to the loading and that can be measured. An example of a typical AA7050 fatigue fracture surface is shown in Figure 1; with crack growth *bands* that can be attributed to the sub-blocks of VA and CA loading. Using optical and electro optical microscopes, post-test Quantitative Fractographic (QF) assessment of the fractures makes it possible to identify and measure these features down to tens of nm (10^{-8}m).

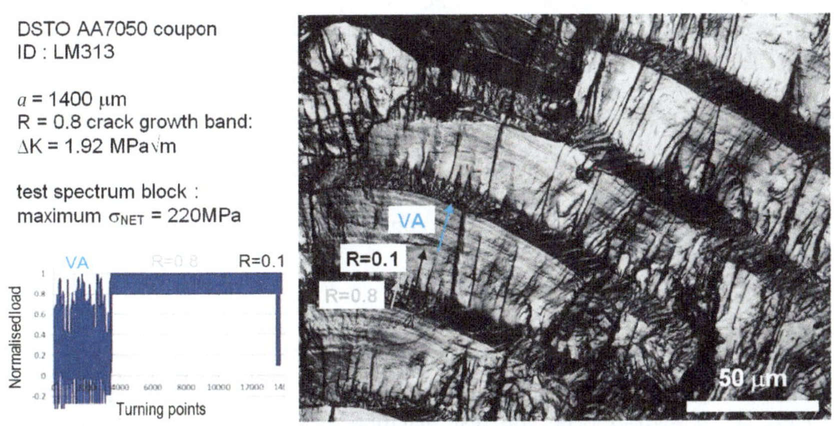

Figure 1 Observed effect of changing load spectrum (insert) on the fracture surface of a fatigue crack at a depth of 1.4mm, in an AA7050 coupon.

The threshold and short crack (TASC) test program has produced fatigue cracks in many AA7050 coupons, with clearly visible marks on the crack surfaces that enabled crack growth increment measurement [8, 9]. The individual CA growth band measurements were then divided by the number of cycles applied to calculate a crack growth per cycle rate (da/dN). For each CA growth band measurement and R, the ΔK was calculated via, $\Delta K = \Delta\sigma\beta\sqrt{(\pi a)}$; where β is the geometry factor for the crack being investigated [15]. In this way, and by using 1000's of cycles (N) to produce a growth band, very small incremental, da/dN data, at very low ΔK values were measured. A summation of this data for two stress ratios (R=0.1 and 0.8) is given in Figure 2, along with a standard material data as modelled in AFGROW, using the Harter-T [6] (labelled HT model) method. For very low ΔK values, higher growths rates were observed during TASC testing compared to standard published data; as well as at ΔK_{th} ~0.5 MPa√m which is significantly less than that typically reported for this material.

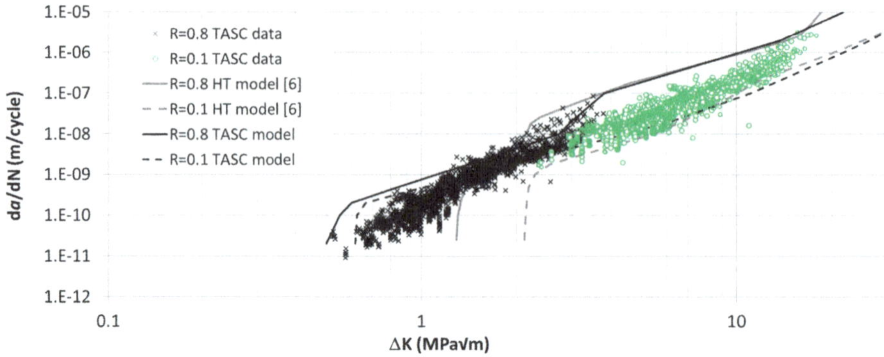

Figure 2 TASC AA7050 short crack growth data compared to modelled standard material, by the Harter-T [6] method.

3 IMPLEMENTING UPDATED DATA IN AFGROW AND FASTRAN

The AA7050 material data model for AFGROW crack growth analyses that is proposed here, is shown in Figure 2, (labelled TASC model) and has been shown to accurately predict crack growth under a fighter aircraft wing root combat spectrum for cracks between 0.1mm and several millimetres depth [8]. The rate curve for the FASTRAN analyses were adjusted using the TASC data and in part informed by long crack testing performed after compression pre-cracking and minimal load reduction [8], a method that is a modification to the standard E647 crack growth collection method. Compliance based crack opening data were also used to determine the effective (ΔK_{eff}) rate curve to account for crack closure. Drawing upon results from recent related research [16-18] into roughness closure an appropriate 3-D constraint factor (α = 2.0) that could be used in the FASTRAN code, was selected.

4 COUPON TEST PROGRAM

Fifteen simple (optimised) hour glass coupons were tested in a 100 kN, computer-controlled, servo hydraulic fatigue test machine at ambient room temperature and humidity. An example of a cracked coupon is shown in Figure 3. The coupons were designed to have a constant stress distribution along the majority of the test section [19], and a net stress concentration factor [Kt_{NET}] of 1.03. The coupons were etched to ensure crack initiation at naturally occurring material discontinuities similar to those observed in typical RAAF aluminium alloy airframe components [12]. The test spectrum consists of 62,442 cycles of mini-TWIST (which represents 4000 flight hours of a transport aircraft's wing loading) followed by a bar-coded CA marker band (CAMB) load sequence between each spectrum block [20,21] to assist post-test QF. The CAMB load sequences were visible on the fatigue fracture surfaces to crack depths <0.1mm. Examples of the CAMB bands are shown in Figure 4 (left view). Five coupons were tested at each of net spectrum peak stress levels: 250, 300 and 350 MPa. The crack growth results for the 250MPa stress level coupons are shown in Figure 4 (right view).

DSTO AA7050 coupons
size: 200 x 40 x 6.35 mm
net test section: 50 mm
Kt_{NET} = 1.03
yield strength = 418 MPa
ultimate strength = 483 MPa

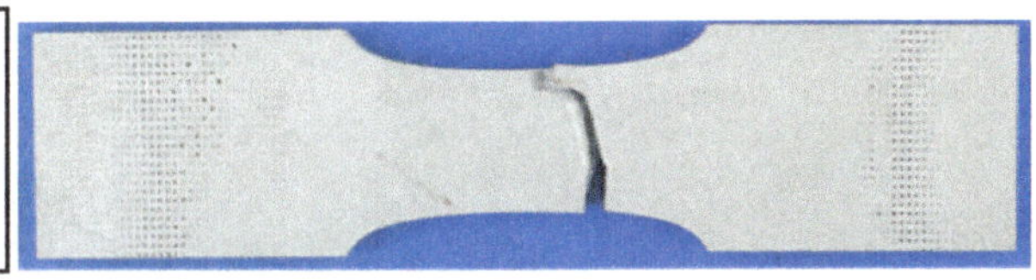

Figure 3 Example of a cracked AA7050 coupon from the study.

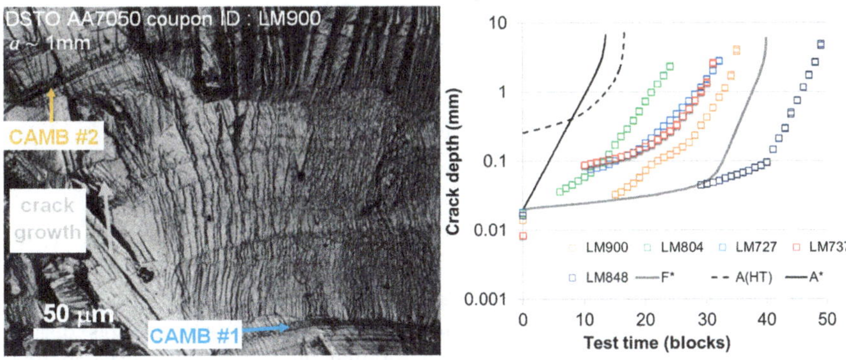

Figure 4 The results from the mini-TWIST coupon tests: (left) fatigue crack fracture surface with CAMB, (right) 250MPa test crack growth data and analytical growth curves.

5 RESULTS AND DISCUSSION

The analytical growth curves using updated material models in AFGROW (A*) and FASTRAN (F*) for the 250MPa stress level are shown in Figure 4, and compare favourably with the coupon data. Crack growth rates increased significantly above about 0.1 mm depth and the FASTRAN model was able to predict that effect. While the AFGROW (A*) model predicted a higher rate, and approximately log-linear growth rates from the initial flaw sizes observed here. The analytical results both assume an initial flaw depth based on the average of all flaws measured in the 15 coupons: 0.02mm. The analytical result for AFGROW with the standard material data and an initial flaw size of 0.02mm produced no crack growth. The analysis shown (labelled A(HT)) is for a fatigue crack growing grown from a 0.25mm initial depth. The 250 MPa crack growth curves from the 250 MPa load case coupons, shown in Figure 4 (for the lead crack that resulted in the failure of each coupon only), were also typical in form for the 300 and 350MPa load cases, although, these curves were, naturally, steeper and the lives generally shorter. (Note that even very short crack growth depths were able to be measured via QF due to the use of CAMB which were apparent on the fracture fatigue surfaces).

A summary of all coupon and analytical results is shown in Figure 5, where the crack growth per block of test loading (mini-TWIST plus CAMB) is compared at increasing crack depths. As expected, at the longer crack depths, all analytical models predict similar crack growth rates per block of loading. However, for short cracks (less than ~0.5mm), standard material data significantly under predicts the measured crack growth. Assuming a typical transport aircraft design goal of 24,000 flights (i.e. 6 blocks of mini-TWIST), fatigue life predictions from the updated models were found to be particularly good and resulted in a conservative (reasonably so) estimate when compared to the coupon results.

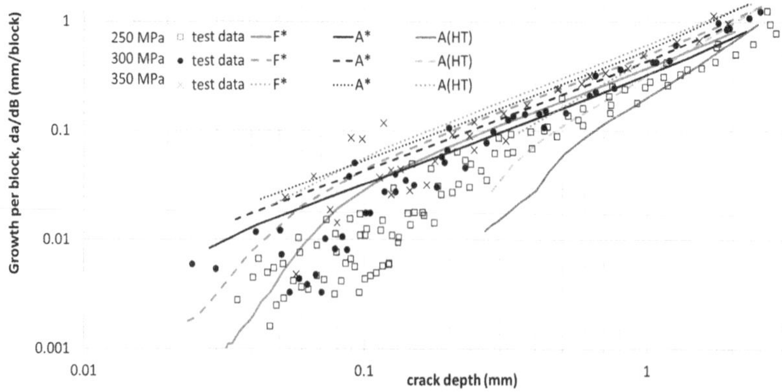

Figure 5 Summary of crack growth rate per block data from test and analyses.

6 CONCLUSION

The coupon data was found to compare favourably to blind predictions from AFGROW and FASTRAN crack growth models using the updated material models discussed. Testing and analysis continues at DSTO in refining the AA7050 material models, in particular for very short cracks, as well as complementary research into developing a better understanding of the growth mechanisms of fatigue cracks, through the examination of crystalline slip and damage accumulation ahead of fatigue crack tips. The aim of this work is to increase understanding and improve prediction capabilities for fatigue damage, in particular for crack growth in RAAF aircraft.

REFERENCES

[1] H. El Haddad, T. H. Topper, and K. N. Smith, Eng. Frac. Mech. **11** (1979).
[2] J. Schijve, Int. Journal of Fatigue **25** (2003).
[3] J. Newman and Y. Yamada, ASTM/ESIS, Vancouver, Canada, 2009.
[4] R. Jones, L. Molent, and S. A. Barter, Int. Journal of Fatigue **55** (2013).
[5] C. Wallbrink, P. Jackson, and W. Hu., ICAF Symposium, Montreal, Canada 2011.
[6] J. A. Harter, AFRL Report AFRL-VA-WP-TR-206, WPAFB, USA, 2006.
[7] Newman, J.C., Jr. NASA report, NASA TM-104159, USA 1992.
[8] K. F. Walker, S.A. Barter, ICAF Symposium, Montreal, Canada 2011.
[9] M. Burchill, S. Barter, and E. Amsterdam, AIAC, Melbourne, Australia, 2013.
[10] H. Lowak, J. B. de Jonge, J. Franz, and D. Schutz, Vol. 146, Laboratorium fur Betriebsfestigkeit, Germany, 1979.
[11] P. K. Sharp, R. Byrnes, and G. Clark, DSTO Report DSTO-TR-0729, Australia, 1998.
[12] L. Molent, Q. Sun, and A. Green, DSTO Report DSTO-TR-624, Australia, 2005.
[13] P. White, S. Barter, and L. Molent, Int. Journal of Fatigue **30** (2007).
[14] ASTM-E647, American Society for Testing and Materials, 2011.
[15] Stress Intensity Factors Handbook, Pergamon press (Murakami, Y., ed.), 1987.
[16] K. F. Walker, Wang, C.H., and Newman,J.C., APCFS/SIF, Sydney, Australia, 2014.
[17] J. C. Newman, Jr., Bigelow, C.A., and Shivakumar, K.N., Eng. Frac. Mech **46** (1993).
[18] J. C. Newman Jr., Crews, J.H., Bigelow, C.A., and Dawicke, D.S., ASTM STP 1244 (M. Kirk and A. Bakker, eds.), (1995).
[19] X. Yu, M. Burchill, R. Kaye, and S. Barter, APCFS/SIF, Sydney, 2014.
[20] M. McDonald, R. Boykett, and M. Jones, ICAS, Brisbane, Australia, 2012.
[21] M. Burchill, S. Barter, and M. McDonald, AIAC, Melbourne, Australia, 2015 (draft).

Optimal coupon design to achieve natural crack start in coupon fatigue tests

X. Yu, M. Burchill, R. Kaye, S. Barter
Aerospace Division, Defence Science and Technology Organisation (DSTO), Australia

ABSTRACT

Fatigue cracks in typical aircraft alloys usually start from statistically distributed small material discontinuities or surface defects. Over a low stress concentration region, the largest crack that dictates fatigue life is often the fastest growing crack that starts from one of the worst discontinuities. This natural crack start cannot be reproduced using traditional dog-bone coupons. In this study, an alternative notch-shaped coupon was designed using the DSTO shape optimisation code. This 'optimal' coupon has a large area of uniform peak stress. Its effectiveness was validated using 54 tests of AA7050-7451 coupons under representative F/A-18 load spectra and stresses.

1 INTRODUCTION

Fatigue test results underpin the aircraft structural integrity advice that the Defence Science and Technology Organisation (DSTO) provides in the management of Australian Defence Force aircraft. Quite often, DSTO will undertake a coupon test program to supplement, or in substitution of, a full-scale fatigue test of a particular critical detail for the sake of time and cost efficiencies. To allow this, coupon fatigue tests must be as representative as possible for the area being considered.

This study focuses on the design of coupons representing typical low stress concentration (low-K_t) features such as shallow fillets or run outs that are common in airframes. It has been observed [1], in both coupon and component fatigue tests, that fatigue cracking in typical aircraft alloys initiates from small material discontinuities or surface defects. The size, shape and orientation of these natural material discontinuities usually follows a statistical distribution. In addition, the growth of a small crack is affected by local material properties, which also follows a statistical distribution. Therefore, over a low-K_t area, fatigue cracks often start from the worst discontinuities, and the crack that grows the fastest develops into the largest crack that dictates the fatigue life. This phenomenon is here referred as natural crack start.

The ASTM dog-bone coupon design [2] has been widely used in low-K_t fatigue tests. However, it is not an ideal coupon design to reproduce natural crack start. This can be understood by examining the stress distribution as well as the starting locations of the largest cracks observed in coupon tests. The plots given in Fig 1 relate to previous work completed at DSTO where 21 flat-plate dog-bone coupons were tested under a range of representative F/A-18 load spectra and stresses. The starting locations of the largest cracks were all confined within a local area associated with the radii run outs at either side of the gauge section. This is the same area where the peak of the maximum principal stress occurs; indicating that the stress concentration is significant enough to dictate the starting positions of the fatigue cracks. In confining the crack starting locations to the area of higher stress, fewer initial discontinuities and local material properties – the latter affecting small crack growth – are sampled. Thus, the largest crack that results in failure may start from a less severe discontinuity or grow at a slower rate. Also, when the largest crack starts from a location that is next to, but not exactly at, the peak stress location, the use of nominal peak stress over-estimates the actual fatigue driving force. Consequently, the fatigue test results from the dog-bone coupon will exhibit more scatter in total fatigue life and any distribution of life data would be skewed to a longer life, when compared to those on a real aircraft structure.

The purpose of this study is therefore to propose an alternative coupon design, referred to here as an optimal coupon shape, which eliminates the local stress concentrations inherent in the dog-bone coupon. The effectiveness of the optimal coupon design will be examined through experimental strain measurement as well as fatigue tests under spectrum loading.

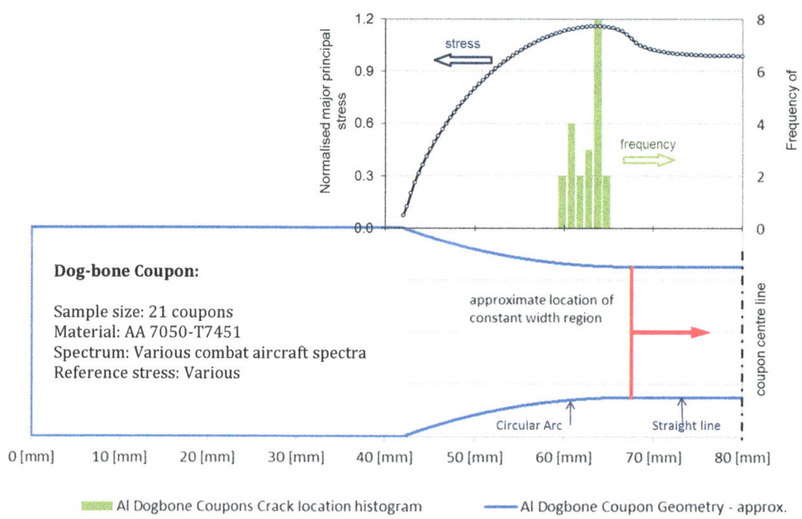

Figure 1. Coincidence between the local stress concentration, and initiating locations of the largest fatigue cracks in twenty-one dog-bone coupons.

2 DESIGN OF OPTIMAL COUPON

In this study, the optimal coupon was designed using a DSTO-developed Shape OPTimisation (SOPT) code. The code was initially developed for the design of optimal rework shapes that seek to minimise peak stresses around a local stress concentration feature. It has been successfully applied to airframe components to extend their fatigue lives [3]. The code works in conjunction with finite element (FE) analyses of stress. The reduction of peak stress is achieved via a gradientless technique [4], which analytically removes material at low-stress areas starting from a kick-off shape that has extra material. The SOPT code is able to consider multiple stress peaks, load uncertainties, and minimum radius of curvature and geometry constraints. In designing an optimal low-K_t coupon, only the minimum radius of curvature constraint needs to be considered. Thus the optimisation algorithm for material removal can be expressed using Eq (1):

$$d_i = \left(\frac{\sigma_i}{\sigma_{th}} - 1 \right) s \quad (1)$$

where d_i is the normal movement at a boundary location i, mimicking material removal, σ_i is the major principal stress at the same boundary location referenced by d_i, σ_{th} is a non-zero threshold stress, and s is a step-size scaling factor. The process of the boundary movement as described by Eq (1) is repeated iteratively, until a uniform σ_i stress distribution is reached along the gauge length, or when another convergence criterion is satisfied. Here, the uniformity of the σ_i distribution is the main feature of the SOPT code that has been relied on for the design of the optimal coupon presented here.

An initial optimal design that has a 160 mm total length is shown in Fig 2. The stresses used in the optimisation process were calculated using a 3-D Nastran FE analysis. The FE model was a half-symmetric one using 6 layers of 8-noded hexagonal elements through the thickness. The grip load was modelled as uniform surface tractions over the grip area. All nodes along the 70 mm notch region were chosen as being part of the movable boundary, and a minimum radius of curvature of 20 mm was enforced for the notch. In the optimisation process, a 2.5-D simplification was applied such that the nodes sharing the same in-plane (x-y) coordinates are treated as a group, and referred as a point. At each boundary point, the major principal stresses are retrieved at all through-thickness nodes and the maximum of these stress values was taken as σ_i for the calculation of boundary movement d_i at the point. After d_i was calculated, all through-thickness nodes related to this point were moved by the same amount. The optimised shape was finalised after 200 iterations as the stress variation was less than 0.01% per iteration. It is noted that the optimal shape is *not* a circular arc, but it can be easily manufactured using numerically-controlled machining methods. As designed, the coupon features a large uniform major principal stress region, and also this uniform stress is the maximum major principal stress within the coupon.

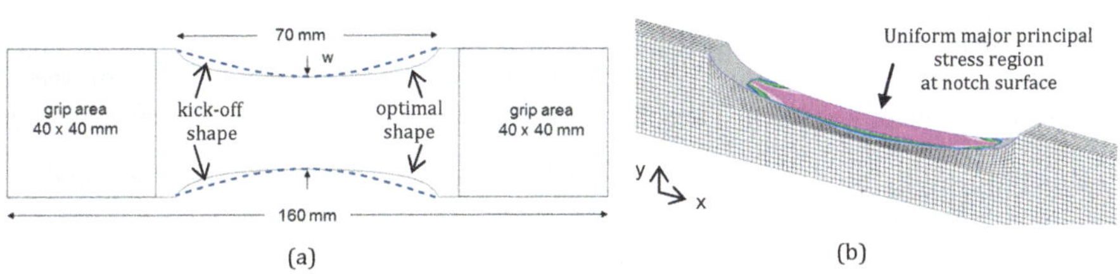

Figure 2. (a) Geometry of the 160 mm optimal coupon (w = 25 mm, thickness = 6.35 mm; kick-off in dashed lines); (b) Contours of major principal stress at a side face of the optimal shape.

To suppress potential effects of grip load uncertainty, a post-optimisation change was introduced to the coupon design such that: the total length of the coupon was increased from 160 mm to 200 mm, meanwhile the 70 mm notch length and the optimal notch shape remain the same. This change allows a longer buffer zone between the notch and the grip area. A sensitivity study using StressCheck [5] *p*-element analyses was performed to investigate the effects of grip length variation. The results are given in Fig 3, showing that the variation of grip length from 30 mm to 55 mm has insignificant effects on the major principal stress at the notch face of the optimal coupon. It is noted that the slight tail-up of stress at the mid-plane is of less concern as it occurs over an area of significant stress decay. Nevertheless, the effect of this tail-up stress will be monitored in the following experimental assessment.

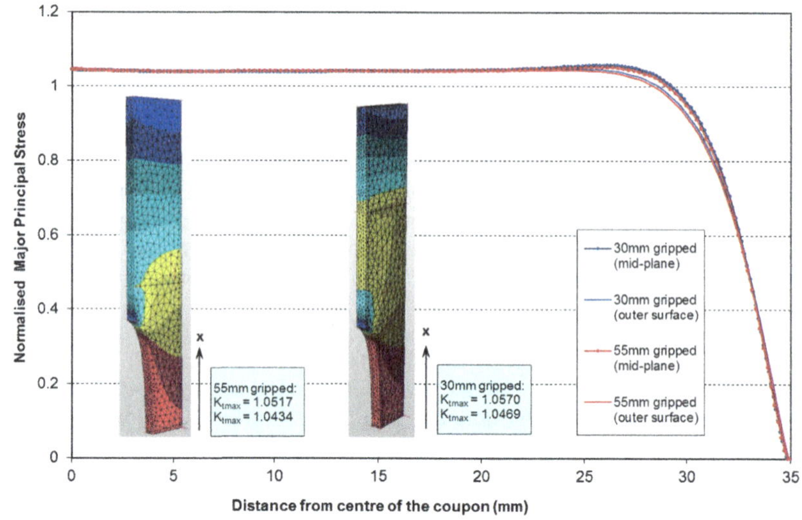

Figure 3. Major principal stress distribution at notch surface of a 200mm optimal coupon.

3 EXPERIMENTAL ASSESSMENT

Experimental assessment of the optimal coupon design was performed in two stages. Firstly, Thermoelastic Stress Analysis (TSA) was performed to assess the stress distribution in a 200 mm optimal coupon under a typical grip configuration. Cyclic loads were applied and the temperature change was measured and compared to FE results, as shown in Fig 4. When the heat dissipation is ignored, the measured temperature change is proportional to the sum of principal stresses [6], which is equivalent to the major principal stress along edge lines. The TSA image indicates that a uniform stress distribution was achieved along a large region of the edge lines, as it was designed to do.

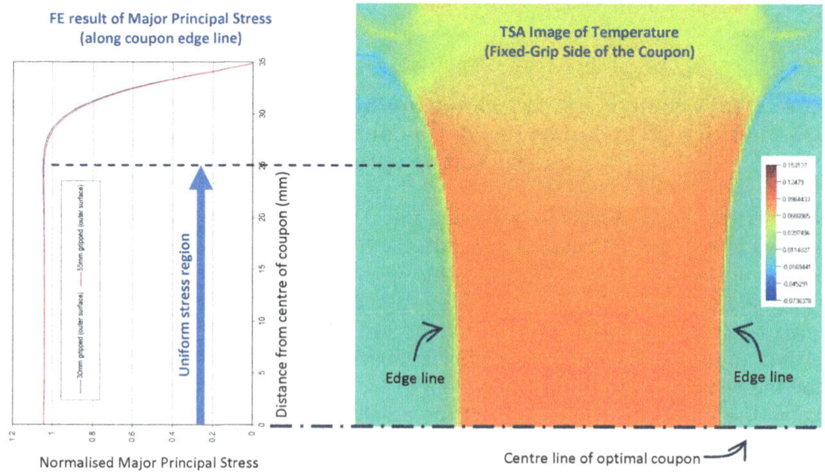

Figure 4. FE result of stress and TSA image of temperature in a 200 mm optimal coupon.

Fatigue tests were then performed using fifty-four 200 mm optimal coupons, tested under a range of representative F/A-18 airframe load spectra and stresses. The results are plotted in Fig 5, and they show that the starting locations of the largest cracks are distributed over a large region within the gauge length. This implies an increased likelihood for a crack to start from one of the worst discontinuities, and to also grow within a local material region that produces the fastest crack growth, *i.e.* an enhanced likelihood to achieve a natural crack start.

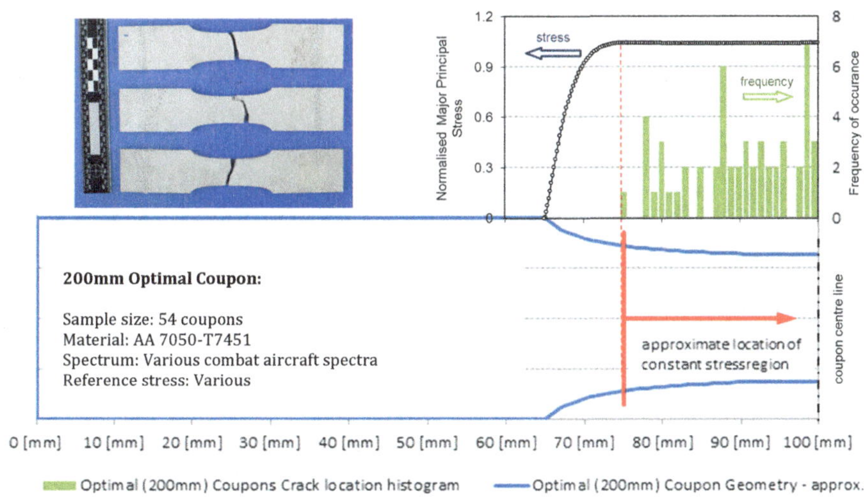

Figure 5. Coincidence between the local stress concentration and the initiating locations of the largest fatigue cracks in twenty-one dog-bone coupons.

4 FINAL REMARKS

A low-K_t coupon was designed using the DSTO-developed shape optimisation code. It features a large region of uniform peak stress at the notch surface. The effectiveness of the optimal coupon design was validated by thermo-elastic stress analysis measurement and 54 coupon fatigue tests. The optimal coupon design enhances the likelihood of achieving fastest crack growth starting from a larger discontinuity, *i.e.* natural crack start, thus providing a better representation of fatigue life for the low-K_t case.

5 ACKNOWLEDGEMENT

The authors would like to thank R Evans for the StressCheck analyses, J Choi for the TSA measurements, and G Sykiotis for performing the fatigue tests and crack measurements.

6 REFERENCES

[1] Barter, S. A., Molent, L. and Wanhill, R. J. H. (2012) *Int. J. Fatigue* **41**, 11–22.
[2] ASTM Standard, E606/E606M-12 (2012).
[3] Heller, M. et al. (2009) *Proc. 25th ICAF Symposium*, 279–299.
[4] Waldman, W., Heller, M. and Chen, G.X. (2001) *Int. J. Fatigue* **23**, *509–523.*
[5] ESRD Inc. (2011) *StressCheck Master Guide, Release 9.2.*
[6] Thomson, W. (Lord Kelvin) (1853) *Trans. R. Soc. (Edinburgh)*, **20**, 261–283.

Small fatigue crack initiation mechanisms and growth behavior of 304 stainless steel at room temperature

G.J. Deng, S.T. Tu, X.C. Zhang, Q.Q. Wang, F.Z. Xuan
Key Laboratory of Pressure Systems and Safety, Ministry of Education, School of Mechanical and Power Engineering, East China University of Science and Technology, P.R. China

ABSTRACT

The aim of this paper was to clarify the initiation and small fatigue crack growth mechanisms of 304 austenitic stainless steel at room temperature. The two-part silicon based replica method, repliset, was employed for measuring surface small crack growth in this research. Results showed that grain boundaries were the crack initiation sites and played an important role on the propagation of small cracks. Once the surface small crack length reached the critical crack size of 0.2 mm, rapid growth of this crack occurred. Stress intensity factor range is not an appropriate parameter to correlate the small crack growth rate.

Keywords: small crack; fatigue; replica; 304 stainless steel

1 INTRODUCTION

In the past decades, small fatigue cracks have been extensively studied for the reason that most of the total fatigue life could be spent in the initiation and small crack growth stages. Various investigations [1-4] have showed that cracks imitated at the very beginning of fatigue life and the propagation of microstructurally small crack could be greatly influenced by the inherent microstructure, resulting in the abnormal propagating behaviour when compared to the long crack behaviour. Therefore, the precise measurement of small crack growth rate as well as the good understanding of small crack growth mechanism was very important in the reliable prediction of the fatigue life of materials.

304 austenitic stainless steel is an alloy that has been widely used in process and power generating industries as piping and structural material. Fatigue fracture is one of the main causes of failure for these components due to the dynamic or alternating stresses they are often subjected to. Despite the fact that there have been numerous investigations of long fatigue crack growth in 304 austenitic stainless steel [5-6], relatively few studies, however, have focused on small fatigue crack growth mechanisms. C.M.Suh et al. [7-9] had made a contribution to the understanding of the small crack growth behaviour of 304 austenitic stainless steel. They obtained quantitative information such as the initiation, growth and coalescence behavior of small cracks in 304 stainless steel via surface replicas and photomicrographs. The

results showed that the accurate determination of these parameters was critical for the application of fracture mechanics to fatigue life assessment. However, experimental photos or data showing a complete small crack propagation process have not yet been available in their researches. The reason was that small crack testing was more difficult to perform than long crack testing.

The aim of this paper was to clarify the initiation and small crack growth mechanisms of 304 austenitic stainless steel at room temperature. The two-part silicon based replica method, which was recently employed for measuring surface small crack growth with good success [10, 11], was used in this research.

2 EXPERIMENTAL PROCEDURE

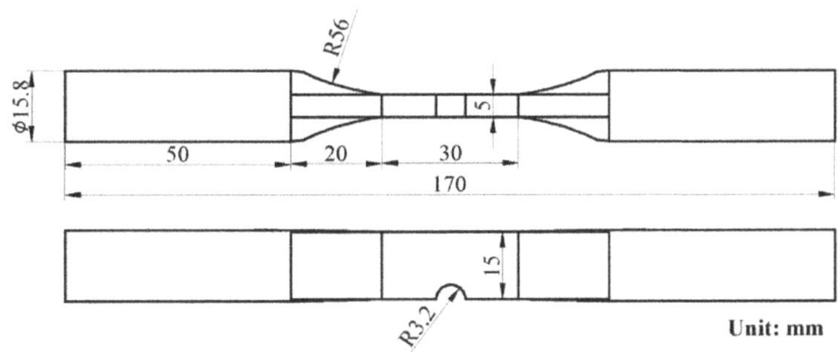

Fig. 1 Shape and dimensions of single edge notch tension specimen used in fatigue tests.

The material used in this study was a 304 austenitic stainless steel supplied in the form of a rolled bar with a diameter of 30 mm, and with the chemical compositions in wt %: 0.015 C; 0.53 Si; 1.74 Mn; 0.0048 S; 0.03 P; 18.37 Cr; 8.16 Ni; 0.019 Mo; 0.059 N; balance Fe. The yield strength and ultimate tensile strength of the material were 297.55 MPa and 668.64MPa respectively. A single edge notch tensile (SENT) specimen, shown in Fig. 1, was used to produce naturally initiated cracks. The semicircular geometry provides easy access to the notch surfaces to allow monitoring of the cracks by replicas. In order to remove residual machining marks and burs, and to identify grain boundaries, the notch of the specimen was polished to a mirror finish and etched with oxalic acid. The microstructure of 304 stainless steel was shown in Fig. 2a. The initial grain structure consists of equiaxed austenitic grains with a few annealing twins. The probability density functions of the grain sizes were determined by Image-Pro Plus software and the results were shown in Fig. 2b. The average grain size of the steel was estimated to be 27μm. The direction of the surface crack growth is perpendicular to the rolling direction of the specimens. Fatigue test was performed under a stress-controlled sine wave loading mode on an Instron 8800 servo-hydraulic test machine at room temperature, air environment. The test was axial tensile fatigue test with the frequency of 8 Hz and the stress ratio of 0.1. A maximum stress of 370 MPa was loaded at the center of specimen. The test was terminated when a single continuous crack formed across the entire notch root. The cyclic test was interrupted sequentially at a given time interval while a tensile load of 80% of σmax was kept when the replica was applied on the specimen surface. A static tip mixing nozzle combined with dispensing gun was used to dispense the RepliSet material onto the notch. The detailed procedure of replicating process can be found in Ref. [11]. After the tests, the major crack that led to the final fracture could be identified from the replicas and the crack initiation can be backtracked through observing the replicas at different time intervals.

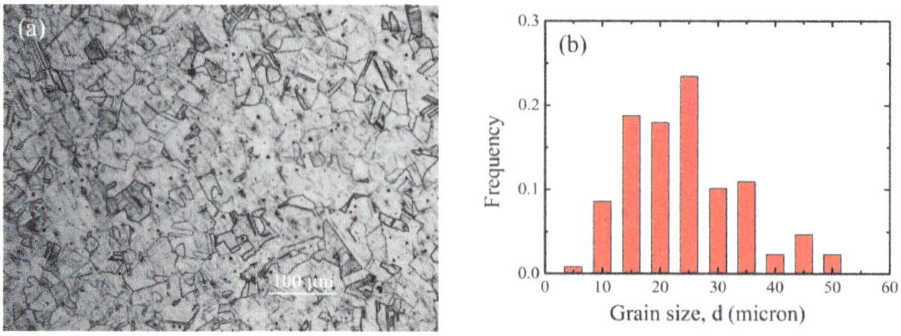

Fig. 2 (a) Optical micrograph of 304 austenitic stainless steel, (b) Probability density function of grain size.

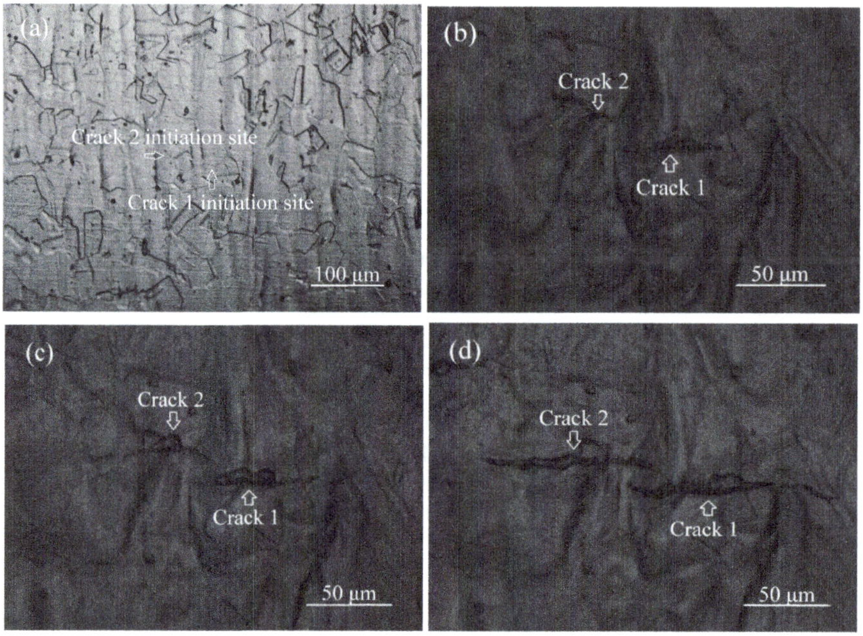

Fig. 3 Images from replicas showing the propagating process of crack 1 and crack 2. Replicas after (a) 0 cycles, (b) 20000 cycles, (c) 25000 cycles, (d) 30000 cycles.

3 RESULT AND DISCUSSION

3.1 Crack initiation

In order to characterize the fatigue crack initiation mechanism of 304 stainless steel, the replicas were observed at different time intervals for the specimen tested at 370MPa. It was found that there were eight major cracks initiated on the specimen. For all the eight cracks, the initiation sites were the grain boundaries. For instance, Fig. 3 showed the crack initiation site and propagating process of crack 1 and crack 2. The cracks followed a slip band path after nucleating at the grain boundaries, and then they propagated in a way perpendicular to the loading axis. Researches by Zhang et al. showed that the intergranular fatigue cracking strongly depended on the interactions of persistent slip bands with grain boundaries [12]. Moreover, crack nucleation might be due to the forward and reverse plastic flow of a slip band [13]. The localization of plastic deformation into a shear band was considered as an instability of plastic flow and a precursor to rupture [14]. Hence, the initiation and propagation of small cracks depended on both the grain boundary and the localization of plastic deformation along the slip bands.

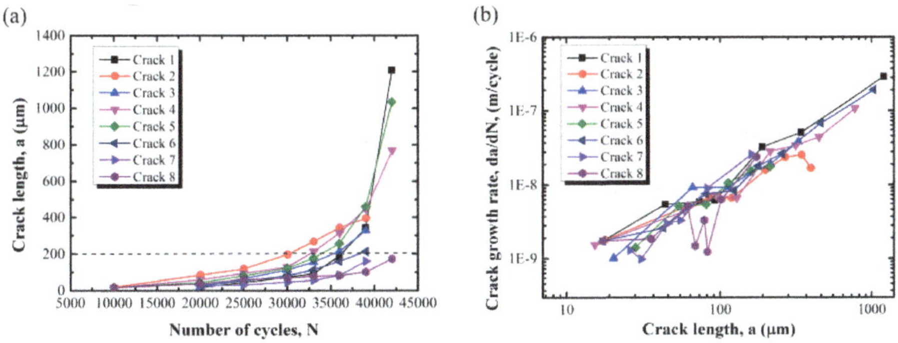

Fig. 4 (a) Major crack lengths against the number of cycles, (b) Small fatigue crack growth rate against crack lengths

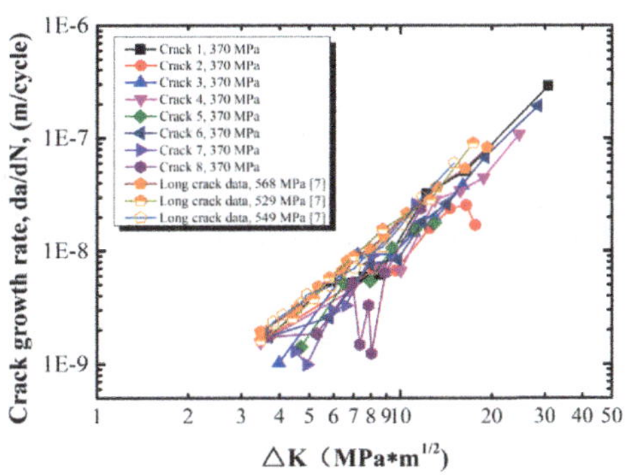

Fig. 5 Comparison of small crack growth rates with long crack data at a stress ratio R = 0.1.

3.2 Small crack propagation behavior

On the basis of the measurement from the replicas, Fig. 4a showed the variation of the lengths of major crack, a, which may cause the ultimate fracture, along with the number of cycles, N. The slope of the major crack length curve almost did not changed too much when the crack length was lower than 0.2 mm. Once the surface small crack length reached around 0.2 mm, the crack will propagate rapidly. Connolley et al. [15] concluded that there was a period of stable crack growth for the fatigue small crack. When the length of the small crack reached a critical size, there would be a transition to rapid crack growth in which the crack growth rate increased with increasing the crack length. From Fig. 4a, it can be deduced that the critical small crack size for the transition to rapid crack growth was around 0.2 mm for 304 stainless steel. Figure 4b showed the variation of the small fatigue crack growth rates along with crack lengths. It was easy to be seen that there was not a general increase in crack growth rate with increasing crack length, as was normally observed for long cracks. Based on the observation of the replicas, the fluctuations of crack growth rate were due to the blocking effect of grain boundaries.In order to characterize the difference between small crack data and long crack data, the nominal stress intensity factor range, ΔK, were calculated by using the equations derived by Newman et al. [16] for the small cracks in the semi-circle notches. The calculation results are plotted in Fig. 5, where the long crack growth rate of the similar 304 material tested at room temperature was used for comparison [7]. Unlike the small crack growth data, the long crack data from different stress level coincide with each other and almost collapse onto a single curve. This indicates that the stress intensity factor range could not be used as an appropriate parameter to correlate the small crack growth rate.

4 CONCLUSIONS

The small crack initiation and growth mechanisms of 304 austenitic stainless steel at room temperature were investigated. The following conclusions can be drawn from this work:

1. Grain boundaries are not only the initiation sites of cracks for austenitic 304 stainless steel, but also are the main factor that influence the growth rates of microstructurally small cracks.
2. Once the surface small crack length reaches the critical crack size of 0.2 mm, rapid growth of this crack occurs, resulting in final specimen fracture.

ACKNOWLEDGEMENTS

The authors are grateful for the support by National Natural Science Foundations of China (51322510, 11172102). X.C. Zhang is also grateful for the support by New Century Excellent Talents in University (NCET-11-0643) and Shanghai Pujiang Program and Key Research Program of Shanghai Science and Technology Committee (12JC1403200).

REFERENCES

[1] Newman JC Jr, Philips EP, (1999) Int J Fatigue; 21: 109-119.
[2] Gao YK, Wu XR., (2011) Acta Mater; 59: 3737-3747.
[3] Ogura KJ, Miyoshi YS., (1986) Eng Fract Mech; 25(1): 31-46.
[4] Li WF, Zhang XP.(2001) Mater Sci Eng A; 318: 129-136.
[5] Ye DY, Xu YD, Xiao X, Cha HB. (2010) Mater Sci Eng A; 527: 4092-4102.
[6] Kelestemur MH, Chaki TK. (2001) Int J Fatigue; 23: 169-174.
[7] Suh CM, Lee JJ, Kang YG. (1990) Fatigue Fract Engng Mater Struct; 13(5): 487-496.
[8] Okazaki M, Endoh T, Koizumi T. (1988) J Eng Mater Tech; 110: 9-16.
[9] Suh CM, Suh MS, Hwang NS. (2011) Fatigue Fract Engng Mater Struct; 35: 22-29.
[10] Newman JA, Willard SA, Smith SW, Piascik RS.(2009) Eng Fract Mech; 76: 898-910.
[11] Jordon JB, Bernard JD, Newman JC Jr. (2012) Int J Fatigue; 36:206-210.
[12] Zhang ZF, Wang ZG.(2003) Acta Mater; 51: 347-364.
[13] Tanaka K, Mura T. (1981) J Appl Mech; 48(1): 97-103.
[14] Rice JR.(1976) Theor Appl Mec; 1: 207-220.
[15] Connolley T, Reed PAS, (2003) Mater Sci Eng A; 340: 139-154.
[16] Newman JC Jr, Wu XR, Venneri SL, Li CG. Small-crack effects in high-strength aluminum alloys. NASA Reference Publication 1309; 1994.

Evaluation on dispersion and degradation of creep rupture property based on Z-parameter

J. Zhao, C.Q. Cheng, T.S. Cao, H. Wang
School of Materials Science and Engineering, Dalian University of Technology, P.R. China

ABSTRACT

The current paper demonstrates the usage of Z-parameter method in evaluating creep rupture property and the deterioration process. Symbol Z_{cr} is used to represent the random deviation of creep rupture data from a "master curve" in stress σ–TTP plot to illustrate the random distribution of creep rupture property and provide a probability evaluation approach. Z_s is used to illustrate the random deviation of operating parameters from the designed condition for a component, and a so-called SCRI interference model is proposed for reliability analysis by considering the intersection distributions of "operating conditions"(Z_s) and "creep strength"(Z_{cr}). Z_d is proposed to represent the deviation of a "deterioration curve" from original "master curve", and it demonstrates a systematically shift or degradation of creep rupture property from original one and can be used for quantitative degradation assessment of creep property.

1 INTRODUCTION

There are generally two basic phenomena developed at components serviced at high temperature: creep rupture under loads by sustaining creep deformation and property deterioration because of microstructure degradation and creep damage [1]. It is seen that dispersion inevitably existed because of uncertainty factors in materials and experiment conditions. The experimental data distributes scattering around the "mean value curve" in stress σ-rupture time t_r plot, as well as inσ-TTP plot. (TTP is abbreviation of time-temperature parameter, such as Larson-Miller and Manson-Haferd parameters [2-3]. The "mean value curve" in σ-TTP plot is called "master curve".). Z parameter, which represents the deviation of creep rupture data from "master curve", was proposed for statistically analysing the dispersion of creep rupture data.

On the other hand, for components serviced at high temperature where microstructural aging and creep damage occurs continuously, the rupture data band exhibits a deviation from "master curve". In our former reports [6-7], we also use the Z-parameter to describe different degradation. Although deviation of data from "master curve" can be observed either from property scattering or from property degradation, it represents quite different characteristic of creep rupture behaviour. The former demonstrates the random distribution of data scattering while the latter is proposed to represent the magnitude of systematically shifting due to deterioration of creep rupture property. In order to distinguish the specific application using Z-parameter method, symbol Z_{cr} is applied to illustrate the random distribution of creep rupture property, while symbol Z_d is used to illustrate deviation of "deterioration curve" from original one.

Besides deviation caused by scattering of creep rupture data and deterioration process, scattering phenomenon also exists when there are fluctuations of serviced condition(temperature or/and load) from the design condition. In order to demonstrate the deviation resulting from operation conditions, symbol Z_s is proposed to distinguish the application from Z_{cr} and Z_d. The purpose of current paper is to clearly describe the meaning and application of these three Z-parameters, and provide applicable method for high temperature property evaluation.

2 Z_{CR} FOR STATISTICS ANALYSIS OF CREEP RUPTURE DATA

Creep rupture data obtained at various stresses and temperatures is generally correlated by using a so-called time-temperature parameter (TTP) method to combine time and temperature into one expression. For example, Larson-Miller and Manson-Haferd parameters [2-3] are among the most common used methods. Figure 1 is an example of 5Cr-0.5Mo steel using NIMS data sheet[8], where all the data can be condensed to a narrow band and a so-called "master curve" is mathematically regressed to represent its creep rupture property.

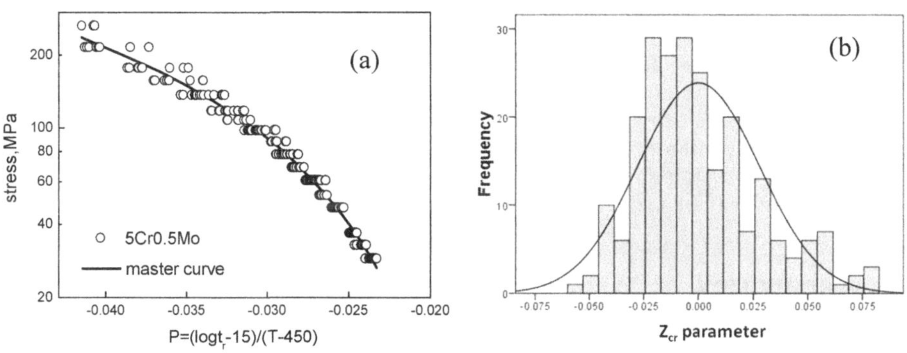

Fig. 1. The master curve and deviation of experimental data from the master curve for 5Cr0.5Mo. a), the master curve; b), the distribution of Z_{cr} parameter

For the "master curve" in Fig.1, stress σ is expressed as a function of TTP parameter P: $\log\sigma = Z_0 + f(P)$. Where Z_0 is a constant; f(P) is a function of TTP parameter P. The deviation of a data point from the "master curve" is supposed to be represented using symbol Z_{cr} that can be expressed as: $Z_{cri} = Z_0 + f(P_i) - \log\sigma_i$, where P_i is the value of calculated TTP parameter for data i (T_i, t_{ri}), Z_{cri} is corresponding to the deviation of data log(stress) to master curve. The distribution of parameter Z_{cr} based on the data shown in Fig. 1(a) is plotted in Fig. 1(b) after probability evaluation. It is verified that the values of Z_{cr} are mostly supported by normal distribution after using $\chi 2$ test. Therefore once the distribution characteristics is obtained, probability evaluation can be performed. The "master curve" in Fig. 1(a) can be extended to σ-TTP-R curves as shown in Fig. 2. The "master curve" is the curve with 50% reliability, and the curves with reliability of 95%, 99% are also plotted. It can be seen that almost all the experimental data are distributed on the right side of the 99% reliability curve.

From σ-TTP-R curves, it is convenient to deduce σ-tr-R curve for given temperatures. Fig.2(b) shows an example of σ-tr-R curves at various temperatures for 5Cr-0.5Mo steel. For each temperature range, the deduced curves at each probability show good tendency with the experimental data. It is supposed that optimum of TTP parameter and appropriate estimation for data distribution can result in good correlation between the curves and experimental results. It is also possible to deduce allowable stress at required probability, i.e. [σ]-T-R curve as shown in Fig.2(c). It is clear that the allowable stress [σ] can be chosen according the selected probability and it provides more flexible choice for strength design of high temperature components.

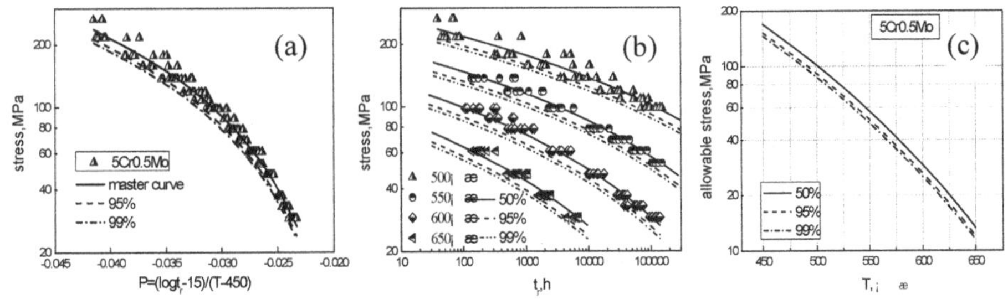

Fig. 2. The reliability curves of 5Cr0.5Mo. a)σ-TTP-R; b)σ-tr-R; c) [σ]-T-R

From σ-TTP-R curves, it is convenient to deduce σ-tr-R curve (Stress-Rupture time-Reliability curve) for given temperatures. Fig.2(b) shows an example of σ-tr-R curves at various temperatures for 5Cr-0.5Mo steel. For each temperature range, the deduced curves at each probability show good tendency with the experimental data. It is supposed that optimum of TTP parameter and appropriate estimation for data distribution can result in good correlation between the curves and experimental results. It is also possible to deduce allowable stress at required probability, i.e. [σ]-T-R curve (allowable stress-temperature-reliability curve) as shown in Fig.2(c). It is clear that the allowable stress [σ] can be chosen according the selected probability and it provides more flexible choice for strength design of high temperature components.

3 Z_s FOR DESCRIBING THE FLUCTUATION OF SERVICE CONDITIONS AND THE PROPOSAL OF SCRI INTERFERENCE MODEL

During evaluation of failure probability for components exposed to high temperature, two factors should be taken into account: one is the scattering of creep rupture property which can be analysed using symbol Z_{cr} as former proposed; another is the fluctuation of operating conditions such as service temperature and service stress. In the case of fatigue fracture, a so-called "Stress-Strength interference model" (SSI model) has been utilized to evaluate fatigue failure probability. Refer to the idea in fatigue life evaluation, Z_{cr} is supposed to be similar to "strength distribution" in SSI model and represents the scattering distribution of creep rupture strength. As for describing fluctuation of operation conditions, symbol Z_s is expressed as: $Z_{Si} = \log(S_{Si}) - [Z_0 + f(P(t_S, T_{Si}))]$, where S_{si} is a value of operating stress and T_{si} is a value of operating temperature, they are assumed to follow some kind of random distribution. t_s is designed service life, it is a certain value (i.e.: 100,000hrs). It can be seen that Z_s and Z_{cr} are two independent random variables. Z_s is similar to "stress distribution" in SSI model and its probability density function is expressed as: $g(Z_S) = g(S_S, \Delta S_S, T_S, \Delta T_S, t_S)$, where S_s is service stress, T_s is service temperature, t_s is designed service time, ΔS_s is the applied stress fluctuation and ΔT_s represents service temperature fluctuation.

By employing symbol Z_{cr} and Z_s, a so-called SCRI model ("Service condition -Creep rupture property" interference model has been established to evaluate the failure possibility in component undergone creep deformation [9]. Fig.3 (a) illustrates the schematic diagram of SCRI model. The interference area where $Z_s > Z_{cr}$ indicates that the creep rupture property of material cannot offer enough capability for service and it reflects the failure probability corresponding to certain operating conditions and servicing time. Only in the interference area, creep fracture can occur.

It is possible to evaluate reliability of a component if the distributions of Z_{cr} and Z_s can be obtained. Generally, Monte-Carlo simulation method is implemented to evaluate the safety reliability in SCRI model. Fig. 3(b) takes 91 steel as an example using NIME data sheet to show the influence of fluctuation of operating conditions on the area of interference region. Fig. 3(c) plots the relationship between failure probability and fluctuating levels, and it is shown that failure probability increases rapidly at higher fluctuation condition.

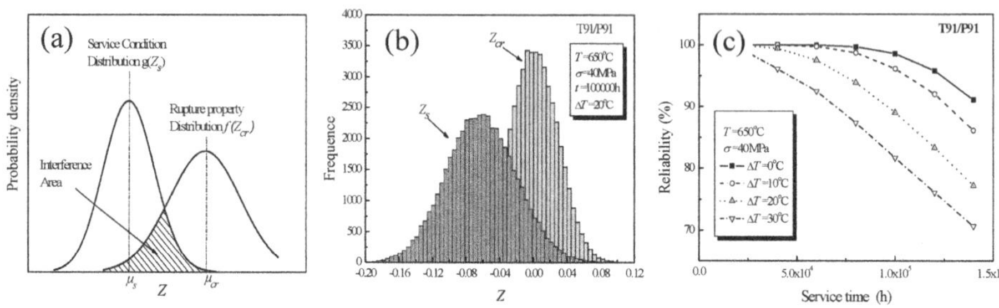

Fig. 3. Illustration of SCRI model. (a) Schematic diagram; (b) Temperature fluctuation on interference region; (c) Influence of temperature fluctuations

4 Z_D FOR REPRESENTING THE DETERIORATION OF CREEP RUPTURE PROPERTY

Long-term exposure of components at high temperature inevitably leads to deterioration of creep rupture property because of microstructure aging and creep damage. Fig. 4(a) shows creep rupture data of 12Cr1MoV steel with different microstructural degradation (i.e. pearlite spheroidization level)[7]. It appears that the deviation of the "data band" from the master curve increases with microstructural degradation. Similar results has also been reported in a austenite steel HK40[6] after different exposure-time of 53,600, 73,000 and 131,000h at high temperature as shown in Fig. 4(b), it is obvious that the longer the service time, the greater the deviation from the original master curve.

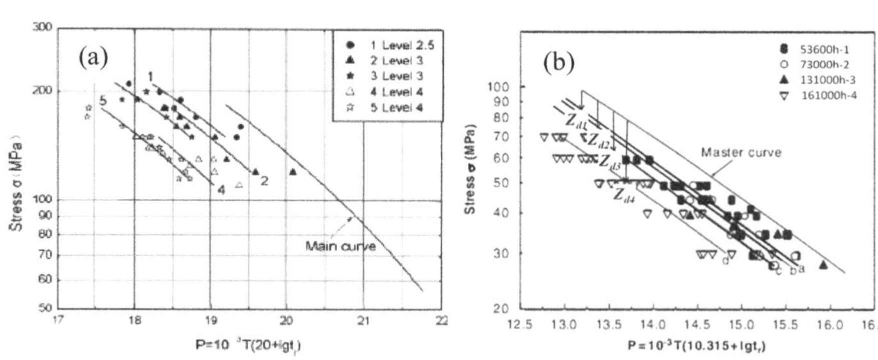

Fig. 4. Deterioration y after long-term exposure: a) 12Cr1MoV; b) HK40 steel

With the deterioration of creep rupture property, the tested data points exhibit systematically deviation from original master curve. This kind of deviations results from not only random scattering of property but also deterioration process. A regressive "deterioration curve" based on the data tested from deterioration samples represents its temporary creep rupture property, and the deviation from original curve exhibits the decrease of creep rupture property. For the purpose to describe the deterioration of creep rupture property after long-term exposure, symbol Zd is proposed to represent the deviation amplitude of a "deterioration curve" from original "master curve".

By using symbol Z_d, a linear relationship between the deterioration of creep rupture property and spheroidization level E in 12Cr1MoV steel is observed as shown in Fig. 5(a): $Z_d = 0.486(1\text{-}E)$. It is therefore possible to quantitatively assess creep property deterioration by analysing the microstructural degradation according to following equation. As for austenite steel HK40, the deterioration evolution using symbol Z_d is demonstrated in Fig.5(b). It is seen that a obvious decrease of Z_d value at initial stage, followed by a relatively stable stage where Z_d develops gradually with service time, and a significant change in Z_d value corresponds to the serious deterioration at the last stage.

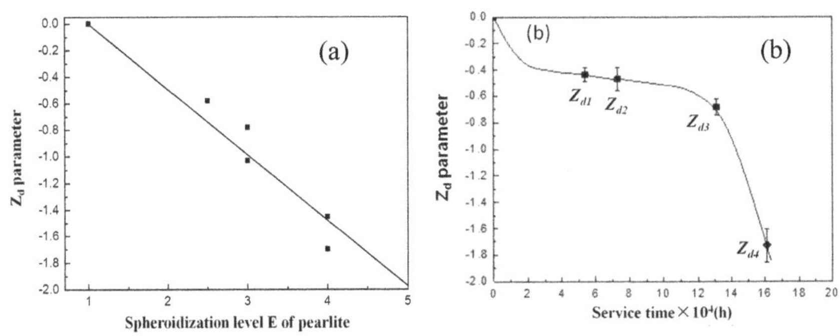

Fig. 5. Plot between deterioration and Z_d values. a) 12Cr1MoV; b) HK40 steel

5 CONCLUSIONS

Three kinds of symbols are proposed to distinguish the specific applications and are summarized below: (1) Symbol Z_{cr} is used to represent the random deviation of creep rupture data from a "master curve" in stress σ–TTP plot, it illustrates the random distribution of creep rupture property resulting from uncertainty factors in materials and/or experiment conditions and provides a method for probability evaluation of creep rupture strength. (2) Symbol Z_s is used to illustrate the random deviation of operating condition from the designed one for a operating component, it represents the fluctuation of operating conditions and a so-called SCRI interference model is proposed for reliability analysis. (3) Symbol Z_d is proposed to represent the deviation of a "deterioration curve" from original "master curve", and it demonstrates a systematically shifting or degradation of creep rupture property from original one and can be used for quantitative degradation assessment of creep property.

ACKNOWLEDGMENT

The authors appreciate the financial support by National Natural Science Foundation of China (NNSFC) under the granted No. 51171037and No.51134013.

REFERENCES

[1] Viswanatan. R. (1993) Damage Mechanisms & Life Assessment of High-Temperature Components. ASM International, Metals Park, Ohio 44073. Second print. 183-264.

[2] Larson, F. R., Miller J. (1998) ASME Transaction. 74(5), 765-775.

[3] Manson S. S., Haferd A M. (1953) NACA TN 2890. 1-49.

[4] Zhao, J., Li, D.M., Fang, Y.Y. (2010) J. Press. Ves. Tech.-Trans. ASME. 132(6), 061209

[5] Xing, L., Zhao, J., Shen, F.Z., J. (2007) Mater. Sci. & Tech. 23, 329-332.

[6] Xing, L., Zhao, J., Shen, F. Z., (2006) Int. J. Pressure Vessels Piping, 83, 730–735.

[7] Zhao, J., Han, S. Q., Gao, H. B., (2004) Int. J. Pressure Vessels Piping, 81, 757–760.

[8] NRIM creep data sheet, (1992) Data sheets on the elevated temperature properties of 5Cr-0.5Mo steel tubes for boilers and heat exchangers(STBA 25), No.12B National Research Institute for Metals, Japan.

[9] Zhao, J., Li, D.-M., Zhang, J.-S., (2009) Int. J. Pressure Vessels Piping, 86, 599–603.

Creep-fatigue life prediction through multiple regression analyses

X.C. Zhang, X.F. Zhang, S.T. Tu, K.L. Zhu, Z.D. Wang
Key Laboratory of Pressure Systems and Safety, Ministry of Education, School of Mechanical and Power Engineering, East China University of Science and Technology, P.R. China

ABSTRACT

The aim of this paper was to propose a creep-fatigue life prediction model based on multiple regression analysis of the existing experimental data. Three different factors including temperature, hold time, and total strain range, were taken into consideration in this model. The prediction capacity of the present model was verified through comparison between predicted results and existing experimental data of different materials with different testing conditions. A good agreement was found between the predicted results and the experimental data.

Keywords: Creep fatigue; Life prediction; Model

1 INTRODUCTION

Creep-fatigue deformation is a dominant damage mechanism for a lot of high-temperature components, which is often depicted microstructurally as the combined effects of fatigue damage and creep damage. The creep-fatigue interaction behavior of materials is frequently simulated in the laboratory by high-temperature low-cycle fatigue tests with incorporation of hold time at constant strain or stress. These tests are often used to determine the effects of different factors, such as hold time, strain range, strain rate, and temperature, on the creep-fatigue resistance of materials and to develop the design codes for materials at some given conditions. In the past decades, more than one-hundred life prediction models have been proposed. Each has enjoyed some degree of success in dealing with a specific set of creep-fatigue data. For instance, the Coffin-Manson plot shows that there is a linear relationship between the number of cycles to failure, N_f, and the plastic strain range, $\Delta\varepsilon_p$, in double logarithmic coordinate. However, the hold time used in the creep-fatigue experiments could have an obvious influence on the slop of the $\Delta\varepsilon_p - N_f$ curve [1].

In this paper, a life prediction model considering three variables, namely, temperature, hold time and total strain range was developed from a phenomenological consideration on the basis of the data from the open literature. For a given material, the constants used in the model could be evaluated by a multiple regression analyses on some experimental results.

2 FRAMEWORK OF PROPOSED MODEL

2.1 Consideration of temperature and total strain rage effects

Zhang et al. derived a simple power relationship to describe the effect of temperature on fatigue life [2], i.e.,

$$N_f=f(\gamma,\dot{\varepsilon})\left(1-\frac{T}{T_m}\right)^b \quad (1)$$

where f is a function of both strain range and temperature, b is the material constant, T and T_m are respectively the experimental temperature and the melting temperature, $\dot{\varepsilon}$ denotes the strain rate, and the parameter γ, which is a function of total strain range, $\Delta\varepsilon_t$, can be expressed as

$$\gamma = 1+\ln(1+\Delta\varepsilon_t/\Delta\varepsilon_0) \quad (2)$$

where $\Delta\varepsilon_0$ is the fatigue limit strain range. Additionally, according to the frequency-modified approach proposed by Coffin [3], the equivalent strain rate for one strain cycle with hold time, $\dot{\varepsilon}$, can be expressed as $2\Delta\varepsilon_t/(t_f+t_h)$, where t_h and t_f denote respectively the hold time and the single continuous cycle time. When the wave shape used in fatigue test is given, the single cycle time, t, which is the sum of t_h and t_f, is constant. Therefore, Eq. (1) can be rewritten as

$$N_f=f(\Delta\varepsilon_t,t)\left(1-\frac{T}{T_m}\right)^b \quad (3).$$

Figure 1 shows the variations of N_f of 304 stainless steel (SS) and grade 91 steel along with $1-T/T_m$ with respect to different strain rates and total strain ranges. The data of 304 SS and grade 91 steel are respectively collected from Ref. [4] and Refs. [5-7]. It can be seen that the value for N_f linearly increases with increasing the value for $1-T/T_m$ in double logarithmic coordinate. For 304 SS, the value of slope, b, can be determined to be 5.811 through regression analysis, which is almost independent of $\Delta\varepsilon_t$ and $\dot{\varepsilon}$, as seen in Fig. 1a. The similar phenomenon can be found for grade 91 steel, as seen in Fig. 1b.

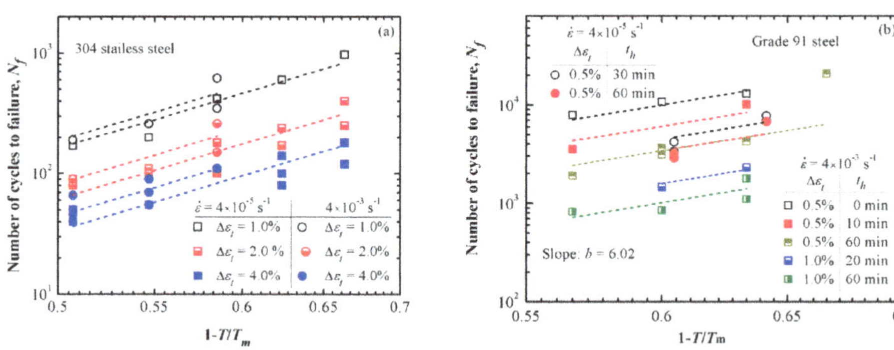

Fig. 1 Effect of temperature on N_f of (a) 304 SS, and (b) grade 91 steel.

2.2 Consideration of hold-time effects

Yoon et al. developed a correlation between hold time, t_h, and N_f [8], i.e.,

$$N_f=C^*\cdot t_h^{\beta} \quad (4)$$

where C^* is a parameter which is related to the total strain range and temperature and β is material constant at a given test condition which is independent of temperature and hold time. The hold time, t_h, is assumed to be the sum of tensile and compression hold times. Equation (4) represents a reasonable relationship between creep-fatigue life and hold time for 304 stainless steel, as shown in Fig. 2a. The data tested at three temperatures, i.e., 593 °C, 648 °C and 704 °C, and at different total strain in the range from 0.3% to 4%, are collected from Ref. [4]. In double logarithmic coordinate, a linear relationship between N_f and t_h with a slop of -0.25 can be plotted. The linear relationship between creep-fatigue life and tensile hold time in double logarithmic coordinate can also be found for grade 91 steel, as seen in Fig. 2b. With increasing the temperature, the intercept C^* decreases at a given total strain range, as seen in Figs. 2. In order to further identify the effects of total strain range and temperature on the magnitude of intercept C^*, the experimental data of 304 stainless steels from Refs. [4,9-11] and those of grade 91 steel from Refs. [5-7] are fitted by using Eq. (4), as seen in Fig. 3. It is interesting to find that there is a liner relationship between $\Delta\varepsilon_t$ and C^* in double logarithmic coordinate and the slop is independent of temperature. Hence, the effect of the total strain range on the intercept, C^*, in Eq. (4) could be expressed as

$$C^*=C'(T)\cdot\Delta\varepsilon_t^{\alpha} \quad (5)$$

where $C'(T)$ is a parameter with is a function of temperature and α is amaterial constant.

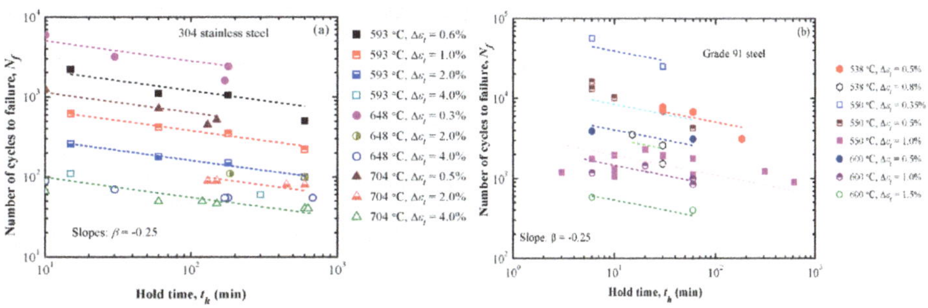

Fig. 2 Effect of t_h on number of N_f of (a) 304 SS, and (b) grade 91 steel.

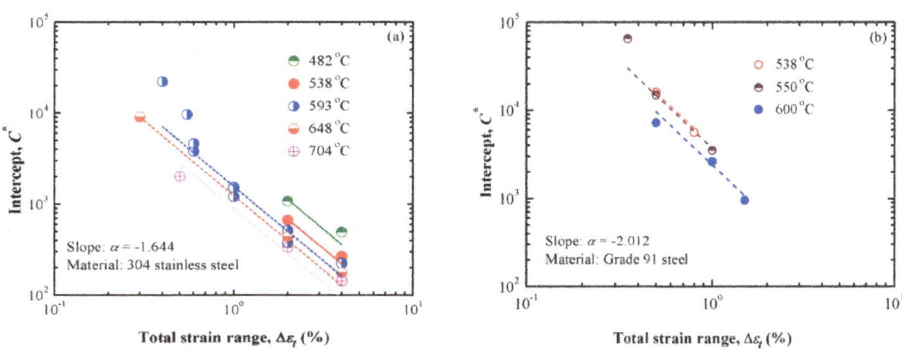

Fig. 3 Linear relationship between C* and $\Delta\varepsilon_t$ of (a) 304 SS, and (b) grade 91 steel.

2. 3 Resulting life expression

Combination of Eqs. (4) and (5), the relationship between the creep-fatigue life and total strain range as well as hold time can be obtained and expressed as

$$N_f=C'(T)\cdot\Delta\varepsilon_t^{\alpha}\cdot t^{\beta} \quad (6)$$

For 304 stainless steel and grade 91 steel, the values of $C'(T)$ at different temperatures can be respectively obtained from Figs. 3a and 3b. Using the similar approach proposed by Zhang et al. [2], the linear relationship between $C'(T)$ and $(1-T/T_m)$ in double logarithmic coordinate can also be observed, as seen in Fig. 4. Moreover, it is interesting to find that the slopes of the lines in Fig. 1 and Fig. 4 are same for a given material. Hence, the creep-fatigue life prediction model can be obtained through combination of Eqs. (1) and (6) and can be expressed as

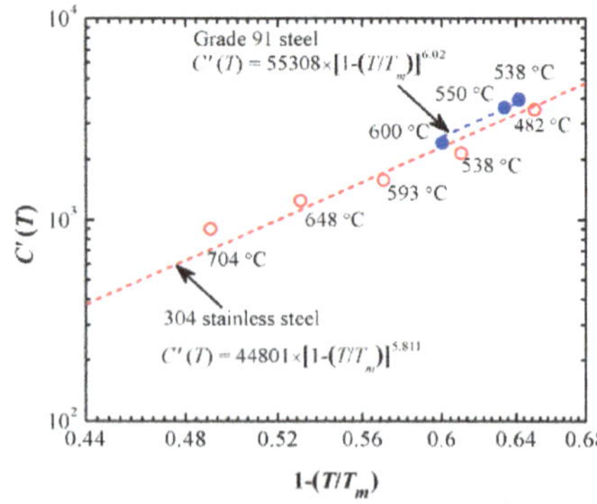

Fig. 4 The relationship between $C'(T)$ and *T* for 304 SS and grade 91 steel.

$$N_f=K\cdot\left(1-\frac{T}{T_m}\right)^b\Delta\varepsilon_t^{\alpha}\cdot t^{\beta} \tag{7}$$

where *K* is the parameter depending on the material. The values of *K*, *b*, α and β for 304 stainless steel and grade 91 steel are listed in Table 1.

Table 1 The constants used for life prediction of 304 stainless and grade 91 steels.

Material	T_m (°C)	β	α	b	ln(K)
304 stainless steel	1400	-0.25	-1.644	5.811	10.71
Grade 91 steel	1500	-0.25	-2.012	6.020	10.92

3 APPLICATION OF THE PROPOSED MODEL

The 304 stainless steel and grade 91 steel are used to illustrate the application of the present phenomenological model. The experimental data of 304 stainless steel and grade 91 steel are respectively collected from Refs. [4,8,10,12-14] and [5-7] to obtain the constants in Eq. (7) through regression analyses and logarithmic transformation. The comparison between the calculated lives obtained from the present approach and the experimental results of 304 stainless steel at total strain range 1.0% and at five different temperatures including hold period cases is shown in Fig. 5. The hold periods are

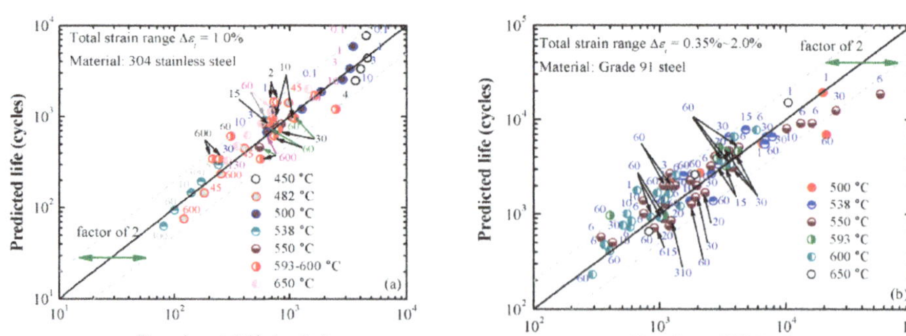

Fig. 5 Comparison between the predicted lives obtained from the present approach and the experimental results of (a) 304 SS and (b) grade 91 steel.

indicated by the number. It can be seen that most of the predicted data tend to fall inside the factor of 2 scatter band on life. The power-law relationship between creep-fatigue life and hold period is often observed for lots of materials. However, experimental results by Asayama [6] indicated that there was no strict linear relationship between t_h and N_f in double logarithmic coordinate for Mod.9Cr-1Mo steel when the hold period is very long. The creep-fatigue life decreased gradually with increasing the hold period. Hence, when the hold period is extremely long, the present model may lead a large error in life prediction. In such a case, Eq. (4) should be modified.

4 CONCLUSIONS

A phenomenological model was proposed to predict the creep-fatigue life. In this model, three different factors, i.e., total strain range, hold time and temperature were included. The 304 stainless steel and grade 91 steel with different load conditions was used to verify the prediction capacity of the proposed model. The deviation of the predicted lives for the material was within a factor of two, indicting the present model is accepted as a more generalized expression.

ACKNOWLEDGEMENTS

The authors are grateful for the support by National Natural Science Foundations of China (51322510, 11172102).

REFERENCES

[1] Zhang, X.C., Tu, S.T., Xuan, F.Z., (2014) Theor. Appl. Fract. Mech. 71, 51-66.
[2] Zhang, A.G., BuiQuoc, T., (1990) Trans. ASME. 112, 422-428.
[3] Coffin, L.J., (1969) Prediction parameters and their application to high temperature low-cycle fatigue, Proceedings of Second International Conference on Fracture, Brighton, London, 643-654.
[4] Esztergar, E.P., (1972) Creep-Fatigue interaction and cumulative damage evaluations for type 304 stainless steel. National Technical Information Service US. Contract No. W-74Q5-eng-26.
[5] Takahashi, Y, (2008) Int. J. Press. Vessels Pip. 85, 406-422.
[6] Asayama, T., (2009) Update and Improve Subsection NH Alternative Simplified Creep-Fatigue Design Methods, JAEA Task 10: Final Report.
[7] Taguchi, K., Kanno, E, Ozaki, S., Uno, T., (1990) Int. J. Press. Vessels Pip. 44, 99-115.
[8] Yoon, Y.C., Nam, S.W., (1994) Mater. Sci. Lett. 13, 1270-1271.
[9] Maiya, P.S., (1978) Creep-Fatigue life prediction for different heats of type 304 stainless steel by linear damage rule, strain-range partitioning method and damage-rate approach. Prepared for the U.S. Department of Energy under contract W-31-109-Eng-38, 1978
[10] Majumadar, S., (1981) Nucl. Eng. Des. 63, 121-135.
[11] Baig, G.C., Nam, S.W., (1996) J. Mater. Sci. 31, 4957-4966.
[12] Esztergar, E.P., (1972) Creep-fatigue interaction and cumulative damage evaluations for type 304 stainless steel, hold-time fatigue test program and review of multiaxial fatigue. Oak Ridge National Lab., ORNL-4757.
[13] Takahashi, Y., Ogata, T., Take, K., (1995) Nucl. Eng. Des. 153, 235-244.
[14] Sauzay, M., Mottot, M., Allais L., et al., (2004) Nucl. Eng. Des. 232, 219-236.

Tensile strength of silica optical fibers for high-temperature sensing applications

Y. Tu, S.-T. Tu
Key Laboratory of Pressure Systems and Safety (Ministry of Education),
School of Mechanical and Power Engineering,
East China University of Science and Technology, P.R. China

ABSTRACT

Using tensile fracture stress testing, the effects of acid stripping, high temperature and high-temperature annealing on the tensile strength of silica optical fibers are investigated. The results confirm careful stripping by hot (~190 °C) concentrated sulfuric acid has a negligible effect on the strength. The statistical strength and scatter are reduced with an increase in temperature. A significant reduction in the strength of the fibers is found after annealing. Higher annealing temperature leads to a larger reduction in the strength. It is concluded that not only the temperature of testing, but also annealing treatment has marked effects on the strength of fibers.

1. INTRODUCTION

Optical fiber sensors have several inherent advantages and provide attractive candidates for a wide range of industrial applications due to their small size, light weight, electrically passive operation, immunity to electromagnetic interference etc. (1). The implementation of optical fiber sensors requires a systematic study of the reliability of the optical fiber in such harsh environments where long-term reliability of critical industrial equipment should be guaranteed. With the increasing interest in optical fiber sensors in which the load is applied to fibers of a brittle material, it thus becomes more and more important to understand the factors which govern the strength of fibers.

The strength of silica optical fibers is known to be dependent on the atmosphere and moisture is a particularly important factor (2, 3). The strength of silica optical fibers has been studied for many years, but until quite recently most work was carried out at room temperature (RT) (4, 5). It has been shown that the effects of temperature on the strength of silica fibers are irreversible in a temperature range above 200 °C (2). Emphasis should be paid on the strength of silica optical fibers for sensing applications at high temperature.

In this paper, we investigate the effects of acid stripping, high temperature and high-temperature annealing on the tensile strength of silica optical fibers commonly used for fabrication of FBG by tensile fracture stress testing to present a clear account of the strength behavior. Some discussion of proposed mechanisms has been introduced for each effect to explain the results described in section 3, before conclusions.

2. EXPERIMENTAL DETAILS

Silica optical fibers used in this work are commercial single-mode fibers (Corning, SMF-28) with 125 μm in diameter and a ~60 μm thick dual acrylate coatings. The 350 mm coatings at the center of its gauge length were stripped by a hot concentrated sulfuric acid (98%) wash followed by a thorough acetone rinse. The coatings were removed by immersion bent optical fiber for 15–20 s in a sulfuric acid bath that was heated to ~190 °C by an oil bath. After acid stripping, acetone was used to clean the acid and residuals away from the surfaces of the fibers with which the stripped samples were suspended in the laboratory to avoid any contact, before tensile testing.

Uniaxial tensile fracture stress tests are used to obtain the tensile strength of optical fiber samples referring to an IEC standard (6). The tests were used a universal testing machine with a load cell of 200 N and two clamping devices which avoid damage and slipping of the sample, as shown in figure 1(a). The vertical tube furnace traversed with three calibrated K-type thermocouples indicated that the test section of the sample lay in a length of 100 mm soaking zone which is within ± 2 °C of the nominal temperature. The optical fiber sample was suspended vertically with a distance of 750 mm between the two axes of the gripping capstans with a diameter of 150 mm, as shown in figure 1(b). The fiber length to be tested was gripped at both ends and the fiber was subjected to tension until fracture occurs in the gauge length section of the fiber. The fiber fracture at the grip could be minimized by providing a surface friction that prevents excessive slippage. The section of the fiber that would not be tested around the capstan was wrapped two times without a crossover and it was secured at the end with an elastic band.

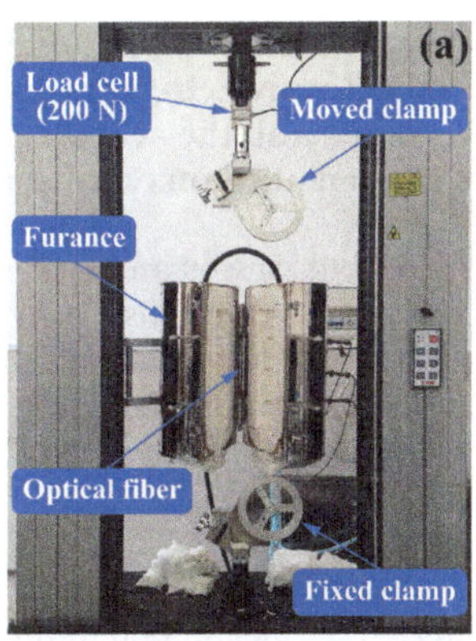

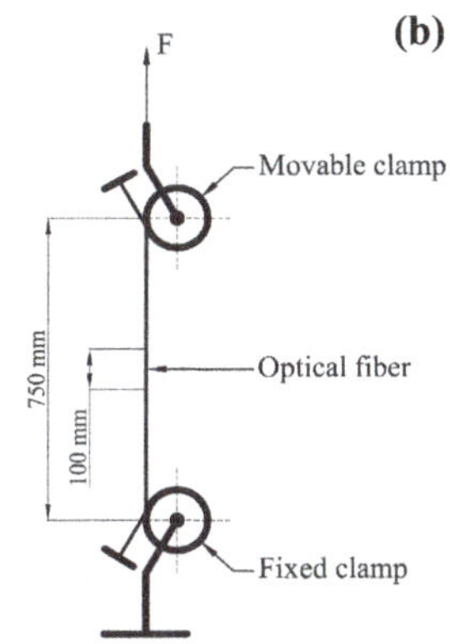

Figure 1. Tensile test apparatus of optical fibers at high temperature.

In our experiment, a strain rate of 4% min^{-1} is selected for the speed control by applying the forces to the fiber via the movable clamp. A minimum of 15 specimens were tested and drop the lowest breaking fracture stress data point for each test condition. All specimens with the same length shall be pre-conditioned in the test environment for a minimum period of 30 min. The tensile tests should be carried out as soon as possible after acid stripping due to the time lapse between the acid stripping and the tensile testing has a major influence on the strength of stripped fibers (7). The ambient temperature is 26 °C with a tolerance of ± 1 °C and the relative humidity (RH) is 63% with a tolerance of 3% for the duration of the tensile tests. Strength measurements were taken in air. Great care was taken not to touch the fiber surfaces in any way before testing since the slightest mechanical abrasion was observed to weaken the silica.

3. RESULTS AND DISCUSSION

3.1 Effects of acid stripping on the strength of silica optical fibers

Optical fibers are almost always coated with a polymer during the draw process to protect their surfaces from mechanical damage introduced by subsequent handling. To investigate the strength of the fibers for high-temperature sensing applications, it is necessary to remove the coatings that allows the strength of the fibers to be obtained by direct contact with the test environment as the coatings would be burned off at high temperature and influence the mechanical behavior of the fibers (8). In this work, the coatings were stripped using hot concentrated sulfuric acid described in section 2. For comparison, the tensile fracture stress tests were performed on a set of as-received fiber samples and a set of stripped fiber samples in ambient conditions at room temperature.

Figure 2 shows the results of the tensile strengths for both as-received and acid stripped optical fiber samples, where a two-parameter Weibull model is employed to analyze the strength distribution of each set of experimental data. The Weibull modulus, *m*, is calculated using a linear regression method. The median fracture stress and the mean strength determined by statistical analyzes of the data for the stripped samples are 4.715 GPa and 4.568 ± 0.369 GPa respectively, corresponding to ~3% and ~6% smaller than those for the as-received samples. The strength distribution of the stripped samples is wider than that of the as-received. The results confirm that careful complete stripping has a negligible effect on the strengths of fibers but causes an increase in the scatter of the strength data, which are entirely consistent with those observed by Matthewson *et al.* (5). There is no evidence that the acid stripping does anything other than slightly change the environment on the fiber surface. A small number of low strengths shown in figure 2 probably indicate poor handling rather than degradation by the stripping process itself.

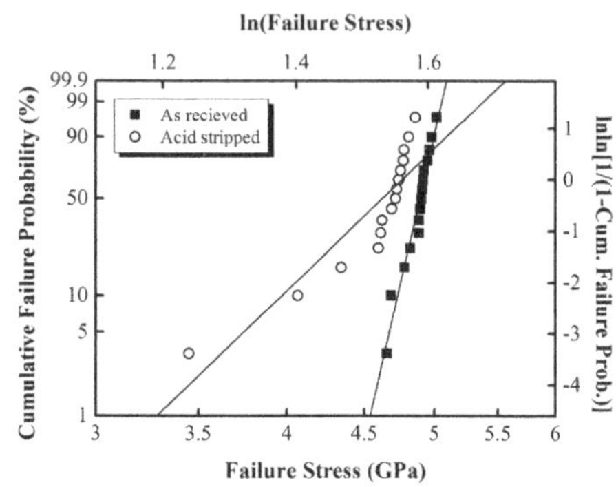

Figure 2. Weibull plot of the tensile strengths for the as-received and the acid stripped fibers tested at room temperature.

3.2 Effects of high temperature on the strength of silica optical fibers

For application in power plant, individual stripped fibers were suspended in the vertical tube furnace and tested at 300 °C and 540 °C in order to obtain strengths of fibers at high temperature in comparison with those at RT. Individual fiber strengths are plotted in figure 3 where the test temperatures are differentiated. It is apparent that the small scatter could be achieved with the fiber samples tested at RT. The scatter is much larger for the samples tested at 300 °C than that for the samples tested at RT, and the scatter is much reduced for the samples tested at 540 °C but still remained greater than that for the samples tested at RT. The strengths and Weibull modulus as a function of temperature are shown in figure 4. In general a statistical decrease in tensile strengths of the fibers is obtained with an increase in temperature even though the differences in upper and lower limit of the confidence interval, as well as Weibull modulus. The results above indicate that the strength decrease at high temperature may be accounted for specific flaws on the fiber surface which are formed and grew at high temperature.

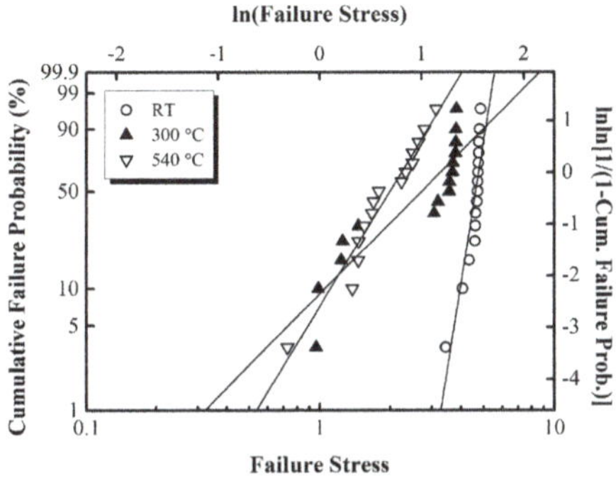

Figure 3. Weibull plot of tensile strengths for the stripped fibers tested at room temperature, 300 °C and 540 °C.

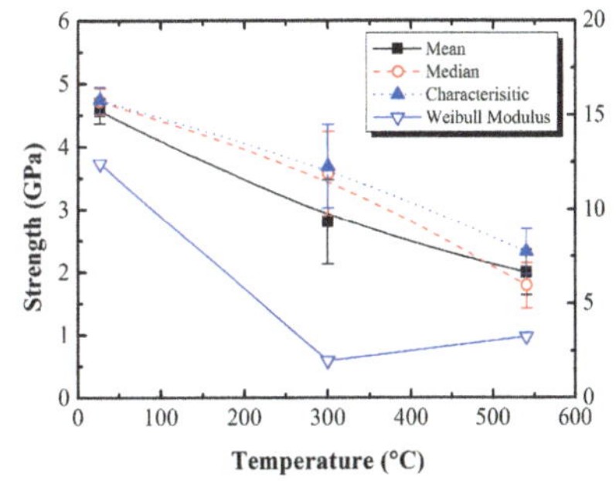

Figure 4. Tensile strength and Weibull modulus for the stripped fibers as a function of temperature.

3.3 Effects of high-temperature annealing on the strength of silica optical fibers

Considering the fabrication process of regenerated fiber Bragg gratings, further investigation of strength was conducted on two sets of stripped optical fibers in air after annealing process described previously (9). Five stripped fiber samples were suspended vertically in the furnace and annealed at 500 °C or 900 °C at one time while mounted on the movable clamp. In order to prevent the fibers being damaged by moisture and accident these samples were subsequently tested at RT as rapidly as possible which occupied at most a further 10 min. Figure 5 shows the tensile strengths of the annealed fiber samples tested at RT, compared with those of the samples tested at RT, 300 °C and 540 °C without annealing. The experimental results demonstrate that annealing process leads to a significant reduction in strengths of the silica optical fibers. The variation of mean strength and Weibull modulus with test condition reveals that the strength of the annealed samples is much reduced and smaller than that of the samples tested at RT, even at 300 °C and 540 °C without annealing, as shown in figure 6. Thus, not only the high temperature introduces the irreversible reduction in the strengths of fibers heated at 300 or 540 °C, but a further weakening occurs on cooling (2). Figure 6 also demonstrates that Higher temperature of annealing could lead to a larger reduction in

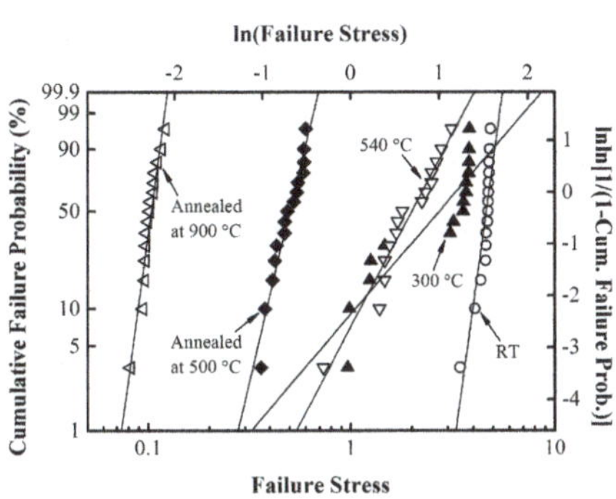

Figure 5. Weibull plot of the tensile strengths for the fibers tested at room temperature (26 °C) after annealing at 500 °C and 900 °C respectively, compared with those for the fibers tested at room temperature, 300 °C and 540 °C.

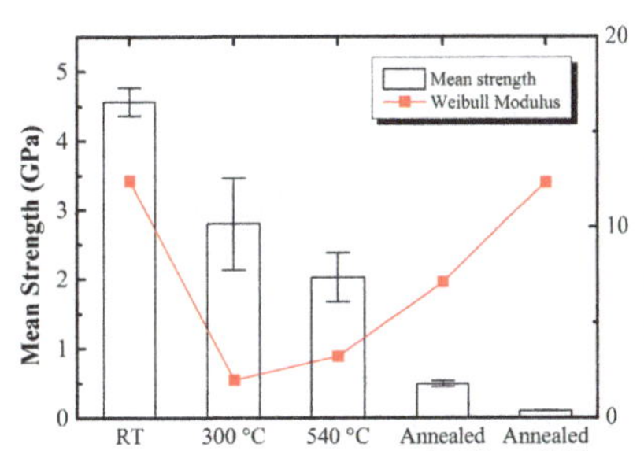

Figure 6. Tensile strengths and Weibull modulus for the fibers tested at room temperature after annealing at 500 °C and 900 °C respectively, compared with those for the fibers tested at room temperature (26 °C), 300 °C and 540 °C.

strength of the fibers. One possible explanation is that sufficient water is absorbed to give the stress-induced water reaction and lower strength during cooling to room temperature in air.

4. CONCLUSIONS

Using tensile fracture stress testing, the effects of acid stripping, high temperature and high-temperature annealing on the tensile strength of silica optical fibers are investigated. Some conclusions obtained from the investigation are as follows:

(1) Stripping the acrylate polymer coatings from the fiber by immersion in hot (~190 °C) concentrated sulfuric acid has a negligible effect on the tensile strength of the fibers, while leading to the relatively high scatter on the strengths, provided the process is conducted with sufficient care.
(2) The temperature of testing has a considerable effect on the strength of the fibers. The fibers tested at higher temperature are weaker than comparable fibers tested at lower temperature, as well as the larger scatter on the strengths.
(3) High-temperature annealing has a significant effect on the strength of the fibers. Higher temperature could lead to a larger reduction in the strength of the fibers.

Such research provides potentially useful experimental values for the design, fabrication and reliability evaluation of optical fiber sensors used at high temperature. Further investigations into the failure mechanisms of silica optical fibers for high-temperature sensing applications should be completed based on analysis of the fracture faces.

ACKNOWLEDGEMENTS

Y. Tu would like to thank Professor Lin Ye from Sydney University for his constructive and useful advises. This work is supported by the Natural Science Foundation of China (Nos. 11172102 and 51175178) and the National High Technology Research and Development Program 863 (No. 2009AA044803).

REFERENCES

[1] Lee B. (2003) *Opt. Fiber. Technol.* **9**, 57-79. doi: 10.1016/S1068-5200(02)00527-8
[2] Proctor B. A., Whitney I. & Johnson J. W. (1967) *Proc. R. Soc. Lond. A.* **297**, 534-557. doi: 10.1098/rspa.1967.0085
[3] Gougeon N., Poulain M. & El Abdi R. (2004) *Opt. Mater.* **27**, 75-79. doi: 10.1016/j.optmat.2004.02.020
[4] Choi S. R. (1991) *J. Mater. Sci. Lett.* **10**, 267-268. doi: 10.1007/bf00735653
[5] Matthewson M. J., Kurkjian C. R. & Hamblin J. R. (1997) *J. Lightwave Technol.* **15**, 490-497. doi: 10.1109/50.557565
[6] IEC (2001) IEC 60793-1-31.
[7] Wei C. Y., James S. W., Ye C.-C., et al. (1999) *Proc. SPIE 3670 Smart Structures and Materials 1999: Sensory Phenomena and Measurement Instrumentation for Smart Structures and Materials.* 164-170. doi: 10.1117/12.349727
[8] Severin I., El Abdi R. & Poulain M. (2008) *Surf. Coat. Technol.* **202**, 2494-2499. doi: 10.1016/j.surfcoat.2007.08.005
[9] Tu Y., Qi Y.-H. & Tu S.-T. (2013) *Smart Mater. Struct.* **22**, 075026. doi: 10.1088/0964-1726/22/7/075026

Numerical modelling of impact response of aluminium foam/FML sandwich panels

C. Liu, Y.X. Zhang
School of Engineering and Information Technology,
The University of New South Wales, Australian Defence Force Academy, Australia

ABSTRACT

Numerically modelling the failure of aluminium foam sandwich panels with fibre metal laminate skins subjected to high velocity impact has not been adequately conducted. A few papers reported the dynamic failure behaviour of aluminium foam sandwich panels subjected to impact, but the perforation of aluminium foam cored sandwich panels was rarely modelled. This paper reports a numerical model that models the perforation of an aluminium foam sandwich panel with fibre metal laminate skins subjected to high velocity impact using the commercial code LS-Dyna. The modelled results are compared to those from the experiment. The residual velocity from the modelling was found extremely close to the test result. The top fibre metal laminate was stuck to the wall of the perforated foam hole and the bottom fibre metal laminate was petaled, both of which matches very well with failure of the tested panel.

1 INTRODUCTION

Sandwich panels with aluminium (AL) foam core and fibre metal laminate (FML) face sheets exhibit superior material properties, such as light weight, high energy absorption and high impact resistance (1). Hybrid sandwich panels of this sort have the potential to support the development of future military, marine and aerospace facilities (2). Impact response of the sandwich panels therefore becomes one of the top concerns and has attracted significant research interests in the past decade.

Finite element modelling as a robust and versatile numerical tool for the analysis and modelling of engineering structures, especially for complicated structures under complex loadings, has been used to model the impact performance of sandwich panels with aluminium foam core successfully. It has been demonstrated to be a very effective method for investigation of the impact response of aluminium foam cored sandwich panels subject to various impact loadings such as low velocity impact (3), bird strike (4), explosion (5) and high velocity impact up to 186 m/s (6). However, these literatures focus on numerical modelling of aluminium cored sandwich structures with monolithic metal skins rather than FML skins.

In this paper, a finite element model and analysis procedures are developed in the commercial finite element code LS-DYNA to model the high velocity impact response of a type of aluminium foam/FML hybrid sandwich panel. The impact response of the panel subjected to an impact from a steel ball bearing at an impact velocity of 210 m/s is modelled using the developed finite element model and procedures. The numerical results are compared to those obtained from experiment to validate the numerical model developed in this paper.

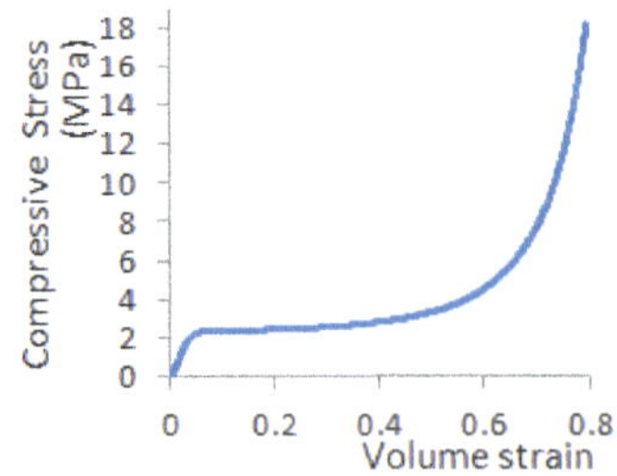

Figure 1 Compressive stress-strain curve of AL foam at 2 mm/min loading rate

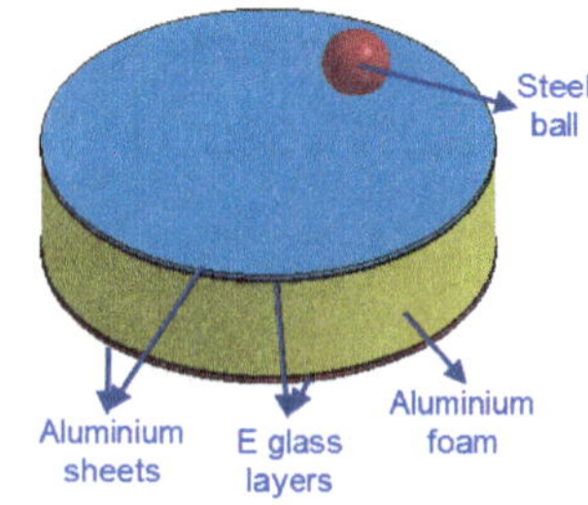

Figure 2 Geometric models of the panel and the projectile

2 FINITE ELEMENT MODEL FOR IMPACT MODELLING

The impact response of an AL foam/FML hybrid sandwich panel fabricated and tested by the authors recently is modelled using the developed numerical model. The tested sandwich panel is of 120 mm × 120 mm × 23 mm and restrained at all the edges. It was seated on a platform with an 80-mm-diameter circular hole when testing. Thus only the circular part of the panel of 80 mm in diameter is modelled with fully fixed boundary conditions. The panel is of a combination of AA/GF/Foam/GF/AA, where 'GF' stands for one ply of plain woven E glass fibre (0°/ 90°) and 'AA' for one layer of AL alloy sheet 5005. The FML face sheets are comprised of a 1 mm thick AA layer and a 0.5 mm thick ply of GF and epoxy. The 20-mm-thick foam core is of 0.3 g/cm^3, and the tested compressive stress strain curve of the AL foam is shown in Figure 1. The material property parameters of the AA, GF and AL foam are given in Table 1, 2 and 3 respectively. The panel was subjected to an impact from a 7.05-gram steel ball bearing of 12 mm in diameter fired from a gas gun facility at a velocity of 210 m/s. The geometry of the panel and the projectile is shown in Figure 2. As seen in Figure 2, the steel ball deviated from the centre of the panel because in the experiment the projectile deviated about 22 mm from the centre of the panel and the platform in the testing. It should be noted that in gas gun test, it is hard to guarantee that the projectile will impact on the centre of the panel.

The finite element modelling is conducted utilizing the commercial code LS-DYNA. The projectile, the AAs and AL foam are meshed using the eight-node hexahedron solid elements, in which transition and rotation are considered, while the two layers of glass fibre/epoxy prepregs are meshed using the Belytschko-Tsay shell elements with four integration points through the thickness. Considering symmetry, a half of the model is built for computation efficiency.

2.1 Material models

The foam material model used is the well-known crushable foam model (7), in which strain rate effect is considered. But in this paper, strain rate effect is not taken into account as rate dependence in low density AL foam is very limited (8, 9). The AA layers are modelled using the simplified Johnson Cook material model (7). Different from the full Johnson Cook material model, temperature and erosion is not considered in the simplified model. The GF plies are modelled using the Composite Damage model based on the Chang-Chang criterion (7). These material models exclude the erosion criteria inside, and the maximum principle strain and maximum shear strain principles based on the erosion criteria are employed to enable damage and fracture modelling. Since the projectile is of much higher level of stiffness than other materials of the sandwich panel, it is regarded as a rigid body.

Table 1: Parameters for the E glass/epoxy (10)

Density (kg/m^3)	Poisson's ratio			Young's modulus(GPa)			Shear Modulus(GPa)		
	Prba	Prca	Prcb	Ea	Eb	Ec	Gab	Gbc	Gca
2000	0.0575	0.0575	0.33	54	9.4	9.4	5.6	5.6	5.6

Shear strength (MPa)	Longitudinal tensile strength (MPa)	Transverse tensile strength (MPa)	Transverse compressive strength (MPa)	Ultimate strain (%)
76	1900	57	285	3.5

Table 2: Parameters for Al 5005 (11)

Density (kg/m^3)	Poisson's ratio	Young's modulus	Static yield limit	Strain hardening modulus	Strain hardening exponent	Strain rate coefficient
2700	0.33	65.04 GPa	369 MPa	684 MPa	0.73	0.0083

Table 3: Parameters for the aluminium foam

Density	Young's Modulus	Poisson's Ratio (12)	Tensile stress cut off (13)	Damping coefficient
300 kg/m^3	0.037 GPa	0	0.012 GPa	0.1

2.2 Contact algorithms

Three types of contact algorithms are employed to define different contact conditions in the model. The Erosion Contact type is selected to define the relation between the projectile and components of the sandwich panel, i.e. the AA layers, GF plies and AL foam. The Tie Break Contact type is adopted to define the bonding of the adjacent components in the sandwich panel. The general 3D contact type is used to define the location relationship of the components having no initial contact in the sandwich panel to correctly model the failure of the sandwich panel.

3 FINITE ELEMENT MESH AND CONVERGENCE STUDY

For computation efficiency a dense mesh is employed in the vicinity of the impact area while the other areas are divided into several regions with each region discretized by $n \times n$ mesh as shown in Figure 3. For convergence study, four values of *n* are considered and tested, namely, 5, 10, 15 and 17, and the foam through the thickness is meshed using 25, 60, 60 and 60 elements respectively. The finite element model is employed to model the impact response of the panel subjected to the impact from the steel projectile and the residual velocity of the projectile is computed. It is found that the computed residual velocity of the projectile at *n*=17 is extremely close to that at *n* =15 (see Figure 4) and this means a converged result can be obtained when *n* =15. The finite element mesh when *n*=15 is shown in Figure 5.

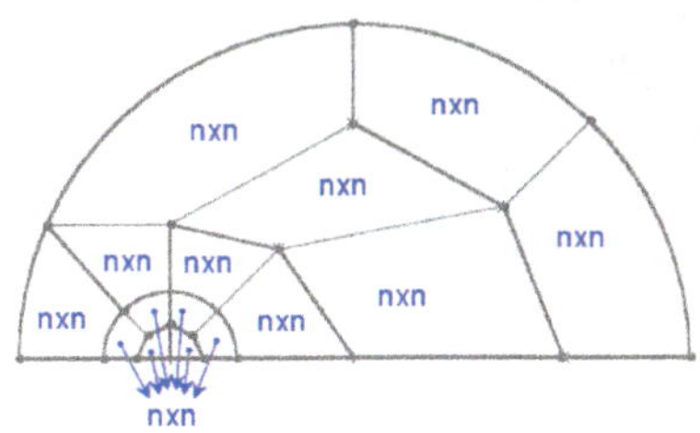

Figure 3 Mesh solution

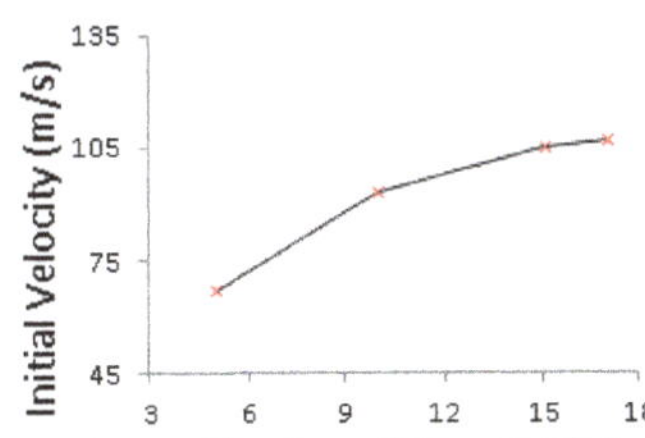

Figure 4 Convergence study

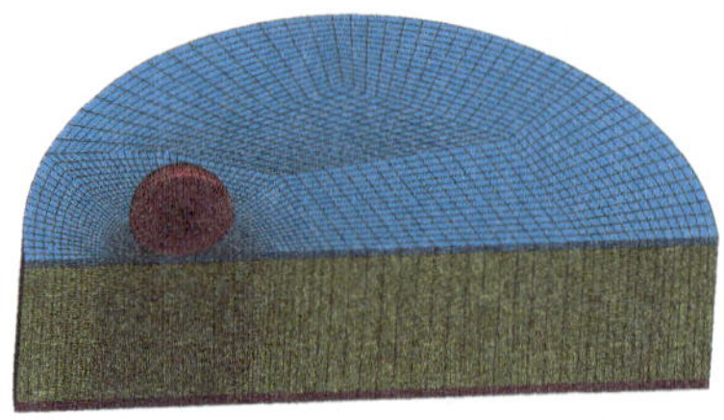

Figure 5 Finite element mesh

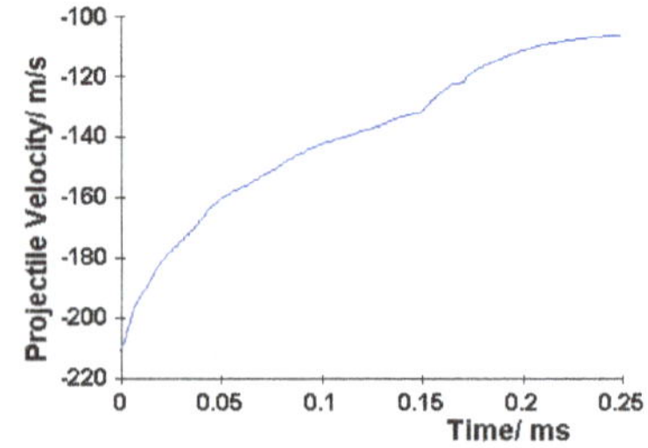

Figure 6 The Computed residual velocity-time curve

4 NUMERICAL RESULTS

The computed residual velocity versus time curve is shown in Figure 6. The computed residual velocity of the steel ball bearing is 106 m/s, and this agrees well with value from the testing of 110 m/s. Figure 7 compares the failure mode of the panel obtained from the numerical modelling and experiment, and it is found that they match each other well. The modelling shows that the top FML skin is fully perforated and tightly stuck to the wall of the perforated foam hole, and this is coincident with that happens in the experiment. At the back face, the AA layer petals and the crack are well modelled and matches well with that in the testing. But the bottom GF layer failure in the test is not exactly the same as that presents in the modelling. This may result from the erosion principles employed. When the maximum principle strain reaches the set value, elements are eroded and deleted. As a result, no elements standing for GF prepreg ply are left in the central area while in the testing the GF prepreg is still in the impact area even though failure occurs. Overall, the numerical model can simulate the failure of the hybrid panel subjected to high velocity steel ball bearing impact very well.

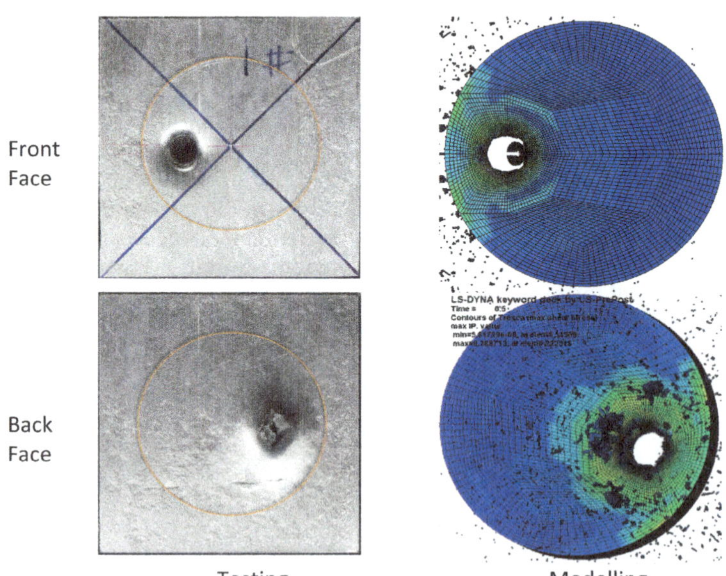

Figure 7 Comparison of panel failure from testing and modelling

5 CONCLUSION

A finite element model and impact modelling procedures are developed to simulate the high velocity impact response of hybrid sandwich panels with AL foam core and FML skins. The developed numerical model is used to model the impact response of a sandwich panel with Al foam core and FML skins comprised of aluminium alloy sheets and E glass fibres. A convergence study is conducted and the finite element mesh can generate convergence result. The computed residual velocity of the projectile and the failure mode of the sandwich panel obtained from the numerical modelling are compared to those obtained from experiment and numerical results agree very well with those obtained from the experiment. It is demonstrated that the developed numerical model can simulate the impact response of the sandwich panel accurately and effectively.

6 REFERENCES

[1] Vlot, A., La, Rocca G.,Frediani, A. (2001) *ICCM 13 proceedings*, ID 1483.

[2] Reyes, G. (2010), *J. Sandw. Struc. Mater.*, **12**, 81-96. doi: 10.1177/1099636209103287

[3] Zou, R., Ruan, D., Lu, G. (2011) Adv. Mater. Res., 261-263, 761-764. doi:10.4028/www.scientific.net/AMR.261-263.761

[4] Hanssen, A.G., Girard, Y., Olovsson, L., Berstad, T., Langseth, M. (2006) *Int. J. Impact Eng.*, **32**, 1127-1144. doi:10.1016/j.ijimpeng.2004.09.004

[5] Zhu, F., Zhao L., Lu G., Wang, Z. (2008) *Adv. Struc. Eng.*, **11**, 525-536. doi: 10.1260/136943308786412005

[6] Liu, C. Zhang Y.X. (2014) *Appl. Mech. Mater.*, **553**, 745-750. doi: 10.4028/www.scientific.net/AMM.553.745

[7] Hallquist, J. LS-DYNA Keyword manual, Livermore Software Technology Corporation, 2006.

[8] Dannemann, A. K., Lankford, J. (2000) *Mat. Sci. Eng. A-Struct.*, **293**, 157–164. doi: 10.1016/S0921-5093(00)01219-3

[9] Nemat-Nasser, S., Kang, W.J., McGee, J.D., Guo, W.-G. and Isaacs, J.B. (2007) *Int. J. Impact Eng.*, **34**, 1119-1146. doi: 10.1016/j.ijimpeng.2006.05.007

[10] Yaghoubi, A. Seyed, Liaw, B. (2012) *Compos. Struct.*, **94**, 2585-2598. doi: 10.1016/j.compstruct.2012.03.004

[11] Kay G., Goto D., Couch R., (2007) Statistical testing of aluminium, titanium, lexan and composites for transport airplane rotor burst fragment shielding. FAA Rep. No. DOT/FAA/AR-07/26, Federal Aviation Administration, Washington, DC.

[12] Shen, J., Lu, G., Wang, Z., Zhao, L. (2010) *Int. J. Impact Eng.*, **37**, 960-970. doi: 10.1016/j.ijimpeng.2010.03.002

[13] Motz, C., Pippan, R. (2001), *Acta Mater.*, **49**, 2463–2470. doi: 10.1016/S1359-6454(01)00152-5

Numerical study on effects of buffer bulbous bow structure in collisions

Y. Liu, L.S. Zhang, L.P. Sun, B. Li
Harbin Engineering University, P.R. China

ABSTRACT

In order to reduce the risk of cargo leakage in case of ship-to-ship collision, in this paper, a novel bulbous bow is introduced. A numerical simulation by using FEM was conducted to validate its effectiveness compared with the conventional one. Collapse mechanism, force-time curves of the novel bulbous bow structure were investigated and compared with those of conventional bow structure. The computed results indicate that the novel bulbous bow structure is expected to be efficient to reduce the risk of cargo leakage, and the approach of modification on the bulbous bow structure in ship collision accidents protection is reasonable.

1. INTRODUCTION

Bulbous bow can help to reduce a ship's resistance and thus to save the fuel consumption up to 15%, however, it is also regarded as a threat to a struck ship in collision accidents because it may generally penetrate the side shell of the vessel, which may cause the leakage of hazardous goods. How to avoid the serious consequence of ship-to-ship collision accidents through the bulbous bow design have become a focus of ship design.

So far many investigators have studied the behavior of bulbous bow structure in collision accidents. Zhang [(1)] used analytical and numerical methods to analyze ship-ice collision, and summarized the regularity of the damage deformation of the ship, the collision force and the energy absorption under different working conditions. Yamada & Endo [(2)] investigated that the crashing mechanisms of the buffer bulbous bow with a transverse stiffening system and verified the effectiveness of a prototype buffer bulbous bow structure in ship-ship collisions as compared with that of standard bulbous bows by conducting a series of large-scale finite element analyses.

The aim of collision protection is to ensure the intact of cargo hold instead of bulbous bow. Therefore, many novel bulbous bow types concerning collision protection have also been proposed. Cheung[(3)] put forward the concept design of buffer bow. Endo et al. [(4)] raised a bulbous bow design by using low yield point steel to accelerate the bow bending process in oblique collision scenarios. Tautzet al. [(5)] estimated removing the longitudinal structural elements in foremost part of the bulbous bow, which would increase the bow's energy absorption ability.

2. MECHANICAL PROPERTY FOR BULBOUS BOWS

2.1 Material properties for steel

High tensile steel like HT690 has a very high yield point and demonstrates a reduced plastic ductility and brittle fracture behavior as compared with mild steel [6]. Figure 1 illustrates the engineering strain and stress for brittle steel (high tensile steel, e.g., HT690, HT460) and ductile steel (mild steel, e.g., MS235).

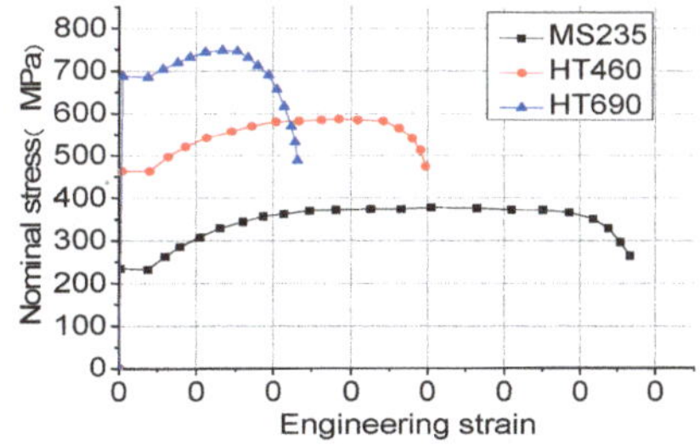

Figure 1 Engineering strain and stress for brittle steel and ductile steel

It is considered that from the right material curves, the higher a material's strength, the lower the fracture strain. That is the source of the novel bulbous bow idea.

2.2 Constitutive models for steel

The true stress-strain relationship material models in the following form were used in this paper (Zhang et al. 2004):

$$\sigma = C\varepsilon^n$$

Where $n = \ln(1 + A_g)$, $C = R_m \cdot (e / n)^n$

Ag is the maximal uniform strain and Rm is the ultimate tensile stress.

2.3 Failure criteria

During the stimulation, when the effective plastic strain reaches the predefined failure stain, the element will be deleted from the FE model. Considering the mesh size effect, the following definition for failure strain was used in this paper.

$$\varepsilon_f(l_e) = A_g + \varepsilon_e \cdot \frac{t}{l_c}$$

where A_g is the uniform strain and εe is the necking strain, t is the plate thickness and L_c is the characteristic length for individual element.

3. COLLISION SIMULATIONS

The collision scenario, as shown in figure 2, is assumed to be that the striking vessel with a bulbous bow collides perpendicularly with a double hull container vessel in full loaded condition, with main dimensions shown in table 1.

Table 1 The principle dimension of two vessels

Item	Value	Unite
Length	230	m
Breadth	32	m
Depth	19	m
Scantling draught	12	m
Length of bulbous bow	7.5	m
Displacement	$5.7*10^7$	kg

The forward speeds of both vessels are taken into account in this simulation as it is assumed to be a more real collision scenario (see table 2).

Table 2 The velocity of the vessels in this simulation

Description		Velocity (m/s)	
		Striking vessel	Struck vessel
case 1	Bow with HT690	4.5	6.5
case 2	Bow with MS235	4.5	6.5

Considering the fact that full structure modeling would spend plenty of time and make the computation almost impossible since a tremendous finite element exist, in other hand, the collision mostly occurred in the a specific, small areas compared to the whole structures, in this paper, the dimension of the plates, stiffeners and holes in this FE model are precisely modeled according to a conversional container vessel using GL Poseidon software (POSEIDON, 2003), as shown in figure 2.

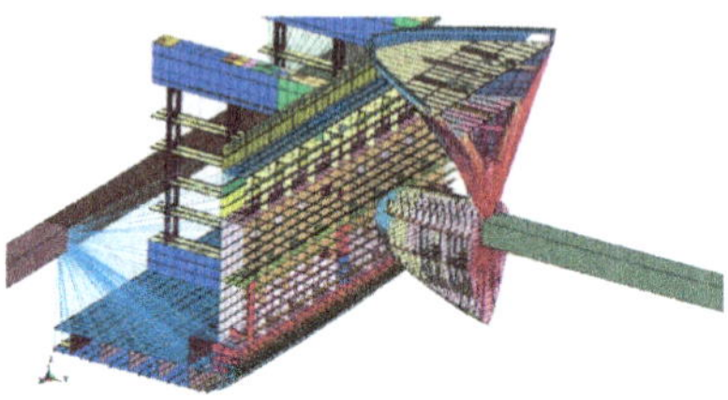

Figure 2 Overview of the collision model with beam elements visible

The idea of novel bulbous bow is to use the brittle fracture behavior of high tensile steel to realize the quickly separation of the bulbous bow in a collision accident. When such a long and sharp bulbous bow collides with another vessel, the plate made of high tensile steel would break quickly under the combination of large compression force and shear force induced by collision. With the motion of the struck vessel, the bow tip can separate, which would keep the bow tip from penetration into another vessel's cargo hold.

4. RESULTS OF THE SIMULATION

4.1 Overall structure deformation

It can be seen in the simulation for both cases that the differences of the overall structure behaviors of the bow and double hull occurred at 0.619s in figure 3-4. Before the differences, the outer shell plate of the struck ship was penetrated at about 0.16s, whereas the bow structure had almost no deformation and damage for both cases.

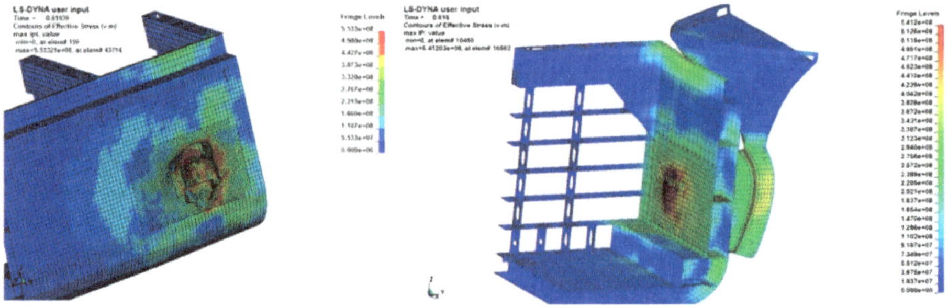

Figure 3-4 Deformation for side shell in case 1(L) and case 2 (R) at time 0.619s

As shown in figure 3, a buckling failure of the strip plate of bulbous occurred at time 0.619, and then 0.2s later the plate ruptured rapidly. Therefore the quickly separation of the bulbous bow will prevent the inner shell plate from penetrating, as shown in figure 5.

In case 2, the bulbous bow is conventional one with mild steel, the inner plate of the struck ship began to break down at 0.619s as shown in Figure 4 and was penetrated seriously at 0.8s as indicated in figure 6. Through comparing the results of both cases, it can be seen that positive separating bulbous bow can be efficient to reduce the risk of cargo leakage in case of ship collisions.

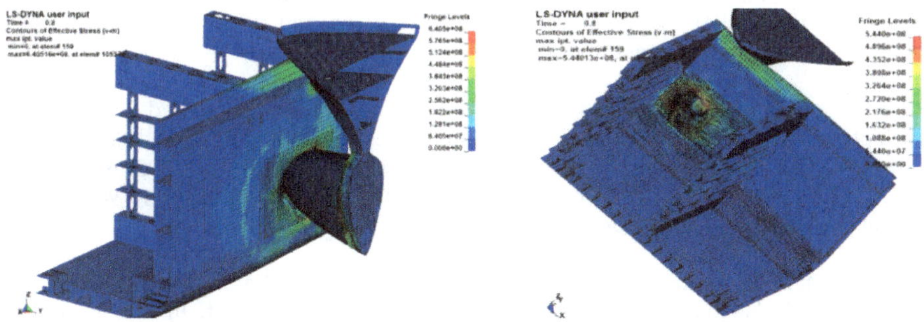

Figure 5-6 Deformation for side shell in case 1 (L) and case 2 (R) at time 0.8s

Through comparing the results of both cases, it can be seen that positive separating bulbous bow can be efficient to reduce the risk of cargo leakage in case of ship collisions.

4.2 Time history of contact force

For both cases, time history of resultant force is presented in figure 7. The contact force fluctuates indicates that the failure of the structure appeared during the collision and the collision was a complicated process.

The curves of resultant force for both cases were nearly the same in first 0.619s, and afterward varied respectively until the end of the simulation, as shown in Figure 7.

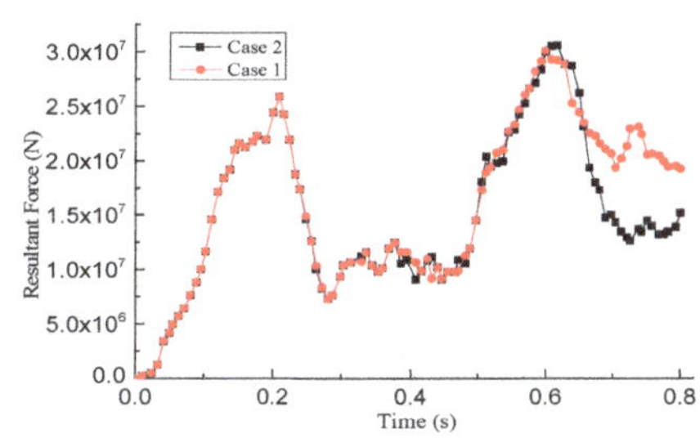

Figure 7 Comparison of the resultant force

4.3 Time history of energy transformation

For both cases, time history of internal energy was almost the same in the whole simulation, as shown in the Figure 8.

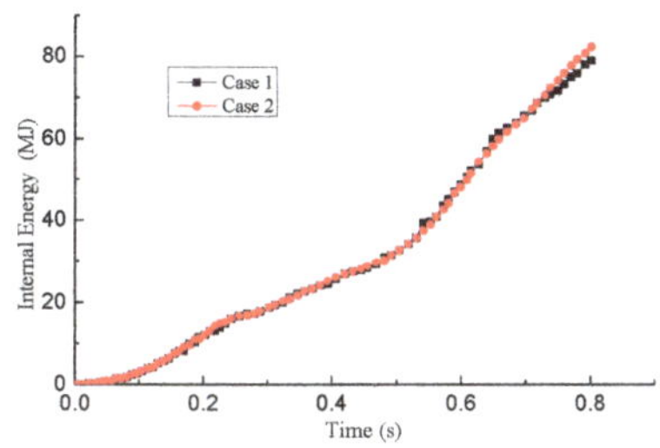

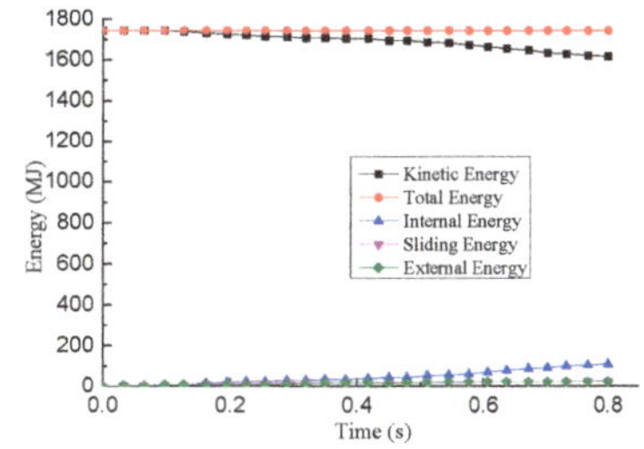

Figure 8 Comparison of internal energy **Figure 9 Energy dissipating distribution**

For case 1, prior to collision, the total energy storage in two vessels behave in form of kinetic energy, as indicated in the Figure 9. During the simulation, the kinetic energy of the structures transformed into internal energy, sliding energy and external energy until declined to 1617Mpa at t=0.8s. The dissipated energy accounted for 7.18% of the total energy.

5. CONCLUSION

An integrated analysis of ship-to-ship collision was presented in this paper. A numerical simulation using FEM was conducted to produce computed results to be used as the reference for the effectiveness of the proposed bulbous bow.

Based on simulation analysis of ship-to-ship collision, it can be concluded that the outer shell plate ruptured quickly after the collision happened, and the proposed bulbous bow can avoid the inner shell plate to be penetrated, which would reduce the risk of cargo leakage. Collapse mechanism, force -time curves were also investigated to validate the availability of this kind of bulbous bow compared to a conventional one.

The finding of this paper can provide useful reference for the study the ship-to-ship collision, and also have a good effect on protecting environment. In the future, the proposed bulbous bow should be further developed to account for varying angles of collision.

6. REFERENCES

[1] Zhang., Wan., Chen, Research on structure dynamic response of bulbous bow in ship-ice collision load. Journal of Ship Mechanics.Volume18, 2014.

[2] Yamada Y., Endo H., & Pedersen PT., Numerical Study on the Effect of Buffer Bow Structure in Ship-to-ship Collisions, 15th International Offshore and Polar Engineering Conference (ISOPE), pp. 604 - 611,2005.

[3] Cheung L., A soft bow for ships. European shipbuilding No.3, 52-53, 1969.

[4] Endo, Hisayoshi, Yamada, & Yasuhei, vessel having lateral bending absorption type bow [P], JP2004314825, 2004.

[5] Tautz I.; Schöttelndreyer M.; Fricke W., Lehmann E., Experimental Investigations on Collision Behavior of Bow Structures, 5th International Conference on Collision and Grounding of Ships (ICCGS), 2010.

[6] Zhang L., Egge E.D., Scharrer M., Evaluation of Thickness Measurements of Drop Tower Tests, Germanischer Lloyd AG, 2003.

Energy absorption of crashworthy structure for rolling stock of a railway

S. Kimura [a], T. Mochida [b], H. Nakamura [b], T. Kawasaki [b], T. Yamaguchi [a]
[a] Hitachi, Ltd, Hitachi Research Laboratory, Japan
[b] Hitachi, Ltd, Industrial Systems, Japan

ABSTRACT

The energy absorptions of crashworthy structures composed of welded aluminium alloys were evaluated using finite element analysis (FEA). In the FEA, a damage-mechanics model was employed as a material model for the parent aluminium alloys and the welding materials. For the welded regions, several combinations of material parameters included in the damage-mechanics model were examined. Using the numerical results, sensitivity of the energy absorption to the material parameters was assessed. Results of the sensitivity study clarified which parameters had the largest effect on the energy absorption.

1 INTRODUCTION

In railway systems, the integrity of areas occupied by passengers and crews, called a survival zone, must be preserved in the case of a collision accident. In order to secure the integrity of the survival zone, a structure for absorbing the collision energy, called a crashworthy structure, is deployed at the end portions of a railway's rolling stock (1).

Numerical simulation using finite element analysis (FEA) plays an important role in evaluating the energy-absorbing ability of crashworthy structures and so many types of crashworthy structures have been designed using FEA (2, 3). For a crashworthy structure composed of welded aluminium alloys, the energy absorption can be accompanied by fractures at the welded regions. In FEA, it is therefore crucial to accurately predict these fractures and to properly evaluate their effect on the energy absorption because they lead to degradation of the structural energy absorption (4).

In this paper, numerical simulations applying FEA were performed to simulate the collapse of two types of crashworthy structures and to evaluate their energy-absorbing abilities. A damage-mechanics model was employed as a material model for the parent aluminium alloys and the welding materials. In the FEA, for the welded regions, a range of material parameters included in the damage-mechanics model was examined. The numerical results were then used for a sensitivity study to clarify which material parameter had the largest effect on the energy absorption.

2 CRASHWORTHY STRUCTURE

Two types of crashworthy structures deployed at the end portions of intermediate vehicles were designed, the schematics of which are shown in Fig. 1. Both structures are composed of welded aluminum alloys. STRUCTURE A consists of a crushable zone in which plastic deformation is allowable and a survival zone representing the part of the passenger's area to be preserved. As shown in Fig. 1(a), it contains part of a door frame in the crushable zone, whereas the whole part of STRUCTURE B is a crushable zone, as indicated in Fig. 1(b). Beneath the floor of STURCTURE B, energy-absorbing blocks are attached by welding. The aluminium alloys used in the crashworthy structures and their static mechanical properties are listed in Table 1.

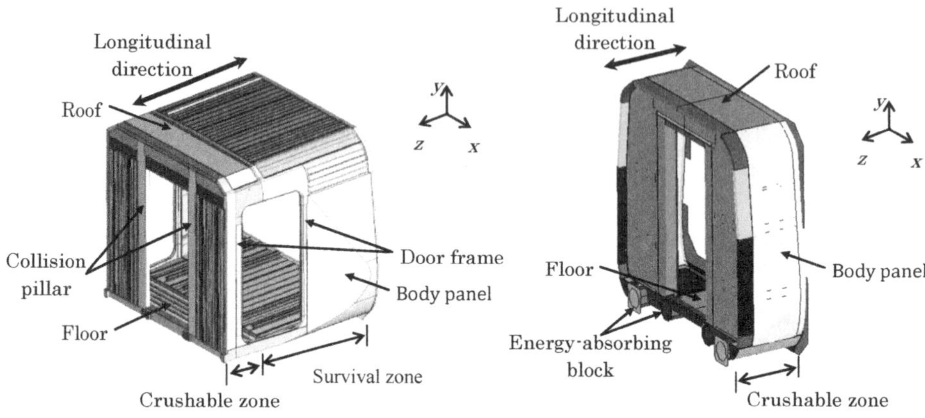

Figure 1 Overview and schematic of designed crashworthy structures.

Table 1 Static mechanical properties of aluminium alloys used in crashworthy structures.

	A5083P-O	A6N01-T5	A7N01-T5	A6063-T5
Application	Body panel, roof	Floor	Pillar, door frame	Energy-absorbing block
Young's modulus [GPa]	70.0	69.0	70.4	67.5
Yielding stress [MPa]	164	235	315	110

3 NUMERICAL SIMULATION

3.1 Material model

A damage-mechanics model was applied as a material model for the base aluminium alloys and the welding material. In this work, we used a Gurson-Tvergaard-Needleman (GTN) model in which ductile fractures are represented by the nucleation, growth, and coalescence of micro-voids (5).

In the GTN model, the yielding condition is provided as

$$\Phi = \left(\frac{\sigma_{\mathrm{eq}}}{\sigma_{\mathrm{M}}}\right)^2 + 2q_1 f^* \cosh\left(\frac{3q_2\sigma_{\mathrm{h}}}{2\sigma_{\mathrm{M}}}\right) - 1 - \left(q_1 f^*\right)^2 = 0\,, \tag{1}$$

where σ_{eq} and σ_{h} are the equivalent and mean stresses of the material containing a void, respectively, and σ_{M} is the equivalent stress of the matrix. q_1 and q_2 are constants and set to central values of 1.5 and 1.0, respectively, as determined by Tvergaard (5). f^* is a function of the volume fraction of void f and defined as

$$f^* = \begin{cases} f & , \text{for} \quad f \le f_C \\ f_C + \dfrac{1/q_1 - f_C}{f_F - f_C}(f - f_C) & , \text{for} \quad f > f_C \end{cases}, \tag{2}$$

where f_C and f_F are the void volume fractions at which the void coalescence starts and the final fracture occurs, respectively. Variations of f with respect to time are expressed as the summation of void nucleation and growth,

$$\dot{f} = \frac{f_N}{S_N\sqrt{2\pi}}\exp\left[-\frac{1}{2}\left(\frac{\varepsilon_M^p - \varepsilon_N}{S_N}\right)^2\right]\dot{\varepsilon}_M^p + (1-f)\dot{\varepsilon}_{kk}^p , \tag{3}$$

where ε_M^p and ε_{kk}^p are effective plastic strain of the matrix and plastic volumetric strain, respectively, and f_N, ε_N, and S_N are respectively the volume fraction of particles causing the void nucleation, the mean value of equivalent strain at which a void nucleates, and the standard deviation of ε_N. At the right-hand side of Eq. (3), the first term represents the void nucleation and the second the void growth.

In the FEA, for the welded regions, several values of the material parameters relating to the fracturing, f_N, ε_N, f_C, and f_F, were examined. As for A6N01-T5, A7N01-T5, and A6063-T5, whose yielding stresses degrade in the vicinity of the welded regions, several values of the yielding stress σ_Y were also examined.

3.2 FE Modelling

Overviews of the FE models for both structures are shown in Fig. 2(a) and (b). Since the designed crashworthy structure was in a plane-symmetrical conformation against the longitudinal axis, the simulation was of a 1/2 model. The crashworthy structure was collapsed by the movable rigid wall from one side while fixing the displacement at the other side. Similar to previous research (4), the normal element size was selected as 10 mm, and for the regions where deformations concentrate, the finer size of 3 mm was applied. Numerical conditions for the FEA are listed in Table 2. A commercially available FE code, LS-DYNA version 971®, was applied.

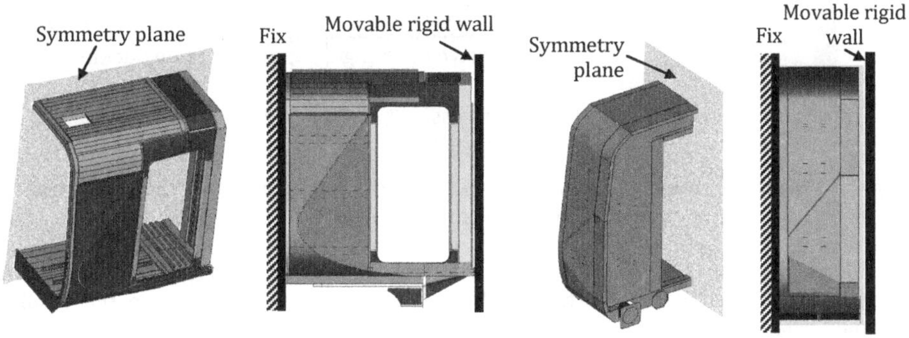

(a) STRUCTURE A (b) STRUCTURE B

Figure 2 Overviews of FE modelling of crashworthy structures.

Table 2 Numerical conditions for FEA.

Element size		Number of elements		Number of nodes	
Normal	Fine	STRUCTURE A	STRUCTURE B	STRUCTURE A	STRUCTURE B
10 mm	3.0 mm	3.9×10^5	3.0×10^5	3.7×10^5	2.4×10^5

4 DISCUSSION

Typical results obtained from the FEA are shown in Fig. 3. Specifically, for the fractures at the welded regions for STRUCTURE A and the progressive plastic buckling of the energy-absorbing block for STRUCTURE B, experimental results obtained from compression tests using full-scale mock-ups of the crashworthy structures (4) are shown.

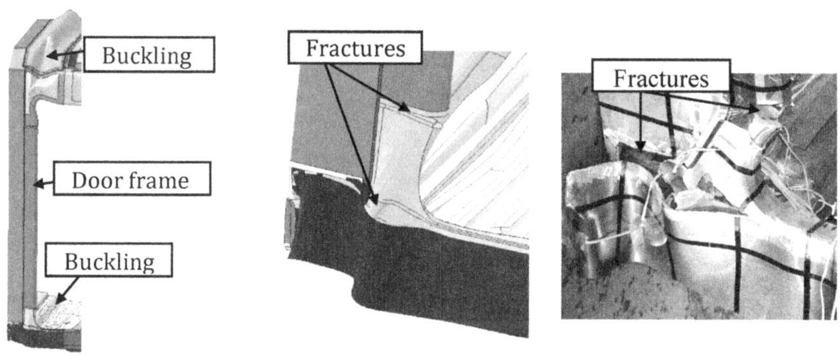

(a-1) Side view (a-2) Fractures at welded region in lower side of door frame

(a) STRUCTURE A

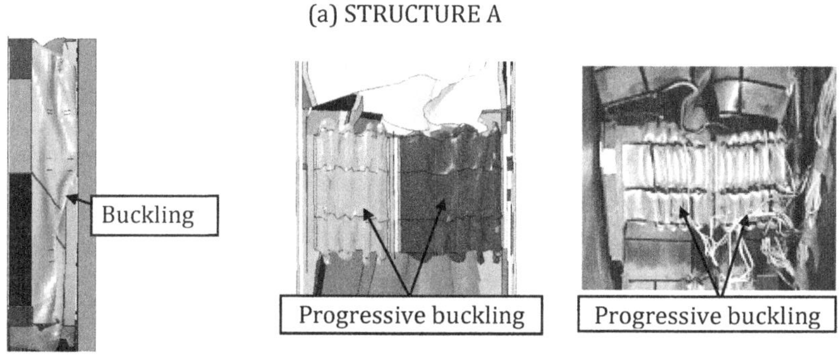

(b-1) Side view (b-2) Progressive buckling of energy-absorbing block

(b) STRUCTURE B

Figure 3 Numerical results obtained from FEA and experimental results obtained from compression test using full-scale mock-up (4).

As indicated in Fig. 3, the FEA using the GTN model can accurately predict both the fracture initiation and the plastic buckling behaviours. Therefore, using the results obtained from the FEA is a worthwhile means of evaluating the sensitivity of the energy absorption to the material parameters relating to the fractures.

In order to conduct the sensitivity study, a sensitivity index was calculated as follows:

$$\alpha_i = \left(\frac{\partial E_\mathrm{a}}{\partial p_i}\right)\left(\frac{\overline{p}_i}{\overline{E}_\mathrm{a}}\right), \tag{4}$$

where E_a is the absorbed energy obtained from the FEA and p_i represents the material parameters in the welded regions normalized by those in the corresponding base aluminium alloys $f_\mathrm{N}/f_\mathrm{N0}$ etc. where the subscript 0 represents the parent aluminium alloy. In Eq. (4), $\overline{E}_\mathrm{a}$ and $\overline{p}_i$ are the average values of E_a and p_i, respectively. We approximated the relationship between p_i and E_a with a method of least squares to calculate the differentiation in Eq. (4).

The sensitivity indices obtained from Eq. (4) are listed in Table 3. As shown, the sensitivity of the energy absorption to f_N is the largest and the sensitivity to σ_Y is smaller than the sensitivity to the parameters relating to the fractures for STRUCTURE A, whereas that to σ_Y is the largest for STRUCTURE B. These results are related to the following phenomena occurring in the crashworthy structures: in STRUCTURE A, the fractures occurred at the welded regions and were measurable as indicated in Fig. 3, and in STRUCTURE B, the energy was mainly absorbed by the progressive plastic buckling of the energy-absorbing blocks and the fracture initiation was negligible (3, 4).

Table 3 List of sensitivity indices.

p_i	$\lvert\alpha_i\rvert$	
	STRUCTURE A	STRUCTURE B
$\sigma_\mathrm{Y}/\sigma_\mathrm{Y0}$	0.0301	**0.126**
$f_\mathrm{N}/f_\mathrm{N0}$	**0.116**	0.103
$\varepsilon_\mathrm{N}/\varepsilon_\mathrm{N0}$	0.0334	0.0175
$f_\mathrm{C}/f_\mathrm{C0}$	0.0601	0.0682
$f_\mathrm{F}/f_\mathrm{F0}$	0.0568	0.0594

From the above, we conclude that the sensitivity study based on FEA using the GTN model is effective for quantitatively evaluating the effect of fractures on the energy absorption of crashworthy structures.

5 CONCLUDING REMARKS

In this paper, the energy absorptions of two types of crashworthy structures were numerically evaluated using FEA. In the FEA, the GTN model was employed as a material model and several combinations of the material parameters included in this model were examined. Using the numerical results, sensitivity indices correlating the energy absorption to the material parameters were calculated. From the sensitivity study, the volume fraction of the particles causing the void nucleation was found to have the largest effect on the energy absorption for STRUCTURE A, whereas the yielding stress of the welded regions in the energy-absorbing blocks was the largest for STRUCTURE B.

REFERENCES

[1] Scholes, A. & Lewis, J. H. (1993) *Proceedings of the Institution of Mechanical Engineers, Part F: Journal of Rail and Rapid Transit* **207**, 1–16.

[2] Sutton, A. (2002) *Proceedings of the Institution of Mechanical Engineers, Part F: Journal of Rail and Rapid Transit* **216**, 97–108.

[3] Kawasaki, T., Yamaguchi, T. & Mochida, T. (2008) *Transactions of the Japan Society of Mechanical Engineers Series A* (in Japanese) **74**, 154–161.

[4] Kimura, S., Mochida, T., Kawasaki, T., Nakamura, H. & Yamaguchi, T. (2013) *Journal of Solid Mechanics and Materials Engineering*, **7**, 102–117.

[5] Tvergaard, V. & Needleman, A. (1984) *Acta Metallurgica* **32**, 157–169.

Research on the strength of air cushion vehicle

H. Zhang, H.L. Ren, N. Liu, S. Li
Harbin Engineering University, P.R. China

ABSTRACT

Due to its unmatchable advantages, the air cushion vehicle (ACV) was widely applied in the field of economy and military. So far, there is no specialized strength assessment of ship structures for the ACV. A reliable assessment method for structural strength is the key to ensure the ship safety. Therefore, it is necessary develop a suitable strength assessment method. One finite element model of a ACV is built in this paper. Then, the strength of ACV of the ship structure is assessed.

1 INTRODUCTION

The ACV suspends on the water or land by the pressure of air cushion. This special theory makes the ACV used rapid widely, such as the cargo transportation on marsh area and island landing campaign (1). First of all, the air cushion makes a big difference between the ACV and general ships on the structure and loading characteristic. In the next place, in order to improve efficiency, the weight of the hull is tend to be reduced as far as possible with the maximum rigidity. The last, there is no specific rules and research about the strength assessment of ACV for the moment. Therefore, it is necessary to estimate the structure strength reasonably for the using and development the ACV.

In the first place, according to the ACV wave loading test results, the loading features of the ACV are analyzed. Then the main loading state was determined and the reasonable design values of loading component in every loading state were calculated and shown in the paper. After that, using the relevant rules for reference, the strength of the sample ship was assessed by the finite element method. So far, it has formed a set of complete and reasonable process of strength assessment of the ACV.

2 TO ESTABLISH THE FINITE ELEMENT MODEL OF THE WHOLE SHIP

In order to estimate the ship's structural strength, a finite element model is necessary to get the structural stress response. In this study, the finite element model of the ACV was modeled by MSC.Patran that is reliable to simulate structures properly proved by massive case histories.

2.1 Mesh generation

Due to the asymmetry of the ACV, the full width finite element model is used. The plate units are used to simulate deck, side shell, buoyancy tank roof, airway deck, bulkhead and bottom. The stiffeners, longitudinal girders, cross girder of suspended framing and webs perpendicular to the panels of main supporting components are represented by beam elements (2).

2.2 The finite element model

The coordinate system of model adopted Cartesian coordinates and the origin located at the intersection of longitudinal section in the zero frame and baseline. The whole ship structure finite element model is shown in figure 1.

Fig.1 The finite element model of ACV

3 THE DETERMINATION OF LOAD

When navigating in the sea, ACV was supported off waves by high-pressure gas. While conventional ships are floating on the waves. Therefore, the ACV no longer suffers water pressures, instead of gas pressure. And a reasonable estimate on the effects of high-pressure is the key to ensure the strength assessment of the structure safely.

3.1 The goods pressure

According to relevant data, the pressure of goods was selected as following. There are five tracked vehicles on the deck of buoyancy tank, each of which full fighting weight is 13300 kg and the pressure is 51274.6 Pa. The ballast distribution is shown in figure 2.

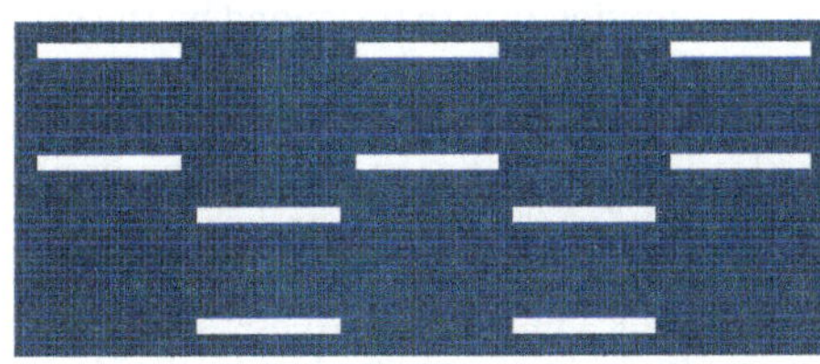

Fig.2 The good mass distribution

3.2 The static air cushion pressure

According to the theoretical calculation of the simplified model, the static air-cushion pressure is 4447.4Pa (the front air chamber) and 5437.78Pa (the back air chamber).

3.3 The wave impact pressure and dynamic air cushion pressure

When waves impact on the bow and stern, there will be sagging bending moment lead by larger hydrodynamic impact, and the moment will cause the most dangerous state for the hull [3]. The wave impact pressure is calculated according to the "Rules For Construction And Classification Of Sea-going High Speed Craft" [4].

When the wavelength is close to the length of the ship, the bow and stern are impacted by the waves at the same time, the wave trough is at the hull medium, wave crest is at the bow and stern, then the pitching could be regarded. Wave impact pressure at stern should be around 18000Pa. The wave loading test shows that the dynamic air cushion pressure of every gas chamber varies well-proportioned and synchronously. According to the test conclusion and d'Alembert principle, base on the rules calculation result, a group of reasonable dynamic pressure distribution was obtained.

3.4 The torsion loading

As the theoretical calculation is immature at this stage, the torsion loading method was referred to "Rules for the Classification of Trimarans" (5) and combined with the characteristics of hull structure. To take the torsion peak loading as an example. According to the rules, the torsion distribution was determined and shown in figure 3.

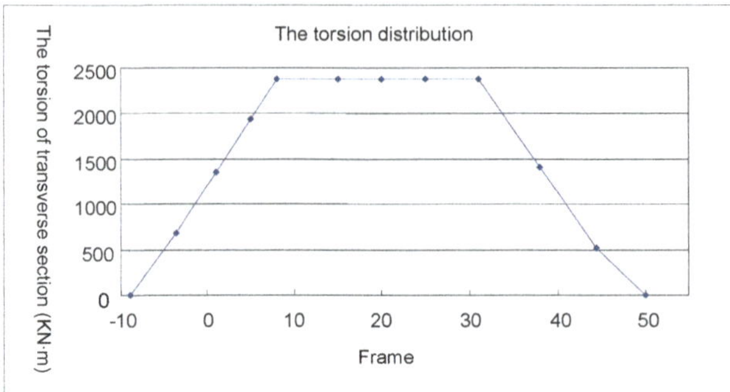

Fig.3 The torsion distribution

According to loading principle by the rules given, combining with the hull structural characteristic, transverse bulkheads or transverse strong frames were selected to load on. The torsion loaded on frame was the torsion difference between the adjacent two sections ($\Delta M_T = M_{T2} - M_{T1}$). The forces in the same numerical value but opposite direction were loaded on both sides of the transverse bulkheads or transverse strong frames. The valley of torsion was to be loaded by the same method.

3.5 The bending moment loading

The loading method of bending moment on the longitudinal section in center plane was also referred to the rules (5) and combined with the characteristics of hull structure.

The bending moment was loaded on both sides of the bottom of buoyancy tank (the lower edge of buoyancy tank side shell), and distributed uniform along the longitudinal line, make sure that the bending moment would reach the design target.

4 THE RESULT AND ANALYSIS OF STRESS CALCULATION

The calculation results are shown in figure 4-5:

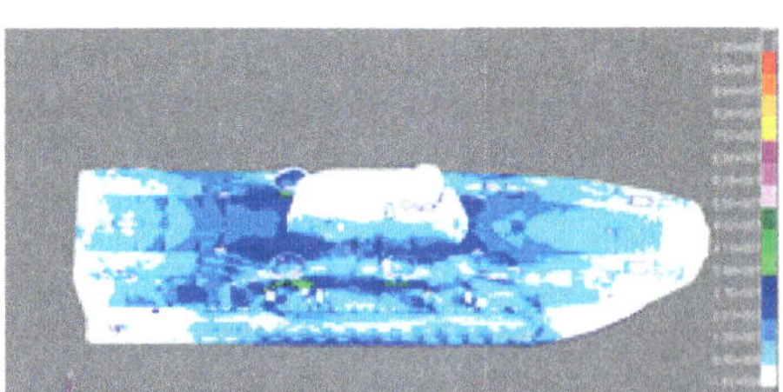

Fig.4 The stress cloud (hogging)

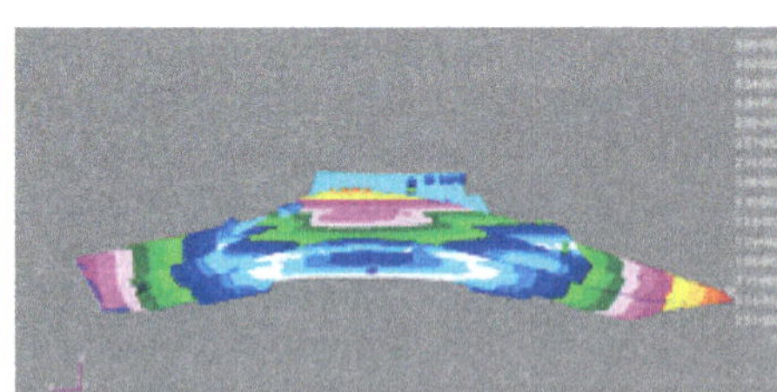

Fig.5 The deformation cloud (hogging)

In order to reduced weight, the aluminium-magnesium alloy is adopted on the main hull. According to the Rules For Construction And Classification Of Sea-going High Speed

Craft[3], the allowable stress is shown as follows: $[\sigma]=0.85\sigma_{sw}$, $[\sigma_x]=0.75\sigma_{sw}$, $[\tau]=0.36\sigma_{sw}$. The stress and check results are shown in table 1-2:

Table 1 The stress results in the case of hogging with ahead sea state

Components	σ	σ_x	τ	Evaluation
deck	54.9	53.7	41.0	Yes
buoyancy tank roof	20.9	-13.1	20.5	yes
airway deck	39.7	-38.3	-23.9	Yes
buoyancy tank bottom	63.4	-38.9	-27.6	Yes
side shell	103.0	59.4	-29.6	Yes
longitudinal bulkhead	56.1	-47.2	-23.6	Yes

Table 2 The stress results in the case of sagging with ahead sea state

Components	σ	σ_x	τ	Evaluation
deck	147.1	-140.7	56.7	Yes
buoyancy tank roof	59.7	34.6	32.2	Yes
airway deck	83.0	70.5	41.2	Yes
buoyancy tank bottom	111.9	102.1	-50.5	Yes
side shell	148.2	-151.2	60.2	Yes
longitudinal bulkhead	206.8	-181.2	-98.4	no

By analyzing the results of stress, it is found that the stress of longitudinal bulkhead exceeded the allowable stress in lifting sagging condition. Therefore, it is necessary to refine the unqualified units and around. According to the Appendix II of (3), the allowable stress should be 1.45 times of the original. The stress results are shown in table 3:

Table 3 The stress result after refined

Refined part	Unit size (mm*mm)	Allowable stress (MPa)	Actual stress (MPa)	Evaluation
longitudinal bulkhead	50*50	269.7	199.8	Yes

The refined model of longitudinal bulkhead is shown in figure 6 and the stress cloud of refined longitudinal bulkhead model is shown in figure 7:

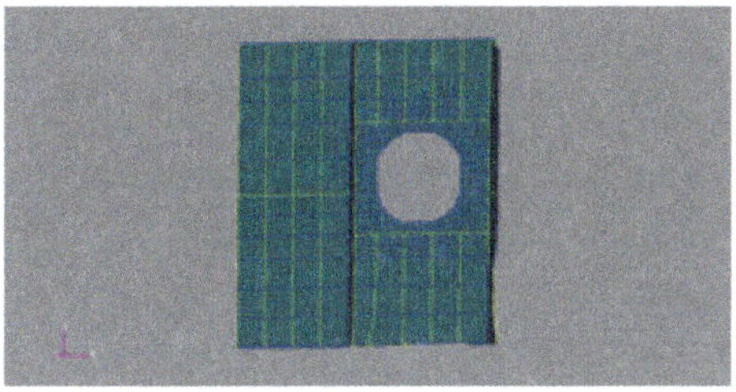

Fig.6 The refined model

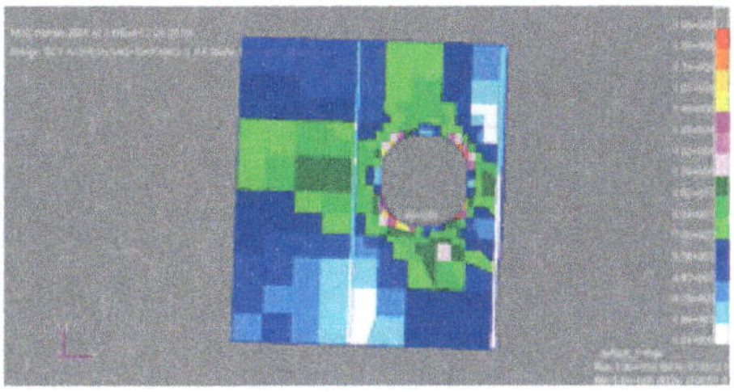

Fig.7 The stress cloud of refined model

It can be known by the tables and figures that the maximum stress was 103MPa in the case of hogging with head sea state, which match the allowable stress of the hull plate. However, the maximum stress was 207MPa in the case of hogging with head sea state, which appeared on the hatch corners of longitudinal bulkhead. Therefore, the refined model was calculated, and it satisfied the rules.

5 CONCLUSION

It can be seen through the above data analysis, the ACV meet the requirements of rules very well. The actual stress of the most structure is lower than the allowable stress relatively in the most conditions. It implied that, to a certain extent, the ship structure can be further optimized, so as to increase the economic benefit by improve the utilization rate of the material of the ship. In this paper, after assessing the strength of the sample ship, it has formed a set of complete and reasonable process of strength assessment of the ACV.

REFERENCES

(1) Zhou Y. H. Air Cushion Vehicle and its Military Purposes. Ship Electronic Engineering, 2008

(2) В.А.Постнов. Application of Finite Element Method in the Calculation of Ship Structure. Beijing: National Defense Industry Press, 1977

(3) CCS. Rules for Construction and Classification of Sea-going High Speed Craft. Beijing. 2005

(4) Yun L. ACV Principle and Design. Beijing: National Defense Industry Press, 1990

(5) Lloyd's Register of shipping. Rules for the Classification of Trimarans. 2011

A new super element for the use in failure assessment techniques utilising FAD concept

W.F. Qiao[1], **D. Xie**[1,2,3]

[1] School of Naval Architecture and Ocean Engineering, Huazhong University of Science and Technology, P.R. China

[2] Hubei Key Laboratory of Naval Architecture & Ocean Engineering Hydrodynamics, P.R. China

[3] Collaborative Innovation Center for Advanced Ship and Deep-Sea Exploration, P.R. China

ABSTRACT

In Failure Assessment Diagram based on failure evaluation techniques, it is important to determine normalised stress intensity and load ratios accurately. In this paper, a new super element is developed and implemented into commercial FEA software ABAQUS® with user defined element for determining normalised stress intensity and load ratios simultaneously with one step linear analysis. The capacity of proposed super element has been assessed by the classic example of cylinder with circumferential through crack under axial tension and bending. The results obtained from the super element are quite consistent with predictions based on analytical models.

1 INTRODUCTION

The concept of Failure assessment diagram (FAD) has been well accepted and has been implemented into failure assessment codes and standards such as BS 7910(1), CEGB R6 (2), API 579(3, 4) etc. At present, FAD has been widely used to perform failure assessment of engineering structures such as pressure vessels, pipes and joint etc (5-9).

A typical FAD is shown in Fig. 1. In this figure, the ordinate, fracture ratio K_r, which is ratio of linear elastic stress intensity factor K_I to the material fracture toughness K_{IC}, indicates the proximity to failure by brittle fracture while the abscissa, load ratio L_r, which is the ratio of reference stress σ_{ref} to the material yield strength σ_y, indicates the proximity to failure by plastic collapse. Therefore, determination of K_r and L_r is key issue to apply FAD. For simple component such as plate and cylinders, analytical formulae are available in regulations and standards above mentioned. For complex structures, finite element analysis (FEA) is

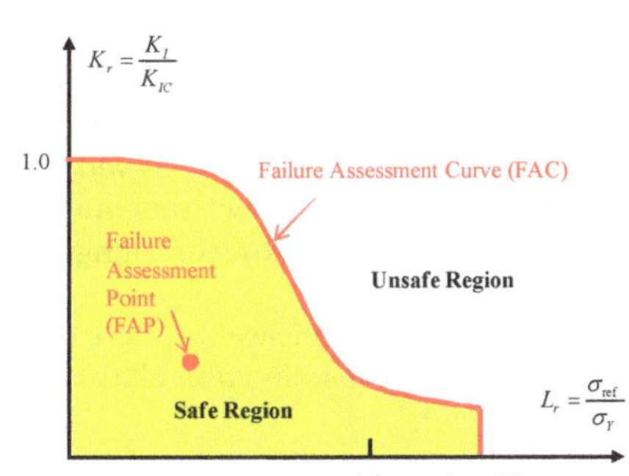

Fig. 1. Typical Failure Assessment Diagram.

applied to obtain fracture ratio with singular elements around crack tip and to obtain load ratio under nonlinear elastic-plastic analysis with crack. Obviously, this process requires much time and effort.

To meet this challenge, the super element has been developed to obtain the K_r and L_r simultaneously in frame of one step linear elastic analysis in this paper. Cylinders with circumferential through crack under axial tension and bending are chosen to be typical examples to assess the capacity of the super element proposed. The results obtained from super element and analytical solutions are quite consistent. Therefore, this super element is expected to be very useful tool to perform failure assessment of offshore structures.

2 METHODOLOGY

In order to determine coordinates (L_r, K_r) of FAP on FAD to assess the integrity, stress intensity factor and reference stress should be known.

2.1 Calculation of stress intensity factors

In this paper, the visual crack closure technique (VCCT) is used to calculate stress intensity factor. Xie and Biggers (10, 11) have introduced interface element with "dummy nodes" to accommodate VCCT into ABAQUS through its user subroutine UEL for 2-D plate element. For 3-D shell element, the computation of stress intensity factor can be shown in Fig. 2. As illustrated in Fig. 2, node 1 and node 2 are assumed to be located at the crack tip while node 3 and node 4 are the nodes next to the crack tip. Therefore, the local coordinate system ($\bar{X}, \bar{Y}, \bar{Z}$) could be easily established based the coordinates of those nodes in the global system (X, Y, Z).

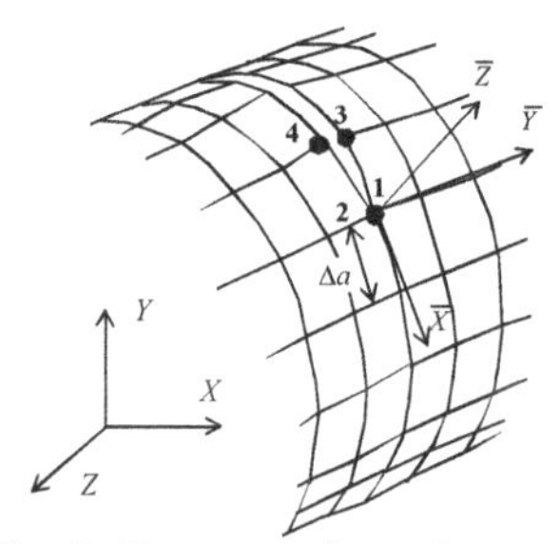

Fig. 2. Computation of stress intensity factors by VCCT.

For four nodes shell elements, the strain energy release rates could be computed based on VCCT in terms of local coordinate system by using following equations:

$$G_I = \frac{1}{2B\Delta a}\left(F_{\bar{y}1,2}\Delta u_{\bar{y}3,4} + M_{\bar{z}1,2}\Delta\theta_{\bar{z}3,4}\right) \tag{1}$$

Where, $F_{\bar{y}1,2}$ is the component of nodal force between node 1 and 2, while, $M_{\bar{z}1,2}$ are the components of nodal moment between node 1 and 2; $\Delta u_{\bar{y}3,4}$ is the component of opening displacement between node 3 and 4 in local coordinate system while $\Delta\theta_{\bar{z}3,4}$ is the component of turn angle between node 3 and 4. B is the thickness of shell element at the crack tip. Δa is virtual extension length of the crack.

Once, strain energy release rates are computed, it is pretty easy to obtain stress intensity factors by using the following relations:

$$K_I = \sqrt{\frac{EG_I}{1-\upsilon^2}} \tag{2}$$

2.2 Calculation of references stress

Through observing the equations to calculate the reference stress in BS7910, the definition of reference stress is the global property of the ligament section. Therefore, the equations which can be summed up in two main kinds:

For tensile load,

$$\sigma_{ref} = \frac{P}{\bar{A}} \tag{3}$$

For bending moment,

$$\sigma_{ref} = \frac{M}{\bar{I}}\bar{d} \tag{4}$$

where, P is the tensile load acting on the ligament section, M is the bending moment acting on the ligament section, $\bar{A}$ is the effective area of the ligament section, $\bar{I}$ is the effective inertia of moment of the ligament section, $\bar{d}$ is the specified distance to the neutral axis of ligament section for special structure.

2.3 Super Element

As a natural extension of interface element with "dummy nodes", super element has been developed to calculate stress intensity factors and the reference stresses simultaneously by one step linear elastic analysis. The typical super element is shown in Fig. 3.

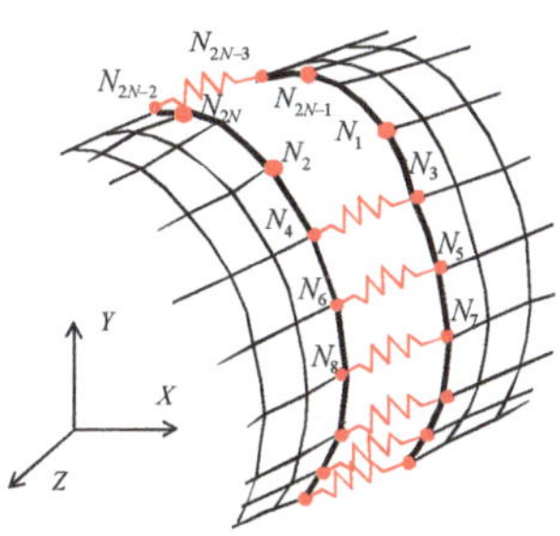

Fig. 3. Definition and node numbering of super element.

The node pair (N_3, N_4) and node pair (N_{2N-3}, N_{2N-2}) corresponding to each initial crack tip, respectively. In order to compute the nodal forces, very stiff spring is placed between node pairs excluding the first and the last node pairs. The first node pair (N_1, N_2) and the last node pair (N_{2N-1}, N_{2N}) next to each crack tip are included only for the purpose to extract the nodal displacements so as to compute strain energy release rates by VCCT. Once the nodal force and nodal displacements are obtained from FEA, stress intensity factors and the reference stresses could be computed simultaneously. When the fracture toughness and the yield strength of materials are known, the fracture ratio and the load ratio can be directly computed within the super element.

3 CASE STUDY

To verify and to validate the new approach, two examples were studied (1) cylinder with circumferential crack under bending, (2) cylinder with circumferential crack under axial loading. As shown in Fig. 4 a cylinder with circumferential crack is under bending (M) and axial loading (P) with principal dimensions of external radius (R_o), internal radius(R_i), mean radius (R), and the thickness of wall (B). θ is the half crack angle.

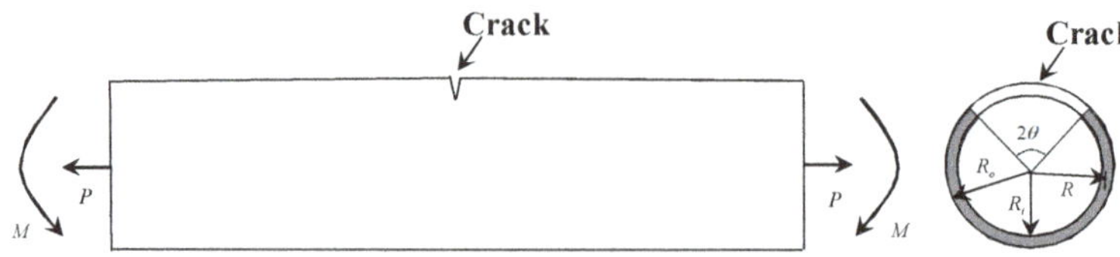

Fig. 4. Cylinder with a circumferential crack.

The analytical solutions for stress intensity factor and reference can be found in Ref. 1 and 12. Fig. 5 show the comparisons between the numerical results obtained from the super element to the analytical solutions for non-dimensional stress intensity factor and non-dimensional reference stress, respectively. The range of half crack angle is taken from $\pi/18$ to $\pi/4$ with the interval of $\pi/72$. The excellent agreement demonstrates the accuracy of the proposed super element to deal with such a problem with curved crack on cylindrical surface under different loading cases.

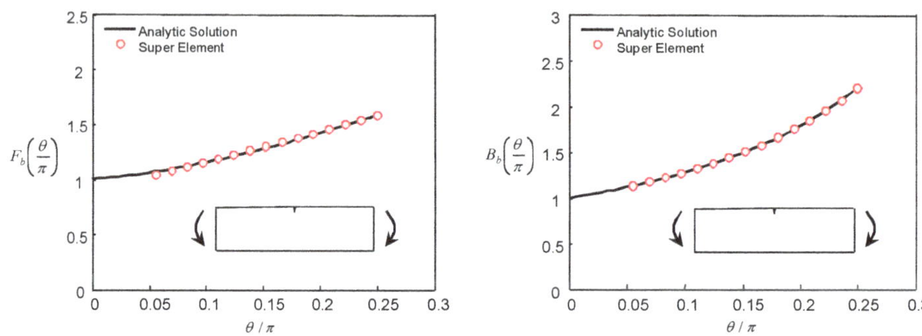

(a) pure bending

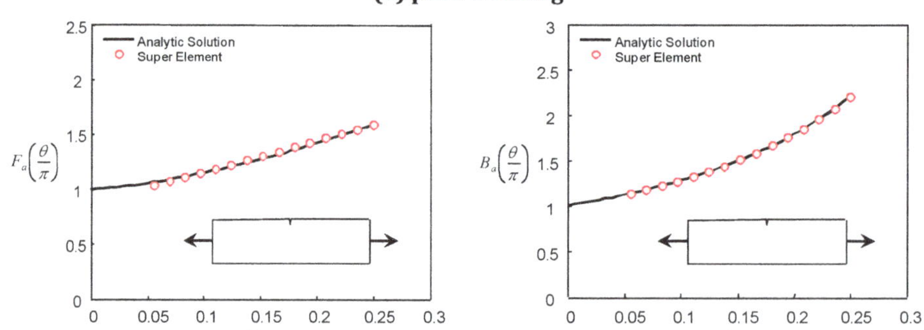

(b) pure axial loading

Fig. 5. Comparison of present results to analytical solution (cylinder with circumferential crack).

4 CONCLUSIONS

This paper proposed a supper element to compute stress intensity factors and reference stresses. With the proposed supper element, fracture ratio and load ratio can be obtained simultaneously with only one step elastic analysis. Two loading cases of cylinder with circumferential crack have been analyzed to demonstrate the accuracy and efficiency of the new formulation. Therefore, the super element developed in this paper is computational efficient and has a great potential to perform failure assessment of offshore structures based on FAD.

REFERENCES

(1) BS7910-Amendment 1. Guide to methods for assessing the acceptability of flaws in metallic structures. London, UK: British Standards Institution; 2005.

(2) R6. Assessment of the integrity of structures containing defects. Revision 4. Gloucester, UK: British Energy; 2001.

(3) API 579-1/ASME FFS-1. Fitness-for-service. New York, USA: American Society of Mechanical Engineers; 2007.

(4) Anderson TL and Osage DA, API 579: a comprehensive fitness-for-service guide, International Journal of Pressure Vessels and Piping, 2000(77): 953-963.

(5) Zerbst U, Hamann R, Beeck F, Scheider I, and Brock W, Engineering failure assessment methods applied to pressurised components, Nuclear Engineering and Design, 2000(200): 371-381.

(6) Kanno S, Hasegawa K, Shimizu T, Saitoh T and Gotoh N, Analysis of leak and break behavior in a failure assessment diagram for carbon steel pipes, Nuclear Engineering and Design, 1992(138): 251-258.

(7) Ainsworth RA, Budden PJ, Oh CY and Kim YJ, The treatment of secondary strains with a strain-based failure assessment diagram, International Journal of Pressure Vessels and Piping, 2013(104): 14-20.

(8) Flom Y, Evaluation of brazed joints using failure assessment diagram, 5th International Brazing and Soldering Conference, 2012.

(9) Iwashita T, Kurobane Y, Azuma K and Makino Y, Prediction of brittle fracture initiating at ends of CJP groove welded joints with defects: study into applicability of failure assessment diagram approach, Engineering Structures, 2003(25): 1815-1826.(10)

(10) Xie D, Biggers Jr SB. Progressive crack growth analysis using interface element based on the virtual crack closure technique. Finite Elements in Analysis and Design, 2006, 42: 977-984.

(11) Xie D, Biggers Jr SB. Strain energy release rate calculation for a moving delamination front of arbitrary shape based on virtual crack closure technique, part I: formulation and validation. Eng Fract Mech 2006;73:771-85.

(12) Zahoor A, Dunctile Fracture Handbook, Volume 1: Circumferential Throughwall cracks, EPRI Report NP-6301-D, Electric Power Research Institute, Palo Alto, CA, 1989.

Real-time signal processing of guided waves acquired on high-speed trains for health monitoring of bogie systems

M. Hong[1], Q. Wang[2], Z. Su[1,3], L.-M. Zhou[1]
[1] Department of Mechanical Engineering, The Hong Kong Polytechnic University, P.R. China
[2] College of Automation, Nanjing University of Posts and Telecommunications, P.R. China
[3] The Hong Kong Polytechnic University Shenzhen Research Institute, P.R. China

ABSTRACT

An in-situ structural integrity monitoring (SIM) technique using guided waves has been developed and deployed through an online diagnosis system, which was recently implemented on a new high-speed train operated between Beijing and Shanghai. The system's hardware and software features are briefly introduced. Focus is placed on real-time signal processing of waves acquired under various working conditions of the train, in order to gain practical insights of continuous SIM that may not be realized in laboratory. Mock-up damage affixed to the bogie was identified and visualized, demonstrating the feasibility and effectiveness of the technique and the system toward field applications.

1 INTRODUCTION

The rapid development in high-speed train industry in the last 20 years has significantly improved the efficiency of public transportation. Yet, with increased operating speed, more intensive use, and more complicated structures, high-speed trains also raise safety concerns and sometimes leave painful lessons to engineers and the public. Defects and damage presented in wheel sets, axles, and bogies may lead to disastrous consequences. To avoid such accidents, nondestructive testing (NDT) techniques have been widely employed, such as rail wheel inspection (1) and hollowed axle testing (2). However, NDT methods usually require interruptions of the normal operations of inspected structures, costing both time and money to a large extent. In addition, they are not condition-based in nature, rendering discontinuity in the inspection process.

Structural integrity monitoring (SIM), in contrast, is a condition-based method aimed at leveling up operating safety of structures while reducing their maintenance costs by achieving online, real-time inspection. An SIM system typically consists of a sensor network for data acquisition and a control unit for health evaluation and knowledge storage. Among a variety of SIM methods, guided waves (GW) have been considered one of the most effective and promising tools, due to their wave-guiding characteristics and sensitivity to damage. However, most investigations now are developed on simple structures in well-controlled laboratory environments. Real-world GW implementations, nevertheless, usually involve complicated structural geometries and ambient conditions.

Therefore, this paper aims to examine GW performance in real-world, continuous SIM. The developed diagnosis system was experimented on the bogie frame of a running high-speed train between Beijing and Shanghai, as reported in (3). While system features and experiment procedures have been accounted in detail there, this paper will briefly reiterate some important aspects for completeness, and will focus on some practical issues in signal acquisition and processing, including GW excitation, impacts of working conditions of the train, signal processing, and mock-up damage diagnosis and imaging.

2 GUIDED WAVE-BASED SIM

Guided waves have a wide variety of superb properties, including fast propagation, strong penetration, omnidirectional dissemination, and most importantly, high sensitivity to damage small or inaccessible in structures. On the other hand, GWs are multimodal at any frequency. For example, for Lamb waves (a type of GWs in thin plates), there are symmetric modes S_0, S_1, S_2, *etc.*, and antisymmetric modes A_0, A_1, A_2, *etc.* The dispersive nature of GWs also leads to different propagation velocities of different modes. These characteristics may create difficulties in extracting desired wave modes for damage detection, especially when the inspected structure varies in thickness and/or shape.

Even so, GWs have been well utilized for structural damage detection and integrity monitoring (3, 4). As a typical GW application, the basic principle of active Lamb wave-based damage detection is shown in Figure 1, where the inspected structure is a panel of a train bogie frame. First, a sensor network is installed for wave actuation and acquisition. Given the multimodal nature of Lamb waves, narrowband signals are usually preferred to excite a single dominant mode (to the extent possible) to simplify structural responses. Then, the structural condition can be evaluated by comparing current responses with stored benchmarks, exploiting various phenomena like wave transmission/reflection, energy intensity, signal nonlinearity, *etc.* In this process, many advanced signal processing techniques and tools, such as the theory of time reversal and phased arrays (5, 6), can be adopted and refined to quantitatively evaluate damage location, size, and severity.

3 ONLINE DAMAGE DETECTION SYSTEM

To realize the guided wave technique, an active sensor network and an integrated online damage diagnosis system have been developed, as detailed in (3). This section recapitulates the important aspects of the system.

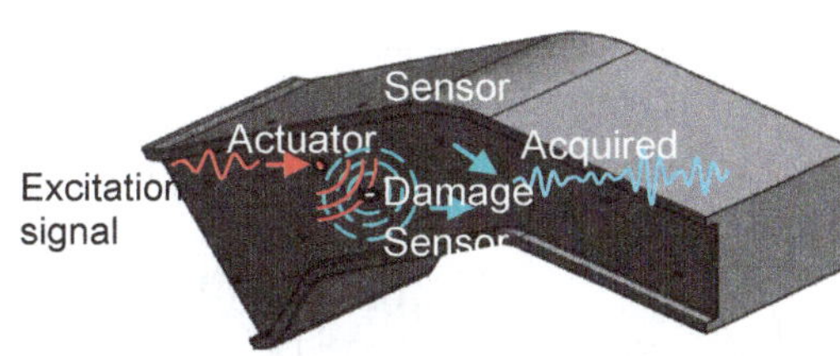

Figure 1. Basic principle of active Lamb wave-based damage detection.

3.1 Sensor network

Lead zirconate titanate (PZT) elements have been widely used as actuators and sensors in GW-based SIM because of their easy manipulation, low cost, and lightweight feature. However, PZT elements are also fragile; their connecting wires can be disordered and unstable, or soldered onto each element inconsistently. Thus, in order to address issues like sensor packaging and performance consistence, the flexible printed circuit (FPC) technique is adopted. A polyimide film is used to embed each PZT element for protection, which is connected to a printed circuit (also embedded) to provide uniform electric connection between individual PZT elements and the system, improving performance consistence of sensors. Meanwhile, time division multiplexing is employed to control the network, so that any two of the sensors can be simultaneously selected (by a switch array) to act as the actuator and sensor respectively, setting up one monitoring path.

3.2 Hardware and software

The damage diagnosis system resides on the virtual instrument technique and the PXI platform (7). The hardware consists of four basic components: the switch controller of active sensor networks, wave generation, multi-channel data acquisition, and central control and signal processing. These four parts are integrated through one PXI Bus, and controlled by the in-house software. Commands to the hardware components are issued by the program planted into the central control unit, which fulfills all the major functions for real-time diagnosis, including management of hardware, man-machine interface (MMI), signal processing, damage detection, and presentation of diagnostic results.

4 REAL-TIME SIGNAL PROCESSING

When the sensor network was installed on the bogie of a running train, several factors may come into play and affect the quality of acquired GW signals, including the choice of excitation frequency and frequency tuning, impacts of different working conditions and environmental noise, and temperature variation through continuous monitoring.

4.1 Excitation frequency

Consider the bogie frame's side panel as a simple plate, a desired Lamb wave mode can be excited and collected using frequency tuning (6). First, the spectrum of a pure noise signal acquired by an arbitrary sensor attached to the panel (without actively generating GWs) was obtained as the train was running. As can be seen in Figure 2(a), a large amount of energy is present in the very low frequency band up to 200 kHz, which is attributed to train vibration. Above 800 kHz, on the other hand, random environmental noise becomes dominant. Thus, a preliminary working frequency band of 200–800 kHz was selected. Then, by comparing the responses at various frequencies within this band, it was found that at 300 kHz and 400 kHz the responses had reasonably large amplitudes and well separated wave modes, as shown in Figure 2(b). Therefore, later tests used these two candidate frequencies for GW excitation.

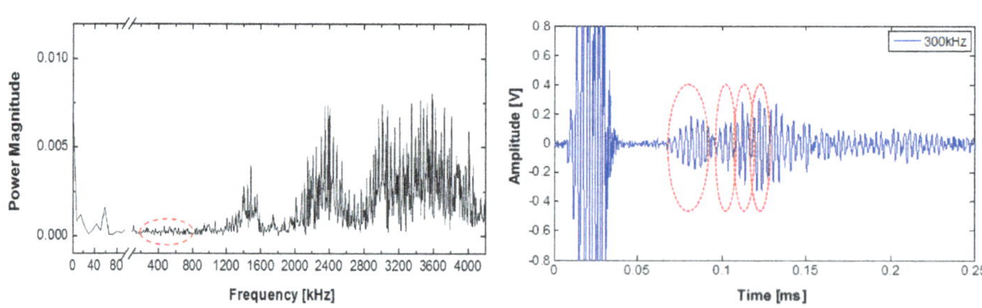

(a) (b)

Figure 2. (a) Spectrum of a pure noise signal with a working frequency band from 200 to 800 kHz (red dashed circle); (b) A typical response at 300 kHz with well-separated wave modes (red dashed circles).

4.2 Various working conditions and noise

Train structures usually bear various loads, both static and dynamic, and their working conditions may change rapidly. At the same time, ambient noises in long-distance operation can also be significant. These facts would bring possible disturbances to the monitoring system and acquired signals. In this study, structural responses under several working conditions (*e.g.*, startup, acceleration, full-speed operation, track change, emergency brake) were recorded and compared to evaluate their impacts. Typical signals are shown in Figure 3(a), and it is found that due to the carefully selected excitation frequency, the impacts from various train events are negligible. After averaging and a low pass digital filter, the signal-to-noise ratio of these signals are further improved, as shown in Figure 3(b). It can thus be concluded that the impacts of different working conditions and environmental noise on the system is negligible if the excitation frequency is carefully chosen and appropriate signal processing is performed.

4.3 Temperature variation and baseline compensation

Temperature variation in continuous measurement can create discrepancies between a current signal Y from its corresponding baseline signal X even when there is no damage in the bogie, manifested by time shifting and/or different amplitude scales, which leads to outdated benchmarking. This is because X is pre-acquired under specific environmental conditions, which may vary significantly from the current conditions when Y is recorded. In order to compensate for continuous changes in temperature, the compensated baseline X_{comp} can be defined as:

$$X_{comp} = a_X X_\tau \tag{1}$$

where $\tau = \tau\big|_{\max\left(\mathbf{E}\left[(Y-\mu_Y)(X_\tau-\mu_X)\right]\right)}$ is the time shift at which the cross correlation of X and Y reaches its maximum, $\mathbf{E}$ signifies the expected value, μ is the mean value, and X_τ hence denotes the lagged baseline signal. $a_X = \mathbf{E}\left[\left(Y-\mu_Y\right)\left(X_\tau-\mu_{X_\tau}\right)\right] \Big/ \mathbf{E}\left[\left(X-\mu_X\right)^2\right]$ here is a scaling ratio between the magnitudes of Y and X_τ, which is also utterly attributed to ambient effects, so that after such an adjustment, the rescaled, time-shifted baseline signal X_{comp} eliminates discrepancies created by the changing temperature and is paired to its corresponding current signal Y as if it were measured under the current condition.

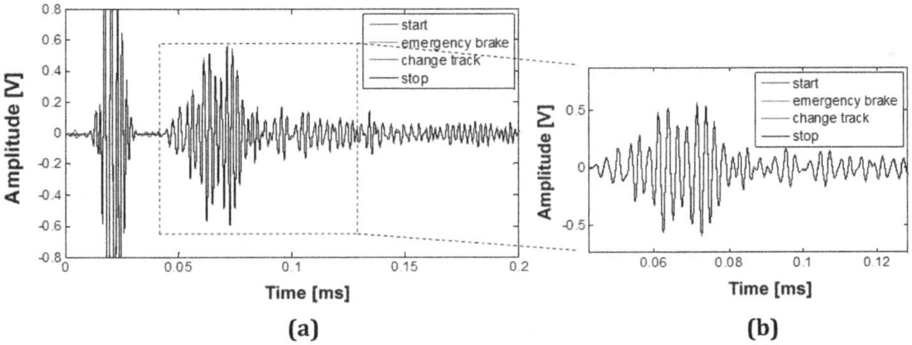

Figure 3. Structural responses under various working conditions: (a) before low-pass filtering; and (b) after filtering, showing a higher signal-to-noise ratio.

5 DAMAGE DETECTION RESULTS

Using the above mentioned techniques and tools for signal processing, continuous monitoring was able to be carried out, with a sensor network layout displayed in Figure 4(a). The mock-up damage attached to an arbitrary spot covered by the sensor network on the side panel of the bogie frame (during one suspension window of the train operation) was successfully identified and visualized as shown in Figure 4(b), using the in-house software in which a variety of detection algorithms were preprogrammed.

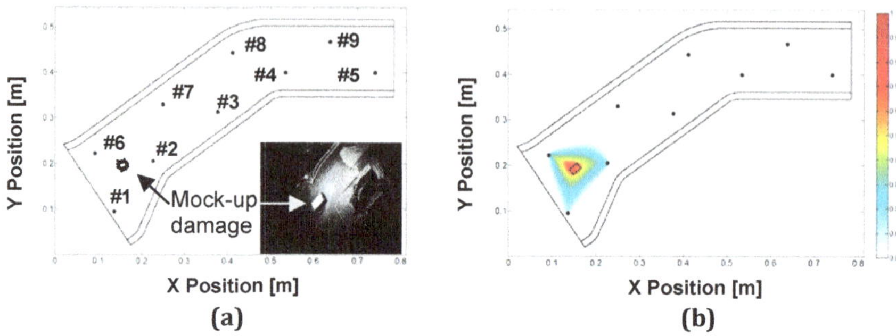

Figure 4. (a) Layout of the sensor network and artificial damage on the side panel of the bogie frame; and (b) real-time damage diagnostic image.

6 CONCLUSIONS

An SIM technique using active GWs is validated on a running high-speed train by monitoring its bogie frames in real time. Features of the diagnosis system are recapitulated. Some important aspects of the processing of wave signals acquired in real-world working conditions are described, and feasibility and effectiveness of GW-based SIM in field applications are demonstrated by producing real-time diagnostic images that accurately identified the mock-up damage on the bogie.

REFERENCES

(1) Salzburger, H. J., Wang, L. & Gao, X. In-motion ultrasonic testing of the tread of high-speed railway wheels using the inspection system AUROPA III. In: 17th World Conference on Nondestructive Testing, 25–28 Oct 2008, Shanghai, China.

(2) Rockstroh, B., Kappes, W., Walte, F. et al. Ultrasonic and eddy-current inspection of rail wheels and wheel set axles. In: 17th World Conference on Nondestructive Testing, 25–28 Oct 2008, Shanghai, China.

(3) Hong, M., Wang, Q., Su, Z. & Cheng, L. (2014) *In situ* health monitoring for bogie systems of CRH380 train on Beijing–Shanghai high-speed railway. Mech. Syst. Signal Process. 45, 378-395.

(4) Raghavan, A. & Cesnik, C. E. S. (2007) Review of guided-wave structural health monitoring. Shock Vib. Digest 39, 91-114.

(5) Wang Q. & Yuan, S. (2009) Baseline-free imaging method based on new PZT sensor arrangements. J. Intell. Mater. Syst. Struct. 20, 1663-1673.

(6) Yu, L. & Giurgiutiu, V. (2008) In situ 2-D piezoelectric wafer active sensors arrays for guided wave damage detection. Ultrasonics 48, 117-134.

(7) Zhou, C., Su. Z. & Cheng, L. (2011) Probability-based diagnostic imaging using hybrid features extracted from ultrasonic Lamb wave signals. Smart Mater. Struct. 20, 125005.

Effects of mistuning patterns on forced response for an integrally bladed disk

G. Chen, J. Hou
Aerospace Division, Defence Science and Technology Organisation, Australia

ABSTRACT

Under the same amount of mistuning strength, different mistuning patterns may have different effects on the forced response amplification of the blades in a blisk. In this research, sensitivity analyses on the effects of different mistuning patterns on forced responses for a military blisk were carried out. Additionally, only the modes with natural frequencies close to the modes of interest were considered, therefore, a relatively shorter computational time is required and also a much smaller model size than a full blisk model is used. The effects of different mistuning patterns on forced responses were examined with the worst case identified.

1 INTRODUCTION

High Cycle Fatigue (HCF) failures have been increasingly considered as one of the major failure mechanisms for conventional aircraft engines, having a significant negative impact on safety, cost of ownership and engine availability. Small variations among blade sectors, termed as mistuning, usually increase the maximum blade amplitude (1) and may detrimentally affect blade HCF lives. The engine blades for modern aircraft, like the JSF, are manufactured as a one-piece integrally bladed disk – termed blisk. As more blisks are used in modern aircraft engines like JSF, HCF related issues are becoming more dominant regarding airworthiness and safety. Therefore, it is essential to build up a capability to predict the forced response of a blisk due to mistuning in an accurate and efficient manner (2-3).

In recent years, there have been two main streams of methods developed to produce Reduced-Order Models (ROM) (2-10) for mistuning analyses: Component Model Synthesis (CMS) Methods and System-Model-Based methods. The CMS approach calculates the modes of each sub-component separately, and then the system model is synthesized by enforcing compatibility conditions at the interfaces between sub-components (4-5). The System-Model-Based method was firstly introduced by Yang and Griffin in 2001 (6) based on their earlier observations. In this method, a new ROM technique was defined as the Subset of Nominal Modes (SNM) method, and the ROM was constructed by selecting a subset of tuned modes. Compared to the CMS methods, a key advantage of the SNM is that there is no need to use sub-component modes to introduce the mistuning to the system. Thus, with only fraction of computational costs, it retains high accuracy relative to the parent FE model.

It is well known that the different mistuning strength may have different effects on the blade response (7). Also it has been reported that even for the same amount of mistuning strength, different mistuning patterns may affect the forced response amplification of the blades in a blisk in a different manner (8, 9). Under even a very small mistuning strength, a particular mistuning pattern may possibly have a very significant effect on blade response (10) and hence it is important to understand the effects of different mistuning patterns on the forced response. In this study, sensitivity analyses of the forced response of a blisk to different mistuning patterns, i.e. same amount of mistuning implemented on different blades were carried out using the SNM method. Because the mode shapes of the mistuned blisk were calculated using a weighted sum of the nominal tuned modes, and only the modes with natural frequencies close to the modes of interest were considered, a relatively shorter computational time, with smaller model size, was used than those for a full blisk model. The various patterns of mistuning strength were examined and the worst case was identified for the studied blisk.

2 THEORY FOR THE MISTUNING ANALYSIS

For a harmonic excitation, in implementing the SNM method (6), the steady-state response for a mistuned blisk assembly can be written as in a modal domain:

$$\{[\omega_r^{\ 2}]+[\Delta\bar{K}]+i\omega[\bar{C}]-\omega^2([I]+[\Delta\bar{M}])\}\{\eta\}=[\phi]^T\{F\} \tag{1}$$

In Equation (1), $[\Delta\bar{K}]$ and $[\Delta\bar{M}]$ are the modification matrices of stiffness and mass for the mistuned system, $[\bar{C}]$ is the viscous damping matrix and ω_r is the natural frequency. $\{\eta\}$ is the modal displacement vector, $\{F\}$ is the vector of external excitation force, ω is the excitation frequency, and i is the unit imaginary number. If only the elastic modulus for an individual sector is changed by a portion from the tuned sector, the modification matrix for i^{th} sector in modal domain can be written as:

$$[\Delta\bar{K}]=\sum_{k=1}^{N}\sum_{i=1}^{N}(1+\delta_i)\left[E_{k,i}\right]^*[\phi]_{i-1}^*\begin{bmatrix}k_s & 0\\ 0 & 0\end{bmatrix}[\phi]_{i-1}\left[E_{i,j}\right] \tag{2}$$

In this equation, $[\phi]_i$ is the mass normalised eigenvector matrix for the *i-th* nodal diameter modes and * is the Hermitian conjugate. Using Equation (2), $[\Delta\bar{K}]$ can be calculated. Therefore, if only the change of stiffness matrix is considered, the eigenvalue equation (1) for the mistuned system becomes:

$$\{[\omega_r^{\ 2}]+[\Delta\bar{K}]-\omega^2[I]\}\{\eta\}=\{0\} \tag{3}$$

Solving Equation (3) gives the natural frequencies and mode shapes of the mistuned blisk. The mode shapes of the mistuned blisk are calculated using a weighted sum of the nominal tuned modes. Furthermore, only the modes with natural frequencies close to the modes of interest are considered. Therefore, it uses relatively much shorter computational time and smaller model size than those for using a full blisk model.

3 MISTUNING ANALYSIS

A full cyclic symmetric FE model with 20 identical sectors for a military engine was created in ANSYS (11), shown in Fig. 1, based on a sector model with1284 elements and 2294 nodes. A cyclic dynamic analysis was performed using the FE model and the stiffness matrix and modal shape results were produced for the tuned blisk. The responses for the tuned blisk were calculated under the first engine order excitation. The calculated results are displayed in Fig. 2.

Figure 1. FE model for the blisk

After the modal analysis for the tuned blisk was carried out, these modal results and model information data were transferred into a mistuning analysis system developed in MATLAB (12). The mistuning analysis system casts the data into the modal domain and adds the information of the mistuning pattern. It then solved an algebraic eigenvalue problem for the mistuned system. For the mistuning analysis, the mistuning patterns were assigned deterministically. A file containing the frequency ratios for each sector of the blisk was used as an input for the mistuning analysis. The frequency ratio is defined as the ratio of the frequency of the mistuned blisk sector to the frequency of the tuned blade sector. For the first mistuned pattern, only one sector is mistuned, with a frequency ratio of 0.90 and all other sectors' frequency ratios being 1. The calculated data under the first forward (1F) engine order excitation are displayed in Fig. 3. The analysis results show that the sectors in the mistuned assembly do not experience the same resonant amplitudes at the same frequency and a small amount of random mistuning significantly changes the response amplitudes. Under the studied condition, the results show that the maximum amplitude for the mistuned case is 19% higher than the one for the tuned system.

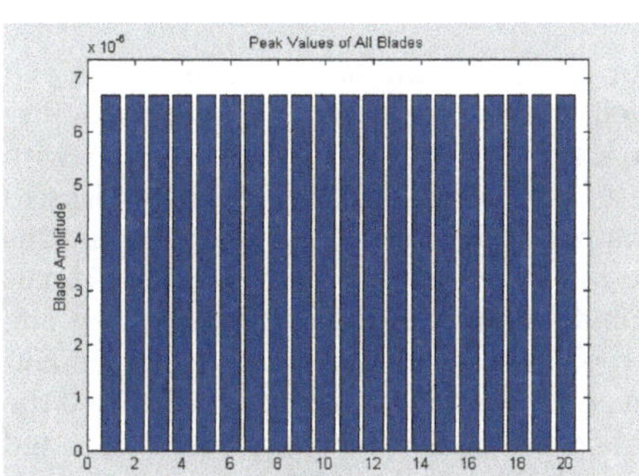

Figure 2. The tuned case

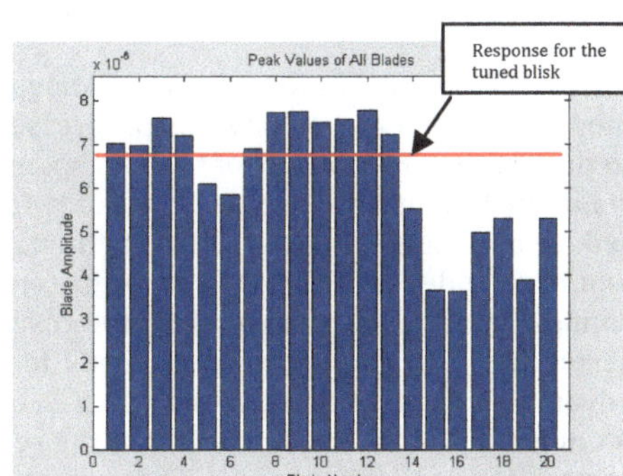

Figure 3. The mistuned case

4 MISTUNING PATTERN ANALYSIS AND THE DISCUSSION

Besides the above mentioned mistuning case, three other mistuning patterns were studied under the 1F, first backward (1B) and second forward (2F) engine order excitations. For these three patterns, the frequency ratio for the mistuned sectors is 0.90 (denoting 10% mistuning strength) with all other sectors' frequency ratios being 1. For Pattern 1, only first sector is assumed mistuned with 10% mistuning strength. With the same mistuning strength, Patterns 2 has two mistuned sectors (Sectors 1 and 11, which was the best case selected from to a sensitivity study for only two mistuned blades), Patten 3 has four mistuned sectors (Sectors 1, 5, 11, and 15) and Pattern 4 has 10 mistuned sectors (Sectors 1, 3, 5, 7, 9, 11, 13, 15, 17, 19). The response results for all the excitations were also obtained and depicted in Figure 4. The analysis results show that the introduction of a mistuned sector changes both the frequencies and the responses of each individual blade; the maximum blade amplitude changes with different mistuning patterns. Pattern 1 with a single mistuned sector represents the worst effect for all the studied cases. There is a big difference between Pattern 1 and Pattern 2, where two particular sectors were mistuned. However, there are no significant differences among Patterns 2-4. This means that the blisk is not sensitive for the mistuning in other sectors with the same mistuning strength if the mistuning is introduced in Sectors 1 and 11.

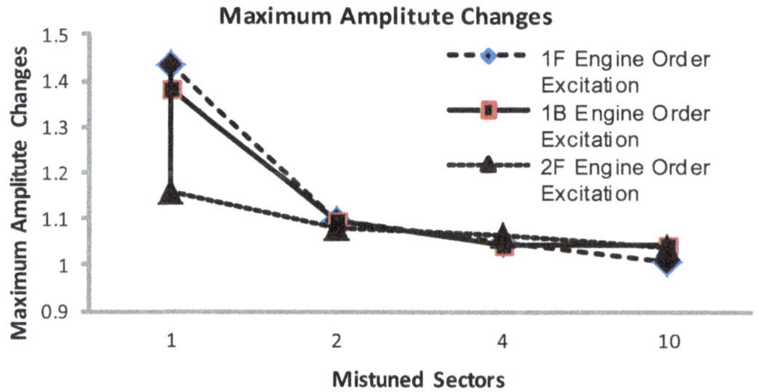

Figure 4. The maximum blade amplitude change vs. mistuned patterns for various excitations

The development of modern methodologies for predicting the natural frequencies and mode shape of a tuned blisk has progressed well. What is still lacking now is the ability to predict the insensitivity of mode response to a specific design or a repaired blisk with some mistuning patterns, and the knowledge of how to design or to repair blisks to reduce response sensitivities to mistuning for a particular disk. The outcome from this work provides an insight on the relative sensitivities of mistuning to repair configurations. For example, from the results obtained in this study for this blisk, if one sector (e.g. Sector 1) in a blisk needs to be repaired, the repair of this sector might adversely introduce a mistuning effect. Such an effect may be significantly reduced if the modification with the same effect of the repair is conducted at another properly located sector, in this case Sector 11. The outcome from this work also offers a useful tool for accurate predictions of the dynamic response required for a realistic fatigue life prediction. This type of prediction is essential for the provision of engine structural integrity support to the ADF, particularly for modern engines with blisks, i.e. F35.

5 CONCLUSIONS

The mistuning analysis of a blisk has been studied using the Subset of Nominal Modes (SNM) method in order to understand the effect of different mistuning patterns on the blisk dynamic response. A perfectly tuned system has been used to create the baseline FE model. Based on the analysis results of mistuning patterns studied, it is observed that the introduction of a mistuned sector changes both the frequencies and the responses of each individual blade. The mistuning pattern with one single mistuned sector represents the worst case; and the blisk is not sensitive for the mistuning in other sectors with the same mistuning strength if the same mistuning is introduced in two particular sectors (Sectors 1 and 11) for the studied case.

Most importantly, as the amplitude magnification exhibits a peak with respect to different mistuning patterns for the blisk, this result provides two important insights. Firstly to design engineers, it presents an idea of implementing intentional mistuning at some locations to keep the nominal design less sensitive to the mistuning. Secondly to maintenance engineers, it can be used as general guidance for blending repairs to avoid the sensitive mistuning patterns and to increase the safety margin during maintenance of engine components. The future work will include further evaluations of the sensitivity of mistuning patterns with different mistuning strengths and the correlations between the predictions and laboratory tests when results are available.

REFERENCES

[1] Srinivasn, H. H, (1997), Flutter and Resonant Vibration Characteristic of Engine Blades, *ASME J. Eng. Gas Turbine Power*, **119**, pp. 742-775.

[2] Chen, G., & Hou, J. (2014) Probabilistic Vibration Analysis of a Mistuned Integrally Bladed Disk, *Advanced Material Research,* **891-892**, p762-731.

[3] Chen, G., & Hou, J. (2013) *T700 1st Stage Blisk Mistuning Analysis Using a Deterministic Approach*, DSTO-TR-2903.

[4] Irretier, H., (1983) Spectral Analysis of Mistuned Bladed Disk Assemblies by Component Mode Synthesis, *Vibrations of Bladed Disk Assemblies*, ASME, New York, pp. 115–125.

[5] Zheng, Z.-C., & Wang, F.-R., (1985), Dynamic Analysis of Blade Groups Using Component Mode Synthesis, *Vibrations of Blades and Bladed Disk Assemblies*, ASME, New York, pp. 97–103

[6] Yang, M.-T., & Griffin, J. H., (2001) A Reduced-Order Model of Mistuning Using a Subset of Nominal System Modes, *Journal of Engineering for Gas Turbine and Power*, **123**, No. 4, pp 893-900.

[7] Whitehead, D. S., (1998) The Maximum Factor by Which Forced Vibration of Blades Can Increase Due to Mistuning, *Journal of Engineering for Gas Turbines and Power*, **120**, No.1, pp. 115-119.

[8] Castainer, M. P. & Pierre, C., (1998) *Investigation of the Combined Effects of Intentional and Random Mistuning on the Forced Response of Bladed Disks*, AIAA-98-3720.

[9] Hou, J. & Cross, C., (2005) Minimising Blade Dynamic Response Through Design Optimisation, *Journal of AIAA*, **43**, No.2, pp 406-412.

[10] Chen, G. & Hou, J, (2013) Analysis of a Mistuning Pattern for an Integrally Bladed Disk, *Proceedings of Australian International Aerospace Congress 15*, 2013.

[11] ANSYS Inc. (2010) *ANSYS V.13 Mechanical User Guide*. Pittsburgh, U.S.A.

[12] Mathworks Inc. (2000) *MATLAB 6 Getting Started Guide*. Massachusetts, U.S.A.

Dynamic buckling tests of cylindrical tubes with and without pellets

N. Morishige[1,2], K. Takahashi[1], K. Inaba[1], K. Kishimoto[1]
[1] Department of Mechanical Sciences and Engineering,
Tokyo Institute of Technology, Japan
[2] Nuclear Fuel and Core Research & Development Department,
Nuclear Development Corporation, Japan

ABSTRACT

In this study, the dynamic buckling tests using cylindrical tubes with and without internal pellets were carried out to investigate the impact behaviour. Various materials of cylindrical tubes and pellets were used to examine the buckling mode and the maximum loads at several impact velocities. The tubes with low stiffness pellets behaved similarly to the empty tubes, but high stiffness pellets caused different behaviours of tubes. At low impact velocity buckling deformations were suppressed by the strengthening effect of the pellets, while at high impact velocity a fracture occurred due to the constraint effect of pellets on plastic deformations.

1 INTRODUCTION

Cylindrical tubes are widely used as transport containers to protect internal products. The containers are required to prevent from collapsing even when they would fall to the ground and be impacted. It is known that buckling modes of a cylindrical tube depend on slenderness (length/diameter) ratio, wall thickness/diameter ratio and impact velocity. Impact velocity also influences impact loads [(1)-(4)]. In addition, fillers such as honeycomb or form change buckling modes and increase impact loads [(2), (5)]. These studies focused on low slenderness ratio and collapse behaviour because these tubes were designed for impact energy absorbers. However, few studies have been reported on tubes with high slenderness ratio regarding the effect of fillers on buckling modes and the maximum load. In this study, the dynamic buckling tests using slender cylindrical tubes with internal pellets were carried out to investigate the impact behaviour. Various materials of cylindrical tubes and pellets were used to examine the buckling mode and the maximum load.

2 EXPERIMENTS

Impact tests of cylindrical tubes with and without pellets in the axial direction were conducted using a gas gun. A test apparatus is consisted of a specimen holder including a load cell transducer, an accelerating tube including trigger sensors for measuring the velocity of a weight, and an air chamber, as shown in Fig. 1. Axial load was applied to specimens by impact of the weight (3 kg) which was accelerated by compressed air in the air chamber. Load histories were measured by the load cell transducer attached at the opposite side of impact. Since the natural frequency of the load cell transducer was about 5 kHz, loads were evaluated by 3 kHz low-pass filtering. Deformation histories of specimens were recorded at 5000 frame per second by two high-speed cameras controlled by a trigger. A 16 mm white circle with a 6 mm black dot was used as a marker. Markers were attached on specimens along axial direction at 20 mm spacing between centers, and their three-dimensional coordinates were calculated by image analyses. This apparatus can synchronize loads, deformation histories, and coordinates of markers by the trigger signal.

Cylindrical tubes (length: 485 mm, outer diameter: 9.5 mm, inner diameter: 8.4 mm) were made of zirconium alloy and SUS304. Young's modulus and yield strength are 97 GPa and 590 MPa for zirconium alloy [(6)], and 193 GPa and 360 MPa for SUS304 from literature [(7)] and mill seat. Two types of pellets (length: 11 mm, diameter: 8.2 mm) were prepared as filler. One was lead as low Young's modulus material and the other was aluminium oxide as high Young's modulus material, compared to property of the tubes. The pellets were inserted in the tube, and both ends of the tube were plugged by the Swagelok. Strain gages were attached on specimens to measure longitudinal strain histories. Tests were carried out at room temperature and pressure. Impact velocities of the weight were set to be 6, 10, 16 m/s.

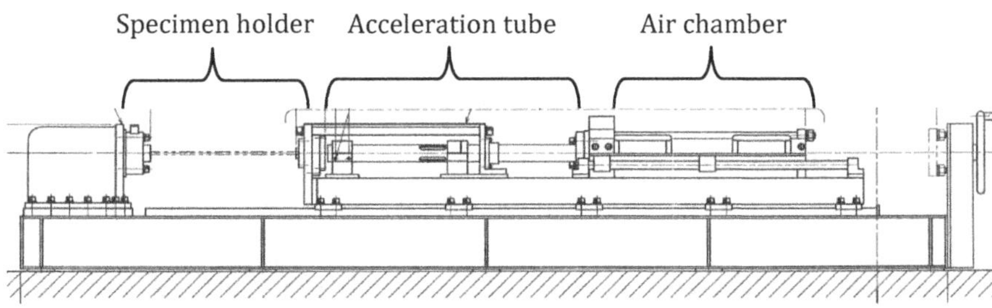

Fig. 1 Test apparatus

3 RESULTS AND DISCUSSION

3.1 Deformation histories

3.1.1 Low impact velocity

At impact velocity of approximately 6 m/s, deformation histories of zirconium alloy tubes with and without pellets are shown in Fig. 2. Deformation history of the empty tube by impact of the weight showed axial compression initially (Fig. 2 (a) (1)) and subsequent overall bending (Fig. 2 (a) (2)). The shape of bending deformation indicated the 3rd order mode ("W" shape). When a tube was filled with lead pellets, deformation history and bending shape were similar to those of the empty tubes (Fig. 2 (b) (1), (2)). On the other hand, the bending deformation was little when the tube was filled with aluminium oxide pellets, and the mode shapes seemed to be the 2nd order mode ("S" shape) (Fig. 2 (c) (2)). Fig. 3 plots shapes of each tube obtained from image analyses and those of the Euler's equation. Deformation and axial position are normalized by the maximum displacement and the initial length of each tube. Bending shapes for the empty tube and the tube with lead pellets are similar to the buckling shape in the 3rd order mode by the Euler's equation under simply supported condition, as shown in Fig. 3 (a). Similar deformation between the experiment and the Euler's equation was confirmed in the 2nd order mode for the tubes with aluminium pellets, as shown in Fig. 3 (b). We expected that the influence of such low stiffness as lead pellets was small on deformation, but that of such high stiffness as aluminium oxide pellets restricted the deformation of the tube and changed the mode shape.

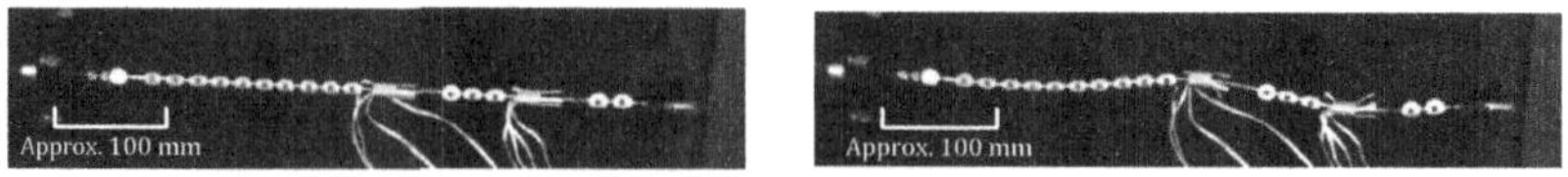

(1) Axial compression (0.0216s) (2) 3rd order mode bending (0.0226s)

(a) Empty tube

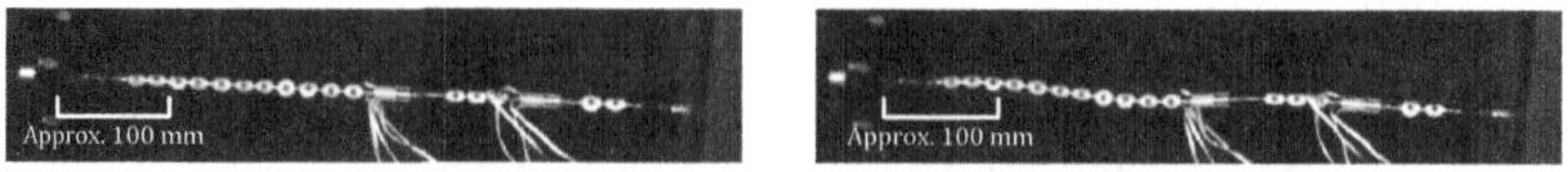

(1) Axial compression (0.0220s) (2) 3rd order mode bending (0.0242s)

(b) Tube with lead pellets

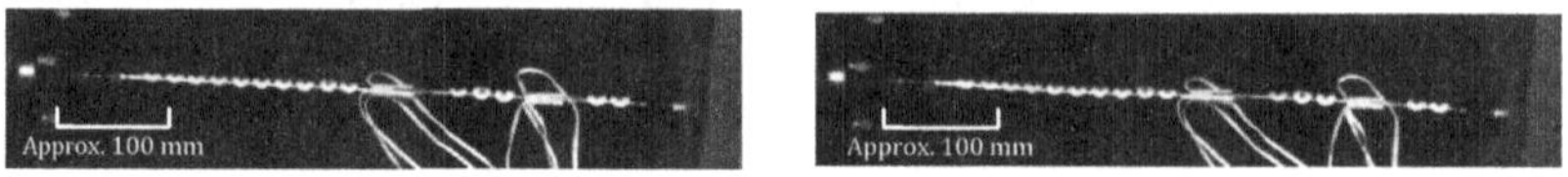

(1) Axial compression (0.0212s) (2) 2nd order mode bending (0.0230s)

(c) Tube with aluminium oxide pellets

Fig. 2 Deformation histories of zirconium alloy tube at low velocity

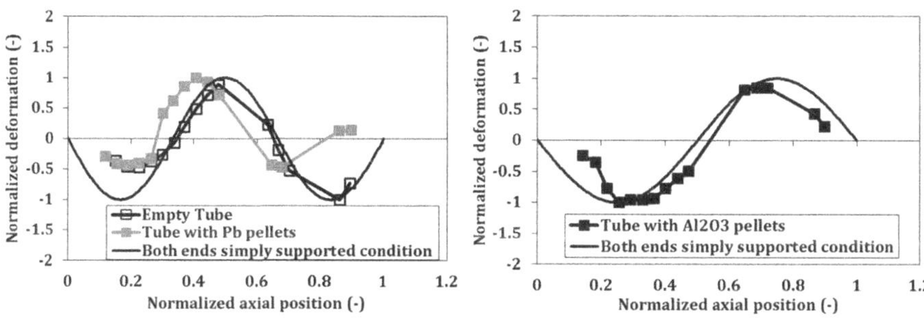

(a) Empty tube and tube with lead pellets (b) Tube with aluminium oxide pellets

Fig. 3 Comparison the shape of specimens with those of Euler's equation

3.1.2 High impact velocity

At impact velocity of approximately 16 m/s, deformation histories of the zirconium alloy tube with and without pellets are shown in Fig. 4. For the empty tube, the tube locally indicated large bending deformation at the impact side and resulted in diametric collapse (Fig. 4 (a) (2)) after axial compression (Fig. 4 (a) (1)). On the other hand, a tube with aluminium oxide pellets caused bending deformation at impact side and fracture (Fig. 4 (b) (2)) after axial compression (Fig. 4 (b) (1)). High-order mode and local deformation at high impact velocity observed in this experiment was also reported on a rectangular bar [(8)]. Shapes before buckling or failure obtained from image analyse were shown in Fig. 5. Shapes of deformation were normalized by the maximum displacement of the empty tube, and axial position was normalized by the same method of Fig. 3. Significant difference is observed in deformation at about 0.8 of normalized axial position between the tube with the aluminium oxide pellets and the empty tube. We considered that the aluminium pellets produced a strengthening effect, and then brittle like fracture occurred because of a constraint effect on plastic deformation due to the interaction between the tube and pellets. Such a local deformation cannot be expressed by the Euler's equation, which assumes overall deformation along the tube's axis. Therefore, FEM analyses are required to simulate the dynamic buckling phenomena, especially at high impact velocity to clarify the effect of pellets and the mechanism of the fracture.

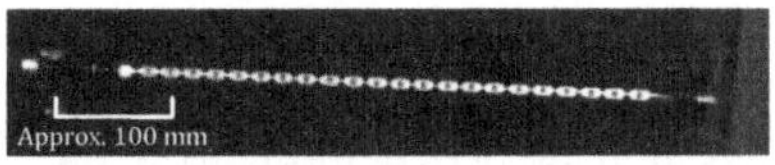

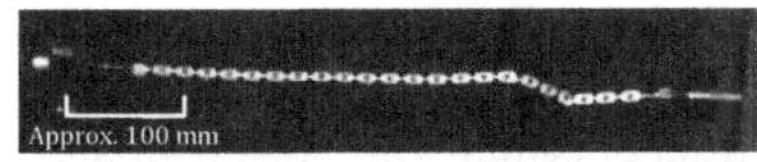

(1) Axial compression (0.0082s) (2) Local bending and collapse (0.0096s)

(a) Empty tube

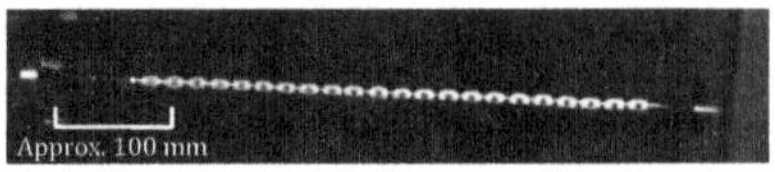

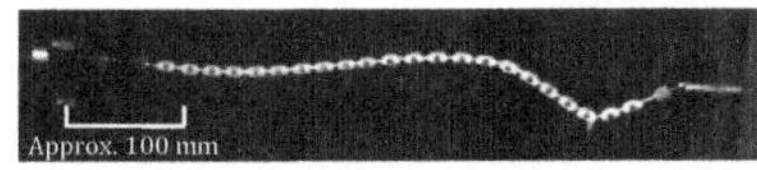

(1) Axial compression (0.0084s) (2) Local bending and failure (0.0106s)

(b) Tube with aluminium oxide pellets

Fig. 4 Deformation histories of zirconium alloy tube at high velocity

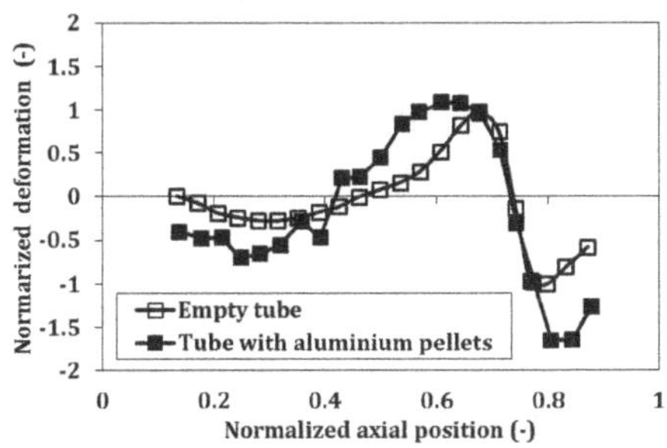

Fig. 5 Deformation shapes before buckling or failure obtained from image analyses

3.2 Maximum impact load

The maximum impact loads are summarized in Fig. 6. The loads are normalized by the measurement of the tube with aluminium oxide pellets at high impact velocity. For empty tubes made of zirconium alloy, the maximum loads slightly increased with higher impact velocity. Since estimated stress of the tubes caused by the impact is greater than yield strength, plastic deformation occurs at impact velocity higher than 6 m/s. The tubes with lead pellets showed almost the same maximum loads with empty tubes. It is confirmed that influence of the lead pellets was small. On the other hand, the tube with the aluminium oxide pellets showed fairly larger loads than empty tubes and those with lead pellets. It is expected that the strengthening effect of the pellets increases axial stiffness of the tubes, resulting in increase of impact load before failure occurs.

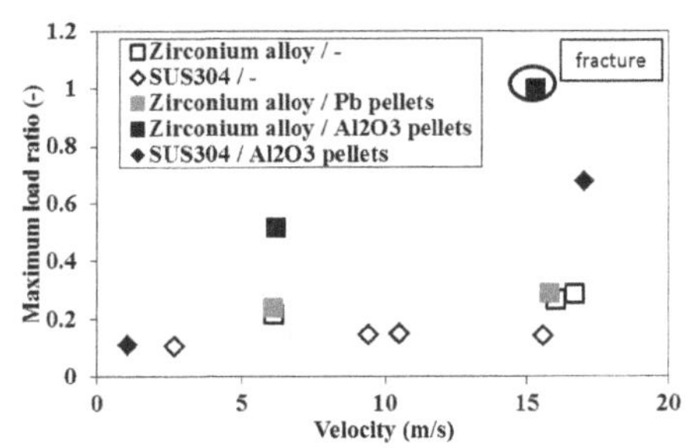

Fig. 6 Relation between maximum load ratio and impact velocity

For SUS304 tubes with and without aluminium oxide pellets, the maximum impact loads were small because of low yield strength, and tendency of loads against impact velocities was similar to that of zirconium alloy tubes. However, no fracture in the tube with aluminium oxide pellets occurred at impact velocity of 16 m/s. Further investigation is required to clarify the occurrence of fracture by considering the stress field and failure strain.

4 CONCLUSION

Impact tests of cylindrical tubes with and without internal pellets in the axial direction were conducted. For zirconium alloy tubes, the tubes with lead pellets indicated almost the same behaviours as empty tubes. In the case of low stiffness pellets, the influence of pellets was small, so deformation shapes and the maximum impact loads were almost the same. On the other hand, the tubes with aluminium oxide pellets indicated different behaviours compared to empty tubes. At low impact velocity, pellets with high stiffness caused the strengthening effect and increased both axial and lateral stiffness for tubes. As a result, the pellets led to increase of impact load, decrease of deformation and change of buckling mode. At high velocity, the strengthening effect of pellets made the tubes stiffer, but led to fracture by the constraint effect of pellets on plastic deformation. The deformation of tubes was compared to Euler's equation, and it was confirmed that the Euler's equation could be applied at the low impact velocity. However, it was not effective for the high velocity impact because of local deformation at the impact side. FEM analyses will be conducted to clarify the deformation shape and the mechanism of fracture at high impact velocity as a further study.

REFERENCES

[1] K. R. F. Andrews, G. L. England and E. Ghani, Classification of the axial collapse of cylindrical tubes under quasi-static loading, International Journal of Mechanical Sciences, Vol.25, No.9-10, pp.687-696, 1983.

[2] S. R. Guillow, G. Lu, R. H. Grzebieta, Quasi-static axial compression of thin-walled circular aluminium tubes, International Journal of Mechanical Sciences, 43, 2103-2123, 2001.

[3] K. Murase, N. Jones, Experiments on the dynamic axial plastic buckling of circular tubes, Journal of the Society of Materials Science, Japan, Vol.42, No.483, 1420-1426, 1993. (in Japanese)

[4] S. Haruyma, M. Takano, K. Ushijima, D. Chen, Effect of impact velocity on circular tubes subjected to axial loading, Transactions of the Japan Society of Mechanical Engineers, Series A, 71, 707 1030-1037, 2005. (in Japanese)

[5] S. Santosa, T. Wierzbicki, Crash behaviour of box columns filled with aluminium honeycomb or form, Computers and Structures, 68, 343-367, 1998.

[6] KEISUIRONENRYOUNOFURUMAI, Nuclear Safety Research Association, 1998. (in Japanese)

[7] SUTENRESUKOUBINRAN -DAI3HAN-, NIKKANKOUGYOUSHINBUNSHA, 1995. (in Japanese)

[8] M. Oda, K. Hukatsu, The longitudinal impact of a bar accompanied by lateral deformation, Transactions of the Japan Society of Mechanical Engineers, Series A, 45, 397, 1098-1103, 1979. (in Japanese)

Effects of leak rate on LOCA probability of pipes in nuclear power plants

J.H. Park[1], Y.K. Cho[1], S.H. Kim[2], J.H. Lee[2]

[1] Dept. of Safety Engineering, Chungbuk National University, Korea

[2] Mechanical and Material Assessment Dept., Korea Institute of Nuclear Safety, Korea

ABSTRACT

In the leak before break (LBB) concept, catastrophic pipe failure can be prevented if leak rate through a crack is more than the minimum detectable leak rate of a leak detection system before a crack grows to the critical crack size. Therefore accurate estimation of leak rate is important to evaluate the validity of LBB concept in pipe line design. Usually LOCA probability analysis is performed with constant crack morphology parameters. Since leak rate is affected by the crack morphology parameters, the LOCA probability may be affected by the parameters also. In this paper the effect of crack morphology parameters on LOCA probability was examined assuming the crack morphology parameters to be normally distributed random variables. A developed probabilistic fracture mechanics program, called P-PIE, was used in the analysis.

1 INTRODUCTION

A probabilistic fracture mechanics program, called P-PIE, was developed for pipes in nuclear power plants. Using the program leak and LOCA (loss of coolant accident) probabilities of pipes can be calculated for initially existing cracks and initiated cracks during operation. Crack growth due to fatigue and SCC (stress corrosion cracking) can be simulated using adequate crack growth rate equations and the effects of residual stresses, inspection period and operation temperature also can be considered in the program.

In the program leak rate through a crack in a pipe is estimated using Henry-Fauske flow model (Henry and Fauske 1970a, 1970b, 1970c) or modified Henry-Fauske flow model (Rahman et al. 1995). It is known that the analytical leak model agrees relatively well with the mean value of the measured leak rate. However, significant scatter was observed in the experimental leak rate data (Collier et al. 1984).

From the study it is found that if crack morphology parameters are treated as random variables in leak rate analysis, scattering distribution characteristics of leak rate can be simulated using the developed program. And the crack morphology parameters give great effect on leak rate.

Usually LOCA probability analysis is performed with constant crack morphology parameters. In this paper the effect of crack morphology parameters on LOCA probability was examined assuming the crack morphology parameters to be normally distributed random variables.

2 THEORETICAL BACKGROUND

2.1 Leak Models

Henry and Fauske (1970a, 1970b, 1970c) proposed the following equation to estimate flow rate through a crack in a pipe.

$$\psi(G_c, p_c) = G_c^2 - \frac{1}{\left[\frac{X_c \upsilon_{gc}}{\gamma_o p_c} - (\upsilon_{gc} - \upsilon_{Lc})N \frac{dX_E}{dp}\right]_c} = 0 \tag{1}$$

$$\Omega(G_c, p_c) = p_c + p_e + p_a + p_f + p_k + p_{aa} - p_o = 0. \tag{2}$$

Here G_c is mass flux at crack exit plane; p_c and p_o are pressure values at crack exit and crack entrance plane respectively; v_{gc} and v_{Lc} are specific volumes of saturated vapour and saturated liquid at exit pressure; γ_o is the isentropic expansion coefficient. In Eq. (2) p_e, p_f, p_k, p_a and p_{aa} are the pressure losses due to entrance effects, friction, bends and protrusions in the flow path, phase change acceleration and area change acceleration respectively. In Eqs. (1) and (2), G_c and p_c are unknowns. After solving the equations leak rate through a crack can be obtained by multiplying G_c by the crack opening area at crack exit plane, A_c. The detail expression and definitions for variables and pressure loss terms in Eqs. (1) and (2) can be found in Harris et al. (1972).

The pressure loss due to friction, p_f is given by

$$p_f = \left(f \frac{L}{D_H}\right) \frac{\overline{G}^2}{2} [(1 - \overline{X})\overline{\upsilon}_L + \overline{X}\overline{\upsilon}_g], \tag{3}$$

where f is the friction factor, L is flow path length, D_H is the hydraulic diameter, X is fluid quality and bar on the variable means the average value in the region. And based on the PRAISE program (Harris et al. 1972) the friction factor f is given by

$$f = \left[C_1 \log\left(\frac{D_H}{2\mu}\right) + C_2\right]^{-2}, \tag{4}$$

where C_1 and C_2 are coefficients and μ is the surface roughness and has the value of 0.0002441 in SCC growth and 0.0015748 in fatigue crack growth.

Rahman et al. (1995) proposed a flow model modifying Henry-Fauske flow model. In their model the surface roughness, number of turns and actual length of flow path are assumed to be a function of COD at the crack center.

The surface roughness is assumed to be a function of COD, δ as follows (Rahman et al. 1995):

$$\mu = \mu_L \qquad \text{for } 0 < \frac{\delta}{\mu_G} \le 0.1$$

$$\mu = \mu_L + \frac{\mu_G - \mu_L}{9.9}\left(\frac{\delta}{\mu_G} - 0.1\right) \quad \text{for } 0.1 < \frac{\delta}{\mu_G} \le 10 \tag{5}$$

$$\mu = \mu_G \qquad \text{for } 10 < \frac{\delta}{\mu_G},$$

where μ_L and μ_G are the local and global surface roughness respectively and δ is COD at the crack center.

The number of 90-degree turns in flow path, n_t is also assumed as a function of δ as follows:

$$n_t = n_{tL} \qquad \text{for } 0 < \frac{\delta}{\mu_G} \le 0.1$$

$$n_t = n_{tL} - \frac{n_{tL}}{11}\left[\frac{\delta}{\mu_G} - 0.1\right] \qquad \text{for } 0.1 < \frac{\delta}{\mu_G} \le 10 \qquad (6)$$

$$n_t = 0.1 n_{tL} \qquad \text{for } 10 < \frac{\delta}{\mu_G},$$

where n_{tL} is the local number of turns in flow path. Because one 90-degree turn corresponds to one velocity head loss.

Since the flow path is not perpendicular to the pipe surface and not straight, the real flow path length is longer than the wall thickness. The real flow path length, L_a can be obtained by multiplying the wall thickness, t by a correction factor K as $L_a = Kt$. The correction factor K is also given as a function of δ. Typical values of crack morphology parameters were given in Table 1 (PRO-LOCA 2009).

2.2 Crack Growth Simulation and LOCA Criterion

In the program fatigue and SCC crack growth simulation was performed based on suitable growth rate equation. When a crack becomes a through-thickness crack leak occurs. And if a growing surface or through-thickness crack satisfies a critical condition LOCA occurs. LOCA can be determined based on the tearing modulus or net-section yielding. When set-section yielding condition is used, the LOCA criterion is expressed as the following (Harris 1992):

$$\sigma_{LC} A_p > \sigma_{flo}(A_p - A_{crack}) \qquad (7)$$

Here σ_{LC} is the load controlled component of the axial stress, σ_{flo} is the flow stress, A_p is the cross-sectional area of the pipe and A_{crack} is the crack surface area.

3 ANALYSIS RESULTS

3.1 Leak Rate

Leak rate was calculated using the modified Henry-Fauske model. In the analysis, pipe material is assumed to be ASME SA351 CF8M. Yield strength and ultimate strength of the material are 201.1 MPa, 529.2 MPa respectively and α=2.22, n=4.84, where α and n are parameters in the Ramberg-Osgood relationship. The outer diameter and thickness of the pipe are 878.8 mm and 71.1 mm respectively. Normal operating pressure is assumed to be 15.51 MPa. The applied axial stress from deadweight is 14.34 MPa and the stress from deadweight and restraint of thermal expansion is 59.2 MPa.

In the calculation crack morphology parameters are treated as random variables. The values of PWSCC-BASE in Table 1 are used in the analysis. The probability density functions of leak rate were given in Figure 1 when the half crack length is 50.8 mm and 101.6 mm. Figure 2 shows leak rate values when 10 values were obtained for each half crack length.

Table 1 Mean and standard deviation of crack morphology parameters (PRO-LOCA 2009)

Crack Morphology Parameter	Corrosion Fatigue		IGSCC		PWSCC - Base		PWSCC - Weld	
	Mean	Stan. Dev	Mean	Stan. Dev.	Mean	Stan. Dev.	Mean	Stan. Dev.
μ_L (μm)	8.814	2.972	4.70	3.937	10.62	9.870	16.86	13.57
μ_G (μm)	40.51	17.65	80.0	39.01	92.67	65.26	113.9	90.97
n_L (mm^{-1})	6.730	8.070	28.2	18.90	8.043	2.043	5.940	4.540
K_G	1.017	0.0163	1.07	0.100	1.060	0.095	1.009	0.011
K_{GL}	1.060	0.0300	1.33	0.170	1.327	0.249	1.243	0.079

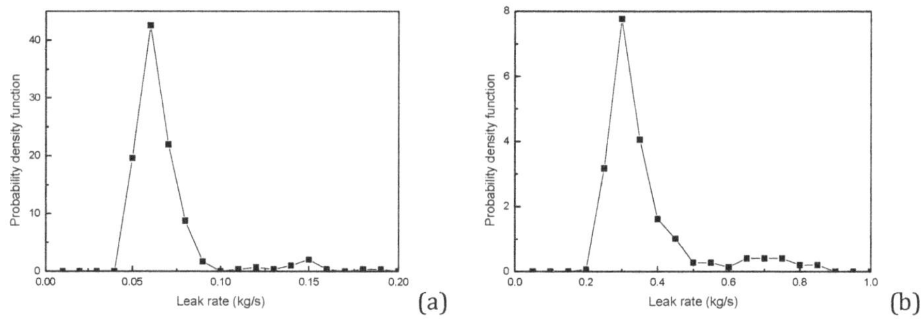

Figure 1 Probability density function of leak rate for crack morphology parameters of PWSCC base when the half crack length is (a) 50.8 mm and (b) 101.6 mm.

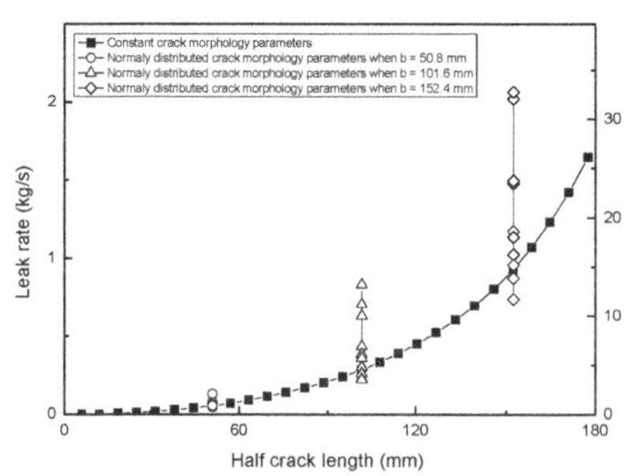

Figure 2 Leak rates for normally distributed crack morphology parameters of PWSCC base.

3.2 LOCA Probability

LOCA probability was calculated using the developed program, P-PIE and the effect of the crack morphology parameters on the LOCA probability was examined. It was found that the crack morphology parameters give very limited effects on the LOCA probability under usual operating condition. Because the critical crack length for LOCA is large, the leak rate at the critical crack length is much larger than the minimum detectable leak rate of a leak detection system even if large n_L value is used in the simulation.

However if the operating condition is changing the crack morphology parameters can give an effect on LOCA probability. For example if the applied axial stress is low and n_L value is very large, the leak rate remains very low until the crack length becomes long. And if very large load is applied because of changing operating condition or earthquake LOCA can occur. In this study several cases were examined to consider the effects of crack morphology parameters on LOCA probability.

4 CONCLUSIONS

1. Scattering characteristics in the measured leak rate data can be generated treating the crack morphology parameters as normal distributed random variables.
2. The crack morphology parameters give very limited effects on the LOCA probability under usual operating condition.
3. Several cases were examined to consider the effects of crack morphology parameters on the LOCA probability.

5 REFERENCES

[1] Collier, R.P., Stulen, F.B., Mayfield, M.E., Pape D.B. & Scott P.M., (1984) EPRI Report, NP-3540-LD.

[2] Henry, R. E. (1970a) *Nuclear Science and Engineering*, 41, 336-342.

[3] Henry, R. E., & Fauske, H. K. (1970b) *Nuclear Science and Engineering*, 41, 79-91.

[4] Henry, R. E. & Fauske, H. K. (1970c) *Nuclear Science and Engineering*, 41, 92-98.

[5] Harris, D. O., Dedhia, D. D. & Lu, S. C. (1992) Theoretical and User's Manual for pc-PRAISE, USNRC Report NUREG/CR-5864.

[6] PRO-LOCA-GUI/PRO-LOCA User's guide (Version 3.5.32), Battelle (2009).

[7] Rahman, S., Ghadiali, N., Paul D. & Wilkowski, G. (1995) Probabilistic Pipe Fracture Evaluations for Leak-Rate-Detection Applications, USNRC Report NUREG/CR-6004.

Investigation on acceleration methods for seismic analysis of through-wall cracked piping

J.S. Kim
Dept. of Mech. Engineering, Sunchon National University, South Korea

ABSTRACT

Structural integrity of the safety class I piping with detected flaws shall be evaluated in accordance with FFS (fitness-for-service) assessment codes such as the ASME B&PV Code Sec.XI. Time history dynamic seismic analysis should be performed considering plastic deformation because the cyclic beyond-design-basis earthquake brings out considerable plastic deformation and cumulative damage. Two acceleration methods, an effective force method (or inertia method) and a large mass method, have been applied in performing time history seismic analysis. The acceleration methods for the no cracked structures have been verified via the previous studies. However, there is no study to identify validity of the acceleration methods for cracked piping. In this study, the acceleration methods for through-wall cracked piping were validated via time history implicit dynamic seismic analysis. As a result, it is found that two acceleration methods have the same results for cracked piping.

1. INTRODUCTION

When flaws are detected on safety class I piping in nuclear power plants, structural integrity of the safety class I piping with the detected flaws shall be evaluated in accordance with FFS assessment codes such as the ASME B&PV Code Sec.XI (1). Level D events, which is assumed to occur once during the lifetime of a nuclear power plant, shall be considered at this assessment as one of DBAs. After the Daiichi Fukushima earthquake, it was found that beyond-design-basis earthquake may occur repeatedly. It is one of post-Fukushima safety issues to identify that structural integrity of the safety class I cracked piping is maintained under this cyclic beyond-design-basis earthquake (2). Time history dynamic seismic analysis should be performed considering plastic deformation because the cyclic beyond-design-basis earthquake brings out considerable plastic deformation and cumulative damage. Two acceleration methods, an effective force method (or inertia method) and a large mass method, have been applied in performing time history seismic analysis. The acceleration methods for the no cracked structures have been verified via the previous studies (3, 4). However, there is no study to identify validity of the acceleration methods for cracked piping.

In this study, validity of the acceleration methods for the through-wall cracked piping was assessed via time history implicit dynamic seismic analysis using the commercial finite element analysis program, ABAQUS (5). And, effect of the analysis variables such as magnitude of added mass and maximum time increment on maximum stress intensity factor was investigated.

2. FINITE ELEMENT SEISMIC ANALYSIS

2.1 Target model

Figure 1 shows an analysis target pipe simplifying a surge line. As shown in Fig. 1, the pipe was modeled as a 1/4 model because of symmetry of geometry and boundary conditions. The pipe was made of austenitic stainless steel, TP347. A through-wall circumferential crack was assumed to be located on center of the pipe. Circumferential crack length considered was 150°. Simplified time history data based on the El Centro seismic data (6) as depicted in Fig. 2, corresponding to maximum acceleration 0.3g, was applied to support nodes.

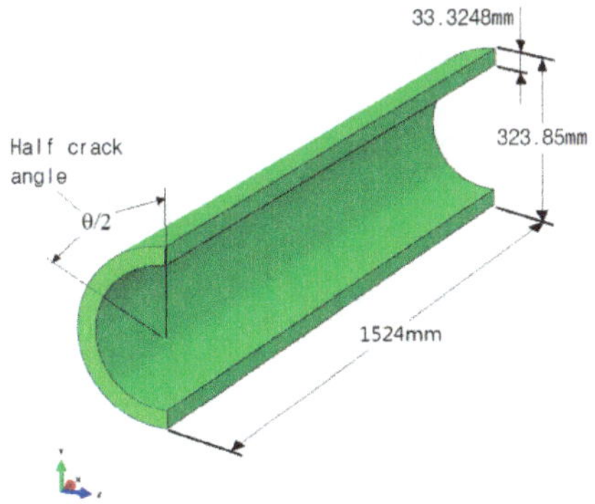

Fig. 1 Configuration of a simplified surge line with the through-wall crack.

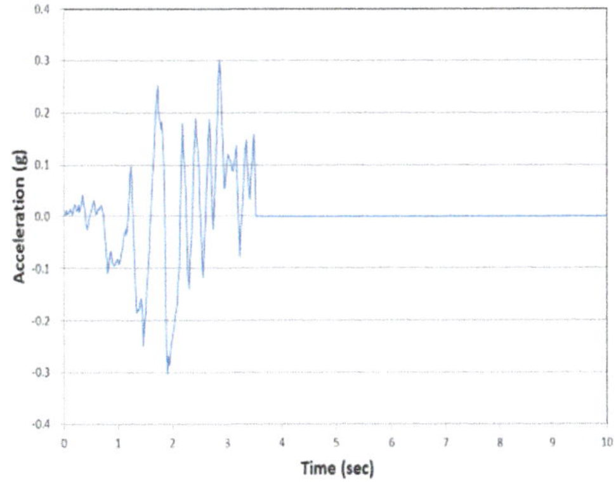

Fig. 2 Time history for simplified seismic acceleration.

2.2 Seismic crack analysis

Time history dynamic seismic analysis for the through-wall cracked piping was performed using ABAQUS. SIF (stress intensity factor) was calculated as average of the values at four integration paths selected near the crack tip. Figure 3 depicts the variation of the SIF history vs. crack angle derived via large mass method (added mass ratio, $R_M=M/m=10^{11}$, maximum time increment, $\Delta t_{max}=0.01sec$). M is added mass, m is mass of the target model. As depicted in Fig. 3, it is found that maximum SIF increases with increasing crack angle.

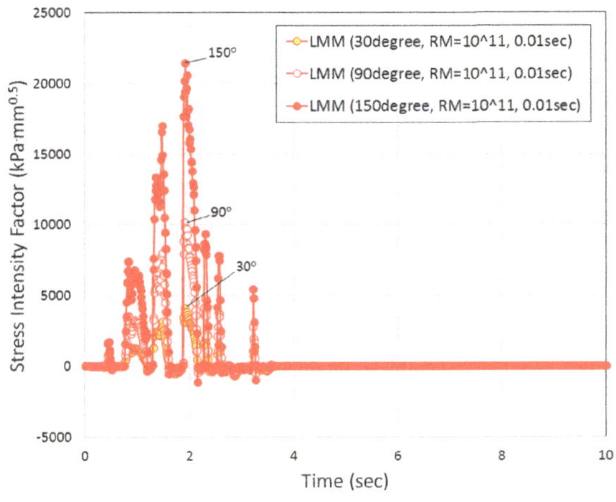

Fig. 3 Variation of SIF history vs. crack angle.

3. REVIEW OF THE ACCELERATION METHODS

Figure 4 compares SIF histories between two methods. From Fig. 4, it is found that the SIF results by the effective force method are a bit larger than those by the large mass method, but difference between two methods are insignificant. Figure 5 shows variations of maximum SIF by two methods vs. maximum time increment. As shown in Fig. 5, for both the methods, it is identified that the maximum SIFs converge at the maximum time increment, 0.01sec. Figure 6 depicts variation of a maximum SIF by the large mass method vs. the additional mass ratio, defined as a ratio of added mass to pipe mass. As depicted in Fig. 6, it is found that constant maximum SIF is derived at the additional mass ratio of more than 10^7. As a result, it is identified that both of two acceleration methods have same results for the cracked piping if the large mass magnitude and maximum time increment are adequately selected.

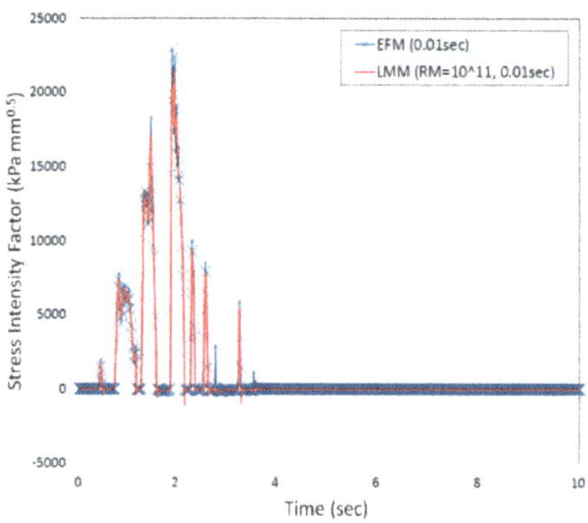

Fig. 4 Comparison of SIF histories between the effective force method and the large mass method.

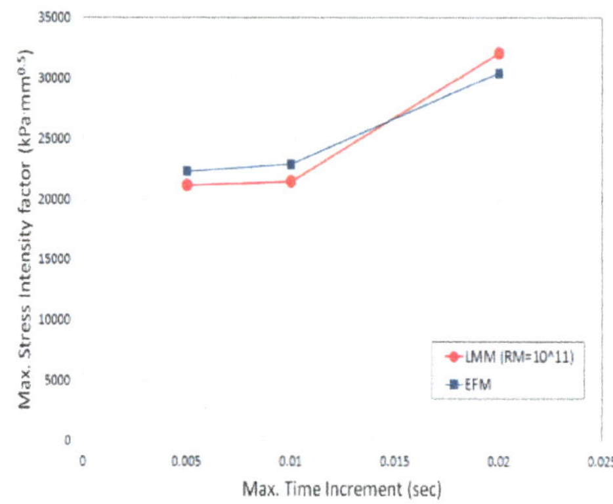

Fig. 5 Variation of maximum SIF for each acceleration method with maximum time increment.

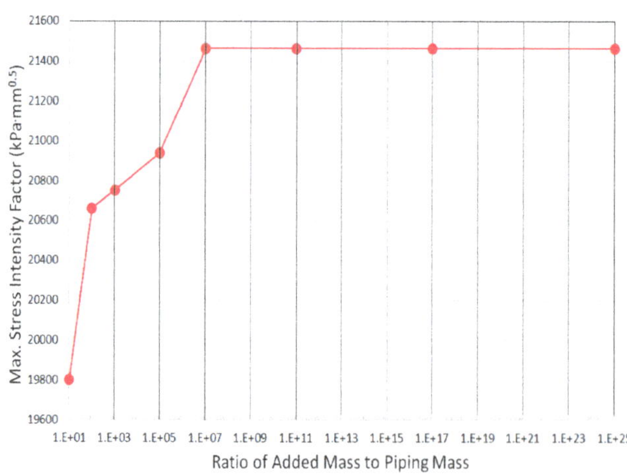

Fig. 6 Variation of maximum SIF with additional mass ratio of the large mass method.

4. CONCLUSIONS

From the time history implicit dynamic seismic analysis for the through-wall cracked piping, the following findings were derived:

1) The SIF results by the effective force method are a bit larger than those by the large mass method, but difference between two methods are insignificant.
2) Constant maximum SIF is derived at the additional mass ratio of more than 10^7.
3) Both of two acceleration methods have same results for the cracked piping if the large mass magnitude and maximum time increment are adequately selected.

ACKNOWLEDGEMENT

This research was supported by National Research Foundation of Korea(NRF) funded by the Ministry of Science, ICT and Future Planning. (NRF-2013M2A8A 1040924)

REFERENCES

(1) ASME B&PV Code Committee (2011) ASME B&PV Code, Sec. XI.
(2) USNRC (2011) *Recommendations for Enhancing Reactor Safety in the 21st Century: The Near Term Task Force Review of Insights from the Fukushima Dai-Ichi Accident.*
(3) Kim, Y.W., Jhung, M.J. (2011) *Nuclear Engineering and Technology.* **43(4)**, 361-374.
(4) Kim, Y.W., Jhung, M.J. (2014) *Journal of Mechanical Science and Technology.* **28**, 25-41.
(5) Dassault Systems (2012) *Simulia, User's Manuals for ABAQUS*, Ver.6.12.
(6) http://www.vibrationdata.com/elcetro.htm.

Simulation and failure analysis of strain clamp failed caused by deterioration of contact resistance

X.H. Li, H.L. Wu, H. Ding, C.L. Zhu
School of Power and Mechanical Engineering, Wuhan University, P.R. China

ABSTRACT

Strain clamp (SC) is a kind of electric fittings compressed with conductor, clamp aluminum tube (CAT), steel strands and steel anchor. Good contact resistance (CR) between aluminum strands and CAT is the SC safety guarantee. Failure mechanism of SC caused by CR deterioration (CRD) was revealed by macro examination, metallograph and FEM through a failure example. Temperature and stress fields (TSFs) of SC with normal and deteriorated CR were compared in order to describe the failure characteristics of SC. Influenced by TSFs, the exposed steel strands fracture first, leading to fracture of conductor at the orifice of CAT.

Keywords: strain clamp; contact resistance; temperature; stress

1 INTRODUCTION

Strain clamp (SC) in high-voltage transmission line is a kind of electric power fittings, which is compressed with conductor, clamp aluminum tube (CAT), steel strands and steel anchor tube, meeting the transmission function of mechanical and electrical loading in stranded conductors across towers [1]. In order to enhance the electrical contact effect, an appropriate thickness of conductive paste (CP) must be coated on the surfaces of aluminum strands before the compression of CAT [2]. Contact resistance (CR) and joint temperature can be reduced effectively by the proper use of CP.

However, the CR may increase on the contrary if the conductive paste is coated uneven, too thick, or even there is no CP [3]. The contact effect between aluminum strands (AS) and CAT will reduce due to oxidation, corrosion, winds vibration loosening, during the long-term operation, leading to abnormal temperature elevation or even failure of SC [4].

In this paper, failure characteristics, morphology and microstructure of steel strands fractures were investigated by macro examination, metallographic and FEM through an example of SC failure accident. And the mechanism and characteristics of SC failure originated from CRD were obtained.

2 STRUCTURE OF STRAIN CLAMP

The structure of strain clamp is shown in figure 1. Section AB is the compression area of conductor and CAT. Section BD is the non-compression area of CAT. Section DE is the compression area of steel anchor grooves and CAT (SAGCAT). Section BC is the area of exposed steel strands (SS), and section CD is the compression area of steel strands and steel anchor tube (SSSAT). Section WA is the free conductor out of strain clamp [1]. The conductive paste is coated at the aluminum strands surfaces within section AB.

3 FAILURE ANALYSIS OF STRAIN CLAMP

3.1 Fracture location

Failure analysis was performed through an example of stranded conductor fracture accident in strain clamp in a 220kV transmission line.

Two fracture spots, fracture of exposed steel strands at section BC, and fracture of conductor at the orifice of CAT (point A), were found after splitting the failed SC (Fig.2). The fracture at section BC should be prior to the fracture at point A. According to logic reasoning, the conductor will drop onto the floor as soon as the fracture of point A happens. The steel strands at section BC will not fracture with little tension and current loading within the SC if the conductor fracture at point A occurs first.

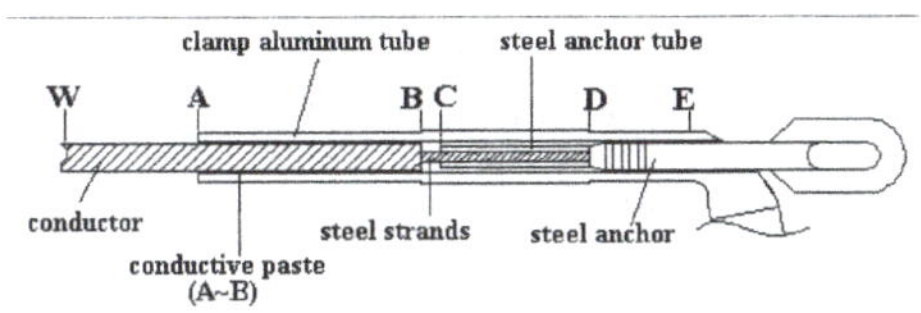

Fig.1 Structure of strain clamp

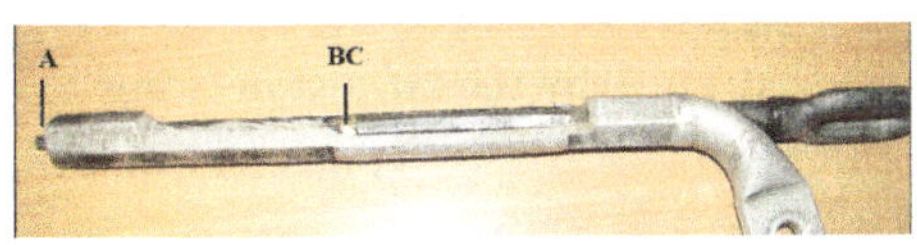

Fig.2 Failed strain clamp

3.2 Experimental analysis

3.2.1 Macro examination

It's found that in macro examination there are obvious neckings at all steel strands fractures and molten phenomenon in some steel strands fracture at point A. The fracture morphology of a steel strand at point B is shown in Fig.3. In order to further analyzing the fracture characteristics of the real fracture, several old steel strands specimen, gained from the free conductor out of the SC, were tensile broken at room temperature and compared with real fractures. The average area reduction of the fracture steel strands and the old steel strands are 42.23% and 32.38%, respectively. The average area reduction of the fracture steel strands is about ten percentages larger than that of the old steel strands. The failed steel strands are preliminarily judged to have suffered from a certain degree of high temperature before fracture.

The inside wall morphology of CAT (section AB) is shown in Fig.4. The CP is extremely uneven and most of the area is not coated by CP. The adherent conductive paste has already become black instead of original deep green. It suggests that the conductor and CAT within section AB also suffered a certain degree of high temperature [5].

Fig.3 Fracture of steel strand

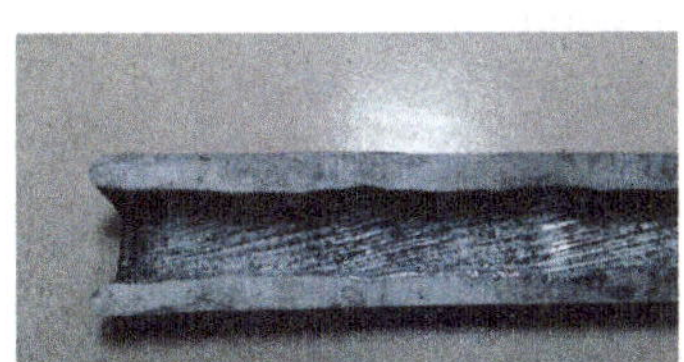

Fig.4 Clamp aluminum tube with CP

3.2.2 *Metallograph*

Longitudinal section samples from fracture steel strands and old steel strands of the free conductor out of the strain clamp were prepared for metallographic test (Fig.5).

Fig.5 shows that the old steel strand material still presents fibrous microstructure characteristics, while the failed steel strands change a lot and equiaxial crystal grains are found to form near the necking place. It gives powerful evidence that the steel strands suffered a certain degree of high temperature before fracture and recovery and recrystallization occurred under the influence of abnormal temperature elevation [6, 7].

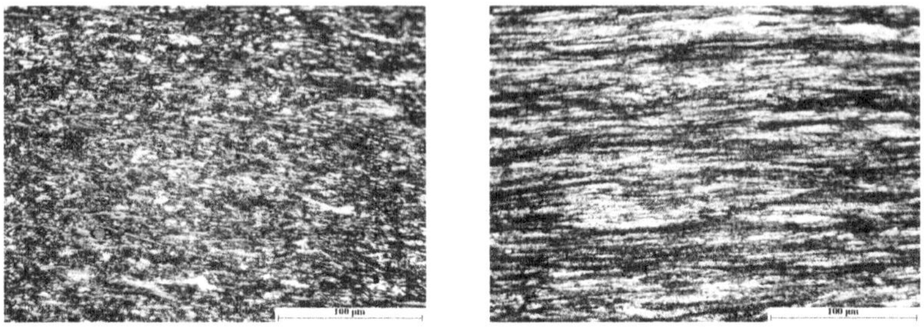

(a) nearby the largest necking (b) old steel strand in free conductor

Fig.5 Metallograph of steel strands

3.3 Simulation analysis

The current path changes when the CR between aluminum strands and CAT increases abnormally causing redistribution of TSFs. The TSFs were analyzed by finite element method (FEM) for investigating the failure mechanism of SC.

3.3.1 *Simulation model*

Simulation FEM model (Fig.6) were built according to the NY-240/40 type strain clamp in the accident example. Only a quarter model of strain clamp is shown in consideration of the symmetrical structure of SC. The steel strands, aluminum strands, CAT and steel anchor were considered as bulk material, respectively, ignoring the gap among the strands. Heat source of the FEM model is the current joule heat. And there is no joule heat at right side of point E without current passing through. The stress loading of the model is the conductor tensile force. The hanging tail of SC is negligible in the FEM model because the loading force by drainage line is much smaller than the conductor tensile force. Simulation of TSFs of SC influenced by CRD was performed.

3.3.2 *Simulation results*

Temperature field (TF, °C) distributions of SC under typical loading condition (500A of current, 25% of ultimate tensile strength [8]) are show in Fig.7~8. TF of strain clamp with normal CR is shown in Fig.7. And Fig.8 is the TF of SC with 50% of CRD.

Temperature within free conductor (section WA) is the highest when the CR is normal (Fig.7). And the temperature within SC (section A~E) is lower than that within free conductor. It is because that the cross section area of CAT within strain clamp, 540.5mm^2, is larger than that of aluminum strands, 238.9mm^2 [8, 9], meaning that the resistance per length of strain clamp is smaller than that of conductor.

The peak temperature transfers to the compression place (SSSAT) (section CD) when the CR deteriorates (Fig.8), because the current path is blocked in the CAT. And temperature in all section is higher with CRD than that with normal CR.

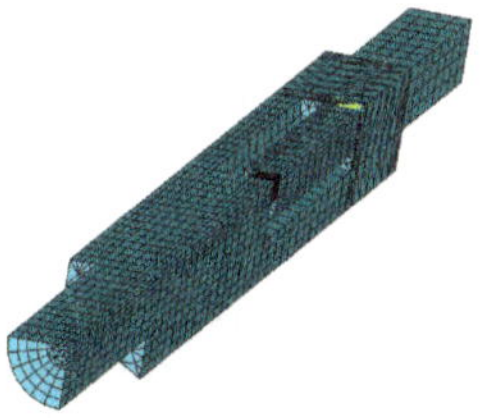

Fig.6 FEM model of strain clamp

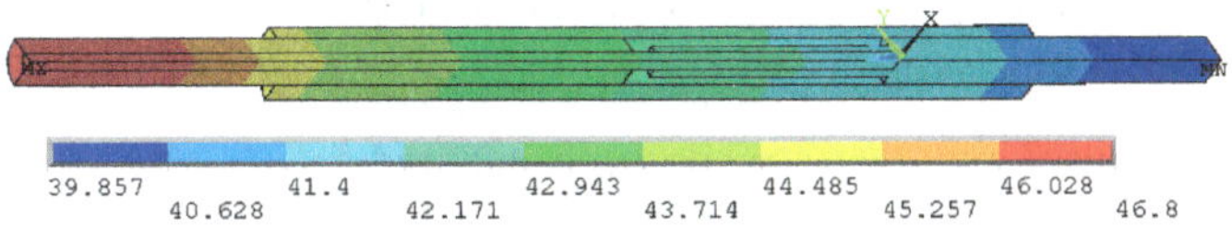

Fig.7 Temperature of SC with normal CR

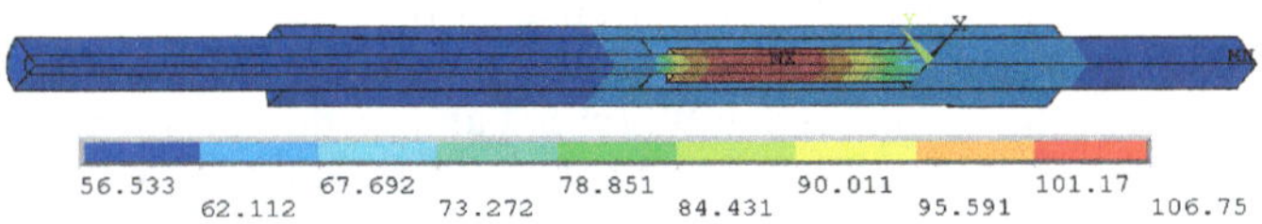

Fig.8 Temperature of SC with 50% deterioration of CR

The TSFs value along the steel strands (SS) axis path extracted under various deterioration degree of CR are shown in Fig.9~10. The origin point, positive axis and negative axis are the orifice of CAT (point A), steel anchor and free conductor, respectively.

Temperature in steel strands reduces from free conductor to steel anchor gradually with normal CR (deterioration degree λ=0). While the temperature in steel strands increases from free conductor to steel anchor gradually when CR deteriorates. There is a bump of steel temperature within section BD (in SS), and the peak temperature occurs at the spot 40mm inside of steel anchor tube orifice (Fig.9).

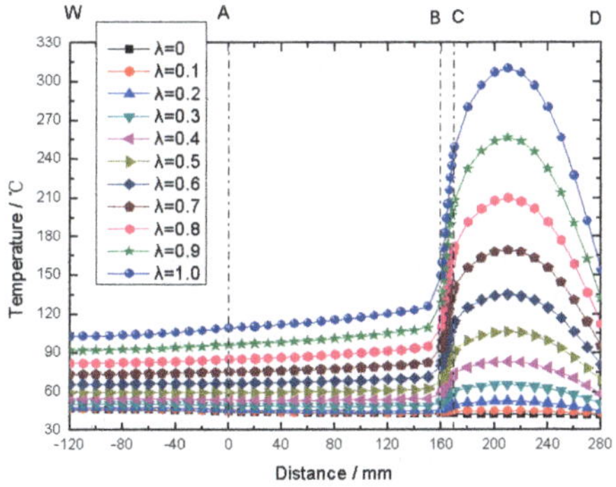

Fig.9 Temperature of steel strands

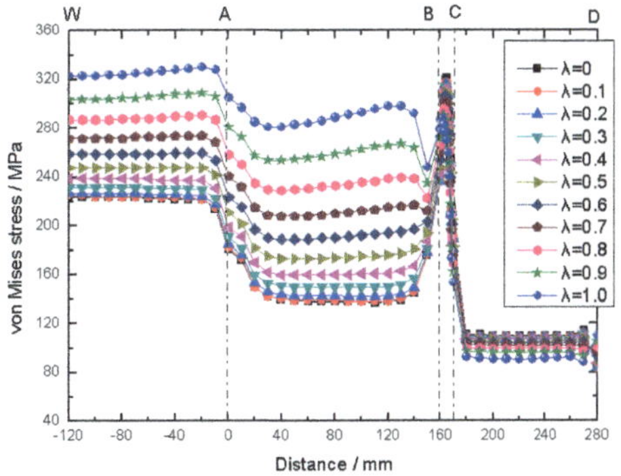

Fig.10 Von Mises stress of steel strands

Fig.10 shows that the ranking of average von Mises stress in various section (in SS) is as follows, section BC>section WA>section AB>section CD. Stress in section WA is slightly higher than that in section BC when deterioration degree λ is 1.0. Stresses in both section WA and section AB increase as the CR deteriorates. Stresses in section WA and section AB increased from 220MPa to 330MPa, and from 139MPa to 298MPa, respectively. Stress in section CD decrease, from 110MPa to 91.5MPa, as the CR deteriorates. Stress in section BC keeps stable between 290MPa and 320MPa.

4 MECHANISM AND CHARACTERISTIC OF FRACTURE

The experimental results show that the failure characteristics of SC, two fracture spots, is caused by steel strands been suffered by abnormal temperature elevation. The simulation results show that the distribution of TSFs is greatly influenced by CRD.

When CR is normal, the majority part of current in aluminum strands (AS) flows through CP (section AB) to CAT, and through section BD (in CAT) to section DE (in CAT) finally. Only a small part of current in steel strands (in SS) flows through section WA (in SS), section AB (in SS), section BC (in SS) and section CD (in SSSAT) to section DE (in steel anchor, in SA) finally. The two parts of current join together at section DE (in SAGCAT) and flow to the next span through the drainage plate.

When the electrical contact effect decreases as CR deteriorates, the current in AS can not flow through compression area totally to the CAT. Thus, the current in section BD of CAT is described by formula (2). Where I_{BD} is the current in section BD of CAT, λ is the deterioration degree of CR, I_{AL} is the current in AS of free conductor.

$$I_{BD} = (1-\lambda)\times I_{AL} \qquad (2)$$

The residual part of current, which can not flow to CAT caused by CRD, has to flow through steel section BC (in SS), CD (in SSSAT) to DE (in SA). Because the steel strand material is not a good conductor for heat, the influx of extra large current into steel strands will certainly cause the abnormal temperature elevation at steel section BC (in SS) and CD (in SSSAT). High stresses occur at steel section WA (in SS) and BC (in SS). And the stress at steel section CD (in SS) is low (Fig.9~10). It suggests that the steel strands at section BC (SS) will fracture first by comparing the TSFs in the three sections, where stress value is large and temperature is lowest in section WA (in SS), stress value is large and temperature is high in section BC (in SS), temperature is highest but stress value is smallest in section CD (in SS).

The fracture of exposed steel strands (section BC) causes the sharp decrease of bearing loading capacity of SC. The total tensile force of conductor transfers to compression portion, between AS and CAT once the steel strands fracture at section BC. Materials soften and stress level rises at the compression portion under the dual influence of CRD and steel strands fracture at section BC. And the AS at the orifice of CAT (point A) fracture fast, leading to steel strands temperature elevation rapidly, fracture and even molten.

5 CONCLUSIONS

(1) The strain clamp failure accident is caused by the CRD. The typical characteristic of strain clamp failure caused by the CRD is the existence of two fracture spots. The first one is the fracture at the exposed steel strands (section BC, SS), the second one is the conductor fracture at the orifice of CAT (point A).

(2) Temperature in steel strands reduces from free conductor to steel anchor gradually with normal CR and the highest temperature is within the section WA. The temperature in steel strands increases from free conductor to the steel strands and steel anchor tube gradually as CR deteriorates and the highest temperature is within the section CD (in SS).

(3) The mechanism of strain clamp failure caused by the CRD is as follows. The redistribution of TSFs in strain clamp occurs when the CR deteriorates. Influenced by the combination influence of temperature and stress, the exposed steel strands (section BC, SS) fracture first, causes sharply decrease of bearing loading capacity. The total tensile force of conductor transfers to compression portion AB between AS and CAT (ASCAT) once steel strands fracture at section BC (SS). Materials soften and stress level rises at the compression portion AB under the dual influence of CRD and steel strands fracture at section BC (SS). And the aluminum strands at the orifice of CAT (point A) fracture fast, leading to temperature elevation, fracture and even molten of steel strands rapidly.

REFERENCES

[1] Dong Ji-E. Electric Power Fittings Manual[M]. Beijing: China Electric Power Press, 2001: 137-142.

[2] The National Energy Bureau. DLT 373-2010 Technical conditions of conductive paste[S]. Beijing: China Electric Power Press, 2010.

[3] Chen Lian-Sheng. The use of electrical compound grease [J]. Electric Engineering, 1999, 05: 58.

[4] Guo Bin, Liu Jiang-Tao. Analysis and research on the heat problem of wires in high voltage transmission line[J]. Industrial control computer, 2011, 12: 117-118.

[5] Qiang Chun-Mei, Wang Guo-Gang, Zhang Miao, et al. Effect of electrical compound grease on saving energy and reducing consumption of electric power fittings[J]. Electrical Manufacturing, 2013, 07: 38-40.

[6] Cui Zhong-Qi, Qin Yao-Chun. Metal Science and heat treatment[M]. Beijing: China Machine Press, 2009: 196-212.

[7] Shi De-Ke. Foundation of material science[M]. Beijing: China Machine Press, 2006: 355-365.

[8] Zhang Zhong-Ting. The design principle of overhead transmission line[M]. Beijing: China Electric Power Press, 2010: 211-225.

[9] Yu Hong-Yun, Yu Cheng-Biao, Li Rui. The application and detection of electric power line equipment[M], Beijing: China Electric Power Press, 2007: 22-23

Applicability of net section collapse load approach to assessment of pipes with multiple circumferential cracks

M.Y. Lee, S.J. Kim, J.Y. Jeon, Y.J. Kim
Dept. of Mechanical Engineering, Korea University, Korea

ABSTRACT

In this paper, an applicability of net section collapse load approach to assessment of pipes with multiple circumferential cracks was investigated. Ductile fracture simulations of pipes with two circumferential cracks are used for verification. The finite element damage analysis that are used for these simulations are based on the element-size-dependent critical damage concept with the stress-modified fracture strain model. Virtual testing for verification is performed on 4-inch diameter schedule 80 pipes with circumferential flaws that have several geometries. The assessment results of net section collapse load approach are in good agreement with virtual test results using finite element damage analysis.

1 INTRODUCTION

Operating nuclear pressurized components often have a multiple cracks due to stress corrosion cracking. Assessment of pipes with multiple cracks can be complex, depending on the proximity of adjacent cracks. If the distance between adjacent clacks is short, these cracks can coalesce. If the distance between adjacent clacks is larger than specific distance, these cracks can propagate without interaction. Due to such complexities, the development of assessment method of pipes with multiple cracks is a difficult task.

One practical method to assess pipes with multiple cracks is to provide crack re-characterization rules. These rules that are based on a proximity are given to decide whether two cracks should be combined or considered separately. In accordance with the ASME Code Section XI, in the case of multiple aligned planar surface cracks that are illustrated in Fig. 1, the distance S between the adjacent cracks is compared with the cracks depths. If distance S is smaller than a half of maximum depth, max(a1,a2), the cracks shall be replaced as a single crack. Re-characterization rules assume that two cracks can propagate without interaction when distance S is larger than maximum depth (1). However, these cracks are interacted in reality.

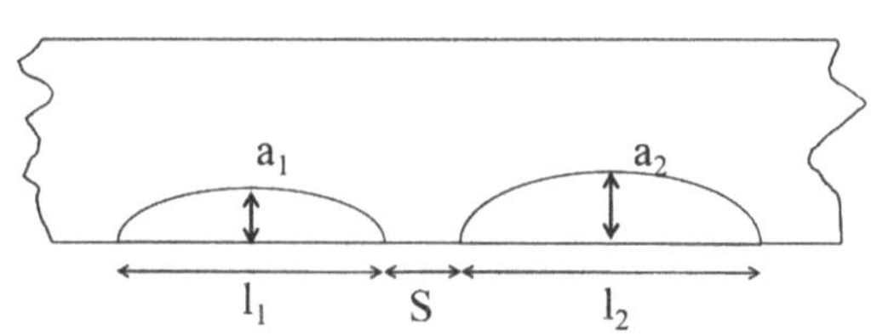

Fig. 1 Two surface cracks.

Hasegawa et al have investigate the assessment method for pipes with multiple circumferential cracks based on net section collapse (NSC) load approach (2). Researchers at Hitachi then performed six pipe tests with two circumferential inner surface cracks (3). Those tests are four-point bending that was performed at pipes made of 304 stainless steels. Pipe test results illustrated in Fig. 2 are compared with theoretical results based on NSC load approach concept. Researchers at Hitachi failed to draw meaningful conclusions on applicability of NSC load approach due to a limited number of tests.

One effective way to covering all possible ranges of variables to draw conclusions is to use finite element (FE) damage analyses. A major concept of the FE analyses is a damage model of ductile fracture. In this paper, element-size-dependent critical damage concept with the stress-modified fracture strain model that was investigated by Kim et al (4) was used to simulate ductile failure. Simulated results illustrated in Fig. 3 are compared with pipe test results for verifying simulation method.

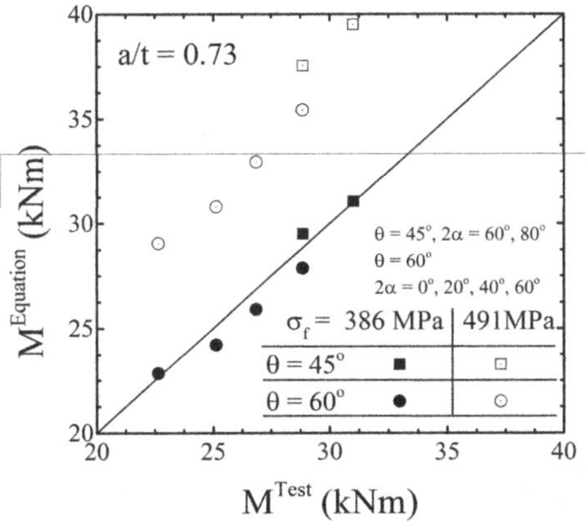

Fig. 2 Comparison of pipe test results and theoretical results based on NSC load approach concept.

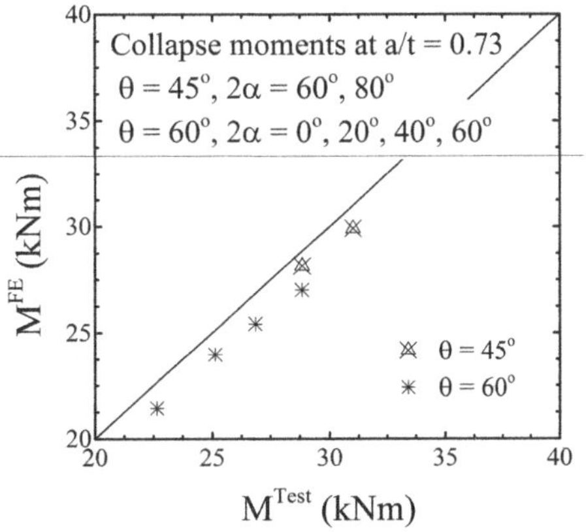

Fig. 3 Comparison of Simulated results and pipe test results.

2 LIMIT LOAD SOLUTION BASED ON NET SECTION COLLAPSE LOAD APPROACH CONCEPT

Fully plastic collapse moment for a pipe with multiple circumferential cracks is easily obtained from the limit load criteria and NSC load approach. β that is the angle to the neutral axis of the pipe cross section containing the flaw is was derived from equilibrium equation. For a pipe with multiple circumferential cracks, β is given by,

$$\beta = \frac{1}{2}\left[\pi - \frac{a}{t}\theta - \pi\frac{\sigma m}{\sigma f}\right] \qquad \text{Eq. (1)}$$

For the situation where $\beta \le (\pi - \theta - \alpha)$, the external bending moment at incipient plastic collapse is given by,

$$M_b^c = 2\sigma f R^2 t\left[2\sin\beta + \frac{a}{t}\sin\alpha - \frac{a}{t}\sin(\alpha+\theta)\right] \qquad \text{Eq. (2)}$$

where σ_f is the flow stress of the pipe material, a is the flaw depth, t is the wall thickness of the pipe, θ is one-half of the angle of the circumferential extent of the flaw, α is one-half of the angle between two flaws and β is given by Eq(1).

3 DUCTILE FRACTURE SIMULATIONS OF PIPES WITH TWO CIRCUMFERENTIAL CRACKS

Pipes with two aligned symmetrical circumferential surface cracks illustrated in Fig. 4(a) were simulated using finite element damage analysis based on element-size-dependent critical damage model. The damage model is based on the concept that fracture strain ε_f for dimple fracture depends on the triaxial stress states, which are quantified by the ratio of the mean stress σ_m and equivalent stress σ_e. Such a model is often referred to as the stress modified fracture strain model. The relationship between ε_f and the stress triaxiality is given, for instance, by,

$$\varepsilon_f = A\exp\left(-C\frac{\sigma_m}{\sigma_e} + B\right) \quad \text{Eq. (3)}$$

where A, B and C are material constants. In previous works (4), each constants were determined based on test data.

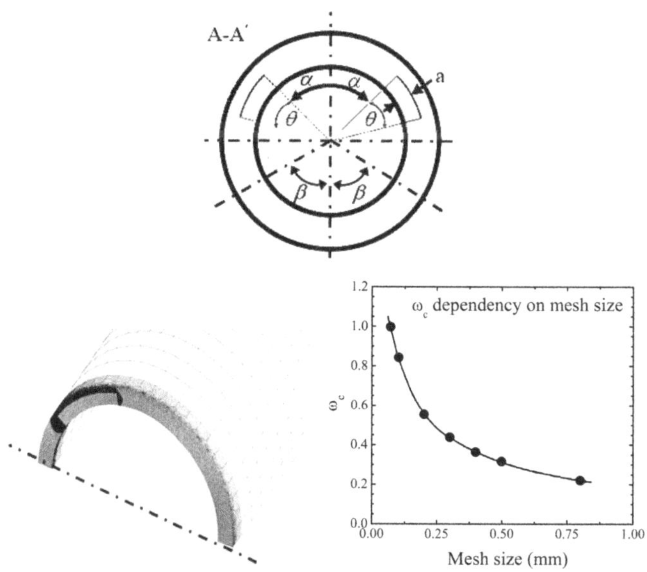

(a) (b)

Fig. 4 (a) Schematic illustration and FE model of pipes with two aligned symmetrical circumferential surface cracks and (b) dependence of the critical accumulated damage with element size.

A concept of the element-size-dependent critical damage is use the calibrated value of the critical accumulated damage. The critical accumulated damage is a sum of incremental damage due to plastic deformation at rupture state. Incremental damage due to plastic deformation, $\Delta\omega$, is defined by the equivalent plastic strain increment, $\Delta\varepsilon_e^p$, and ε_f:

$$\Delta\omega = \frac{\Delta\varepsilon_e^p}{\varepsilon_f} \qquad \text{Eq. (4)}$$

When the accumulated damage becomes unity, $\sum \Delta\omega = \omega_C = 1$, generally, crack growth is simulated. In this paper, the calibrated value of the critical accumulated damage that depends on a mesh size is used. The dependence of the critical accumulated damage with element size is shown in Fig .4(b). Conventional 3-D elastic-plastic FE analysis was performed to simulate ductile fracture of pipes .The analyses were conducted by using ABAQUS Standard 6.13 (5).

4 ASSESSMENT RESULTS AND CONCLUSION

In this paper, an applicability of net section collapse load approach to assessment of pipes with multiple circumferential cracks was investigated based on virtual testing using finite element damage analysis. Finite element damage analysis are validated by Hitachi's full-scale pipe bending test (3). Hitachi test pipes (4-inch diameter schedule 80) were made by a 304 stainless steel and contained two circumferential surface cracks. The comparison of assessment results between NSC load approach and virtual testing is described in Fig. 5. A Collapse moment ratio is compared depending on angle between two flaws, each flaw angle, and flaw depth. The assessment results of NSC load approach are in good agreement with virtual test results. So, we can conclude that net section collapse load approach can be available to assessment of pipes with multiple cracks.

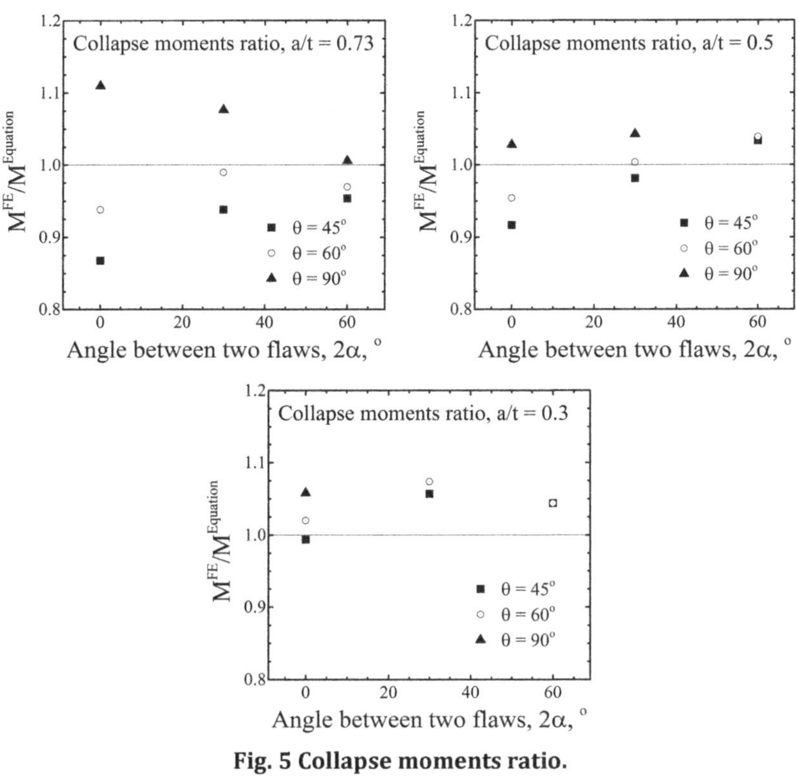

Fig. 5 Collapse moments ratio.

5 ACKNOWLEDGMENTS

This research was financially supported by the Ministry of Trade, Industry & Energy (MOTIV), Korea Institute for Advancement of Technology (KIAT) and Dongnam Institute for Regional Program Evaluation through the Leading Industry Development for Economic Region (No. 20131520202170).

This research was supported by National Research Foundation of Korea (NRF) funded by the Ministry of Science, ICT and Future Planning. (NRF- 2013M2A7A1076396)

This research was supported by Basic Science Research Program through the National Research Foundation of Korea (NRF) funded by the Ministry of Science, ICT & Future Planning (MSIP) (No. 2007-0056094).

6 REFERENCES

(1) ASME, (2011), *ASME Boiler and Pressure Vessel Code,* **XI**, 17. ISSN: 0517-5321

(2) Li Y., Hasegawa, K., Onizawa K. and Masayoshi S., (2009), *Proceedings of ASME 2009 PVP Division Conference*, **1**, 287-295. doi:10.1115/PVP2009-77061

(3) Hasegawa, K., Saito, K., Iwamatsu, F. and Miyazaki K., (2009), *J. Pressure Vessel Technol.*, **131**, 021209-021206. doi:10.1115/1.3066967

(4) J. H. Kim, N. H. Kim, Y. J. Kim, K. Hasegawa and K. Miyazaki, (2013), *Fatigue Fract. Engng. Mater. Struct.*, **00**, 1-13. doi: 10.1111/ffe.12072

(5) Dassault Systems, (2013), ABAQUS version 6.13. User's manual. http://50.16.176.52/v6.13/

Thermal aging effect on tensile and ratcheting behaviour of nuclear power pipeline steel

T. Liang, H.C. Cheng, G. Chen, X. Chen
School of Chemical Engineering and Technology, Tianjin University, P.R. China

ABSTRACT

The uniaxial tensile and the ratcheting-fatigue behaviors of the nuclear power pipeline steel (Z2CND18.12N) were studied with different thermal aging times (from 1 hour to 500 hours) at two thermal aging temperatures (500℃ and 700℃). The aged specimens conducted slight but consistent decrease of yield stress and Young's modulus than those of the original specimen. The ultimate strength and the elongation were not strongly affected by any thermal aging conditions. The thermal aging process resulted in the apparent changes of the ratcheting behavior and the ratcheting-fatigue life.

1 INTRODUCTION

The aging effect on the mechanical properties of stainless steels exposed to the elevated temperature range has been studied since several decades ago. For the duplex stainless steel (the amount of ferrite phase above 15wt% in the austenitic matrix) exposed to the intermediate temperature range, the thermal aging usually promotes an increase in hardness, yield stress and ultimate tensile stress, as well as a decrease in ductility, toughness and impact properties (Xue, 2009). The cyclic deformation is also influenced in the intermediate range and large range of the plastic strain (Llanes et al. 1996). The piping under bending, torsion and pressure loads in nuclear power plant will produce ratcheting effect which due to vibration and the ratcheting deformation will influence their lives. As the important significance of the ratcheting effect, there are large quantity of research on the ratcheting and ratcheting-fatigue behavior and the influence factors such as the mean stress, stress amplitude, stress rate, chemical composition, temperature, pre-strain/pre-corroded and etc. (Hassan et al. 1994, Kang et al. 2002, Chen et al. 2005, Wang et al. 2013).

In this study, the Z2CND8.12N austenitic stainless steel specimens with different thermal aging conditions (different aging times from 1 hour to 500 hours at two temperatures of 500℃ and 700℃) were tested under cyclic stress. The ratcheting behaviors were analyzed in detail on the hysteresis loop, ratcheting strain and the ratcheting strain rate. The ratcheting-fatigue lives under mean stress of 75MPa and stress amplitude of 275MPa were also obtained for the piping material of the nuclear power plant.

2 MATERIALS AND EXPERIMENTS

Nitrogen controlled austenitic stainless steel Z2CND18.12N was used as the test material with the chemical composition shown in Table 1. The original form of the material was a pipe with the outer and inner diameter of 76mm and 67mm, respectively. Specimens with the gauge length of 20mm and gauge width of 5mm were cut longitudinally from this pipe. Holding parts at both ends of the specimen were designed long enough to fit for the plate clamps. In the thermal aging process, the specimens were heat treated at 500℃ and 700℃ respectively for various time durations of 1 hour, 10 hours, 50 hours, 100 hours and 500 hours in the heating furnace. All the uniaxial tensile tests and cyclic tests were conducted on a 20-kN servohydraulic tension-compression fatigue testing machine with a closed-loop digital controller. The gauge length of the extensometer was 12.5mm. All the experimental equipment was developed by CARE lab, Tianjin University.

Table 1 The chemical composition of Z2CND18.12N (in wt%).

C	Si	Mn	P	S	Ni	Cr	Mo	N	Cu	Co
0.025	0.43	1.211	0.021	0.003	12.073	17.517	2.388	0.07	0.075	0.035

In this study, the uniaxial tensile tests were conducted under strain rate of 0.1%/s. A series of stress-cycling tests were conducted under stress amplitude of 275MPa and mean stress of 75MPa with the frequency of 0.2Hz. The cyclic test mode was stress-controlling with typical triangular waves.

The ratcheting strain is defined as:

$$\varepsilon_r = \frac{\varepsilon_{\max} + \varepsilon_{\min}}{2} \tag{1}$$

where $\varepsilon_{\max}$ and $\varepsilon_{\min}$ are the maximum strain and the minimum strain respectively in a cycle.

The ratcheting strain rate is defined as:

$$\dot{\varepsilon}_r = d\varepsilon_r / dN \tag{2}$$

where N is the number of cycle.

3 RESULTS AND DISCUSSION

3.1 Uniaxial tensile properties

The Young's modulus, 0.2% yield strength (YS), ultimate tensile strength (UTS) and the elongation of both the original specimens and the thermal-aged specimens were obtained from the uniaxial tensile tests. The tensile properties, the Young's modulus, yield strength, ultimate tensile strength and elongation, with different thermal aging time were listed in the Table 2. The results show that compared with the original specimens, there is a slight but consistent decrease in both Young's modulus and the yield strength of the aged specimens. The decreases of the Young's modulus and the yield strength of the specimens after aging at 700℃ are more apparently than that of 500℃.

Table 2 Tensile properties of specimens in different aged conditions.

Aging condition	Young's modulus (GPa)	YS (MPa)	UTS (MPa)	Elongation (%)
Original	195.1	290.7	603	52.0
500℃/1h	175.7	254.3	612	47.0
500℃/10h	159.8	236.0	631	52.3
500℃/50h	177.2	247.0	614	51.7
500℃/100h	185.3	250.0	612	52.0
500℃/500h	181.9	241.3	612	51.2
700℃/1h	155.2	237.7	593	54.8
700℃/10h	160.3	221.9	600	54.2
700℃/50h	157.4	224.1	601	52.9
700℃/100h	154.7	222.7	593	53.2
700℃/500h	172.7	224.7	605	53.0

3.2 Effect of thermal aging on the cyclic behavior

Figure. 1 shows the evolutions of the initial 50 hysteresis loops and the ratcheting strain of the original specimen under mean stress of 75MPa and stress amplitude of 275MPa at room temperature. As shown in Fig. 1(a), the plastic strain accumulation was apparent in the initial several cycles, and then declined because of the cyclic hardening effect. The evolutions of the ratcheting strain defined by Eq. (1) with the increasing number of cycles for the original specimen under mean stress of 75MPa and stress amplitude of 275MPa are shown in Fig. 1(b). The curve of the ratcheting strain could be divided into three stages. In the primary stage, the ratcheting strain increased rapidly since the plastic strain accumulated fast. In the second stage, the ratcheting strain increased steadily with a constant rate. In the last stage, the ratcheting strain increased rapidly because of the crack initiation. Finally, the specimen failed as a result of crack propagation.

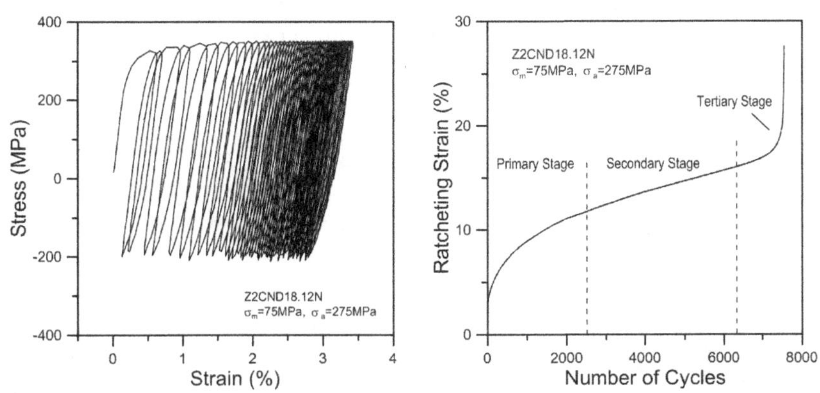

Fig. 1. The evolutions of the (a) initial 50 hysteresis loops and (b) ratcheting strain of the typical ratcheting behavior of the Z2CND18.12N without thermal aging.

Figure 2 shows the first hysteresis loop of both original specimen and thermal-aged specimens under mean stress of 75MPa and stress amplitude of 275MPa with aging temperatures of 500℃ and 700℃ respectively. The yield stresses of the aged specimens were lower than that of the original specimen, which was coincidence with the uniaxial tensile results. Since the decrease of the yield stress of the aged specimen, the plastic strains of the aged specimens in the first hysteresis loop were larger than that of the original specimen. The widths of the hysteresis loops increased with the increasing thermal aging time for both thermal aging temperatures. As shown in the Fig. 2(a) and Fig. 2(b), the ratcheting strains in the first cycle after aging at 700℃ were larger than the ratcheting strains after aging at 500℃. Figure 3 shows the evolutions of the ratcheting strains with cycles after thermal aging at both 500℃ and 700℃ under mean stress of 75MPa and stress amplitude of 275MPa in the initial 1000 cycles. All the ratcheting strains increased with the number of cycles since the plastic strain accumulation. After aging at 500℃, the ratcheting strains increased with the increasing thermal aging time. For the aging time of 50h, 100h and 500h, the ratcheting strains increased faster significantly than the aging time of 1h, 10h and the original specimen. After aging at 700℃, the ratcheting strain also increased with the increasing thermal aging time. The ratcheting strains after thermal aging increased faster apparently than the original specimen.

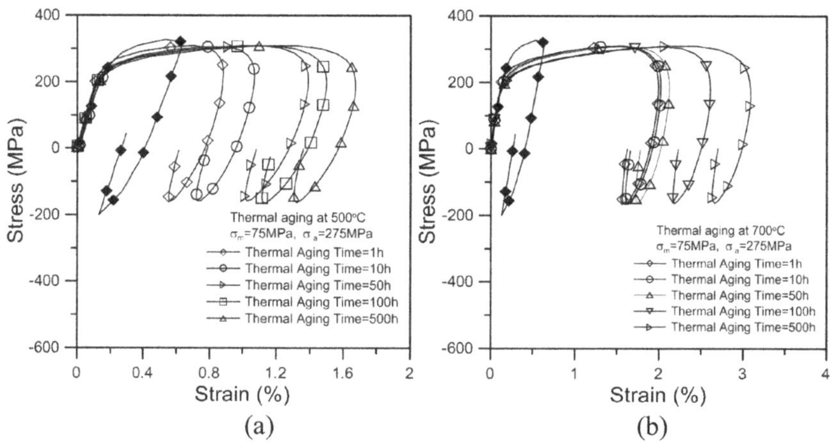

Fig. 2. The hysteresis loops of the first cycle of both the original specimen and the thermal-aged specimens after thermal aging at (a) 500℃ and (b) 700℃.

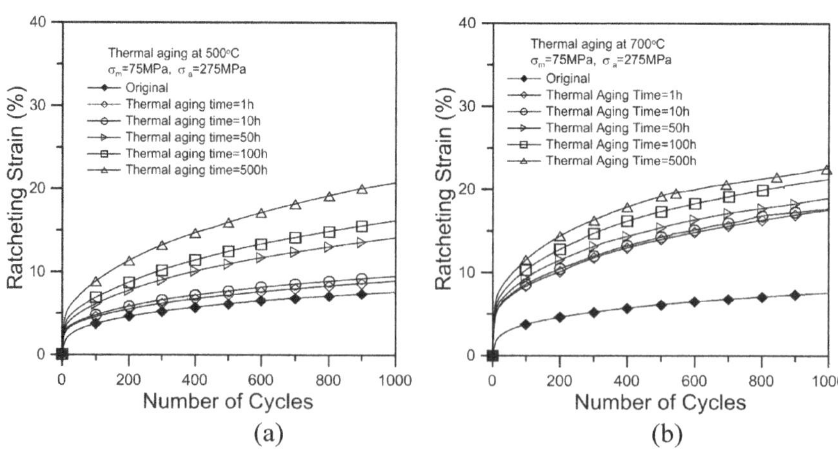

Fig. 3. Ratcheting strains for both the original specimen and the thermal-aged specimens after thermal aging at (a) 500℃ and (b) 700℃.

3.3 Effect of thermal aging on the ratcheting-fatigue life

Figure 4 shows the effect of thermal aging with different conditions on the ratcheting-fatigue life. Under the both aging temperatures of 500℃ and 700℃, the ratcheting-fatigue lives decreased with the increasing aging time apparently. The curve with 500℃ shows the constant decreasing trend while the curve with 700℃ shows the rapid decrease at the point of 50h and then saturated trend at the point of 500h. As shown in Fig. 4, the ratcheting-fatigue lives also decreased with the increasing aging temperature with the same aging time. With the short aging time of 1h and 10h, the differences of the fatigue lives between two aging temperatures were small. With the aging time of 50h and 100h, the differences of the fatigue lives increased since the rapid decrease of the lives under 700℃. Generally, the ratcheting-fatigue life decreased with the increasing aging temperature and the increasing aging time.

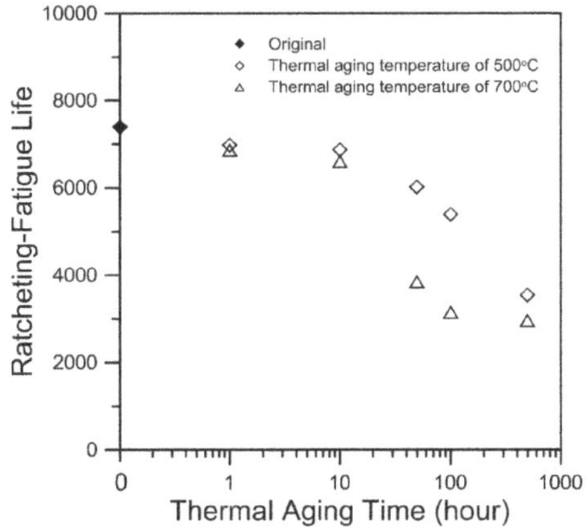

Fig. 4. Ratcheting-fatigue lives for original and thermal aged specimens with various aging time.

4 CONCLUSIONS

In the present work, the thermal aging effect on the mechanical properties of Z2CND18.12 stainless steel was studied after aging for 1h, 10h, 50h, 100h and 500h at aging temperature of 500℃ and 700℃ respectively. The mechanical properties of the aged material were analyzed in uniaxial tensile properties, cyclic behaviors and the ratcheting-fatigue lives. The following conclusions can be drawn:

(1) For the tensile properties, the aged specimens conducted slight but consistent decrease of yield stress and Young's modulus than those of the original specimen. The ultimate strength and the elongation were not strongly affected by any thermal aging conditions.

(2) For the cyclic behaviors, the widths of the first hysteresis loops increased with the thermal aging time for both aging temperatures, and all the widths of the aged specimens were larger than that of the original specimen.

(3) The ratcheting strains in the initial 1000 cycles also increased with the thermal aging time for both aging temperatures. The ratcheting strain rates increased apparently after aging at 500℃ and almost unchanged after aging at 700℃.

(4) Generally, the ratcheting-fatigue lives decreased with the increasing thermal aging time and the increasing thermal aging temperature.

ACKNOWLEDGEMENTS

The authors are grateful for the financial support from Ph.D. Program Foundation of Education Ministry of China(No. 20130032110018).

REFERENCES

[1] Xue, F., Wang, Z-X., Shu, G., Yu, W., Shi, H-J. & Ti, W. (2009) Nucl Eng Des. **239**, 2217-23.
[2] Llanes, L., Mateo, A., Iturgoyen, L. & Anglada, M. (1996) Acta Mater. **44**, 3967-78.
[3] Hassan, T. & Kyriakides, S. (1994) Int J Plast. **10**, 149-84.
[4] Kang, G., Gao, Q. & Yang, X. (2002) Mech Mater. **34**, 521-31.
[5] Chen, X., Jiao, R. & Kim, K. S. (2005) Int J Plast. 21, 161-84.
[6] Wang, Y., Yu, D., Chen, G. & Chen, X. (2013) Int J Fatigue. **52**, 106-13.

Effect of models and derivation methods for initial flaw size distribution on probability of failure of airframes

R.F. Torregosa, W. Hu
Aerospace Division, Defence Science and Technology Organisation, Australia

ABSTRACT

The distribution of initial flaw sizes is one of the key inputs for the calculation of the probability of failure of aircraft structures. In practice, an equivalent flaw size is used instead of the physical flaw size, to avoid numerical difficulties and to provide consistencies. The equivalent flaw size is derived, ultimately, from crack-time data obtained from teardown inspections or purpose-conducted experiments, but different methods may be used to derive the distribution of the equivalent flaw size. In this paper, two such methods are examined, and their effects on probability of failure are assessed.

1 INTRODUCTION

Aircraft structural integrity assessment is a major activity for aircraft owners and operators, during acquisition and the subsequent operation, and the analysis of probability of failure is an integral part of this assessment. MIL-STD 1530C [1] specifies that a probability of catastrophic failure at or below 10^{-7} per flight for aircraft structure is adequate to ensure safety for long term military operations. The standard also specifies that probabilities of catastrophic failure exceeding 10^{-5} per flight should be considered unacceptable, and mitigation measures should be taken if the risk is between 10^{-7} and 10^{-5}, in order to ensure the safety of the personnel and the aircraft. The probability of failure as defined in MIL-STD1530C refers to the probability of happening within the next flight, referred to as single flight probability of failure (SFPoF).

To calculate SFPoF, a minimum of four categories of inputs are required, namely a) the distribution of the sizes of initial flaws that characterises the build condition; b) the crack growth curve that governs the evolution of these flaws under typical service load spectrum; c) the residual strength that indicates the load bearing capacity of the structure and d) the peak stress exceedance that characterises extreme loads experienced by the structure, in a probabilistic sense. In practice, an equivalent initial flaw size (EIFS [2]) is used instead of the physical flaw size, to avoid numerical difficulties and to provide consistencies. References [2, 3] show that the EIFS distribution has the most significant effect on the probability of failure. However, these studies did not investigate the sensitivity of the SFPoF values to the assumed distribution models for the EIFS and the methods used to derive them from crack-time data obtained from teardown inspections or purpose-conducted experiments.

In this paper, two methods used for EIFS distribution derivation are examined to evaluation their effect on SFPoF. The two methods are referred to as the Direct EIFS method and the Time to Crack Size (TTCS) [2, 4] method. For the TTCS method, different baseline crack sizes were assumed and their effects on the SFPoF values are compared. The outcome of this study will provide guideline on the selection of most suitable models and method of derivation for the EIFS distribution.

2 SINGLE FLIGHT PROBABILITY OF FAILURE (SFPOF)

Probability of failure (PoF) is the likelihood that the applied stress, S, exceeds the residual strength, S_{cr} based on fracture toughness of the material. $S_{cr}(a)$ is the load bearing capacity of the structure containing a crack of size a. PoF, therefore, is a function of crack size and is defined as,

$$PoF(a) = P(S \geq S_{cr}(a)) \tag{1}$$

The probability that an aircraft will fail during the next flight, SFPoF, is given by [5, 6],

$$SFPoF(t) = \int_0^{\infty} f_a(a_t)\left(1 - \int_0^{S_{cr}(a_t)} f_s(s)ds\right)da, \tag{2}$$

where $f_a(a_t)$ is the probability density function of crack size, and the subscript t indicates the dependence of crack size on time. $f_s(s)$ is the peak stress probability density function for a single flight. At t = 0, $f_a(a_0)$ refers to the EIFS distribution. In Equation (2), without the integral of $f(s)$ term the equation would equal to 1 as the integral of any probability density function is a cumulative density function, but the integral of $f_s(s)$ reduces the integral at each *da* interval and therefore reduces the integral to a value less than one, the SFPoF.

3 EIFS DISTRIBUTION DERIVATION METHODOLOGIES

For a given set of crack data (a_i, t_i), the EIFS distribution may be derived using two methods, the Time to Crack Size (TTCS) and the Direct EIFS methods.

3.1 Time to Crack Size (TTCS) Method

In the TTCS method, the EIFS distribution is derived by first projecting observed crack data to a specific crack size, referred to as the baseline crack size in this paper, using a master crack growth curve, and then determining a life distribution at that crack size. The life distribution is then back-extrapolated to time zero to get the EIFS distribution. The method is illustrated in Figure 1a. The steps are as follows:

1. Set the baseline crack size (a_{TTCS});
2. Use the crack growth curve to project each in-service crack size data to a_{TTCS} and determine its corresponding fatigue life;
3. Determine the life distribution at a_{TTCS} using an appropriate distribution model, e.g., Lognormal distribution or Weibull distribution;
4. Use the same crack growth curve to back-extrapolate each point in the life distribution to get the EIFS distribution.

Clearly the EIFS distribution is not directly derived from the crack data; rather, it is obtained by projecting a fatigue life distribution at a selected crack size. As such, the derived EIFS distribution is not unique; it depends on the assumed baseline crack size.

3.2 Direct EIFS Method

The Direct EIFS method is a simple approach in determining the EIFS distribution. The EIFS corresponding to each pair of crack data is directly regressed to time zero and a probabilistic distribution is constructed using the regressed data, see Figure 1b.

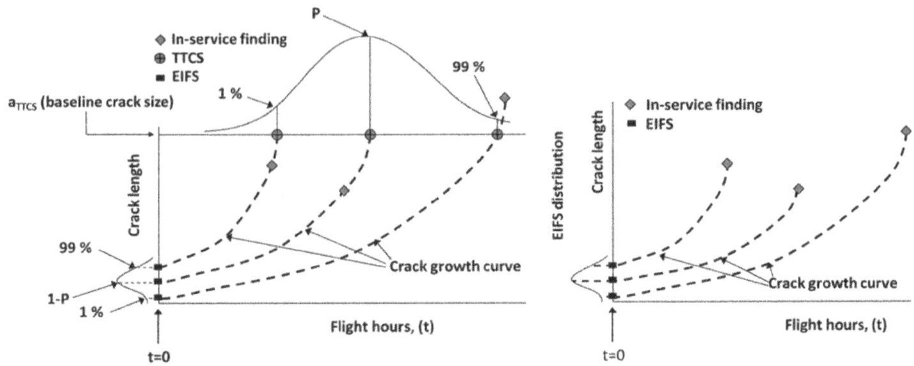

Figure 1 Two most common EIFS derivation methods, a) TTCS Method and b) Direct EIFS Method

3.3 Derived EIFS Distributions and Probabilistic Risk Analysis of Fracture

A total of 145 crack data obtained from teardown inspection of C-130H aircraft were compiled to be used for EIFS distribution development. The number was reduced to 64 by excluding data relating to multi-site damage. Both the Direct EIFS method and the TTCS method were used to derive the EIFS distribution. In Direct EIFS method, the lognormal distribution model was used to fit the regressed flaw sizes. In the TTCS method, the lognormal distribution model was used to represent the life distribution at the baseline crack size. Three baseline crack sizes (i.e., 0.10, 0.20 and 0.30 inches) were assumed. The resulting EIFS distributions are shown in Figure 2a, and the right tails of the distributions are shown in the enlarged view in Figure 2b, as cumulative distribution functions (CDF). The mean EIFS values from the Direct method and the TTCS method with baseline sizes of 0.10, 0.20, 0.30 inches are 0.00111, 0.00211, 0.00161 and 0.00157 inches, respectively. The CDF of the EIFS distribution from the Direct EIFS method shows a curve extending indefinitely and asymptotes to F(x)=1.0. For the three distributions from the TTCS method, F(x)=1.0 is reached at x values corresponding to each baseline crack size. It should be noted that in Figure 2, the maximum horizontal coordinate is fixed at 0.08 inches to emphasize the differences in the right tail of the distributions. The derived EIFS distributions were used to analyse the SFPoF of C-130H using the risk input parameters from reference [6].

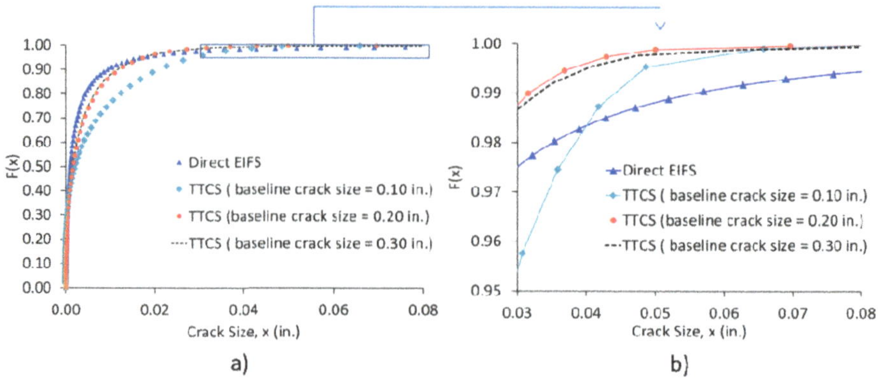

Figure 2 EIFS distributions derived using the Direct EIFS method and the TTCS method corresponding to baseline crack sizes of 0.10, 0.20 and 0.30 inches, a) showing CDF curve and b) showing the right tail of the distribution

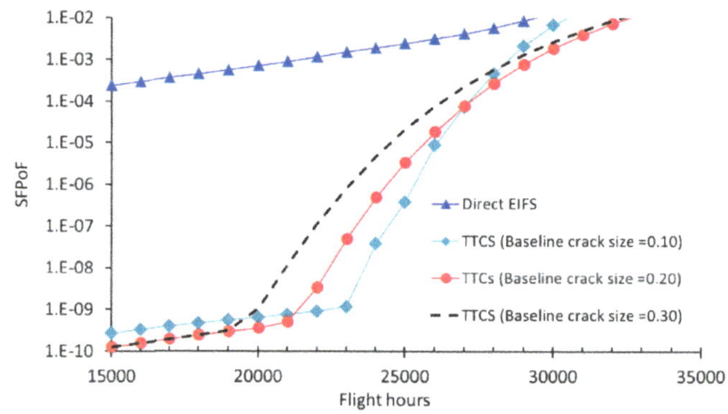

Figure 3 Comparison of SFPoF for different EIFS methods and baseline crack sizes

4 RESULTS AND DISCUSSION

Comparison of SFPoF curves of the four EIFS distributions are shown in Figure 3. It can be seen that the Direct method gives the highest risk value for the data considered. This is an interesting observation because the mean of the EIFS distribution from the Direct method is the smallest yet the method leads to the highest probability of failure. The reason for this is that the right tail of the distribution is asymptotic to 1.0, which means that very high EIFS values are also possible. In the TTCS distributions, the EIFS has an upper bound limit which is the baseline crack size. Since the probability of failure is sensitive to the right tail of the EIFS distribution, the Direct EIFS method gives the highest risk values.

The SFPoF curves from the TTCS method also show significant differences for different baseline crack sizes. At acceptable SFPoF value of 10^{-7} [1], the difference in time between baseline crack size of 0.10 and 0.30 is about 2400 flight hours, which is a significant usage difference and involves huge operating cost savings if aircraft life can be safely extended. The result in this study showed that assumed baseline crack size has significant effect on the SFPoF. The results also show that unbounded distribution models such as the lognormal distribution may not be appropriate for the Direct EIFS method, and the TTCS method may be a better option since the resulting EIFS distribution has an upper bound limit. However, when the TTCS method is used, an appropriate upper bound relevant to the structural location being analysed must be used.

5 CONCLUSIONS AND RECOMMENDATION

The sensitivity of SFPoF to EIFS distribution models is presented in this paper. It has been shown that the risk analysis of aircraft structures may be refined by selecting the most appropriated EIFS distribution model and derivation methods. Based on the example presented here, the following may be concluded:

1) The EIFS distribution derived by the Direct Method may lead to very high probability of failure, especially when an unbounded distribution model is used.

2) The EIFS distribution derived by the TTCS Method give probability of failure which are sensitive to the assumed baseline crack size. The baseline size should correspond to the actual condition of the location being analysed so as not to produce unrealistic EIFS values.

3) The mean of the EIFS distribution has very little influence on the SFPoF values whereas the right tail of the EIFS distribution has the most significant influence.

4) More accurate and realistic assumption of the upper bound of the EIFS distribution is necessary in analysing the fatigue of aircraft structures.

The authors recommend that distribution models that have finite upper bound limit should be explored further, especially when Direct EIFS distribution is used.

ACKNOWLEDGMENT

The authors wish to acknowledge Dr. Colin Pickthall for reviewing the manuscript.

REFERENCES

1. US Department of Defence, *MIL-STD-1530C, Aircraft Structural Integirty Program (ASIP)*. 2005.
2. Liao, M., Bombardier, Y., and Renaud, G., Quantitative Risk Assessment for the CC-130 Centre Wing Structure (Phase II), National Research Council Canada, 2009.
3. White, P., L. Molent, and S. Barter, *Interpreting fatigue test results using a probabilistic fracture approach.* International Journal of Fatigue, 2005. **27**(7): p. 752-767.
4. Rudd, J., Yang, J., Manning, S. and Yee, B., *Probabilistic Fracture Mechanics Analysis Method for Structural Durability*, in *55th Meeting AGARD Structures and Materials Panel.* 1982. Toronto, Canada.
5. Hovey, P.W., Berens, A.P., and Loomis, J.S., Update of the PROF Computer Program for Aging Aircraft Risk Analysis, 1998.
6. Torregosa, R.F., and Hu, W., C-130H Centre Wing Lower Surface Panel Number 3 Probabilistic Risk Analysis, Defence Science and Technology Organisation, DSTO-RR-0371, 2011.

A life prediction method for aircraft structure based on enveloping life surface

Y.T. He, T. Zhang, L.M. Wu, C.F. Li
Aeronautics and Astronautics Engineering College, Air Force Engineering University, P.R. China

ABSTRACT

A new concept named aircraft structure enveloping life surface (ASELS) is presented. The ASELS describes the interrelationships between stress level, fatigue life, and calendar life. This may be used to predict the residual life of aircraft structures. The development of the ASELS is described. Furthermore, by combining ASELS with Miner linear cumulative damage theory, a residual life prediction method for aircraft structures under typical environment is established.

1 INTRODUCTION

The structural life of an aircraft is affected by its in-service environment, which relates specifically to issues of corrosion and cyclic loading. Some US Military Standards, such as MIL-A-8860B and MIL-STD-1573, incorporate corrosion into aircraft production and usage process protocols. In China, there are two indices to monitor aircraft structural life, which include fatigue life and calendar life.

Previous studies have presented the concept of an aircraft structure life enveloping curve (ASELC) based on an analysis of aircraft structural life indices and considering the interaction between fatigue life and calendar life (1-3). These studies also presented applications of this theory.

In this paper, a new concept named aircraft structure enveloping life surface (ASELS) is presented based on the ASELC. Furthermore, an approach to establishing the ASELS is developed and a residual life prediction method for aircraft structures is presented based on the ASELS.

2 CONCEPT OF AIRCRAFT STRUCTURE ENVELOPING LIFE SURFACE

ASELS describes the safe and reliable life scope for aircraft structures in service. It reflects the interrelationships between stress level (***S***, in MPa), fatigue life ($\boldsymbol{N_f}$, in flight hours), and calendar life ($\boldsymbol{N_y}$, in years). When an aircraft is used so heavily that it exceeds the limits of ASELS, the structural state is considered to be unsafe. As shown in Fig. 1, the ASELS can be considered as an extension of the ASELC (Sight B) along the coordinate direction of stress level or an extension of the S-N curve (Sight A) along the coordinate direction of calendar life.

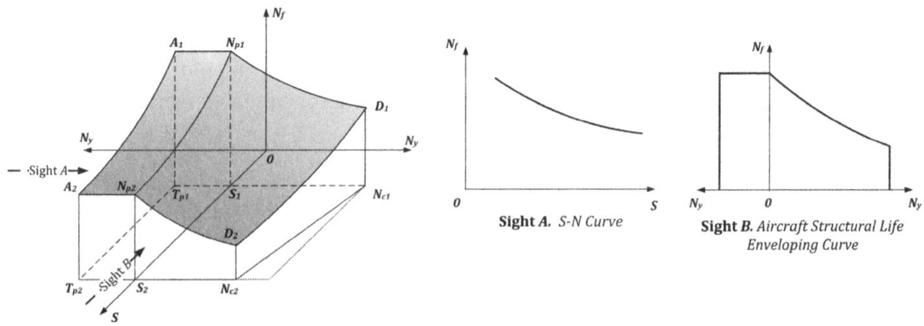

Fig. 1 Diagram of an aircraft structure enveloping life surface

Fig.1 shows the ASELS in a typical environment. This can be used to predict the structural residual life of an aircraft grounded in the environment. If an aircraft is grounded in different environments in full life circle (the corrosion properties relevant for high altitudes can be ignored), different ASELSs corresponding to these environments should be used for the prediction of residual life. In Fig.1, both directions of the abscissa axis (***N$_y$***) are positive; they are calendar lives under different states of protective coating. The abscissa of T_{p1}/T_{p2} is the effective period of protective coating; structures can be considered to be suffering from pure fatigue damage within this period. The points N_{p1} and N_{p2} represent safety lives with high reliability under the stress levels S_1 and S_2, respectively. These parameters can be obtained through component or full-scale fatigue testing and through reliability analyses. The surface N_{p1}-D_1-D_2-N_{p2} reflects change laws of the fatigue lives of structures under different stress levels in the corrosive environment without protective coating. The surface D_1-N_{c1}-N_{c2}-D_2 is a boundary limit designed to prevent unexpected fracture of a structure due to corrosion fatigue damage.

3 ESTABLISHMENT OF AIRCRAFT STRUCTURE ENVELOPING LIFE SURFACE

To consider the effect of the corrosive environment on the life quality of an aircraft structure and following the approach to establishing the ASELC described in the literature (2, 4), the ASELS can be established as follows. A schematic diagram illustrating the establishment of the ASELS is shown in Fig.2. It shows the ASELS determined by only three stress levels. In reality, the more tests that are carried out under more stress levels, the more accuracy is achieved with the ASELS.

In Fig.2, the effective period of protective coating (the abscissa of T_{pi}) can be determined by equivalent corrosion testing of coating specimens in laboratory experiments with conditions simulating actual service environment. The N_{p1} to N_{p3}, which are fatigue safe lives with high reliability, can be determined by performing fatigue tests under several stress levels (S_1 to S_3 in Fig.2). The N_{p1}-N_{p3} curve is actually a *p-S-N* curve with a high reliability determined by reliability analysis.

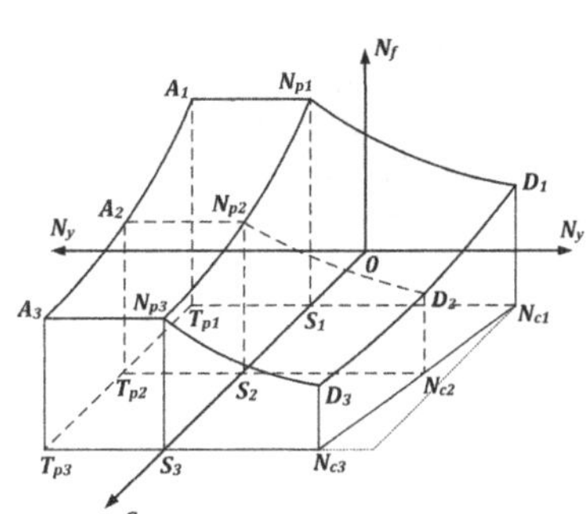

Fig. 2 Schematic diagram of the establishment of the ASELS

The $C(T)$ curve, which is also called the corrosion effect coefficient curve (4, 5), represents the effect of corrosion on fatigue life. It can be determined by pre-corrosion fatigue tests or by alternative corrosion fatigue tests of unprotected specimens. Any point on the $C_i(T)$ curve under the stress level S_i can be determined using Eq. (1)

$$C_i(T) = \frac{N_{99.9i}(T)}{N_{pi}} \tag{1}$$

where N_{pi} is the fatigue safe life under stress level S_i, $N_{99.9i}(T)$ is a fatigue safe life with a reliability of 0.999 and a confidence of 0.95 (when the fatigue lives follow a log-normal distribution) after an equivalent of T years of corrosion. The $C_i(T)$ curve can be fitted as

$$C_i(T)=1-aT^b \tag{2}$$

where T is the equivalent calendar years and a and b are fitting parameters.

With an increase in service time, the life quality of an aircraft structure decreases. In order to prevent unexpected fractures, the boundary limit (the surface D_1-N_{c1}-N_{c3}-D_3 in Fig.2) must be determined. It relates to parameters such as the demands of static strength and fracture characteristics in corrosive environment, the financial demands associated with the repair of structures, and to the tactical capabilities of the aircraft.

The right part of the ASELS can be fitted by multiplying the fatigue safe lives (N_{pi}) by the $C_i(T)$ curves and by cutting off the curve at the boundary limit. Finally, the whole ASELS of a typical environment is produced.

4 APPLICATION OF AIRCRAFT STRUCTURE ENVELOPING LIFE SURFACE

The ASELS can be used to predict the residual life of an aircraft structure. Aircraft structural life is a dynamic variable in actual service processes, due to dynamic changes in loading and in the service environment. In the prediction process of aircraft structures, damage quantity is a bridge between the consumed life and the residual life. Combining with flight damage and Miner cumulative damage theory, when the overall damage reaches 1, the structure has reached its life limit and it should retire. The life prediction method for aircraft structures based on the ASELS is described below.

When the protective coating is in the effective period (the abscissa of T_{pi}), the structure may be considered to be suffering from fatigue damage only. The cumulative fatigue damage of a structure is related to the flight strength, which includes flight load level and flight density. The flight load level can be calculated using data obtained from the flight data recorder and the critical structure sensors. The flight density relates to flight hours per year. According to the Miner theory, the cumulative damage of an aircraft structure in the effective period of coating (d_A) can be calculated as

$$d_A = \sum_{i=1}^{n} \sum_{T=0}^{T_{pi}} \frac{T_{Ti}}{N_{pi}} \tag{3}$$

where i is the sequence number of the stress level, n is the total number of stress levels, T is the calendar time of the aircraft, T_{pi} is the effective period of the protective coating, N_{pi} is the fatigue safe life under the ith stress level, S_i, and T_{Ti} is the total flight time under the S_i.

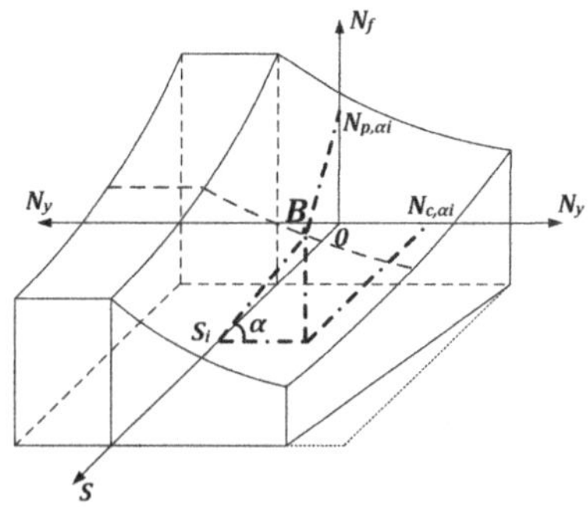

Fig. 3 Schematic diagram of an application of the ASELS

Since the protective coating loses integrity, the base material of the aircraft structure suffers interactive damage of fatigue and corrosion. The right side of the ASELS can be used to calculate the cumulative damage in this period. Any position on the ASELS corresponds to a state of flight strength. As shown in Fig.3, point B ($N_{c,\alpha i}$, S_i, $N_{p,\alpha i}$) on the surface N_{c1}-N_{c3}-D_3-D_1 corresponds to a stress level S_i and flight density $I_{\alpha i}$ ($I_{\alpha i}=N_{p,\alpha i}/N_{c,\alpha i}=\tan\alpha$). This indicates that an aircraft can be used for at most $N_{c,\alpha i}$ years if it is in service for $I_{\alpha i}$ hours per year under the flight load level S_i without the protection of coating. Consequently, if a structure is used in $I_{\alpha i}$ hours per year under the flight load level S_i, since the protective coating loses integrity with service time, the structural damage per year (d_B) can be calculated as

$$d_B = \frac{I_{\alpha i}}{N_{p,\alpha i}} = \frac{1}{N_{c,\alpha i}} \tag{4}$$

where $I_{\alpha i}$ is the flight density and $N_{p,\alpha i}$ and $N_{c,\alpha i}$ are the fatigue life limit and calendar life limit of point B, respectively. The $N_{p,\alpha i}$ and $N_{c,\alpha i}$ can be calculated using the following equation

$$N_{p,\alpha i} = I_{\alpha i} \times N_{c,\alpha i} = C_i(N_{c,\alpha i}) \times N_{pi} \tag{5}$$

where N_{pi} is the fatigue safe life under stress level S_i and $C_i(N_{c,\alpha i})$ can be determined using Eq. (2).

If an aircraft is only grounded in a typical environment, the residual life of the aircraft structure can be predicated using the ASELS corresponding to these service conditions. For instance, the service conditions of an aircraft are shown in Table 1.

Table 1 Service conditions of an aircraft

Conditions of protective coating	Stress level	Flight density	Service time
Effective	S_1	I_1	T_1
	S_2	I_2	T_2
Loses integrity	S_1	I_3	T_3
	S_3	I_4	T_4

The cumulative damage of the aircraft structure (d_C) can therefore be calculated as follows

$$d_C = \frac{I_1 \cdot T_1}{N_{p1}} + \frac{I_2 \cdot T_2}{N_{p2}} + \frac{I_3 \cdot T_3}{N_{p,1}} + \frac{I_4 \cdot T_4}{N_{p,3}} \tag{6}$$

In Eq. (6), N_{p1} and N_{p2} can be determined using the ASELS according to the stress levels $N_{p,1}$ and $N_{p,3}$, which can be calculated using Eq. (5). The residual damage of the structure (d_R) is

$$d_R = 1 - d_C \tag{7}$$

If the aircraft will services in I_5 hours per year under the flight load level S_4, the residual fatigue life ($N_{p,R}$) is

$$N_{p,R} = d_R \times N_{p,5} \tag{8}$$

where $N_{p,5}$ can be calculated using Eq. (5). The residual calendar life ($N_{c,R}$) is

$$N_{c,R} = d_R \times N_{c,5} = N_{p,R} / I_5 \tag{9}$$

Due to the space limitations, the test and application results are not presented in the present paper. Some correlative examples can be found in the literature (4).

5 SUMMARY

A concept named aircraft structure enveloping life surface (ASELS) is presented and its development is described. The $C(T)$ curve is used to simulate the interactive processes of corrosion and fatigue on the aircraft structure in actual service environment and the Miner theory is used to predict the residual life of the aircraft structure based on the ASELS. The establishment of ASELS will pave the way for supervision of structural life for individual aircraft. It also can be used to predict residual lives of other mechanical equipment and structures.

ACKNOWLEDGEMENTS

Financial supports for this study were granted by National Natural Science Foundation of China (51201182) and Nature Science Foundation of Shanxi Province (2014JQ7256).

REFERENCES

[1] Y. T. He (2005) Journal of Air Force Engineering University **6**, 4-6.

[2] Y. T. He, C. H. Fan (2006) Journal of Air Force Engineering University **7**, 1-3.

[3] Y. T. He (2006) Engineering Science **8**, 23-27.

[4] H. W. Zhang, Y. T. He, L. M. Wu, Q. Shao, J. Q. Du (2011) Advanced Materials Research **299-300**, 1198-1202.

[5] X. F He, W. T. Liu, D. B. Jiang (2003) Journal of Beijing University of Aeronautics and Astronautics **29**, 229-232.

Fuzzy modeling to predict the adhesion strength of TiN ceramic thin film coating on aerospace AL7075-T6 alloy

E. Zalnezhad, A.A.D. Sarhan
Advanced Manufacturing and Material Processing Centre, Department of Mechanical Engineering, Faculty of Engineering, University of Malaya, Malaysia

ABSTRACT

In this research work, predicting of titanium nitride (TiN) coating adhesion on AL7075-T6 is presented. First TiN was coated on Al7075-T6 in different conditions and the surfaces adhesion of TiN coated specimens were measured using micro scratch force machine. Second a fuzzy logic model was established to predict the TiN coating adhesion with respect to changes in input process parameters, DC power, DC bias voltage, and nitrogen flow rate. Finally, new five experimental tests were carried out to verify the predicted results achieved via fuzzy model. The result indicated settlement between the fuzzy model and experimental results with 95.534% accuracy.

Keywords: AL7075-T6 alloy, TiN coating, PVD, Adhesion, fuzzy logic model

1. INTRODUCTION

Aerospace applications and energy saving strategies in general raised the interest and study in the field of lightweight materials, especially on aluminum alloys. Aluminum 7075-T6 alloy which is used in this research work is widely used in industry and in particular in aircraft structure [1]. This alloy has low specific weight and high strength to weight ratio but itself does not have suitable wear resistance. Therefore, improvements of surface properties are required in practical applications [1]. With the advent of new technologies, such as vacuum processing and advances in materials, such as ceramics and composites, the surface modification techniques based on new technologies have attracted more attention with respect to the traditional surface modifications ranging from glazing and painting to gas carburizing and electroplating over past decade [2].

Physical Vapor Deposition (PVD) is one of the vacuum coating processes in which the film of coating material is usually deposited atom by atom on a substrate by condensation from the vapor phase to the solid phase. This technology improved durability, higher surface hardness and increased service temperatures can be achieved from less expensive [3]. In this research work, TiN ceramic PVD-coating is used to improve the surface properties of Al7075-T6 for higher wear resistance [4]. The most critical issue in application of coated elements is the adhesion between a coating and a substrate. Adhesion is usually evaluated by scratch test technique [5]. The critical load, the load at which the coated film is removed from the substrate, is influenced by many different coating parameters that can be setup in advance, such as; DC power, temperature, nitrogen flow rate, and substrate DC bias voltage. Hence, a reliable systematic approach to investigate and predict the surface adhesion strength at different TiN coating parameters is thus required.

Soft computing techniques are useful when exact mathematical information is not available. Fuzzy logic is one of the soft computing techniques that play an important role in input-output parameter relationship modeling [6]. Fuzzy logic is preferred in predicting coating adhesion based on the input variables due to nonlinear condition in coating process [7].

In this present work, TiN was coated on Al7075-T6 substrate in different parameters conditions. Each parameter has four levels which, these parameters include; the DC power, substrate temperature, the nitrogen flow rate and DC substrate biases voltage. Fuzzy modelling was proposed to predict adhesion of TiN coating on AL7075-T6 alloy.

2. DESIGN OF EXPERIMENTS

The most important stage in the design of an experiment lies in the selection of parameters and identifying the experimental array. In this experiment with four parameters and four levels each, the experimental array used is L_{16} (4^4). The parameters and levels are shown in Table 1.

Table 1. Factors and levels

Control factors		Experimental levels			
		1	2	3	4
A	DC Power (w)	300	350	400	500
B	Temperature(°C)	150	180	200	220
C	Ni flow rate (%)	3	4	5	6
D	Biases voltage (v)	25	50	75	100

3. TEST SPECIMENS AND COATING PREPARATION

Aluminum alloy 7075-T6 was used in this research work. The surface of all samples were polished with SiC papers grit 800-2000, after that all samples were surface mirroring by diamond liquid and the substrate were ultrasonically cleaned in acetone for 14 min, thoroughly rinsed with distil water and dried using nitrogen gas to avoid contamination. A SG Control Engineering Pte Ltd series magnetron sputtering system was used to experimentally deposit thin films of metal. This system contained 600W RF and 1200 W DC generators with 4" × 12" electrodes 15 cm away from the target. To easily sputter metals we designed DC generators. The substrate carrier was circular and was rotatable at various speeds for required co-sputtering deposition. The chamber was evacuated to below 2×10^{-5} Torr before the argon gas for sputtering was introduced. Here, we used constant sputtering pressure 5.2×10^{-3} Torr. The substrate temperature, DC bias voltage, nitrogen flow rate and DC power as coating parameters are arranged according to the experimental array shown in Table 1 to learn how to improve the adhesion of the sputtered TiN thin film. A pure titanium 99.995% target was selected for investigating the sputtering conditions for Al 7075-T6 alloy. The layers were characterized using scanning electron microscopy (FE/SEM-FEG), focused ion beam techniques (Quanta FEG250). Adhesion of surfaces was determined using micro scratch force equipment (Micro Material Ltd, Wrexham, U.K.).

4. EXPERIMENTAL RESULT

Figure 1, shows a typical example of a TiN coating, it can be seen under SEM that the coating structure is Laminar. Figure 1 is also shown the diffusion rate of Ti and Nitrogen, chemical composition of AL 7075-T6, the interfacial layer of Titanium, TiN and Aluminum. The adhesion of TiN coating was measured using micro scratch force equipment. Each measurement repeated three times and the averages are calculated. A diamond indenter (Rockwell type) of 25 µm radius applied an initial load zero onto a sample. The sliding velocity was 5µm/sec. The load was increased gradually by 9.2mN/sec. The scratch's length during scratch test was 1000µm. In the scratch test, critical load, Lc, is used to calculate the adhesion strength. In order to obtain the magnitude of the critical load, acoustic signal, friction curve and microscope observation were utilized. Figure 2 shows an example of TiN coating failure on AL7075-T6. The selection of the deposition conditions is essential for fabricating composite thin films.

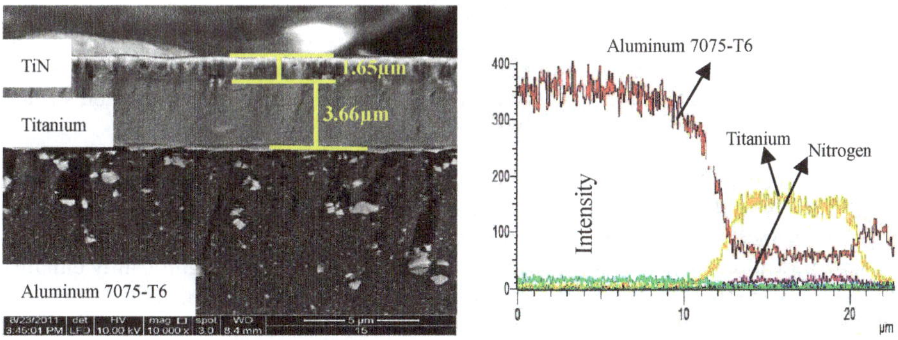

Figure 1. A typical example of a TiN coating

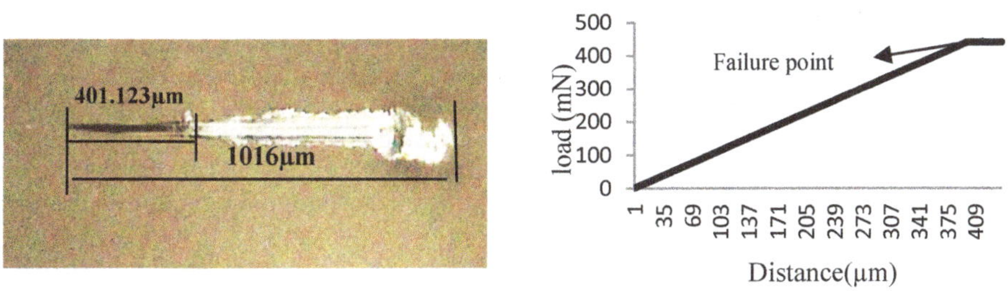

Figure 2. Scratch adhesion testing on a coated sample and the critical load

5. FUZZY LOGIC MODELLING

Fuzzy linguistic variables and fuzzy expression for input and output parameters are shown in Table 2. For each variable, four membership functions were used which are Low (L), Medium (M), High (H), and Very High (VH) for inputs. The output variable (Adhesion) also used four membership functions, ranging from Bad (B), Average (A), Good (G), and Excellent (E), shown in Table 2.

Table 2. Fuzzy linguistic

Inputs		Range
Par.	Ling. Var.	
A	(L), (M), (H), (VH)	300-500
B		150-220
C		3-6
D		25-100
Output		
Adh. (mN)	(E), (G), (A), (B)	248-2314

5.1. Membership functions and fuzzy rules structure

In choosing the membership functions for fuzzification, the event and type of membership functions are mainly dependent upon the relevant event. In this model,

Gauss shape of membership function was employed to describe the fuzzy sets for input variables, while triangular shape of membership function is used for the output variable. These membership functions are generally used and have gradually increasing and decreasing characteristics with only one definite value [6, 8]. As for the fuzzy rules structure, a set of 16 rules have been constructed based on the actual experimental surface adhesion of TiN coating on AL7075-T6. Experimental results were simulated in the Matlab software on the basis of Mamdani Fuzzy Logic.

5.2. Defuzzification

Defuzzification is the conversion of a fuzzy quantity to a precise value. In this method, the resultant membership functions are developed by considering the union of the output of each rule, which means that the overlapping area of fuzzy output set is counted as one, providing more result. Figure 3 (a) and (b) shows the predicted surface adhesion by fuzzy logic in relation to parameters change.

As can be seen in Fig.3, the surface adhesion of TiN coated specimens is increased significantly with the increasing of DC power from 300 to 500 (w). It could be attributed to that with increasing the DC power the sputtering rate is also increases [9]. For the nitrogen flow rate, it can be seen that the nitrogen gas mixture can significantly enhance the TiN-AL7075-T6 adhesion, with 5%. With more increasing in nitrogen flow rate up to 6% the adhesion of surface is decreased slightly. This might be happen because of the brittleness of the surface at higher nitrogen flow rate.

Figure 3 also shows that the DC bias voltage is another important factor affecting the adhesion of TiN coating. The adhesion of coated samples is increased with increasing substrate bias voltage from 25 to 85 (v), while it decreases with the further increase of bias voltage from 85 to 100 (v). It is happening because at higher substrate bias the incoming ions onto the substrate obtained excessive energy to damage the bounding between atoms in the film and substrate, leading to weakening the adhesion of the film to the substrate. In addition, temperature parameter does not have significant affect on adhesion of TiN coating [10].

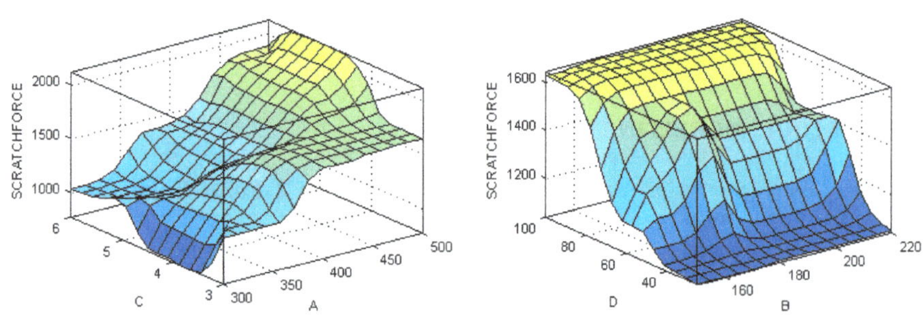

(a) **(b)**

Figure 3. The predicted surface adhesion by fuzzy logic in relation to parameters change

5.3. Investigate the fuzzy model error

New five experimental tests were carried out. Surface adhesions are measured and compared with the predicted values obtained using the proposed fuzzy model, shown in Fig. 4. As can be seen in Fig.4, the close assent between the measured and predicted surface adhesion obviously display that the fuzzy logic model can be used to predict adhesion of TiN coating. The average prediction error is found to be 6.729%.

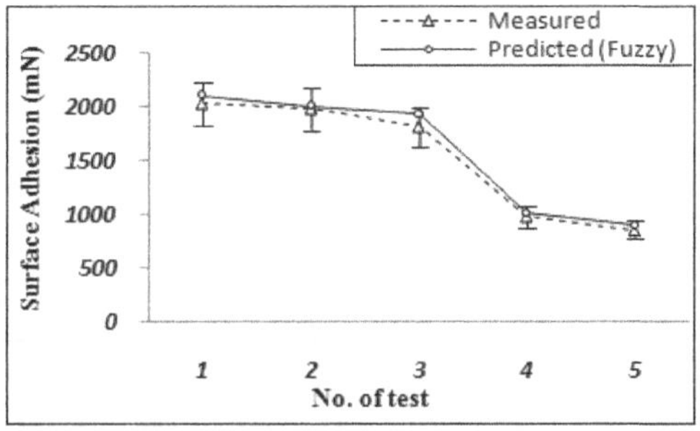

Figure 4. Comparison of measured and predicted adhesion

6. CONCLUSION

In this research work, predicting of TiN coating surface adhesion on aerospace AL7075-T6 alloy using fuzzy system was carried out. The combination parameters of 500(w) DC power, 200 (°C), 5% nitrogen flow rate, and 85(v) substrate DC bias voltage are recommended to obtain the highest surface adhesion. In addition, the fuzzy model prediction error were found to be 6.729% indicating that the fuzzy logic prediction model could be used to predict the surface adhesion of the coated thin film of TiN on AL7075-T6 alloy in a very accurate manner.

ACKNOWLEDGEMENTS

The authors acknowledge the financial support under the University Malaya Research Grant (UMRG Grant No.: RP001C-13AET) from the University of Malaya, Malaysia.

REFERENCES

[1] R. Sadeler, S. Atasoy, A. Arici, Y. Totic., The fretting fatigue of commercial hard anodized aluminum alloys, J. Mat. Eng. Perf. 1282-volume 18(9), (2009) 1280-1284.

[2] G.H. Majzoobi, M. Jaleh, Duplex surface treatment on AL7075-T6 alloy against fretting fatigue behaivior by application of titanium coating plus nitriding. Mat. Sci. Eng. A 452-453(2007) 673-681.

[3] J. Wagner, C. Mitterer, M. Penoy, C. Michotte, W. Wallgram, M. Kathrein, The effect of deposition temperature on microstructure and properties of thermal CVD TiN coatings, J. Refractory Met. Hard Mat. 26 (2008) 120–126.

[4] B. Subramanian, M. Jayachandran, characterization of reactive magnetron sputtered nanocrystalline titanium nitride (TiN) thin films with brush plated Ni interlayer, J. Appl. Electrochem 37 (2007) 1069-1075.

[5] D. Kakas, B. Skoric, A. Miletic, P. Terek, L. Kovacevic, M. Vilotic, Influence of substrate roughness on adhesion strength of hard TiN films, international conference on production engineering 28 - 30 September 2011, Niš, Serbia University of Niš, Faculty of Mechanical Engineering (2011).

[6] R. W. K. Leung, H. C. W. Lau, C. K. Kwong, An expert system to support the optimization of ion plating process: an OLAP- based fuzzy-cum-GA approach, Expert. Sys. Appl. 25 (2003) 313-330.

[7] H. Oktem, T. Erzurumlu, F. Erzinchanli. Prediction of minimum surface roughness in end milling mold part using neural network and genetic algorithms J. Mat. Design, 27 (2006) 735-744.
[8] M. Chandrasekaran, M. Muralidhar, C. Murali Krishna, U.S. Dixit, "Application of soft computing techniques in machining performance prediction and optimization: a literature review", J. Adv Manuf. Tech. 46 (2009) 445-464.
[9] Y. Chunyan, T. Linhai, W. Yinghui, W. Shebin, L. Tianbao, X. Bingshe, The effect of substrate bias voltages on impact resistance of CrAlN coatings deposited by modified ion beam enhanced magnetron sputtering, Surf. Sci. 255 (2009) 4033–4038.
[10] N. Panich, P. Wangyao, S. Hannongbua, P. Sricharoenchai, Y. Sun, effect of argon-nitrogen mixing gas during magnetron sputtering on titanium interlayer deposition with TiB2 coatings on high speed steel Rev. Adv. Mat. Sci. 16 (2007) 80-87.

Solidification simulation of copper-iron alloy for lead frame by phase-field method

T. Hara*, Y. Kanegae*, T. Tonogi**
* Hitachi, Ltd., Hitachi Research Laboratory, Japan
** SH Copper Products Co., Ltd, Japan

ABSTRACT

We analyzed the solidification process and microstructure of a copper-iron alloy by using the phase-field method and found that the crystal grain size of the iron is significantly affected by the cooling rate and composition of the alloy system. By using this method, we can easily understand the formation process of the alloy microstructure and evaluate various process conditions.

1 INTRODUCTION

Copper alloys are generally used as lead-frame material because of their high electronic conductivity, thermal transfer property, and corrosion resistance. During the manufacturing process of lead frame, copper alloys are bent, coated, and etched. Therefore, precipitates included in copper alloys should be fine enough not to disturb uniform treated surfaces.

One of the candidate materials for this type of alloy is the copper-iron alloy. For this type of alloy, however, care must be taken to prevent iron crystal grains, which have a harmful effect on lead frame quality. The market is currently demanding an etching process for lead frame manufacturing and a high quality copper alloy strip for LED use, so we need to find out how to reduce inclusions such as crystal grains, precipitates, and oxides formed during the casting process.

Directly observing what happens in the microstructure during the casting process is very difficult. Therefore, when we try to find the best casting process via experimentation, we have to observe a microstructure made under a certain condition, and then repeat the same experiment under a different condition. This style of trial and error investigation has never been regarded as an efficient process.

Our response to this issue is to attempt an analysis of the solidification process and microstructure of copper-iron alloy in order to suppress the coarsening of iron grains by using a numerical simulation technique. In this study, we used a phase-field method(1) that can treat the time-dependent concentration distribution of alloy elements and the specific morphology of each phase. We examined how alloy composition, cooling rate and the mechanism behind iron grain formation affect the solidification microstructure of copper-iron alloy, with a particular focus on the size of iron grains.

2 ANALYSIS OBJECT

We treated copper-iron alloy C194, a type of alloy often used for lead frame, as an analysis object. The alloy composition of C194 is shown in Table 1. In this study, our target is the major components of the copper-iron binary system.

3 SIMULATION RESULTS

We used Thermo-Calc software[2] as the numerical simulation technique to calculate a copper-iron system phase diagram. The multi-phase multi-component phase-field code MICRESS[3] was used to simulate the microstructure formation of the non-equilibrium solidification process.

3.1 Equilibrium diagram

The copper-iron system phase diagram is shown in Fig. 1. In this calculation, we used SSOL4[4] as a thermodynamic database. Figure 1(b) shows an expanded view of the region around the range of the C194 iron composition. As shown, in the cooling process fcc Fe phase is precipitated from the fcc Cu phase when the iron composition is lower than 2.8 wt%. In contrast, under an iron rich condition, iron grain is crystallized directly from the liquid phase. Generally speaking, crystallized grains tend to get bigger than precipitated grains, and so we conclude that the iron grain coarsening is due to crystallization. However, we cannot determine how much iron grains grow in the practical cooling process only from the phase diagram. We therefore tried to calculate the iron grain formation process by using a phase-field method to analyze solidification microstructure.

Table 1 Chemical composition of C194 (wt%).

Fe	Zn	P	Cu
2.1–2.6	0.05–0.20	0.015–0.15	>97

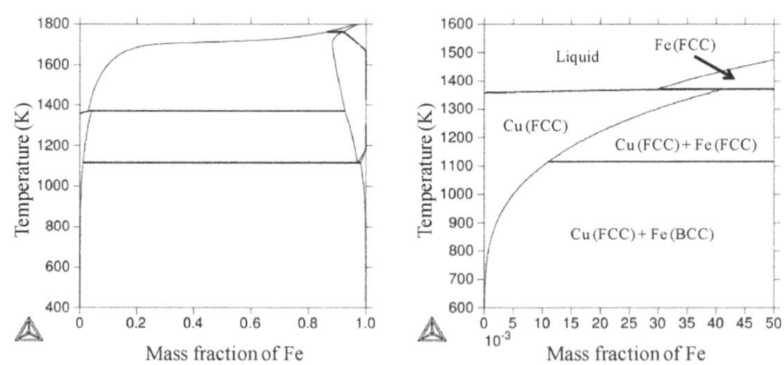

Fig. 1 Copper-iron equilibrium phase diagram.

3.2 Precipitation of iron grains

First, we examined the case of the formation of iron grains as precipitates. In our calculation, we considered liquid phase, fcc Fe phase, bcc Fe phase, and fcc Cu phase. Because precipitated iron grains are expected to be very small compared to fcc Cu grains, we have to set the grid cell size of the analysis model small enough, which leads to a long computation time.

In this study, we shortened the calculation time by introducing a larger iron diffusion coefficient. That is why the qualitative evaluation of iron grain size toward alloy compositions and cooling conditions is possible. The microscopic simulation domain was 300 phase-field grid cells high and the same number of cells wide. The grid size was $\Delta x = 0.5$ μm. Figure 2 shows the solidification process of the copper-iron alloy when the iron composition was 2.7 wt%, heat extraction rate was −2 J/(s·cm^3), and the initial temperature was 1370 K. We found that copper grain formation, fcc Fe phase precipitation, phase transformation, and Ostwald growth of precipitates were all visible by using the phase field method.

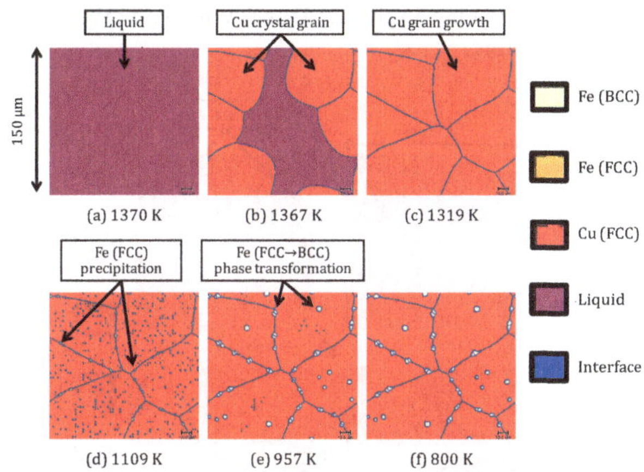

Fig. 2 Solidification process of copper-iron alloy (concentration: 2.7 wt%Fe, heat extraction rate: −2 J/(s·cm^3)).

The distribution of iron precipitates under different heat extraction rates from −20 to −1 J/(s·cm^3) are shown in Fig. 3. In the case of a rapid cooling condition like that in Fig. 3(a), each iron precipitate is very small and its number is large. However, with a decreasing heat extraction rate, iron precipitates get bigger and their numbers decline.

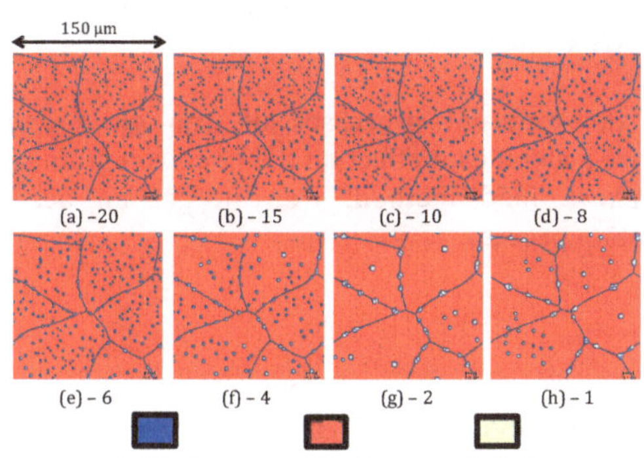

Fig. 3 Solidification microstructure of copper-iron alloy of each heat extraction rate written below each figure (J/(s·cm^3)) (concentration: 2.7 wt%Fe, temperature: 800 K).

Reproducing the various cooling conditions in experiments is very costly in all aspects, and using phase field method to calculate material microstructure is therefore a very attractive approach.

3.3 Crystallization of iron grains

Next, we studied the case of iron grain formation as crystal from the liquid phase. Iron composition was 3.5 wt% and initial temperature was 1410K. The iron diffusion coefficient in this alloy was set on the basis of previously reported values(5-8). Three iron grain nuclei were randomly distributed in the calculation domain and the heat extraction rate was set to -20 J/(s·cm^3). The solidification process under this condition is shown in Fig. 4. The crystallization of the fcc Fe phase is identified from distributed nuclei at the beginning, after which copper grains were formed. We found that the iron component of the fcc Cu phase was supplied both directly from the liquid phase and from the iron crystal grains via liquid phase. This is why iron crystal grains in the liquid phase start shrinking with time.

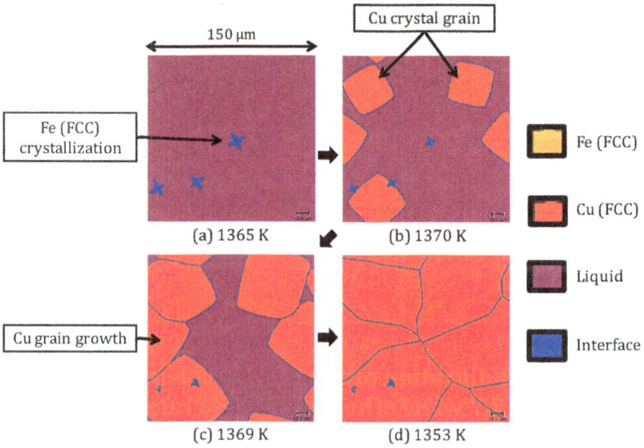

Fig. 4 Solidification process of copper-iron alloy (concentration: 3.5 wt%Fe, heat extraction rate: -20 J/(s·cm^3)).

The surviving iron grains in the copper phase had a maximum size of 7.5 μm. As discussed in section 3.2, iron precipitates had at most a size of 2μm under the same heat extraction rate of -20 J/(s·cm^3).

This indicates that the different processes of iron grain formation due to iron composition value have a significant effect on the final iron grain size. Using a numerical simulation of the phase field method, we can get an accurate picture of the microstructure formation process.

4 CONCLUSION

We took a phase field method approach to calculating the solidification process of copper-iron alloy for lead frame and drew the following conclusions.

(1) Different iron grain distribution patterns of the solidification microstructure are obtained from different cooling conditions.

(2) Simulation by the phase-field method showed that the different iron grain formation mechanism, precipitation, and crystallization due to iron composition significantly affect the iron grain size after complete solidification (2 μm for precipitation, 7.5 μm for crystallization).
(3) We found that we can visualize how the microstructure of the copper-iron alloy changing during the solidification process, which is difficult to directly observe. In particular, by using a numerical simulation such as the phase-field method, we can experiment with many different cooling conditions and alloy compositions economically. This approach is expected to accelerate the process of searching for the appropriate manufacturing condition.

5 REFERENCES

(1) R. Kobayashi, (1993) Phys. D **63**, 410–423.
(2) Andersson, J.O., Helander, T., Höglund, L., Shi, P.F., and Sundman, B., (2002) Calphad **26** 273–312.
(3) MICRESS®, http://web.access.rwth-aachen.de/MICRESS/
(4) SSOL4 SGTE Solutions database version 4 (assessed 23 Aug. 2013).
(5) T. Ejima and M. Kameda, (1969) J. Japan Inst. Metals **33** 96.
(6) S.K. Sen, M.B. Dutt, and A.K. Barua, (1978) Phys. Stat. Sol. (a) **45 (2)** 657.
(7) B. Million, J. Ruzickova, J. Velisek, and J. Vrestal, (1981) Mater. Sci. Technol. **50 (1)** 43.
(8) L.G. Ivantsov and A.M. Blinkin, (1965) Phys. Met. Metall. **19 (2)** 112.

Return mapping considerations for tangential inelastic effect on the subloading surface model

S. Tsutsumi, R. Fincato
Osaka University, Joining and Welding Research Institute (JWRI), Japan

ABSTRACT

The assumption of an associative flow rule in the classical elastoplastic models generates a plastic stretch which is not influenced by the tangential component of the stress rate. As a consequence, a significant underestimation of the inelastic strains is predicted whenever a non-proportional loading is take into account. The work presented in this paper aims to correct the excessive stiffness estimated by the traditional elastoplastic theories using the formulation proposed by [1] and at the same time it combines the subloading surface model constitutive equations with the return mapping technique ([2], [3]) for a faster but more accurate computation.

1 THE CONSTITUTIVE EQUATIONS

The plasticity model adopted in the current work defines an additional surface (i.e. subloading surface) within the conventional yield one with the purpose of driving the stress and realizing a smooth transition from elastic to the elastoplastic regime. Since the model has been widely developed and explained, here it will be briefly introduced, referring the reader to [4] for a more detailed explanation. According to [1] the stretch is decomposed into an elastic and inelastic part, where the latter is further subdivided in plastic and tangential as shown in the following equation:

$$\mathbf{D} = \mathbf{E}^{-1}\overset{\circ}{\boldsymbol{\sigma}} + \frac{tr(\mathbf{N}\overset{\circ}{\boldsymbol{\sigma}})}{M_p} + \frac{\overset{\circ}{\boldsymbol{\sigma}}{}^*_t}{T} \tag{1}$$

where $\overset{\circ}{\boldsymbol{\sigma}}{}^*_t$ is the deviatoric tangential stress rate, which can be defined starting from the corotational stress rate, $\mathbf{N}$ is the vector normal to the yield surface and T is a stress function that will be explained later:

$$\overset{\circ}{\boldsymbol{\sigma}}{}^*_t = \overset{\circ}{\boldsymbol{\sigma}}{}^* - \overset{\circ}{\boldsymbol{\sigma}}{}^*_n = \left(\mathbf{I} - \mathbf{N}\otimes\mathbf{N}\right)\left(\overset{\circ}{\boldsymbol{\sigma}} - \frac{1}{3}\sigma_{ii}\mathbf{I}\right) \tag{2}$$

Through some mathematical passages, and introducing some simplifications proper of plastically-incompressible materials, is possible to invert Eq. (1) obtaining the stress rate as function of the strain rate (i.e. symmetric part of the velocity gradient $\mathbf{L}$).

$$\overset{\circ}{\boldsymbol{\sigma}} = \frac{1}{(1+2GA)}\left[\mathbf{E} - \frac{\mathbf{EN}\otimes\mathbf{EN}}{M_p + tr\mathbf{NEN}} + \frac{2GA}{3}\mathbf{E}\otimes\mathbf{I} + 2GA\frac{\mathbf{EN}\otimes\mathbf{N}}{M_p + tr\mathbf{NEN}}M_p\right]\mathbf{D} \tag{3}$$

where A is an arbitrary function which depends on the stress state. Since the tangential inelastic strain rate does not affect the hardening behaviour it is possible to consider separately the return mapping algorithm and the tangential relaxation in the computation of the stress rate by using a specific form for the function A, that allows to separate the two contributes of Eq. (3).

$$A = \left(\frac{T}{2G - 2GT}\right); \qquad T = \xi R^b$$

$$\overset{\circ}{\boldsymbol{\sigma}} = \overset{\circ}{\boldsymbol{\sigma}}{}^*_e - \overset{\circ}{\boldsymbol{\sigma}}{}^*_p - \overset{\circ}{\boldsymbol{\sigma}}{}^*_t = \left\{ \mathbf{E} - \frac{\mathbf{EN} \otimes \mathbf{EN}}{M_p + tr\mathbf{NEN}} \right\} \mathbf{D} - \overset{\circ}{\boldsymbol{\sigma}}{}^*_t \tag{4}$$

$$\overset{\circ}{\boldsymbol{\sigma}}{}^*_t = T\left(\mathbf{E} - \frac{\mathbf{EN} \otimes \mathbf{EN}}{M_p + tr\mathbf{NEN}} - \frac{1}{3}\mathbf{E} \otimes \mathbf{I} - \frac{\mathbf{EN} \otimes \mathbf{N}}{M_p + tr\mathbf{NEN}} M_p \right) \mathbf{D}$$

ξ and b are material parameters regulating the percentage of tangential relaxation, whereas R is the similarity ratio introduced in the subloading surface model [4], which allows a smooth development of irreversible deformations for every change of the stress state (i.e. [3]).

The starting point for the evaluation of the plastic strain is that of freezing all the internal variables (i.e. isotropic hardening, back-stress) between a generic n step and the subsequent $n+1$, and perform a so called 'elastic trial', assuming a perfect elastic material. This can be an admissible stress state only if it satisfies the first condition of Eq. (5), in the second case it must be corrected as shown by Hashiguchi in [3] and [5].

$$\begin{aligned} &\mathrm{f}(\boldsymbol{\sigma}_{n+1}^{(trial)}) \leq R_n F(H_n) \;\; \text{then } \boldsymbol{\varepsilon}_{n+1}^p = 0 \text{ and } \boldsymbol{\sigma}_{n+1}^{(trial)} = \boldsymbol{\sigma}_{n+1}^{(final)} \\ &\mathrm{f}(\boldsymbol{\sigma}_{n+1}^{(trial)}) > R_n F(H_n) \;\; \text{then } \boldsymbol{\varepsilon}_{n+1}^p \neq 0 \text{ and } \boldsymbol{\sigma}_{n+1}^{(trial)} \neq \boldsymbol{\sigma}_{n+1}^{(final)} \end{aligned} \tag{5}$$

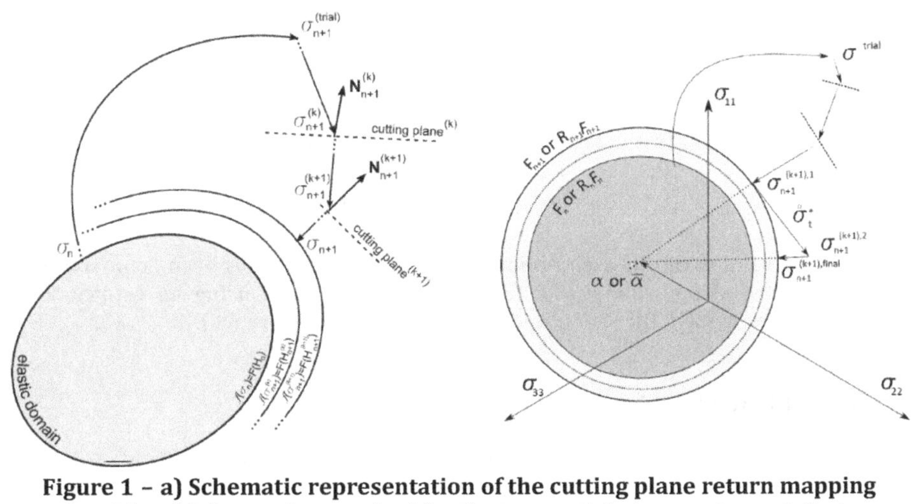

Figure 1 - a) Schematic representation of the cutting plane return mapping algorithm. b) Correction method for the tangential stress rate contribute.

Briefly, the plastic relaxation is carried out by computing the correct amount of plastic deformation λ through an iterative procedure, which ends when the stress is brought back to lay on the correct updated plastic surface (the magnitude of the tolerance *toll* in Eq. (7) should be defined by the user for the numerical algorithm):

$$\boldsymbol{\sigma}_{n+1}^{(k+1)} = \boldsymbol{\sigma}_{n+1}^{(k)} + d\boldsymbol{\sigma}_{n+1}^{p(k+1)} = \boldsymbol{\sigma}_{n+1}^{(k)} - \mathbf{E} : \lambda_{n+1}^{(k+1)}\mathbf{N}_{n+1}^{(k)} \tag{6}$$

$$\left(f(\overline{\boldsymbol{\sigma}}_{n+1}^{(k+1)}) - R_{n+1}^{(k+1)}F_{n+1}^{(k+1)}\right) / F_{n+1}^{(k+1)} \leq \text{toll} \tag{7}$$

The apex *k* stands for the sub-iterations that should be performed for every macro *n* step, ***N*** is the vector normal to the plastic surface (i.e. associative flow rule), *F* is the isotropic hardening function exhibiting the size of the yield surface and ***E*** is the tensor of the elastic constants.

Once the correct elastoplastic stress is computed it is possible to subtract the tangential one, evaluated by means of the last of Eq. (4), updating Eq. (6) as follows:

$$\boldsymbol{\sigma}_{n+1} = \boldsymbol{\sigma}_{n+1}^{(\mathrm{k})} - \mathbf{E} : \lambda_{n+1}^{(k+1)}\mathbf{N}_{n+1}^{(k)} - \overset{\circ}{\boldsymbol{\sigma}}{}_t^{*} \tag{8}$$

The problem of computing a *n+1* step using two separate procedures lays on the fact that, when the elastoplastic equilibrium is fulfilled, a unique set of values for *R* and *F* is possible.

The subtraction of the tangential stress rate will bring the point to lay outside the correct plastic surface, with a loss of local equilibrium. This aspect is enhanced whenever a large step simulation is carried out, since the entity of the deviation is directly proportional to the magnitude of the deviatoric tangential stress rate. A simple solution adopted is that of performing a further correction of the stress state after the tangential relaxation has been counted. In detail, looking at Figure 1b, once estimated the $\boldsymbol{\sigma}_{n+1}^{(k+1),2}$ through Eq. (8) a sort of radial return mapping is performed to bring back the stress to the current subloading surface $R_{n+1}F_{n+1}$ or normal yield surface F_{n+1} (depending on the magnitude of the similarity ratio):

$$\begin{aligned}
&1)\quad \boldsymbol{\sigma}_{n+1}^{(k+1),2} = \boldsymbol{\sigma}_{n+1}^{(k+1),1} - \overset{\circ}{\boldsymbol{\sigma}}{}_t^{*} \\
&2)\quad \mathbf{N}^{,2} = \partial f(\overline{\boldsymbol{\sigma}}_{n+1}^{(k+1),2}) / \left|\partial f(\overline{\boldsymbol{\sigma}}_{n+1}^{(k+1),2})\right| \\
&3)\quad \lambda_{n+1}^{\text{tangential}} = \sqrt{3/2}\left|\overline{\boldsymbol{\sigma}}_{n+1}^{(k+1),2}\right| - \sqrt{3/2}\left|\overline{\boldsymbol{\sigma}}_{n+1}^{(k+1),1}\right| \\
&4)\quad \boldsymbol{\sigma}_{n+1}^{(k+1),\text{final}} = \boldsymbol{\sigma}_{n+1}^{(k+1),2} - \lambda_{n+1}^{\text{tangential}}\mathbf{N}^{,2}
\end{aligned} \tag{9}$$

To prove the validity of this assumption the numerical model has been compared with the results carried out solving Eq. (3), without splitting it, using an explicit Euler formulation (i.e. named 'Direct Method' D.M. in the following graphs).

2 NUMERICAL RESULTS

A simple two steps test, under plain strain hypothesis, has been analysed for a four node plate element (CPE4) using a user subroutines in ABAQUS v13.4. During the first step the plate is pulled from the top (the base has been constrained to give symmetrical boundaries condition respect the pulling axis) up to 10%, whereas in the second one a pure shear test, with 20% angular distortion, is realized. A similar numerical example has also been investigated by Yu and Yuan [6] for a granular material with a rounded Hyperbolic Mohr-Coulomb surface.

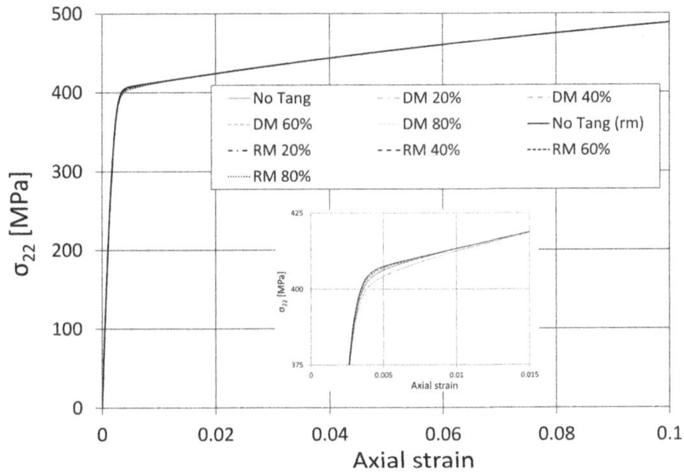

Figure 2 - Tangential effect during the pull up step.

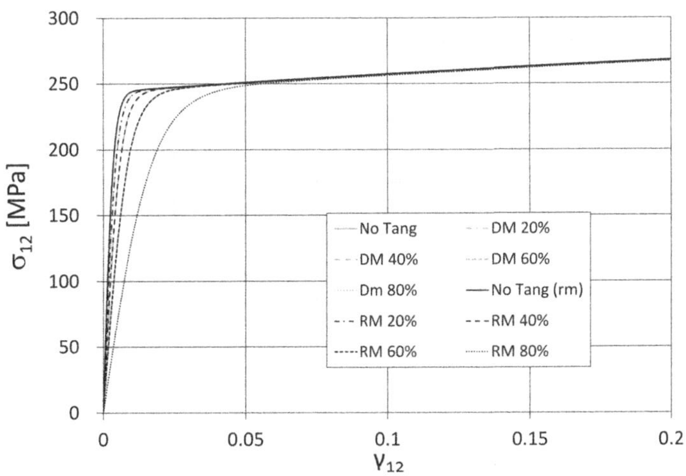

Figure 3 - Tangential effect during the shear step (DM - direct method; RM - return mapping).

As it can be seen in Figure *2* the effect of the tangential reduction becomes not negligible during the transition between the elastic and fully plastic state due to the non-constant ratio of the stress components under the plane strain assumption.

After the plate has reached 10% deformation along the vertical axis it is subjected to a simple shear test which causes a sudden change in the direction of the stress rate along the yield surface, inducing a strong non-coaxiality between principal plastic stress and strain axes. Figure 3 shows the decrease of the shear component *xy* for different percentages of the tangential reduction. This effect tends to disappear around 5% of angular distortion, when large strains lead the axes to coincide once again, as experimentally confirmed by Roscoe [7].

The influence of the tangential relaxation is pointed out tracing the evolution of the shear modulus computed at the beginning of the second step of test for different value of the T function. It has to be noticed that, even with 0% reduction, the shear stiffness (G) appears slightly affected by the elastoplastic deformation developed in the first part of the analysis.

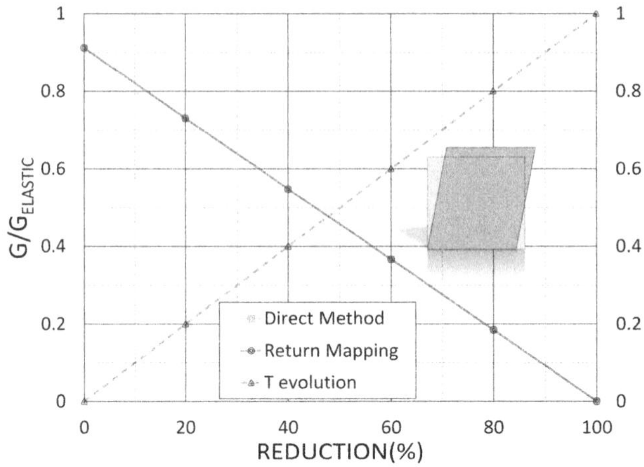

Figure 4 – Tangential effect during the shear step.

CONCLUSIONS

The present work introduced a new approach for the evaluation of the inelastic behaviour of metals when non-proportional loading is involved. The innovative aspect lays on the use of the return mapping technique, with an associative plastic flow rule, in combination with the tangential relaxation. Preliminary results have been compared with the solution carried out adopting an Explicit Euler method, proving the reliability of the implementation.

REFERENCES

[1] Hashiguchi, K., Tsutsumi, S., (2001) Elastoplastic constitutive equation with tangential stress rate effect, *Int. J. Plasticity*, **17**, 117-145.

[2] Simo, J. C., Ortiz, M. (1985) A unified approach to finite deformation elastoplasticity based on the use of hyperelastic constitutive equations, *Compt. Meth. Appl. Mech. Eng.*, **49**, 221-245.

[3] Hashiguchi, K., Mase, T., (2001) Subloading surface Model and its Return-mapping Formulation, *ACTA 60th Nat. Congr. Of Theoretical & Applied Mechanics.*

[4] Hashiguchi, K., Elastoplasticity theory. In: F Pfeiffer, P Wriggers (Eds.), Lecture notes in applied and computational mechanics, Springer: Berlin 2009; 42.

[5] Hashiguchi, K., Suzuki, N., Ueno, M. (2014) Elastoplastic analysis by return-mapping and consistent tangent modulus tensor based on subloading surface model (1st report, formulation of return-mapping), *Transaction of the JSME*, Vol.80, No.81.

[6] Yu, H. S., Netherton, M. D. (2000) Performance of displacement finite elements for modelling incompressible materials, *Int. J. Numer. Anal. Meth.Geomech.*, **24**, 627-653.

[7] Roscoe, K. H. (1970) The influence of strains in soil mechanichs, *Geotechnique*, **20**, 129-170.

Evaluation of guided wave propagation in steel pipes

T. Wang[1], C. Yang[1,2], L. Ye[3], D. Spray[1], Y. Xiang[1]

[1] School of Computing, Engineering and Mathematics, The University of Western Sydney, Australia

[2] Institute for Infrastructure Engineering (IIE), The University of Western Sydney, Australia

[3] Centre for Advanced Materials Technology (CAMT), School of Aerospace, Mechanical and Mechatronic Engineering, The University of Sydney, Australia

ABSTRACT

In this study, an active structural health monitoring (SHM) system is developed for non-destructive defect identification in steel pipes, which is based on guided waves generated and collected using piezoelectric ceramics patches. Finite element models of the elastic wave propagation in steel pipes are devised to determine their characteristics and used to optimise wave mode and frequency. An experimental study is also conducted for the validation of the proposed structural health monitoring system and the numerical modelling. The obtained results can be employed to optimally design the online structural health monitoring system with an integrated piezoelectric actuator-sensor network for steel pipes.

1 INTRODUCTION

Guided wave-based structural health monitoring (SHM) technique, which is a process of implementing an in-situ damage identification and detection, has been developed in several decades and it brings significant influences on non-destructive defect detection of engineering structures (1). Many researchers contributed to this field of research for the applications of the guided waves which can propagate for a long distance in an engineering structure and detect defects of small sizes in the structure based on the collected wave signals. However, there are several technical problems on developing such an SHM system for pipes, which are typical shell-like structures, including dispersion relations, wave mode selection, wave propagation characteristics and wave mode conversion in complex structures [3], etc. When piezoelectric materials are used to act as actuators and/or sensors to generate and collect signals of elastic waves in pipes, their performances need to be evaluated too. As the most useful and powerful numerical tool, finite element method-based dynamic analysis has been widely employed to capture guide waves' characteristics in complex engineering structures. To develop robust and cost-effective finite element-based models, researchers (2-4) have developed effective or equivalent models for piezoelectric actuators and/or sensors using different theories.

In this study, the evaluation of elastic wave propagation in steel pipes is conducted both numerically and experimentally. In the finite element models of the pipes, the enhanced effective PZT actuator model [4] developed in the authors' previous work is further applied and the performance of piezoelectric actuators used including the shape effect is investigated. For validation purpose a steel pipe is tested using a 16-actuator/sensor network to understand the elastic wave propagations. Furthermore, the relationship between the defects and the elastic wave propagation is identified.

2 FUNDAMENTALS OF ELASTIC WAVE PROPAGATION IN CYLINDRICAL STRUCTURES

Considering a solid medium, the governing equation for the propagation of free harmonic waves along a hollow circular cylinder is the Navier's equation of motion which can be given in a vector form as (6),

$$\mu\nabla^2\mathbf{u}+(\lambda+\mu)\nabla\nabla\cdot\mathbf{u}=\rho\left(\partial^2\mathbf{u}/\partial t^2\right) \tag{1}$$

where $\boldsymbol{u}$ is the displacement vector, ρ is the density, μ and λ are Lame's constants, and ∇ is the three-dimensional Laplace operator, respectively. For an anisotropic cylinder hollow pipe, the roots of Eq. (1) can be assumed as follows:

$$u_r=U_r(r)\cos(n\theta)\cos(\omega t+\varsigma z) \tag{2}$$

$$u_\theta=U_\theta(r)\sin(n\theta)\cos(\omega t+\varsigma z) \tag{3}$$

$$u_z=U_z(r)\cos(n\theta)\sin(\omega t+\varsigma z) \tag{4}$$

where u_r, u_θ, and u_z are radial, circumferential and axial displacements, and U_r, U_θ and U_z are amplitudes described using Bessel functions, respectively. There are three modes for guided wave propagation in pipes: longitudinal mode L(0, m), torsional mode T(0, m) and flexural mode F(n, m). Only the dispersion characteristics of the elastic waves are of research interests and hereby the characteristic or dispersion equation is listed without reproducing its complicated algebraic expressions for the sake of brevity,

$$D=\begin{vmatrix} C_{11} & C_{12} & \cdots & C_{16} \\ C_{21} & C_{22} & \cdots & C_{26} \\ \vdots & \vdots & \vdots & \vdots \\ C_{61} & C_{62} & \cdots & C_{66} \end{vmatrix}=0 \tag{5}$$

where C_{ij} are constants, and i, j = 1, 2, 3, 4, 5, and 6. Considering guided waves have a zero wave number, $\varsigma = 0$, Eq. (5) can be broken into the product of two sub determinants as

$$D_1D_2=0 \tag{6}$$

Hence, either D_1 or D_2 is equal to zero to satisfy Eq. (5). The case of $D_1 = 0$ corresponds to plane-strain vibrations L(0, m) whereas the case of $D_2 = 0$ is for longitudinal shear vibrations T(0, m) (6). The dispersion curves of a steel pipe with an outer diameter 323mm and a wall thickness 6.4 mm based on Eq. (7) can be obtained as given in Fig. 1.

As can be seen from Fig. 1, it is clear that the group velocity and phase velocity will change when the frequency increases. It is necessary to select a suitable wave mode and frequency for appropriate guided waves used in the proposed structural health monitoring system and in current study, and the longitudinal L(0, 2) wave mode at 20 kHz was chosen due to a low attenuation (5).

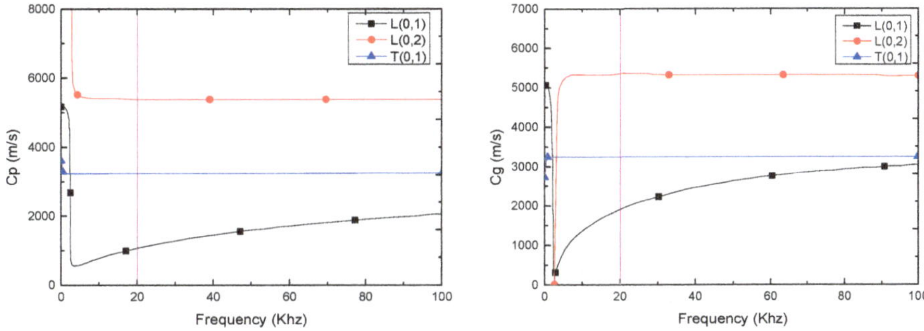

Figure 1 Disperse curves of a pipe with an outer diameter of 323mm and a wall thickness of 6.4 mm

3 FINITE ELEMENT MODELLING AND EXPERIMENTAL SETUP

3.1 Finite element modelling

In this study, both numerical modelling and experimental test were conducted for steel pipes, which are of 1500 mm in length, 323 mm in outer radius and 6.4 mm in wall thickness and are widely used for both liquid transport and piling in reality. The steel has a Young's modulus of 210 GPa and a Poisson's ratio of 0.27. A sensor network with 16 PZT actuators/sensors was devised. Eight actuators/sensors were installed at each end of the pipe and they functioned as either generating or collecting the signals of elastic waves. The waveform used for the testing was selected as 40 kHz toneburst in a 5-cycle Hanning window and the Hilbert-transform (HT) was utilised to decompose collected signal into different frequency bands, thereby making the useful wave components appeared after screening broadband noise.

As for the finite element modelling, the steel pipes were treated as a 3D deformable solid and meshed with 684,512 C3D8R solid elements. Since piezoelectric materials are used to act as actuators and/or sensors to generate and collect signals of elastic waves, the enhanced effective PZT actuators model [4] was applied in the finite element models of steel pipes. In the FE models, A1 to A8 are standing for actuators while S1 to S8 are for sensors, as shown in Fig. 2. The defects added to the pipe model are a hole of diameter 12 mm and depth 5 mm, a crack of length 120 mm and width 2 mm and a notch of length 120 mm, width 12 mm and depth 5 mm as shown in Fig. 2. Those defects were designed to locate at the middle of the pipeline and kept aligned with the actuator-sensor pair: A1 and S1.

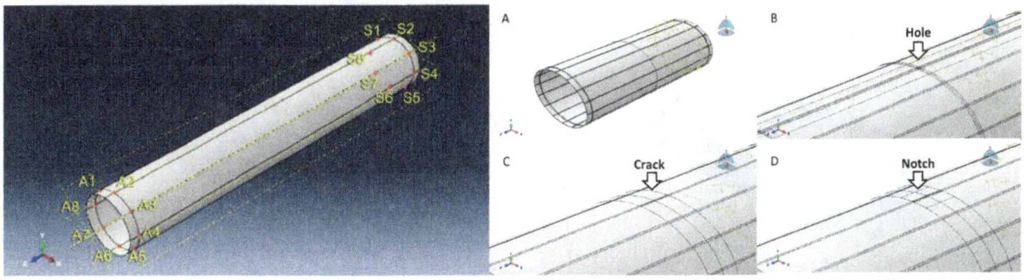

Figure 2 FE models of pipes with/without defect: A) no defect; B) hole; C) crack; and D) notch 3D solid model

3.2 Experimental setup

Being similar to the FE modelling, in the experimental study, two identical pipes were employed using two types of PZT patches. Both PZT patches were glued to the pipes using LOCTIGHT 401 and distributed evenly around the pipes' circumference, ensuring that each actuator was in line with a sensor. The experiment followed the same setups used in the FE modelling in both pipes which were tested one by one in a four-stage procedure for four cases: a) no defect; b) a hole defect; c) a crack defect; and d) a notch defect as shown in Fig. 3. As the same as designed in the FE modelling, eight PZT actuators were installed at the related positions as actuators at one end of the pipe and the other eight PZTs were installed as sensors on the other end to collect data. The defects were added to the pipes right in their middle at different stages.

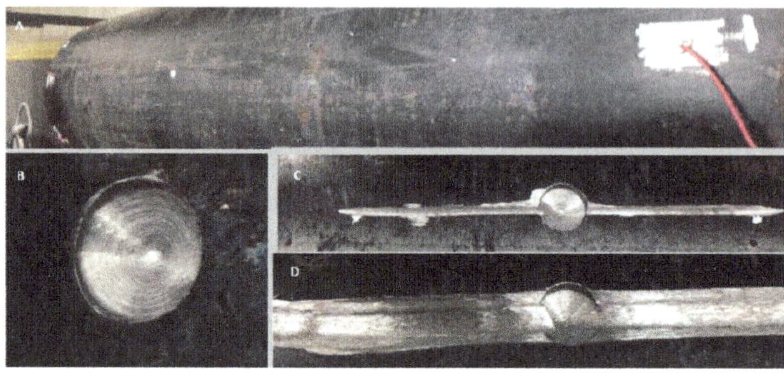

Figure 3 the pipe with piezoelectric actuators/sensors with a defect in forms of B) hole C) crack and D) notch

4 RESULTS AND DISCUSSION

Fig. 4 shows guided wave propagations along the pipes, which is without any defect for benchmarking. It can be clearly seen that there are two kinds of wave modes generated in pipes - L(0, m) wave modes and F(n, m) wave modes, respectively.

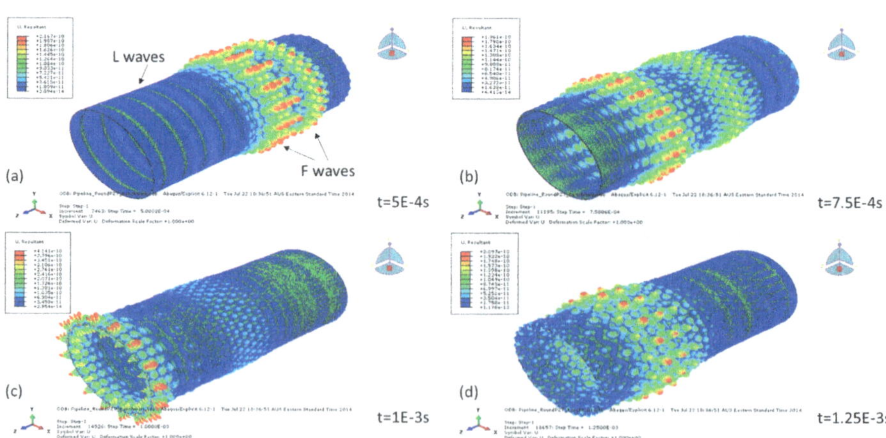

Figure 4 Displacement outputs from FE dynamic simulations along the longitudinal direction of the pipe

To investigate the relationship between the guided wave propagation and defects, wave modes were identified based on their group velocities. According to Fig. 1 (b), the group velocity for L(0, m) mode is the fastest. Then, the wave packet I in Fig. 5 (b) should be longitudinal axial symmetric modes L(0, m). According to simulation results, the group velocity was calculated as 5278 m/s. According to Fig. 1 (b) (Cg = 5400 m/s), it is evident that wave packet I is L(0, 2). The result is consistent with the experimental result of 5300 m/s. The group velocity which we calculated for wave packet II is 2107.90 m/s. This value is very closed to the group velocity of L(0, 1) at 20 kHz in Fig. 1 (a). Then, the wave packet II is L(0, 1). In addition, the wave packet III in Fig. 5 (b) must be F(n, m) modes. For the wave packet IV, it could be an F(n, m) mode combined with L(0, m) mode. This is because the amplitude of VI wave packet is the largest one. Therefore the basic relationship between the guided wave propagation and defects can be determined. First of all, L(0, 2) mode is hard to detect the longitudinal defects, which means the results from the crack structure are almost identical to the benchmark ones in wave packet I. Rose [7] had the same opinion in his works. L(0,1) can be used to detect the circumferential cracks because there are differences between the results from the model with a notch defect and those from the benchmark model in the time instants of 0.0006 s to 0.0008 s. Moreover, the F(n, m) is useful to detect some large defects. According to wave packet III depicted in Fig. 5 (b), it is clearly indicated that the results from the model with a hole defect show there is no F(n, m) mode compared to those from the benchmark model. Lastly, according to the comparison on the results from the model with a notch defect and the ones from the benchmark, it can be found that the notch can lead to waveform transformation, especially for the period between 0.0008 s to 0.001 s.

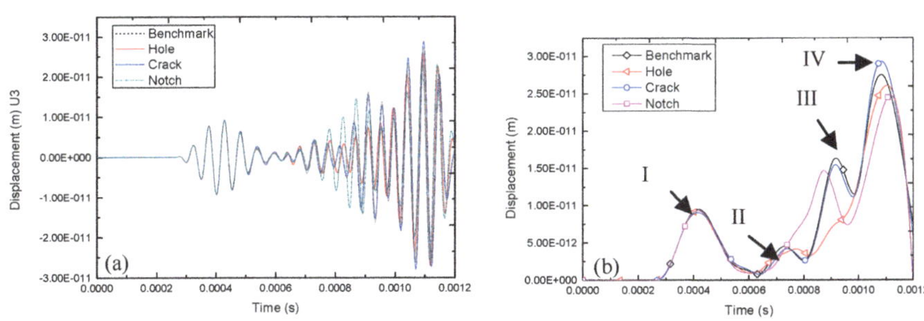

Figure 5 Displacement signals from FE dynamic simulations of steel pipes with and without a defect, a) Before HT-processed and b) after HT-processed

5 CONCLUDING REMARKS

Guide wave propagation in steel pipes and the relationship between the guided wave propagation and defects have been evaluated and investigated mainly using finite element modelling and simulations in the present study. The enhanced effective piezoelectric actuator model has been further applied in the FE models and the results show their effectiveness for generating elastic waves although a shape effect has been observed. Experimental results have primarily been used for verification on wave modes only at this stage and the on-going experimental work will be conducted considering the complexity of the elastic wave propagation in steel pipes. Based on the numerical results, the longitudinal axially symmetric modes L(0, m) cannot be used on the detection of longitudinal defects while the non-axially symmetric modes F(n, m) are sensitive to different kinds of defects, which can be focused in future research for more comprehensive understanding on them.

REFERENCES

[1] D. Alleyne, M. Lowe, and P. Cawley (1998) *Journal of Applied mechanics*, **65**, 635-641.

[2] C.Y.K. Chee, L. Tong, and G.P. Steven (1998) *Journal of Intelligent Material Systems and Structures*, **9**, 3-19.

[3] C. Yang, L. Ye, Z. Su, and M. Bannister (2006) *Composite Structures*, **75**, 267-275.

[4] T. Wang and C. Yang (2014) *Applied Mechanics and Materials*, **553**, 705-710.

[5] D.N. Alleyne, P. Cawley (1997) *Materials Evaluation*, **55(4)**, 504-508.

[6] D.C. Gazis (1959) *The Journal of the Acoustical Society of America*, **31**, 568-573.

[7] J. L. Rose (2004) *Ultrasonic Waves in Solid Media*, Cambridge University Press.

Nonlinear response of a closed defect to Lamb waves: perturbation analysis using hybrid finite element method

S. Biwa, N. Ito
Department of Aeronautics and Astronautics, Graduate School of Engineering, Kyoto University, Japan

ABSTRACT

The nonlinear interaction of Lamb waves with a closed subsurface defect lying parallel in a thin elastic plate is studied numerically. The closed defect is modelled as a nonlinear spring interface with quadratic nonlinearity. Assuming weak nonlinearity, the governing equations are linearized using a perturbation technique and solved by the hybrid finite element method in the frequency domain. As a numerical example, the second-harmonic generation at the defect for the lowest-order antisymmetric Lamb mode incidence is examined. It is demonstrated that the generation of the second-harmonic Lamb mode is most significant at the lowest resonant frequency of the defect.

1 INTRODUCTION

Understanding the interaction of ultrasound with different types of defects in structural members is of paramount importance for advanced non-destructive testing and structural health monitoring. Traditional linear ultrasonic methods are often difficult to apply when defects are closed and their surfaces are in partial contact since the reflection or scattering of ultrasound can be significantly reduced. As an alternative and more sensitive measure of the state of defects and interfaces, their nonlinear ultrasonic responses, often manifested as higher harmonic generation, have been attracting increasing attention.(1,2) In plate structures, ultrasonic waves propagate as Lamb modes, and their interaction with closed defects is a subject of high interest from both fundamental and practical points of view.(3) Experimentally, Solodov et al.(4) recently demonstrated an interesting idea of enhancing the nonlinear response of a defect due to its local resonance.

In this study, the nonlinear response of a closed subsurface defect lying parallel in a thin elastic plate to Lamb waves is studied numerically. The closed defect with contacting surfaces is modeled as a nonlinear spring interface with quadratic nonlinearity. Assuming weak nonlinearity, the governing equations are linearized using a perturbation technique(5) and solved by the hybrid finite element method in the frequency domain.(6) Linear resonance characteristics of the defect as well as second-harmonic generation behaviour to the incidence of the lowest-order antisymmetric Lamb wave are demonstrated as a numerical example.

2 FORMULATION

The two-dimensional plane-strain motion of a thin isotropic linearly elastic plate of thickness d is considered in the x_1-x_2 plane as shown in Fig. 1. In the frequency domain, the governing equations for the displace components u_α (α = 1, 2) are

$$(c_L{}^2 - c_T{}^2)\frac{\partial^2 u_\gamma}{\partial x_\alpha \partial x_\gamma} + c_T{}^2 \frac{\partial^2 u_\alpha}{\partial x_\gamma \partial x_\gamma} + \omega^2 u_\alpha = 0, \quad \alpha = 1, 2, \tag{1}$$

where $\omega = 2\pi f$ is the angular frequency (f is the frequency), and c_L and c_T are the longitudinal and shear wave speeds. The two-dimensional summation convention is adopted here. The plate has a parallel crack-like defect with length a and is subjected to the incidence of a Lamb mode with frequency f. The upper and lower surfaces of the plate are assumed to be traction-free, and the defect is modelled as a nonlinear spring interface where the stress components obey

$$\sigma_{22} = K_n(1 - \beta u_n) u_n, \quad \sigma_{12} = K_t u_t, \quad u_n = u_2{}^+ - u_2{}^-, \quad u_t = u_1{}^+ - u_1{}^-, \tag{2}$$

where K_n and K_t are the normal and tangential linear stiffnesses of the closed defect, and β is a parameter representing its quadratic nonlinearity. It is noted that the above model of the defect only incorporates the nonlinearity in the normal direction to the defect, which represents the variation of the normal stiffness due to its opening/closure.

In the present analysis, the solution to the above problem is decomposed into two contributions. One is the linear (fundamental) response $\bar{u}_\alpha(\boldsymbol{x})$ with frequency f which is solved for Eq. (1) with the linearized condition

$$\bar{\sigma}_{22} = K_n \bar{u}_n, \quad \bar{\sigma}_{12} = K_t \bar{u}_t, \quad \bar{u}_n = \bar{u}_2{}^+ - \bar{u}_2{}^-, \quad \bar{u}_t = \bar{u}_1{}^+ - \bar{u}_1{}^-, \tag{3}$$

instead of Eq. (2). The other is the perturbation contribution due to the nonlinearity consisting of the zero-frequency component as well as the double-frequency component $\tilde{u}_\alpha(\boldsymbol{x})$. The latter is solved for Eq. (1) but with 2ω instead of ω, and

$$\tilde{\sigma}_{22} = K_n\{\tilde{u}_n + \beta F(\bar{u}_n)\}, \quad \tilde{\sigma}_{12} = K_t \tilde{u}_t, \quad \tilde{u}_n = \tilde{u}_2{}^+ - \tilde{u}_2{}^-, \quad \tilde{u}_t = \tilde{u}_1{}^+ - \tilde{u}_1{}^-, \tag{4}$$

instead of Eq. (2). The function $F(\bar{u}_n)$ represents the wave excitation source for the second-harmonic component due to the quadratic nonlinearity of the defect. Due to the limit of space the details of the formulation will be presented elsewhere.

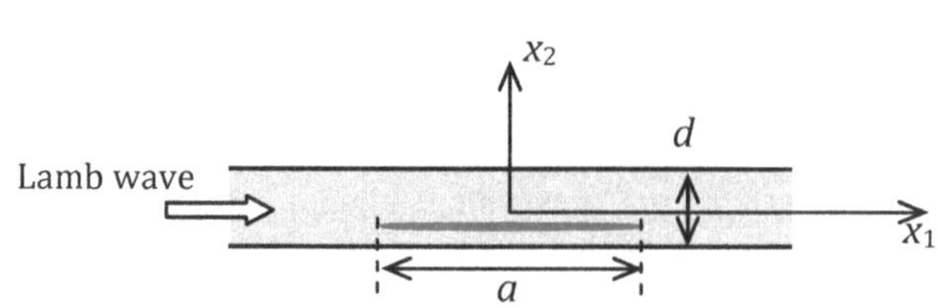

Fig. 1 A thin plate with a closed defect

3 NUMERICAL ANALYSIS

The fundamental as well as the double-frequency responses are solved using the hybrid finite element method. Namely, a central region of the plate $-L/2 < x_1 < L/2$ containing the defect is divided by finite elements with nodal displacements as unknown variables. For the fundamental response, the displacement fields in the outer semi-infinite regions are expressed as the modal expansion of Lamb waves

$$\overline{u}_\alpha(\boldsymbol{x}) = \begin{cases} u_\alpha^{inc}(\boldsymbol{x}) + \sum_n R_{\mathrm{S}n}\, \overline{u}_\alpha^{Sn(-)}(\boldsymbol{x}) + \sum_n R_{\mathrm{A}n}\, \overline{u}_\alpha^{An(-)}(\boldsymbol{x}), & x_1 < L/2, \\ \sum_n T_{\mathrm{S}n}\, \overline{u}_\alpha^{Sn(+)}(\boldsymbol{x}) + \sum_n T_{\mathrm{A}n}\, \overline{u}_\alpha^{An(+)}(\boldsymbol{x}), & x_1 > L/2, \end{cases} \tag{5}$$

where $R_{\mathrm{S}n}$ ($T_{\mathrm{S}n}$) and $R_{\mathrm{A}n}$ ($T_{\mathrm{A}n}$) are the coefficients of the nth-order symmetric (Sn) and antisymmetric (An) Lamb modes for the left (right) regions, and the corresponding modal displacement functions are denoted by e.g. $\overline{u}_\alpha^{Sn(-)}(\boldsymbol{x})$, where the superscripts (+) and (-) distinguish the propagation direction of each mode. The continuity of the displacements and tractions at the boundaries of the inner and outer regions is applied to build up a set of linear equations for the unknown modal expansion coefficients of the outer regions and the nodal displacements of the inner region, which are solved numerically. The double-frequency wave field is solved in a similar fashion, in the absence of the incident field but with the wave excitation by the quadratic nonlinearity of the defect.

4 RESULTS AND DISCUSSION

Numerical results are shown for the case when a crack-like defect of length a = 4d lies parallel in an aluminium plate at distance 0.1d from the lower surface. The incident mode is assumed to be the lowest-order antisymmetric (flexural) Lamb mode (A0 mode). The linear stiffnesses of the defect are set as $K_{\mathrm{N}}\, d/(\rho\, c_{\mathrm{T}}^2) = 0.01$ and $K_{\mathrm{T}}/K_{\mathrm{N}} = 0.4$, where ρ is the density of the solid. The amplitude U_0 of the incident A0 mode is chosen so that β $U_0 = 0.01$. The inner region is taken for the length L = 6d and is divided into 2,400 finite elements. In the outer regions, only four Lamb modes (A0 to A3 and S0 to S3) are considered in the modal expansion. For the frequency range $0 < fd/c_{\mathrm{T}} < 0.2$ considered here, only the lowest-order Lamb modes (A0 and S0) are the propagating modes which travel away from the defect: the other modes are localized near the defect.

As the linear response, the reflection and transmission coefficients of the A0 and S0 modes are shown as functions of the normalized fundamental frequency in Fig. 2. In Fig. 2, the coefficients for the S0 mode remain small for the entire frequency range. The reflection coefficient of the A0 mode takes a high value at certain discrete frequencies, where the corresponding transmission coefficient shows sharp drops. This phenomenon is due to the resonance of the defect to the incident Lamb wave. The resonant frequencies of the defect depends on its interfacial stiffnesses in addition to its length and location. When $K_{\mathrm{N}} = K_{\mathrm{T}} = 0$, the present solution procedure accurately reproduces the resonance spectrum of a traction-free crack in a plate as analyzed by Rokhlin.(7)

The nonlinear response of the defect leads to the generation of second-harmonic Lamb modes. In Fig. 3, the coefficients of the double-frequency A0 and S0 Lamb modes (M_{A0} and M_{S0} for the left region and P_{A0} and P_{S0} for the right region) are shown as functions of the normalized frequency of the incident wave. The second-harmonic A0 mode is generated significantly at the frequencies of linear resonance of the defect. This is due to the feature that the opening displacement of the defect, which gives the source for second-harmonic generation, is amplified at resonance. The coefficient of the second-harmonic A0 mode is particularly high at the lowest resonant frequency of the defect.

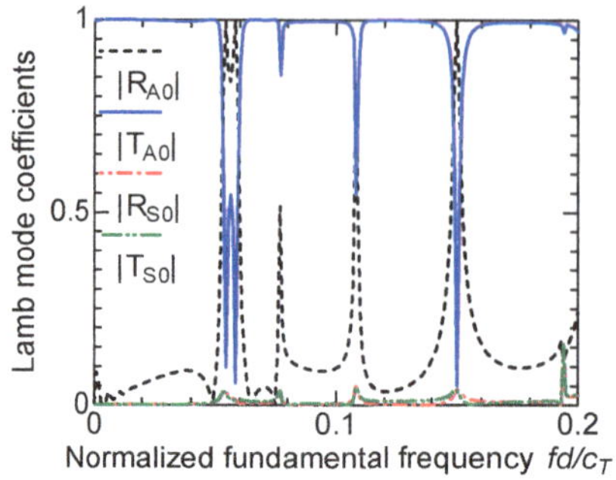

Fig. 2 Magnitude of the fundamental Lamb mode coefficients

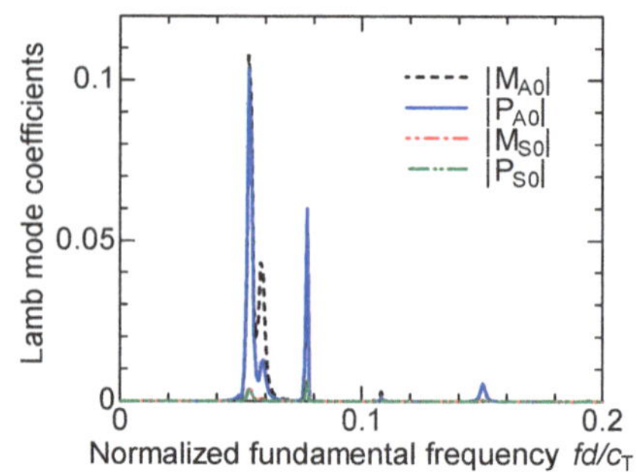

Fig. 3 Magnitude of the second-harmonic Lamb mode coefficients

5 SUMMARY

The resonance of the lowest-order antisymmetric Lamb mode as well as the second-harmonic generation at a closed defect in a thin plate has been demonstrated by numerical analysis based on a perturbation technique and the hybrid finite element method. The second-harmonic Lamb mode can be generated significantly when the frequency of the incident Lamb mode matches the resonant frequency of the defect. The resonant frequency and the second-harmonic mode amplitude can thus be used to characterize the state of closed defects.

ACKNOWLEDGEMENT

This work has been supported by JSPS KAKENHI Grant Number 25289005.

REFERENCES

(1) Solodov, I. Y. (1998) *Ultrasonics* 36, 383-390. doi: 10.1016/S0041-624X(97)00041-3
(2) Biwa, S., Hiraiwa, S. & Matsumoto, E. (2006) *Ultrasonics* 44, e1319-e1322. doi: 10.1016/j.ultras.2006.05.010
(3) Shkerdin, G. & Glorieux, C. (2008) *J. Acoust. Soc. Am.* 124, 3397-3409. doi: 10.1121/1.2996301
(4) Solodov, I., Bai, J., Bekgulyan, S. & Busse, G. (2011) *Appl. Phys. Lett.* 99, 211911. doi: 10.1063/1.3663872
(5) Biwa, S., Nakajima, S. & Ohno, N. (2004) *Trans. ASME J. Appl. Mech.* 71, 508-515. doi: 10.1115/1.1767169
(6) Mori, N., Biwa, S. & Hayashi, T. (2013) *J. Appl. Phys.* 113, 074901. doi: 10.1063/1.4791711
(7) Rokhlin, S. I. (1981) *J. Acoust. Soc. Am.* 69, 922-928. doi: 10.1121/1.385614

Laser shock peening simulation of mitigation on residual stress in Alloy 600

J.S. Kim [1], H.S. Nam [1], Y.J. Kim [1], J.H. Kim [2]
[1] Dept. of Mechanical Engineering, Korea University, Korea
[2] Dept. of Weapon and Mechanical Engineering, Korea Military Academy, Korea

ABSTRACT

Laser shock peening simulation method especially for Alloy 600 penetration nozzles in pressurized water reactor is proposed. In order to simulate multiple laser peening process, the strain rate dependent stress-strain relationship with isotropic hardening model was considered. In this paper, we present analytical results about the effect of multiple laser peening on mitigation of tensile residual stress in Alloy 600.

1 INTRODUCTION

In recent years, primary water stress corrosion cracking (PWSCC) on Alloy 600 penetration nozzles has been reported in worldwide pressurized water reactors (PWR). To solve these problem, various mitigation technologies for PWSCC are also developed (1-2). Laser shock peening (LSP) process, which is one of the techniques, produce compressive residual stress on the surface of target metal using laser-induced shock wave (3). LSP process has been applied to eliminate welding residual stress in Alloy 600 nozzles at reactor vessel head penetrations and bottom-mounted instruments in Japan (4).

In this study we carried out a numerical simulation of the LSP process for Alloy 600 penetration nozzles. By comparing the analytical results with experimental data, we could confirm the validity of the simulation. And we present results of mitigation on welding residual stresses by LSP process.

2 ALLOY 600 LSP SIMULATION

The finite element analysis was performed using the commercial FEM code ABAQUS v6.12. To simulate generation and propagation of laser-induced pressure wave in function of time, the analysis was carried out using the ABAQUS/Explicit code, which is efficient for the analysis of relative short dynamic event.

The analysis model and input variables are shown in Figure 1. The model consisted of finite elements for calculating residual stresses induced by laser shock and infinite elements for removing the influence of the reflected wave.

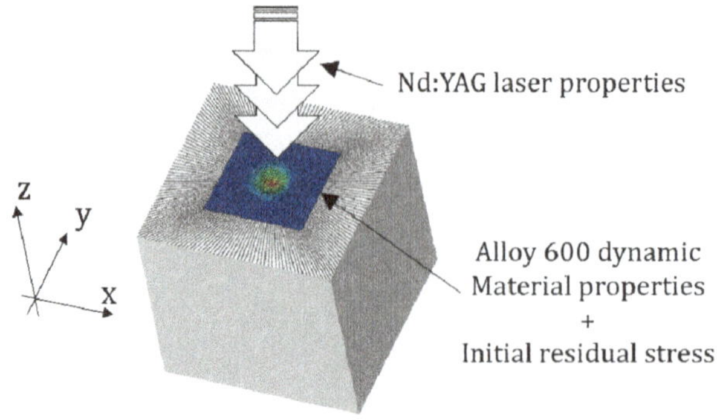

Figure 1. Finite element model for LSP simulation and input variables

2.1 Material variables

In order to consider the strain rate dependent stress-strain relationship, a Johnson-Cook plastic law with isotropic hardening was used (5). Neglecting temperature effects in this study, the stress is given as a function of plastic strain and strain rate as follows:

$$\sigma_{eq} = \left(A + B(\varepsilon_{eq}^{pl})^n\right)\left(1 + C\ln(\frac{\dot{\varepsilon}}{\dot{\varepsilon}_o})\right) \quad (1)$$

where $\varepsilon^{pl}{}_{eq}$ is the equivalent plastic strain, $\dot{\varepsilon}$ is the plastic strain rate and $\dot{\varepsilon}_o$ is a reference strain rate. The parameter A, B, and n are material constant to be determined by tensile test at low strain rate. The parameter C represents strain rate effect to be determined by Hopkinson bar test.

The hydrodynamic behaviour in material was considered by a Mie – Gruneisen equation of state in which the material's volumetric strength (6).

2.2 Laser variables

The laser system used for the Alloy 600 penetration nozzles consists of a Q-switched Nd:YAG laser with wavelength of 532nm (4). The laser variables such as maximum pulse pressure, pulse duration and laser geometry are determined based on the properties of the Nd:YAG laser shown in Table 1. The maximum pulse pressure generated by laser is given by the following equation (3):

$$P = 0.01\sqrt{\frac{\alpha}{2\alpha+3}}\sqrt{Z(g/cm^2\cdot s^2)}\sqrt{I_0(GW/cm^2)} \quad (2)$$

where α is the efficiency of the interaction, I_o is the laser power density and Z is the reduced shock impedance between the target metal and the confining water.

Table 1. Q-switched Nd:YAG laser properties

Pulse energy	Spot Size	Pulse duration	Pulse Interval time
40~250 (mJ)	ϕ 0.4~1 (mm)	4~10 (ns)	2~5 (ms)

2.3 Initial residual stress

In order to evaluate the effect of welding residual stress in the Alloy 600 penetration nozzles, the initial condition option is used in the simulation. Corresponding to the general welding residual stress in Alloy 600 penetration nozzle, initial residual stresses is assumed to be about σ_{xx}=0~200MPa when σ_{yy}=0MPa.

3 VERIFICATION OF SIMULATION

To verify the LSP simulation method, the analytical results were compared with the experimental data on Alloy 600 flat plate performed by Toshiba (1). The Nd:YAG laser was used in the experiment. The laser pulses were irradiated in zigzags with pulse energy of 60mJ, focal spot diameter of 0.7mm, and pulse density of 4500 pulses/cm^2. And residual stress measurements were performed by EPRI and Toshiba using X-ray diffraction.

Figure 2 shows the depth profile for the residual stress after LSP process. A comparison between the square symbols representing test data and line representing FE result shows that the simulation can predict well the residual stress generated by LSP.

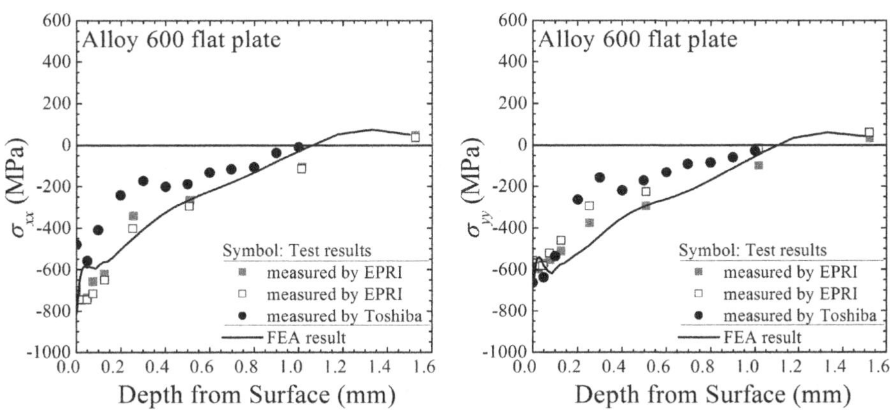

Figure 2. Verification with LSP test on Alloy 600 flat plate

4 SIMULATION RESULTS

When the laser is irradiated 1~10 times at same place in Alloy 600, a variation of the residual stresses is shown Figure 3. Figure 3(a) shows a mitigation on initial residual stress, σ_i =200MPa with increasing the number of laser shots in multiple laser peening process. And Figure 3(b) shows a change in compressive stress applied depth, D_o with increasing the number of laser shots at σ_i=0MPa, 100MPa, 200MPa.

The results show that the mitigation on initial residual stress converges gradually as the number of laser shots is increased. And the final compressive stress applied depth by LSP process deceases as the initial residual stress increase.

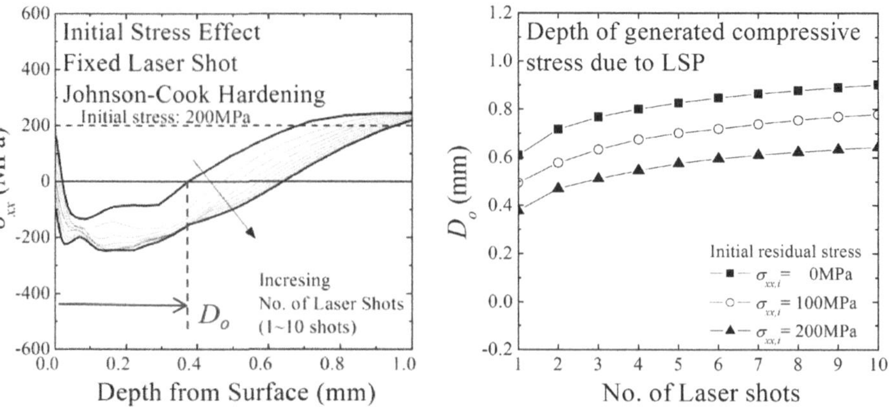

(a) Mitigation of residual stress as number of laser shots increases

(b) Variation of Compressive stress applied depth as number of laser shots increases

Figure 3. Effect of multiple laser peening

ACKNOWLEDGMENTS

This research was supported by National Research Foundation of Korea(NRF) funded by the Ministry of Science, ICT and Future Planning. (NRF-2013M2A8A1040924)

This work was supported by Nuclear Power Core Technology Development Program of the Korea Institute of Energy Technology Evaluation and Planning(KETEP) granted financial resource from the Ministry of Trade, Industry & Energy, Republic of Korea. (No. 20131520000140)

REFERENCES

(1) Material Reliability Program (MRP-162) (2006) *Electric power research institute*

(2) Material Reliability Program (MRP-267) (2012) *Electric power research institute*

(3) Fabbro, R., Fournier, J., Ballard, P., Devaux, D., & Virmont, J. (1990) *J. Appl. Phys.*, **68(2)**, 775-784. http://dx.doi.org/10.1063/1.346783

(4) Yoda, M., et al. (2004) *12th international conference on nuclear engineering.* **1**, 36-49. doi: 10.1115/ICONE12-49238

(5) Johnson, G. R., & Cook, W. H. (1985) *Engineering Fracture Mechanics*, **21**, 31~48. doi:10.1016/0013-7944(85)90052-9

(6) Peyre, P., Sollier, A., Chaieb, I., Berthe, L., Bartnicki, E., Braham, C. & Fabbro, R. (2003) *The European Physical Journal Applied Physics.* **23**, 83~88. doi:10.1051/epjap:2003037

(7) Kim, J. H., Kim, Y. J. & Kim, J. S. (2013) *Journal of Mechanical Science and Technology*, **27**, 2025-2034. doi:10.1007/s12206-012-1263-0

Deformation and behaviour of membrane structure by large deformation and contact simulation

T. Kinoshita [1)], T. Kawakami [1)], T. Sugiura [1)], T. Matsuda [2)], Y. Oura [2)]
[1)] Mechanical Systems Enginerring, Toyama Prefectural University, Japan
[2)] Sankyo Tateyama, Inc., Japan

ABSTRACT

Large deformation and contact simulations were performed to study deformation behaviour and stresses of membrane structure at conditions of snow weight by using a large scale simulator ADVENTURECluster® which was based on Finite Element Method (FEM). The initial state of the membrane structure was reproduced by loading gravity on membrane. Pressure was applied on surface of membrane as snow weight. In this study, it is suggested that applying gravity on membrane is reasonable method for reproducing initial state of membrane structure. Deformation, stresses and behaviour of membrane were quantitatively discussed at condition of snow weight.

1 INTRODUCTION

Recently, membrane structure is extensively studied (1)-(6) and used as permanent structure because of improvement of endurance and resistance fire. It has advantages that large space is easily and economically constructed due to that membrane materials are light and thin sheet materials. On the other hand, membrane materials have low resistance to compressive force, out-of-plane shearing force and out-of-plane bending moment. Irregular tensional force makes wrinkle and warp on membrane materials. Shapes of membrane structure would be limited due to above reasons and it has possibility that expected curvature of membrane could not be suitable curvature. Generally, larger curvature would be formed in membrane and sheet materials in a construction because that the materials have low stiffness for out-of-plane force and moment. In case of discussion and analysis for membrane materials, geometric non-linearity, large deformation, should be considered. If shape of membrane materials in completed products is understood at the planning or the design phases, much cost of products will be reduced.

In this study, we performed large deformation and contact simulation by using finite element method (FEM) to quantitatively discuss deformation, stresses and behaviour membrane at condition of snow weight.

2 SIMULATION MODEL AND DIMENSIONS

Configuration of membrane structure was shown in Fig. 1. The membrane structure was constructed by Al alloy member and membrane. Only Al alloy members were shown in Fig. 2. The membrane was made of polyester foundation cloth and of vinyl chloride resin. The members were common Al alloy material. The dimensions of whole member structure were 2486.6 mm and 2836.6 mm, and height was 313.5 mm. The dimensions of membrane were 2700 mm and 2836.6 mm, and thickness is 0.52 mm. Aluminium alloy is used as structural members, and Young's modulus, Poisson's ratio and mass density were assumed to be 70.0 GPa, 0.350 and 2,700 kg/m^3, respectively. Young's modulus, Poisson's ratio and mass density of membrane were assumed to be 0.719 GPa, 0.141 and 1,330 kg/m^3, respectively. Material properties in this study were shown in Table 1. Both materials were assumed to be isotropic.

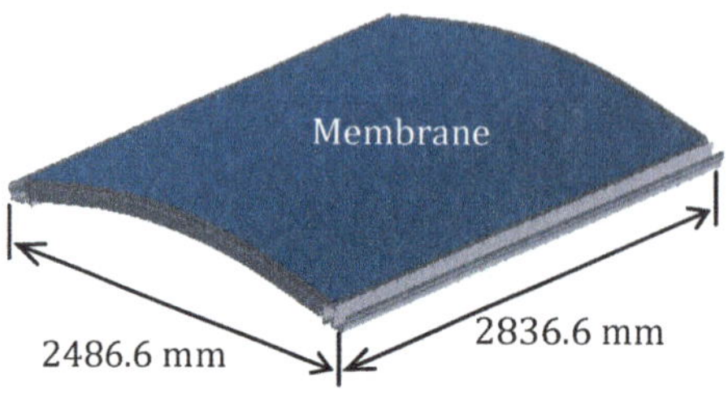

Fig. 1 Membrane Structure

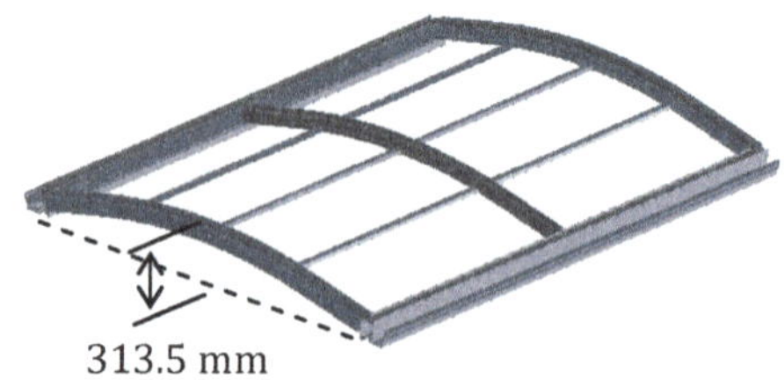

Fig. 2 Aluminium Alloy Members

Table 1 Material Properties

	Young's Modulus [GPa]	Poisson's Ratio [-]	Mass Density [kg/m^3]
Al alloy	70.0	0.350	2,700
Membrane	0.719	0.141	1,330

3 FEM MODEL

FEM model at starting state was shown in Fig. 3. Detail shapes of Al alloy member where were not concerned about analysis were modeled as simplified shapes. However, curvatures and corner shapes of Al alloy member were modeled in truly because that the above factors were influential on results of our simulation. Primary hexahedral elements were adopted and the elements were generated by mesh generator Hyper Mesh (7). The areas where were expected to be contacted between membrane and Al alloy member were divided by sufficiently small size elements to estimate accurate displacements and stresses. Almost uniformly-sized elements were adopted on areas where were not expected to be contacted between the membrane and the member. Small size elements provided accurate displacements and stresses. Number of elements and degree of freedom were 106,763 and 597,723, respectively.

In this study, FEM simulator of ADVENTURECluster (ver. 4. 8) (8) is adopted to perform large deformation and contact simulation. The simulator is for a large scale and parallel computing.

4 BOUNDARY CONDITION

Large deformation and contact simulation were performed by assumption under the condition of snow weight. In our study, initial state of membrane structure was created by gravity process and tension process. Gravity process means the process that gravity was applied to planar configuration of membrane to fit between membrane and the member. Gravity process was shown in Fig. 4. Starting state of membrane structure was shown in Fig. 4(a). After applying gravity, terminal of the membrane were lie down on Al alloy plates which were built as a part of member structure as shown in Fig. 4(b). The plate is not a part of actual product, but the plate was modeled to avoid spread of calculation due to fluttering motion of the terminal of membrane. After then, initial tension, 1kN/m, was applied to terminal of membrane as shown in Fig. 4(c) to create initial state of membrane structure.

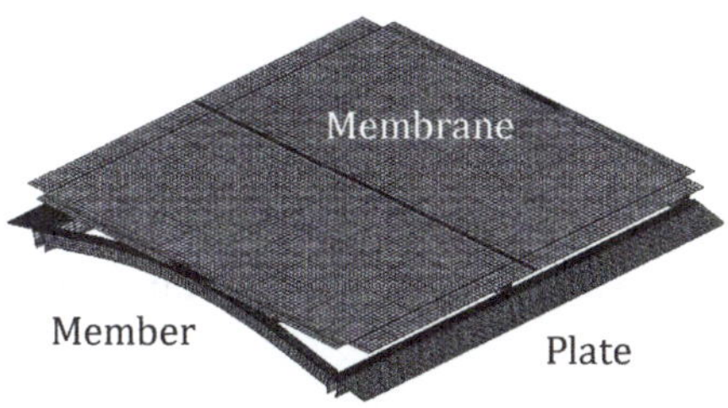

Fig. 3 FEM Model at Starting State

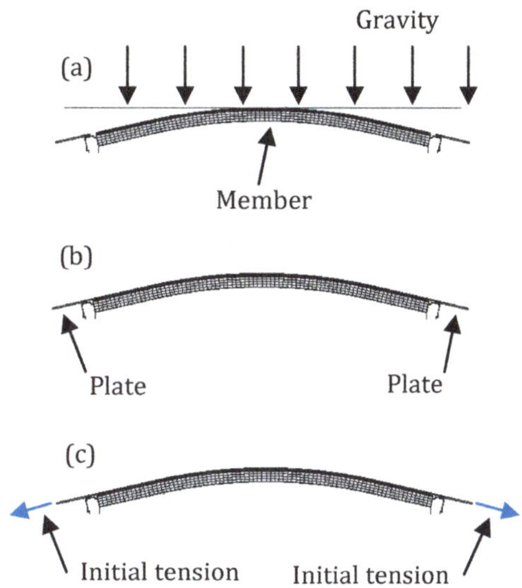

Fig. 4 Gravity Process for Reproducing Initial State of Membrane

5 SIMULATION RESULTS

Deformation behaviours of membrane structure were compared between initial state before snow loading and after snow loading. Initial state before snow loading was shown in Fig. 5. Initial tensions have been already applied on the terminal of membrane. Alphabet and arrows in the figure showed origin of data sampling and direction of data sampling, respectively. Relationships between distance from the origin and displacement to membrane thickness direction were shown in Fig. 6. Zero level of displacement means the positions on the top of the curved Al alloy member. In both cases along direction of arrow from A and B, maximum displacements along thickness direction were about 20 mm.

Deformation behaviour of membrane structure under snow weight loading was shown in Fig. 7. Pressure, 600 Pa, has been applied on the surface of membrane as snow weight. Relationships between distance from the origin and displacement to membrane thickness direction were shown in Fig. 8. Origin and direction along data sampling were same in case of initial state. Zero level of displacement was also same meaning. In the cases along direction of arrow from point A, maximum displacements were about 33 mm. In the cases along direction of arrow from point B, maximum displacements were about 56 mm. Equivalent stresses of membrane at initial state and snow weight condition were shown in Fig. 9.

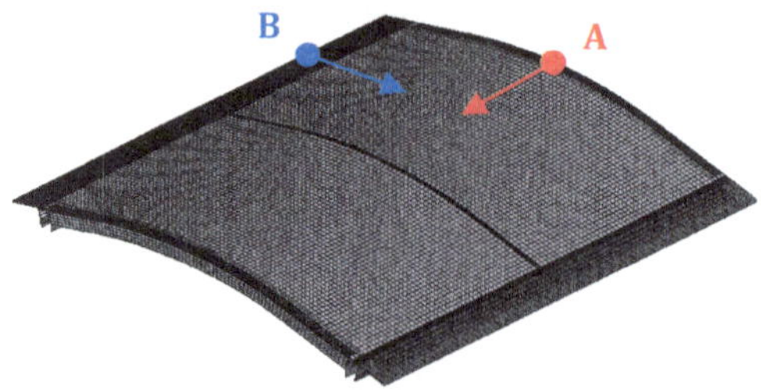

Fig. 5 Deformation Behaviour at Initial State

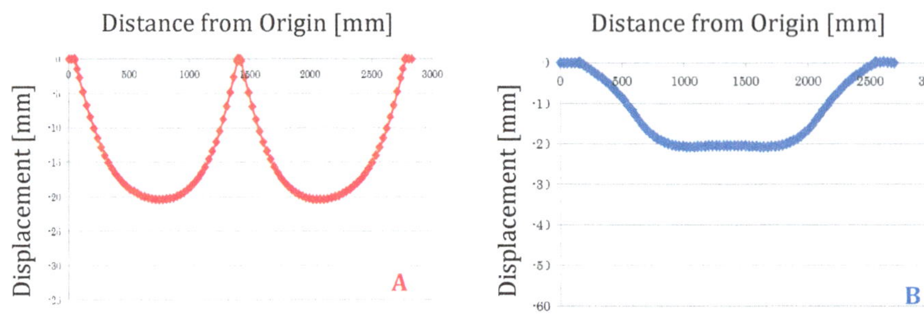

Fig. 6 Relationship between Distance from Origin and Displacement at Initial State

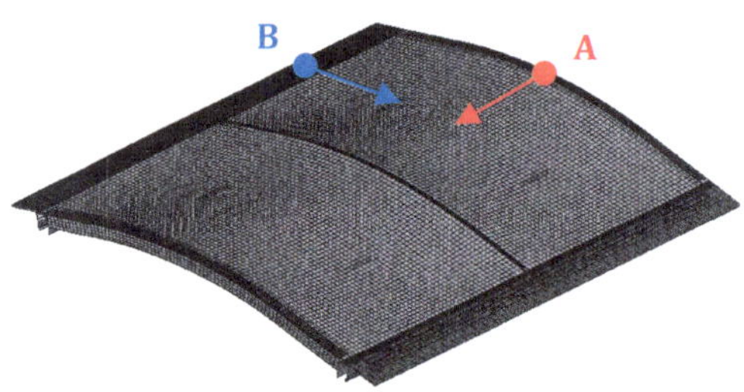

Fig. 7 Deformation Behaviour at Snow Weight

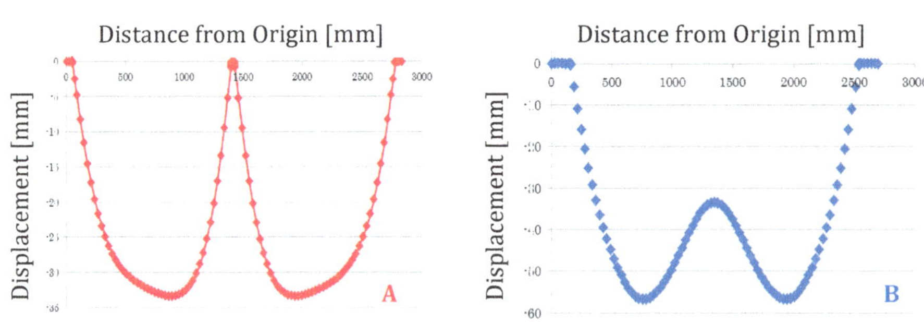

Fig. 8 Relationship between Distance from Origin and Displacement at Snow Weight

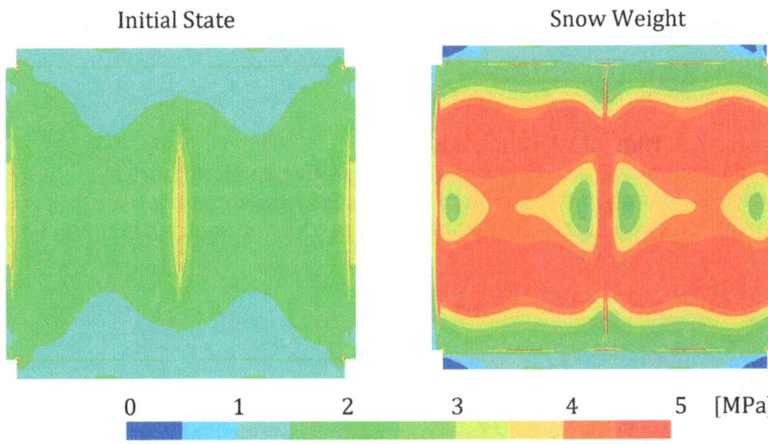

Fig. 9 Equivalent Stress at Conditions of Initial State and Snow Weight

6 CONCLUSIONS

In this study, large deformation and contact simulation were performed to discuss on deformation, stresses and behaviour of membrane structure of membrane and aluminium alloy members. at conditions of initial state and snow weight. It is suggested that applying gravity on membrane is reasonable method for contacting between membrane and aluminium alloy members, and also reasonable method for reproducing initial state of membrane structure.

In case at initial state of membrane, maximum displacement was estimated about 20 mm and higher equivalent stress were generated at the membrane where has been contacted on aluminium alloy member. In case at snow weight condition, displacement was increased up to 60 mm and higher equivalent stresses were generated at membrane where has been contacted on the member and where has been shown maximum displacement of membrane. It insisted that it had possibility of fracture of membrane where has been contacted on the member, and also insisted surface ponding at position of maximum displacement of the membrane.

REFERENCES

[1] E. Jeong, S. Yasutake & M. Kawabata (2012) Research Report on Membrane Structures 2012. **26**, 1-12.

[2] T. Muranaka, I. Shinohara, R. Takaki, O. Mori, R. Funase, Y. Shirasawa & G. Ono (2014) Trans. JSASS Aerospace Tech. Japan. **12**, 33-38.

[3] K. Kawaguchi, W. Ke & M. Miki (2008) J. Struct. Constr. Eng. **73**, 1773-1777.

[4] T. Ogawa, T. Kumagai, Y. Kurokawa, K. Tabuchi, M. Nakayama & K. Masuda (2012) J. Struct. Constr. Eng. **77**, 585-591.

[5] Y. Nagai, A. Okada, M. Kanda, N. Miyasato & M. Saitoh (2012) J. Struct. Constr. Eng. **77**, 211-219.

[6] T. Yamashita, T. Kumagai, T. Ogawa, T. Miyamura & M. Ohsaki (2009) J. Struct. Constr. Eng. **74**, 297-304.

[7] Altair Engineering Corporation, (http://www.altairjp.co.jp/Default.aspx)

[8] Allied Engineering Corporation, (http://www.alde.co.jp/)

Burst simulations of steam generator tubes using FE damage analyses

J.Y. Jeon[*], Y.J. Kim[*], J.W. Kim[**], J.S. Kim[***]
* Korea University, Department of Mechanical Engineering, Korea
** Chosun University, Department of Nuclear Engineering, Korea
*** Sunchon National University, Department of Mechanical Engineering, Korea

ABSTRACT

This paper compares FE analysis results with burst test data using Alloy 690TT tube specimens. The tube specimens (the outer diameter of 19.05 mm and the thickness of 1.07 mm), commonly used in steam generator (SG) of nuclear power plants, were tested. To obtain material properties of Alloy 690TT, tensile tests were conducted, according to ASTM standard, using the tube specimens. Burst tests were also performed using tube specimens with single axial crack at outer surface of tubes. Through-wall cracks and surface cracks with several depths and lengths were considered in the tests. Burst pressures for single cracked tube specimens were predicted using finite element (FE) analyses. FE analysis method to simulate ductile failure is based on 'the multi-axial fracture strain model', recently proposed by the authors. Predicted results were compared with test data. Crack propagation behaviours were observed from the FE results.

1 INTRODUCTION

The steam generator (SG) is one of the most important part in the pressurized water reactor (PWR). Heat from the primary reactor coolant system is transferred to the secondary non-radioactive water to produce steam. The steam generator tubes are the barrier between the primary system and the secondary system. Thus, the steam generator tube has to be within enough safety margin to avoid leakage of radioactive water to the secondary system. For this reason, the defects in the steam generator tubes should be assessed by proper evaluation methods.

Several researches to assess the defects in the steam generator tube have been carried out by both of experiments and FE analyses. The experimental ways have limitation of cost and time. For this reason, FE analysis methods have been used to consider various cases that might be occur in the actual steam generator tubes. There are FE analysis results using damage mechanics in the simulations by the authors (1-2). The damage mechanics can be applied to FE analysis to describe ductile fracture. Yuuki Miyajima et. al. (1) used the GTN model and Xaver Schuler et. al. (2) used Rousselier model as damage mechanics model in their studies. For using these models, several material constants should be determined by a number of tests.

In this paper, burst simulations of steam generator tubes with single axial cracks were conducted using the damage mechanics model that is relatively simple but shows accurate results. Simulated results were compared with test data and were observed in detail.

2 EXPERIMENT

2.1 Material and Tube Tensile Test

Alloy 690TT steam generator tubes (the outer diameter (D_o) of 19.05mm and the thickness (t) of 1.07mm), commonly used in pressurized water nuclear reactors, were tested. To measure tensile properties, tests were conducted at room temperature using tube specimens according to ASTM E08-09 (3). The displacement rate was 0.5mm/min, and the axial displacement of specimen was measured using extensometer with the gage length of 25mm. Yield (0.2% proof) strength and tensile strength of the material were about 306.9MPa and 691.9MPa, respectively. Resulting engineering and true stress-strain curves are shown in Fig. 1.

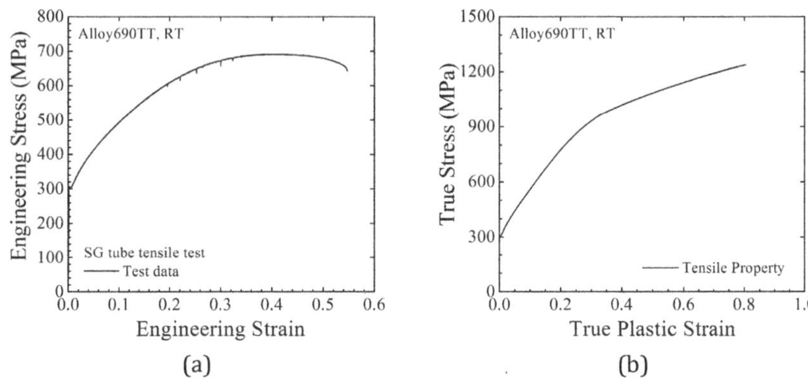

Fig. 1 (a) Tube tensile test result and (b) true stress-strain curves of Alloy690TT

2.2 Burst Test of Axial Cracked Tubes

Burst tests were performed for Alloy690TT tubes with single axial through-wall cracks and surface cracks. For the single axial surface crack, six tubes with different notch lengths (L) and depths (a) were tested. Tested tubes with axial surface crack are schematically shown in Fig. 2.

Fig. 2 Geometry of single axial cracked tube specimen

3 FINITE ELEMENT DAMAGE ANALYSIS

3.1 Multi-Axial Fracture Strain

In this study, the damage model used to simulate ductile failure that lead to crack propagation is well known 'the multi-axial fracture strain' model. This model based on the theory that fracture strain of ductile materials depends on triaxial stress states. The dependence of the multi-axial fracture strain, ε_f, on the stress triaxiality (defined by the ratio of the mean normal stress σ_m and equivalent stress σ_e) can be represented by the formula below (4).

$$\varepsilon_f = A \exp\left(-C\frac{\sigma_m}{\sigma_e}\right) + B \tag{1}$$

where *A*, *B* and *C* are material constants that can be determined from notched bar tensile tests data with different notch radii (5-7).

The test specimens used in this study to obtain material properties were only Alloy690TT tube specimens. Unlike notched bar, these types of thin-walled specimens could not be tested under high triaxial stress states. For this reason, Eq. (1) is simplified as follows. According to Rice and Tracey (4), the constant C in Eq. (1) can be assumed to be 1.5. Furthermore the multi-axial fracture strain can be conservatively estimated by assuming B=0 in Eq. (1), leading to simplified form

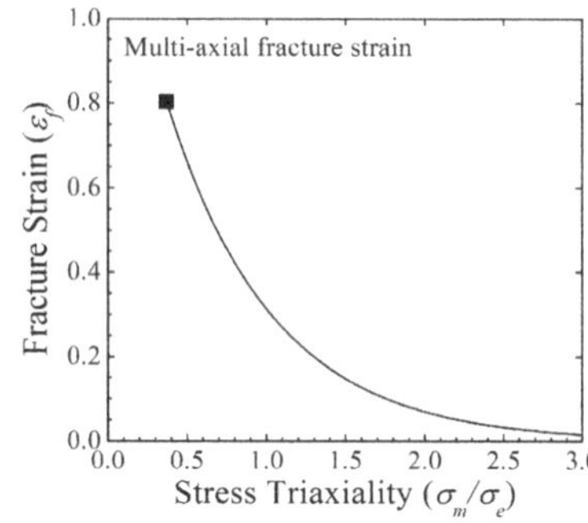

Fig. 3 Multi-axial fracture strain for Alloy690TT

$$\varepsilon_f = A \exp\left(-1.5\frac{\sigma_m}{\sigma_e}\right) \tag{2}$$

where *A* can be determined from the tube tensile test data. By the same process with Ref. (5-7), the constant *A* for Alloy690TT studied in this paper was determined to be 1.402. Resulting fracture strain as a function of the stress triaxiality is shown in Fig. 3 (one point indicated as 'tube tensile specimen').

3.2 Burst Simulation and Comparison with Experimental data

To simulate ductile failure and crack propagation using FE analysis, damage accumulation and ductility exhaustion concept was introduced. In FE analysis, incremental damage, $\Delta\omega$, defined by the ratio of the equivalent plastic strain increment to the multi-axial fracture strain is calculated. When accumulated incremental damage becomes unity, $\omega=\Sigma\Delta\omega=1$, ductile failure is assumed at the gauss point and all stress components are reduced forcibly using user-subroutine USDFLD and UHARD in ABAQUS (8). More detailed information about crack propagation simulation technique can be found in Ref. (5-7).

$$\Delta\omega = \frac{\Delta\varepsilon_e^p}{\varepsilon_f} \tag{3}$$

Using the ductile failure simulation technique, bursting of single cracked Alloy690TT tubes were simulated. All of the tested cases were simulated by FE analyses, and additionally the short crack cases (crack length, *L*, is 3mm) were also simulated without testing. FE mesh used in the simulations is shown in Fig. 4. FE mesh was modelled by first order solid element (C3D8 within ABAQUS)

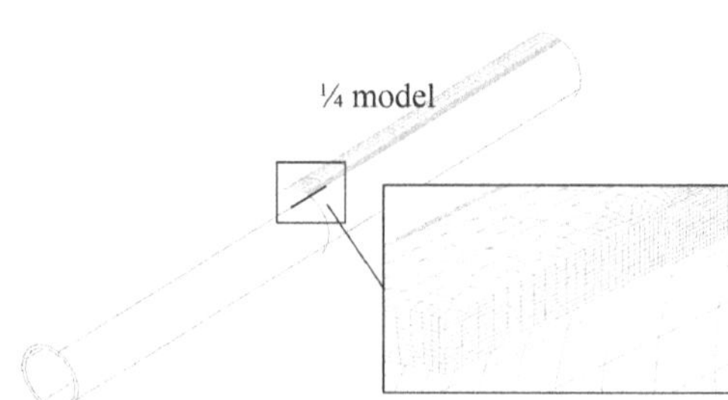

Fig. 4 FE mesh used in the simulation

and the FE analyses were conducted with large geometry (nonlinear) change option. Element size in the crack tip region is arranged to 0.1mm that was found to be proper in the previous study. Internal pressure and axial-tension load were applied to inner surface and end-cap of the tube, respectively. To consider non-linear behaviour after bursting, RIKS (8) option supported by ABAQUS was applied.

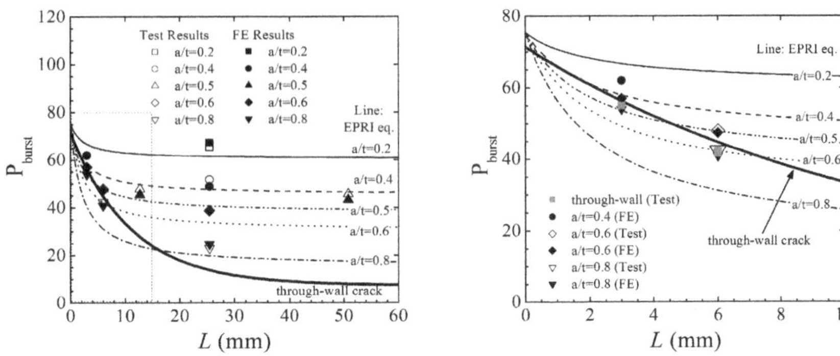

(a) (b)

Fig. 5 Comparison of burst test data with simulated one using FE analysis: (a) for all cases and (b) for only short crack cases

Predicted burst pressures are compared with each test data in Fig. 5. In the figure, closed symbols and open symbols represent FE results and test data, respectively. Figure 5 also includes burst pressure prediction for tubes with through-wall crack and surface crack (suggested by EPRI (9)) by solid and dotted lines. FE results agree well with test data within 7% of error range. Both of the FE results and test data are slightly higher than EPRI predictions. The square dotted box in Fig. 5(a) is expanded in Fig. 5(b). In the figure, the short and deep cracked tubes have a same burst pressures with the values for through-wall cracked tubes. To study about this phenomenon, internal pressure variations followed by increasing crack mouth displacement (δ) are shown in Fig. 6. In the short crack case (as shown in Fig. 6(a)), internal pressure increases even after the crack penetration, and decreases by plastic collapse. In contrast with Fig. 6(a), relatively long crack case (see Fig. 6(b)) shows bursting before showing crack penetration. Therefore, a burst pressures for short cracked tubes are affected by plastic collapse rather than crack penetration. For this reason, short and deep cracked tubes behave like a through-wall cracked tube.

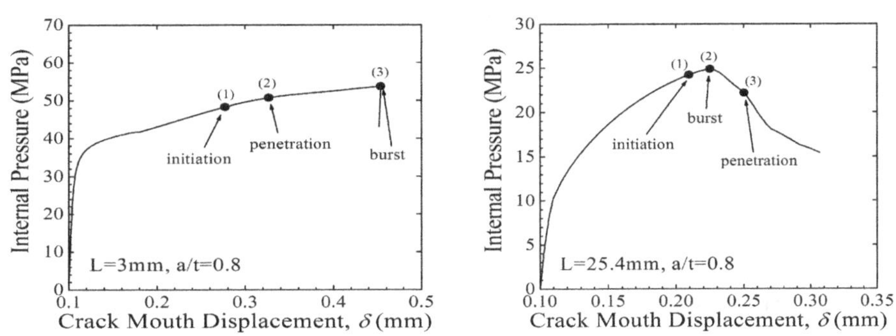

(a) (b)

Fig. 6 Observed crack behaviour with increasing of crack mouth displacement

4 CONCLUSIONS

In this study, bursting of single axial cracked Alloy690TT tubes were tested experimentally and simulated using FE damage analysis. To consider the material properties of Alloy690TT in the simulations, tensile tests were performed using the tube specimens and the tensile property was obtained from the test data. Within the multi-axial fracture strain model in the FE damage analysis, ductile fracture and crack propagations were implemented. In the simulation, tested cases and additional cases could be considered with low cost, and the behaviours of the cracks in the tubes could be observed in detail. In the future, burst pressures and crack behaviours of various cracks (position, direction and length) might be existed in actual steam generator tubes could be considered using the FE method suggested in this paper.

ACKNOWLEDGEMENTS

This work was supported by the Nuclear Energy Technology Innovation Program of the Korea Institute of Energy Technology Evaluation and Planning (KETEP) grant funded by the Korea government Ministry of Trade, Industry and Energy. (No. 2012T100100443) and by the Ministry of Trade, Industry & Energy(MOTIV), Korea Institute for Advancement of Technology(KIAT) and Dongnam Institute for Regional Program Evaluation through the Leading Industry Development for Economic Region (No. 20131520202170).

REFERENCES

(1) Miyajima, Y., Kikuchi, M., Takahashi, A., "Ductile fracture simulation of a pipe of steam generator in PWR", Proceeding of the ASME Pressure Vessels and Piping Conference, Paris, France, 2013.

(2) Schuler, X. Seidenfuss, M., Gehrlicher, S., Mahlke, J., "Damage modelling of the load-bearing capacity of damaged steam generator tubes in nuclear power plants", Proceeding of the ASME Pressure Vessels and Piping Conference, Paris, France, 2013.

(3) ASTM E08-09 2006, "Standard test method for measurement of fracture toughness. In: Annual book of ASTM standards", American Society for Testing and Materials, Philadelphia; USA.

(4) Rice, J. R., Tracey, D. M., "On the ductile enlargement of voids in triaxial stress fields", Journal of the Mechanics and Physics of Solids. Vol. 17, pp. 201-217, 1969.

(5) Oh, C. K., Kim, Y. J., Baek, J. H., Kim, Y. P., Kim, W., "A phenomenological model of ductile fracture for API X65 steel", International Journal of Mechanical Sciences. Vol. 49, pp. 1399-1412, 2007.

(6) Oh, C. S., Kim, N. H., Kim, Y. J., Baek, J. H., Kim, Y. P., Kim, W. S., "A finite element ductile failure simulation method using stress-modified fracture strain model", Engineering Fracture Mechanics. Vol. 78, pp. 124-137, 2011.

(7) Kim, N. H., Oh, C. S., Kim, Y. J., Yoon, K. B., Ma, Y. H., "Comparison of fracture strain based ductile failure simulation with experimental results", International Journal of Pressure Vessels and Piping. Vol. 88, pp. 434-447, 2011.

(8) ABAQUS Ver. 6.13 User's manual. Dassault Systemes Simulia, Inc. USA.

(9) Electric Power Research Institute, EPRI 1019037, Steam Generator Degradation Specific Management Flaw Handbook, Rev. 1, 2009.

Effect of pile-ups of dislocations in numerical analysis of fatigue crack propagation using discrete dislocations method

T. Fukihara [1], T. Kakiuchi [2], Y. Uematsu [2], Y. Motoyashiki-Besel [3]
[1] Mechanical and Systems Engineering Division, Graduate School of Gifu University, Japan
[2] Department of Mechanical Engineering, Gifu University, Japan
[3] German Aerospace Center, Germany

ABSTRACT

In this study, the numerical simulation of fatigue crack propagation in plane strain state was conducted using the discrete dislocations method and the effect of pile-ups of edge dislocations in this method is discussed. In the numerical simulation, the nucleation and the motion of dislocations are constrained by the pile-ups of dislocations which are nucleated in the past cycles, thus, the crack propagation rate after the second cycle is decreased compared with the first cycle.

1 INTRODUCTION

A fatigue crack in metal forms a plastic zone and a stress field in front and it propagates in the stress field formed by itself. The fatigue crack propagation behavior is affected by the stress field significantly. From the microscopic viewpoint, a stress field in front of a crack is attributed to the pile-ups of dislocations after a number of fatigue cycles. Thus the dislocations behavior, such as the nucleation, the motion and the annihilation, and the stress field formed by the dislocations are required to be considered to evaluate the fatigue crack propagation. The discrete dislocations method is one of the methods to consider the discrete dislocations behavior in which the stress field is represented by the superposition of that by the external load, which is solved by the equations of continuum body, and that formed by the discrete dislocations. The problems on the crack have been discussed using this method (1) - (5).

In this study, the numerical simulation of fatigue crack propagation in plane strain state is conducted using the discrete dislocations method in which the fatigue crack propagation is modeled by the motion of dislocations near the crack tip and the effect of pile-ups of edge dislocations on the fatigue crack propagation in the early stage of fatigue crack propagation in this method is discussed.

2 ANALYTICAL PROCEDURE

2.1 Model of analytical target

Fig. 1 shows the analytical target, which is a Mode I crack in a plate. The sizes of a plate are W = 400 μm, a/W = 0.5 and L/W = 2. A crack is modeled as a slit of 10 nm clearance with a round tip of 5 nm radius. The coordinate system is constructed as the the origin is set at the crack tip, the x-axis is the macroscopic crack propagation direction and the y-axis is the vertical direction to the x-axis as shown in Fig. 1.

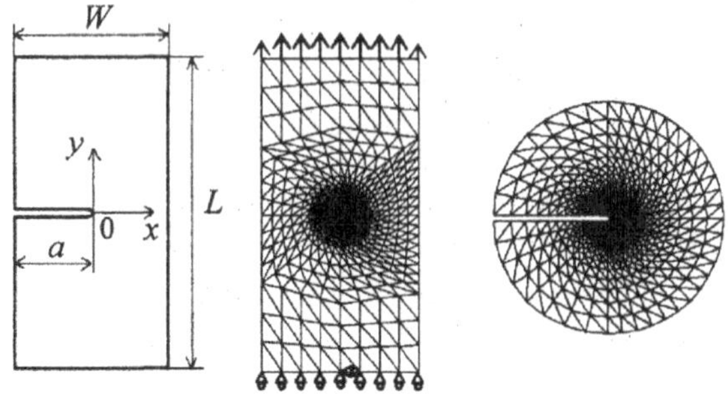

(a) Model (b) Mesh (c) Mesh near crack tip

Fig. 1 Model of analysis.

The applied fatigue stress σ, which is depending on time t, is given in the following equation (1) at the upper and the lower side, $y = \pm L/2$:

$$\sigma = \sigma_0 \left(1.1 - \cos \frac{2\pi t}{T} \right) \quad (1)$$

where the time period of cycle T = 480 ns and σ_0 = 33.8, 56.3 and 78.8 MPa, which corresponds to the stress intensity factor range ΔK = 4.8, 8.0 and 11.1 MPa√m, respectively.

The material is assumed to be a bcc-structured iron with two preferred slip planes of ±54.7 ° to the x-axis, which is the macroscopic crack propagation direction. The elastic moduli are the Young's modulus E = 210 GPa and the Poisson's ratio ν = 0.3. In the analysis, the plane strain state is assumed and the edge dislocation, which is nucleated and moves in the preferred slip planes whose Burgers vector is also parallel to them, is considered. The absolute value of a Burgers vector is b = 0.25 nm. Fig. 2 shows the schematic illustration of dislocation and slip. A dislocation which has a Burgers vector in the direction away from the crack surface is denoted as a "positive" dislocation and that which has a Burgers vector in the opposite direction is denoted as a "negative" dislocation. The motion of a positive dislocation away from the crack surface and that of a negative dislocation close to the crack surface is denoted as a "forward" slip and the motion in the opposite direction is denoted as a "reverse" slip. It is modeled that a negative dislocation disappeared when it approaches to the crack surface within a given range in a positive slip. The disappearance of a negative dislocation near the crack tip is regarded as the crack propagation. The stress field, which is represented by the superposition of that formed by the external load and that formed by the dislocations, affect the behavior of dislocations and the crack propagation. In the numerical simulation, a set of dislocations are distributed in the state of equilibrium as the initial condition.

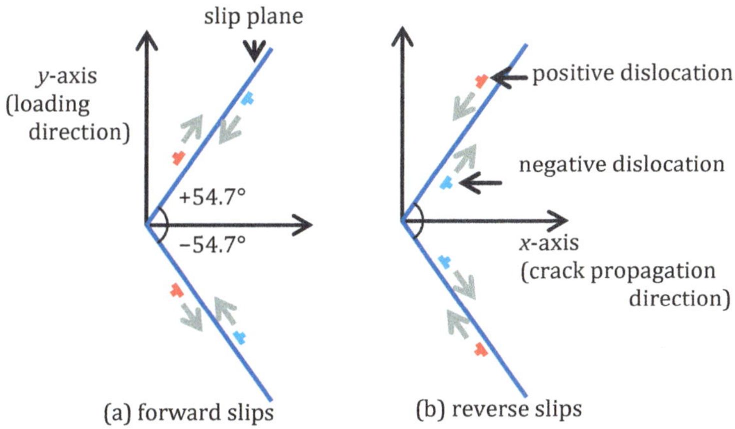

Fig. 2 Schematic illustration of dislocation and slip.

2.2 Motion of dislocation

It is modeled that a dislocation moves according to the speed law in the following equation (2) as a function of the resolved shear stress τ_{rss}, which is the shear stress component in the direction of the preferred slip planes,

$$v = v_0 \left(\frac{\tau_{rss} - \tau_p}{\tau_0} \right)^m \quad (2)$$

when τ_{rss} exceeds τ_p, which is corresponding Peierls stress. The constants are τ_p = 75 MPa, m = 35, v_0 = 0.01 m/s and τ_0 = 200 MPa (6), (7).

2.3 Nucleation and annihilation of dislocation dipole

A dipole of dislocations, which is a set of a positive and a negative dislocation in the distance of 15.625 nm, is nucleated when τ_{rss} exceeds a critical value τ_{nuc} = 800 MPa over the time period t_{nuc} = 1 ns, while a new dipole is not nucleated when the evaluation point is within the distance of 25 nm (= 100 b) from the other dislocations or the crack tip (8). A set of a positive and a negative dislocation is annihilated when they approaches close to each other. In addition, a dislocation disappeared when it passes cross the crack or free surface.

2.4 Crack propagation

It is considered that a small step is formed when a dislocation goes through a surface and the motion of a dislocation to the crack tip is attributed to the mechanism of crack propagation. In this study, it is modeled that a crack propagates when a negative dislocation approaches near the crack tip by a forward slip and the dislocation disappears with the crack propagation. The amount of crack propagation is evaluated based on the number of negative dislocations which disappear near the crack tip and the Burgers vector. In this study, the coordinate system is moved to the negative direction of the x-axis when a crack propagates to represent the crack propagation, i.e., a crack tip is moved relatively forward and the pile-ups of dislocations are moved relatively backward along the crack wake.

3 RESULTS AND DISCUSSION

Fig. 3 shows the number of dislocations which are nucleated and disappeared. The number of nucleated and disappeared dislocations is increased with the increase of ΔK. The number of dislocations is the largest in the first cycle and it decreased in the second cycle regardless of ΔK. However the change of number is small after the second cycle. It is considered that the change of number of dislocations is attributed to the stress field by the pile-ups of the residual dislocations.

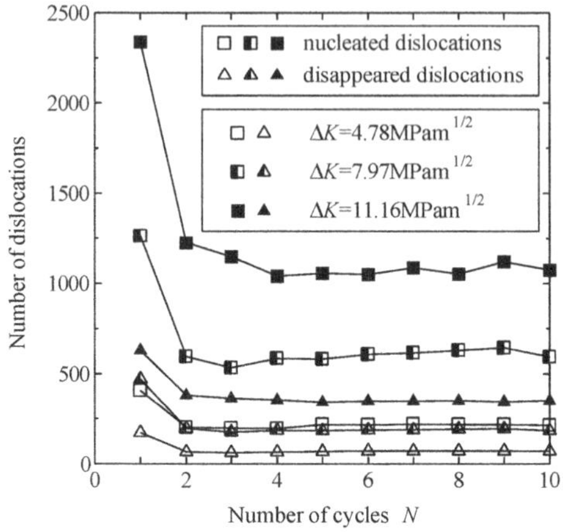

Fig. 3 Change of number of dislocations.

Fig. 4 shows the crack propagation rate da/dN. The dependence of da/dN on ΔK is clearly distinguished. da/dN is also dependent on the number of cycle as is the same with the number of dislocations, which is attributed to the pile-ups of the dislocations.

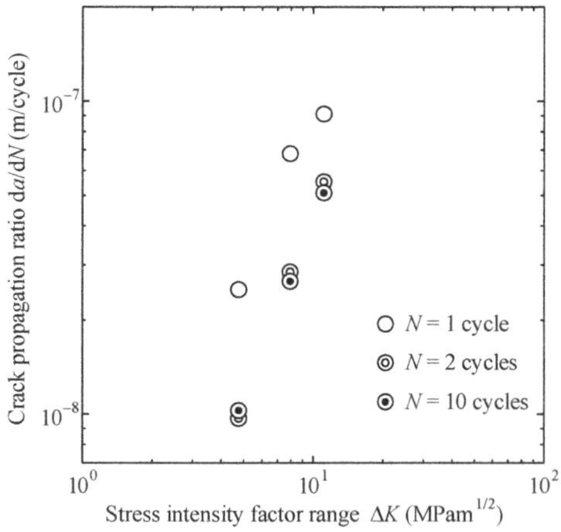

Fig. 4 Relationship between da/dN-ΔK.

Fig. 5 shows the distribution of dislocations in the condition of ΔK = 4.8 MPa√m. ((a): after 1cycle, (b): after 50 cycles). Dislocations indicated by the pink symbols are newly nucleated in the last cycle and dislocations indicated by the black symbols are nucleated in the past cycles except for the last cycle and left in the crack wake. In the end of the first cycle, the negative dislocations are piled up near the crack tip and the positive dislocations are piled up in the distance. The piled-up dislocations produce the negative shear stress field toward the crack tip direction, which relaxes the positive shear stress field by the external stress. As a result, the forward slip and the nucleation of dislocations are constrained.

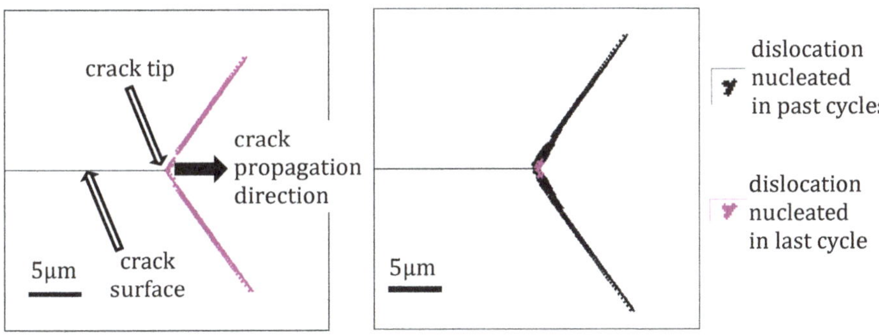

(a) After 1 cycle (b) After 50 cycles

Fig. 5 Distribution of dislocations.(ΔK=4.8 MPa$\sqrt{m}$)

4 CONCLUSION

In this study, the numerical simulation of fatigue crack propagation in plane strain state was conducted using the discrete dislocations method and the effect of pile-ups of edge dislocations in this method is discussed. In the numerical simulation, the piled-up dislocations produce the negative shear stress field toward the crack tip direction, which relaxes the positive shear stress field by the external stress. As a result, the forward slip and the nucleation of dislocations are constrained and the crack propagation rate after the second cycle is decreased compared with the first cycle.

REFERENCES

[1] Clegeringa, H.H.M., Van der Giessen, E. & Needleman, A. (2000) *Journal of the Mechanics and Physics of Solids*, **48**, 1133-1157.

[2] Van der Giessen, E., Deshpande, V.S., Clegeringa, H.H.M. & Needleman, A. (2001) *Journal of the Mechanics and Physics of Solids*, **49**, 2133-2153.

[3] Deshpande, V.S., Needleman, A. & Van der Giessen, E. (2001) *Acta Materialia*, **49**, 3189-3203.

[4] Motoyashiki, Y., Sugeta, A., Uematsu, Y. & Jono, M. (2003) *Journal of the Society of Materials Science, Japan*, **52**, 1217-1224 (in Japanese).

[5] Sugeta, A., Motoyashiki, Y., Uematsu, Y. Moriyama, T. & Jono, M. (2005) *Journal of the Society of Materials Science, Japan*, **54**, 540-545 (in Japanese).

[6] Stein, D.F., & Low Jr., J.R. (1960) *Journal of applied physics*, **31**, 362-369.

[7] Wang, J.N. (1996) *Materials Science and Engineering A*, **206**, 259-269.

[8] Essmann, U., Gösele, U. & Mughrabi, H. (1981) *Philosophical Magazine A*, **44**, 405-426.

Effects of mesh size and specimen configuration in simulating ductile fracture of metals by GTN model

Y.R. Oh, J.S. Kim, Y.J. Kim
Mechanical Engineering, Korea University, Korea

ABSTRACT

This work simulates the ductile fracture behaviour of metals by GTN model. The GTN model has a few issues need to be resolved in practical application of these methods. The first one is that depend on mesh size. For confirmation, the simulations has been tested with three different sizes of 0.05, 0.1, 0.2 mm characteristic length. The second issue is that, results vary according to the type of specimen configuration. So, simulated results using the GTN model are compared with experimental data for CT tests and SET tests.

1 INTRODUCTION

A number of works have been reported in the literature up to present on finite element ductile failure analysis. Depending on the model employed for damage simulating, the Gurson-Tvergaard-Needleman model can be broadly used. That model is using a micro-mechanical model for ductile fracture, incorporating void nucleation, growth and coalescence.

As many researchers have already published a number of papers using these methods, applicability and validity of these methods have been well discussed in the literature. From the author's point of view, it is felt that a few issues need to be resolved in practical application of these methods. The first one is that depend on mesh size. For instance, finite element analysis using by GTN model is hard to resolve of large and complex geometry. Because GTN model has a small mesh size. So, an alteration in model's mesh size require to change of GTN analysis parameters. The second issue is that, Results varies according to the type of specimen configuration. This paper identify that effects of mesh size and specimen configuration in simulating ductile fracture of Metals by GTN model.

2 GURSON-TVERGAARD-NEEDLEMAN (GTN) MODEL

The GTN model to simulate ductile damage and failure is given by [1, 2, 3]

$$\Phi = \left(\frac{\sigma_e}{\sigma_0}\right)^2 + 2q_1 f \cosh\left(-q_2 \frac{3p}{2\sigma_0}\right) - \left(1 + q_1^2 f^2\right) = 0 \tag{1}$$

where σ_e is the equivalent stress, p the hydrostatic pressure and σ_0 the yield stress of void-free material. The material parameters q_1 and q_2 depend on the hardening exponent n and on the ratio E/σ_0, where E is the Young's modulus.

The void growth rate can be expressed in terms of the current value of f (the void volume fraction) and the plastic strain rate tensor $\dot{\varepsilon}^p$ as

$$\dot{f}_{gr} = (1-f)\dot{\varepsilon}^p : \mathrm{I} \quad (2)$$

where I is the second order unit tensor.

3 NUMERICAL SIMULATION OF DUCTILE FRACTURE

3.1 Effect of mesh size

The material data compiled in this paper were obtained from tests conducted on weld samples.[4] The values of E and Poisson's ratio v were taken as 210 GPa and 0.3, respectively. And yield stress is 462 MPa. The fracture tests have been conducted on standard CT specimens with thickness of 25 mm and initial crack length/specimen width ratio of a_0/W=0.5. Values for the void growth parameters in the Gurson model, q_1=1.44 and q2=0.94. The magnitude of f_0, f_c and f_f were taken as 0.003, 0.11 and 0.16. The length of mesh sizes were 0.05, 0.1 and 0.2 mm. ABAQUS explicit was used for the analyses, with the standard porous plasticity model, without void nucleation. A typical mesh used in the computations is shown in Fig. 1. Plane strain conditions and finite deformations were assumed in the analyses.

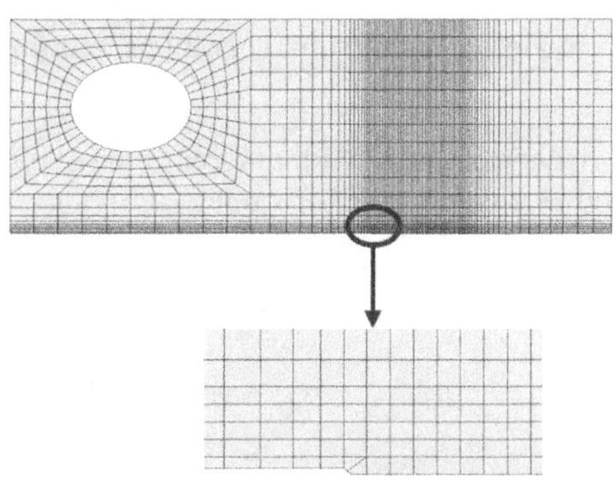

Fig. 1 Finite element mesh for the CT specimens

The computational predictions obtained with different values of mesh sizes are shown in Fig. 2-(a). These results show that a mesh size 0.05 mm best describes the experimental trend. Increasing mesh sizes predict conservative of J-resistance curves. To accurate describes the experimental trend from mesh sizes 0.1 and 0.2 mm need to change values of f_c and f_f. For example, FE results obtained with different values of f_c and f_f are shown in Fig. 2-(b).

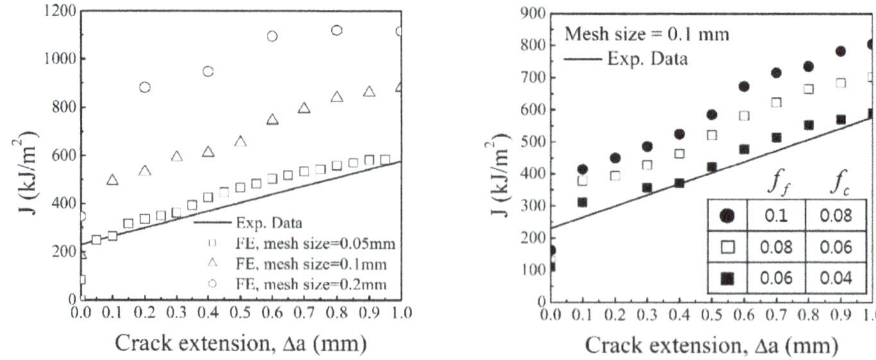

(a) **(b)**

Fig. 2 J-resistance curves (a) for various mesh sizes and (b) for various f_c and f_f

3.2 Effect of specimen configuration

The material data compiled in this paper were obtained from tests conducted on API X60 steel.[5] The values of E and Poisson's ratio v were taken as 210 GPa and 0.3, respectively. And yield stress is 483 MPa. The fracture tests have been conducted on standard CT specimens with thickness of 13 mm and initial crack length/specimen width ratio of a_0/W=0.5 and SET specimens with thickness of 12.5 mm and initial crack length/specimen width ratio of a_0/W=0.2. Values for the void growth parameters in the Gurson model, q_1=1.43 and q2=0.83. The magnitude of f_0, f_c and f_f were taken as 0.008, 0.04 and 0.20. The length of mesh size was 0.1 mm. ABAQUS standard was used for the analyses, with subroutine, without void nucleation. A typical mesh used in the computations is shown in Fig. 1. Plane strain conditions and finite deformations were assumed in the analyses.

The computational predictions obtained with different type of tests are shown in Fig. 4-(a). These results show that simulation of CT test best describes the experimental trend. Simulations of SET test best describes the Load-CMOD and Load-Δa curves in Fig. 3. However, simulation of SET test predict non-conservative of J-resistance curves in Fig. 4-(b). In case of SET tests, the determination of J values by tests is not clear.

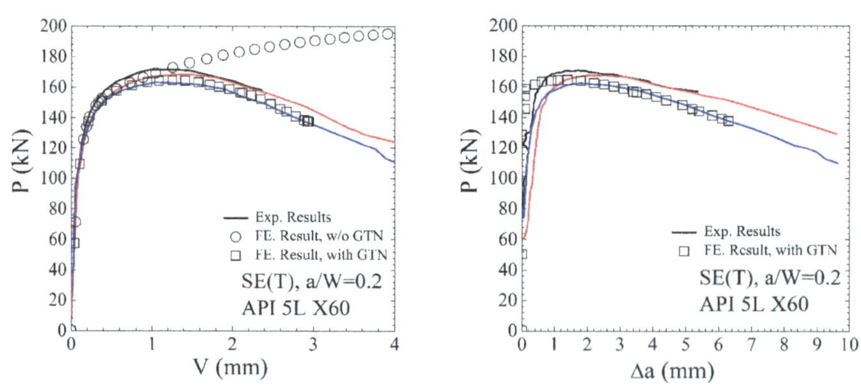

(a) **(b)**

Fig. 3 (a) Load-CMOD (b) Load-Δa curves

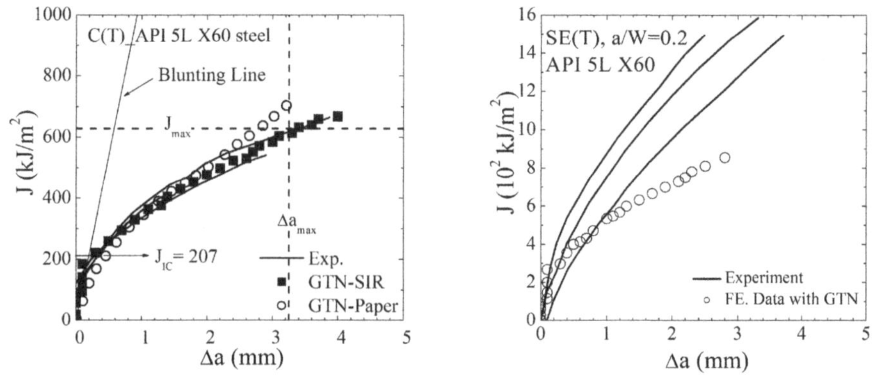

(a) **(b)**

Fig. 4 J-resistance curves for (a) CT tests, (b) SET tests

4 CONCLUSIONS

In this paper, confirm that effect of mesh size, specimen configuration in GTN model. The simulations has been tested with three different sizes of 0.05, 0.1, 0.2 mm characteristic length. They do show different J-resistance curves. To accurate descrives the experimental trend from different mesh sizes need to change values of f_c and f_f. Also GTN model parameters of specific material predict well various type of tests.

ACKNOWLEDGMENTS

This research was supported by National Research Foundation of Korea(NRF) funded by the Ministry of Science, ICT and Future Planning(NRF-2013M2B2B1075733) and Engineering Research Center Program through the National Research Foundation of Korea(NRF) funded by the Ministry of Science, ICT and Future Planning. (NRF-2007-0056094)

REFERENCES

[1] Gurson AL. "Continuum theory of ductile rupture by void nucleation and growth. Part 1-yield criteria and flow rules for porous ductile media", J Engng Mater Rech, 99:2-15, 1977

[2] Chu C, Needleman A. "Void nucleation effects in biaxially stretched sheets", J Engng Mater Tech, 102:249-56, 1980

[3] Tvergaard V, Needleman A. "Analysis of the cup-cone fracture in a round tensile bar", Acta Metall, 32:157-69, 1982

[4] Moskovic R. "Fracture toughness and crack growth resistance of pressure vessel plate and weld metal steels", Engng Fract Mech, 30:839-61, 1988

[5] Roberto Piovatto, Eduardo Hippert, Sebastian Cravero "Experimental measurements of crack growth resistance curves for a pipeline steel using SE(T) fracture specimens", Int. Congress Mech Engng 19, 2007

Modelling localisation and spatial scaling of constitutive behaviour: a kinematically enriched continuum approach

G.D. Nguyen, C.T. Nguyen, V.P. Nguyen
School of Civil, Environmental and Mining Engineering, The University of Adelaide, Australia

ABSTRACT

It is well known that classical constitutive models fail to capture the post-peak material behaviour, due to localisation of deformation. In such cases the concept of Representative Volume Element (RVE) on which classical continuum models rest ceases to exist and hence the smearing out of local inhomogeneities over the whole RVE is no longer correct. This paper presents a new approach to capturing localised failure in quasi-brittle materials, focusing on the kinematic enrichment of the constitutive model to describe correctly the behaviour of a volume element with an embedded localisation band. The resulting models possess an intrinsic length scale which in this case is the width of the embedded localisation band. The behaviour therefore scales with both the width of the localisation band and the size of the volume on which the model is defined. As a consequence, size effects are automatically captured in addition to the model capability in capturing behaviour at the scale of the localisation zone.

1 INTRODUCTION

It is well known that failure in several engineering materials starts from a diffuse deformation process and gradually transforms to a localised deformation stage followed by material separation. Fracture Process Zone (FPZ) in quasi-brittle failure, shear bands in metals, and shear/compaction bands in soils are typical examples of localised failure. Constitutive models therefore have to correctly describe both diffuse and localised stages of failure. While it is straightforward to assume homogeneous deformation in diffuse stage of failure, such an assumption is no longer valid for a volume of material with an embedded localisation zone. In such cases, the homogeneity might be valid only at different scales of behaviour: homogenous deformation inside the localisation band, and in the region outside this band. In the literature, several different regularisation approaches have been proposed to tackle the modelling of localised failure. They range from simple (e.g. smeared crack/deformation [1]) to more complicated (nonlocal/ gradient regularisation; [2, 3]), or focus on enriching the spatial discretisation schemes (e.g. eXtended Finite Element (XFEM [4]) and Enhanced Assumed Strain (EAS [5, 6]) to capture the kinematics of localised deformation.

Despite the success in describing localised failure, the above regularisations are either too simple to correctly capture the kinematics of localised deformation (smeared deformation), too complicated and computationally expensive to be practical applicable (nonlocal/gradient) or dependent on the discretisations (XFEM, EAS). In this study, the focus is on the constitutive modelling aspects of localised failure. This brings adequate enhancements to the kinematics of localised deformation while not making the proposed approach complicated or expensive and also facilitating its implementation in any spatial discretisation schemes.

The paper starts with a numerical illustration of failure in quasi-brittle materials. This shows the evolution of failure from diffuse to localised, and addresses associated issues in continuum modelling to capture this evolution. It is then followed by the introduction of a kinematically enriched continuum framework with different constitutive behaviour for the material inside and outside the localisation zone and their connections. Key features will be addressed, together with a numerical example to illustrate the potentials of the new approach.

2 LOCALISED FAILURE IN QUASI-BRITTLE MATERIALS

The development of micro-cracks together with the diffuse to localized failure of a RVE made of cement matrix composite is illustrated here. The RVE size is 25mm; plane stress condition (thickness 10mm) is assumed and the uniform mesh sizes are 0.25mm and 0.125mm for the coarser and finer meshes, respectively. The generation of sample is facilitated by the use of the Material Point Method (MPM [7]) and all three phases of the composite material including cement matrix (Young modulus E=45000MPa; uniaxial tensile strength f_t'=2MPa, fracture energy G_F=1.51Nmm/mm^2), inclusions (E=45000MPa, f_t'=3.0MPa, G_F=3.71Nmm/mm^2) and their weak interfacial transition zones (E=30000MPa, f_t'=1.5MPa, G_F=0.65Nmm/mm^2) can be generated using material points (Fig. 1). Further details on the modelling and constitutive models can be found in [8].

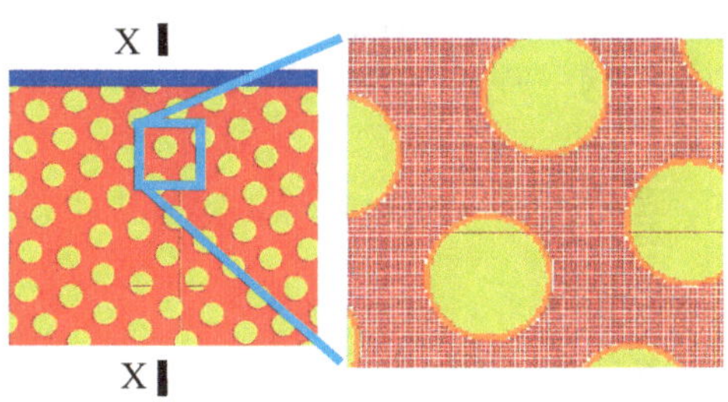

Figure 1: RVE made of cement matrix composite.

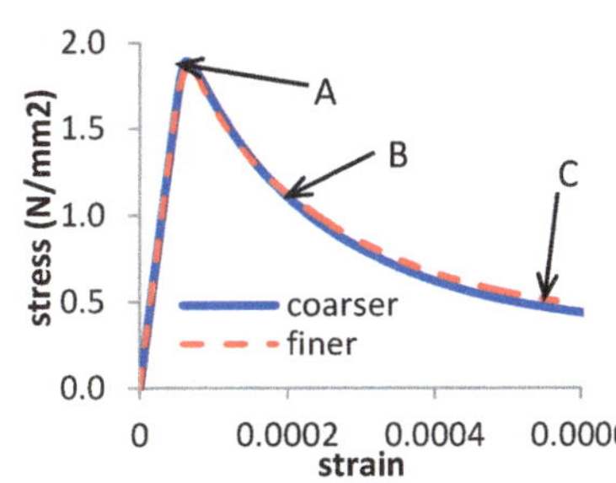

Figure 2: Mesh independent stress strain response.

The response of the RVE, in terms of macroscopic stress-strain relationship is plotted in Fig. 2, showing the softening branch associated with localised deformation inside the RVE. The micromechanical analysis in this case provides us with details on the distributions of stresses and strains during the failure process. For illustration purpose, only tensile loading in the vertical direction is considered.

In relation to the loading condition, the distribution of strain ε_{yy} and Von Mises stress during failure are plotted in Figs. 3 & 4. It is clear that the distributions of both stress and strain are strongly localised towards the end of the failure process where macro crack and separation are expected. In this case continuum assumption breaks down as soon as the localisation of deformation is clearly visible, e.g. close to macro peak stress.

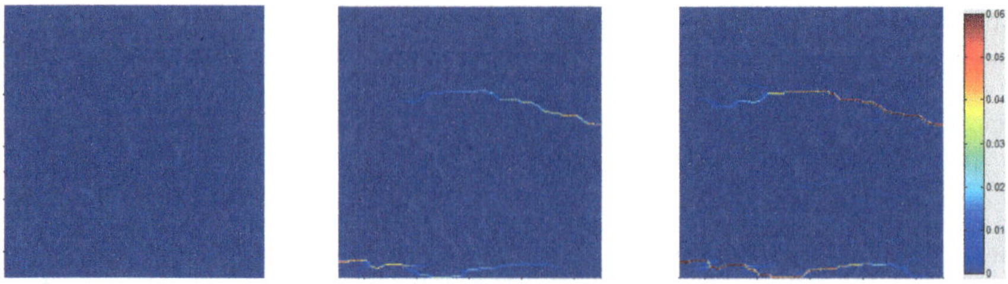

Figure 3: Contour of strain ε_{yy} corresponding to points A, B, C in Fig. 2.

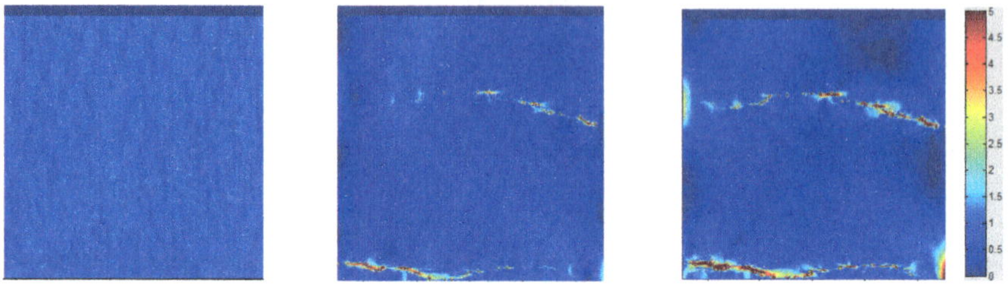

Figure 4: Contour of Von Mises stress corresponding to points A, B, C in Fig. 2.

3 ENRICHING THE CONTINUUM KINEMATICS

From the above micromechanical analysis, it is clear that only a very small fraction of the RVE undergoes inelastic deformation during loading, while the rest just remain elastic all the time, or unloads elastically after peak. It is then desirable to capture as much as possible the micromechanical details and their evolution, while maintaining a balance between accuracy and practicality. In this sense a mechanism should be devised for mapping essential micromechanical details to a continuum model.

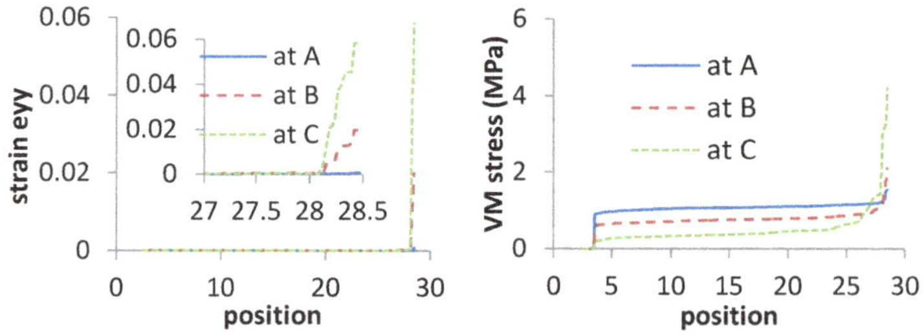

Figure 5: Strain and stress profiles on line X-X at different stages of failure in Figure 2.

Given the spatial distribution of elastic and inelastic behaviour associated with different material points, two separate regions corresponding to elastic and inelastic behaviours in the RVE are needed in the continuum models. For illustration purpose, the Von Mises stresses, strains ε_{yy} on the line X-X (Fig. 1) are sorted in ascending orders to give an indication of the evolution from diffuse to localised failure. This is presented in Fig. 5.

Due to the strong localisation of deformation, the use of a single-stress strain relationship is not appropriate, as an average strain profile is not a good representation of the observed behaviour (Fig. 6). A mechanism of localised failure is therefore needed, with the introduction of both stress $\boldsymbol{\sigma}_i$ and strain $\boldsymbol{\varepsilon}_i$ inside the localisation band to represent inelastic behaviour (Fig. 6). The elastic behaviour outside the localisation band is represented by a different stress-strain relationship $\boldsymbol{\sigma}_o$-$\boldsymbol{\varepsilon}_o$.

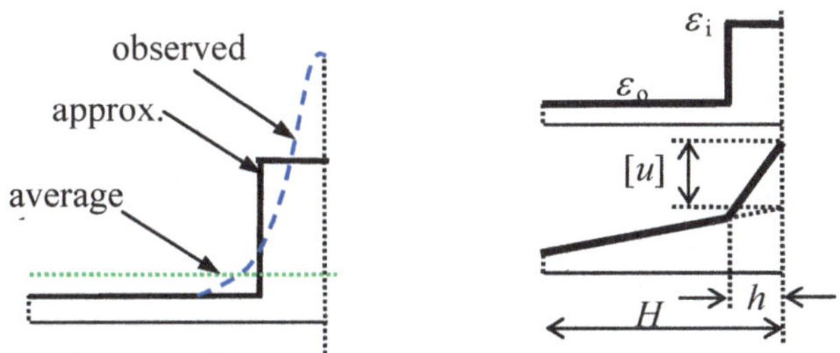

Strain profiles Approximate strain and disp. profiles

Figure 6: Strain profiles and approximation.

Homogeneity assumption is therefore valid only in each region (inside or outside the band; Fig. 6), and the macroscopic strain tensor is obtained from a simple homogenisation:

$$\dot{\boldsymbol{\varepsilon}} = f\dot{\boldsymbol{\varepsilon}}_k + (1-f)\dot{\boldsymbol{\varepsilon}}_o \tag{1}$$

where f is the volume fraction of the inelastic region [9]. Given the above strain, the macroscopic stress is obtained from the virtual work equation, so that:

$$\boldsymbol{\sigma} : \dot{\boldsymbol{\varepsilon}} = f\boldsymbol{\sigma}_i : \dot{\boldsymbol{\varepsilon}}_i + (1-f)\boldsymbol{\sigma}_o : \dot{\boldsymbol{\varepsilon}}_o \tag{2}$$

For a narrow localisation band, e.g. $h<<H$ (Figs. 5 & 6), the localisation strain $\boldsymbol{\varepsilon}_i$ can be assumed to take the following form (Kolymbas, 2009):

$$\dot{\boldsymbol{\varepsilon}}_i = \frac{1}{h}\left(\mathbf{n} \otimes [\dot{\mathbf{u}}]\right)^{\text{sym}} = \frac{1}{2h}\left(\mathbf{n} \otimes [\dot{\mathbf{u}}] + [\dot{\mathbf{u}}] \otimes \mathbf{n}\right) \tag{3}$$

where $[\dot{\mathbf{u}}]$ is the relative velocity between opposite sides of the localisation band. In this case, (1-3) lead to $\boldsymbol{\sigma} \equiv \boldsymbol{\sigma}_o$ [9], and the constitutive equation describing the RVE behaviour becomes:

$$\dot{\boldsymbol{\sigma}} = \dot{\boldsymbol{\sigma}}_o = \mathbf{a}_o : \dot{\boldsymbol{\varepsilon}}_o = \frac{1}{1-f}\mathbf{a}_o : (\dot{\boldsymbol{\varepsilon}} - f\dot{\boldsymbol{\varepsilon}}_i) \tag{4}$$

in which $\mathbf{a}_o$ is the elastic stiffness of the elastic region. Denoting $\mathbf{a}_i$ the tangent stiffness, the inelastic behaviour inside the localisation band is therefore:

$$\dot{\boldsymbol{\sigma}}_i = \mathbf{a}_i : \dot{\boldsymbol{\varepsilon}}_i = \frac{1}{h}\mathbf{a}_i : \left(\mathbf{n} \otimes [\dot{\mathbf{u}}]\right)^{\text{sym}} \tag{5}$$

The internal equilibrium across the localisation band with normal vector $\mathbf{n}$ then completes the model definition:

$$(\dot{\boldsymbol{\sigma}} - \dot{\boldsymbol{\sigma}}_i) \cdot \mathbf{n} = 0 \tag{6}$$

Equations (4-6) are generic structures of the constitutive equations describing the behaviour of a RVE crossed by a localisation band. The computational, energetic aspects, numerical implementation and examples demonstrating promising features of the kinematic enrichment of this type have been documented in some recent papers [9].

4 ONE-DIMENSIONAL NUMERICAL EXAMPLE

This numerical example shows the response of a bar of length *H*, with unit cross sectional area and an embedded localisation zone of size *h*. The material is assumed to be elastic-softening with linear softening law and specific fracture energy g_F is taken as 20 times the elastic strain energy at peak, $g_F = 20f_t^2/(2E)$. The critical bar length L_c at which snap back starts to occur (e.g. $H>L_c$ for snap back) is therefore L_c=20*h*. Fig. 7a shows the effects of varying the bar length *H* on the overall response, while keeping the size *h* of the localisation band fixed. The scaling of response with respect to size *H* can be seen, while the energy dissipation produced by inelastic behaviour inside the localisation zone is insensitive to *H*, indicated by the same area under all three curves corresponding to three values of *H*. In all three cases, the constitutive response inside the localisation zone (Fig. 7b) is independent of the structural size *H*, thanks to the use of two separate constitutive laws for the localisation band and elastic region outside that band. In contrast, the single stress-strain law in the traditional smeared crack approach requires the variation of the constitutive behaviour with respect to the discretisation to meet the requirement on the dissipation (Fig. 7c).

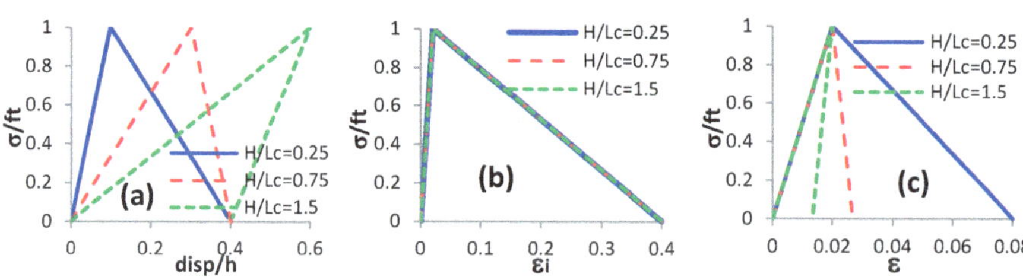

Figure 7: (a) normalised stress-normalised displacement; (b) normalised stress-strain response in the localisation zone of the proposed approach, (c) normalised stress-strain response of a smeared crack model.

The coupled stresses in the proposed constitutive modelling framework allow us to keep a meaningful constitutive response inside the localisation zone that is invariant with the discretisation (Fig. 7b). Essentially, all parameters of the model remain unchanged with respect to the resolution of the discretisation, a property that is missing in traditional smeared crack approach.

5 CONCLUSIONS

We used micromechanical analysis to identify key issues in constitutive modelling of materials that exhibit localised failure. An approach to enhance the constitutive kinematics was therefore proposed to capture correctly the material failure in both diffuse and localised stages. The resulting structure of constitutive models possesses both sizes and behaviours of the elastic and inelastic regions of a RVE crossed by a localisation band. The response of such models then scales with both the RVE size *H*, and the width *h* of localisation band, while dissipation is dependent on *h* and its associated inelastic behaviour only. Further work is on-going to apply this approach to the modelling of localised failure in geomaterials.

6 ACKNOWLEDGEMENTS

Support from the Australian Research Council (project DP140100945) is acknowledged.

7 REFERENCES

[1] Bažant, Z.P., Oh, B. (1983) *Materials and Structures* **16**, 155-77.
[2] Pijaudier-Cabot, G., Bazant, Z.P. (1987) *J Engrg Mech.* **113**(10), 1512-1533.
[3] Chen, Z., Schreyer, H.L. (1987) *J Engrg Mech.* **113**, 1665-1677.
[4] Samaniego, E., Belytschko, T. (2005) *Int J Numer Meth Engng.* **62**, 1857-1872. Oliver, J. (1996) *Int J Numer Meth Engng.* **39**, 3575-3600.
[5] Borja, R.I. (2000) *Comput Methods Appl Mech Engrg* **190**, 1529-1549.
[6] Sulsky, D., Chen, Z., Schreyer, H.L. (1994) *Comput Methods Appl Mech Engrg* **118**(1-2), 179-196.
[7] Nguyen, G.D. (2014) *Applied Mechanics and Materials* **553**, 731-736.
[8] Nguyen, G.D., Korsunsky, A.M., Einav, I. (2014) *Eng Fract Mech.* **115**, 221-240.

Natural frequency analysis of rat whiskers

W. Yan[a], Q. Kan[b], R. Rajan[c]
[a] Department of Mechanical & Aerospace Engineering, Monash University, Australia
[b] School of Mechanics and Engineering, Southwest Jiaotong University, P.R. China
[c] Department of Physiology, Monash University, Australia

ABSTRACT

Rat whiskers, and their associated neural substrate, comprise a highly evolved bio-sensory system used to perceive the world with an acuity that matches human vision or hearing. Study of the whisker sensory system is becoming an attractive research topic due to its fundamental importance as a fine sensor in neuroscience and due to the potential to develop biomimetic robotics. In this paper, a truncated conical beam model is used to study the natural frequency of a rat whisker. In addition to the proof of two generic natural frequency relationships for a truncated conical beam, our numerical results show that there exists a window where the natural frequencies of a rat whisker are very sensitive to the change of the rotational constraint at the base. This has significance for the function of the sensor since the rat follicle constraints can be manipulated by a behaving rat, thereby altering the natural frequencies of the whisker.

1 INTRODUCTION

A Chinese idiom says that "the eyes of a rat can see only an inch of light", which vividly describes rat's poor sight. However, rats, as a ubiquitous and highly successful mammalian species that carry out most of their activities in nocturnal poor-light conditions, use their smell and their large face whiskers (also called macrovibrissae or vibrissae) to perceive the world. The vibrissae are long and thick hairs on the side of the rat's snout, arranged in a characteristic pattern of rows and columns (arcs) that is consistent between rats (Fig. 1)(1). Each hair arises from a follicle with sensory receptors innervated by some 100 myelinated fibres and a number of unmyelinated fibres, travelling in the vibrissal nerve. The region around the follicle, the whisker pad, is innervated by striated muscles that the rat can contract to actively move the whiskers. The rat whiskers have discrimination acuity that easily matches that of human hearing and vision (2).

During a typical exploration, rats sweep the large face whiskers against and over an object to obtain information such as its position, shape and surface properties, and identity. Mechanical interactions between the moving whiskers and objects can cause whisker vibration or even resonance which multiplies motion amplitude at the whisker's natural frequency (3, 4). The whiskers vary systematically in length and thickness across the face (long and thin at the back, short and stubby at the front, see Fig. 1) and in turn whisker natural frequency varies systematically with whisker position across the face, allowing for encoding of a range of different textures (3, 4). Whisker resonance theory is considered as one of three existing theories to explain how a rat discriminates surface texture, i.e., to distinguish a rough surface from a smooth surface (5). Therefore, study of whisker resonance and the natural frequencies of whiskers can assist the understanding of this exquisite tactile sensory system.

In this paper, we report our recent theoretical study on the natural frequency analysis of rat whiskers (6) and then outline our thoughts on the future work. Based on experimental measurements, a truncated cone is applied to simulate the geometry of rat whiskers. Considering that the base of the rat whisker is attached underneath the skin in a follicle-sinus complex in an animal, a more realistic constraint of a transverse and a rotational spring is applied in our model to simulate such an elastic body support at the base. To cover all real situations for the rat whisker tip, similar constraints are applied to the tip in our model.

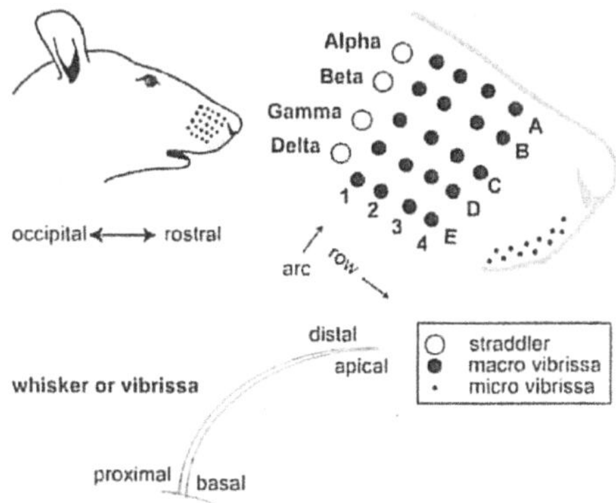

Figure 1 Morphological designation of facial whiskers (1).

2 THEORETICAL MODEL

As noted above, a truncated conical beam illustrated in Fig. 2 is used to simulate a rat whisker. The radius of the beam cross-section evolves linearly along the axial direction. All possible boundary conditions at the two ends of the whisker are represented by spring elements with translational springs to constrain the transverse deflection and rotational springs to constrain the angular movement. k_t and T_t are the stiffness constants for the translational and rotational springs at the whisker tip, and k_b and T_b are the stiffness constants at the whisker base (see Fig. 2b). The classical free, pinned, sliding and fixed conditions can be considered as special cases from this generic treatment.

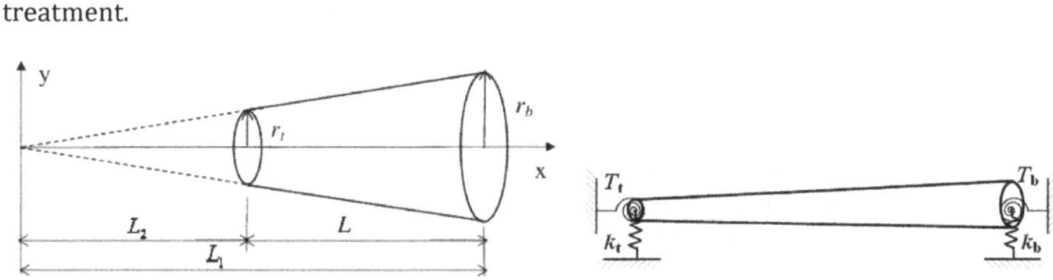

(a) (b)

Figure 2 (a) The truncated cone for modelling rat whiskers with specified geometrical parameters and (b) illustration of the boundary conditions at the two ends of the truncated beam model to simulate ex vivo or in vivo rat whisker vibration (6).

Under the assumption of small deformation and small rotation for a Bernoulli-Euler beam, applying the dynamic equilibrium condition for the forces and the moment equilibrium equation, the following partial differential equation can be obtained,

$$\frac{\partial^2}{\partial x^2}\left[EI(x)\frac{\partial^2 v(x,t)}{\partial x^2}\right]+\rho A(x)\frac{\partial^2 v(x,t)}{\partial t^2}=0 \tag{1}$$

where $I(x)$ and $A(x)$ are the moment of inertia and the cross-section area at location x, respectively. $v(x,t)$ is the transverse displacement. E and ρ are the elastic modulus and mass density of the rat whisker material, respectively. Here, the rat whisker material is assumed as isotropic and homogenous. When the rat whisker vibrates transversely in one of its natural modes n, the transverse displacement at any location $v(x,t)$ varies harmonically with time, i.e.,

$$v(x,t)=y(x)\sin(\omega_n t+\phi) \tag{2}$$

where ω_n is the circular (angular) frequency for natural mode n. The natural frequency f is related to its circular frequency ω by,

$$f=\frac{\omega}{2\pi} \tag{3}$$

Substituting Eq. (2) into Eq. (1) results in

$$\frac{d^2}{dx^2}\left(EI(x)\frac{d^2 y(x)}{dx^2}\right)-\rho A(x)\omega^2 y(x)=0 \tag{4}$$

The general solution of Eq. (4) can be expressed in terms of Bessel functions [7],

$$y(x)=\frac{1}{x}\left[C_1J_2\left(\lambda\sqrt{x}\right)+C_2Y_2\left(\lambda\sqrt{x}\right)+C_3I_2\left(\lambda\sqrt{x}\right)+C_4K_2\left(\lambda\sqrt{x}\right)\right] \tag{5}$$

where C_1, C_2, C_3 and C_4 are constant coefficients. J_2 and Y_2 are Bessel functions of the first and second kind, respectively, of order 2. I_2 and K_2 are the modified Bessel functions of the first and second kind, respectively, of order 2. The numerical values of Bessel functions can be obtained easily from mathematical software such as Matlab. Parameter λ in Eq. (5) is defined as

$$\lambda^4=\frac{64\rho L_1^2\omega^2}{Er_b^2} \tag{6}$$

Under the assumption of small deformation and small rotation (since such fine movements appear to be more than adequate for the rat to be able to discriminate fine detail such as texture [2], the slope $\theta(x)$ of the transverse displacement curve, bending moment $M(x)$ and shear force $S(x)$ at any location can be derived as

$$
\begin{cases}
\theta(x) = \dfrac{dy}{dx} = -\dfrac{\lambda}{2x^{3/2}}\left[C_1 J_3\left(\lambda\sqrt{x}\right) + C_2 Y_3\left(\lambda\sqrt{x}\right) - C_3 I_3\left(\lambda\sqrt{x}\right) + C_4 K_3\left(\lambda\sqrt{x}\right)\right] \\[2ex]
M(x) = EI\dfrac{d^2 y}{dx^2} = \dfrac{\pi r_b^4 \lambda^2 x^2 E}{16L_1^4}\left[C_1 J_4\left(\lambda\sqrt{x}\right) + C_2 Y_4\left(\lambda\sqrt{x}\right) + C_3 I_4\left(\lambda\sqrt{x}\right) + C_4 K_4\left(\lambda\sqrt{x}\right)\right] \\[2ex]
S(x) = \dfrac{d}{dx}\left(EI\dfrac{d^2 y}{dx^2}\right) = \dfrac{\pi r_b^4 \lambda^3 x^{3/2} E}{32L_1^4}\left[C_1 J_3\left(\lambda\sqrt{x}\right) + C_2 Y_3\left(\lambda\sqrt{x}\right) + C_3 I_3\left(\lambda\sqrt{x}\right) - C_4 K_3\left(\lambda\sqrt{x}\right)\right]
\end{cases}
\tag{7}
$$

The spring constraints illustrated in Fig. 1(b) can be described as follows:

At the tip, $x = L_2$ and

$$S = -k_t y(x) \quad \text{and} \quad M = T_t \theta \tag{8}$$

At the base, $x = L_1$ and

$$S = -k_b y(x) \quad \text{and} \quad M = T_b \theta \tag{9}$$

Applying boundary conditions (8) and (9) to Eqs. (5) and (7) results in a set of simultaneous equations with respect to C_1, C_2, C_3 and C_4

$$
\begin{cases}
\left[\varepsilon_1 \lambda^3 J_3\left(\lambda\sqrt{L_2}\right) + J_2\left(\lambda\sqrt{L_2}\right)\right]C_1 + \left[\varepsilon_1 \lambda^3 Y_3\left(\lambda\sqrt{L_2}\right) + Y_2\left(\lambda\sqrt{L_2}\right)\right]C_2 + \left[\varepsilon_1 \lambda^3 I_3\left(\lambda\sqrt{L_2}\right) + I_2\left(\lambda\sqrt{L_2}\right)\right]C_3 - \left[\varepsilon_1 \lambda^3 K_3\left(\lambda\sqrt{L_2}\right) - K_2\left(\lambda\sqrt{L_2}\right)\right]C_4 = 0 \\
\left[\varepsilon_2 \lambda J_4\left(\lambda\sqrt{L_2}\right) + J_3\left(\lambda\sqrt{L_2}\right)\right]C_1 + \left[\varepsilon_2 \lambda Y_4\left(\lambda\sqrt{L_2}\right) + Y_3\left(\lambda\sqrt{L_2}\right)\right]C_2 + \left[\varepsilon_2 \lambda I_4\left(\lambda\sqrt{L_2}\right) - I_3\left(\lambda\sqrt{L_2}\right)\right]C_3 + \left[\varepsilon_2 K_4 \lambda\left(\lambda\sqrt{L_2}\right) + K_3\left(\lambda\sqrt{L_2}\right)\right]C_4 = 0 \\
\left[\varepsilon_3 \lambda^3 J_3\left(\lambda\sqrt{L_1}\right) + J_2\left(\lambda\sqrt{L_1}\right)\right]C_1 + \left[\varepsilon_3 \lambda^3 Y_3\left(\lambda\sqrt{L_1}\right) + Y_2\left(\lambda\sqrt{L_1}\right)\right]C_2 + \left[\varepsilon_3 \lambda^3 I_3\left(\lambda\sqrt{L_1}\right) + I_2\left(\lambda\sqrt{L_1}\right)\right]C_3 - \left[\varepsilon_3 \lambda^3 K_3\left(\lambda\sqrt{L_1}\right) - K_2\left(\lambda\sqrt{L_1}\right)\right]C_4 = 0 \\
\left[\varepsilon_4 \lambda J_4\left(\lambda\sqrt{L_1}\right) + J_3\left(\lambda\sqrt{L_1}\right)\right]C_1 + \left[\varepsilon_4 \lambda Y_4\left(\lambda\sqrt{L_1}\right) + Y_3\left(\lambda\sqrt{L_1}\right)\right]C_2 + \left[\varepsilon_4 \lambda I_4\left(\lambda\sqrt{L_1}\right) - I_3\left(\lambda\sqrt{L_1}\right)\right]C_3 + \left[\varepsilon_4 K_4 \lambda\left(\lambda\sqrt{L_1}\right) + K_3\left(\lambda\sqrt{L_1}\right)\right]C_4 = 0
\end{cases}
\tag{10}
$$

where

$$\varepsilon_1 = \frac{\pi r_b^4 L_2^{5/2} E}{32k_t L_1^4}, \varepsilon_2 = \frac{\pi r_b^4 L_2^{7/2} E}{8T_t L_1^4}, \ \varepsilon_3 = \frac{\pi r_b^4 E}{32k_b L_1^{3/2}}, \varepsilon_4 = \frac{\pi r_b^4 E}{8T_b L_1^{1/2}} \tag{11}$$

According to mathematics, to have non-zero solutions for C_1, C_2, C_3 and C_4, the determinant consisted of the coefficients of Eq. (10) must be zero, i.e.,

$$
\begin{vmatrix}
\varepsilon_1 \lambda^3 J_3\left(\lambda\sqrt{L_2}\right) + J_2\left(\lambda\sqrt{L_2}\right) & \varepsilon_1 \lambda^3 Y_3\left(\lambda\sqrt{L_2}\right) + Y_2\left(\lambda\sqrt{L_2}\right) & \varepsilon_1 \lambda^3 I_3\left(\lambda\sqrt{L_2}\right) + I_2\left(\lambda\sqrt{L_2}\right) & -\left[\varepsilon_1 \lambda^3 K_3\left(\lambda\sqrt{L_2}\right) - K_2\left(\lambda\sqrt{L_2}\right)\right] \\
\varepsilon_2 \lambda J_4\left(\lambda\sqrt{L_2}\right) + J_3\left(\lambda\sqrt{L_2}\right) & \varepsilon_2 \lambda Y_4\left(\lambda\sqrt{L_2}\right) + Y_3\left(\lambda\sqrt{L_2}\right) & \varepsilon_2 \lambda I_4\left(\lambda\sqrt{L_2}\right) - I_3\left(\lambda\sqrt{L_2}\right) & \varepsilon_2 K_4 \lambda\left(\lambda\sqrt{L_2}\right) + K_3\left(\lambda\sqrt{L_2}\right) \\
\varepsilon_3 \lambda^3 J_3\left(\lambda\sqrt{L_1}\right) + J_2\left(\lambda\sqrt{L_1}\right) & \varepsilon_3 \lambda^3 Y_3\left(\lambda\sqrt{L_1}\right) + Y_2\left(\lambda\sqrt{L_1}\right) & \varepsilon_3 \lambda^3 I_3\left(\lambda\sqrt{L_1}\right) + I_2\left(\lambda\sqrt{L_1}\right) & -\left[\varepsilon_3 \lambda^3 K_3\left(\lambda\sqrt{L_1}\right) - K_2\left(\lambda\sqrt{L_1}\right)\right] \\
\varepsilon_4 \lambda J_4\left(\lambda\sqrt{L_1}\right) + J_3\left(\lambda\sqrt{L_1}\right) & \varepsilon_4 \lambda Y_4\left(\lambda\sqrt{L_1}\right) + Y_3\left(\lambda\sqrt{L_1}\right) & \varepsilon_4 \lambda I_4\left(\lambda\sqrt{L_1}\right) - I_3\left(\lambda\sqrt{L_1}\right) & \varepsilon_4 K_4 \lambda\left(\lambda\sqrt{L_1}\right) + K_3\left(\lambda\sqrt{L_1}\right)
\end{vmatrix} = 0
\tag{12}
$$

For a given problem, λ can be numerically obtained from Eq. (12). Then, by applying Eqs (3) and (6), the natural frequency f can be determined.

3 RESULTS

Our dimensional analysis proves that the natural frequency f can be expressed by the dimensionless function Π_f as [6]

$$f = \frac{1}{L}\sqrt{\frac{E}{\rho}}\Pi_f\left(\frac{r_t}{L}, \frac{r_b}{L}, \frac{k_t}{EL}, \frac{k_b}{EL}, \frac{T_t}{EL^3}, \frac{T_b}{EL^3}\right) \tag{13}$$

Eq. (13) shows that the natural frequency f of the rat whisker is always inversely proportional to the square root of the density of the whisker material because the dimensionless function Π_f is independent of the density.

Under any classical free, pinned, fixed or sliding constraints at the ends of a truncated conical beam, it has been proved that the natural frequency can be expressed as [6]

$$f = \alpha\frac{r_b}{L^2}\sqrt{\frac{E}{\rho}} \tag{14}$$

where α only depends on r_t / r_b.

Applying general constraints at the whisker ends, a prototypical rat whisker was generated by taking the mean values of the 18 macrovibrissae of a Sprague Dawley rat measured by [3]. These values are: $L = 34.1\,\text{mm}$, $r_b = 0.062\,\text{mm}$, $r_t = 0.007$ mm and $\rho = 1.4\,\text{mg/mm}^3$. The elastic modulus of the rat whisker material was taken as 3 GPa, which is within the range of reported data from 1.4 GPa to 7.8 GPa (3,4,8). Figures 3(a) and 3(b) shows numerical results of the first natural frequency when freely whisking in air and pinned at the tip, respectively.

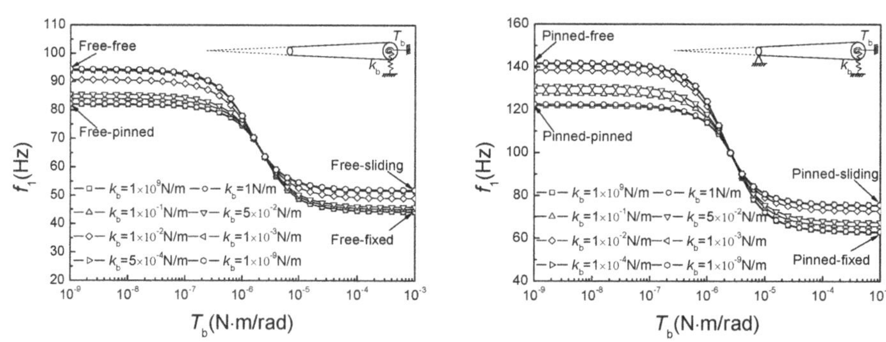

(a) (b)

Figure 3 The first natural frequency of a representative rat whisker (a) freely whisking in air and (b) with the tip touching an object under different elastic constraining stiffness k_b and T_b at the base (6).

The most interesting finding from Fig. 3 is that the first natural frequencies in both situations are very sensitive to the rotational stiffness T_b within a certain range. The sensitive ranges of T_b in both situations are almost the same, which is $5\times10^{-7} \sim 10^{-5}$ Nm. Within this range, the corresponding natural frequency decreases quickly with the increase in T_b and the frequency is generally insensitive to the change of k_b. This means that the rat could easily adjust the natural frequency of individual whiskers by increasing or decreasing the rotational constraint at the whisker base. According to the physiology, the whiskers are supported by the follicle sinus complexes which act as controllable bearing through control by surrounding intrinsic muscles [9]. Therefore, it is practically feasible for the rat to control the natural frequency of individual whiskers over a wide range by adjusting the rotational constraint at the base. This finding provides support for the theory that whisker resonance could be used to discriminate textures [3, 4].

Current studies of the mechanics of rat whiskers treats them as straight beams with a single phase elastic material in the same way as we have done here.. Future work to construct more realistic beam models with curved shape, a three-layered structure and incorporating viso-elasticity will be carried out to further confirm that current finding can be applied to real rat whiskers.

REFERENCES

[1] Voges, D., Carl, K., Klauer, G., Uhlig, R., Schilling, C., Behn, C. & Witte, H. (2010) *Sensors* 4718.

[2] Carvell, G. E. & Simons, D. J. (1990). *J Neurosci* **10**, 2638–2648.

[3] Hartmann, M. J., Johnson, N. J., Towal, R. B. & Assad, C. (2003) *J. Neurosci.* **23**, 6510–6519.

[4] Neimark, M. A., Andermann, M. L., Hopfield, J. J. & Moore, C. I. (2003) *J Neurosci* **23**, 6499–6509.

[5] Jadhav, S. P. & Feldman, D. E. (2010). *Curr. Opin. Neurobiol.* **20**, 313–318.

[6] Yan, W., Kan, Q., Kergrene, K., Kang, G., Feng, X.-Q. & Rajan, R. (2013) *J. Biomech.* **46**, 1987-1995.

[7] Conway, H. D., Becker, E. C. H. & Dubil, J. F. (1964) *J. Appl. Mech.* **31**, 329-331.

[8] Kan, Q., Rajan, R., Fu, J., Kang, G. & Yan, W. (2013) *Mater. Res. Bull.* **48**, 5026-5032.

[9] Szwed, M. & Bagdasarian, K. (2003) *Neuron* **40**, 621-630.

Age-related degradation of mouse cortical bone: implications for the *α-klotho* gene responsible for bone mechanical integrity in a series of nanoindentation experiments

N. Maruyama[1], Y. Shibata[2], A. Mochizuki[3], T. Miyazaki[2], T. Inoue[3], K. Maki[1]
[1] Department of Orthodontics, Showa University School of Dentistry, Japan
[2] Department of Conservative Dentistry, Division of Biomaterials and Engineering, Japan
[3] Department of Oral Physiology, Showa University School of Dentistry, Japan

ABSTRACT

This study examined the bone mechanical properties of *α-klotho*-deficient mice using a series of nanoindentation tests, to obtain novel findings on age-related bone degradation at the material level. Similar extents of elastic moduli were found in cortical bone regions of wild-type and *α-klotho*-deficient mice using depth-dependent nanoindentation tests. Despite the reduced viscoelasticity, normal calvarial bone exhibited instantaneously enhanced moduli against higher stain-rate indentations. Such strain-rate stiffening against large deformation is of particular importance for maintenance of tissue integrity, and hence the reduced stiffening behavior of aged bone under large strain might result in unexpected fractures.

1. INTRODUCTION

Bone degradation becomes more common with age, and is related to increased porosity and excessive local mineralization at the macroscale and microscale, respectively [1, 2]. Owing to the structural complexity of bone, namely its intrinsic inhomogeneity and multiscale tissue structures, the age-related changes in bone mechanical properties have not been fully elucidated at the material level. In this respect, nanoscale mechanical testing technologies, such as nanoindentation, may resolve specific features of bone mechanical properties through precise measurement protocols [3]. In addition to the difficulties associated with the bone hierarchical structure, the development of representative aging animal models needs to be addressed. Although age-related changes have often been considered in young-to-old animal models, inherent deviation along with aging is an unavoidable event. The *α-klotho* mutant (α-*kl*) mouse has been expected to serve as a "representative" aging model. The α-*kl* mice were originally described as a short-lived model that displays a variety of aging-related phenotypes, including (i) arteriosclerosis, (ii) ectopic calcification in various soft tissues such as the lung, kidney, stomach, heart, and skin, (iii) decreased bone mineral density, (iv) emphysema, (v) uncoordinated movement, (vi) atrophy of the skin, and (vii) severe hyperphosphatemia in association with increased concentrations of $1{,}25(OH)_2D_3$ [4]. Thus, nanoindentation tests on bone obtained from α-*kl* mice comprise an attractive alternative protocol to illustrate age-related bone degradation at the material level.

Bone mineralization is considered to be responsible for the bone stiffness, whereas the bone durability partially associated with the time-dependent viscoelasticity of matrix proteins remains poorly elucidated. The present nanoindentation experiments invariably and simultaneously captured both the elastic and viscoelastic properties of bone using quasi-static and dynamic nanoindentation tests at varied frequencies (strain-rates). The partition of the corrected elastic modulus (storage modulus) and viscoelastic response (loss modulus) needs to be clarified, so that strain-rate-dependent moduli and viscoelastic energy dissipation with time can be discussed in parallel. We anticipate that such strain-rate responses of bone will be highly appreciated for bone durability at the material level.

2. MATERIALS AND METHODS

2.1 Sample preparation

Male wild-type (n=4)and *α-kl* (n=4) mice (Japan Clea, Tokyo, Japan) at 5 weeks of age were used in this study. The animal care and protocol followed the Guiding Principles for the Care and Use of Animals, as approved by Showa University (Tokyo, Japan; ref. #1372). The calvarial bone and tibia were harvested, and the surrounding soft tissue was removed. The bones were then cross-sectioned in the longitudinal direction using a cryostat, so that a fine surface could be obtained without manual polishing. The samples were ultrasonically cleaned with distilled water for 10 s between each step.

2.2 Nanoindentation

The samples were briefly submerged in Hanks' balanced salt solution prior to testing. Nanoindentation experiments were performed on the cortical regions of the bone samples (Fig. 1) using a diamond Berkovich tip and a quantitative nanomechanical test instrument interfaced with a scanning probe microscope (TI950; Hysitron, Minneapolis, MN).

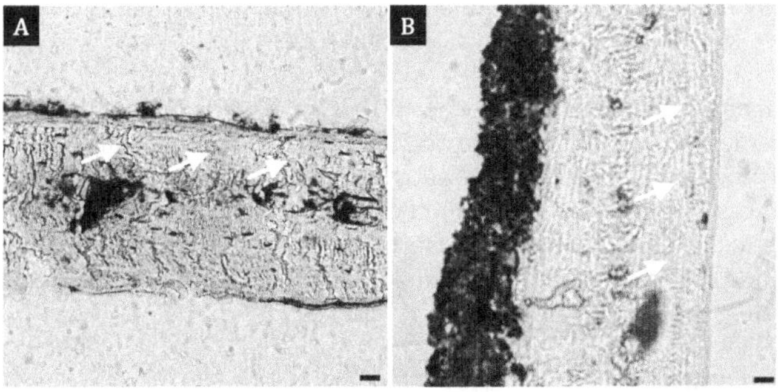

Figure 1- Light microscopy images of cross-sectioned normal mouse calvarial (A) and tibia bone (B) (scale bars=20 μm). Nanoindentation mechanical testing regions are marked with arrows.

A high bandwidth transducer (nano-DMA III; Hysitron) that enables nanoscale dynamic testing was used throughout the study. Fused quartz was used as the standard calibration material to determine the indenter area function [5]. The measured indentation data for each sample in the contact depth range of 20–200 nm were used to ensure that the effective contact area of the Berkovich indenter always dominated the contribution to the measured mechanical properties of the samples. To determine the effective range of nanoindentation depths for each sample, the elastic moduli of all samples were measured as a function of depth using a partial unloading technique [6] with a load function containing a total of 33 partial unloading cycles, each comprising a 1-s loading segment, a 1-s hold segment, and a 1-s unloading segment. For the current tests, each loading portion was followed by a 1-s hold period to minimize the effect of creep on the unloading stiffness. The elastic moduli of biological tissues may also vary as a function of the unloading strain-rate [3]. Material homogeneity of the specimens was confirmed by the observation of near-constant elastic moduli within the effective measurement area. The following dynamic or quasi-static nanoindentation tests were performed within the range of the constant elastic moduli for each sample.

2.2.1 *Dynamic mechanical analysis*

Dynamic oscillations were superimposed at a maximum loading force of 500 μN on a force-displacement curve. The applied strain amplitudes for the oscillations were approximately 2 nm at frequencies of 0.1–200 Hz. During superimposed dynamic indentations, a phase lag between the applied stress and measured strain signal reveals viscoelastic behaviors. The tangent of the phase lag (tan δ) is defined as the ratio of the storage modulus (E') to the loss modulus (E'') as described below:

$$\tan\delta = \frac{E''}{E'} \tag{1}$$

$$E' = \frac{S_e\sqrt{\pi}}{2\sqrt{A}} \tag{2}$$

$$S_e = \frac{\sigma_0}{\varepsilon_0}\cos\delta \tag{3}$$

where A is the projected contact area obtained by manual curve fitting of the slope at the onset of unloading following sinusoidal indentation, σ_0 is the stress amplitude, ε_0 is the strain amplitude, and δ is the phase lag. A correction for elastic modulus determination can be made by equation (2) with the effective contact stiffness, S_e, defined as equation (3).

2.2.2 *Displacement control*

Force-displacement curves were recorded using a 20-nm/s loading rate to a maximum depth of 200 nm determined by the effective tip contact area (see section 2.2). The loading slope was followed by a 60-s hold at 200-nm contact depth. During this period, the force decreased as expected because of the stress relaxation associated with the viscoelastic (and/or poroelastic) response. The stress relaxation ratio (%) to the maximum loading force was calculated.

2.3 Statistical analysis

Three indentations were performed in the preliminary loading/partial unloading tests. At least five regions for each sample with five indentations were evaluated. The results are expressed as the mean ± standard deviation for each experiment. The normal distribution of each dataset was confirmed using the Kolmogorov–Smirnov test. The validity of the homogeneous variances assumption was investigated by Bartlett's test. Data were analyzed by ANOVA followed by a post-hoc t-test. A P-value of less than 0.05 was considered significant. The P-value less than significance level rejected the null hypothesis as we expect to see the observed outcome only 5% of the time if the null hypothesis was true.

3. RESULTS AND DISCUSSION

The elastic moduli of all bone samples were approximately constant at 15 GPa ($P>0.05$) between 50–200 nm in indenter contact depth. Bone is a complex hierarchical tissue with different structural levels, namely cortical and trabecular bone at the macroscale, haversian osteons and lamellae at the microscale, and hydroxyapatite crystals and collagen fibers at the nanoscale. Using the estimated tip contact area and the constant elastic modulus of each bone sample (contact depth: 50–200 nm), a series of nanoindentation tests within this target measurement range revealed homogeneous bone nanomechanical properties. Similar extents of elastic moduli ($P>0.05$) were found in cortical bone regions of normal (wild-type) and *α-kl* bones. The loss tangents of calvarial bone in the α-*kl* samples revealed an estimated higher viscoelasticity ($P<0.05$) compared with normal bone. In particular, the normal bone enabled extremely lower loss tangents ($P<0.05$) at higher frequencies. The loss tangents of tibial bones were higher than those of calvarial bones.

An enhanced storage modulus at 10 Hz compared with 0.1 Hz revealed a strain-rate stiffening that was only observed for normal calvarial bone. Despite the intrinsically lower viscoelasticity, normal calvarial bones enabled large time-dependent stress relaxation responses as well as the *α-kl* bone samples during constant strain. Bone is often considered to be a fluid-saturated porous network [7]. Based on a poroelastic theory [8], the strain-rate stiffening of normal bone seemed to be an intrinsic pileup response against rapid load indentation, while surface softening was dominant during time-dependent stress relaxation without large viscoelasticity that corresponded to a theoretically zero strain-rate during constant displacement indentation. As a consequence of the pileup response against higher strain-rate dynamic indentation, it can be assumed that normal calvarial bone enabled extremely lower loss tangents regardless of the viscoelastic responses.

Such strain responses are highly dependent on the physical contacts in each apatite crystal within a collagen matrix. The estimated porosity and collagen polymerization linearity content of *α-kl* calvarial bone and tibial bones in all samples rendered them less able to generate sufficient resistance to rapid strain than normal bone at least at the material level. Besides the given macrostructural and microstructural disorders, age-related degradation might result in a great reduction of the nanomechanical durability of calvarial cortical bone on the basis of its lower mineral homeostasis partially related to *α-klotho* gene deficiency. Strain-rate stiffening and following stress relaxation may be vital responses of normal bone to excessive deformation during the initial application of large stresses. A series of nanoindentation experiments can provide a precise protocol for measuring the mechanical properties of mineralized tissues, as reported here.

REFERENCES

[1] McCalden RW, McGlough JA, Barker MB, Court-Brown CM. Age-related changes in the tensile properties of cortical bone. The relative importance of changes in porosity, mineralization and microstructure. Journal of Bone and Joint Surgery - Series A. 1993;75:1193-205.

[2] Zioupos P, Currey JD. Changes in the stiffness, strength, and toughness of human cortical bone with age. Bone. 1998;22:57-66.

[3] Ebenstein DM, Pruitt LA. Nanoindentation of biological materials. Nano Today. 2006;1:26-33.

[4] Kuro-o M, Matsumura Y, Aizawa H, Kawaguchi H, Suga T, Utsugi T, et al. Mutation of the mouse klotho gene leads to a syndrome resembling ageing. Nature. 1997;390:45-51.
[5] Oliver WC, Pharr GM. Improved technique for determining hardness and elastic modulus using load and displacement sensing indentation experiments. Journal of Materials Research. 1992;7:1564-80.
[6] Field JS, Swain MV. Determining the mechanical properties of small volumes of material from submicrometer spherical indentations. J Mater Res. 1995;10:101-12.
[7] Jast J, Jasiuk I. Age-related changes in the 3D hierarchical structure of rat tibia cortical bone characterized by high-resolution micro-CT. Journal of Applied Physiology. 2013;114:923-33.
[8] Oyen ML. Poroelastic nanoindentation responses of hydrated bone. Journal of Materials Research. 2008;23:1307-14.

Mechanical analysis of the influence of the change in the height of gravitational center on body sway properties for postural control in the human body

H. Nakagawa[1], T. Yoshida[2], K. Yamada[3]

[1] Department of Physical Therapy, Toyohashi Sozo University, Japan
[2] Osaki Orthopedic Rehabilitation Clinic, Japan
[3] Toshiwakai Medical Corporations, Showaku Care Facilities, Japan

ABSTRACT

To mechanically elucidate the influence of the difference in the height of gravitational center on postural balance, we analyzed the body sway while standing in both young and elderly subjects. Body sway of young subjects was evaluated with and without weights of 50N attached to the knees and shoulder regions respectively. Although the body sway was not influenced by the difference in the height of gravitational center in young subjects, it increased significantly when loaded with weights. In elderly subjects, however, it became clear that the body sway was affected by the height of gravitational center even without any load.

1 INTRODUCTION

If we anatomically observe our posture when we are standing erect, we can see that it is brought about by a complex framework of numerous bones connected at the joints by countless muscles, ligaments, soft tissues and cartilage attached in a specific way so that such posture is sustained when the muscles are tensed. At a glance, our upright posture which is affected only by gravity looks stationary. However, our body is actually swaying constantly to keep our balance and maintain such posture. If such system to maintain our balance and posture is impaired even in the slightest way, it will be detected and determined as an abnormality and affect our body sway and balance. In the past, there were many attempts to check for such abnormalities in body sway and balance and detect symptoms of diseases for clinical diagnosis and medical treatment (1-2). A time sequential tracing of the gravity line moving along with the body sway, we believe, may enable us to monitor the parameter of equilibrium and the apparatus to visually record such parameter is the stabilometer. This device is composed of a pressure sensory board attached with numerous pressure sensors and a computerized calculating apparatus and is presently being widely used in the fields such as neurology, otolaryngology, orthopedics, rehabilitation and psychiatrics as well as sports medicine.

However, there are yet many unknown facts about our ability to maintain our equilibrium as it is complicated and intricately evolving, adapting to various external elements such as environment, age, etc. Therefore, it is necessary that we identify the various elements affecting our postural stability and the character of our body sway.

It is also thought that the height of gravitational center greatly affects the body sway as well. The gravitational center in humans is located way above the leg joints and it is thought that this difference in the height of gravitational center affects the body sway but no study has been done on the matter yet. The first persons to record the height of gravitational center were M.I. Croskey, et al (3) and the first in Japan were Y. Akita (4) and E. Isomae (5). However, the effect of gravitational center on body sway has yet to be clarified.

Thus we have embarked on this study to clarify the effect of the height of gravitational center on body sway from a mechanical point of view by investigating the change in body sway when the center of gravity was intentionally moved up and down the body along the body center by attaching weighted cuffs on the shoulders and patellae of the subjects.

2 MATERIALS AND METHOD

2.1 Subjects

We have studied 20 healthy university students between the ages of 20-21 (20.5±0.5). The average height was 163.7±7.5cm and the average weight was 540±108N. Furthermore, we have studied elderly people between the ages of 68-80 (74.1±3.8). Their average height was 156.3±6.9cm and their average weight was 553±63N. We have conducted this study upon informing the subjects thoroughly about the details of this study and receiving the subjects' written consent and upon approval of the board of ethics of the Toyohashi SOZO University.

2.2 Measurement of the height of gravitational center

We have measured the height of gravitational center using an apparatus of our own creation shown in Fig. 1. We have taken a flat board (length 200 cm × width 50 cm × depth 3 cm) and have attached one end of the board by hinge to a support so as to establish a free edge on the sides and attached the other end to a spring scale which can measure up to 500N so as the board is parallel to the base. The subjects were asked to lie down on the board in a supine position and measurements were taken and recorded. The formula to find the height of gravitational center was:

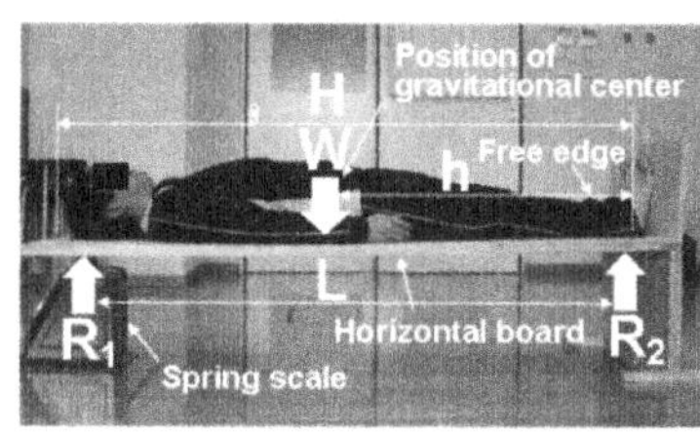

Fig. 1 Measuring device of height of gravitational center using principle of leverage

$$h = R_1 L/W \qquad \text{.........} (1)$$

h=height of gravitational center, R_1=total weight applied on spring scale, W=body weight, L=distance between the point of load. In order to test the accuracy of our apparatus, we placed a 100N weight in the middle of the board so that the gravitational center was precisely known and gradually moved the weight on the board away from the gravitational center and measured the values. A linear relationship could be found between the 2 values and when tested using the least -square method applying the linear regression equation, the coefficient of determination (R^2) was found to be close to 0.999 or 1 which indicated the appropriate accuracy was high. Next, two 25N weighted cuffs (total of 50N) were attached around the subjects' shoulders and patellae to change the gravitational center vertically along the body axis as shown in Fig. 2.

In Fig. 3 we have plotted the height of gravitational center without any load as abscissa and the height of gravitational center with a load of 50N on the shoulders and patellae as ordinate to show the relationship between the height of gravitational center with and without weights using the least-square method applying the linear regression equation. The line in the middle indicates the measurement taken without loading. The 3 lines show some dispersion but their inclination show a similar tendency proving that the displacement in the Y direction is similar and that the weighted cuffs have been placed on the subjects correctly.

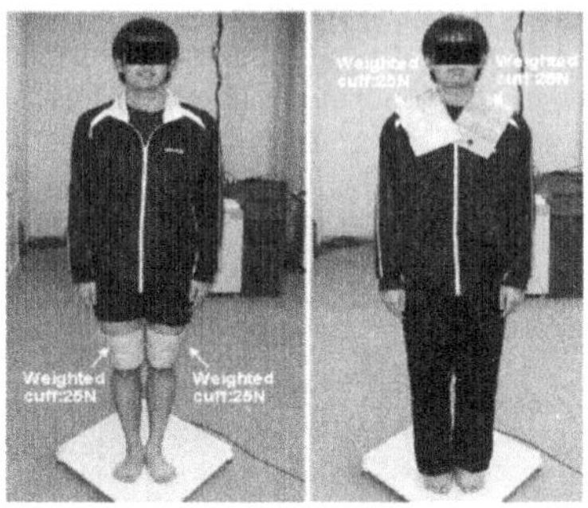

Patellar region Shoulder region

Fig. 2 Measuring body sway when weights were attached on patellar and shoulder regions

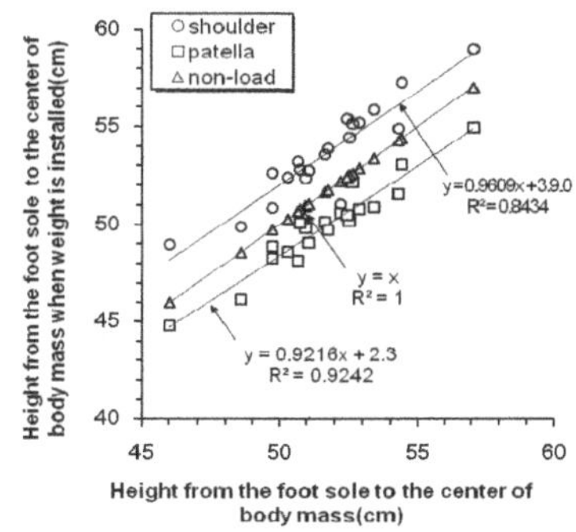

Fig. 3 Relationship between the height of gravitational center with weights loaded on patellar and shoulder regions as compared to the height of gravitational center when the subjects were unloaded

2.3 Analysis of body sway using force platform

For our study, we have utilized a gravicoder manufactured by Anima Corp. and as shown in Fig. 4, our apparatus is a thick triangular board with load cells attached under it in 3 places. When the subjects stood in the middle of the plate, the electronic wave from the load cells picked up the body sway. Such electronic data was then recorded and processed by a computer so that we could understand the center of foot pressure displacement following the movement of the gravitational center on an x-y plane. We calculated the center of plantar pressure as follows:

$$x = m(R_1 - R_3)/(R_1 + R_2 + R_3) \quad ....(2)$$

$$y = \{R_1 l - (R_2 + R_3) l_0\}/(R_1 + R_2 + R_3) \quad ...(3)$$

R_1, R_2, R_3 =load on load cells, m= distance between the points of loads in both left and right directions from the center coordinate, l and l_0= the distance between the center coordinates and the point of load in both fore and aft directions.

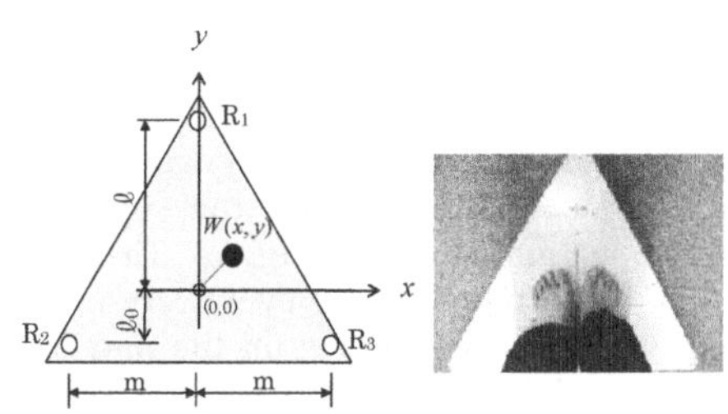

Fig. 4 Force platform measuring apparatus to visualize body sway by transforming the displacement of plantar pressure center on an x-y plane

In measuring the actual body sway, the subjects were asked to take the Romberg position and stand on both feet. We put a visual marker at a point 2m in front of them and a stabilometric recording of their body sway was plotted at a sampling frequency of 20 Hz.

Taking into consideration the influence of visual information on the result, we have also done the test with the eyes closed. We have evaluated the parameters of the body sway by calculating the TSL (total sway length) and TSA (total sway area) from the stabilometric recording of body sway using exclusive analysis software developed by a certain manufacturer.

3 RESULTS AND DISCUSSION

3.1 Relationship between stature and height of gravitational center

We have been unable to correctly assess the height of gravitational center as there is a great difference in body size between children and adults. Therefore in our study, we have divided the height of gravitational center (h) by stature (H) and multiplied it by 100 to standardize the values and have stipulated this value as proportional height of gravity center (ρ). We have obtained the value by the following formula 4:

$$\rho = h/H \times 100(\%) \quad \text{......... (4)}$$

Fig. 5 shows the relationship between the stature and the height of gravitational center and ρ of 20 subjects. As the height, indicated horizontally in the figure, increased, the height of gravitational center also gradually increased, but on the other hand, the ρ value remained stable proving that the value is not related to stature.

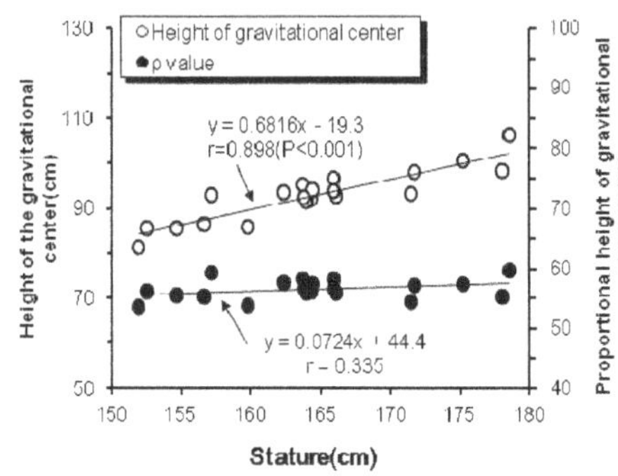

Fig. 5 Relationship between stature and height of gravitational center and proportional height of gravitational center (ρ)

3.2 Effect of the change in the height of gravitational center on body sway by loading with weighted cuffs

We have placed weighted cuffs on the shoulders and patellae of the young subjects and have recorded both the TSA values of when the height of gravitational center was shifted vertically along the body axis by loading with the subjects' eyes open and closed in Fig. 6. The TSA value when the subjects' eyes were open and weights were loaded on the patellae was 3.1±1.1 cm^2; the TSA value of when the subjects' eyes were open and no weights were loaded was 3.3±1.2 cm^2; and the TSA value when the subjects' eyes were open and weights were loaded on the shoulder region was 3.9±1.3 cm^2 and the TSA value when weights were loaded on the shoulder region, 3.9 cm^2, was the highest. The TSA value when the subjects' eyes were closed and weights were loaded

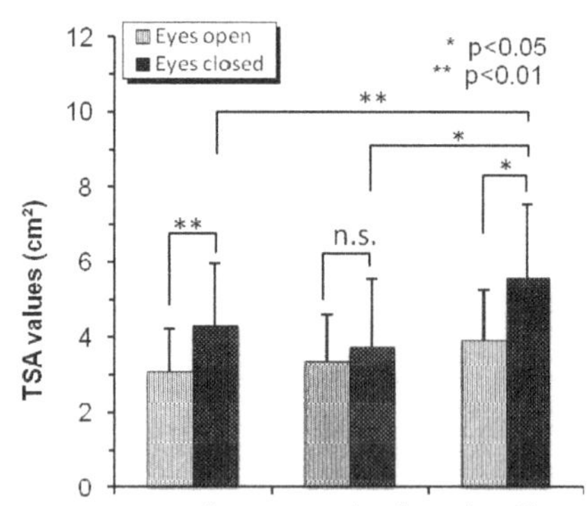

Fig. 6 Comparison between the TSA values with weights on patellar and shoulder regions and with no weights when eyes were open and closed

onto the patellae was 4.3 ± 1.9 cm^2; 3.7 ± 1.6 cm^2 when there were no weights; and 5.6 ± 1.8cm^2 when weights were loaded on the shoulder area. As in the closed-eye test, the TSA value when weights were loaded onto the shoulder region was the highest. Also in the open-eye test, there was a tendency for the TSA value to increase gradually according to the height of the gravitational center, but in the closed-eye test, the TSA value without any weights was lower than that when weights were loaded onto the shoulder region and the TSA value when weights were loaded on the patellae, 5.6 cm^2, was significantly higher than the other 2 TSA values ($P<0.05$). With the eyes open, there was no prominent change in both the TSL and TSA values when the height of gravitational center was intentionally changed. However, when the eyes were closed, there was change in both values as the height of gravitational center was changed and the highest value was seen when weights were loaded onto the shoulder region of the subjects, which was the highest point of gravitational center.

3.3 Association of body sway and height of gravitational center in elderly persons

We elucidated the relationship between body sway and the height of gravitational center in elderly persons by asking them to stand erect for 30 seconds with their eyes open and closed. Fig. 7 shows the relationship of the traces of the centers of pressure (COP) and the proportional height of gravitational center with eyes closed. The body sway increased according to the increase in proportional height of gravitational center when the eyes were open and closed. In particular, a significant correlation between the traces of COP and the height of gravitational center ($r=0.521$, $p<0.05$) could be seen indicating that as the height of gravitational center was elevated, their body sway increased and the postural balance became unstable. Therefore when assessing postural balance in elderly persons, it would be better if we take into consideration the height of gravitational center.

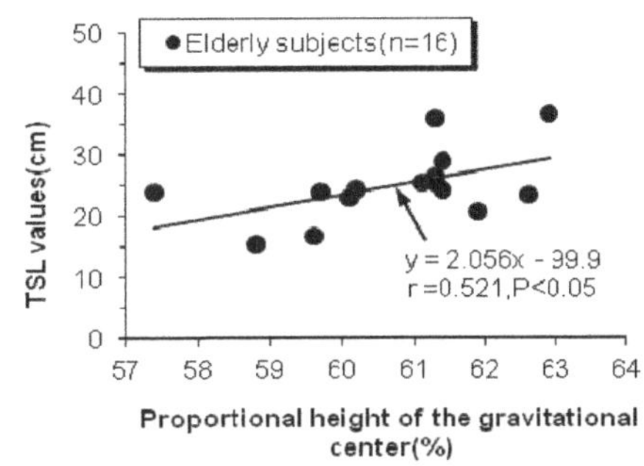

Fig. 7 Correlation between proportional height of gravitational center and TSL values in elderly people without weights on body

4 CONCLUSIONS

We conclude from this study as follows: 1) there was no relationship between the proportional height of gravitational center and body sway when the young subjects were not loaded with weights. 2) When the young subjects' height of gravitational center was moved vertically along the body axis by loading with weights and their eyes were open, there was no effect on body sway because visual information acted as a compensational function for postural control. 3) When the young subjects' eyes were closed, the change in the height of gravitational center affected the body sway and the body sway increased the most when weights were loaded on the shoulder region, which was the highest point of gravitational center. 4) Finally, it was found that the height of gravitational center greatly affects postural stability in elderly persons and that it is necessary to consider the height of gravitational center when assessing their postural ability.

REFERENCES

(1) M.P. Murray, A.A. Seireg, S.B. Sepic. Normal postural stability and steadiness: quantitative assessment. J Bone Joint Surg Am. 1975; 57(4):510-6

(2) J. Cernacek, J. Jagr, B. Harman, et al. Stabilographic findings in central vestibular disturbances. Agressologie 1973; 14(Spec No D):21-6

(3) M.I. Croskey, P.M. Dawson, A.C. Luessen, et al. The height of the center of gravity in man, American Journal of Physiology, 1922; 61.171

(4) Y. Akita. Proportional body mass of Japanese people (In Japanese): J. Tokyo Medical Society. 1929; 43,149-207

(5) E. Isomae. Study on center of gravity of the living body (1) A Simple binary simultaneous measurement method (In Japanese): The Keio Medical Society, 1956; 33, 87

Evaluation of mechanical properties of porcine sclera

K. Tanaka, K. Yamamoto, T. Katayama
Department of Biomedical Engineering, Doshisha University, Japan

ABSTRACT

Scleral buckling surgery is a common procedure in the management of rhegmatogenous retinal detachments. However, this method requires extensive experiences and advanced techniques of the surgeons. To support to have these experiences and techniques, mechanical behaviour analysis of the eyeball and studies of the relevant numerical simulation have been advanced. However, evaluation methods of the mechanical properties of the sclera have not been established yet. In this study, the effects on the mechanical properties of porcine sclera were evaluated for specimen shape as well as the location and direction in which the specimen was harvested.

1 INTRODUCTION

Rhegmatogenous retinal detachment is an ophthalmologic disease caused by various factors such as aging and external injuries (1). Recently, scleral buckling surgery is utilised as a common procedure in the management of rhegmatogenous retinal detachments due to the low risk of intraocular infection and the relatively smaller burden on the patients. However, this method requires extensive experiences and advanced techniques of the surgeons (2). To support to have these experiences and techniques, mechanical behaviour analysis of the eyeball and studies of the relevant numerical simulation have been advanced (3-6). Evaluation methods of the mechanical properties of the sclera have not been standardized yet because of various issues such as standardizing the preparation of the desired shape of specimens as well as the handling of the specimens (7). In this study, the effects of various specimen shapes on the mechanical properties of the porcine sclera are examined in order to propose the most appropriate tensile testing method for the porcine sclera. The location and direction of the porcine sclera in which the specimen was harvested were evaluated for their effects on the mechanical properties.

2 MATERIAL AND EXPERIMENTAL PROCEDURE

2.1 Material and specimens

Porcine scleras with an average age of 180 days were harvested within 24 hours of the death of the test animal. Each sclera was detached from the cornea, the muscle tissue, the retina and the choroid of the eye and then immediately stored in normal saline at 4 °C until performing experiments. Four types of specimens - temporal-meridian, temporal-equator, nasal-meridian and nasal-equator directions - were harvested as shown in Fig. 1. These specimens were designated as "TM", "TE", "NM", and "NE", respectively (T=temporal, N=nasal, M=meridian, and E=equator). These specimens were harvested using a cutting tool as shown in Fig. 2. The dimensions of the scleras were measured with an optical microscope. Polyester thin film as shown in Fig. 3 was used for a tab. The porcine sclera was fixed to the tab by using carbon tape and adhesive glue (cyanoacrylate system) as shown in Fig. 4.

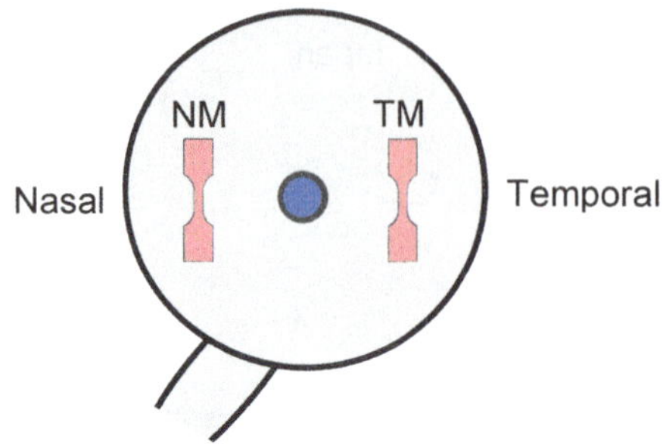

(a) Specimens cut in meridian direction

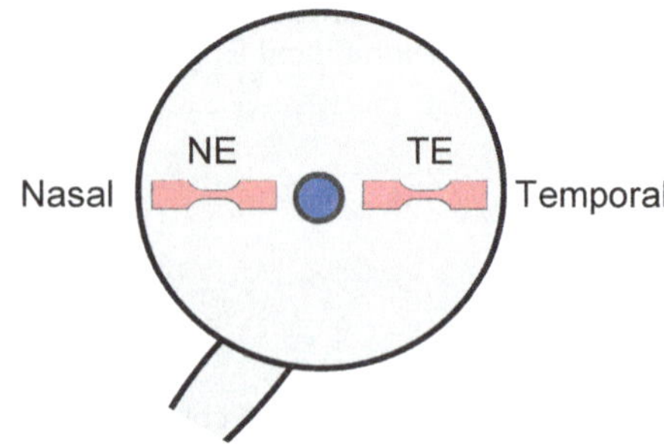

(b) Specimens cut in equator direction

Fig. 1 Harvested location and direction

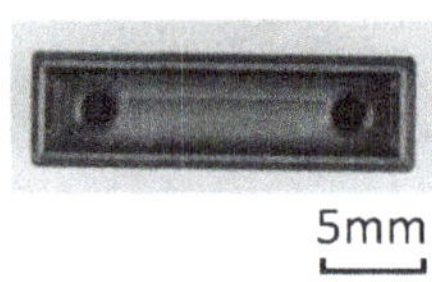

(a) Rectangle shape

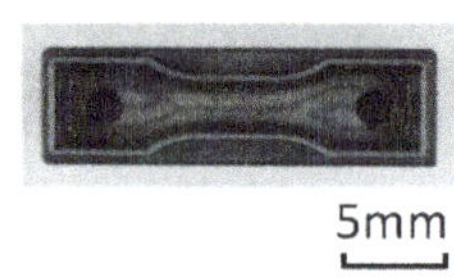

(b) Dumbbell shape

Fig. 2 Shape of cutting tool

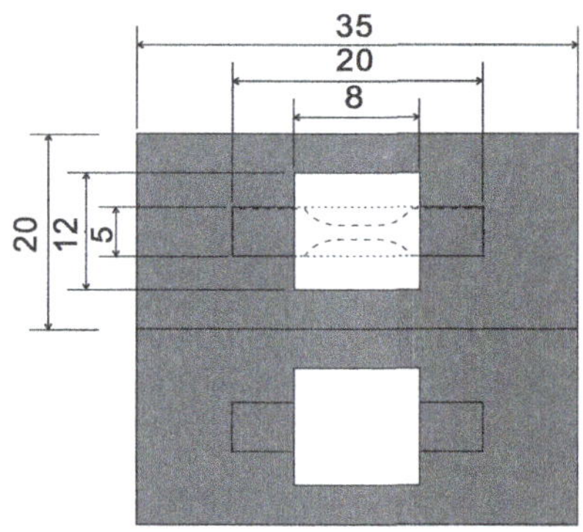

Fig. 3 Tab made of polyester thin film

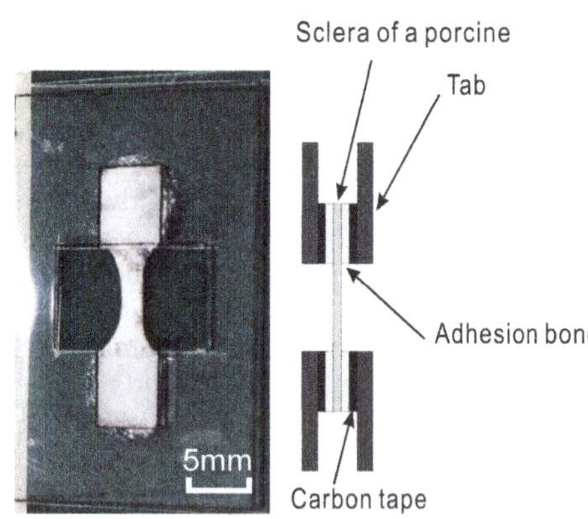

Fig. 4 The fixation method of the sclera

2.2 Experimental procedure

Tensile tests were conducted at room temperature by a universal testing machine equipped with a 100 N load cell with displacement rate of 0.17 mm/sec. Both sides of the tab were cut after the specimen was fixed to the testing machine. To evaluate the effect of specimen shape on the mechanical properties, two types of specimens - rectangular specimens and dumbbell-shaped specimens as shown in Fig. 5 - were harvested for NE specimens. Only dumbbell-shaped specimens were used to evaluate the effects of location and direction on the mechanical properties of porcine sclera.

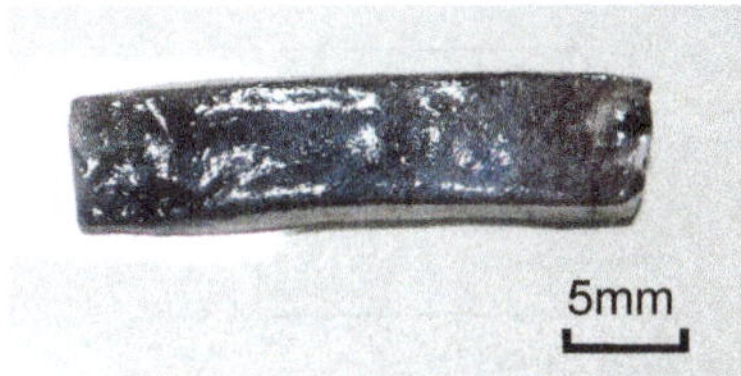

(a) Rectangle shape

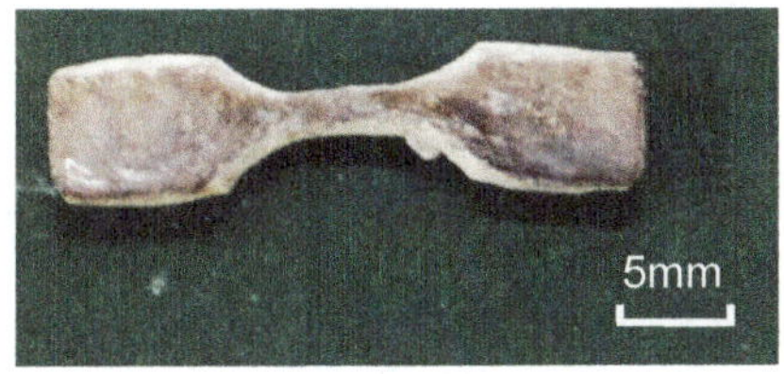

(b) Dumbbell shape

Fig. 5 Specimen shape

3 RESULT AND DISCUSSION

3.1 Effects of specimen shape

Figure 6 shows typical tensile fractured specimens just after tensile tests. Typical load-displacement curves and tensile strengths for the two types of specimens are shown in Figs. 7 and 8, respectively. The rectangular specimen was fractured at the end of the gripping jig, while the dumbbell-shaped specimen was fractured within the straight parallel part of the specimen. Tensile strengths of the rectangular specimen and dumbbell-shaped specimen were 2.38±0.48 MPa and 4.59±0.26 MPa, respectively. The dumbbell-shaped specimens had higher tensile strength (by 51.8 %) than the rectangular specimens. For the dumbbell-shaped specimens, fracture occurred within the centre parallel part instead of the stress concentration point at the end of the R as shown in Fig. 9. According to the above result, the dumbbell-shaped specimen should be used to evaluate the mechanical properties of porcine sclera.

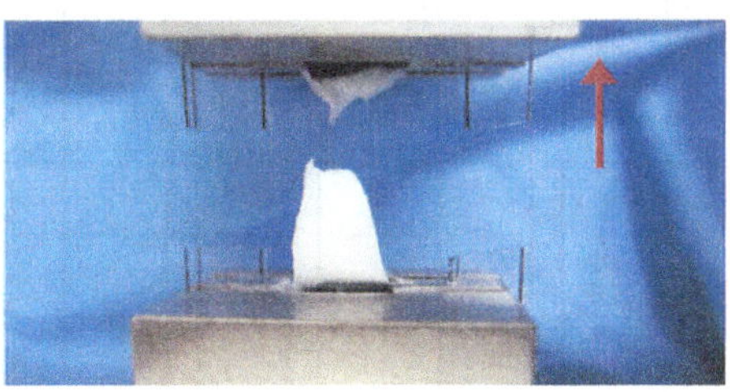

(a) Rectangle shape

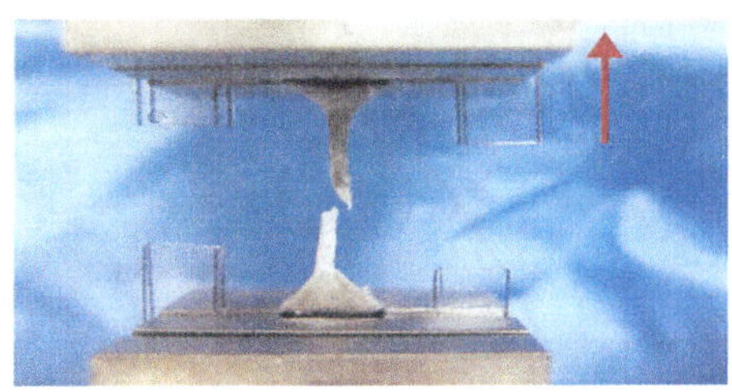

(b) Dumbbell shape

Fig. 6 Tensile fractured specimen

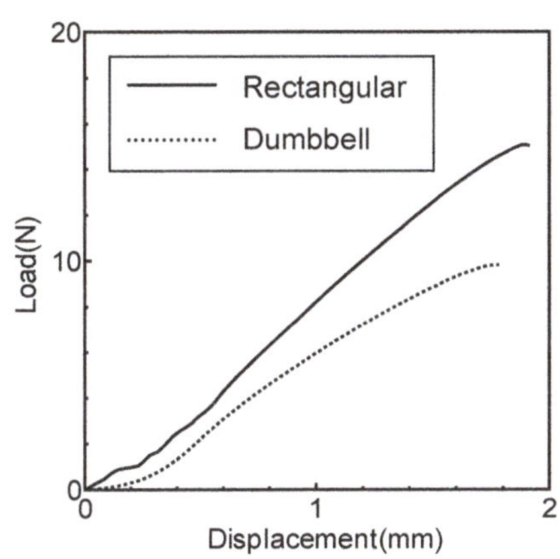

Fig. 7 Load-displacement curves

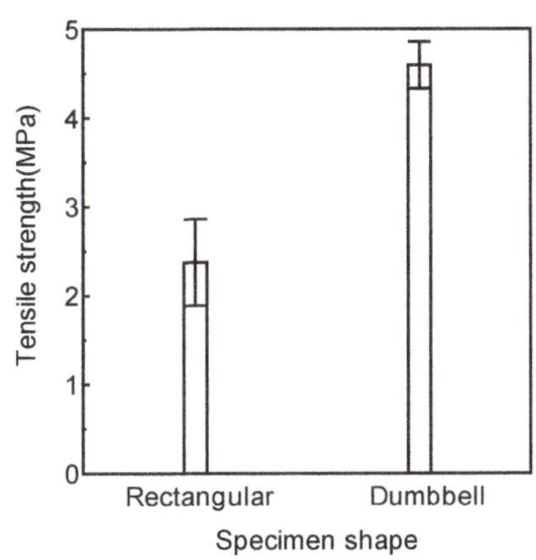

Fig. 8 Tensile strength

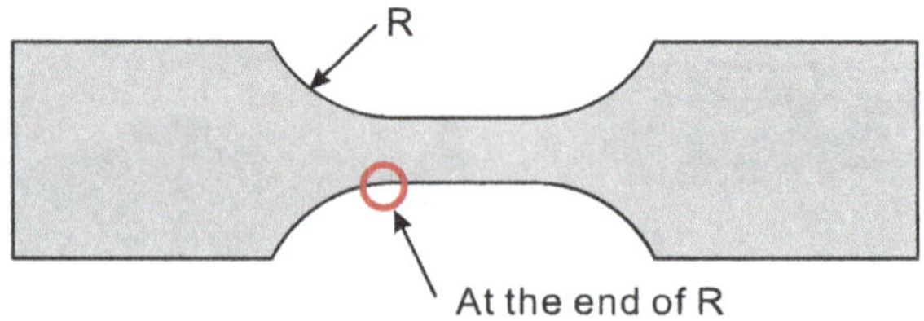

Fig. 9 Dumbbell-shape specimen

3.2 Effects of location and direction of specimens

The load-displacement curves and tensile strengths for the four types of specimens with different harvested locations and directions are shown in Figs. 10 and 11, respectively. Comparing the tensile strength of specimens for the meridian direction and the equator direction, the tensile strength of the equator direction specimen was reduced by 27% as compared to the tensile strength of the meridian direction specimen in the nasal side. Similarly, in the temporal side, it was reduced by 30%. Several studies have reported that the porcine sclera has isotropic property (8), however, the results of this study indicate the anisotropy of the porcine sclera in both the nasal and temporal side, and the tensile strength of the meridian direction was higher than that of the equator direction regardless of the harvested locations.

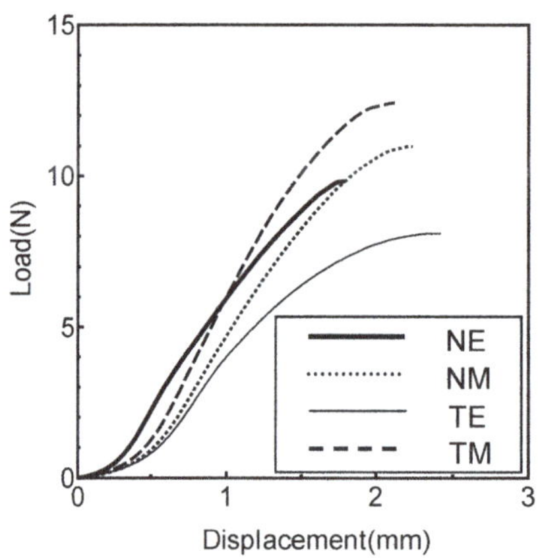

Fig. 10 Load-displacement curves

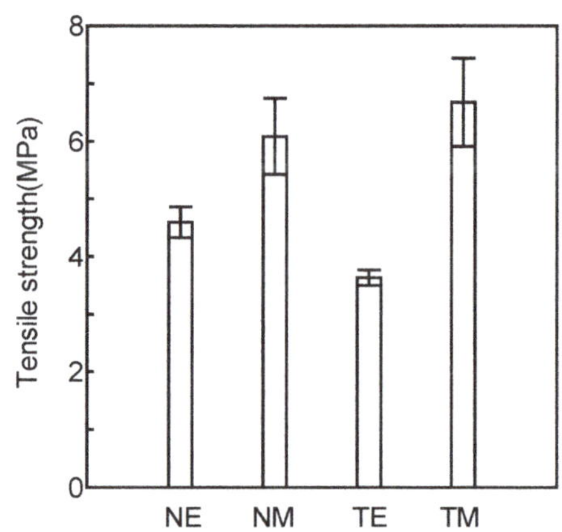

Fig. 11 Tensile strength

4 CONCLUSION

The effects on the mechanical properties of the porcine sclera were clarified for specimen shape as well as the location and direction in which the specimen was harvested. The following conclusions can be drawn from the conducted study:

1. The use of the dumbbell-shaped specimen is appropriate for the tensile test of the sclera.
2. The tensile strength of the meridian direction was higher than that of the equator direction regardless of the harvested locations. This result suggests that the porcine sclera has anisotropy.

5 REFERENCES

[1] Ito, T., Nishio, M., Ise, T. & Chikuda, M. (2010) Dokkyo journal of medical sciences 37, 63-68.

[2] Avitabile, T., Ortisi, E., Scott, I. U., Russo, V., Gagliano, C. & Reibaldi, A. (2010) Can J Ophthalmol 45, 161-165.

[3] Sun, Z. G. & Makinouchi, A. (2002) Transactions of the Japan Society of Mechanical Engineering 68, 357-363.

[4] Uchio, E., Ohno, S., Kudoh, J., Aoki, K. & Kisielewicz, L. T. (1999) Br J Ophthalmol 83, 1106-1111.

[5] Sun, Z. G., Makinouchi, A. & Yabe, H. (2003) Transactions of the Japan Society of Mechanical Engineering 69, 1775-1781.

[6] Tanaka, K., Nagayama, T. & Katayama, T. (2010) High performance structures and materials V, 197-206.

[7] Schultz, D. S., Lotz, J. C., Lee, S. M., Trinidad, M. L. & Stewart, J. M. (2008) Investigative ophthalmology & visual science 49, 4232-4236.

[8] Sunaga, J., Yokota, H., Nakamura, S., Sato, K., Himeno, R. & Makinouchi, A. (2004) The Japan of Mechanical Engineers 16, 275-276.

Tensile behaviour of a sustainable fibre reinforced cementitious composite under different strain rates

H. Tian, Y.X. Zhang
School of Engineering and Information Technology,
The University of New South Wales, Australian Defence Force Academy, Australia

ABSTRACT

In this paper the tensile and cracking behaviour of a newly developed green cementitious composite containing high volume fly ash and 0.7% steel fibre by volume and 3% bagasse fibres by volume is investigated experimentally. Direct tensile test of dog-bone specimens under different strain rates of 10^{-6}/s, 10^{-4}/s and 10^{-2}/s were conducted so as to determine the different mechanical behaviour of the new material under different loading situations, such as creep, static loading and earthquake. The influence of the stain rate on the tensile behaviour of the new composite is investigated, analysed and reported in this paper. The cracking behaviour of the fibre reinforced cementitious composite under tension is also studied and analysed.

1 INTRODUCTION

Fibre reinforced cementitious composites (FRCC) are of a number of superior material properties which make them attractive for use in structural applications, such as high tensile strength and significantly enhanced ductility compared to the conventional concrete. In order to achieve the distinct ductility characteristics, a large proportion of cement has to be utilized in the manufacturing of FRCC. As a result, the manufacturing of the FRCC is hazardous to the environment due to the emission of large amount of CO_2 from the process of manufacturing of cement (1). Therefore, partly replacing cement with other industrial by-products provides a possible prospect for developing of sustainable, green and environment friendly fibre reinforced cementitious composites.

As a by-product of coal fired power station, fly ash has been proved to be an effective pozzolanic material that could be used together with cement as a component of concrete matrix. Experimental studies found that fibre reinforced cementitious composite containing high volume fraction of fly ash could exhibit outstanding short and long term mechanical behavior with reduced material cost (1, 2). Another promising way to improve the sustainability of the fibre reinforced composite is to use natural fibres (3, 4). Bagasse fibre, which is the by-product of the cane sugar industry, has been found to be able to change the setting behavior and improve the basic mechanical properties of the cementitious composites (5, 6). However, the influence of bagasse fibre and fly ash on the mechanical behaviour is of uncertainty when they are used together since the existence of each component may affect the hydration environment and the mechanical properties of the matrix. In fact, the combined influence of high volume of fly ash and natural fibre on the tensile behavior of cementitious composite has not been studied systematically yet and need to be further investigated. Strain rate is an important factor that influences the mechanical behaviour of the composites. It is recognized that material deformation and failure processes are greatly influenced by loading conditions. Furthermore, as a construction material, it is supposed to be confronted with different loading conditions of different strain rates, such as creep (strain rate below 10^{-6}/s), static loading (strain rate about 10^{-5}/s) and earthquake (strain rate between 10^{-3}/s and 10^{-2}/s). Therefore, researches on the influence of strain rate on the tensile behaviour may reflect diverse mechanical behavior of the material under different loading conditions.

In this paper, direct tensile tests were conducted to study the tensile behavior of a newly developed green hybrid bagasse fibre (3% by volume) and steel fibre (0.7% by volume) reinforced cementitious composite by the authors (7, 8). Direct tensile test of dog-bone specimens with fibre volume ratio of 3% under different strain rates of 10^{-6}/s, 10^{-4}/s and 10^{-2}/s were conducted. The influence of the stain rate on the tensile behaviour of the composite is investigated and reported in this paper. The cracking behaviour of the fibre reinforced cementitious composite under tension is studied and analysed.

2 A GREEN FIBRE REINFORCED CEMENTITIOUS COMPOSITE

A green fibre reinforced cementitious composite with cement, fly ash, sand, water, reduce agent, bagasse fibre and steel fibre was recently developed by the authors (7, 8).The composite with a sand to cement ratio of 0.9 by mass, a fly ash to cement ratio of 1.6, 3% bagasse fibre by volume and 0.7% steel fibre by volume was tested in this paper to investigate the tensile and crack behaviour of the material. ASTM Type I OPC and a low-calcium Class F fly ash were used. The high-range water reduce agent (HRWR) used was Grace ADVA 142. The average size of sand used was 250 μm, with the maximum size of 300 μm. The steel fibre used was coppered micro straight steel fibre with a length of 13 mm, and a diameter of 0.2 mm, and the tensile strength and Young's modulus of the steel fibre is 2.2 GPa and 180 GPa, respectively. Original bagasse fibres were provided by Australian Prime Fibre Company, and they were modified with surface treatment to improve its mechanical and chemical properties before the mixing. After the modification, the bagasse fibre had a mean length of 10.1 mm, and mean diameter of 0.39 mm. The tensile strength of the bagasse fibre is around 290 MPa and the Young's modulus is about 17 GPa (9). The mix design of the composite is listed in Table 1. The compressive strength, Young's modules, and modulus of rupture of the composite were determined experimentally to be 36.67 MPa, 15.25 GPa and 5.22 MPa, respectively. (8). In this paper, direct tensile tests under different strain rate, i.e. 10^{-6}/s, 10^{-4}/s and 10^{-2}/s, were conducted on the dog-bone specimen of the composite to study the effect of strain rate on the tensile mechanical behavior.

Table 1 Mixture proportions of the new composite with different bagasse fibre content (by volume)

Cement [kg/m^3]	Fly ash [kg/m^3]	Sand [kg/m^3]	Water [kg/m^3]	Reduce agent [kg/m^3]
487.78	780.45	439	329	6.34

3 TENSION TESTING AND RESULTS ANALYSIS

Dog-bone shaped specimens as shown in Fig. 1 were prepared in the same way as described in previous research (8). As there is not a definite standard for the direct tensile test of fibre reinforced concrete, the dog-bone shaped specimens and tensile test set up from literature (10) (shown in Fig. 2) are utilized herein. Three specimens were tested to get a mean value of the material behavior.

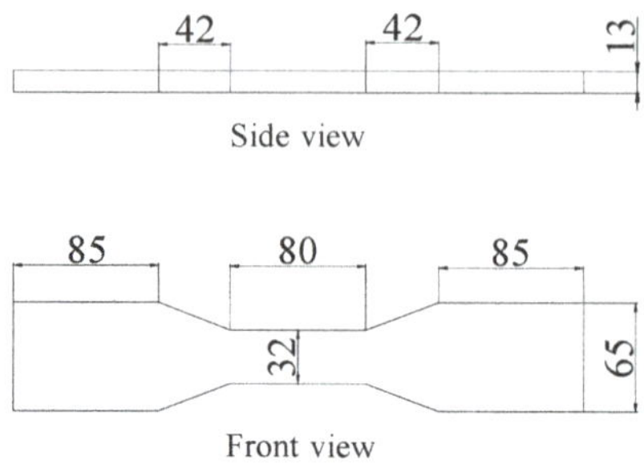

Fig. 1 Dog-bone shaped tensile specimen (mm)

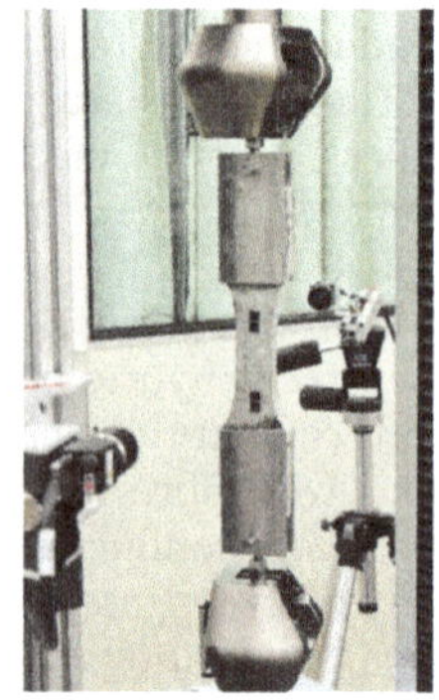

Fig. 2 Direct tensile test setup

Direct tensile tests for the specimens were conducted on the 100 kN Shimadzu AG-X testing machine at the age of 28 days under various strain rates of 10^{-6}/s, 10^{-4}/s, and 10^{-2}/s, and the experimental data was recorded every 20 ms. Non-contact extensometer was used to detect the displacement of the gauge. The obtained tensile strength is listed in Table 2. It is found that the tensile strength of the cementitious composite increases with the increase of the strain rate significantly. When the strain rate was 10^{-6}/s, the tensile strength was 3.58 MPa. As strain rate increased to 10^{-4}/s and 10^{-2}/s, the tensile strength increased to 4.32 MPa and 5.07 MPa, which are 120.7% and 141.6% of the tensile strength under the strain rate of 10^{-6}/s.

Table 2 Direct tensile strength of the composite under different strain rate

Mechanical Properties	Strain Rate		
	10^{-6}/s	10^{-4}/s	10^{-2}/s
Tensile Strength (MPa)	2.95	3.82	5.06
Standard Deviations (MPa)	0.12	0.15	0.08

The tensile stress-strain behavior of the composite tested under different strain rate is shown in Fig. 3. When the strain rate was 10^{-2}/s, the tensile stress of the composite dropped suddenly to around 1.3 MPa from 5.1 MPa as crack happened. Then the tensile stress decreased slightly until the specimen fractured completely. For the tensile tests under the strain rate of 10^{-4}/s, although the tensile stress decreased suddenly when cracks occurred, the residual tensile stress dropped to around 2 MPa from 4.3 MPa. After that, the tensile stress increased slightly with the increase of the strain, and declined

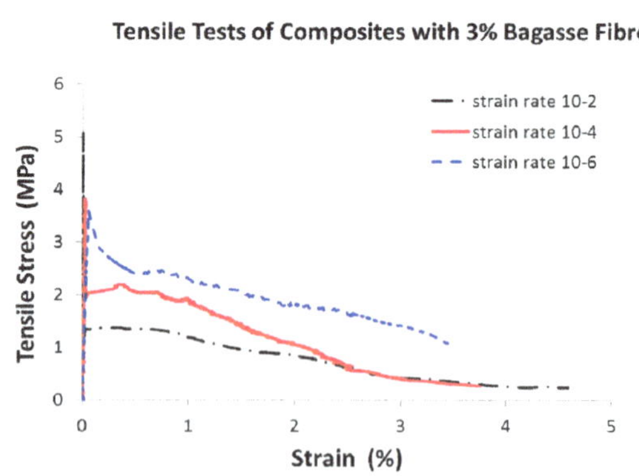

Fig. 3 Direct tensile stress-strain curves of 3% bagasse fibre reinforced composite under different strain rates

gradually as the strain was over 0.6%. The increase in ductility could be found more obviously from the curve of the composite specimen tested under the strain rate of 10^{-6}/s. Although the tensile strength of the composite under this strain rate (3.4 MPa) was much lower than that of the composite tested under the strain rates of 10^{-4}/s and 10^{-2}/s, the residual tensile stress declined much slowly after the specimen cracked and there was no sudden drop of the tensile stress.

4 FRACTURE MECHANISM OF THE DOG-BONE SPECIMENS

It could be observed from the test result that the uniaxial tensile behaviour of the developed composite containing hybrid 3% bagasse fibre and 0.7% steel fibre exhibited increase in tensile strength and ductility as strain rate increased. A general trend of strength enhancement with strain rates was very similar as demonstrated in the literature (11).

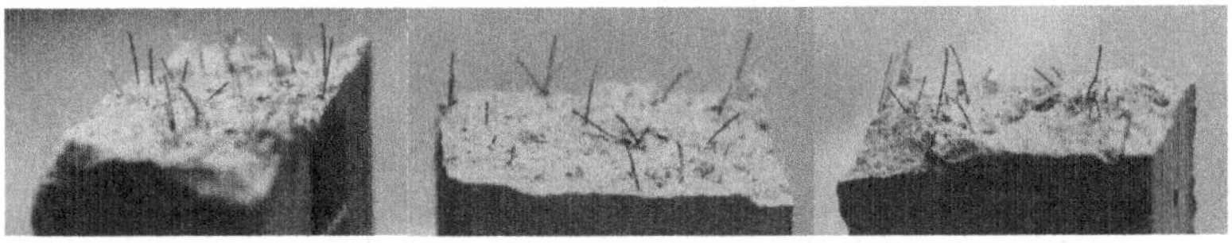

(a) 10^{-6}/s (b) 10^{-4}/s (c) 10^{-2}/s

Fig. 4 Fracture surface of composites under different strain rates

The difference of the tensile behaviour of the composites under different strain rates could be explained with the fracture mechanism associated with the fracture surfaces under specific strain rate. For the tensile tests with the strain rate of 10^{-6}/s, as crack developed slowly and the specimen undertook a long-term and slow cracking progress before specimen fractured, the energy released from the broken cementitious matrix could be reasonably distributed to the fibres which bridged across the cracks. Therefore, the pulled-out bagasse fibres on the fracture surface under the low strain rate tensile test (10^{-6}/s) were long and thick (shown in Fig. 4a), and they could play a role in improving the tensile ductility that reflected from the larger area under the tensile stress-strain curve of strain rate 10^{-6}/s. As the strain rate increased to 10^{-4}/s, the uniaxial tensile strength increased, and this meant more energy released from the broken cementitious matrix was passed to the bridging fibres. In this process, the bridging bagasse fibres had to be subjected to more bridging stress than those in the specimens under a lower strain rate of 10^{-6}/s. As a result, the bridging bagasse fibres began to break from the surface during pull-out process and the pulled-out fibres (shown in Fig. 4b) from the fractured specimen were much thinner and shorter than the ones from the tensile test under strain rate of 10^{-6}/s. Due to the breaking behavior of the bridging fibres during the pull-out process, the reinforcing role of bridging bagasse fibre became partially or completed unsuccessful, and the ductility of the tensile test was not as good as that under the strain rate of 10^{-4}, which was associated with the smaller area under the stress-strain curve. As the strain rate went up to 10^{-2}/s, most bagasse fibres got fractured and very less pulled-out bagasse fibres could found on the fracture surface (shown in Fig. 4c). Therefore, the ductility of the composite showed a further decrease since the broken bagasse fibres could not reinforce the composite effectively. It is imperative to state that the developed cementitious composite shows comparative tensile strength with conventional concrete but much better ductility since the bridging fibres constrained the development of crack after crack happened and, therefore, improved the loading capacity of the tested specimen.

5 CONCLUSIONS

Direct tensile behaviour and the influence of strain rate on the direct tensile behavior of the newly developed hybrid bagasse fibre and steel fibre reinforced cementitious composite are studied in this paper. The cracking and fracture mechanism of the fibre reinforced composites are also explained in the paper.

The tensile strength of the composites under direct tensile test increases with the increase of strain rate. However, the residual tensile strength through the tensile tests ascends with the increase of the strain rate. The tensile strength of the composite with 3% bagasse fibre increases from 2.95 MPa to 3.82 MPa and 5.06 MPa as strain rate increases from 10^{-6}/s to 10^{-4}/s and 10^{-2}/s. Fracture mechanism of the tensile tests under different strain rates is studied associated with the analysis of fractured fibre from the fracture surface.

REFERENCES

[1] Yang, E.H., Yang, Y.Z., Li Victor C. (2007) *ACI Materials Journal.* **104**, 303-311. doi: 10.14359/18966

[2] Mehta, P.K. *Proceedings of the International Workshop on Sustainable Development and Concrete Technology*, University of California, Berkeley, USA, 2004.

[3] Savastano, H., Turner, A., Mercer, C., Soboyejo, W.O. (2006) *Journal of Materials Science.* **41**, 6938-6948. doi: 10.1007/s10853-006-0218-1

[4] Flávio de Andrade, S., Romildo Dias Toledo, F., João de Almeida Melo, F., Eduardo de Moraes Rego, F. (2010) *Constr Build Mater.* **24**, 777-785. doi: 10.1016/j.conbuildmat.2009.10.030

[5] Bilba, K., Arsene, M.-A., Ouensanga, A. (2003), *Cement & Concrete Composites.* **25**, 91-96. doi: 10.1016/S0958-9465(02)00003-3

[6] Aggarwal, L.K. (1995) *Cement & Concrete Composites.* **17**, 107-112. doi: 10.1016/0958-9465(95)00008-Z

[7] Tian, H., Zhang, Y.X. *Applied Mechanics and Materials.* **438**, 275-279.

[8] Tian, H., Zhang, Y.X. *Proceedings of the 2nd International Conference on Australasia and South East Asia Structural Engineering and Construction,* Bangkok, Thailand, Nov 8-10, 2014, accepted.

[9] John, M.J., Anandjiwala, R.D. (2008) *Polymer Composites.* **29**, 187-207. doi: 10.1002/pc.20461

[10] Soe, K.T., Zhang, Y.X., Zhang, L.C. (2013) *Constr Build Mater.* **43**, 399-407. doi: 10.1016/j.conbuildmat.2013.02.021

[11] Yan, D.M., Lin, G. (2006), *Cement and Concrete Research.* **36**, 1371-1378. doi: 10.1016/j.cemconres.2006.03.003

Deformation behavior of polymeric hollow fiber membranes for water purification

S. Iio, T. Itonaga, A. Yonezu, H. Yamamura
Department of Precision Mechanics, Chuo University, Japan

ABSTRACT

The tensile deformation behavior of polytetrafluoroethylene (PTFE) hollow fiber membranes is studied. The present PTFE membrane has sub-micron pores with open cell structure, which plays a critical role of water purification. Using uniaxial tension experiments, we established a fundamental discrete model to describe the deformation behavior of porous structure using finite element method (FEM). The present model enables the prediction of the macroscopic deformation behavior of the membrane, by taking into account the changes of pore structure. The insight may be useful for porous membrane fabrication and provide insights for reliable operation of water purification.

1. INTRODUCTION

With the advance of high packing density, hollow fiber membrane is often used for a water purification process (1). The major drawback of the membrane filtration is membrane fouling. In practice, membrane fouling is mitigated by physical cleaning. The physical cleaning including air scrubbing or backwashing is routinely applied in place and removes the contaminants accumulating on the surface of membrane. Although thin and slim fiber is easy to be swung widely and frequently, such fiber is easily fractured, where the repeated physical cleaning leads to shorten the membrane lifetime. To achieve both physical strength and cleaning efficiency, we need to optimize the macroscopic and microscopic dimensions of fiber on the basis of the deeper understanding of the mechanical properties of hollow fiber membrane. However, there is still lack of information on the relationship between porous structure and mechanical characteristics (i.e. macroscopic deformation and fracture strength of entire fiber membrane). To improve the physical property of the PTFE membrane, this study investigated the deformation behavior of PTFE hollow fiber membrane under uniaxial tensile loading. Based on the experimental results, we proposed the fundamental mechanics of deformation behavior using finite element method (FEM) with porous structure. The predictive model may be useful for designing material structure for a given polymer and porosity. The present study shed some lights on pore-structural design for physical cleaning as well as the reliability for water purification.

2. MATERIAL AND EXPERIMENTAL SETUP

2.1 Materials

The materials used in this study are Polytetrafluoroethylene (PTFE) hollow fiber membrane for water purification. PTFE of crystalline polymer is stretched to create a sub-micron pore structure. For the sample configuration in Fig.1(a), the outer diameter is 2.43 mm and inner one is 1.14 mm. As shown in Fig.1(b), the outer surface of the membrane was observed by atomic force microscope (AFM). We also used a scanning electron microscope (SEM). It is found that the present membrane has an open-cell porous structure. The pore shape is stretched elliptical and the direction of longer axis coincides with the longitudinal direction of the hollow membrane. From the observations of Figs.1, the pore size and shape were measured. The pore size as true circle shape was roughly estimated to be range from 0.1 to 1 μm, and the averaged value is 0.5 μm (Figs.2(a)). Subsequently, we measured the pore shape, i.e. elliptical pore axis ratio (the length ratio of short and long axes). This ranges between 1.5 and 4 with an average value of 3 (Figs.2(b)). The overall porosity is estimated to be 75% based on the density measurement.

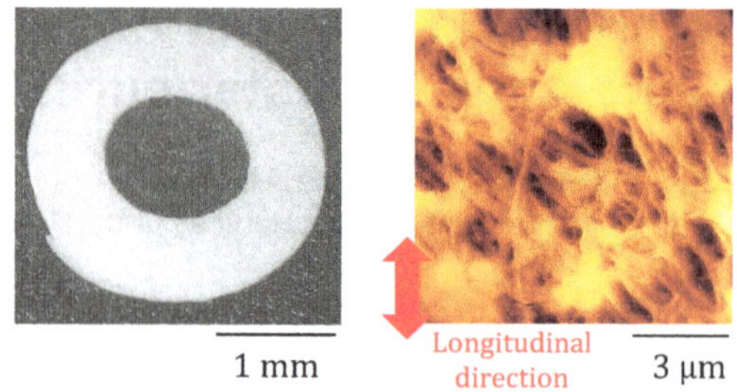

(a) (b)

Fig.1 Configuration and microstructure of hollow fiber membrane

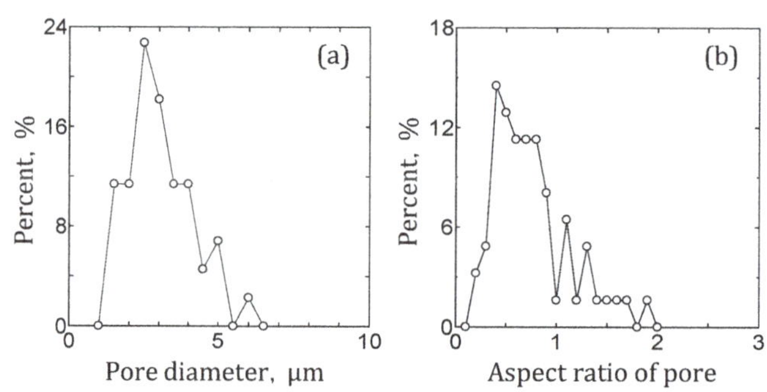

Fig.2 Distribution of pore size and aspect ratio in porous polymer

2.2 Experimental setup

Uniaxial tensile tests were performed using a universal testing machine. The membrane fiber was winded on the testing fixture, so that it applies uniform tensile stress to the fiber membrane. The clamps distance, corresponding to a gage length, is about 195 mm. The tests were conducted under displacement control with the different strain rate of 8.55×10^{-5}, 4.27×10^{-4}, and 2.56×10^{-3} 1/s at room temperature. For each test, the average data of at least three measurements was taken.

3. EXPERIMENTAL RESULTS

Fig.3 shows the nominal stress-strain curve of each test with different strain rate. With the increase in tensile strain, the nominal stress increases gradually and reaches the maximum stress. Subsequently, the stress value was almost constant. This is called "plateau" region. The maximum peek stress (plateau stress) strongly depended on the strain rate $\dot{\varepsilon}$. Finally, it reaches the final fracture. We next conducted the loading and unloading test to investigate how the fiber membrane deforms elastically and plastically. Fig.4(a) shows the nominal stress-strain curve of loading and unloading test. At a certain tensile load, the loading was stopped and released down to zero. Several tests with different peak loads were conducted. For all tests, the strain rate was set to 4.27 x 10^{-4} 1/s. The peak loads were chosen to be (A) (σ=1.51 MPa, ε=0.034), (B) (σ=9.99 MPa, ε=0.13), (C) (σ=15.0 MPa, ε=0.15), (D) (σ=18.0 MPa, ε=0.29) and (E) (σ=18.2 MPa, ε =0.45). After the completely unloading, we carefully observed the changes in the total strain as shown in Fig.4(b). The test of (A),(B) and (C) showed no plastic strain, suggesting the complete deformation recovery. However, (D) and (E) tests remained the plastic strain. This means that the specimen (D) and (E) underwent the plastic deformation, remaining the plastic strain.

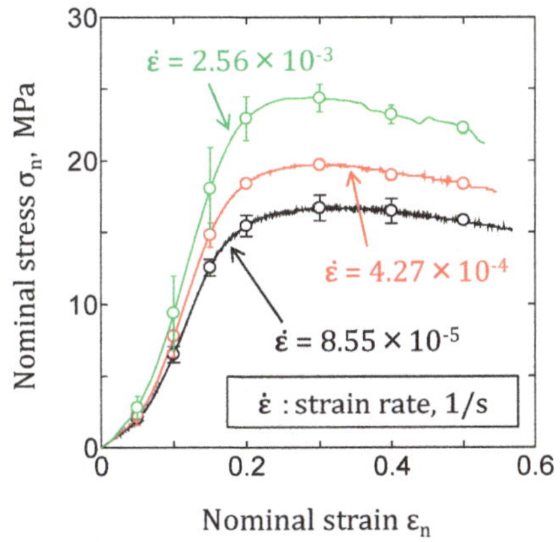

Fig.3 Experimental nominal stress - strain curves at different strain rate

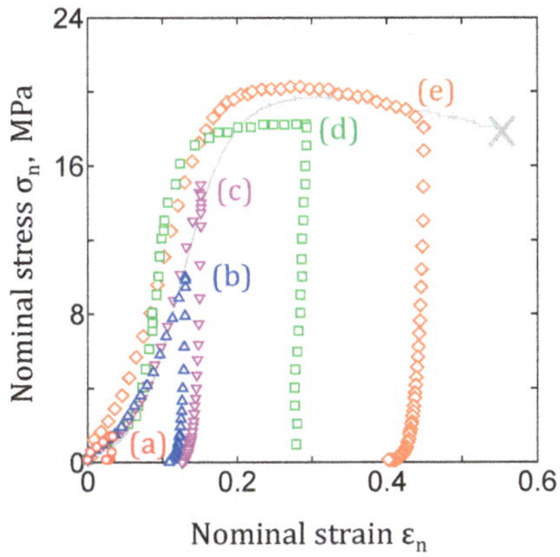

(a)

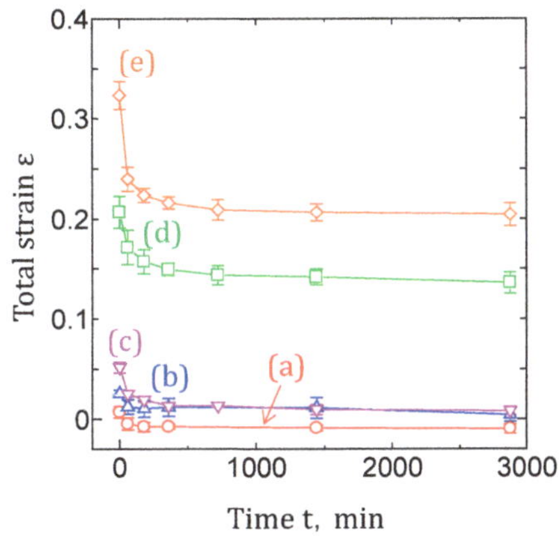

(b)

Fig.4 Nominal stress-strain curve of repeated loading-unloading test (a), and changes in total strain as a function of time after unloading (b)

4. DISCUSSION

4.1 Finite element simulation

In order to establish the computational approach to describe macroscopic deformation of the present membrane, finite element method (FEM) incorporating pore structure was employed. Fig.5 shows the FEM model of the present hollow fiber membrane. As mentioned above, the porosity of the present membrane was 75%. For the representative element of pore structure, the effective pore diameter (as if the pore is circular) was 0.5 μm (the area was about 0.15 μm^2), and the aspect ratio was about 3. The representative element structure of FEM model was given in Fig.5. For simplicity, the pore shape was assumed to semi-periodic along the longitudinal direction. The plane stress condition was assumed for the present 2D model. The present study conducted 1) elastic analysis and 2) elastoplastic analysis, respectively. We employed the elastic parameters where the Young's modulus of PTFE matrix E was 580 MPa and Poisson's

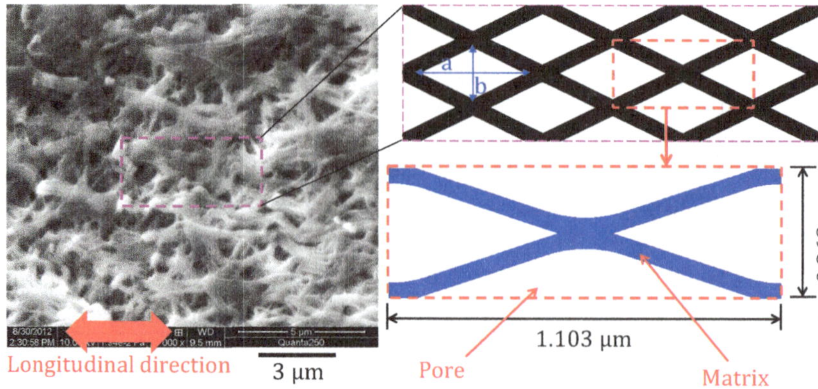

Fig.5 Simplified FEM model of porous structure

ratio ν was 0.4 (2). For the plastic property, there were several constitutive equations for polymers developed in the previous studies. Among them, the following simple phenomenological equation can be applied for large tensile strain (3).

$$\sigma = \sigma_Y + K(\varepsilon)^m \qquad \text{Eq. (1)}$$

Here, σ_Y is the yield stress. K and m are the materials constant to describe the non-linear plastic deformation and ε is the plastic strain. Nunes et al. reported that when the tensile load was applied to PTFE solid material, σ_Y was dependent on the strain rate $\dot{\varepsilon}$, while K and m were independent of $\dot{\varepsilon}$ (3). Note that the material constants (σ_Y, K and m in Eq. (1) for the present PTFE matrix) were unclear in this study. This is due to the changes of the property in matrix microstructure (such as aligned molecular chains), since the present PTFE matrix was significantly stretched in order to create porous structure. Such a fabrication method seems to induce work hardening nature. In order to estimate the properties, the computed stress - strain curve is matched with the experimental one by repeatedly changing the materials constant (σ_Y, K and m) of the constituent material in Eq. (1).

4.2 Deformation behavior

Fig.6 shows the stress - strain curve of experiment and computational one. The computed result (as shown by open circle marks) was obtained by elastic analysis. Similar to the experimental data, the FEM result also showed nonlinear deformation in the first portion, owing to the structural deformation of pore geometry. After the non-linear deformation, the stress linearly increased up to the maximum stress. The stress increasing part showed relatively good agreement with experimental curves.

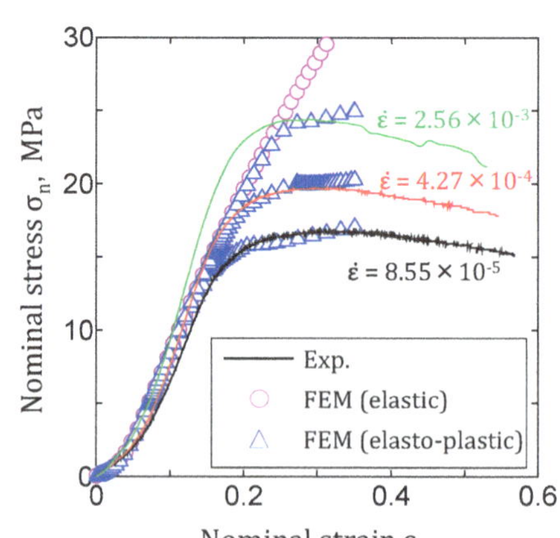

Fig.6 Nominal stress – strain curve of experiment and FEM

We next investigated the plastic deformation with elastoplastic analysis, using Eq. (1) to describe the plastic deformation. We calibrated the material parameters in the FEM model, such that they may deduce the similar strain rate effect measured in experimental stress-strain curves. In order to match the FEM analysis with experimental ones, the materials constants (σ_Y, K and m in Eq. (1)) were optimized at given strain rates through error minimization. As shown in triangle marks (in Fig.6), all tests indicated a good agreement, when the proper material constants were selected. Table 1 shows the material constants of the mechanical property of PTFE matrix. Similar to the experimental data, the computed result showed that the nominal stress became nonlinear with the increase of strain, and developed the plateau region. From these parametric FEM study, the material constants of Eq. (1) were obtained as shown in Table 1. The yield stress σ_Y is dependent on $\dot{\varepsilon}$. On the contrary, the constants of K and m are 330 MPa and 1.4, which is independent of $\dot{\varepsilon}$. Their constants (σ_Y, K and m) correspond to the parameter of PTFE fiber matrix. This result is similar to that of PTFE solid material (3).

Table 1 Mechanical property of PTFE matrix which is identified by FEM

Strain rate, 1/s	σ_Y, MPa	K, MPa	m
8.55×10^{-5}	65	330	1.4
4.27×10^{-4}	85	330	1.4
2.56×10^{-3}	115	330	1.4

K, m : const.

5. CONCLUSION

This study proposed a simple model to predict the tensile deformation behavior of polytetrafluoroethylene (PTFE) hollow fiber membranes. First, we investigated the deformation behavior of entire fiber membrane using uniaxial tensile experiments. The specimen deformed elastically and plastically, at small and large strains, respectively. During the plastic deformation, a plateau region in the stress – strain curve was observed, which corresponded to the maximum nominal stress (i.e. tensile strength), and also strongly depending on the strain rate. Based on the experimental results, we established a mechanical model and simulated the overall behavior using finite element method (FEM), incorporating the pore structure measured by SEM and AFM. A parametric FEM study was conducted to extract the rate-dependent material constants of constitutive equation of the PTFE matrix. This model enables the prediction of the macroscopic tensile deformation behavior at different strain rates. However, more advanced model may be required such as accurate discrete FEM structure and constitutive equation (i.e. visco elastoplastic phenomena). Such an issue will be addressed in the future.

This work is supported by JSPS KAKENHI (Grant No. 26420025) and JGC-S Scholarship Foundation.

REFERENCES

(1) Tang Y, Li N, Liu A, Ding S, Yi C, Liu H. (2012) *Desalination.* 287:326–339. DOI: 10.1016/j.desal.2011.11.045

(2) Rae PJ, Brown EN. (2005) *Polymer.* 46:8128–8140. DOI: 10.1016/j.polymer.2005.06.120

(3) Nunes LCS, Dias FWR, Mattos HSdC. (2011) *Polymer Testing.* 30:791–796. DOI: 10.1016/j.polymertesting.2011.07.004

Effect of in-plane tension to blank on formability of carbon fiber non-crimp fabric

K. Tanaka, Y. Tanaka, T. Katayama
Department of Biomedical Engineering, Doshisha University, Japan

ABSTRACT

Non-Crimp Fabric (NCF) has attracted much attention in the composite industry and its usage is growing rapidly. As NCF are deformed mainly due to shear deformation, they are difficult to be formed in three dimensional shapes. One of methods of formability improvement of metals and fabrics is to introduce in-plane tension to blank. Although the effect of in-plane tension to blank on formability of woven fabrics has been clarified in previous studies, that of NCF has not been clarified yet. In this study, effect of in-plane tension to blank on formability of carbon fiber non-crimp fabric is clarified.

1 INTRODUCTION

For intermediate materials, Non crimp Fabric, which consists of unidirectional plies that are kept together by stitching yarns arranged in a number of different orientations relative to the fabric production direction (1), has attracted much attention due to the high material properties compared to woven fabrics. As NCF are deformed mainly due to shear deformation, they are difficult to be formed in three dimensional shapes (2). One of methods of formability improvement of metals and fabrics is to induce in-plane tension to blank. Although the effect of in-plane tension to blank on formability of woven fabrics has been clarified in previous studies (3), that of NCF has not been clarified yet. In this study, effect of in-plane tension to blank on formability of carbon fiber non-crimp fabric is evaluated by using non-contact 3D deformation measurement system.

2 EXPERIMENTAL PROCEDURE

2.1 Materials

Non-crimp fabric (NCF, 300g/m^2, [0°/90°]), using PAN-based carbon fibers, being stitched together by polyester sewing thread, was used in this study. Stitch type of NCF used was tricot stitch (Fig.1). Table 1 shows the condition of the NCF. The size of the specimen is 300 mm×300 mm.

2.2 Experimental procedure

Drape test by hemisphere punch (Fig.2) of 100 mm in diameter was conducted by universal testing machine (Autograph AG-100kNX, Shimadzu Co. Ltd.) with the displacement rate of 10mm/min and shear distribution during drape test was measured by non-contact 3D deformation measurement system (ARAMIS®, GOM mbH). In-plane tension to blank in the direction of ±45° of NCF was applied by using springs and clips as shown in Fig.3. Three different springs, test A, test B, and test C, as shown in Table 2 were used to apply three different magnitudes of in-plane tension. Test free means the drape test without any in-plane tension to blank. Fig.4 shows the definition of the position for the hemisphere punch and NCF. Fig.4 (a) shows the distance from the top of the punch before and after the drape test.

Table 1 Specimens condition

Specimen	Stitch type	Thickness [mm]	Area density [g/m^2]
NCF	Tricot	0.40	300

Table 2 Condition of in-plane tension

	Spring type	Initial tension [N]	Spring constant
Test free	no spring	0	0
Test A	A	2.94	0.16
Test B	B	3.73	0.16
Test C	C	20.59	1.96

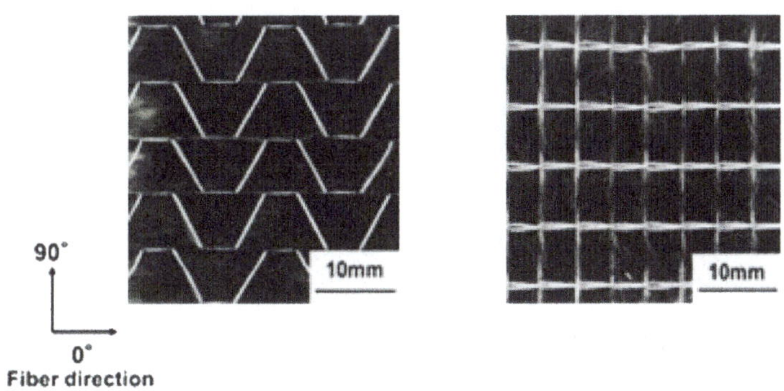

Fig.1 NCF with tricot stitch

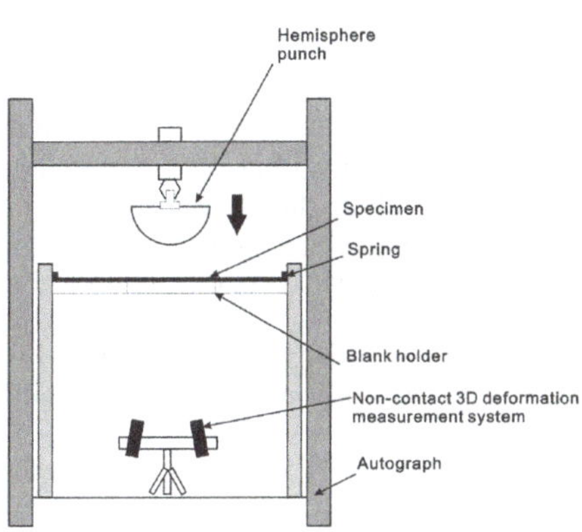

Fig.2 Schematic drawing of drape test

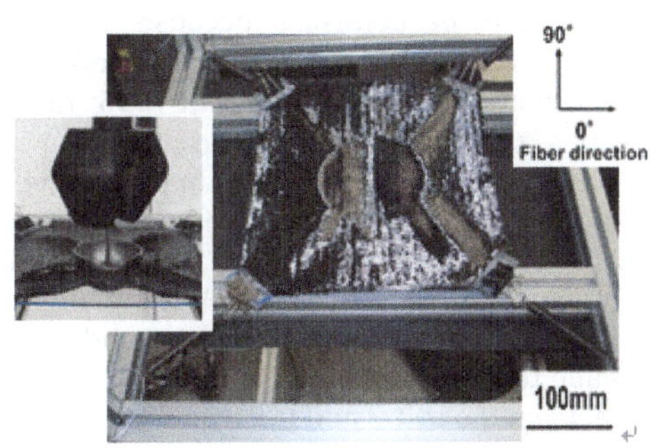

Fig.3 Top view of the blank with in-plane tension applied by using springs and clips during drape test

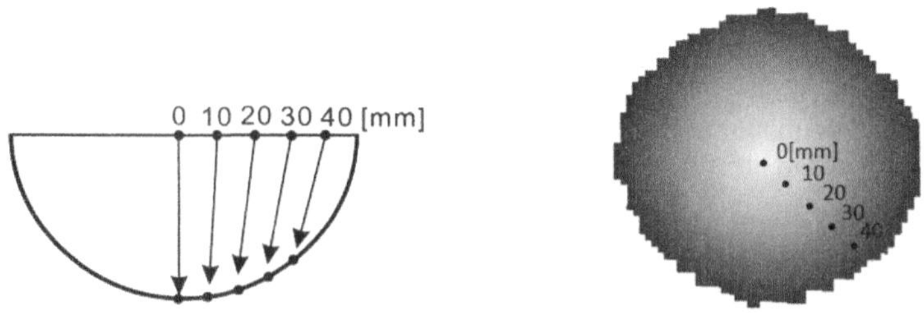

(a)Side view of the hemisphere punch (b)Top view of the hemisphere punch

Fig.4 Definition of the position for hemisphere punch and NCF

Non-contact 3D deformation measurement system is a measuring device which can evaluate non-uniform deformation of the measured object by using a principle of the digital image correlation. In this study, deformation of facets was measured as shown in Fig.5. Shear angle θ_{ARAMIS} is calculated from equation (1). Here, α is the angle [deg] between two directional fiber bundles after deformation as shown in Fig.5.

$$\theta_{ARAMIS} = 90° - \alpha \cdots (1)$$

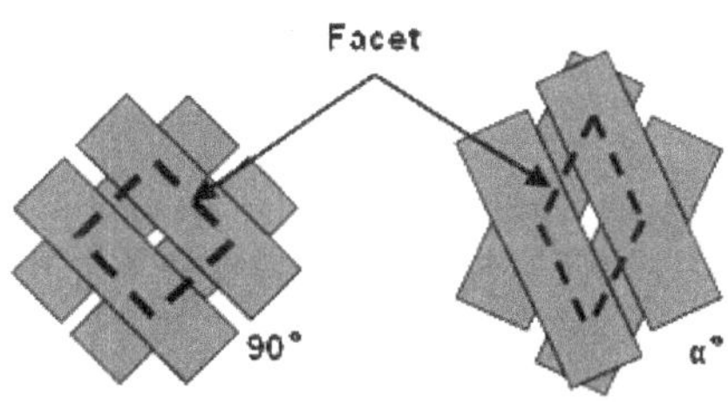

Fig.5 Schematic drawing of facet deformation

3 RESULT AND DISCUSSION

Fig.6 shows a shear deformation of NCF obtained by drape test at the punch displacement of 25mm for four different conditions. Shear deformation was observed in ±45° directions to the fiber direction. 0°/90° directions to the fiber directions are difficult to be deformed because fibers are oriented to circumferential direction at 0°/90° directions. On the other hand, ±45° directions is deformed due to the shear deformation of fiber bundles at ±45° directions. Comparing the results of test free and test A, shear deformation became larger when in-plane tension to blank was induced. Fig.7 shows the relationship between the shear angle of NCF at the position of 40mm and punch displacement during the drape test for test B. The shear angle at the point where the slope of the line changed is defined as the angle of shear deformable limit, the indicator of the formability: locking angle. Fig.8 shows the locking angles obtained by the drape test for four different conditions. The locking angle became larger when in-plane tension to blank was induced. However, no significant difference of locking angle between the test B and the test C was observed.

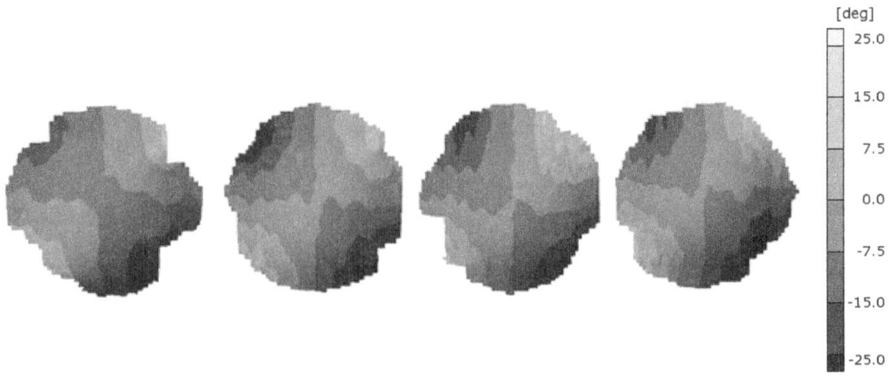

Fig.6 Shear deformation of NCF

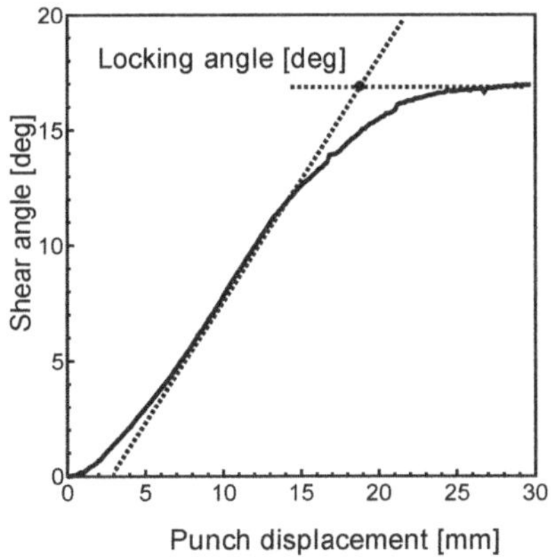

Fig.7 Relationship between shear angle at 40mm and punch displacement (Test B)

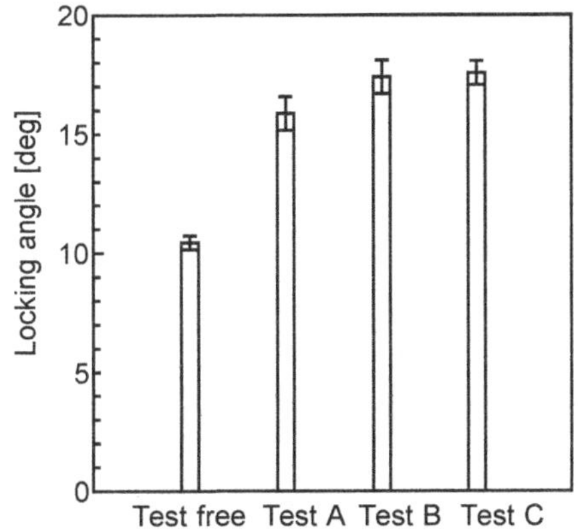

Fig.8 Locking angle obtained by drape test of hemisphere punch

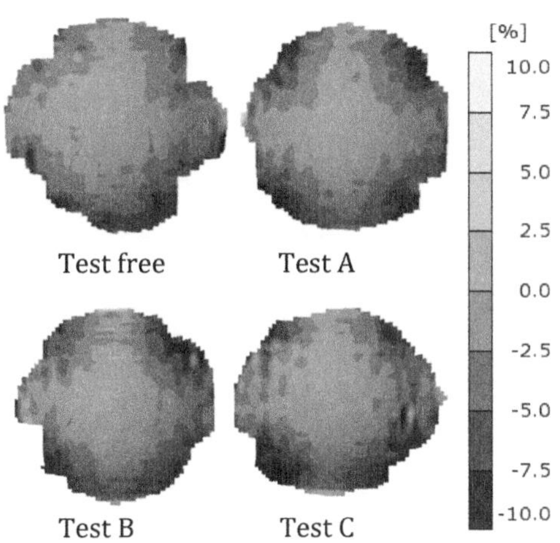

Fig.9 Thickness reduction rate

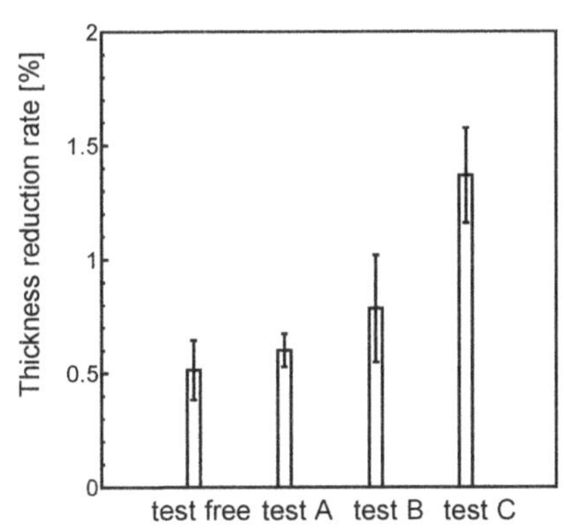

Fig.10 Thickness reduction rate at 40mm

Thickness reduction rate was reported to become the indicator of the formability of metal sheets (4). Fig.9 shows the distribution of thickness reduction rate of NCF obtained by drape test at the punch displacement of 25mm. Thickness reduction was observed in 0°/90° directions to the fiber direction. Comparing the results of test free and test C, the area with large thickness reduction late became more wide when in-plane tension to blank was induced. Fig.10 shows the thickness reduction rate at the position of 40mm from the top of punch in 0°/90° direction of NCF for four different conditions. The thickness reduction rate became larger when in-plane tension to blank was induced.

4 CONCLUSION

Drape test by hemisphere punch was conducted to evaluate the effect of in-plane tension to blank on formability of carbon fiber non-crimp fabric and deformation was measured by non-contact 3D strain measurement system. The investigation yielded the following conclusion:

(1) The locking angle was improved when the in-plane tension force to NCF was applied. To obtain better formability of NCF, in-plane tension force to NCF was revealed to be an important factor.

(2) Excessive in-plan tension force enhances thickness reduction.

5 REFERENCES

[1] Bibo, G. A., Hogg, J. P., Kemp, M., (1997) Composite Science and Technology. **57** 1221-1241.

[2] Boisse, P. H., Deluycker, E., Vanclooster, K., Morestin, F., Lomov, S. V., Verpoest, I., Vandepitte, D. (2008) Composites. Part A. **39**, 1232-1244.

[3] Harrison, P., Gomes, R., Curado-Corrreia, N. (2013) Composites. Part A. **54**, 56-69.

[4] Hatanaka, N., Iizuka, T., Takakura, N., Yamaguchi, K., and Hassan, M. A. (2010) Journal of Japan Institute of Light Metals. **60,** 315-320.

Effect of stitch tension of non-crimp fabric on the mechanical properties of CFRTP

K. Tanaka, D. Tokura, T. Katayama
Department of Biomedical Engineering, Doshisha University, Japan

ABSTRACT

Non-Crimp Fabrics (NCF), whose structure leads to a synergetic effect of high material properties and excellent drape performance, has received much attention over recent years. Up to now, several researchers have reported that the influence of stitch parameters has exerted on the drape characteristic of dry NCF. However, the influence of stitch tension on the mechanical properties of carbon fiber reinforced plastics (CFRTP) has not been clarified yet. In this study, the effect of stitch tension on the mechanical properties of CFRTP laminates using NCF is clarified.

1 INTRODUCTION

Non-Crimp Fabrics (NCF, Fig.1) consists of unidirectional plies which are kept together by stitching yarns arranged in a number of different orientations relative to the fabric production direction (1). The structure of NCF leads to a synergetic effect of high material properties and excellent drape performance, thus it is also suitable for molding three-dimensional-shaped composite component (2). Due to the upsurge in use of textile preform in manufacturing of composites parts, the draping of textile preform like NCF onto surfaces of varying geometry has received much attention over recent years (3). Up to now, several researchers have reported that the influence of stitching parameters has exerted on the drape characteristic of dry fabrics (2, 3). However, the influence of stitchparameters, such as stitch type, stitch tension and so on, on the mechanical properties of carbon fiber reinforced plastics (CFRTP) has not been clarified yet. In this study, CFRP laminates based on three different stitch tension (Normal, Tight and Loose) of NCF were molded and the influence of stitch tension on shear deformation and mechanical properties were evaluated.

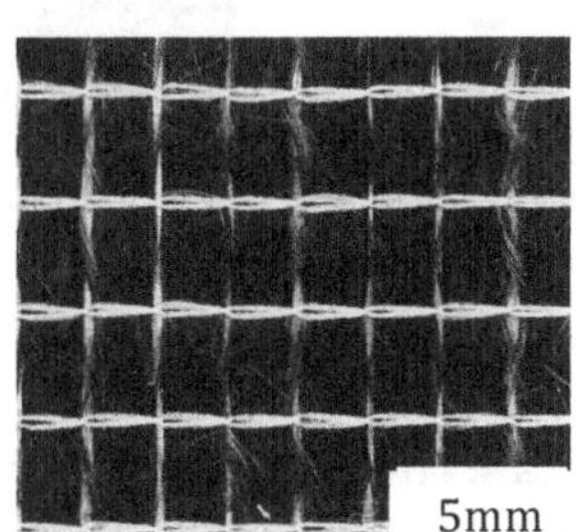

Fig.1 Non Crimp Fabrics (NCF) with tricot stitch.

2 EXPERIMENTAL PROCEDURE

2.1 Materials

Stitch type used in this study was tricot stitch. Non-crimp stitched carbon fabric (NCF, 300g/m^2 [0°/90°]), using PAN-based carbon fibers, being stitched together by polyester sewing thread, was used as the reinforcing fabric. Non-woven fabric (45g/m^2, Prototype, Kuraray Co., Japan) for Polyamide 12 (3014U, PA12, UBE Ltd., Japan) was used as the matrix. NCFs using polyester sewing thread of three different stitch tensions (Normal, Loose, Tight) were used. The fiber volume content of the specimen is set to 50%.

Laminated composites plates were molded by using a high-speed compression molding process (4). Molding pressure, molding temperatures, and maximum temperature holding time were set to 5MPa, 220°C, and 180s, respectively.

Cross sections of the laminated composite plates were observed by confocal microscope (OPTELICS H1200 Lasertec Ltd. Japan) for evaluating resin impregnation.

Although the fiber volume content of the specimen was set to 50% in this study, considering the possibility that it could be varied depending on molding conditions, fiber volume contents were measured by following the recommended testing procedures as described in JIS-K7075. Specimens were burned in the electric furnace (MMF-2, AS ONE Ltd., Japan) and the fiber volume fraction, V_f, was calculated by using the density of resin(ρ_m =1.02) and fiber(ρ_f =1.75).

2.2 Tensile test

Specimens for tensile tests were cut out in fiber direction (0° specimen) and 45° direction to fiber direction (45° specimen) as shown in Fig.2. The dimension of the specimen is 200 mm × 10 mm × 2 mm for 0° specimens and 130mm × 40mm × 2mm for 45° specimen. The tensile tests (Fig.3) were conducted by following the recommended testing procedures as described in JIS-K7164 using a universal testing machine (Autograph AG-100kNX, Shimadzu Co., Japan) with a constant displacement rate of 2mm/min and the extension of specimen was measured by non-contact 3D deformation analysis system (ARAMIS®, GOM mbH).

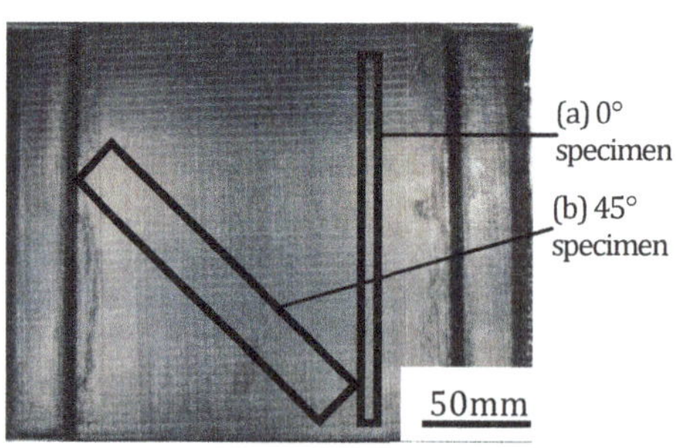

Fig.2 Cutting direction of specimen.

Fig.3 Tensile test setup.

3 RESULT AND DISCUSSION

3.1 Observation of the cross sections of the laminated composite plates

A typical cross section of the laminated composite plate observed by confocal microscope is shown in Fig.4. Regardless of the difference of stitch tension, good resin impregnation was observed from the cross section observation of the laminated composite plates. Fig.5 shows the surface of the laminated composite plates. Stitching yarns were observed on the surface of the laminated composite plates. While the melting point of the stitching yarn of the polyester used in this study was 256°C, molding with a high-speed press molding machine was conducted at 220°C, therefore stitch yarn was not dissolved and remained after molding. While, carbon fiber and stitching yarn are remained straight for the laminated composite plate of Tight specimens, waved stitching yarns are observed for those of Normal and Loose specimens.

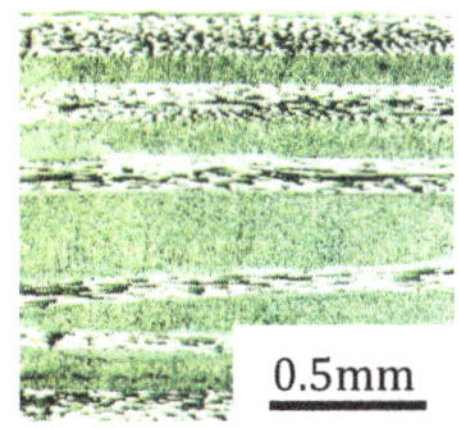

Fig.4 Observation of CF/PA12 cross section of Tight specimen.

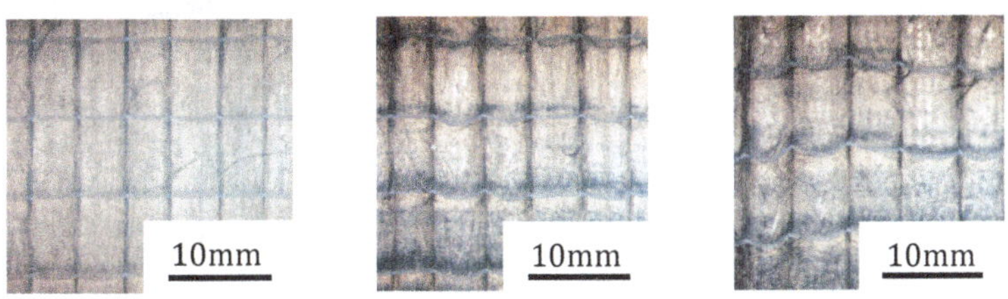

Fig.5 Surface observation of molded laminated composite plates.

3.2 Measurement of fiber volume fraction

Measured fiber volume fraction of CFRTP laminates for Tight, Normal and Loose specimens were 56%, 55% and 56% respectively, and had no significant difference. All specimens showed a higher value than the designed fiber volume fraction. During the press molding process, the resin was considered to flow out from the edge of the laminated composites plates.

3.3 Tensile properties

Tensile strengths for 0° specimens of Tight, Normal and Loose were 751±49MPa, 776±21MPa and 763±21MPa respectively. The tensile strengths for 0° specimens had no significant difference. Fig.6 shows the typical stress-strain curves for 45° specimens. The strain calculated from crosshead displacement (CH) was calculated as dividing crosshead displacement by length of the parallel part between chucking jig. Fig. 7 shows the typical strain distribution measured by non-contact 3D deformation analysis system. The strain measured by this system (ARAMIS) was obtained from the displacement between two points (50mm in distance), which are plotted in Fig.6. Tensile strength for 45° specimens are shown in Fig.8. Tensile strengths of 45° specimens of Tight, Normal and Loose were 219±4MPa, 180±5MPa and 184±15MPa respectively. The 45° specimens with Tight stitch tension had highest tensile strength. The tensile modulus of 45° specimens of Tight, Normal and Loose were 6.4±0.3Gpa, 6.0±0.4Gpa, and 6.1±0.1GPa respectively as shown in Fig.9. The tensile modulus of 45° specimens had no significant difference. Fig.10 shows the tensile fractured specimens of Tight, Normal and Loose specimens. The rotation angle of 45° fibers was measured by using image software. Angles of the fiber to tensile direction of Tight, Normal and Loose were 22°, 26° and 28°

respectively. Tight specimen has the smallest angle, showing the largest shear deformation. According to a previous study, in the bias extension tests of NCF performed on tensile direction to 45°, typical three different shear deformation areas were reported as shown in Fig.11: Areas C exhibits pure shear, Areas A very little or no shear and Areas B with partial shear(5). As shown in Fig.10, the Tight specimen had largest shear deformation in Area C. In the tensile test of 45° specimens, the slope of the stress-strain curves becomes steeper where the strain reaches around 0.3 as shown in Fig.5. Because of the steepest slope of the stress-strain curve for Tight specimens after reaching strain of 0.3, the tensile strength for specimen of Tight had the highest tensile strength.

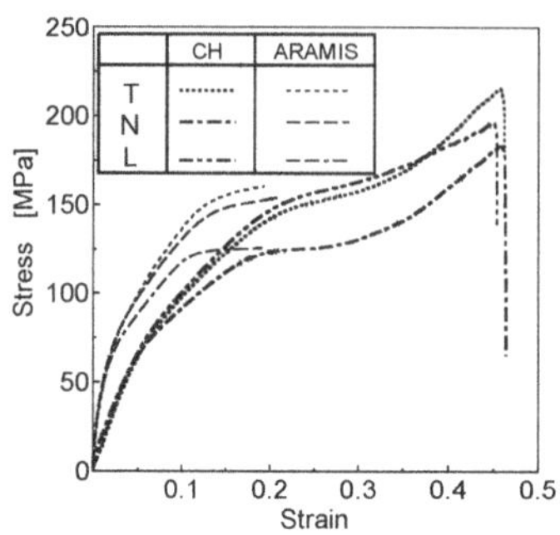

Fig.6 Stress-strain curve of 45°specimens.

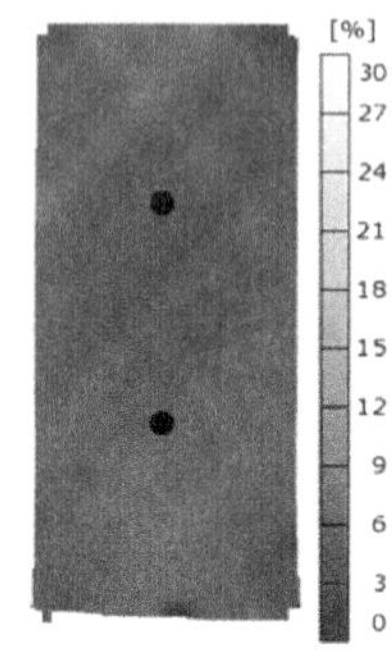

Fig.7 Contour of strain of Tight specimen.

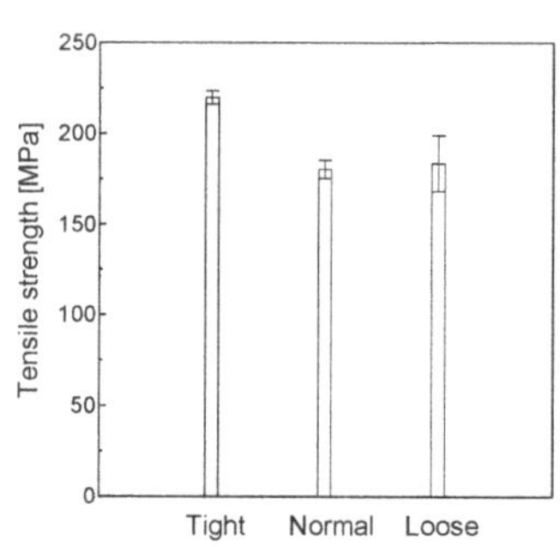

Fig.8 Tensile strength of 45°specimens.

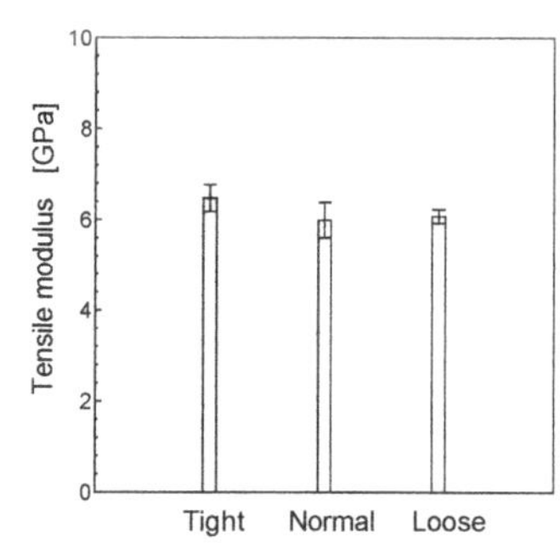

Fig.9 Tensile modulus of 45°specimens.

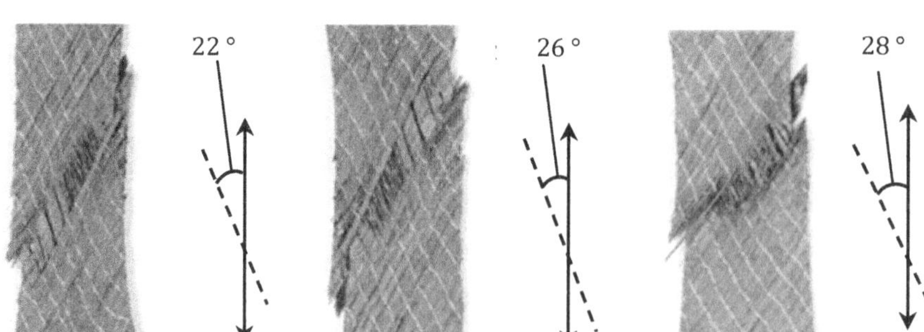

(a) Tight **(b) Normal** **(c) Loose**

Fig.10 Rotation of fiber for each specimen.

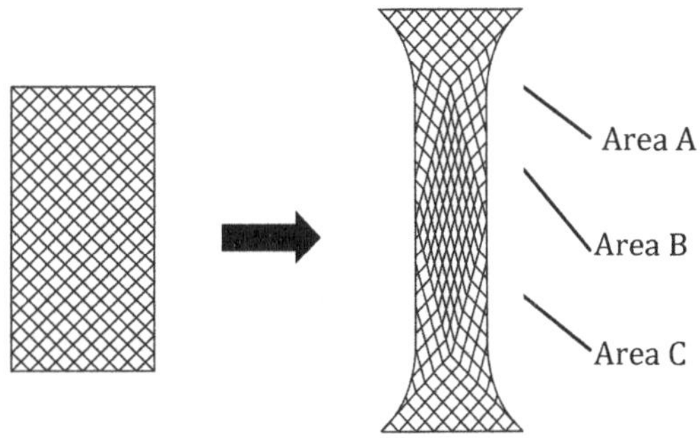

Fig.11 Schematic drawing of 45° specimen (5).

4 CONCLUSION

CFRTP laminates based on three different types of stitch tension (Normal, Tight and Loose) of NCF were molded and mechanical properties were evaluated. The investigation yielded the following conclusion.

(1) 0° specimens had almost same tensile strength regardless to the stitch tension.

(2) Because of the larger shear deformation of 45° fibers to the tensile direction, 45° specimens with Tight stitch tension had highest tensile strength.

5 REFERENCES

[1] Bibo, G. A., Hogg, J. P., Kemp, M. (1997) Composite Science and Technology **57**, 1221-1241.

[2] Kong, H., Mouritz, P. A., (2004) Composites Structures. **66**, 249-259.

[3] Kanazawa, K., Tanaka, K., Enoki, S., Katayama, T., (2013) Key Engineering Materials. **493**, 525-526

[4] Tanaka, K., Katayama, T., Uno K., (2008) WIT Transactions on the Built Environment. **97** 203-210.

[5] Sirtautas, J., Pickett, K.A., Lepicer, P., (2013) Composites Part B. **47** 48-57.

Temperature effect on strength of aluminum based high thermal conductive composites containing VGCF-CNT filler

K. Sasaki[1]**, K. Fukuchi**[2]**, T. Imanishi**[3,4]**, A. Kakitsuji**[5]

[1] Department of Mechanical Engineering, Hokkaido University, Japan
[2] Department of Mechanical Engineering, Kushiro National College of Technology, Japan
[3] Technology Research Association for Single Wall Carbon Nanotubes, Japan
[4] Sumitomo Precision Products Co., Ltd, Japan
[5] Technology Research Institute of Osaka Prefecture, Japan

ABSTRACT

Aluminum (Al) based composites containing vapor-growth carbon fibers (VGCFs) and carbon nanotubes (CNTs) has been developed by authors for a decade using spark plasma sintering (SPS). Temperature dependences of both thermal conductivities and strength properties of the composite are investigated in this paper. Pure tensile tests and measurements of thermal conductivity are conducted at high temperatures.

1 INTRODUCTION

Carbon nanotubes (CNTs) (1) are one of effective materials to increase the thermal conductivity of materials used in radiators and fins, because CNTs have characteristics such as high thermal conductivity, high electrical conductivity, and high strength. The thermal conductivity and the strength of composites containing CNTs have been widely reported (2-4). For example, Li et al. (2) fabricated titanium metal matrix composites reinforced by CNTs and graphite. They showed the effect of the CNTs reinforcement on the strength of the composites. Liu et al. (3) fabricated CNTs/Al composites using friction stir processing. They discussed the relationship between strength properties and aspect ratio of CNTs. Tsai et al. (4) described strength properties of CNTs/Cu composite using both experiments and molecular dynamics. They obtained the relationship between the length of CNTs and local buckling of CNTs in the composite.

The authors have fabricated Al matrix composites containing CNTs and vapor-growth carbon fibers (VGCFs) using spark plasma sintering (SPS). They showed that the composites had high thermal conductivities and further suggested that the high thermal conductivities arise from VGCF-CNT networks in the composite (5, 6). In this paper, temperature dependences of both thermal conductivities and strength properties of the composite are investigated. Pure tensile tests and measurements of thermal conductivity by laser flash methods are conducted at several high temperatures.

2 ALUMINUM BASED VGCF-CNT COMPOSITES (5)

The VGCF-CNT/Al composites were fabricated by powder metallurgy using a spark plasma sintering (SPS) machine. VGCFs used in this work have a thermal conductivity of 2000W/mK and a tensile strength of 7GPa. Figure 1 shows the procedure for fabrication of the composites. The composites were fabricated with the aligned VGCFs. The composites were manufactured by interbedding the aligned VGCF, matrix powder, and foils. For the pilling up VGCFs, matrix powder and foils are used to obtain good alignment of VGCFs in the composite. The CNTs are dispersed into aligned VGCF fillers and the fillers are piled up with the matrix powder and foils alternately. Pure Al containing 10vol% of Al/12Si was used as a matrix. The mean sizes of the powder of pure Al and Al/12Si are 35μm. The foil thickness is 10um.

Figure 2 shows the SEM observation of the VGCF filler sheet dispersed CNTs. It can be seen that the VGCFs align, the VGCFs and CNTs are blended, and there are the VGCF-CNT networks. Lestriez et al. (7) studied the network between carbon fibers and CNTs to improve the electric conductivity of electrode for lithium batteries. They showed that the network between carbon fibers and CNTs is effective to improve the electric conductivity. Figure 3 shows the coordinate system of the composites. X is the VGCFs direction, Y the radial direction of VGCFs, and Z the sintering direction, respectively. The coordinate is used to explain the following experiments.

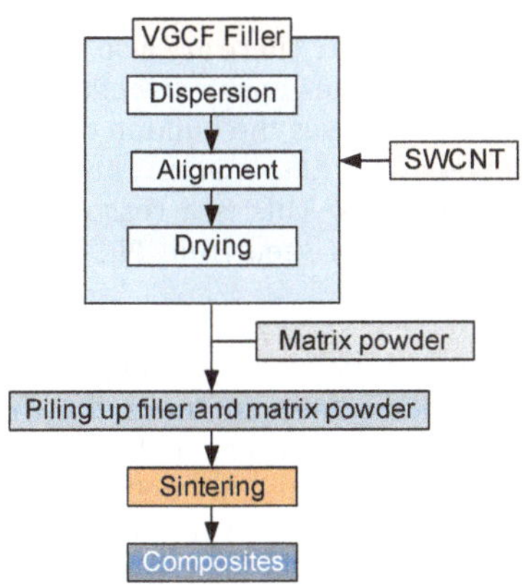

Figure 1 Procedure for the fabrication of the composite

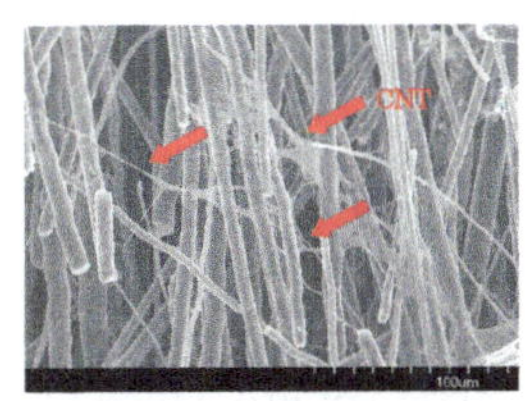

Figure 2 SEM image of VGCF filler (volume fraction of VGCF: 60%)

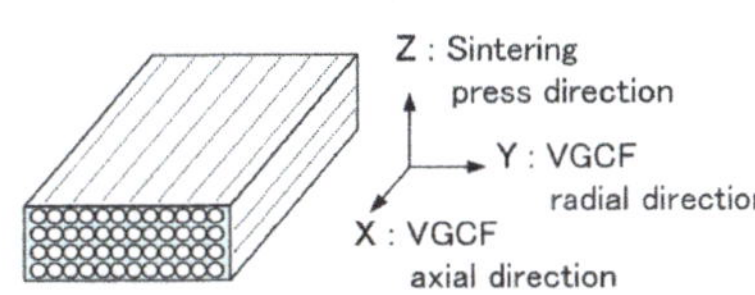

Figure 3 Coordinate of composites

3 EXPERIMENTAL METHODS

Figure 4 shows the geometry of specimens to measure the strength. Fig. 4(a) shows the geometry of the test specimens to measure the strength of X and Y directions of the composite, and Fig. 4(b) Z direction. The specimen for the measurement of Z direction has the composite at the center area, whose length is 2mm of the specimen. This is because that the strength of Z direction is expected to be quite low and brake under the low stress. The volume fractions of VGCF of the specimens are 0, 30, and 60% and that of CNT is 0.2%, respectively. Five specimens for each test condition were prepared. The tensile tests were conducted at five temperatures of 295K, 373K, 473K, 573K, and 673K using an electrical furnace.

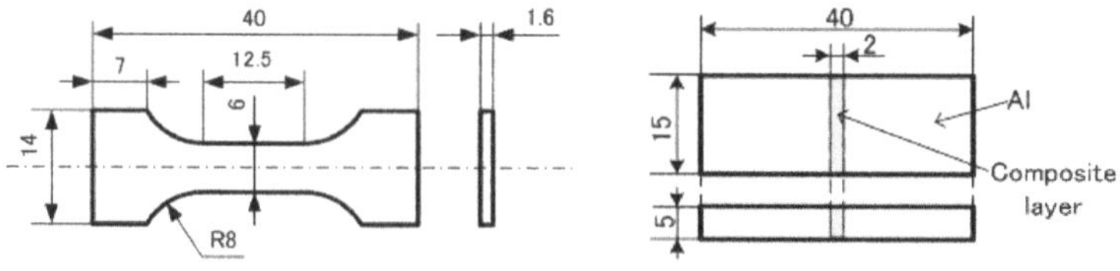

(a) X and Y direction specimen **(b) Z direction specimen**

Figure 4 Coordinate of composites

The thermal conductivity was measured by the laser-flash method using ULVAC, TC-7000 at five temperatures of 295K, 373K, 473K, 573K, and 673K. The geometry of sample for the measurements of X and Y direction are 25mm square and thickness of 3mm, and the measurements of Z direction a disc of diameter of 10mm and thickness of 3mm, respectively. The thermal conductivity was measured five times at each temperature.

4 RESULTS AND DISCUSSIONS

Figure 5 shows the relationship between the tensile strength and the volume fraction of VGCF at 295K. In Fig. 5, each plot shows the mean value of each property. The tensile strength in X direction, the axial direction of VGCF, is larger than those of Y and Z directions, which were along the radial directions of the VGCF. There are also differences in the tensile strength between the Y and Z directions. The tensile strength in the Z direction is smaller than that in the Y direction. Z direction was the buildup direction of VGCF-CNT filler and the matrix layer exists between the filler parts. Therefore, the interfacial strength between fiber and matrix is weakness. This is a reason why the strength in Z direction is significantly weak. It can be also seen from Fig. 5 that the volume fraction of VGCF hugely affects the tensile strength.

Figure 6 shows the thermal conductivity of each direction of the composites. The composites have thermal anisotropy though the matrix does not have the thermal anisotropy, i.e. the thermal conductivity of X direction is much higher than those of Y and Z directions. There is also the difference in thermal conductivity between Y and Z directions. The thermal conductivity of Z direction is smaller than that of Y direction. The thermal conductivity of Z direction is especially smaller than that of the aluminum matrix. This is because that Z direction is the buildup direction of the VGCF-CNT filers and the aluminum layer exists between the filers.

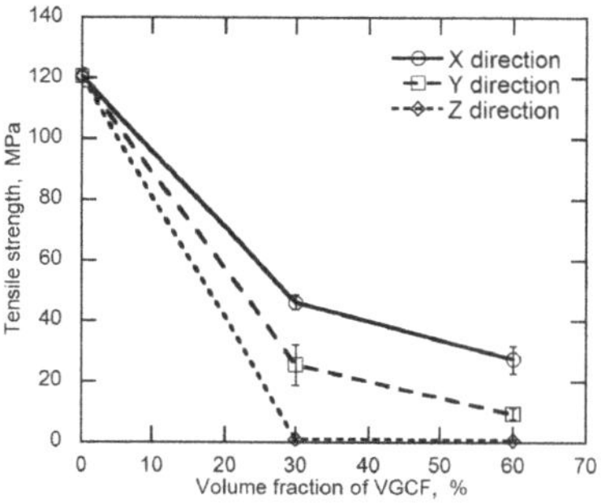

Figure 5 Tensile strength of the composites in each direction and volume fraction of VGCF at 295K

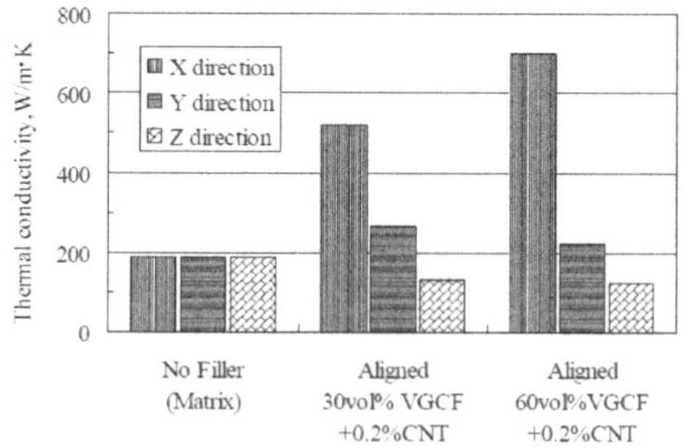

Figure 6 Thermal conductivity of the composite: effect of volume fraction and axial direction of VGCF

Figure 7 shows the relationship between the testing temperature and the tensile strength in X, Y, and Z direction of the composite. In Fig. 7, solid line shows the tensile strength of Al matrix without fibers, and broken lines are the tensile strength of composites. It can be seen from Fig. 7(a) that the tensile strength linearly decreases with increase in the testing temperature. However, the decrease ratios of tensile strength of the composites are smaller than that of Al matrix. It can be also found that the tensile strength of the composite including 30% VGCF in X direction is almost the same as the value of Al matrix at 673K. Figure 7(b) shows the tensile strength of Y direction. The effect of temperature on tensile strength of the composite in Y direction is smaller than that of Al matrix. However the value of tensile strength of the composite of 30% VGCF at 673K becomes near to the Al matrix. Figure 7(c) shows the tensile strength of Z direction. In Z direction, the tensile strength is dramatically small and the temperature dependency of tensile strength cannot be confirmed.

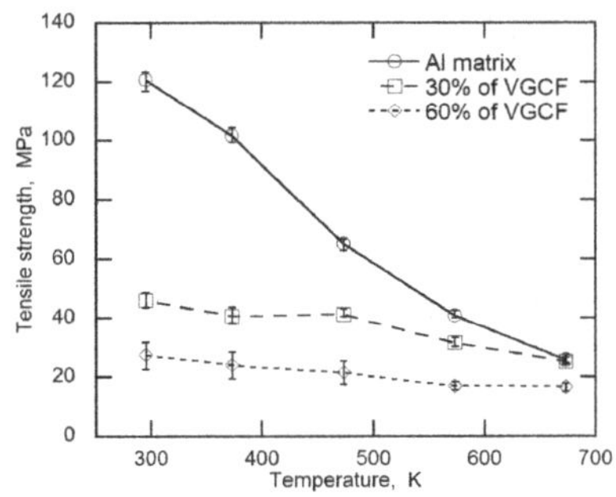

(a) X direction

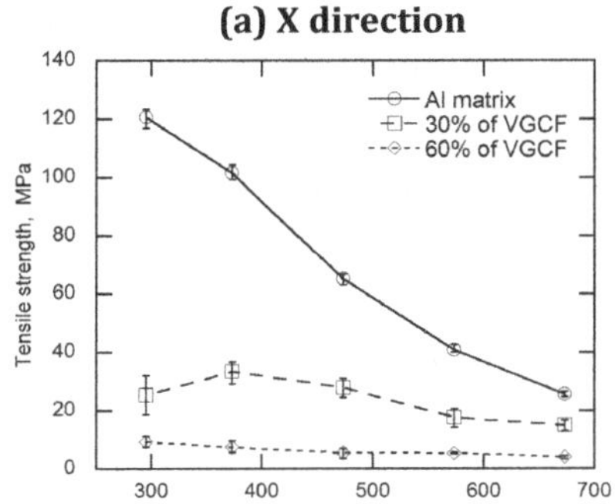

(b) Y direction

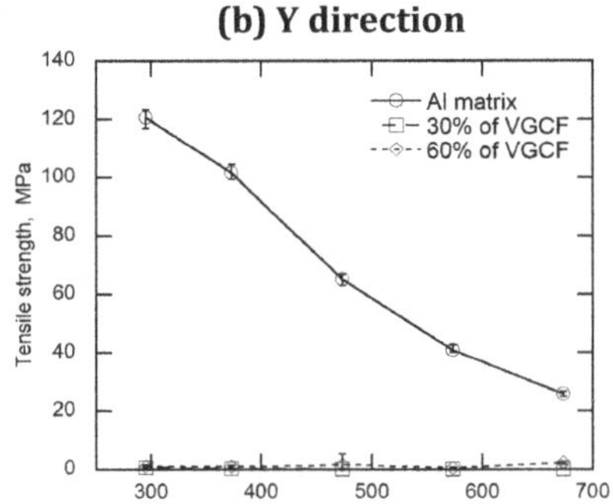

(c) Z direction

Figure 7 Relationship between tensile strength and temperature

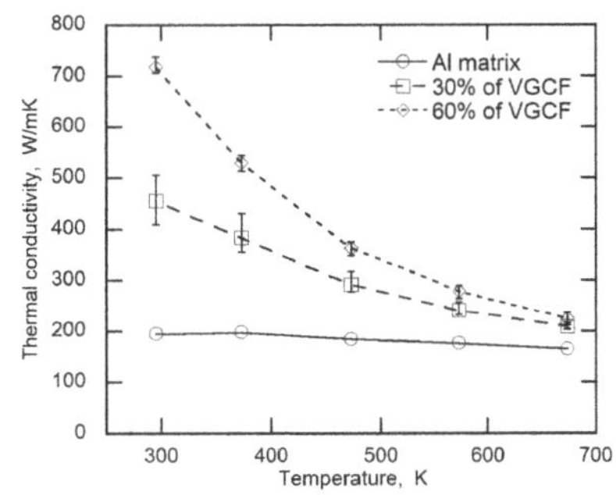

(a) X direction

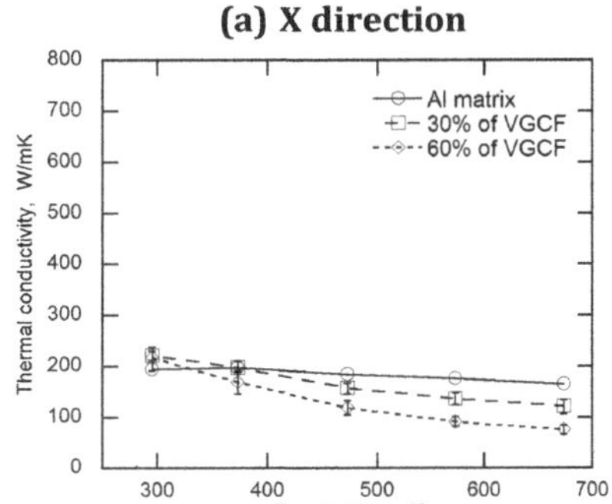

(b) Y direction

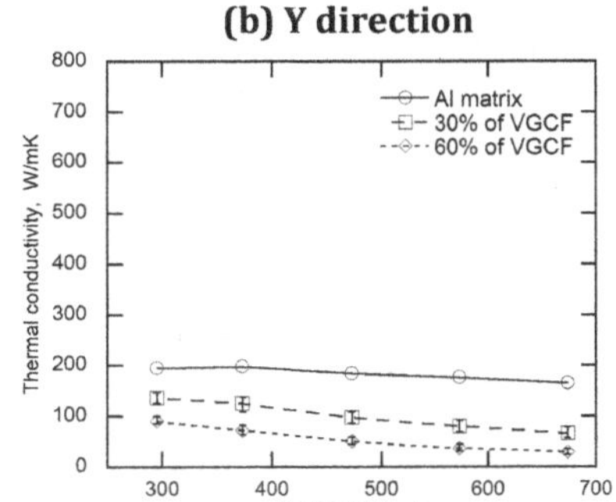

(c) Z direction

Figure 8 Relationship between thermal conductivity and temperature

Figure 8 shows the relationship between the testing temperature and the thermal conductivity in X, Y, and Z direction of the composite. In Fig. 8, the solid line shows the thermal conductivity of Al matrix without VGCF, and the broken lines the thermal conductivities of composites. It can be seen from Fig. 8(a) that the effect of testing temperature on thermal conductivity of Al matrix is a little. The thermal conductivity of the Al matrix is about 200 W/mK at the all temperatures. On the other hand, the thermal conductivities of the composites decrease with increase in temperature. Especially, the larger decrease in the thermal conductivity can be seen in X direction. The reason of the decrease in the thermal conductivity is that the thermal conductivity of CF (carbon fiber) including CNT has lower thermal conductivity at high temperature. However, it should be notified that the thermal conductivity of the composite in X direction at 673K is still higher than that of Al matrix. Figures 8(b) and (c) show the thermal conductivity of Y and Z directions, i.e., radial direction of VGCF and sintering direction. The thermal conductivities in both Y and Z direction decrease with increase in temperature as well as X direction. However, the thermal conductivities at the all temperatures in Y direction are a little higher than that in Z direction because Z direction is the piled up direction.

5 CONCLUSIONS

This paper treated the temperature dependency of the thermal and strength properties of the VGCF-CNT/Al composite having the high thermal conductivity. The results suggested the following conclusions; 1) The composite has the strength anisotropy. The tensile strength in VGCF axial direction is the highest; 2) The tensile strength decreases with increase in the volume fraction of VGCF; 3) The temperature dependency of tensile strength in each direction is clarified. The effect of the temperature on tensile strength of the composite is smaller than Al matrix. At 673K, the tensile strength of the composite in X direction is almost same as Al matrix; 4) The temperature dependency of thermal conductivities in each direction is clarified. The effect of the temperature on thermal conductivities of the composite is larger than that of Al matrix. However, at 673K the thermal conductivities of the composite in X direction are higher than that of Al matrix.

ACKNOWLEDGMENTS

This work is financially supported by Ministry of Economy, Trade & Industry, Japan (MEIR), New Energy & Industrial Technology Development Organization, Japan (NEDO), and Japan Aerospace Exploration Agency (JAXA). The authors gratefully acknowledge their supports.

REFERENCES

(1) Iijima, S. (1991) Nature, 354, 56-58.
(2) Li, S., Sun, B., Imai, H., Mimoto, T. & Kondoh, K. (2013) Composites Part A, 48, 57–66.
(3) Liu, Z. Y., Xiao, B. L., Wang, W. G. & Ma, Z. Y. (2014) Carbon, 69, 264-274.
(4) Tsai, P. C. & Jeng, Y. R. (2013) Composites Science and Technology, 79, 28-34.
(5) Imanishi, T., Sasaki, K., Katagiri, K. & Kakitsuji, A. (2008) Trans. JSME, 74, 651-655.
(6) Fukuchi, K., Sasaki, K., Imanishi, T., Katagiri, K., Kakitsuji, A. & Shimizu, A. (2010) Journal of Solid Mechanics and Materials Engineering, 4, 1273-1281.
(7) Lestriez, B., Desaever, S., Danet, J., Moreau, P., Plée, D. & Guyomard D., (2009) Electrochemical and Solid-State Letters, 12, A76-A80.

Mesoscale analysis of CFRP pressure vessel

N. Yoshikawa, K. Hariya
Institute of Industrial Science, The University of Tokyo, Japan

ABSTRACT

Carbon fiber reinforced plastic (CFRP) pressure vessels have been intensely developed as crucial components for both fuel cell vehicles and hydrogen gas stations. A mesoscopic simulation methodology is promising for accurate strength evaluation by taking account of local stress concentrations caused by the mesostructure of a fiber bundle/resin system. A high-performance computer infrastructure represented by "K" has enabled mesoscale finite element analysis in which the fiber bundle/resin structure is explicitly handled. We demonstrate the validity of this analysis, focusing on the fiber bundle stress enhancement caused by using a large tow.

1 INTRODUCTION

The commercial introduction of fuel cell vehicles in Japan will begin in 2015. Reliable and economical high-pressure hydrogen management technology is indispensable for its popularization. Carbon fiber reinforced plastic (CFRP) pressure vessels for both automobile tanks and hydrogen station accumulators are key components of the management technology. The CFRP pressure vessels are manufactured by the filament winding method with a complicated stacking sequence of winding angles and thicknesses. The design methodology of the vessel is not sophisticated enough to establish a stress criterion defined by material tests. In the track of conventional metallic pressure vessel design, finite element analysis based on anisotropic continuum modeling of CFRP has been performed, but it is not reliable enough to evaluate the actual strength of the vessel. Therefore, all CFRP pressure vessels are designed by trial and error, which requires laborious and destructive examinations using actual vessels.

The reason for the lack of confidence in the finite element analysis comes from the local stress concentration caused by the fiber bundle/resin structure. Macroscopic stress evaluated by the rule of mixtures(1) is not able to indicate the real stress state of the carbon fiber and resin system, which is decisive for the failure of CFRP. We have investigated a strength evaluation methodology based on the mesoscopic modeling of the CFRP vessel and developed simulation software(2) for the finite element modeling and analysis. The filament winding path is determined by following a geodesic line(3). A three-dimensional model of the fiber bundle/resin structure with a liner is constituted by accounting for the fiber bundle crossover topology. We can analyze the stress in the fiber bundle accurately.

We established a demonstration problem concerning the production efficiency. The use of a large tow(4) has been attempted with the aim of reducing the filament winding time and effort, that is, the number of winding revolutions. At the same time, we are apprehensive about burst pressure reduction due to the large tow. Setting three types of mesomodels with different fiber bundle widths, we examine the stress state in the fiber bundle and investigate the mechanism of stress enhancement at the crossover. The proposed methodology based on mesoscopic modeling seems promising for a quantitative discussion on the trade-off between strength and production efficiency.

2 MESOSCALE MODELING OF CFRP PRESSURE VESSEL

The analyzed vessel consists of two layers: an aluminum liner and a CFRP helical layer. The liner dimensions are indicated in Figure 1. We set three types of helical windings, as shown in Figure 2. The numbers of windings are (a) 36, (b) 18, and (c) 12. The widths of the fiber bundles are set as (a) 5.0 mm, (b) 10.0 mm, and (c) 15.0 mm to maintain a constant amount of carbon fiber. The helical winding paths are set as geodesic lines successively overlapped with a regular difference in longitude of (a) 110 deg, (b) 100 deg, and (c) 150 deg. The paths round at the boss neck. Near the crossover position, a sine curve undulation is added to the upper path. The volume data of the bundle are constructed by setting a rectangular cross-section of 0.25 mm thickness.

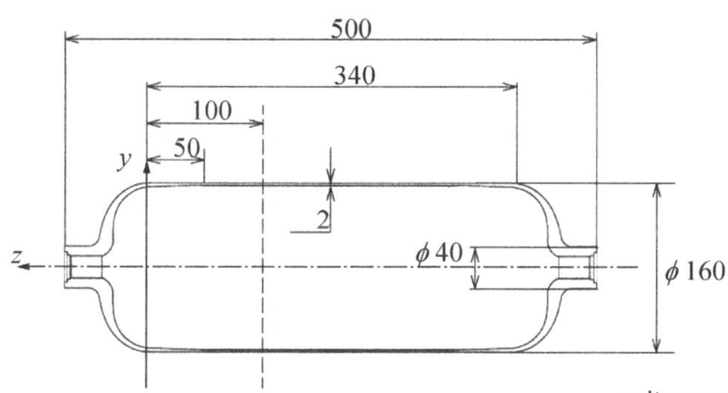

Figure 1 Aluminum liner

(a) 36 windings of 5.0 mm bundle width

(b) 18 windings of 10.0 mm bundle width

(c) 12 windings of 15.0 mm bundle width

Figure 2 Mesoscale model of pressure vessel

The liner, fiber bundle, and resin are characterized by a linear-elastic body with Young's moduli of 68.3, 230, and 3.0 GPa and Poisson's ratios of 0.3, 0.28, and 0.35, respectively. Tetrahedral elements with four nodes are employed for discretization. The total number of nodes is about 2.5 million. The use of high-performance computers represented by "K"[(5)] is necessary to handle the enormous amount of data.

3 STRESS CONCENTRATION IN FIBER BUNDLE

The first principal stress distributions in the bundles are indicated in Figure 3. The internal pressure is set to 1.0 MPa. Local stress concentrations caused by crossover are expected. The maximum first principal stresses increase with the fiber bundle width as (a) 60.4 MPa, (b) 70.4 MPa, and (c) 77.7 MPa, even though the total amounts of carbon fiber are equivalent. The locations of the maximum stresses are marked by solid circles in the figures. An increase in width induces an increase in stress at the crossover on the cylinder part.

Details of the stress enhancement are indicated with a finite element mesh in Figure 4. The stresses at the nodes are indicated by small colored points. The color legend is identical to that in Figure 3. The node location is shifted by a displacement of a hundredfold. The maximum stress regions marked by blank squares are located beside the crossovers for all three cases. The direction of the first principal stress is evaluated at the maximum stress nodes. The direction deviates from the fiber direction. The deviation angles are (a) 28 deg, (b) 39 deg, and (c) 44 deg. The increase in bundle width increases the deviation angle and the maximum stress, accordingly.

In the cylinder part of the thin-walled metal pressure vessel, the hoop stress is twice the axial stress. It is legitimate that the dominant deformation in the hoop direction affects the deviation of the first principal stress. The mechanical constraint caused by the crossover of a wider bundle seems stronger. Therefore, higher stress occurs in a wider bundle. This result implies a burst pressure reduction in large tows together with a high possibility of transverse fracture of the bundle.

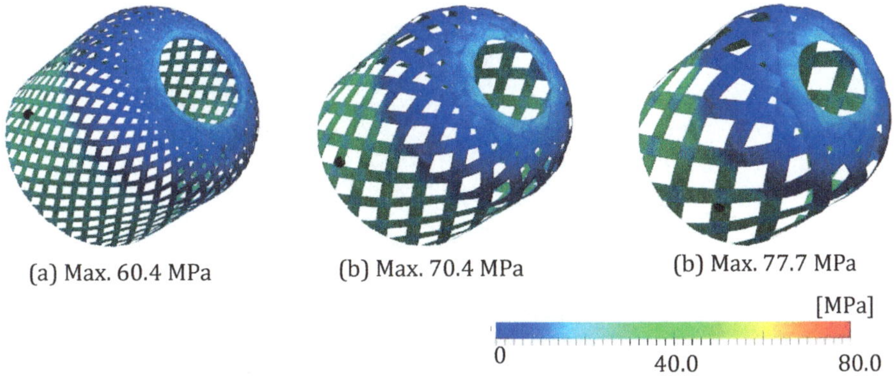

Figure 3 First principal stress distribution in fiber bundle

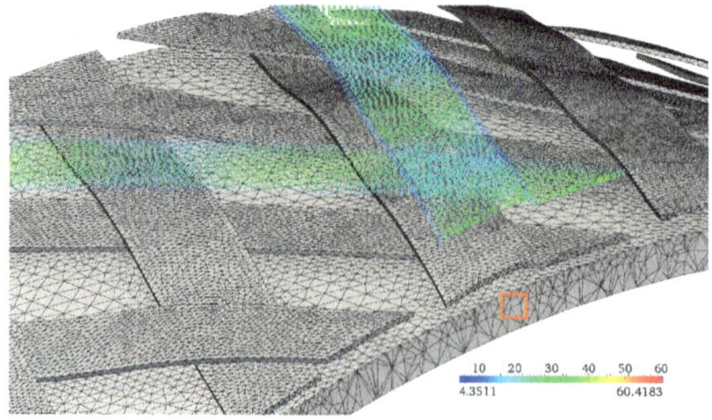

(a)

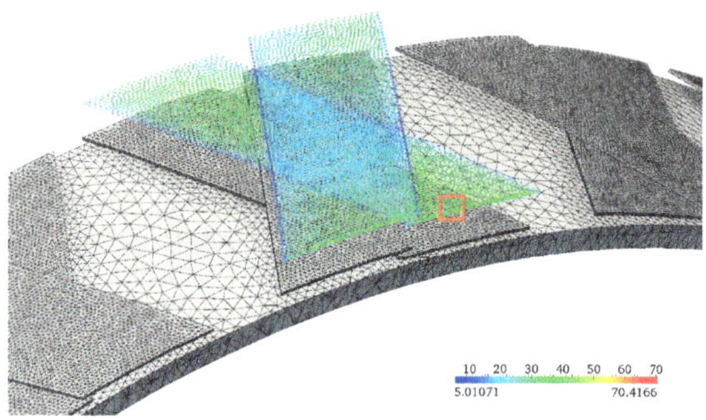

(b)

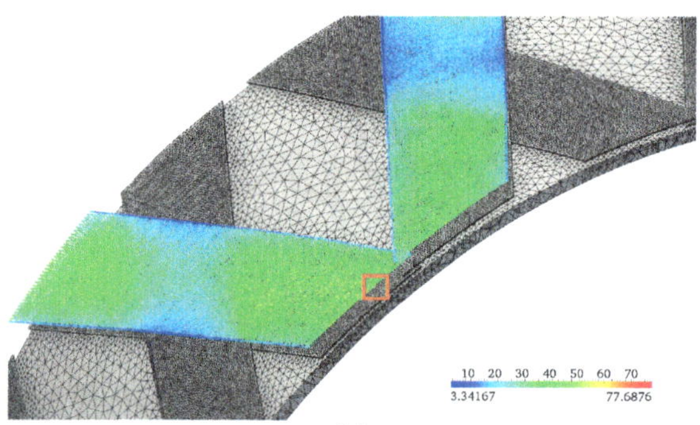

(c)

Figure 4 Detailed inspection of maximum stress region beside crossover

4 CONCLUSIONS

We have developed a mesoscopic simulation based methodology for the strength evaluation of CFRP pressure vessels, which are indispensable for high-pressure hydrogen management in a fuel cell vehicle system. The fiber bundle and resin are strictly separated in the modeling of the vessel. Complicated stress concentrations can be revealed through finite element analysis. A sample demonstration was performed by employing three types of CFRP pressure vessel models in which the bundle width and number of windings were systematically changed. We demonstrated the potential of the proposed methodology for quantitatively solving the trade-off problem between strength and production efficiency.

ACKNOWLEDGEMENT

This research was supported by the grant for "Strategic Programs for Innovative Research" Field No. 4: Industrial Innovations from the Ministry of Education, Culture, Sports, Science, and Technology (MEXT)'s "Development and Use of Advanced, High-Performance, General-Purpose Supercomputers Project."

REFERENCES

(1) Whitney, J. M., and Riley, M. B., AIAA Journal, 4(9): 1537–1542, 1966.
(2) http://www.ciss.iis.u-tokyo.ac.jp/project/rss/software/10_info.html
(3) Vasiliev, V. V., and Morozov, E. V., Mechanics and Analysis of Composite Materials, Elsevier, 2001.
(4) http://web.ornl.gov/~webworks/cpr/pres/105602.pdf
(5) http://www.aics.riken.jp/en/

Low temperature stress estimation of fiber reinforced material

M. Nishida[1], M. Refai Muslih[2], T. Doi[3], T. Matsue[4], T. Hanabusa[5]
[1] Kobe City College of Technology, Mechanical Engineering, Japan
[2] National Nuclear Energy Agency, Indonesia
[3] Kobe City College of Technology, Mechanical System Course, Japan
[4] Niihama National College of Technology, Environmental Materials Engineering, Japan
[5] The University of Tokushima, Japan

ABSTRACT

The tungsten fiber reinforced titanium composite was produced by the spot welding method. The internal stress alteration of W/Ti composite was measured by the neutron diffractometer, DN1, which had been installed at neutron beamport #6 of Indonesian multipurpose reactor RSG-GAS in Serpong. The residual stress alterations were measured by the in-situ neutron stress measurement technique under the cooling cycles from 300K to 10K. Both residual stresses in titanium-matrix and tungsten-fiber were investigated at several temperatures. Furthermore, residual stress alterations were estimated by x-ray stress measurement after cyclic thermal shock testing from 77K to 373K generated by liquid nitrogen and boiling water.

1 INTRODUCTION

Fiber reinforced materials have a large difference in thermal expansion coefficients between the matrix and the fiber. This mismatch causes the thermo-reduce residual stresses in every composite system (1, 2, 3). In this study, the tungsten fiber reinforced titanium composite (W/Ti) was prepared to investigate the residual stresses. This W/Ti composite consisted of the continuous tungsten fiber and the pure titanium matrix. Thermal residual stresses in the composite were examined by the neutron stress measurement method and the x-ray stress measurement. Stress alterations during cryogenic cycling were estimated by the combination of the neutron diffraction and the cryostat cooling systems (4). Temperature was changed from 300K to 10K periodically, and thermal stresses in the tungsten-fiber and the titanium-matrix were measured at each temperature. On the other hand, residual stress alterations were estimated by x-ray stress measurement after cyclic thermal shock testing from 77K to 373K generated by liquid nitrogen and boiling water.

2 PREPARATION OF FIBER REINFORCED MATERIAL

The W/Ti composite was produced for the present investigation. In this W/Ti composite, 99.99 % purity tungsten fiber with 100 μm diameter and 99.9 % purity titanium plate of thickness 0.5mm and 0.2mm were used for the fiber phase and the matrix phase, respectively. In this study, the W/Ti composite was produced by the spot welding method (5). Fig.1 (a) and (b) show the photographs of the W/Ti composite. The final dimension of the W/Ti sample for the thermal stress measurement was 12mm × 12mm, thickness 7.0mm. The volume fraction of the tungsten fiber is about 5 % in this sample.

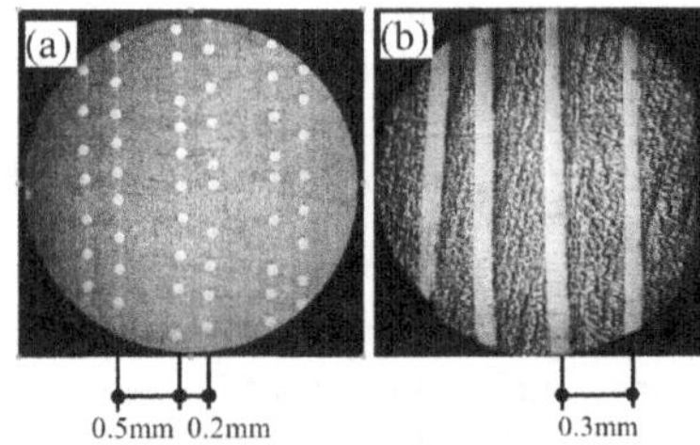

Fig.1 Photographs of the W/Ti composite; (a) a microphotograph of the cross section of the W/Ti composite, (b) the specimen surface polished by emery paper.

3 IN-SITU THERMAL STRESS MEASUREMENT

The long-fiber reinforced composite considered in this investigation is schematically shown in Fig.2. The x_1 axis is defined as parallel to the longitudinal direction of the tungsten-fiber. The x_2 and x_3 axes are normal to the fiber direction. When stresses are calculated by the neutron stress measurement, the strains in the three directions of x_1, x_2 and x_3 are required. The thermal residual stresses σ_1 parallel to the longitudinal direction and σ_2 normal to the longitudinal direction were calculated by the following Hooke's equation;

$$\begin{cases} \sigma_1 = \dfrac{E}{1+\nu}\{\varepsilon_1 + \dfrac{\nu}{1-2\nu}(\varepsilon_1+\varepsilon_2+\varepsilon_3)\} \\ \sigma_2 = \dfrac{E}{1+\nu}\{\varepsilon_2 + \dfrac{\nu}{1-2\nu}(\varepsilon_1+\varepsilon_2+\varepsilon_3)\} \\ \sigma_3 = \dfrac{E}{1+\nu}\{\varepsilon_3 + \dfrac{\nu}{1-2\nu}(\varepsilon_1+\varepsilon_2+\varepsilon_3)\}. \end{cases} \quad (1)$$

In this study, stress alterations caused from low temperature cycling were measured by the in-situ neutron stress measurement technique. The cryostat cooling system was mounted on the neutron diffractometer DN1 with the two-dimensional detector system. Controls of the diffractometer, the two-dimensional detector and the temperature in the cryostat system were managed by the main computer system.

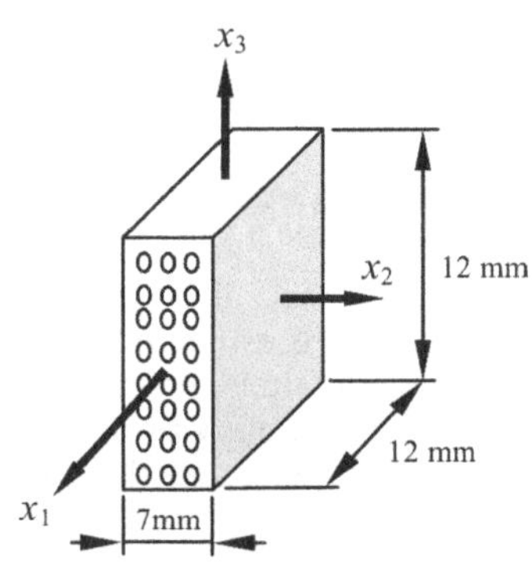

Fig.2 Coordinate system of the W/Ti composite.

Fig.3 shows the schematic diagram of the temperature vs. time program for the in-situ neutron stress measurement of the W/Ti composite. The rate of temperature change is about 2K/min. in the cool down stage and 1.5K/min. in the heat up stage. When the temperature came to a target position, it held for about 20 minutes in order to stabilize, and after that the stress measurement started in every temperature case. One cycle consists of the cool down stage and the heat up stage, and three cycles were repeated in this measurement. Table 1 shows the conditions of the in-situ neutron stress measurement. Young's modulus: E and Poisson's ratio: ν depending on the *hkl* diffraction plane were calculated from Kroener model. In this measurement, E, ν and thermal expansion coefficient α were considered the temperature dependency (6). The measurement time for one diffraction profile was 90 minutes for the titanium-matrix and 75 minutes for the tungsten-fiber. Finally, the measuring time of one cycle in this measurement became about 30 hours.

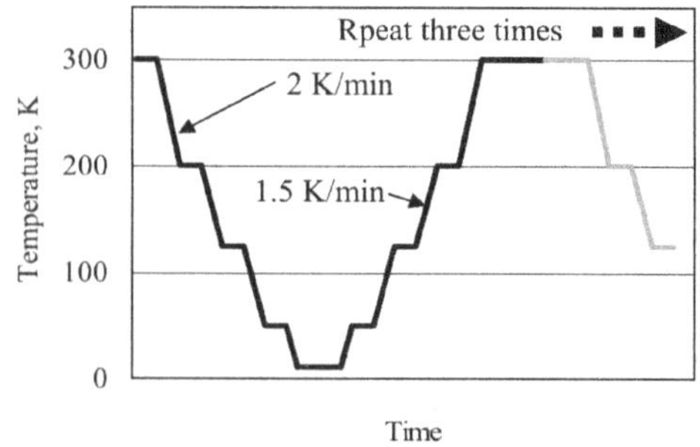

Fig.3 Time program of the in-situ thermal stress measurement.

Table 1 Conditions of neutron stress measurement.

Wave length	λ = 0.18301 nm
Detector	2D detector, 256x256 pixel
Resolution	0.0365 deg./pix.
Reactor power	15 MW
Measurement material	W/Ti composite
h k l plane, diffraction angle	Ti 103, 2θ= 86.5° W211, 2θ= 91.0°, W 220, 2θ= 111.0°
Young's modulus E, at R.T. Poisson's ratio ν, at R.T. Coefficient of thermal expansion α, at R.T.	Ti: E=120.7GPa, ν=0.31, α=8.04x10^{-6} /K W: E=409.2GPa, ν=0.28, α=4.43 x10^{-6}/K
Slit system	Incident slit : 3mm x 10 mm Receiving slit : radial collimator
Measurement time	Ti: 90 min./profile W: 60 min./profile

4 THERMAL SHOCK TESTING

Residual stress alterations in the W/Ti composite were estimated by the x-ray stress measurement after the cyclic thermal shock testing. The sample was dipped alternately in the liquid nitrogen and the boiling water. The temperature change was from 77K to 373K. The residual stresses σ_1 parallel to the longitudinal direction and σ_2 normal to the σ_1 direction were measured by the x-ray stress measurement with the $\sin^2\psi$ method (7). Fig.4 shows the schematic diagram of the temperature vs. time program for the cyclic thermal shock testing. The thermal sock from 77K to 375K were repeated 20 times. After that, residual stresses were measured by x-ray measurement and compared with initial stress values. Table 2 shows conditions of the x-ray stress measurement.

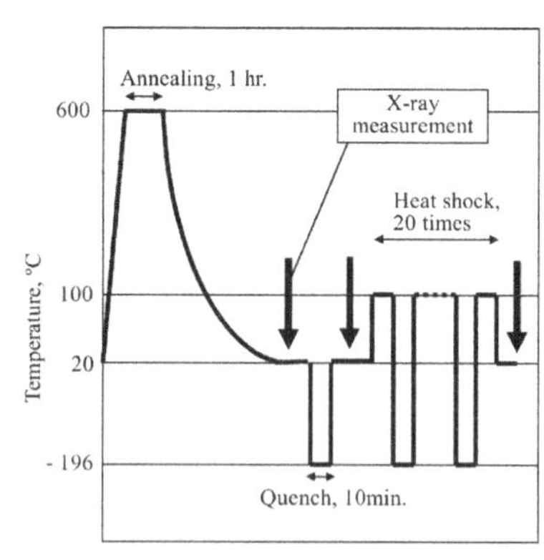

Fig.4 Time program of the thermal shock testing.

Table 2 Conditions of x-ray stress measurement.

Characteristic x-rays	CuKα	hkl plane & Diffraction angle	W 321, 2θ=131.1° Ti 213, 2θ=139.3°
Method	Ω-diffractometer method	2θ step angle & fixed time	W: 0.1°, 30sec. Ti: 0.1°, 60sec.
X-ray optics	Parallel beam	Filter	Nickel
Tube power	40 kV, 25 mA	Irradiated area	10×5 mm
$\sin^2\psi$	0 ~ 0.5, 6 points	Peak deciding method	FWHM method

5 RESULTS AND DISCUSSIONS

Fig.5 (a) and (b) show results of the in-situ thermal stress measurement by the neutron diffraction. In this figure, the up side region in Fig.5 (a) shows the results of the titanium-matrix, and the down side region in Fig.5 (b) is the tungsten-fiber results. The longitudinal direction stress σ_1, the normal direction stress σ_2 are plotted in these figures. These measurement results are assembled as the average in the each temperature. From Fig.5 (a), in the case of the titanium-matrix, the initial stress σ_1 in the fiber longitudinal direction was a tensile state of about 75MPa. It is assumed that this initial small tensile residual stress was caused during the manufacturing process. This initial stress was changed from 75MPa to a tensile state of 150MPa at 10K with the temperature down. In the heat up stage, thermal stresses gradually return to 75MPa with the temperature rise. Furthermore, stress σ_2 in the normal of the fiber direction showed the almost the same tendency of the stress σ_1. However, these stress alterations were relatively small. Therefore, it needs a lot of data to improve the reliability of the measurement results. From the result of tungsten-fiber in Fig.5 (b), the initial residual stress in the fiber longitudinal direction σ_1 increased gradually with the temperature down, and it changed from -780MPa to about -1000MPa in 10K. In the heat up stage, thermal stresses return to -780MPa with the temperature rise. On the other hand, in the case of stresses σ_2 in the normal of the fiber direction, stress values of the σ_2 were close to 0MPa, and this initial stress was changed 100MPa at 10K with the temperature down. These alterations of the thermal residual stress were generated form the thermal expansion mismatch between the tungsten-fiber and titanium-matrix. Furthermore, it is assumed that the stresses in the fiber longitudinal direction σ_1 are the dominant stresses in the stress condition of the W/Ti composite. Table 3 shows results of the thermal shock testing. From these results, the initial stresses generated in the manufacturing process were relaxed by the annealing treatment. Especially, residual stresses σ_1 in the titanium-matrix changed

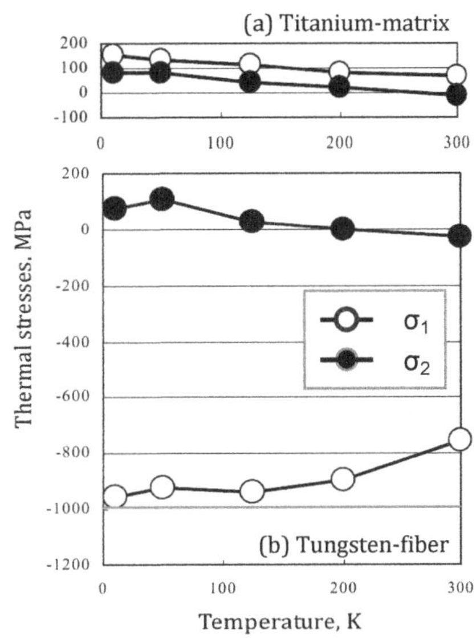

Fig.5 Results of the in-situ neutron stress measurement; (a) titanium-matrix, (b) tungsten-fiber.

from a compressive residual stress state to a tensile state. Residual stress values after the cooling quench and after the cyclic thermal shock show little change in all cases. It is seemed from these results that residual stresses in the W/Ti composite have not yet led to induce the plastic deformation by the thermal shock testing.

Table 3 Results of the thermal shock testing.

Material & direction	Residual stresses, MPa			
	Initial stresses by manufacturing	After annealing	Cooling quench	Cyclic thermal shock
Ti-matrix, σ_1	-155	28	55	36
Ti-matrix, σ_2	-174	-102	-47	-64
W-fiber, σ_1	-894	-829	-876	-777
W-fiber, σ_2	-553	-124	-171	-153

6 CONCLUSIONS

1) Regarding to the W/Ti composite, it was success to measure the inside thermal residual stresses alterations under the cryogenic temperature cycling by the in-situ neutron stress measurement technique.
2) Thermal residual stresses in tungsten fiber and titanium matrix were changed to other states depending on temperature changes.
3) The main factor of thermal stress alteration is the difference of thermal expansion between tungsten-fiber and titanium-matrix.
4) In the thermal shock testing, residual stresses in the W/Ti composite have not yet led to induce the plastic deformation by the thermal shock testing.

ACKNOWLEDGMENT

This work has been supported by Grant-in-Aid for Scientific Research(C), No.25420037 of Japan Society for the Promotion of Science (JSPS) from 2013 to 2017. We are grateful for their support.

REFERENCES

[1] Y. Ikeuchi, T. Matsue and T. Hanabusa: Proc. 5th Int. Conf. on Residual Stresses (1997), **2**, 958-963.
[2] Y.Ikeuchi, K. Iga and T. Hanabusa: J. Soc. Mat. Sci., Japan (1994), **43**, 792-798.
[3] Masayuki Nishid, Rifai M.Muslih, et al.: International Journal of Modern Physics B (2006), **20**(25-27), 3668-3673.
[4] Masayuki Nishida, M Refai Muslih, Yasukazu Ikeuchi, Nobuaki Minakawa and Takao Hanabusa: Materials Science Forum (2005) **490-491,** 239-244.
[5] MASAYUKI Nishida, MASASHI Haneoka, TATSUYA Matsue, TIAN Jing and TAKAO Hanabusa: Materials Science Forum (2014), **768-769,** 335-342.
[6] Information on http://riodb.ibase.aist.go.jp/TPDB/DBGVsupport/index.html
[7] The Society of Materials Science Japan ed. Standard Method for X-Ray Stress Measurement (2005), ISBN 4-901381-27-X, The Society of Materials Science Japan.

Fracture of fiber reinforced materials under off-axis loading

C. Marotzke, T. Feldmann
BAM - Federal Institute for Materials Research & Testing,
Division 5.3 – Mechanical Behaviour of Polymers, Germany

ABSTRACT

The failure behaviour of unidirectional plies tested under off-axis loading is studied. The stress distribution occurring in standard off-axis tests is briefly discussed. The different failure modes, this is, pure fiber breakage, pure inter fiber failure as well as a superposition of both is analysed based on test data and fracture patterns. The large difference of the failure load depending on the failure mode at small off-axis angles is studied by comparison of a short and a long 3° off-axis angle specimen.

1 INTRODUCTION

In structural applications of fiber reinforced materials the composite plies inside a laminate are loaded by complex stress distributions. Composites are very sensible against the loading transverse to the fiber direction because of their highly anisotropic mechanical properties. Even small off-axis angles lead to a significant reduction of the mechanical properties such as stiffness and strength. Especially the resistance against failure strongly depends on the loading angle. The strength parallel to the fiber direction is several times higher than the strength perpendicular to the fibers.

The off-axis test is regarded as the anisotropic pendant to the standard uniaxial tensile test for isotropic materials. However, this is true only at first glance. Off-axis tests suffer from some inherent drawbacks. They usually are performed using standard testing machines equipped with fixed, this is, non rotating clamps. Since the material is anisotropic a non-homogeneous stress state develops due to a superposition of tension and bending. Near the clamps the bending effect leads to a notable disturbance of the stress field. This was first studied by Pagano and Halpin using reinforced rubber strips (1). Additional stresses emerge due to the usual clamping effects, this is, the constraining of the lateral contraction an compressive stresses perpendicular to the plane of the specimen. As a result the likelihood of failure starting near the clamps is enlarged. The specific problems of measuring strength and stiffness of anisotropic materials was studied by several authors, e. g. (2-4).

The longitudinal stresses (stresses in loading direction) exhibit a significant variation along the specimen length. The longitudinal stresses for the cross section at various positions along the specimen length are calculated by a finite element analysis for a 15° off-axis specimen. The specimen has a thickness of 1mm. It is 12mm in width and 70mm long (between the tabs). High maxima occur along the edges in the vicinity of the clamps (fig. 1, path 5). The disturbance due to the clamping decreases with growing distance to the clamps (path 6, path 8). But even in the middle of the specimen (path 10) the stresses differ by about 13% along the width. The bending effect varies with the off-axis angle. For small off-axis angles it increases with growing angle. After a maximum around 30° it decreases again at vanishes almost completely at 60°. For lager angles it changes its direction and reaches a small maximum before it finally vanishes for 90°. This means that the disturbance of the stress field varies with off-axis angle.

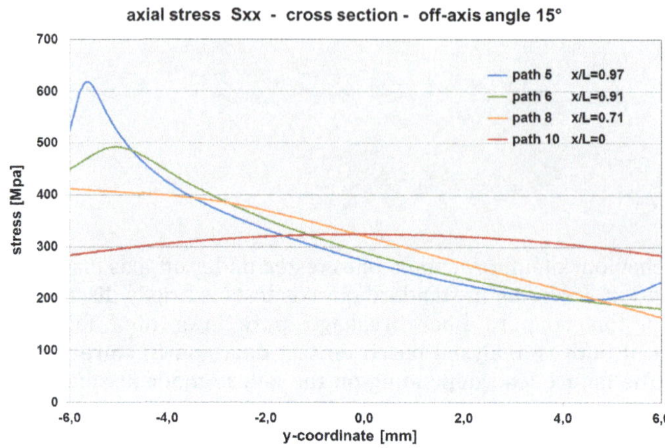

Fig. 1: Longitudinal stresses in 15° off-axis specimen at different positions along the specimen length

In off-axis tests two different failure modes can appear depending on the specimen geometry and the off-axis angle, this is, fiber breakage and inter fiber failure. If the geometry is such that a crack path parallel to the fibers can develop connecting the two edges of the specimen inter fiber failure occurs. If any crack path starting at on edge runs into the clamped ends of the specimen fiber breakage occurs. The fiber breakage may take place at the ends of the tabs or, depending on local imperfections, anywhere in between. Fiber breakage is finally accompanied by additional inter fiber failure.

Except of the two limit cases, this is, pure fiber fracture or pure inter fiber failure the question arises which failure mode takes place in a structural composite part. A single ply inside a laminate cannot fail without delamination or simultaneous failure of the neighbouring plies. This means that the failure of a single ply is constrained by the adherent neighbouring plies. In an off-axis test of an unidirectional ply, on the other hand, the ply is only constrained at the ends of the specimen where it is clamped. As a matter of fact the failure mode taking place within a laminate cannot be perfectly mapped by off-axis tests. However, the measured failure stresses must be related to the same failure mode. Strictly speaking all specimens except of the 0° specimen must be long enough to enable inter fiber failure without fiber breakage.

A further factor decreasing the failure load is the machining of the specimens. If the edges are not polished after cutting significant damage on microscale is present, e. g. debonding of the fiber ends. The fiber ends itself cause stress concentration on microscale due to the mismatch of the elastic constants. It may, however, be doubted that this is a substantial factor in comparison with the others mentioned above. As a result, the strength measured by off axis tests can be regarded as an lower limit of the actual strength as long as the specimen fails by inter fiber failure.

2 EXPERIMENTAL

A series of off-axis tests is performed for a carbon fiber reinforced epoxy resin with a fiber volume fraction of 55%. The specimens are cut out of 50x50mm plates under the respective angles. The plates are manufactured by an RTM process using five layers of non crimp carbon fabrics. The thickness of the plates is 1mm. The ends of the specimens are reinforced by multiaxial glass fiber tabs of 1mm in thickness.

The pure axial fiber loading (off-axis angle 0°) and pure transverse loading, (off-axis angle 90°) define the limits of the strength. Since at small off-axis angles the failure stress strongly depends on the angle small steps of 3° are chosen between off-axis angles between 0° and 15°. Between 15° and 30° steps of 5° are taken and between 30° and 90° the step size is 15°. The specimen geometry and the off-axis angle are chosen in order to enable inter fiber failure, except of the 3° and 0° off-axis angle. In order to show the influence of the different failure mechanisms a second batch of specimens was used which allow inter fiber failure even at an off-axis angle of 3°.

2.1 Off-axis test for off-axis angle 0°

If the fibers are aligned in loading direction the failure occurs by instantaneous fiber fracture followed by additional inter fiber failure. No significant damage occurs before the limit load is reached. The load/displacement curves show only very little indication of damage. Except of one specimen which exhibits a slight drop of the load at an early phase small load drops occur just before final failure (fig. 2a). One specimen fails without any premature failure. A large amount of fibers break directly at the ends of the tabs. Due to the high amount of energy stored the specimens are completely damaged in the moment of final failure (fig. 2b). During this highly dynamic process a large number of inter fiber cracks develops.

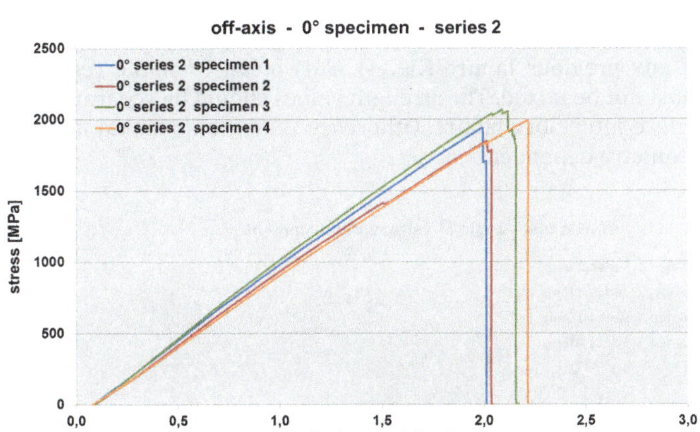

Fig. 2a: Load/displacement curves for 0° specimens - fiber breakage

Fig. 2b: 0° specimens after failure

2.2 Off-axis test for 3° off-axis angle

In case of a short specimen with an off-axis angle of 3° the failure stress is significantly decreased and reaches only less than one half of the respective failure stress of the 0° specimen. This is due to the fact that the failure processes are completely different. The first difference is the course of failure during the loading process. While for the 0° specimen the failure occurs almost instantaneously the failure in case of a small off-axis angle accumulates during loading. This is indicated by several load drops occurring over a large range during the loading process (fig. 3a). The load drops are either caused by inter fiber cracks or by breakage of a number of fiber bundles. The second difference is that in case of a longitudinal fiber orientation necessarily all fibers have to break. In case of a 3° off-axis angle about 30% of the fibers remain unbroken because they are shielded by in an inter fiber crack (fig. 3b). This means that a combination of two failure modes takes place. As a result the values are not directly comparable.

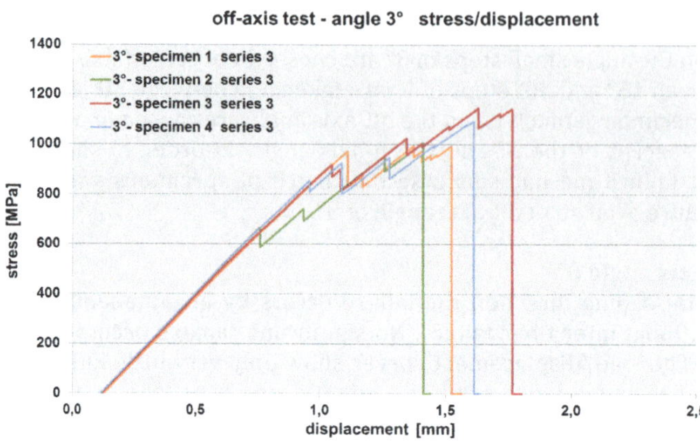

Fig. 3a: Load/displacement curves for 3° specimens

Fig. 3b: 3° specimens after failure

If the specimen length to width ratio allows an inter fiber crack to run from one edge to the other without meeting the clamps (fig. 5) the failure load is significantly decreased compared to the short specimen. The failure occurs instantaneously without any drop of the load, this is, without any previous failure (fig. 4). This means that the results of different failure modes must not be mixed. The strength values should be separated into pure fiber breakage and pure inter fiber failure. Otherwise they are no actual material properties but they are geometry dependent.

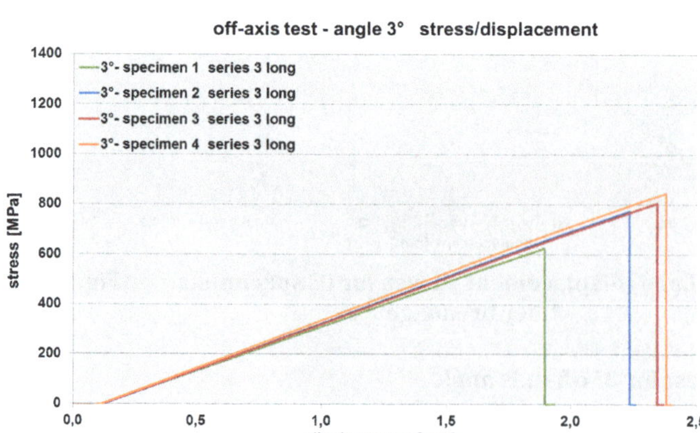

Fig. 4: Load/displacement curves for long 3° off-axis angle specimens – inter fiber failure

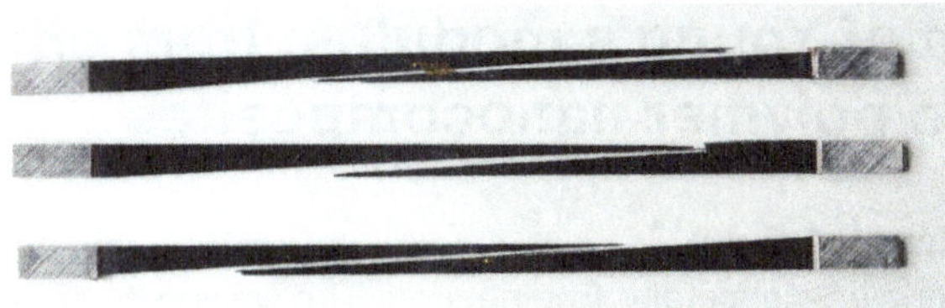

Fig. 5: Long 3° off-axis specimens

2.3 Strength versus off-axis angle

The strong decrease of the strength at small angles (fig. 6) is caused by the weakness of unidirectional plies against stresses perpendicular to the fibers. In an ideal 0° specimen the fibers carry the load and the interface is loaded only by secondary loads. With the inclination of the fibers the interface is loaded directly by the external load even though the stresses are small due to the large fracture surface. In fig. 6 the failure stresses for both failure modes - fiber fracture and pure inter fiber failure - are shown. The change of the failure mode results in a decrease of the failure stress by almost 30%. For an off-axis angle of 30° only 5% of the the maximum failure stress are left. The strength perpendicular to the fibers is only about 2% of the strength in fiber direction.

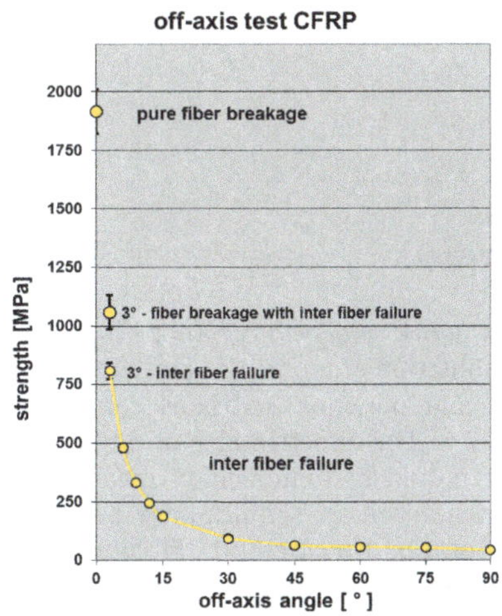

Fig. 6: Strength vs off-axis angle; error bars (small for off-axis angles above 3°)

CONCLUSIONS

Off-axis tests especially when performed with non-rotating clamps suffer from several drawbacks. The measured failure stresses must be divided into two categories: pure fiber breakage and inter fiber failure. This means that the specimens for all off-axis angles except of 0° must be long enough to allow pure inter fiber failure.

REFERENCES

(1) N. J. Pagano, J. C. Halpin, Journal of Composite Materials 2 (1968) 18
(2) J. C. Marin, J. Canas, F. Paris, J. Morton, Composites Part A 33 (2002) 87
(3) G. Odegard, M. Kumosa, Composites Science and Technology 60 (2000) 2917
(4) C. Marotzke, T. Feldmann, Proc. of SAMPE Europe Conference, Paris (2014)

Prediction of Young's modulus: from effective clay clusters to polymer nanocomposites

W. Xu[1,2], Q.H. Zeng[3], A.B. Yu[1,4]

[1] School of Materials Science and Engineering, The University of New South Wales, Australia

[2] College of Materials and Mineral Resources, Xi'an University of Architecture and Technology, P.R. China

[3] The University of Western Sydney, Australia

[4] Department of Chemical Engineering, Monash University, Australia

ABSTRACT

This work aims to predict the properties of clay-polymer nanocomposites by a combination of molecular modelling and micromechanical model. The interface between clay and polymer matrix is considered when determining the effective size of clay clusters. The effective size of different clusters in nylon 6 and their Young's moduli are determined from molecular dynamics simulation. Then, the clay-nylon 6 nanocomposite is considered as a multiphase composite and its Young's modulus is predicted by the rule-of-mixture method. Finally, it is demonstrated that the Young's moduli of nanocomposites with different fractions of effective clay clusters can be predicted.

1 INTRODUCTION

Polymers reinforced with nanoparticles have received extensive attention in the past two decades. Clay minerals have a unique layered structure and properties (e.g., cationic exchangeability), which makes them an ideal reinforcing nanofillers for polymer matrix.[1] Clay clusters (e.g., single-layer clay platelet or partially exfoliated clay cluster) have a high aspect ratio and equivalent length scale to polymer chain structure. Yet, the reinforcement effect depends largely on the degree of clay exfoliation and the dispersion of clay clusters in polymer matrix. Although significant work has been done, the full potential of structural design and property prediction of clay-based polymer nanocomposites has yet to be achieved.[2] The present work aims to extend our capability toward the prediction of Young's modulus of such polymer nanocomposites.

2 SIMULATION METHODS

2.1 Model construction

Molecular dynamics (MD) models were constructed for some representative clay (i.e., montmorillonite) clusters, e.g., fully exfoliated single-layer clay platelet and partially exfoliated two-layer or three-layer clay platelet embedded into surfactants and polymer chains. The MD unit (Figure 1) consists of a clay cluster, ten surfactants of octadecyltrimethyl ammonium (ODTMA) and ninety nylon 6 chains. It has a dimension of a = 25.959 Å, b = 27.0459 Å, c = 175 Å (single clay platelet), 130 Å (two-layer clay cluster), and 150 Å (three-layer clay cluster), $\alpha = \beta = \gamma = 90°$, and a basal spacing of 22.3 Å. The clay model is based on experimental structure and has a formula of $Na_{0.333}$ $[Si_4O_8][Al_{1.667}Mg_{0.333}O_2(OH)_2]$ and cationic exchangeable capacity (CEC) of 90 mmol/100 g. The polymer used is nylon 6, and the surfactant is octadecyltrimethyl ammonium (ODTMA). Figure 2 shows a schematic structure that consists of randomly-oriented effective clay clusters with different size in polymer matrix.

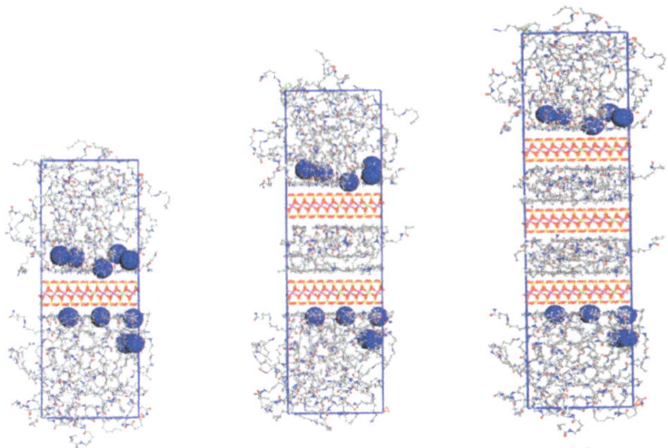

Figure 1. Molecular models of representative clay clusters surrounded by surfactants and polymer matrix: (a) a single-layer clay platelet (N=1); (b) a two-layer clay cluster (N=2); and (c) a three-layer clay cluster (N=3).

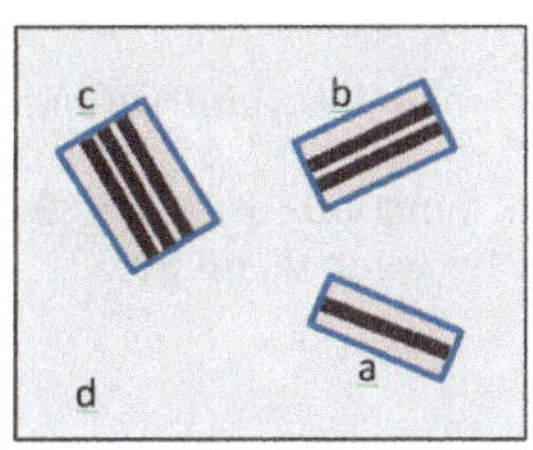

Figure 2. Schematic structures of randomly dispersed effective clay clusters in polymer matrix, where (a) represents fully exfoliated single-layer effective clay cluster, (b) is a partially exfoliated two-layer effective clay cluster, (c) a partially exfoliated three-layer effective clay cluster, and (d) polymer matrix.

2.2 MD Simulation

MD simulations were carried out by Discover module and Forcite module in MATERIALS STUDIO 4.3 (Accelrys Co.). A modified CVFF (Consistent Valence Force Field) force field was employed. For each representative clay cluster, a canonical (constant atom number, volume and temperature (NVT)) MD simulation was initially performed at elevated temperature of 513 K for 10 ps, and then 50 ps at 298 K. The equilibrium state has then been achieved and data were collected to estimate the effective interface thickness and size of clay clusters. Then, using MD simulation, the Young's moduli for pure single-layer clay platelet and effective clay clusters were obtained from stress-strain curves by applying an external tensile stress along one axis direction. Further MD simulations were performed to determine the Young's moduli of effective clay clusters along the longitudinal (z) direction.

2.3 Prediction of Young's modulus of polymer nanocomposites

The Young's moduli of the effective clay clusters obtained above were then used to predict the overall Young's modulus of clay-based polymer nanocomposites by a traditional micromechanical model (i.e., the rules-of-mixtures).

$$E = E_e f_e + E_m f_m \qquad (1)$$

where E_e is the Young's modulus of effective clay clusters. E_m is the Young's modulus of polymer matrix (e.g., nylon 6 of 2.3 GPa). f_e and f_m are the volume fractions of effective clay clusters and polymer matrix, respectively.

Upon the validation, predictions of Young's moduli were made to polymer nanocomposites with either one type of effective clay clusters or a mixture of different types of effective clay clusters under randomly orientation. When a tensile stress is applied to the polymer nanocomposites with an angle θ between the applied force and effective clay clusters. According to the classical Newtonian mechanics, the Young's modulus of effective clay clusters along the direction of applied force E(θ) can be determined by classical Newtonian mechanics as follows:

$$E(\theta) = E_{11} cos\theta + E_{33} sin\theta \qquad (2)$$

where E_{11} is the modulus of fully exfoliated single-layer effective clay cluster along lateral direction, and E_{33} is the Young's modulus of both fully exfoliated and partially exfoliated effective clay clusters along longitudinal direction.

Under a random distribution of the orientation, Young's moduli of fully and partially exfoliated effective clay clusters can be calculated by the following equation:

$$E(\theta) = \int_0^{\frac{\pi}{2}} (E_{11} cos\theta + E_{33} sin\theta) g(\theta) d\theta \qquad (3)$$

where, $g(\theta)$ is the distribution function of angle θ which can be obtained from experimental data and is usually between 0 and 90°.

3. RESULTS AND DISCUSSION

3.1 Effective thickness of the interface

The effective thickness of clay-polymer interface is defined as the distance from which the atomic properties (e.g., atomic density profile, atomic mobility) are different from those in bulk polymer.[3] Based on the atomic concentration profiles of surfactant, polymer and all molecules as well as the mean squared displacement at equilibrium structures, the interface is estimated to be 3 nm on both sides of a clay platelet.

3.2 Young's moduli of clay platelet and effective clay clusters

Based on the MD simulation, the resulting Young's moduli of pure single-layer clay platelet along three directions are E_{11}=277.0 GPa, E_{22}=277.8 GPa, and E_{33}=271.4 GPa ($E_{11}$and E_{22} are the Young's moduli in the lateral directions perpendicular to z axis, E_{33} is the Young's modulus along z axis), respectively. Moreover, for the effective clay clusters with single-layer, two-layer, and three-layer clay platelets, the calculated Young's moduli along the longitudinal direction E_{33} are 36.8±3.22 GPa, 51.7±2.26 GPa, and 60±0.98 GPa, respectively.

3.3 Young's moduli of polymer nanocomposites with one type of clay clusters

In the present calculation, we assumed that there is only one type of effective clusters in the polymer nanocomposites, that is, either fully exfoliated single-layer effective clay cluster (N=1), partially exfoliated two-layer effective clay cluster (N=2), or partially exfoliated three-layer effective clay cluster (N=3). It was shown (Figure 3) that there is a good agreement between the calculated and measured results by Anoukou et al.[4] The Young's modulus increases with the volume fraction of silicate (i.e., montmorillonite). Moreover, the calculated Young's modili of polymer with partially exfoliated clay clusters are closer to the measurements.

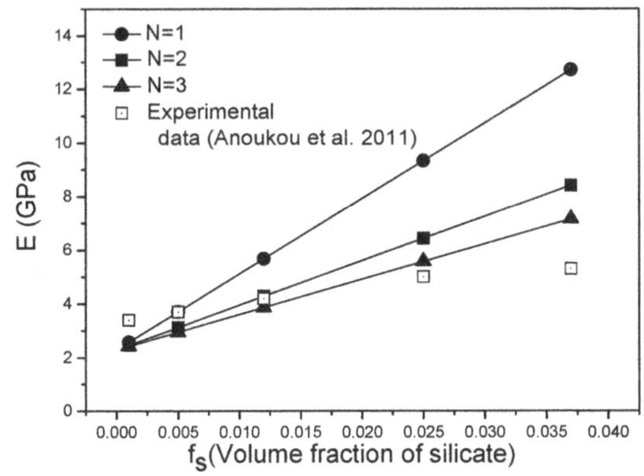

Figure 3. Calculated and measured Young's moduli of clay-nylon 6 nanocomposites with one type of clay cluster under the experimentally observed orientation distribution of angle θ. N=1: fully exfoliated single-layer effective clay clusters; N=2: partially exfoliated two-layer effective clay cluster; N=3: partially exfoliated three-layer effective clay cluster.

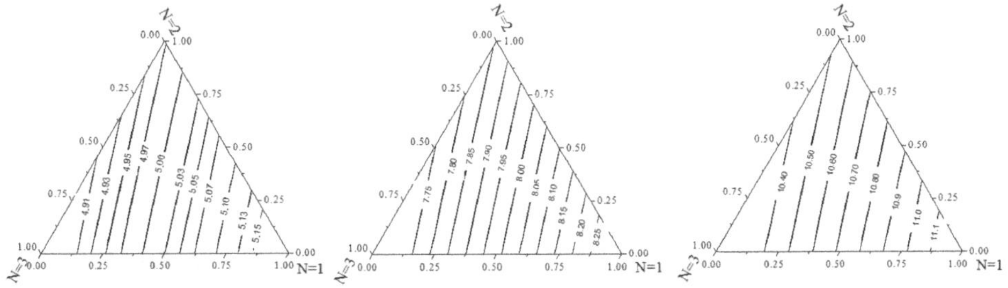

Figure 4. Young's moduli of clay-nylon 6 nanocomposites with 3 types of effective clay clusters and a volume fraction of (from left to right): 1.2%, 2.5%, 3.7%.

3.4 Young's moduli of polymer nanocomposites with a mixture of different effective clay clusters

Furthermore, we assume both fully exfoliated and partially exfoliated morphologies exist with a total volume fraction of clay as 1.2%, 2.5% or 3.7%. The overall Young's moduli of clay-nylon 6 nanocomposites are different under different proportions of these three effective clay clusters. From Figure 4, we can draw the conclusion that the Young's moduli increase with the clay volume fractions from 1.2 vol%, 2.5 vol% to 3.7 vol%. It is also observed that with the increase in fully exfoliated single-layer effective clay clusters, the Young's moduli of nanocomposites increase and reach a maximum value if the polymer nanocomposites consist of fully exfoliated clay clusters, which is in well agreement with many studies.[5, 6]

4 CONCLUSIONS

We have reported herein the use of molecular dynamics simulation to determine the effective size of some representative clay clusters in polymer nanocomposites and their corresponding Young's moduli, and the resulting information is further used to calculate the overall Young's moduli of clay-based polymer nanocomposites. The effective thickness of the interface between clay platelet and polymer matrix in nylon 6 nanocomposites is 3 nm. The calculated Young's moduli of single-layer clay platelet along lateral directions are 277.0 GPa and 277.8 GPa. The Young's moduli along the longitudinal direction are 271.4 GPa, 36.8 GPa, 51.7 GPa, and 60.0 GPa for the single-layer clay platelet, fully exfoliated effective clay cluster, partially exfoliated effective clay clusters with two-layer and three-layer clay platelets, respectively. Such Young's moduli are used to predict the overall Young's moduli of clay-based polymer nanocomposites by the rule-of-mixture method. The predicted Young's moduli of the nanocomposites with randomly dispersed effective clay clusters agree well with the measured ones. This approach can be applied to predict the mechanical properties of other types of nanoparticle-polymer nanocomposites.

ACKNOWLEDGEMENT

The authors are grateful to the ARC and the DIISR for the financial support through the Australia-India collaborative research program. The authors would like to thank Professor Donald Paul for his encouragement and helpful suggestion.

REFERENCES

[1] Zeng, Q.H., et al., Clay-based polymer nanocomposites: research and commercial development. Journal of Nanoscience and Nanotechnology, 2005. 5(10): p. 1574-1592.

[2] Zeng, Q.H., A.B. Yu, and G.Q. Lu, Multiscale modeling and simulation of polymer nanocomposites. Progress in Polymer Science, 2008. 33(2): p. 191-269.

[3] Xu, W., Q.H. Zeng, and A.B. Yu, Young's modulus of effective clay clusters in polymer nanocomposites. Polymer, 2012. 53(17): p. 3735-3740.

[4] Anoukou, K., et al., On the overall elastic moduli of polymer-clay nanocomposite materials using a self-consistent approach. Part II: Experimental verification. Composites Science and Technology, 2011. 71(2): p. 206-215.

[5] Lan, T. and T.J. Pinnavaia, Clay-Reinforced Epoxy Nanocomposites. Chemistry of Materials, 1994. 6(12): p. 2216-2219.

[6] Ray, S.S. and M. Okamoto, Polymer/layered silicate nanocomposites: a review from preparation to processing. Progress in Polymer Science, 2003. 28(11): p. 1539-1641.

Effects of temperature and strain rate on fracture toughness of nano-rubber modified epoxies

F. Xu[1,2], H.Y. Liu[1], X.S. Du[1], Y.W. Mai[1], W.G. Guo[2]
[1] Center for Advanced Materials Technology (CAMT), School of Aeronautics, Mechanical & Mechatronic Engineering J07, The University of Sydney, Australia
[2] School of Aeronautics, Northwestern Polytechnical University, P.R. China

ABSTRACT

An investigation was conducted to characterize the toughening effects of nano-sized rubber particles on epoxies under three temperatures and two loading rates. The results obtained showed that addition of nano-rubber particles significantly toughened epoxies at 20°C and -50°C under quasistatic loading (1 mm/min), but this effect was diminished at 80°C. With a higher loading rate at 4 m/s, the toughness of rubber modified epoxies was even lower than that of neat epoxy, contrary to what was obtained from the low loading rate tests (at 1 mm/min).

Keywords: Impact fracture, Single edge-notched beam (SENB), Temperature effect, Nano -rubber particles

1 INTRODUCTION

As a widely used thermosetting polymer, epoxy resins have continuously attracted great attention and are broadly used in many fields of engineering due to their high modulus and strength. However, because of their three-dimensional molecular network structure, most epoxies display poor fracture resistances for crack initiation and growth. For many years, an effective technology for toughening epoxies has been the addition of a second phase particle [1-3]. Among these, micro-/nano-sized rubber particles have been proven effective tougheners through different established toughening mechanisms [4]. Owing to the increasing applications of epoxies under some extreme conditions, the influences of temperature and impact loading for particle modified epoxies become important. Low and Mai [5] examined the failure mechanisms of micro-sized rubber particles modified epoxy over a wide range of strain rates (10^{-6} s^{-1} -10^{-2} s^{-1}) and temperatures (-80 to 60 °C) and indicted that thermal blunting accounted for toughness improvement under high strain rate (113 s^{-1}). Cardwell and Yee [6] studied temperature effects on the fracture toughness of epoxies modified by 300 nm rubber particles at a given low testing rate. Their results showed that both α-relaxation andβ-relaxation processes can significantly affect the fracture toughness enhancement. Recently, Brown *et al.* [7] also investigated different contents of 250 nm CSR (core-shell rubber) on the increase of Charpy fracture strength under cryogenic and room temperatures. They found that 5 wt% CSR improved the Charpy impact strength of epoxy; however, the Charpy impact strength was reduced by adding only 1 and 3 wt% CSR. Nonetheless, they did not address the issue of nano- rubber toughening of epoxies owing to different loading rates and temperatures. Hence, in this work, we chose 100 nm rubber particles as a toughener for epoxy to examine its effects on fracture toughness under three different temperatures and two loading rates.

2 EXPERIMENTAL METHOD

2.1 Specimen preparation

The materials for the samples were based on a single-part epoxy formulation. The epoxy resin was standard diglycidyl ether of bisphenol A (DGEBA) with an epoxide equivalent weight (EEW) of 185g/mol, Araldite-F, supplied by Sigma-Aldrich Australia. The curing agent used was the cycloaliphatic secondary amine piperidine from Sigm-Aldrich. The spherical rubber particles averaged ~100 nm in diameter and 25 wt% concentration in bisphenol A resin was supplied in master batch by Kaneka Corporation, Japan. The nano-rubber modified epoxy resin was prepared by mixing plain DGEBA resin with 6 wt% nano-rubber. The prepared materials were heated to 60 ^{0}C and mixed continuously by magnetic stirring until a clear mixture was obtained, which was then degassed at 100 ^{0}C by evacuating for 1 h to ensure no bubble was trapped in the solution. Subsequently, a stoichiometric curing agent with a ratio of 100:5 was added and gently stirred. The final mixture was poured into the preheated mould for curing at 120 ^{0}C for 16 h. Following ASTM D5045-99 [8] single edge-notched beams (SENB) measuring 62(length) x 12.5(width) x 5.5(thickness) mm3 were used for the fracture toughness tests. A natural crack was introduced to each SENB specimen by tapping a razor blade to minimize the effects of residual stress and plastic deformation around the crack tip. Before testing, all specimens were annealed at 100^{0}C for 3 h to remove any residual stress which might be introduced during the fabrication process.

2.2 Fracture toughness measurements

To study temperature effects (at -50^{0}C, 20^{0}C and 80 ^{0}C) on the fracture toughness of nano- rubber/epoxies, an environmental chamber was used. Two loading rates of 1 mm/min (using an Instron 5567) and 4m/s (using an impact tester ITR-2000) at 20^{0}C were also studied. These loading rates correspond to equivalent strain rates of $5\times10^{-4}s^{-1}$ and $120s^{-1}$ respectively, based on the SENB dimensions [9]. K_Q can be determined from [8]:

$$K_Q = \frac{P}{B\sqrt{W}} f(a/W) \tag{1}$$

$$f(a/W) = f(\alpha) = 6\alpha^{1/2} \frac{[1.99 - \alpha(1-\alpha)(2.15 - 3.93\alpha + 2.7\alpha^2)]}{(1+2\alpha)(1-\alpha)^{3/2}} \tag{2}$$

where P is maximum load at crack initiation, W width of specimen, a initial crack length and f (α) is non-dimensional shape factor. If both specimen thickness and crack length satisfy the required plane strain conditions [8], $K_Q = K_{IC}$. Otherwise, K_Q stands. In the test results presented below, only those at 80^{0}C are K_Q values; all others are valid K_{IC} values. After testing, the fracture surfaces were also examined with optical microscopy.

3 RESULTS AND DISCUSSION

The effect of two loading rates on toughness and relevant typical load-displacement curves of E and R6 at 20 ^{0}C are shown in Fig. 2(b) and Fig.1 respectively. The opposite trends observed are interesting. For E, the toughness increases, but for R6 it decreases, from $5\times10^{-4}\,s^{-1}$ to $120s^{-1}$. The high rate induced temperature rise [5] is insufficient to cause crack-tip thermal blunting (e.g., 20 ^{0}C in E) and the rate effect becomes dominant. In neat epoxy, both Young's modulus and yield strength increase with loading rate, which explains why the toughness is increased (by ~30%) using a critical crack opening displacement fracture criterion. However, there is little difference in the fracture surfaces shown in Fig. 3(a) and 3(c). In rubber-modified epoxy, the distinctly different failure mechanisms are displayed in Fig. 3(d) and 3(f). At $5x10^{-4}$ s^{-1}, a stress-whitened fracture surface owing to rubber cavitations and matrix shear bands is displayed yielding a high toughness; but at 120 s^{-1}, the fracture surface is brittle because the high strain rate has suppressed these deformation mechanisms.

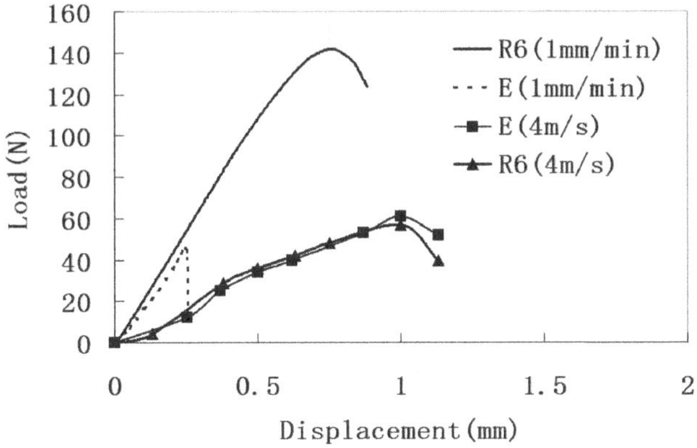

Figure 1. Typical load-displacement curves under two loading speeds of 1mm/min (strain rate: $5x10^{-4}$ s^{-1}) and 4m/s (strain rate:120 s^{-1})

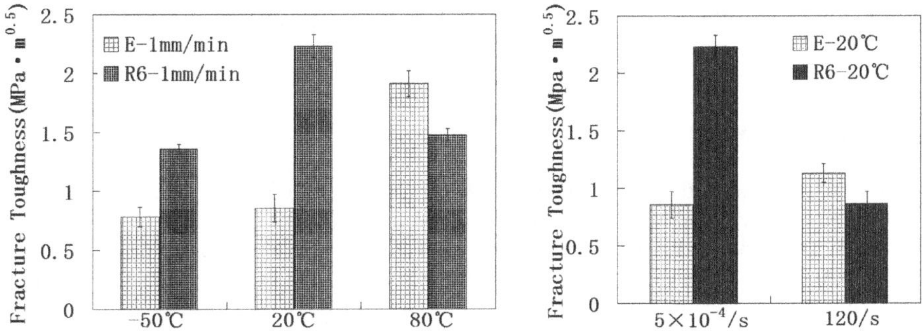

(a) Effect of temperature **(b) Effect of strain rate**

Figure 2. Effect of temperature and strain rate on fracture toughness

The temperature effect on the fracture toughness of neat epoxy (E) and rubber modified epoxy (R6) tested at $5x10^{-4}$ s^{-1} is shown in Fig. 2(a). It can be seen that the toughness of E increases slightly when the temperature changes from -50 ^{0}C to 20^{0}C but it rises sharply at 80^{0}C. In contrast, R6 displays higher toughness values at -50^{0}C and 20^{0}C compared to E. A very high toughness (2.6-fold increase of E) is achieved at 20^{0}C, which is consistent with our early work using compact tension samples [1]. At 80^{0}C, the toughness of R6, however, drops substantially and is even less than that of E.

At -50^{0}C and 20^{0}C, the fracture surfaces of E (Fig. 3(a)) are smooth and brittle with many short ridges at the initial crack-tip. But at 80^{0}C, obvious river pattern formed from these ridges is widespread on the fracture surface (Fig. 3(b)) with a distinct thickness reduction. These fractographs are consistent with the toughness results given in Fig. 2(a). For R6 with 6 wt% nano-rubber in epoxy, stress-whitening (indicative of high toughness) occurs at 20°C (Fig. 3(d)) but it disappears at 80°C and is replaced by the river lines (Fig. 3(e)). Again, there is a thickness reduction in R6 but its toughness is lower than at 20°C.

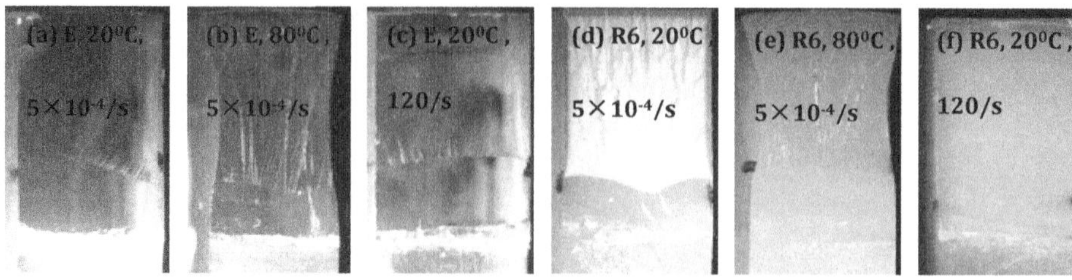

Figure 3. The overall fracture surface morphology of tested specimens

4 SUMMARY AND CONCLUSIONS

The toughening effect of nano-rubber particles under different temperature and loading rate were examined in terms of fracture toughness. Two major conclusions can be made below:

(a) Nano-rubber toughens neat epoxy at a low strain rate of $5x10^{-4}/s$, but it deteriorates the toughness of epoxy at the high strain rate of 120 $s^{-1}$of 4m/s.

(b) Nano-rubber offers significant toughness increase to epoxy at 20°C and -50°C; however, it deteriorates the toughness of epoxy at 80°C.

ACKNOWLEDGEMENT

Feng Xu would like to thank the Chinese Scholarship Council for supporting this work at the Center for Advanced Materials Technology, University of Sydney. H-Y Liu also wishes to thank the University of Sydney for financial support via a Bridging Support Fellowship. Finally, the authors acknowledge Kaneka Corporation, Japan for supply of nano-rubber particles for this study.

REFERENCES

[1] Liu, H.-Y., Wang, G.T., Mai, Y.-W. & Zeng, Y. (2011) Composites: Part B, 42:2170-2175.
[2] Kinloch, A.J., Lee, J.H., Taylor, A.C., Sprenger, S., Eger, C. & Egan, D. (2003) J Adhesion, 79:867-873.
[3] Deng, S.Q., Ye, L. & Friedrich, K. (2007) J Mater Sci, 42:2766-2774.
[4] Garg, A.C. & Mai, Y.- W. (1998) Compos Sci Technol, 31:179-223.
[5] Low, I.M. & Mai, Y.-W. (1989) J Mater Sci, 24:1634-1644.
[6] Cardwell, B.J. & Yee, A.F. (1993) Polymer, 34:1695-1701.
[7] Brown, H.R., Schneider, J.A. & Murphy T.L. (2014) J Mater Sci, 48: 1279-1296.
[8] ASTM D5045-99. Standard test methods for plane-strain fracture toughness and strain energy release rate of plastic materials: 1999.
[9] Birch, M.W. & Williams, J.G. (1978), Int. J Fracture, 14:69-84.

Effect of silane coupling agent and air plasma treatment on interfacial shear strength of carbon fiber/polyphenylene sulfide composites

B.Y. Liu [a,b], X.J. Wang [c,d], H.Y. Liu [d], G. Zhang [c], S.R. Long [c], J. Yang [c,d], Q. Yang [a]
[a] College of Polymer Science & Engineering, Sichuan University, P.R. China
[b] College of Chemistry and Chemical Engineering, Henan University, P.R. China
[c] Institute of Materials Science & Technology, Sichuan University, P.R. China
[d] Centre for Advanced Materials Technology, School of Aerospace, Mechanical and Mechatronic Engineering, The University of Sydney, Australia

ABSTRACT

Interface modification of carbon fiber reinforced polyphenylene sulphide (PPS) composites was studied by microbond test. The effects of silane coupling agent treatment and atmospheric plasma treatment on the interfacial adhesion and micromechanics of PPS/CF composites were investigated. The results showed that the apparent interfacial shear strength (τ_{app}) of CF/PPS with coupling agent treated carbon fiber was increased by 13.2% compared to the untreated samples. Meanwhile, the corresponding macro tensile and bending strength and modulus of PPS/CF composites were also increased. The τ_{app} of samples with plasma treated carbon fiber reduced from 40.9±3.9 to 35.3±5.0MPa, however, which of the micro-composites with plasma treated PPS increased by 17.1%.

1 INTRODUCTION

Polyphenylene sulfide (PPS) is a kind of semi-crystalline engineering thermoplastic with high chemical resistance, excellent thermal stability, flame retardance and good mechanical properties, etc. Therefore, PPS is an ideal high performance matrix for thermoplastic composite [1]. Carbon fiber (CF) has been widely used as a reinforcement of PPS due to its excellent comprehensive performance. However, the strength of CF reinforced PPS composite is always much lower than the theoretically predicted value due to the weak interface [2]. Thus, the research on interface modification of carbon fiber reinforced PPS composites is very important in the composites applications.

Plasma and silane coupling agent treatments are widely used in surface modification of reinforcing fiber and resin fiber to improve the interfacial adhesion. However, the quantitative research on what content the plasma and silane treatment can improve is still limited. In this paper, we present our recent work on the interfacial adhesion behaviour of PPS/CF composites. Either air dielectric barrier discharge (DBD) plasma or silane coupling agent treatment was employed to modify the PPS/CF interface adhesion.

2 EXPERIMENTAL

Poly(phenylene sulfide) resin, injection grade for silane coupling agent modification (Philip Petroleum Co., USA) and fiber grade for plasma modification (Poly Plastics, Japan), and commercial carbon fiber (T700SC-12000, Toray Co, ltd, diameter of 7μm) were employed in this study. Microbond test was used to measure the interfacial adhesion between carbon fiber and PPS resin. A novel, simple and efficient microdroplet preparation procedure has been established for thermoplastic composite (As shown in Figure 1) [3, 4]. Variations of surface morphology and chemical composition were analysed by scanning electron microscopy (SEM) and X-ray photoelectron spectroscopy (XPS).

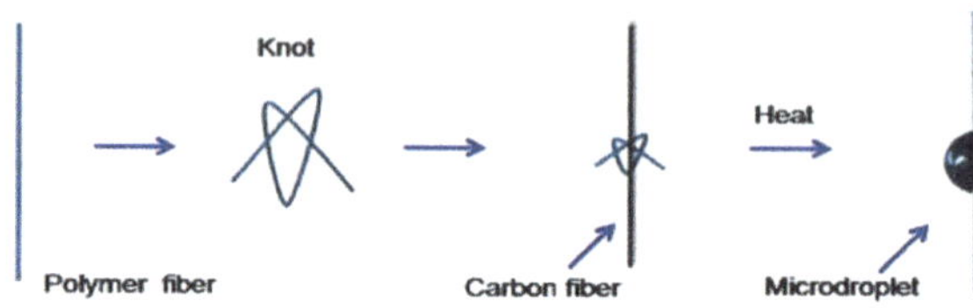

Figure 1. Forming procedure of thermoplastic resin droplet on carbon fiber monofilament and an optical microscope photograph of one microdroplet

The apparent interfacial shear strength (τ_{app}), calculated from Equation 1[5], was traditionally used to estimate the efficiency of matrix and/or fiber surface modification.

$$\tau_{app} = \frac{F_{max}}{\pi d_f l_e} \tag{1}$$

Where d_f is the fiber diameter, l_e is the embedded length.

3 EXPERIMENTAL RESULTS AND DISCUSSION

Generally, there are three main factors in determining adhesion between fiber and matrix: (1) covalent bonds between the fiber and the matrix (chemical bonding), (2) wettability of the fiber by the matrix (physical bonding), and (3) micro-mechanical interlocking [6]. For thermoplastic polymer composite, there may often be little or no covalent bonds formed across the interface, physical bonding and micromechanical interlocking are the primary factors influenced the interfacial strength of fiber/matrix.

3.1 Effect of coupling agent on τ_{app}

Silane coupling agent bears alkoxysilane groups and a large number of functional groups which ensure a good compatibility between the reinforcing element and the polymer matrix. There are only a few articles relating to the influences of silane coupling agent on the interfacial shear strength and mechanical performance of carbon fiber reinforced polymer composites, especially for the CF/PPS composites. In this study, CF was first modified by surface coating of silane coupling agent to improve the interfacial adhesion between carbon fiber and PPS resin. Microbond, tensile and flexural tests were conducted to verify the effects of silane coupling agent on the CF/PPS interfacial adhesion and mechanical properties.

The effect of silane coupling agent on τ_{app} values of the CF/PPS composites is presented in Figure 2. The average shear strength ($\bar{\tau}_{app}$) of CF/PPS with coupling agent treated carbon fiber is 35.6±4.9MPa, which is 13.2% higher than that of the untreated samples which illustrates that the positive effect of the coupling agent on the interface bonding between carbon fiber and PPS resin. Meanwhile, the corresponding tensile and flextural properties of PPS/CF composites were tested by a universal testing machine (AG-10TA, Seiko), according to Chinese Standard GB/T1040-1992 and GB/T9341-2000, respectively, at 20°C. As shown in figure 3(a), for the 20wt% CF reinforced composites, the tensile strength was improved from 129.1MPa to 174.7 MPa when 3% coupling agent was added. Meanwhile, the tensile modulus was also increased from 3.4GPa to 4.3GPa. Figure 3(b) shows that the flexural strength and modulus were also increases by 23.2% and 15.7%, respectively, at the presence of the coupling agent.

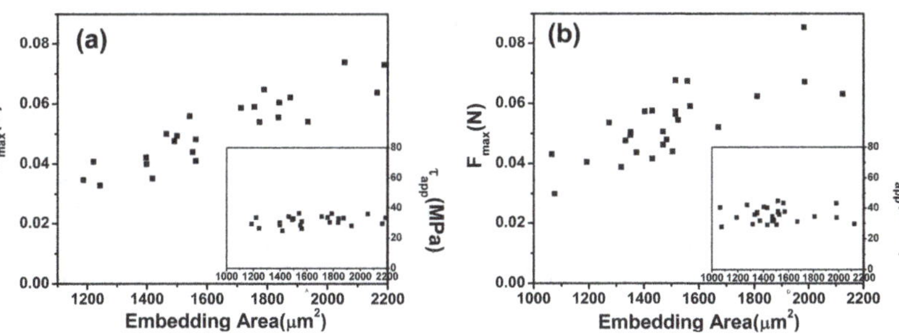

Figure 2. Microbond test results of CF/PPS composites with carbon fiber with (a) and without (b) KH560 coating

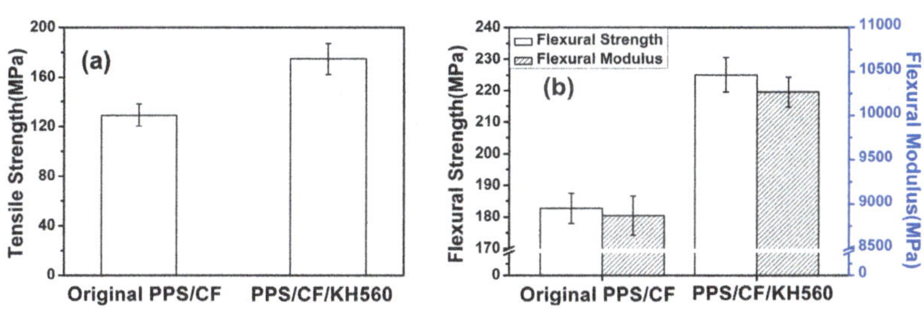

Figure 3. Tensile(a) and flexural properties(b) of PPS/CF composite with/without KH560

As shown in figure 4, coupling agent could form chemical bonding interface layer (as white arrows marked) between fiber and matrix which may be a reason for the modification of interfacial strength and improvement of the mechanical properties.

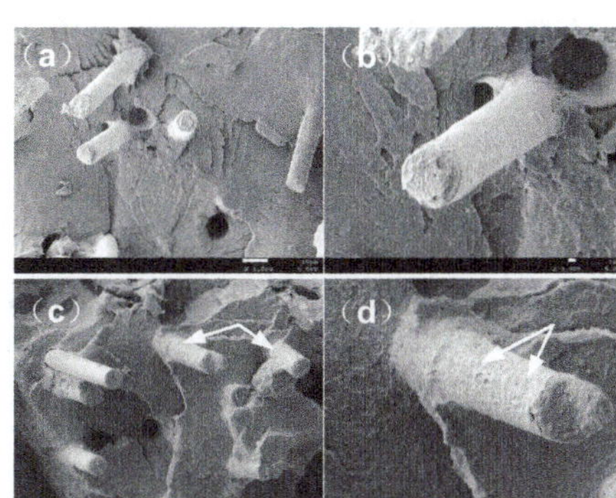

Figure 4. SEM photographs of tensile fractured surfaces of CF/PPS composites (a),(b)-original PPS, (c),(d)-PPS with KH560

3.2 Effect of plasma treatment on τ_{app}

For many types of fibres, such as organic fibers and glass fiber, plasma treatment improves fiber-matrix adhesion largely by introducing polar, excited groups, and even a new polymer layer. In some cases, plasma treatment can roughen the surface of fiber to increase mechanical interlocks between the fiber and the matrix. Nevertheless, for the case of CF, the situation is more complicated [7]. CF and PPS fiber were treated by air dielectric barrier discharge (DBD) plasma at atmospheric pressure in this study. The discharge voltage and discharge time for CF and PPS fiber were 1kW/60s and 1kW/ 30s, respectively. Comparative tests of PPS/CF with treated carbon fiber and treated PPS fiber samples were made to investigate the influence of plasma treatment on the interfacial shear strength of CF/PPS micro-composite.

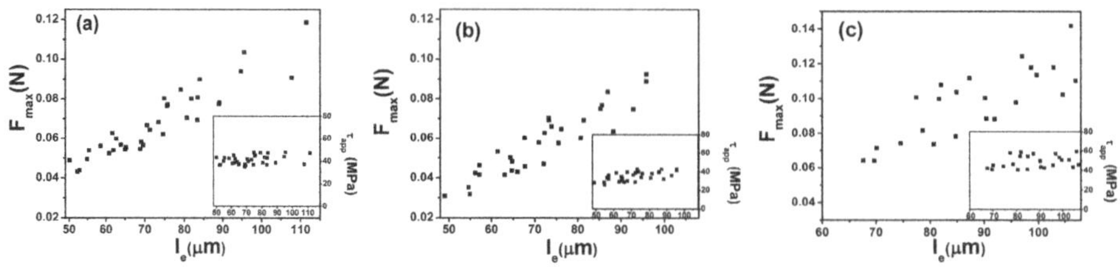

Figure 5. Comparison of maximum pull-out force(F_{max}) and the corresponding interfacial shear strength (τ_{app}) as a function of embedding length(l_e) for microbond specimens with different materials plasma treated (a)-untreated; (b)-plasma treated CF; (c)-plasma treated PPS

The values of $\bar{\tau}_{app}$ of three groups are given by Figure 5 (a)-(c), which are 40.9±3.9, 35.3±5.0 and 47.9±6.7MPa, respectively. It is clear that composites reinforced by plasma treated fibers exhibited lower values of interfacial shear strength compared to that reinforced by untreated fibers. This result is different from most previous studies, in which some possible reasons for the enhancement of interfacial shear strength by plasma treatment were summarized, such as high surface roughness of fiber, introduction of chemical composition on the fiber surface, modification of fiber surface wettability. To figure out the reason of this reduction in interface strength, SEM and XPS were conducted to obtain the variations of morphology and chemical composition on the surface of carbon fiber before and after plasma treatment.

The XPS results of C1s spectra of CF with or without plasma treated were shown in Table 1. The O/C atomic ratio on the surface of CF was increased from 20.2% to 28.5% after plasma treatment. Some reactive oxygen groups, ester for example, were increased obviously. All of these lead to an increase of specific surface area and surface wettability. However, PPS is a kind of nonpolar thermoplastic polymer with an ordered alternating arrangement of phenylene and sulfide atoms. There is little chemical bonding between the matrix and CF with reactive groups after plasma treated, which leads to worse compatibility and the decrease of τ_{app}. In addition, part of sizing agent coated on the surface of CF, which is benefit to the interfacial bonding between CF and PPS, was removed by plasma etch, which may be another reason for the decrease of τ_{app}.

Table 1. XPS C1s spectra of carbon fiber surface with and without plasma treatment and correlative carbon-containing groups on carbon fiber surface

CF	Content of functional group (%) and O/C ratio				
	C-C,C-H (284.8eV)	-C-OH,-C-O-C (286.5eV)	-C=O (286.9-287.3eV)	-COOH,-COOR (289.4eV)	O/C
Un-treated	56.9	39.3	3.8	—	20.2
Plasma-treated	57.8	31.2	12.6	8.4	28.5

However, the average interfacial shear strength of samples with plasma treated PPS fiber increased by 17.1% (as shown in Figure 5). From XPS analysis (Table 2) we can see that the plasma treatment successfully increased the concentration of sulfide oxides (S=O and O=S=O) on the surface of polyphenylene sulfide fiber, which could change the wettability of the resin on the carbon fiber surface and finally improve interfacial mechanical properties of the system. Contrast with the research mentioned above, it can be concluded that the nonpolarity of PPS molecular structure is the reason responsible to the poor wettability and poor interfacial adhesion between carbon fiber and PPS. In light of this finding, some new methods for the modification of CF/PPS composites could be put forward.

Table 2. Concentration of different sulphur-containing groups of PPS fiber surface before and after plasma treatment

PPS	sulfur-containing groups(%)		
	-S-(163.5eV)	S=O(165.6eV)	O=S=O(168.9eV)
Un-treated	100	0	0
Plasma-treated	44.7	8.9	46.4

4 CONCLUSIONS

Interface modification by air plasma treatment and coating silane coupling agent was carried out to improve the interfacial adhesion between carbon fiber and PPS resin. The existence of coupling agent can improve the interface bonding performance between carbon fiber and the PPS resin and improve tensile and bending performance of PPS/CF composites as well. It was found that the interfacial shear strength of samples with carbon fiber plasma treated reduced by 13.7%, due to the remove of the size layer coated on the surface of CF, though the surface roughness and O/C atomic ratio of carbon fiber were increased after plasma treatment. However, the interfacial shear strength of samples with plasma treated PPS fiber increased by 17.1%. The nonpolarity of PPS molecular structure should be responsible for the poor interfacial adhesion between carbon fiber and PPS.

REFERENCES

(1) Yang, J., Wang, X.J. (2006) Published by Chemical Industry Press of China.
(2) Zhang, K., Zhang, G., Liu, B.Y., Wang, X.J., Long, S.R., Yang, J. (2014) Compos. Sci. Technol. 98, 57-63.
(3) Liu, B.Y., Wang, X.J., Long, S.R., Yang, J. (2014). Compos. Interfaces 21(4), 359-369.
(4) Liu, B.Y., Liu, Z., Wang, X.J., Zhang, G., Long, S.R., Yang, J. (2013) Polym. Test. 32,724-730.
(5) Zhandarov, S., Mäder, E. (2005) Compos. Sci. Technol. 65, 149-160.
(6) Coffey, A. B., O'Bradaigh, C. M., Young, R. J. (2007) J. Mater. Sci. 42(19), 8053-8061.
(7) Li, R.Z., Ye, L., Mai, Y.W. (1997) Composites Part A, 28, 73-86.

Electrical conductivity and mechanical performance of polymer/graphene composites developed by two compounding methods

S. Araby, Q. Meng, J. Ma
School of Engineering, University of South Australia, Australia

ABSTRACT

In this work we demonstrated how to utilise graphene platelets by melt compounding and solution mixing to develop polymer composites based on a commercial elastomer, styrene-butadiene rubber (SBR) which has inherently low stiffness and strength. The electrical volume resistivity decreased abruptly by several orders of magnitude at ~5.3 vol% in the case of composites made by solution mixing and at ~17 vol% for the melt-compounding composites, confirming the formation of conductive networks in the matrices. Nearly all the mechanical measurements for the solution-mixing composites except the fracture strain demonstrated higher improvements than those for the melt-compounding composites.

1. INTRODUCTION

Polymers have witnessed the most rapid increase in industrial applications over the past decades due to their high specific strength and low manufacturing cost. Elastomers (rubbers) are a type of polymers that can elongate 1000%, but they are inherently limited by low tensile strength and Young's modulus and lack of electrical and thermal conductivity. Conducting elastomers are vital for aerospace and biomedical, micro-electromechanical systems [1]. Compounding elastomers with fillers is a widely used method to improve the mechanical and physical properties and also to provide new functionalities such as electrical and thermal conductivity.

Graphene—a flat sheet of carbon just one atom thick—shows promise for functional nanocomposites: it is the stiffest and strongest materials measured to-date (Young's modulus 1 TPa and intrinsic strength 130 GPa) while upon loading it can elongate 25% of its original length. Graphene carries higher electrical/thermal conductivity and provides higher reinforcement [2-5] than multi-walled carbon nanotubes; it also carries non-toxicity, isotropic electrical/thermal conductivity on the graphene plane, and low viscosity when compounded with a polymer.

There are three methods for compounding nanofillers with elastomers including *in-situ* polymerisation, solution mixing and melt compounding. Because the 1st method does not suit well industrial applications, we focus on the other two processes in this study. Most of the work in this field used graphene oxide as a precursor, which takes much time to produce large amount, produces much acids and solutions as side products impacting on the environment while adding more costs to the fabrication process. Our group has developed a facile route to fabricate graphene platelets (GnPs, each platelet comprising 2–4 graphene layers), by thermal shocking a graphite intercalation compound and then sonicating the expanded product in solvent such as acetone.

In this article, we will study the effect of fabrication methods on the structure and properties of the SBR composites, which includes: (i) employing a common furnace with an ultrasonication bath to produce GnPs, (ii) fabricating SBR/GnP composites by melt compounding and solution mixing, and (iii) investigating the effect of fabrication on the morphology and the structure–property relations of the composites.

2. EXPERIMENTAL SECTION

2.1. Materials

Styrene-butadiene rubber (SBR 1502, 23.5% styrene and 76.5% butadiene) was provided by Jilin Petrochemical Limited, China, with a Mooney viscosity of ML(1+4) at 100 °C= 52. A commercial graphite intercalation compound (GIC, Asbury 3494) was kindly provided by Asbury Carbons (Asbury, NJ). Curing chemicals were purchased from market, for the vulcanization process. Tetrahydrofuran was purchased from Sigma-Aldrich (Australia).

2.2. Preparations

Graphene platelets (GnPs). 1 g of a graphite intercalation compound (GIC) was carefully moved into a crucible preheated in a common furnace at 700 °C, and treated for 1 min, followed by cooling. Safety procedures, such as placing the furnace in a fume cupboard to prevent nanoparticle inhaling hazard and wearing safety glasses, respirator, and heat resistant gloves, are required in this process. The expanded product was immersed in tetrahydrofuran (THF, 1 wt %) in a metallic container and treated in an ultrasonication bath (200 W at 42 kHz) for 1 hr below 30 °C. Upon sonication, the expanded product in suspension was able to split into platelets of 2–4 nm in thickness, as reported in our previous works [6, 7]. This suspension would be used for the following preparation by solution mixing. For melt compounding, GnPs have been filtered, collected and then dried in a ventilated oven at 120 °C overnight.

Elastomer/GnP nanocomposites: To produce a solution-mixed composites, a given weight of SBR was dissolved in THF (4 wt %) and heated to 90 °C by an oil bath and a round-bottom flask equipped with a condenser. To prevent SBR sticking to the flask bottom, we split it into many small pieces, which were added one by one with magnetic stirring. The SBR solution was mixed with the aforementioned suspension under mechanical stirring at 200 rpm, followed by one-hour sonication below 30 °C. Then ~60% THF was evaporated by heating up to 60 °C with mechanical stirring. Using ethanol, the mixture was precipitated, collected, washed and dried at 60 °C using a ventilated oven for 12 hrs.

Melt compounding was conducted by a two-roll mill. GnPs were added to SBR sheet gradually to avoid loss. Curing agents were mixed with the blend using a compounding formula, where 100 g of SBR is mixed with 2 g sulfur, 0.5 g of Dibensothiazole disulfide (DM), 0.5 g of Diphenylguanidine (D) and 0.2 g of Tetramethylthiuram disulfide (TT). Vulcanization was carried out at 150 °C for 23 min under 3 MPa. In this study, the nanocomposites prepared by solution mixing are denoted s-nanocomposites and those by melt compounding are named m-nanocomposites. The density of graphene was taken as that of graphite, 2.26 g/cm^3, and the matrix density was assumed to be 1.043 g/cm^3; thus we were able to convert wt% to vol% using eq.1:

$$V_f = \frac{\rho_m W_f}{\rho_f(1 - W_f) + \rho_m W_f} \quad \text{(Eq.1)}$$

2.3. Characterization

The characterization techniques are presented elsewhere [8].

3. RESULTS AND DISCUSSION

3.1. Morphology

Figure 1 shows X-ray diffraction patterns of the graphite intercalated compound (GIC), expanded product, neat SBR and its composites at 2.4 vol%. SBR is amorphous and thus shows no diffraction. All other materials show a diffraction at 2θ=26.5° corresponding to the graphitic structure at (002) plane. It also identifies the discrepancy of internal structure of the composites prepared by both methods. Although a diffraction at 2θ = 26.5° is seen for both composites, a higher diffraction intensity is seen for the composite fabricated by solution mixing (s-nanocomposites). We attribute it to the change of coherence of the layered structure of GnPs in the matrix as shown in Figure 1 inset. Solution mixing retained the lateral dimension of GnPs, but they broke into smaller GnPs in melt compounding. It is known that larger lateral dimension of nanolayers should have more coherence and hence higher diffraction intensity.

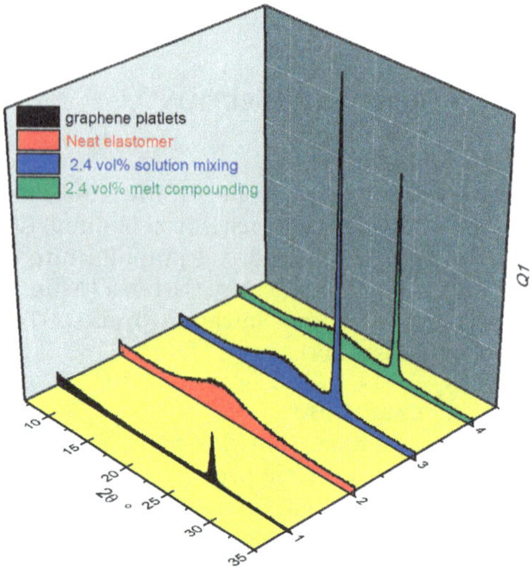

Figure 1. XRD patterns of neat SBR, graphene platelets and their nanocomposites.

Figure 2 contains representative TEM micrographs of the 2.4 vol% SBR/GnP nanocomposite prepared by solution mixing (Figure 2a) and melt compounding (Figure 2b). In Figure 2a, many sites are lightly grey in color as pointed out by blue arrows, and we can also see dark sites by red arrows. The grey areas could be monolayer or few-layer graphene exfoliated in the matrix during preparation. The dark sites may refer to stacked graphene platelets during preparation. On the other hand, graphene stacks can be readily found from the matrix in Figure 2b. In brief, GnPs exfoliate better in the solution-prepared nanocomposites, corresponding to the following analysis on the functional and mechanical properties.

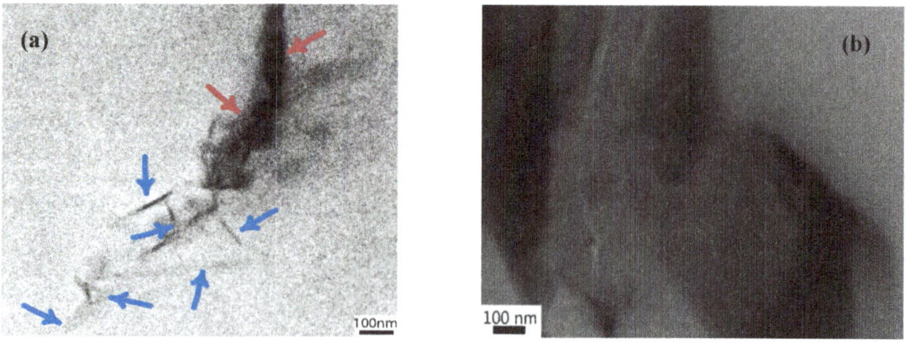

Figure 2. TEM micrographs of 2.4 vol% nanocomposite prepared by: (a) solution mixing [9] and (b) melt compounding [10].

3.2. Electrical conductivity

Figure 3 shows the electrical volume conductivity of these two groups of composites. The electrical volume conductivity values increase abruptly by several orders of magnitude at ~5.3 vol% in the case of solution mixing and ~17 vol% for melt compounding composites, confirming the formation of a conductive network inside the elastomer matrix. The divergence of percolation thresholds between the two methods is mainly because solution mixing is more effective to exfoliate GnPs and also retain their aspect ratio.

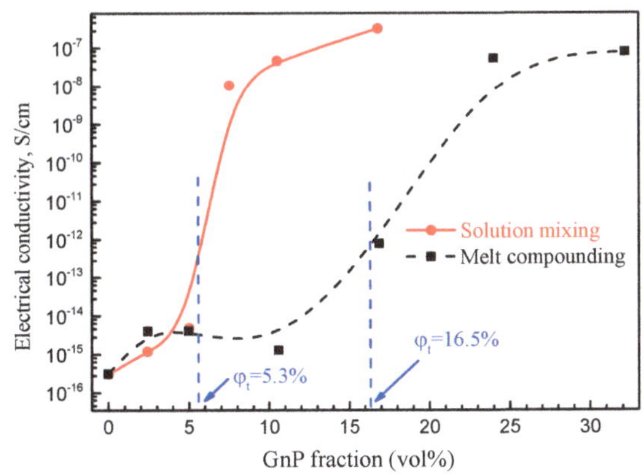

Figure 3. Electrical volume conductivity of SBR nanocomposites prepared by solution mixing and melt compounding.

3.3. Mechanical properties

Figure 4 contains the mechanical properties of neat SBR and its composites prepared by the two methods. Nearly all the measurements except fracture strain demonstrate higher improvements than m-nanocomposites; at 10.5 vol%, tensile strength, Young's modulus and tear strength for s-nanocomposite are respectively improved 390%, 580% and 500%, compared to 137%, 145% and 193% for m-nanocomposite.

The improvements in mechanical performance are attributed to the exceptionally high mechanical performance of graphene — Young's modulus ~1 TPa, ultimate tensile strength ~130 GPa and an ability to elongate a quarter of its original length upon loading, which are far higher than those of elastomers (~1 MPa in Young's modulus).

The key factors for reinforcement include a uniform dispersion of fillers and their possible orientation in the matrix, a large interfacial area and strong interactions between different phases, and a high aspect ratio (or the shape factor). These conditions are met better *via* solution mixing, since solution mixing provides more interlayer spacing for intercalation of macromolecules.

4. CONCLUSIONS

Layered materials, such as clay and graphite, have attracted increasing interest over the past decades. Elastomers can be deformed to large deformations and then elastically spring back to their original form, but they have inherently low stiffness and strength and thus are in need of reinforcement for practical applications. In this work we demonstrated how to utilise graphene platelets by melt compounding and solution mixing to develop polymer composites based on a commercial elastomer, styrene-butadiene rubber (SBR).The comparison between the two processes demonstrated that solution mixing is more effective in promoting the reinforcing effect of GnPs, since it provides more interlayer spacing for elastomer molecules to intercalate and retains the high aspect ratio of GnPs leading to filler–filler network at a low volume fraction. It shows that the structure and various properties elastomer/graphene composites are overwhelmingly dependent upon the processing history and morphology.

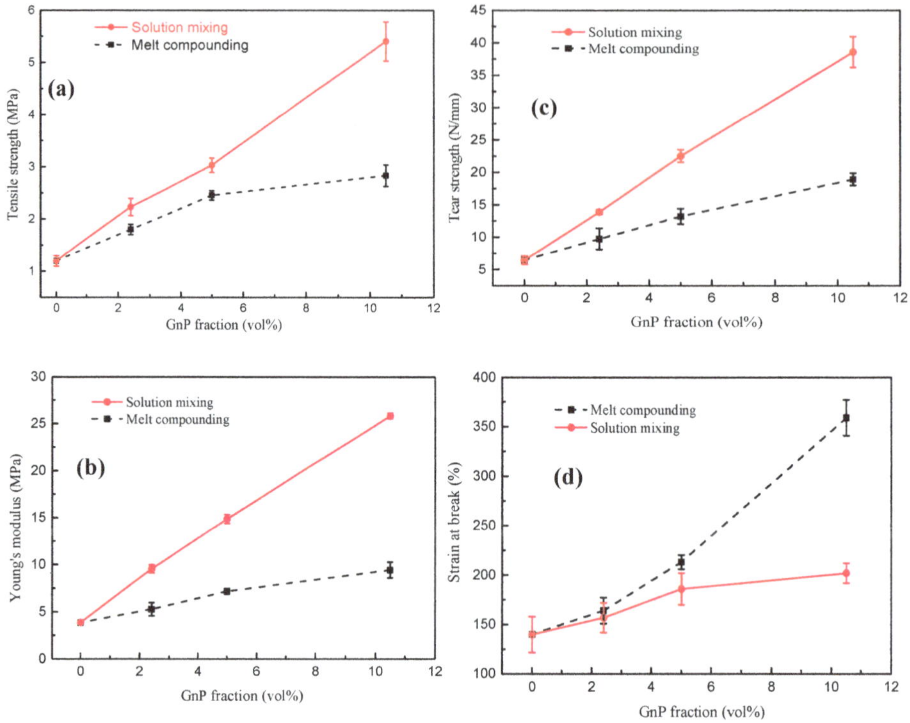

Figure 4. Solution mixing versus melt compounding: (a) tensile strength, (b) Young's modulus, (c) tear strength and (d) strain at break.

5. REFERENCES

1. Yang J, Tian M, *et al.* Acta Materialia 2007;55(18):6372-6382.
2. Paul Podsiadlo AKK, Ellen M. Arruda, *et al.* Science 2007;318 (5847):80-83.
3. Paul DR and Robeson LM. Polymer 2008;49(15):3187-3204.
4. Goettler LA, Lee KY, and Thakkar H. Polymer Reviews 2007;47(2):291-317.
5. Zeng QH, Yu AB, *et al.* Journal of Nanoscience and Nanotechnology 2005;5(10): 1574-1592.
6. Zaman I, Ma J. *et al.* Advanced Functional Materials 2012;22(13):2735-2743.
7. Zaman I, Ma J. *et al.* Polymer 2011;52(7):1603-1611.
8. Sherif A, Jun M, *et al.* Nanotechnology 2013;24(16):165601.
9. Araby S, Ma J., *et al.* Polymer 2014;55(1):201-210.
10. Araby S, Ma J., *et al.* Polymer 2013;54(14):3663-3670.

Effect of rare earth on the characterization of corrosion of low carbon steel in CSP

T. Li[1,2], M. Qin[1]
[1] Key Lab of Integrated Exploitation of Bayan Obo Multi-Metal Resources, Inner Mongolia University of Science & Technology, P.R. China
[2] School of Material and Metallurgy, Inner Mongolia University of Science & Technology, P.R. China

ABSTRACT

The periodic immersion corrosion test of low carbon steel in CSP with and without RE was carried out in simulated marine environment. The surface morphology, corrosion dynamic law and components of the corrosion products and electrochemical properties of samples were investigated by digital camera, XRD, Raman and electrochemical station. The results show that the corrosion of low carbon steel in CSP is mitigated after 240 hours immersion with RE, the corrosion resistance increases too. Compared with the sample without RE, RE promotes the formation of compact α-FeOOH and Fe_3O_4 within the rust layer, thereby mitigates the corrosion of the low carbon steel in CSP.

1 INTRODUCTION

Low carbon CSP steels, as important structural materials, are widely used in buildings, bridges, pressure vessels, etc [1]. With the increase of application, the economic loss caused by the atmospheric corrosion is becoming more and more serious, especially in the marine atmosphere with high humidity and chloride ion [2].

So, the investigation of atmospheric corrosion behavior of low carbon steels is essential to ensure integrity and safety of structural materials. Previous studies[3-5] have shown that RE can improve the corrosion resistance of steels. Due to rich reserves of RE in China, it is significant to carry out the research on the effects of RE on corrosion behavior of low carbon steels in CSP to the development of new RE steels with Chinese characteristics.

2 EXPERIMENTAL

2.1 Experimental material

The experimental materials were both processed in one of the iron and steel companies in China. During the continuous casting process, the rare earth wires were fed into the mold of CC slab. The compositions of two types of steels in CSP were shown in table 1. Sample 1# and sample 2# are low carbon steels without RE and with RE, respectively.

Table 1 The alloying elements of the experimental steels / (mass%)

Number	C	Si	Mn	P	S	Ca	RE
1#	0.044	0.013	0.14	0.008	0.007	0.0018	0
2#	0.037	0.014	0.14	0.014	0.005	0.0018	0.009

2.2 Accelerated corrosion test

An EA-08 type equipment of cyclic wet-dry immersion was used to simulate the corrosion process in marine atmosphere environments, and 3.5% NaCl was used as immersion agents. All experiments were performed at the ambient temperature 45±2℃ and the solution temperature 35℃ as long as 480h. Each wet/dry cycle consists of 10 min immersing period and 50 min a drying period. The specimens for weight loss measured with a dimension of 30mm×20mm×5mm, were degreased by acetone and grounded with SiC papers sequentially to No. 800 grit, then followed by cleaning in ethanol. After drying and weighting, the specimens were hang up in the cyclic wet-dry immersion equipment. Similarly, the electrochemical samples with a dimension 10mm×10mm were polished by abrasive papers from No. 400 to No. 1000, and were inlaid with epoxy resin, then, were hanged up in the equipment.

2.3 Surface analysis and electrochemical measurement

The samples were taken out after 72h, 120h, 240h, 360h, 480h. Right after that, corrosion-section morphologies of corrosion products on the surfaces of samples were examined by a digital camera. The phase analysis of corrosion products were characterized by XRD and Raman. The Potentiodynamic polarization measurements of the samples corroded for different time were carried out by 273A electrochemical station in 3.5 wt.% NaCl electrolyte. During the test, the applied potential ranged from -0.25 to +0.5 V SCE, the scan rate was 0.5 mV/s.

3 RESULTS AND DISCUSSION

3.1 Surface Morphologies of Corrosion Product

The surface morphologies of corrosion production on the low carbon steel with and without RE after different wet-dry immersion time were shown in figure 1. As can be seen that the corrosion products covered all over the substrate after 72h immersion and both samples were most heavily corroded after 240h.

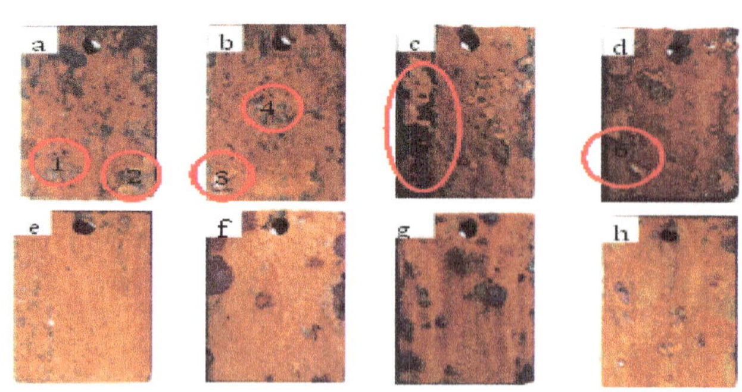

Fig. 1 Surface morphologies of samples after 72h, 120h, 240h and 480h
a, b, c, d-1#; e, f, g, h-2#

Obviously difference between two types steels is that the rust layer of 1# samples was quite rough and uneven. Without RE, the corrosion product was cluster and bump in some area in initial stage, such as sites 1,2,3 and 4 mark in Fig. 1. With the increase of immersion time, local corrosion products drop from the substrate easily, just as sites 5 and 6 marked in Fig1. In contrast, it is flat, uniform of rust layer on low carbon steels with RE, which has stronger adhesion with substrates. As the chemical characteristics of RE elements was very active, La and Ce reacted preferentially with oxygen and sulphur to produce stable oxides and sulphides. The dispersion spherical RE oxides and/or oxy-sulfides inclusions were niform distributed into substrate and might act as corrosion active sites, hindered the cluster and bump of corrosion product, resulting in flat morphology and uniform corrosion.

3.2 Rust composition and structure

Figure 2 shows the XRD pattern of rust layer formed on two type samples after 72h, and 480h wet-dry immersion. In marine atmosphere, Crystallization effect of corrosion product on low carbon steels is poorer. We can see from fig.2 that both rust layers are mainly made up of Fe_3O_4 and γ-Fe_2O_3, with a small amount α-FeOOH and γ-FeOOH in the early stage. Corrosion reaction and formation process of the initial corrosion products progressed very rapidly and those initial corrosion products were subsequently transformed to α-FeOOH, γ-FeOOH. Correspondingly, the percentage composition of Fe_3O_4 and γ-Fe_2O_3 decreased gradually, as shown in table 2.

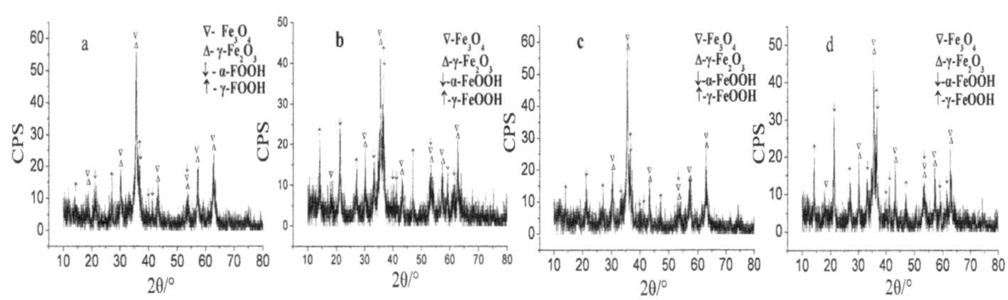

Fig. 2 The XRD of corroded sample after 72h, 240h, 480h a, b -1#; c, d -2#

Table 2 XRD analysis Results/%

Sample	Corrosion time/h	Fe_3O_4, γ-Fe_2O_3	α-FeOOH	γ-FeOOH
1#	72	63	17	18
	480	43	23	30
2#	72	64	17	16
	480	40.6	27.2	30.6

Some researchers[6-8] suggested that α-FeOOH is the stable insulators and the most stable hydroxyl iron oxide. It provides better protection to the substrate. But γ-FeOOH has much less protection than α-FeOOH [9]. It is easy to fall off from the surface. In some cases, reaction of the Fe^{2+} with γ-FeOOH forms the Fe_4O_3. It is more stable than oxyhydroxide and therefore more protective to steel. The Raman results of phase composition of rust layer on two type steels after 360h test also evidenced the transformation from γ-FeOOH to Fe_4O_3, as shown in figure 3. So, It is reasonably accepted that the initial corrosion product of the steel mainly transforms to the mixtures of α-FeOOH and Fe_3O_4.

According to the value in Table 2, the content of α-FeOOH increased much slower and γ-FeOOH increased much faster of 1# samples than that of 2# samples. Therefore, we can deduce that RE could promote the formation of compact phases Fe_3O_4 and could transformate from amorphous ferric oxyhydroxide into the more stable α-FeOOH. Consequently, RE enhanced the formation of uniform, compact rust layer and further improved the protection of the low carbon steel substrate.

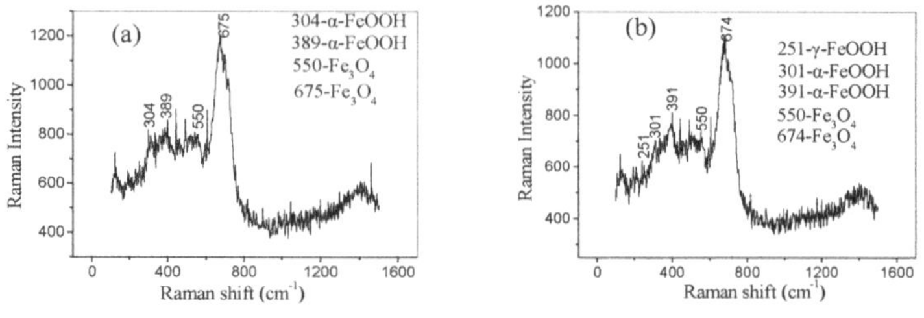

Fig. 3 The Raman results of rust composition of samples after 360h wet-dry immersion (a)-1#; (b)-2#

3.3 Weight loss and Corrosion rate

Fig. 4a shows weight loss with immersion time of two types of samples. It can be seen that corrosion weight losses of both materials increased monotonically with the prolonging of time. Compared to 1# sample, the weight loss of 2# samples was distinctly smaller after 240h immersion. From 240h to 480h, corrosion weight loss of 2# samples was about 5-12 percent lower than that of 1#samples. From Fig.4, it can be seen that the corrosion rate of both steels increased onset and then decreased gradually, as well as the study result of Yuantai Ma[10]. Compared to 1# samples, the corrosion rate of 2# samples was slightly higher on early phase but became very lower in the late stage. After 480h test, the corrosion rate of 2# samples was 1.2 ($gm^{-2}h^{-1}$), about 28% percentage lower than that of 1# sample which was 1.67($gm^{-2}h^{-1}$).

In the initial stage, corrosion active point increased by the addition of RE, promoting the electrochemical reaction occurred on the surface of electrode. As a result, the corrosion rate of 2# samples was higher than that of 1# samples. Meanwhile, rust layers were thin and loose, have no obvious effect on electrochemical reaction of both steels.

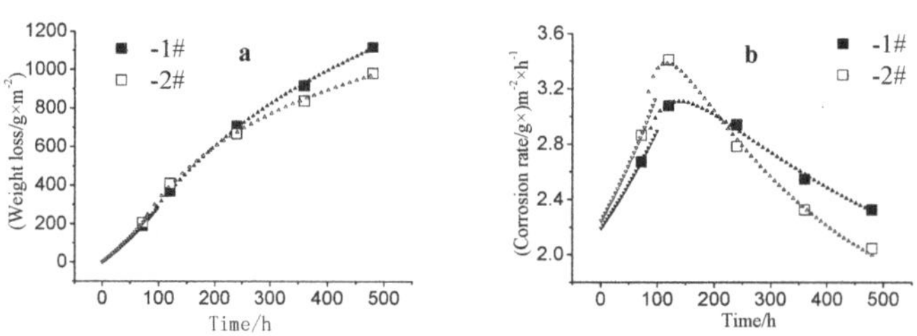

Fig. 4 Fitting curves of Corrosion weight loss and corrosion rate varying with time a- corrosion weight loss: b- corrosion rate

With the increase of time, corrosion products played an important role in the atmospheric corrosion of steels. The compact rust layer can act as a diffusion barrier and thus delay the oxidation and corrosion processes. Thereby, the corrosion rate was decreased. RE promoted the formation of compact rust layer, therefore, the protection of corrosion products is relatively strong. Moreover, the formation of RE compounds possible restricted the availability of aggressive species and electrolyte to the metal surface, thereby reducing the rate of corrosion.

3.4 Potentiodynamic polarization

Potentiodynamic polarization curves of two samples at different time were shown in Fig.5. At early stage, the zero current potential of samples with RE was negative to that of the samples without RE. It perhaps because that RE increased the corrosion active site and accelerated the electrochemical reactions. However, after 240h, the zero current potential which was associated with the oxidation of the steel surface became more positive, indicating that the corrosion product became thick and firm. Consistently, the current density of 2# samples was smaller than that of 1# samples which was possible due to that the RE compounds in rust layer and compact products formation on the surface blocked the anodic dissolution of the steel. Once the compact rust layer was formed and developed by addition of RE, the diffusion process and the electrochemical reaction on the surface of steels substrate would be slower. Combined with results shown in Fig.1, RE reduced the cluster of Fe_3O_4 and enhanced the protection for substrate. The corrosion resistance was improved.

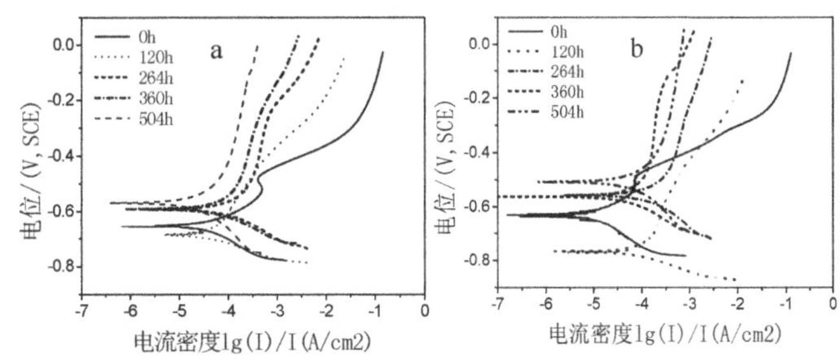

Fig. 5 Polarization curve of samples varying with time a-1#; b-2#

4 CONCLUSIONS

In simulated marine atmosphere, RE can improve the uniformity of rust layer in the early stage and promotes the formation of compact phases α-FeOOH and Fe_3O_4 in the late stage of low carbon steel in CSP. After a certain time, the zero current potential improved and the current density decreased as well as the corrosion rate decreased with the addition of RE with the extension of time. RE promotes the formation of compact and adhesion rust layer is the major reason for the corrosion resistance improvement in certain extent.

REFERENCES

[1] Q. X. Li, Z.Y. Wang and W. Han. (2008) Corros. Sci, 50:365

[2] Robert Jeffrey, Robert E. (2009) Corros. Sci, 51:2291

[3] Wang L M, Du T, Wang Y K. J (2003) China Rare Earth Soc., 21:491

[4] Chen J Q, Chen B W, Hu M, Song Y L. (2003) Res Iron Steel, 31-5:49

[5] Li C Y, Wang X D, Jiang S M, (2005) Chin Rare Earths , 26-6:23

[6] Ph. Dillmann, F Mazaudier, S Hoerrle. (2004) Corros. Sci, 46(6):1401

[7] T. Kamimura, S. Hara, H. Miyuki,etc. (2006) Corros. Sci, 48(9):2799

[8] Desmond C. Cook.. (2005) Corros. Sci, 47:2550

[9] Hara S, Kamimura T, Miyuki H, et al. (2007) Corros. Sci, 49:1131

[10] Yuantai Ma, Ying Li, Fuhui Wang. (2008) Materials Chemistry and Physics, 15(112):844

Effect of the Ce on behavior of corrosion resistance and mechanical properties of A36 plate steel for shipbuilding

J.C. Yang, H.C. Yu, X.Y. Wang
School of Material and Metallurgy, Inner Mongolia University of Science and Technology, P.R. China

ABSTRACT

In order to improve corrosion-resistance and mechanical properties of A36 plate steel in the analogy seawater (3.5% NaCl solution), a rare earth (RE) metal, Ce, was added into the steel during the course of manufacturing. The corrosion products were collected and analyzed by using SEM, EDS and XRD. Meanwhile, the impact toughness of A36 steel plate was measured at 273K. The experimental results indicated that A36 plate steel with the addition of Ce showed better corrosion-resistance than the steel without the element being added. Corrosion products were found mainly being oxides (Fe_2O_3 and Fe_3O_4.) and FeOOH. The mechanical properties of A36 experimental steel were improved, with its tensile strength being increased by 6%, yield strength by 8%. When the Ce content reaches up to 0.009%, the impact toughness of steel was increased by 9%.

1. INTRODUCTION

Nowadays, low alloy steel still remains the main metallic material in ship and ocean engineering for its low price and good general mechanical performance. However, the steel has a poor corrosion resistance in seawater, so, it is very necessary to develop new type of ship plate steel which can endure marine atmospheric corrosion because traditional carbon steel can't form a protective rust layer in marine atmosphere that has a rather high NaCl content [1]. Some researchers both at home and abroad have found that RE can reduce the diffusion coefficient of hydrogen in steel and inhibit beneficiation of hydrogen in the crack tip plastic zone. Therefore, hydrogen's reduction greatly improves the service life of the material [2]. In this paper, A36 ship plate steel is added with RE Ce. Room temperature stretch, impact at 0°C and weekly immersion accelerated corrosion experiments have been carried out. The corrosion behaviors of A36 Ce ship plate steel were studied with simulation of immersion in seawater solution and comparison method. Meanwhile, it aims at improving the service life of material and providing theoretical foundation for industrial manufacture.

2. EXPERIMENTAL PROCEDURE

This experiment selected a A36 ship plate manufactured by Baotou Iron and Steel (Group) Co as the experimental material and pure metal Ce as annexing agent. The chemical compositions of A36 experimental steel are listed in Table 1.

There were three parallel samples in corrosion experiments. The specimens were inserted in the bracket vertically and were immersed in the simulated seawater containing 3.5% NaCl. Experimental cycles were chosen with 2, 9, 16, 21 and 28 days, respectively. After removal of the rust from the specimens, macro morphology of corrosion products was filmed by high digital camera, microstructure of corrosion surface was observed by SEM, elements of corrosion products were analyzed by EDS and XRD. Weight loss was measured to calculate the corrosion rate. Self-corrosion potential, polarization curve and alternating-current impedance were also measured in Solartron1280B electrochemical workstation. Potential sweep frequency of 1×10^{-2} - 1×10^{5}Hz was selected in the range of -0.25-0.5mV.

Table 1 The chemical compositions of A36 experimental steel mass %

No.	C	S	Si	Mn	P	Al_t	Cr	Ce
1#	0.13	0.0037	0.21	0.79	0.015	0.023	0.32	0
2#	0.14	0.0035	0.20	0.78	0.011	0.022	0.34	0.002
3#	0.12	0.0033	0.21	0.77	0.013	0.025	0.33	0.003
4#	0.14	0.0032	0.23	0.78	0.014	0.026	0.30	0.009

Impact test sample should be processed into standard v-notch, sample's size was 10mm×10mm×55mm. The test was carried out in 0°C water. The instantaneous impact velocity of the tester was 5.0-5.5 m/s. Tensile property at room temperature was measured through CSS-88500 electronic universal testing machine with a rate of 1.50mm/min.

3. EFFECT OF RE Ce ON THE MECHANICAL PROPERTIES OF A36 SHIP PLATE STEEL

3.1 Effect of RE Ce on the tensile test of A36 ship plate steel at room temperature

Through tensile test at room temperature, steels tensile strength, yield strength, elongation, reduction of area affected by RE are obtained. The results are listed in table 2.

Table 2 Results of tensile test at room temperature

No.	R_m/MPa	R_e/MPa	A/%	R_e/R_m
1#	551.635	358.563	30.885	0.65
2#	585.310	386.358	42.221	0.66
3#	597.231	406.117	45.281	0.68

After adding RE into the steel, yield strength R_e has improved 8% to 386.35MPa, tensile strength has improved 6% to 585.310MPa and elongation A has a little change while yield ratio R_e/R_m has a little rise. Tensile strength and yield strength of A36RE ship plate steel are both improved while elongation and reduction of area both have a modest increase. These properties change because Ce makes sundry metamorphism and modification in the steel. Long strip inclusions along rolling direction have refined pearlitic structure. Metamorphous inclusions and smaller ones have both reduced the generation of crack source in substrate. All the reasons above have improved the tensile strength, ductility and toughness of the steel.

Macro and micro morphology of tensile fracture from 1# steel and 3# steel at room temperature are as shown in Fig.1. According to macro morphology, there is obvious radiation fiber area at 1# steel's fracture, whose characteristic is ductile fracture mixed with brittle fracture and the direction of crack is the same as that of radiation. There is plastic fracture in 3# steel. Fiber area is made up of numerous fibrous "small peaks" whose small cant form a certain angle among themselves[3]. According to microstructure, in 1# steel, overall appearance of fracture is not homogeneous and there are big holes, dimples are shallow and the number of distribution is uneven. There are also irregular flake inclusions beside the dimples. In 3# steel, overall appearance of fracture is close to isometric dimples. Dimples are well-distributed. The quantity, size and depth of dimples have significantly increased compared with 1# steel. Besides, small globular inclusions that distributed at the bottom of dimples appear in the depth of dimples, which explains that 3# steel has an obvious plastic deformation before fracture by tensile stress.

In 1# steel, its plasticity and toughness is relatively poor because of the existence of radiation area which results in smaller size of fiber area compared with 3# steel. Adding RE Ce into A36 ship plate steel turns tensile fracture from ductile fracture mixed with brittle fracture into ductile fracture, which therefore improves mechanical properties of A36 ship plate steel at room temperature.

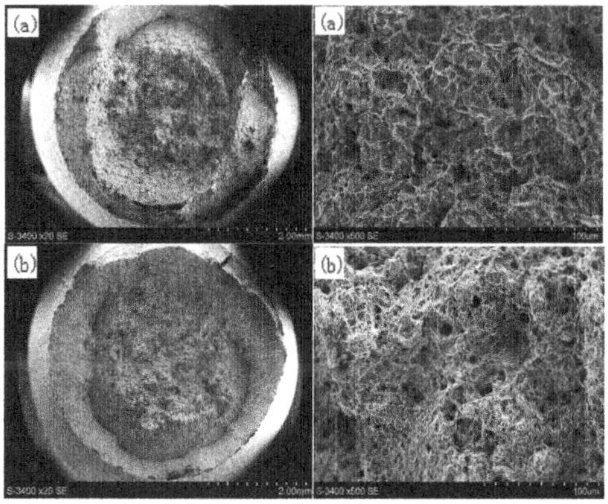

Fig.1 Macro and micro morphology of tensile fracture (a) 1# steel; (b) 3# steel

3.2 Effect of RE Ce on the impact toughness of A36 ship plant steel at 273K

RE Ce can improve the low temperature impact toughness of A36 ship plate steel. Experimental results are listed in table 3. When Ce content is up to 0.003%, impact toughness starts to rise. When Ce content is up to 0.009%, impact toughness has improved by 9%. After adding RE, grain is refined and organization is well-distributed, which contributes to the improvement of impact toughness. By scanning electron microscope, Ce has formed small RE inclusions in steel. The RE inclusions are affirmed to be RE cerium oxide sulfur through analysis of EDS energy spectrum. inclusions in steel. The RE inclusions are affirmed to be RE cerium oxide sulfur through analysis of EDS energy spectrum.

Table 3 The impact toughness of A36 ship plate steel at 273K

No.	1#	2#	3#	4#
Impact toughness /AK/J	257.6	249.4	267.1	281.7
Ce content /%	0	0.0020	0.0030	0.009

4. EFFECT OF RE Ce ON THE CORROSION PROPERTIES OF A36 SHIP PLATE STEEL

4.1 Effect of RE Ce on the corrosion of A36 ship plate steel

As can be seen from Fig.2, there are point corrosion pits on the surface of 1# steel. The corrosion hole is deep and there are more holes than 2# steel. In 2# steel, its surface is relatively flat overall. Few shallow and wide corrosion pits are distributed on local area. The results above illustrate that adding RE into experimental steel has improved pitting corrosion resistance of the ship plate steel.

Fig.3(a) and (b) are respectively rust layer profiles of A36 ship plate steel and A36RE ship plate steel after 28d corrosion. Rust layer thickness of A36 ship plate steel is significantly thicker than that of A36RE ship plate steel. Rust layer of A36 has a poor continuity while the rust falls off seriously. Rust layer of A36RE is compact with good continuity. Because the rust layer of A36RE is connected to the substrate closely, it can prevent the oxygen from diffusing into the substrate effectively, which enhances cathodic polarization. The surface between rust layer and substrate of A36 ship plate steel is curve which means there are serious pitting corrosions on local surface of the steel. On the contrary, the surface between rust layer and substrate of A36RE ship plate steel is smooth which means corrosion in the steel is homogeneous.

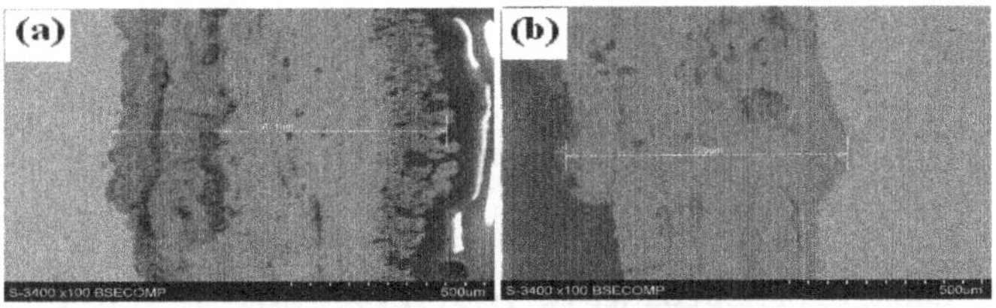

Fig.2 Micro morphology of the surface after 23d corrosion (a)1# steel (b)2# steel

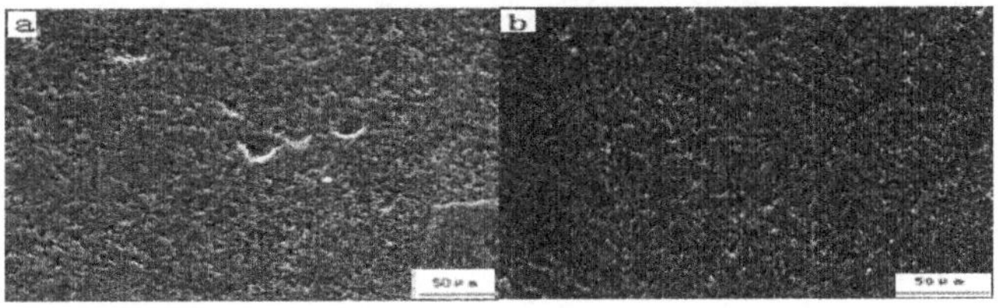

Fig.3 Rust layer of experimental steel after 28d corrosion (a)1# steel; (b)3# steel

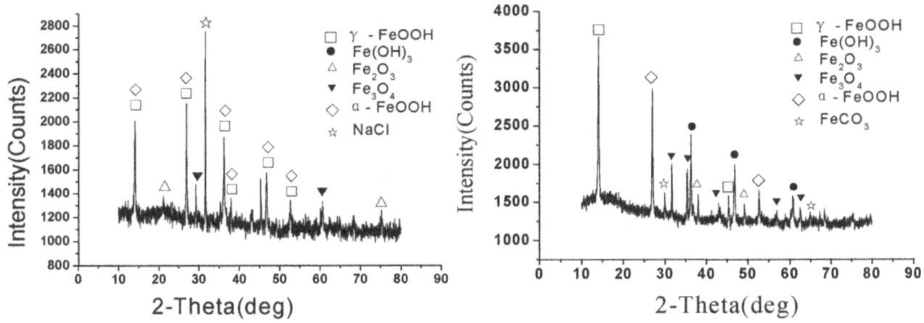

Fig.4 XRD analyses of corrosion products for 1# and 2# steel

At the beginning of the immersion test, sample surface has a direct contact with the corrosive medium. Under this situation, a large number of γ-FeOOH and $Fe(OH)_3$ which have a mass of flocculent shape are generated. The two corrosion products both have unstable structure. As time goes by, corrosion products gradually transform into stable α-FeOOH and Fe_3O_4. Surface corrosion products of 1# steel and 2# steel are scraped to be analyzed by XRD. The results are shown in Fig.4. In two kinds of experimental steels, corrosion products are all $Fe(OH)_3$, α—FeOOH, Fe_2O_3, γ—FeOOH, Fe_3O_4 and $FeCO_3$.

4.2 Electrochemical corrosion impedance measurement

As can be seen from Fig.5(a) and (b), the arcs of capacitive reactance about 30d samples for 1# steel and 2# steel are the longest with maximum value on impedance modulus while arcs of 9d are the shortest. The corrosion rate of A36 steel is slowest on 30d while fastest on 9d. Because from 9d on, inner rust layer gradually increases as corrosion time goes by. Therefore, corrosion products are covered on the surface interfering further corrosion to protect substrate from being corrosive.

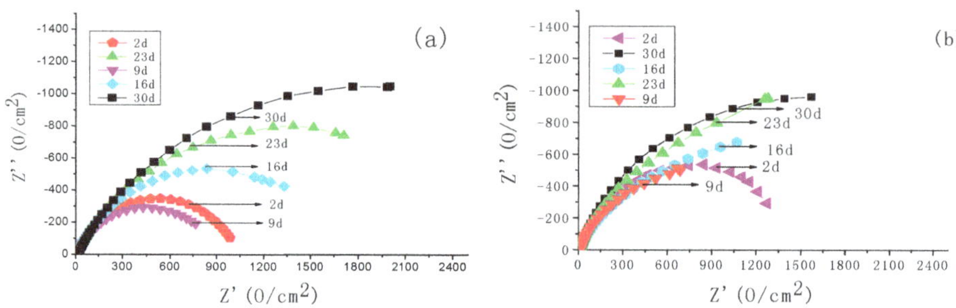

Fig.5 Changes of impedance curve over time (a)1# steel;(b)2# steel

Polarization curves of 1# steel and 2# steel in simulated seawater solution are shown in Fig.6. As can be seen from Fig.6, anode polarization curves of 1# steel and 2# steel are both smooth. There's no phenomenon of passivation around anode area. Oxygen has participated in the electrode reaction process as a depolarizing agent in the cathodic reduction reaction. Under this situation, the electrode reaction is believed to be affected by the diffusion process of oxygen as depolarizing agent. Corrosion process will be diffusion controlled process. 2# steel's polarization resistance is the biggest while corrosion is the lightest.

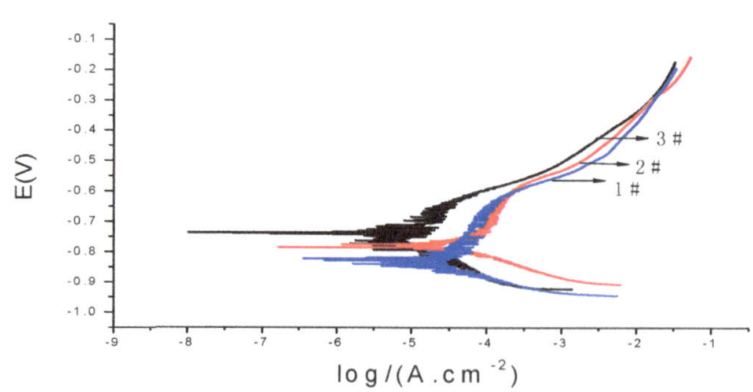

Fig.6 Comparison diagram of polarization curve for 1# and 2# steel

5. CONCLUSIONS

1) There's a very obvious metamorphism effect by adding RE into A36 ship plate steel. Ce has formed small RE inclusions in steel. Appearance of fracture has been improved. Distribution of dimples is more homogeneous than before.
2) When Ce content reaches 0.009%, yield strength increases 8%, tensile strength Rm increases 6% and impact toughness at 0°C increases 9%.
3) The addition of RE has increased protective α-FeOOH in ship plate steel, which improves the protection ability of the rust layer for substrate. For the same corrosion same, A36 ship plate steel is corroded worse than A36RE ship plate steel.

ACKNOWLEDGEMENT

This work was supported by the National Natural Science Foundation of China (51174114).

REFERENCES

[1] Ri-zhang ZHU. Metal corrosion science [M]. Beijing: Metallurgical industry press, 1989:203.
[2] Bai Shan, Zi Meng. The vitamin in steel – rare earth [J]. Metal World, 2004, 2(1):42-43.
[3] Transformation Microstructure of Low Alloy Steels [C]. 2002 International Symposium on Microalloyed Steels, 2002, 134-14. Lou D, Cui K, Grong O. Effects of rare earth metals and calcium on solid phase 75.

Stress corrosion cracking of sensitized stainless steel type 304 in high-temperature, high-purity water environment

T. Fujii, K. Tohgo, A. Kenmochi, Y. Shimamura
Department of Mechanical Engineering, Shizuoka University, Japan

ABSTRACT

This paper deals with experimental investigation of stress corrosion cracking (SCC) in sensitized stainless steel under an accelerated simulated environment in boiling water reactors. Constant load tests are conducted under high-temperature and high-purity water. The number and maximum length of cracks increase with increasing stress and testing time. The length of micro cracks is normally-distributed with the mean of 54μm and standard deviation of 20μm, which are close to those of grain diameter.

1 INTRODUCTION

Stress corrosion cracking (SCC) is a phenomenon related to crack initiation and propagation due to local anodic dissolution on smooth surface and at crack tip (1), and occurs under specific combinations of material, environment, and stress condition (2). In nuclear power industries, much attention has been paid to SCC as a degradation phenomenon of materials and structures in nuclear power plants. Many studies on SCC in materials under high-temperature and high-purity water environment have been carried out (3). According to the fitness-for-service assessments for nuclear power plants in U.S. (4) and Japan (5), SCC remaining life is estimated by crack growth from the size of a crack detected by inspection to final failure. However, crack formation processes are not taken into account in the fitness-for-service assessments. In order to predict SCC remaining life, it is necessary to consider SCC processes from micro crack initiation to macro crack growth, simultaneously (6).

In this study, constant load tests in high-temperature and high-purity water containing dissolved oxygen as an accelerated test of SCC in boiling water reactors were conducted and micro crack initiation behaviour by SCC was investigated. The number of cracks and crack length were measured, and the influence of tensile load and loading time on SCC behaviour was examined.

2 MATERIAL AND EXPERIMENTAL PROCEDURE

2.1 Material and specimen

Austenitic stainless steel type 304 was used. Table 1 shows the chemical composition. The solution heat treatment was performed at 1100°C for 1hour, and the sensitization heat treatment was performed at 750°C for 2hours and 500°C for 24hours. The reactivation ratio of the sensitized material was 29.3%. The 0.2% offset stress obtained by tensile test at 288°C was 140MPa.

Table 1 Chemical composition

C	Si	Mn	P	S	Ni	Cr	Fe
0.06	0.47	0.82	0.03	0.003	8.05	18.16	Bal.

Fig. 1 shows the dimensions of a specimen. The specimens were machined from bulk material after the heat treatment and their surfaces were polished by waterproof papers. The specimens have three parts with different cross-section to realize three stress levels by one specimen. The specimen was wrapped by graphite fibre wool to form crevices, as shown in Fig. 2.

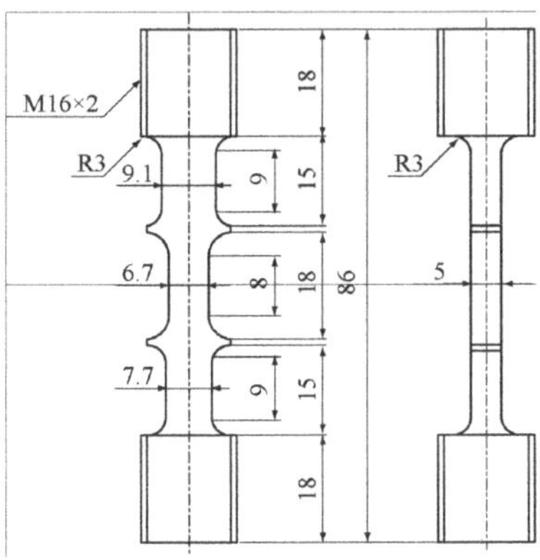

Fig. 1 Specimen configuration

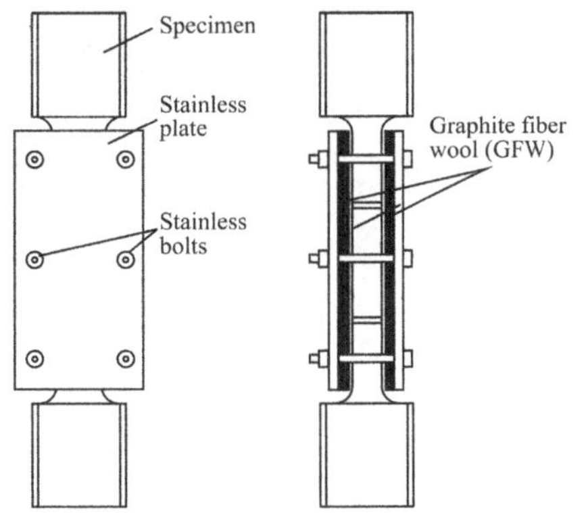

Fig. 2 Specimen setting

2.2 Constant load test

The constant tensile load tests were conducted under high temperature and high-purity water environment. The temperature, pressure, conductivity and dissolved oxygen concentration were 288°C, 8MPa, 0.1μS and 8ppm, respectively.

At first, the specimens were pre-immersed in the corrosive environment for 48hours to create oxide film on specimen surface. After that, the load corresponding to applied stresses of 175, 210 and 250MPa for three test parts of the specimen was applied. The experiments were terminated at 24, 48, 96 or 120hours. For each condition, four specimens were prepared. After the experiments, the specimens were etched by 10% oxalic acid.

3 EXPERIMENTAL RESULTS AND DISCUSSION

3.1 SCC behaviour

Fig. 3 shows variation in crack distribution with testing time of 24, 48, 96 and 120hours under applied stress of 250MPa. Fig. 4 shows schematic drawing of the crack profile corresponding to Figs. 3. In all specimens, a number of micro cracks initiated at grain boundary and relatively large cracks perpendicular to the loading direction created by coalescence and growth are observed. As shown in Fig. 4(d), the crack initiation does not occur in uniform, but the crack initiation and succeeding coalescence and growth are concentrated in some small areas. This localization of SCC behaviour is more or less observed in all specimens, and can be explained from the viewpoint of creviced corrosion. In the present tests, GFW was put on the specimen surface so as to form uniform crevice through the specimen surface as far as possible. However, from an electro-chemical viewpoint of corrosion, small difference in gap of the crevice would create inhomogeneous environment on the specimen surface. In such a case, a local battery was formed and the electro-chemical reaction of corrosion occurred on local areas of the specimen surface. That is, the anodic reaction and SCC behaviour were localized in some areas of the specimen surface.

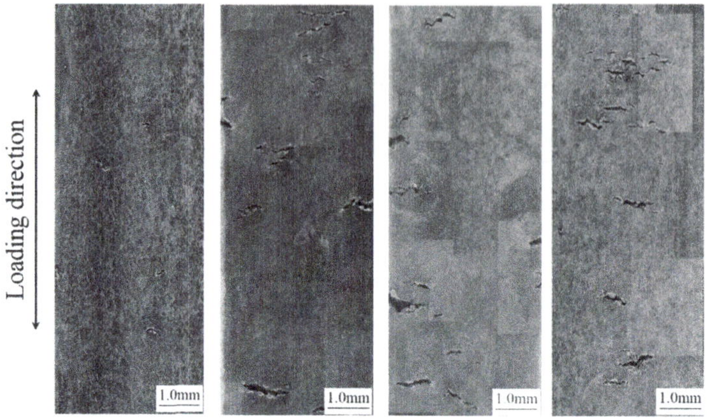

(a) 24hours (b) 48 hours (c) 96hours (d) 120hours

Fig. 3 Crack profile on specimen (applied stress, 250MPa)

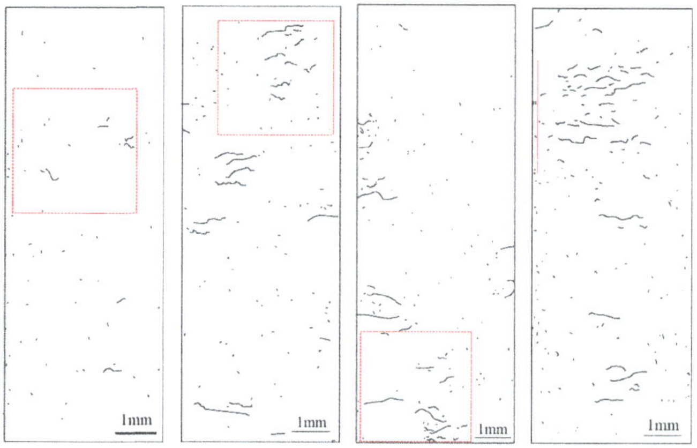

(a) 24hours (b) 48 hours (c) 96hours (d) 120hours

Fig. 4 Schematic illustration of crack profile in Fig. 3

3.2 Influence of tensile stress and testing time on SCC

The 3mm x 3mm square region containing the localization area is assigned on all specimen surfaces for the applied stress and testing time as shown by dotted frames in Fig. 4. In this section, the SCC behaviour within the localized 3mm x 3mm square area will be discussed. Fig. 5 shows the number of cracks per 100mm^2 as a function of testing time. The plots and scatter bars denote the average, maximum and minimum value in four specimens, respectively. The number of cracks increases with increasing testing time and applied stress.

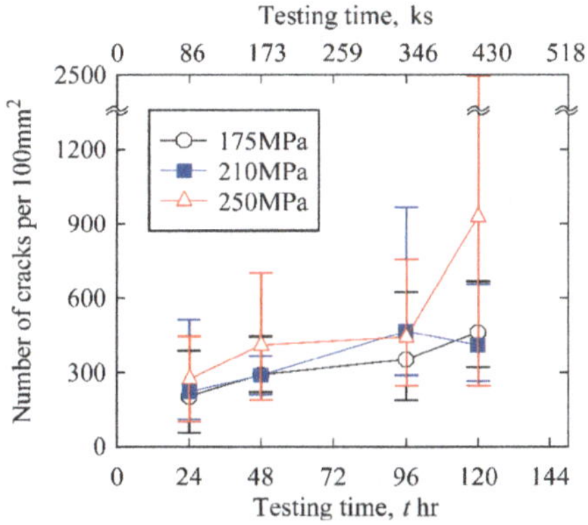

Fig. 5 The number of cracks as a function of testing time.

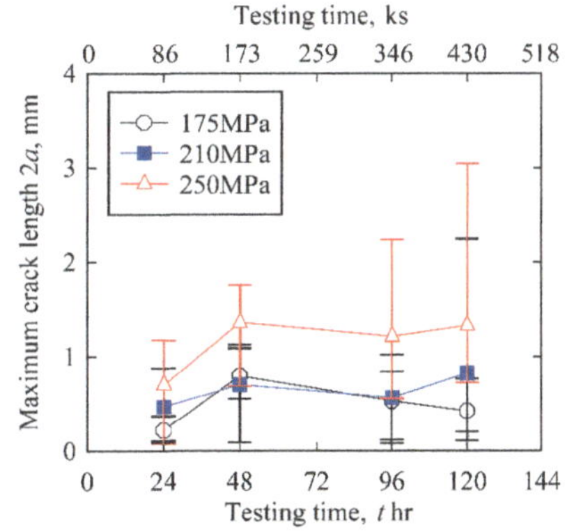

Fig. 6 Maximum crack length as a function of testing time.

Fig. 6 shows the maximum crack length as a function of testing time. The crack length was defined by distance between both ends of crack tips. The scatter bars also denote the maximum and minimum value of maximum length in four specimens. The crack length tends to increase with increasing testing time and applied stress.

Fig. 7 shows the histogram of the number of cracks against the crack length for three applied stresses and testing time of 96 hour. The interval of crack length in the histogram is set to be 0.1mm. For every condition, most of the crack length is shorter than 0.1mm. A micro crack seems to initiate in a single grain boundary because crack length is less than or equal to grain diameter of this material. The number of cracks over 0.1mm in length increases with increasing applied stress. Because long cracks are readily formed since the number of crack initiation and crack coalescence increases and the crack growth is faster when the applied stress is high.

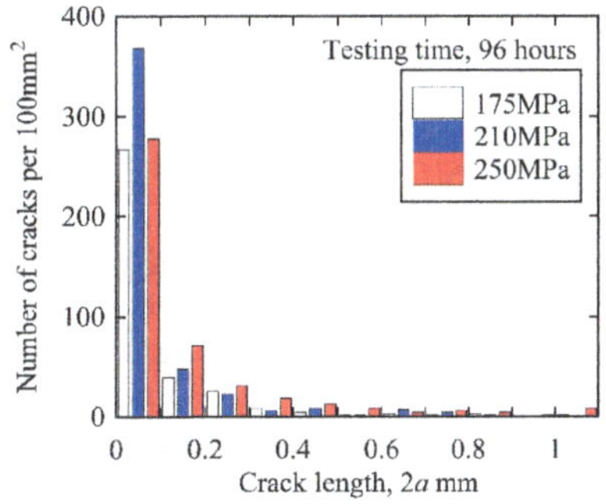

Fig. 7 Histogram of the number of cracks against crack length (96hours)

In order to investigate the initiation behaviour of grain-sized micro cracks, the micro cracks shorter than 100μm were focused on. Fig. 8 shows the histogram of the frequency of cracks against the crack length for three applied stresses and testing time of 96hour. The interval of crack length is set to be 10μm. For every condition, most of the crack length ranges from 30 to 70μm, and the distribution of crack length are bell-shaped, irrespective of applied stress.

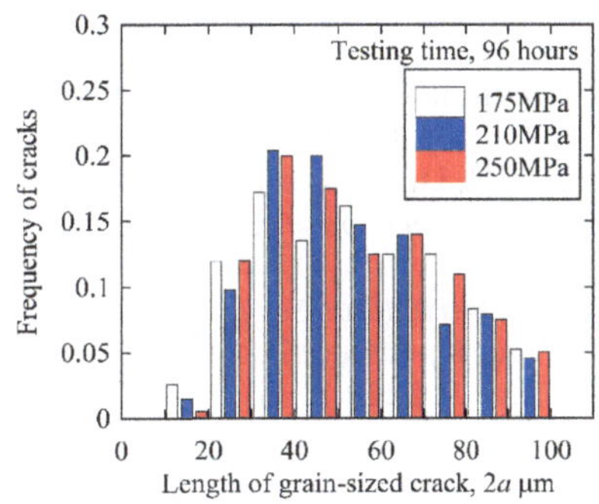

Fig. 8 Histogram of the number of cracks against grain-sized micro crack length (96hours)

The mean and standard deviation of grain diameter are 66.7μm and 16.7μm on the basis on the line-intercept method by the Japanese Industrial Standard (JIS G 0552). Fig. 9 shows the summary of crack length and grain diameter, and the error bars denotes the standard deviation. The mean and standard deviation of micro cracks are 54μm and 20μm, respectively, irrespective of applied stress and testing time. Moreover, the crack length is equal to or a little smaller than the grain diameter.

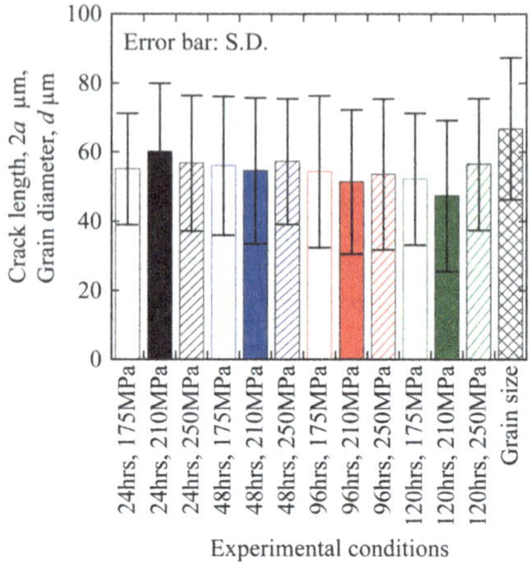

Fig. 9 Comparison of crack length and grain size.

4 CONCLUSIONS

Constant load tests of sensitized austenitic stainless steel type 304 under an accelerated simulated environment of BWRs are conducted. A number of intergranular cracks by SCC initiate on smooth surface. The number and maximum length of cracks increases with increasing applied stress and testing time. The length of grain-sized micro cracks is normally-distributed with the mean of 54μm and standard deviation of 20μm, irrespective of experimental condition and the values of normal distribution are close to those of grain diameter.

REFERENCES

[1] Parkins, R.N. (1992) *J. Miner. Met. Mater. Soc.*, **44**, 12, 12-19.

[2] Trethewey, K.R. (2008) *Mater. Des.*, **29**, 501-507.

[3] Shoji, T., Lu, Z., Murakami, H. (2010) *Corros. Sci.*, **52**, 769-779.

[4] ASME, *Boiler and Pressure Vessel Code Section XI 2003 Addenda*, ASME, New York (2003)

[5] JSME, *Fitness-For-Service Code S NA1-2004*, JSME, Tokyo (2004).

[6] Tohgo, K., Suzuki, H., Shimamura, Y., Nakayama, G. and Hirano, T. (2009) *Corrosi. Sci.*, **51**, 9, 2208-2217.

Hydrogen absorption amount of magnesium formate at room temperature

S. Hirotaki[1], **M. Notomi**[2]
[1] Department of Mechanical Engineering, Graduate School of Science and Technology, Meiji University, Japan
[2] Department of Mechanical Engineering, School of Science and Technology, Meiji University, Japan

ABSTRACT

In this paper, we quantitatively evaluate hydrogen absorption amount of four different particle sizes of Mg-Formate with the pressure-composition-temperature (PCT) measurement at room temperature (296K) and also conduct the PCT measurement of $LaNi_5$ under the same conditions for the comparison. Each powder sample was measured for PCT curve between 0.1 MPa and 2.0 MPa of hydrogen gas pressure. As a result, four kinds of the Mg-Formate samples that have the particle sizes 63~125, 125~250, 250~1000 and 1000~5000 μm averagely absorb hydrogen about 0.513±0.091 wt.% at 296 K and 2.0 MPa.

1 INTRODUCTION

In recently years, global environment problems have been increased by carbon dioxide emissions from the burning of fossil fuels. The world's increasing energy demands have required attractive energy. Although hydrogen energy which doesn't emit carbon dioxide and is renewable energy can have replace fossil fuel energy in the future, hydrogen is a gas at room temperature and ordinary pressure and that is a problem for storage in practical use. Among some hydrogen storage methods, hydrogen storage materials that can absorb and desorb reversibly have attracted attention. The most practicable material is $LaNi_5$, one of the hydrogen storage alloys, because of its hydrogen absorption / desorption properties. Specifically $LaNi_5$ absorbs 1.6 wt.% of hydrogen at 296 K and 2.0 MPa and desorbs 1.4 wt.% of hydrogen at 0.1 MPa(1). Study for hydrogen storage requires the further hydrogen storage materials such that it can absorb hydrogen more than 1.6 wt.% and desorb all the absorbed hydrogen at room temperature. Metal-organic frameworks (MOFs) for hydrogen storage have been extensively studied during the past decade(2). MOFs that are assembled by coordinate bond between metal ions and organic ligands are porous materials having uniform meso pore. MOFs physically absorb hydrogen molecular in pores with van der waals interaction and quickly desorb it when heated. Having a large surface area, MOFs have been considered as a potential candidate material for hydrogen storage. Mg-based MOF i.e. Mg-Formate ($Mg(COOH)_2$), which is assembled by connection of Mg ions and formate ligands, has especially great potential to become a promising material in MOFs due to high porosity, light weight and low cost(3)(4).

In this research, we presume that the hydrogen absorption / desorption property is attributed to particle size difference because it affects storage and quantitatively evaluate hydrogen absorption property of four kinds of particle size of Mg-Formate with the PCT measurement at room temperature (296K) and also conduct the PCT measurement of $LaNi_5$ under the same conditions for the comparison.

2 EXPERIMENTS

2.1 Sample preparation

2.1.1 $LaNi_5$

$LaNi_5$ powder sample (particle size is 45μm) was purchased from Japan Metals & Chemicals Co. Ltd. Because the surface of sample had attached oxygen and other active gas in the air, we pretreat the sample by hydrogenation and evaporation before PCT measurement. Pretreatment parameter is shown in Table 1. Since alternating activation increases reactivity between the surface of the sample and hydrogen(5), the evaporation and the hydrogenation are repeated alternately three times in order to activate it. After that, the sample is evaporated for 6.5 h at 333 K for dehydrogenation.

2.1.2 Mg-Formate

Mg-Formate, that was purchased from Aldrich, have distributed particle size and is separated for four groups according to the size using some sieves. The photos of the samples are shown in Fig.1, Mg-For No.1 : 63~125 μm, Mg-For No.2 : 125~250μm, Mg-For No.3 : 250~1000 μm and Mg-For No.4 : 1000~5000 μm. All samples are evaporated for 12 h at 333 K in order to remove any absorbent molecules before PCT measurement. Pretreatment parameter is shown in Table 1. There is no need to hydrogenate for Mg-Formate by reason that Mg-Formate is physisorption material, doesn't form hydride.

Table 1 Pretreatment parameters for samples

Sample	Evaporation				Hydrogenation		
	Sample mass [g]	Temperature [K]	Time [h]	Cycle	Temperature [K]	Time [h]	Cycle
$LaNi_5$	0.400	333	1	3	333	0.5	3
Mg-Formate	0.400	333	12	1			

Fig.1 Sample images of (a)Mg-For No.1, (b)Mg-For No.2, (c)Mg-For No.3, (d)Mg-For No.4.

2.2 Absorption measurement

Hydrogen absorption / desorption measurement were conducted at 296 K using PCT curve measurement apparatus (see Fig.2) by volumetric method complying with JIS H 7201. The pressure of hydrogen gas in the sample chamber was measured with a pressure gauge consisting of pressure transducer (PTU-S-MG6-13AD, Sweagelok) and digital indicator (PTU-DT-32, Sweagelok). The absorption/desorption amount was measured from a change in pressure of it and we calculated amount plot upon PCT curve.

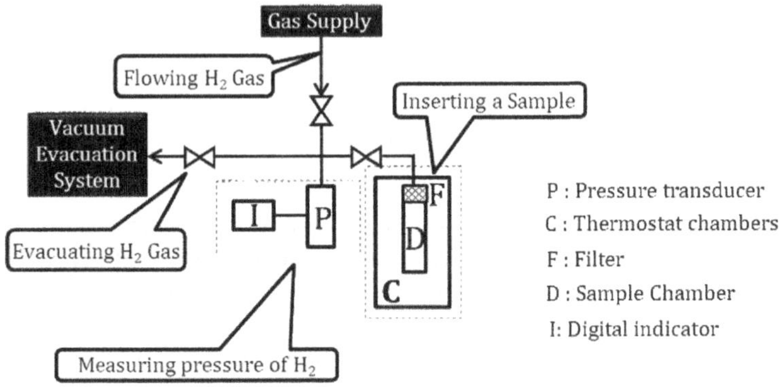

Fig.2 Schematic illustration of PCT curve measurement apparatus.

3 RESULTS

3.1 $LaNi_5$

The PCT curve of $LaNi_5$ is shown in Fig.3. Increasing the pressure of hydrogen gas, $LaNi_5$ absorbed only 0.100 wt.% of it until 0.7 MPa, and that, the reaction between sample and hydrogen gas is promoted at 0.8 MPa. Hydrogen amount increased at constant pressure of it, there is called plateau of absorption. The curve was straight through until absorbing 1.475 wt.% of it at 1.0MPa. After the plateau of absorption, hydrogen amount gradually rise with increasing the pressure of it again. Decreasing hydrogen pressure, the curve shows little hydrogen desorption until 0.15 MPa, and that, $LaNi_5$ desorbed it rapidly at 0.14 MPa, there is plateau of desorption. Table 2 shows that Maximum hydrogen amount at 2.0 MPa and residual hydrogen amount at 0.1 MPa after measurement. This PCT curve tendency of hydrogen absorption / desorption behavior is quantitatively similar to results of An's report[1].

3.2 Mg-Formate

The PCT curves of Mg-Formate are shown in Fig.3. For the curve of Mg-For No.1, hydrogen amount gradually rise with increasing the pressure of it and descent with decreasing the pressure of it, and desorption curve approximates absorption one. Other curves of three Mg-Formate samples were same each other. Table 2 shows that maximum hydrogen amounts at 2.0 MPa and residual hydrogen amounts at 0.1 MPa after measurement. Mg-Formate averagely absorbed hydrogen about 0.513±0.091 wt.% and desorbed 0.377±0.017 wt.% at room temperature. Hydrogen absorption property does not depend on particle size.

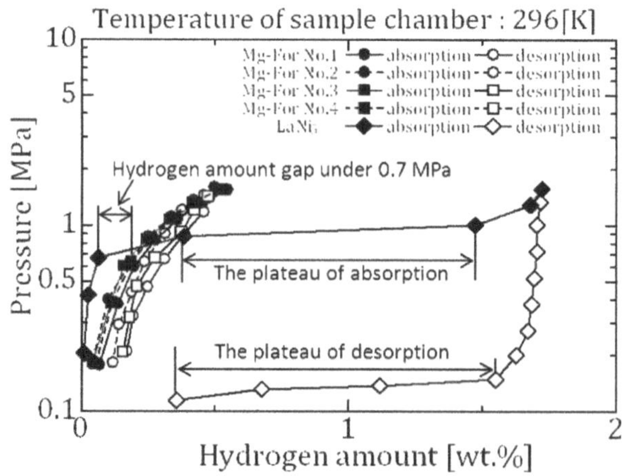

Fig.3 The PCT curves for powder of Mg-Formates and $LaNi_5$ at 296 K. Experimental absorption process are shown as close symbols and desorption process are shown as open symbols.

Table 2 Maximum hydrogen amounts and residual hydrogen amounts at room temperature for all samples

Sample	Maximum hydrogen amount [wt.%]	Residual hydrogen amount [wt.%]
Mg-For No.1	0.536	0.172
Mg-For No.2	0.422	0.136
Mg-For No.3	0.556	0.162
Mg-For No.4	0.540	0.177
$LaNi_5$	1.691	0.314

4 DISCUSSION

As has been explained above, we have demonstrated Mg-Formate can averagely absorb hydrogen about 0.513±0.091 wt.% and desorb 0.377±0.017 wt.% at room temperature. Barbara's report shows that Mg-Foramte absorbs 0.05 wt.% at 298 K and 2.0 MPa, 1.17 wt.% at 77 K and 2.0 MPa(5) and also absorption rate in this study is about 10 times than that report. It is possible that Mg-Formate at 77 K has more hydrogen absorption amount than $LaNi_5$.

Considering the above, we evaluate the hydrogen absorption / desorption property of Mg-Formate. Hydrogen amount of it is still not applicative value for serving hydrogen energy at room temperature. Compared to $LaNi_5$, maximum hydrogen amount is less 1.0 wt.% than $LaNi_5$. For Mg-Formate, there is not plateau at PCT curve and hydrogen amount has gradually rise with increasing the pressure of it. The hydrogen absorption amounts of all samples have increased in almost proportion to increasing the pressure of it under 0.7MPa and Mg-Formate absorbs it more than $LaNi_5$ under this pressure, because of absorption method difference. $LaNi_5$ absorbs hydrogen by forming a hydride and Mg-Formate absorbs hydrogen by physically reaction with van der waals interaction.

Forming a hydride is slowly and not consistently reaction, physically absorption is smoothly and almost consistently reaction in proportion to the pressure of it, it is easy to operate hydrogen absorption / desorption reaction.

Since van der waals interaction is very weak reaction, it has high absorption amount at cryogenic region about decreasing thermal behavior(6). Mg-Formate that has more hydrogen absorption amount than $LaNi_5$ at cryogenic region could be improved increasing more the amount of it than $LaNi_5$ at room temperature. For example, addition of catalyst metal dissociating into hydrogen atoms enhance more storage reaction to hydrogen because van der waals interaction is higher for hydrogen molecular than for hydrogen atoms. Mg-Formate has especially great potential to become a hydrogen absorption material by addition of catalyst metal because Mg-Formate is very light weight among MOF.

5 CONCLUSION

Mg-Formate has high reactivity of hydrogen absorption / desorption and it does not depend on particle size. Particle size is not important factor for this property for hydrogen gas pressure of 0~2.0 MPa at room temperature. Although this property of Mg-Formate is low at room temperature, it has smoothly absorption / desorption reaction and can absorb similar hydrogen amount as $LaNi_5$ at cryogenic temperature, and it has potential to be storage materials exceeding the property of $LaNi_5$ in the near future.

REFERENCES

(1) X.H. An, Q.F. Gu, J.Y. Zhang, S.L. Chen, X.B. Yu and Q. Li, "Experimental investigation and thermodynamic reassessment of La-Ni and $LaNi_5$-H systems", CALPHAD., 40 (2013), pp. 48–55

(2) N.L. Rosi, J. Eckert, M. Eddaoudi, D.T. Vodak, J. Kim, M. O'Keeffe and O.M. Yaghi, "Hydrogen Storage in Microporous Metal-Organic Frameworks", Science, (2003), 300, pp1127-1129

(3) A. Mallick, S. Saha, P. Pachfule, S. Roy and R. Banerjee, "Structure and Gas Sorption Behavior of a New Three Dimensional Porous Magnesium Formate", American Chemical Society, Inorg. Chem., (2011), 50, pp1392–1401

(4) Michael Fisher, Frank Hoffman, and Michael Froba, "Metal-organic frameworks and related materials for hydrogen purification : Interplay of pore size and pore wall polarity", ChemPhysChem, 11 (2010), p. 2220

(5) R.L. Cohen, K.W. West, J.H. Wernick, "Degradation of $LaNi_5$ hydrogen-absorbing material by cycling", J Less-Common Met, 70 (1980), pp. 229–241

(6) Barbara Streppel and Schütz, Gisela, "Hydrogen Adsorption on Metal-Organic Frameworks", Max-Planck-Institut fur Metallforschung, Stuttgart,(2011), pp15-57

The influence of Ti layer on the hydrogen desorption properties of Mg in multi-layer

T. Hashimoto[1], **M. Notomi**[2]

[1] Department of Mechanical Engineering, Graduate School of Science and Technology, Meiji University, Japan

[2] Department of Mechanical Engineering, School of Science and Technology, Meiji University, Japan

ABSTRACT

We evaluated the effect of Ti layer to Mg layer for hydrogen desorption properties in multi-layered. Two types of multi-layer films that are Mg/Pd and Mg/Ti/Pd were prepared with Pulsed Laser Deposition. The thicknesses of Mg and Ti layers were 270±5nm and 35±5nm, respectively and the ultrathin Pd layer could play the role of a catalyst. The hydrogen desorption temperature of these samples were measured with Differential Scanning Calorimetry. The desorption temperature of Mg/Pd was about 430°C, while that of Mg/Ti/Pd was about 150°C. This research revealed that the hydrogen desorption temperature of Mg was decreased by Ti layer.

1 INTRODUCTION

In recent years, with the increasing of the fossil fuel consumption, exhaustion of resources and environmental problems are concerned. The hydrogen energy has attracted attention as an alternative energy to fossil energy. It is required for practical applications to establish the effective and reliable technologies of hydrogen storage and transportation. Mg based alloys are one of the most promising materials among all hydrogen storage alloys because Mg is a low cost material and can absorb a large amount of hydrogen up to 7.6wt.%, as MgH_2 (1). Due to the structure of MgH_2 is a rutile-type (Fig.1), which stores hydrogen stably, the hydrogen absorption/desorption kinetics need high temperature about 300°C and are too high to use in service. Much effort has been spent in alloying with transition metals (Ni, Fe, Cu etc.) and producing thin film to destabilize MgH_2 for improving its reactivity and it is one of the different approaches from above ways that the crystal structure of MgH_2 will be modified by means of multi-layering some films made of these elements. Vermeulen et al. have shown the possibility of improving Mg reaction kinetics for decreasing the desorption temperature in the system of Mg/Ti multi-layer films (2)(3). And Tao et al. have shown 20at.%Ti-Mg effected changing the structure of MgH_2 and then the hydrogen absorption (4)(5).

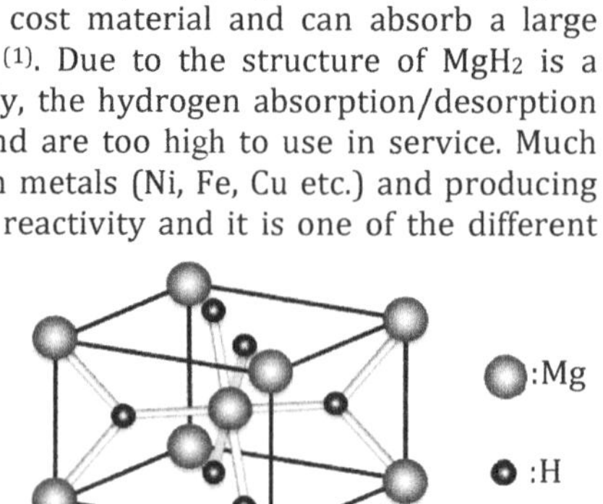

Fig. 1 Crystal structure of rutile type

In this research, we prepared two types of multi-layer films that are Mg/Pd and Mg/Ti/Pd with Pulsed Laser Deposition and measured the hydrogen desorption temperatures of these samples with differential scanning calorimetry. And we evaluate the effect of Ti layer to Mg layer for hydrogen desorption property in multi-layered.

2 EXPERIMENTS

2.1 Preparation

Two types of multi-layer films, sample I: Mg/Pd and sample II: Mg/Ti/Pd, were prepared with pulsed laser deposition designed by ourselves. The disks of Mg (bulk, 99.9% purity, Nilaco corp., φ20×6mm) and Ti (bulk, 99.9% purity, Nilaco corp., φ15×6mm) and Pd (bulk, 99.95% purity, Nilaco corp., φ5×6mm) were the evaporation sources, and a polyimide film (7.5μm thickness, Kapton, DUPONT-TORAY CO.,LTD) was a substrate (50×50mm). The above sources were evaporated by irradiating the Pulsed Nd:YAG laser forth harmonic generation (wavelength:λ=266nm) (Quanta-Ray INDI, Spectra-Physics corp.) The vacuum chamber was evacuated to around 5.0×10^{-4} Pa and the evaporated materials were piled on the substrate in order. Two samples, sample I: Polyimide/Mg/Pd and sample II: Polyimide/Mg/Ti/Pd just after multi-layered, are called sample I as-prepared and sample II as-prepared, respectively.

2.2 Activation and hydrogenation

Both the as-prepared samples were activated and hydrogenated under the series of conditions shown in Table 1. The samples hydrogenated with the process of Table 1 are called sample I after hydrogenation and sample II after hydrogenation, respectively.

2.3 Dehydrogenation temperature measurement

The hydrogen desorption temperatures of the samples after hydrogenation were measured with a differential scanning calorimeter (DSC-60, Shimazu Co., Ltd.). The measurement conditions are shown in Table 2. For the comparison of hydrogen desorption temperature, the dehydrogenation temperature of MgH_2 powder produced by Bio Coke Lab Co., Ltd. was also measured under the same conditions.

Table 1 Conditions of activation and hydrogenation

	Pressure [MPa]	Time [min]	Temperature [°C]
Vacuum dealring	-	60	200
Hydrogenation	2.5 (H_2)	30	200
Vacuum dealring	-	60	200
Hydrogenation	2.5 (H_2)	30	200
Vacuum dealring	-	120	200
Hydrogenation	2.5 (H_2)	60	200

Table 2 Conditions of dehydrogenation temperature measurement

Cell	-	Aluminium
Reference material	-	α-Al_2O_3
Atmosphere	-	Nitrogen
Gas flow rate	[ml/min]	50
Temperature increase rate	[°C/min]	10
Measurement start temperature	[°C]	100
Measurement finish temperature	[°C]	500
Holding time	[min]	1

3 RESULTS

3.1 Sample properties

Fig.2 shows the schematic drawings of the layer structure in sample I and sample II. Table 3 shows the weights of Mg and Ti and the thicknesses of them. The deposited volumes of Mg and Ti were obtained the weights of them divided by their densities, and the thicknesses of each material were those volumes divided by the deposition area (45×45mm).

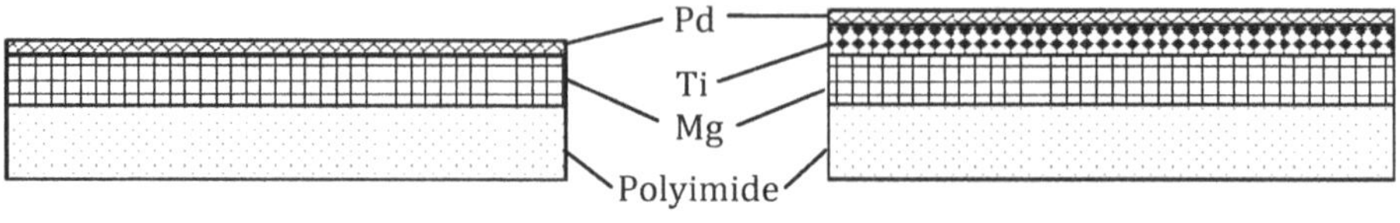

(a) Sample I **(b) Sample II**

Fig. 2 Schematic drawings of sample I and sample II in multi-layer

Table 3 Appearances of samples

Sample No. (sample type)		I (Polyimide/Mg/Pd)	II (Polyimide/Mg/Ti/Pd)
Material mass of Mg	[mg]	0.965	1.14
Thickness of Mg	[nm]	274	323
Material mass of Ti	[mg]	-	0.568
Thickness of Ti	[nm]	-	62.2

3.2 Hydrogen desorption temperature

Fig.3 shows the results of DSC measurement and so that the hydrogen desorption temperatures of sample I after hydrogenation, sample II after hydrogenation and MgH_2 powder sample.

Firstly, the curves of sample I and sample II were declining downward to the right with increasing temperature shown in Fig.3 (b) and (c). And the cause of this behavior would be the change of the specific heat with the softening of the polyimide film using as the substrates due to the elevated temperature.

Secondly, the dehydrogenation temperatures of both MgH_2 powder for comparison and sample I were around 440°C shown in Fig.3 (a) and (b). And that means the hydrogen desorption temperature did not decrease due to Pd layered in sample I. Although Pd has the effect to promote the hydrogen absorption reaction by the dissociation of hydrogen molecules [6], Pd does not have the effect to promote the hydrogen desorption reaction.

Finally, (c) and (d) of Fig.3 show that sample II desorbed hydrogen approximately 150°C. From the results, the hydrogen desorption temperature of sample II was improved about 300°C in comparison with sample I and MgH_2 powder sample. The difference of layer structure between sample I and sample II was the existence of Ti layer. As the result, the remarkable reduction of hydrogen desorption temperature would be caused by Ti layer between Mg layer and Pd one.

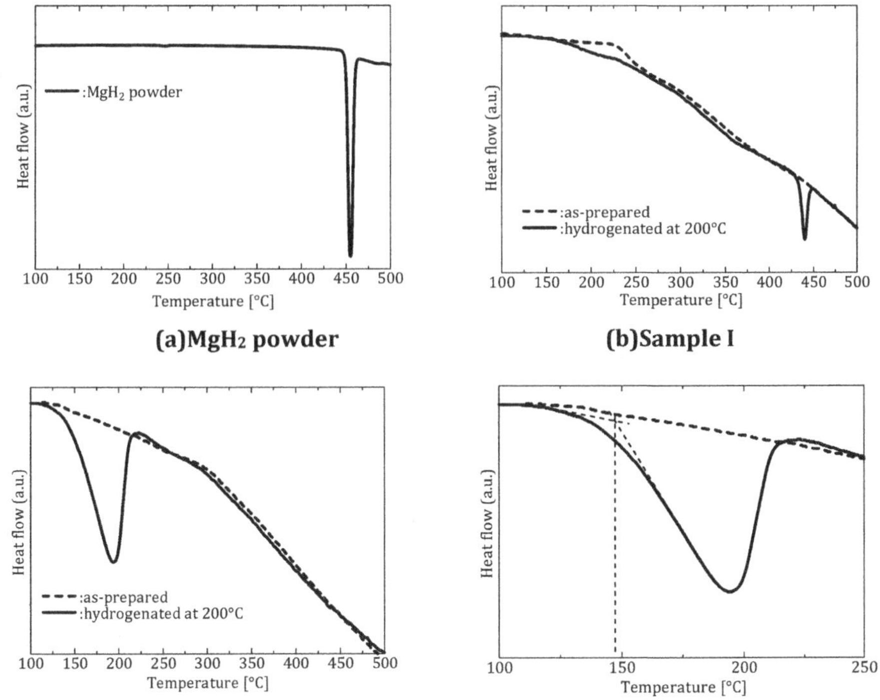

(a)MgH_2 powder **(b)Sample I**

(c)Sample II **(d)Sample II (100-250°C)**

Fig. 3 The hydrogen desorption temperature of samples

4 DISCUSSION

Table 4 shows the results that have reported so far the hydrogenation and dehydrogenation properties of Mg based alloys. Song et al. prepared the sample (Mg-10wt.%Al_2O_3), and it absorbed hydrogen at 300°C (1.1MPa) and desorbed hydrogen at 300°C (0.05MPa). Wang et al. prepared the sample (Mg-20wt.%TiO_2), and it absorbed hydrogen at 350°C (2.0MPa) and desorbed hydrogen at 330-350°C (0.1MPa). Liang et al. prepared the sample (MgH_2-5at.%Ni), and it absorbed hydrogen at 200°C (1.0MPa) and desorbed hydrogen at 300°C (0.015MPa). These results were obtained using powder samples prepared with ball milling. The sample II showed the remarkable dehydrogenation property in film. Vermeulen et al. showed that Mg-20at.%Ti thin film has a fcc structure in the after hydrogenation state and the Mg-H bonds in that structure were weaker than that of rutile structure (2)(3). Although we omit to show the results of

Table 4 The study results of Mg based alloy

Material	Method	Temperature [°C]	Pressure [MPa]	Reference
Mg-10wt.%Al_2O_3	Ball Milling	T_{abs}:300 T_{des}:300	P_{abs}:1.1 P_{des}:0.050	(7)
Mg-20wt.%TiO_2	Ball Milling	T_{abs}:350 T_{des}:330-350	P_{abs}:2.0 P_{des}:0.1	(8)
MgH_2-5at.%Ni	Ball Milling	T_{abs}:200 T_{des}:300	P_{abs}:1.0 P_{des}:0.015	(9)

X-ray diffraction because of limitations of the pages, it is possible that the fcc structure might exist in the part of Mg layer after hydrogenation.

In this study, the weaker Mg-H bonds would be caused due to the laminating of Ti and the rearrangement of the metal lattice.

5 CONCLUSION

In this paper, we prepared two samples (Mg/Pd and Mg/Ti/Pd) and evaluated their hydrogen desorption properties. Mg/Pd laminate desorbed hydrogen about 440°C and Mg/Ti/Pd laminate desorbed about 150°C. The remarkable improvement of dehydrogenation property would be caused by the structural change of Mg lattice with Ti addition and the transition of MgH_2 structure due to the hydrogenation.

REFERENCES

(1) I.P. Jain, C. Lal, A. Jain, Hydrogen storage in Mg: A most promising material, INTERNATIONAL JOURNAL OF HYDROGEN ENERGY 35 (2010) 5133-5144

(2) P. Vermeulen, P.C.J. Graat, H.J. Wondergem, P.H.L. Notten, Crystal structure of Mg_yTi_{100-y} thin film alloys in as-deposited and hydrogenated state, INTERNATIONAL JOURNAL OF HYDROGEN ENERGY 33 (2008) 5646-5650

(3) P. Vermeulen, H.J. Wondergem, P.C.J. Graat, D.M. Borsa, H. Schreuders, B. Dam, R. Grissen, P.H.L. Notten, In situ electrochemical XRD study of (de)hydrogenation of Mg_yTi_{100-y} thin films, The Royal Society of Chemistry (2008) 3680-3687

(4) S.X. Tao, P.H.L Notten, R.A. van Santen, A.P.J. Jansen, Density functional theory studies of the hydrogenation properties of Mg and Ti, The American Physical Society (2009) 14121 1-7

(5) S.X. Tao, P.H.L Notten, R.A. van Santen, A.P.J. Jansen, Dehydrogenation properties of epitaxial (100) MgH_2/TiH_2 multilayers – A DFT study, Computational Materials Science 50 (2011) 2960-2966

(6) A.Krozer, B.Kasemo, Equilibrium hydrogen uptake and associated kinetics for the Mg-H_2 system at low pressures, J Phys. Condens. Matter 1 (1989):1533-1538

(7) Wang P, Wang AM, Zhang HF, Ding BZ, Hu ZQ, Hydrogenation characteristics of Mg-TiO (rutile) composite, J Alloy Compds (2000) 313:218-223

(8) Liang G, Huot J, Boily S, Nestea AV, Schulz R, Catalytic effect of transition metals on hydrogen sorption in nanocrystalline ball milled MgH_2-Tm (Tm_5Ti, V, Mn, Fe and Ni) systems, J Alloys Compds (1999) 292(1-2):247-252

(9) Song MY, Bobet J-L, Darriet B, Improvement in hydrogen sorption properties of Mg by reactive mechanical grinding with Cr_2O_3, Al_2O_3, and CeO_2, J Alloys Compds (2002) 340:256-262

Reduction of hydrogen embrittlement cracking of stainless steel SUS316L by cavitation peening

H. Soyama, O. Takakuwa
Department of Nanomechanics, Tohoku University, Japan

ABSTRACT

In order to demonstrate reduction of hydrogen embrittlement cracking by improving residual stress using cavitation peening, crack growth of stainless steel SUS316L exposed to hydrogen was investigated comparing with and without cavitation peening by a plate bending fatigue test with a notched specimen. The tensile residual stress was introduced into specimen surface by an angle grinder. It was revealed that the residual stress of the specimen surface was improved from tension to compression by cavitation peening using a cavitating jet in air, and the crack growth rate of the specimen exposed to hydrogen was supressed by cavitation peening.

1 INTRODUCTION

In order to realize sustainable society using hydrogen as main energy source, hydrogen embrittlement cracking should be investigated and it is necessary to suppress the hydrogen embrittlement cracking. The hydrogen embrittlement cracking was accelerated, when tensile stress was introduced into surface. Then the hydrogen embrittlement cracking might be reduced by introducing compressive residual stress.

A conventional method to introduce compressive residual stress is shot peening. It was reported that shot peening enhanced the hydrogen embrittlement cracking (1), and it might be caused by the increasing dislocation density by shot peening. A peening method using cavitation impact, which normally caused severe damage in hydraulic machineries such as pumps and valves, was developed by authors (2), (3). It is called "cavitation shotless peening" as shots are not required (4)-(10), or simply called as "cavitation peening" (11)-(15). In the case of cavitation peening, micro strain was reduced with introducing compressive residual stress (10), as there were no solid collisions. Thus, the cavitation peening might reduce hydrogen embrittlement cracking by improving residual stress from tension to compression. Although it was already reported by the authors that the hydrogen embrittlement cracking arisen in thermal refining stainless steel was supressed by cavitation peening (16), it was not clear whether the hydrogen embrittlement cracking in tensile residual stress would be reduced or not by introducing compressive residual stress using cavitation peening.

In the case of conventional cavitation peening, cavitation bubbles were generated by injecting a high speed water jet into a water filled chamber. It is called a cavitating jet in water. Soyama et al. successfully realized a cavitating jet in air without a water filled chamber, by injecting a high speed water jet into a low speed water jet which was injecting into air using a concentric nozzle (17)-(19). The optimized cavitating jet in air can introduce large compressive residual stress into sub-surface comparing with the cavitating jet in water (20). As the cavitating jet in air can treat without a water filled chamber, applications would expand.

In the present paper, in order to demonstrate the reduction of the hydrogen embrittlement cracking by cavitation peening improving residual stress from tension to compression, the crack growth of hydrogen charged specimen was investigated with and without cavitation peening using a displacement controlled plate bending fatigue test with a notched specimen. The tested material was stainless steel Japanese Industrial Standards JIS SUS316L. In order to introduce tensile residual stress into the surface, the tested stainless steel was grounded by an angle grinder, then specimen was treated by the cavitating jet in air to introduce compressive residual stress.

2 EXPERIMENTAL FACILITIES AND PROCEDURES

Figure 1 illustrates a cavitation peening system using a cavitating jet in air. A low speed water jet was injected into tank A without filling a water in the tank and a high speed water jet was injected into the low speed water jet using a concentric nozzle as shown in Fig. 2. The throat diameter for the high speed water jet, d_H, was 1 mm and the inner diameter of the nozzle for the low speed water jet, d_L was 30 mm. As the aggressive intensity of the cavitating jet was affected by the injection pressure of the low speed water jet, p_L, at constant injection pressure of the high speed water jet, p_H, p_H and p_L were chosen as 30 MPa and 0.05 MPa considering previous report (19). The standoff distance from the nozzle to the surface of the specimen, s, was set as 45 mm.

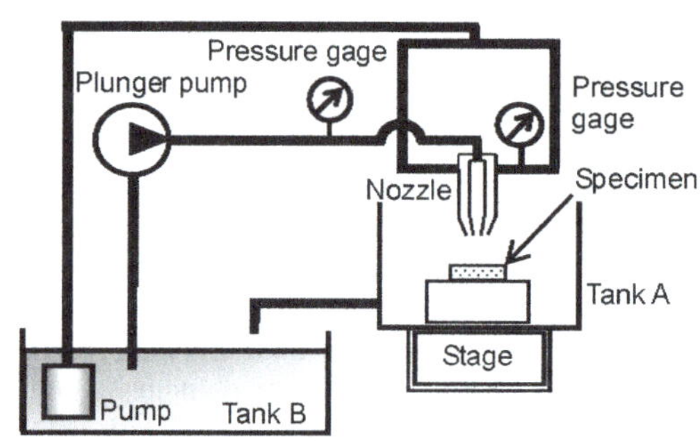

Fig. 1 Schematic diagram of cavitation peening system

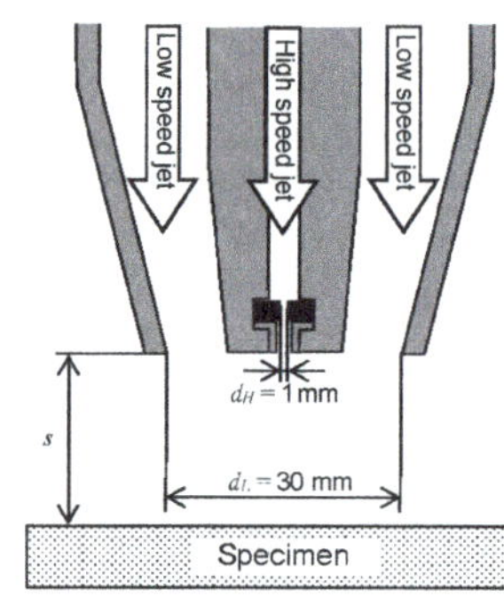

Fig. 2 Nozzle geometry

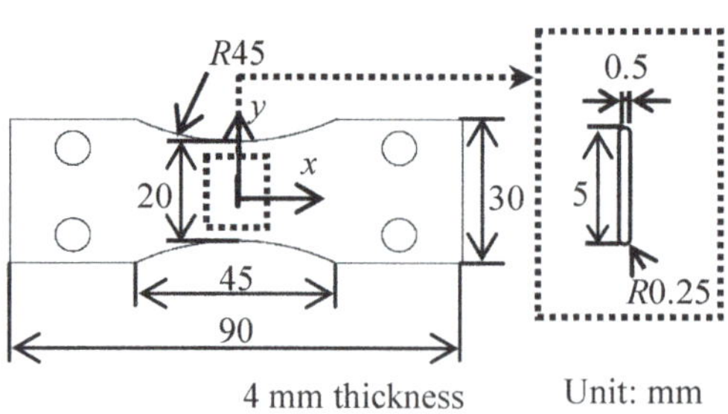

Fig. 3 Geometry of test specimen

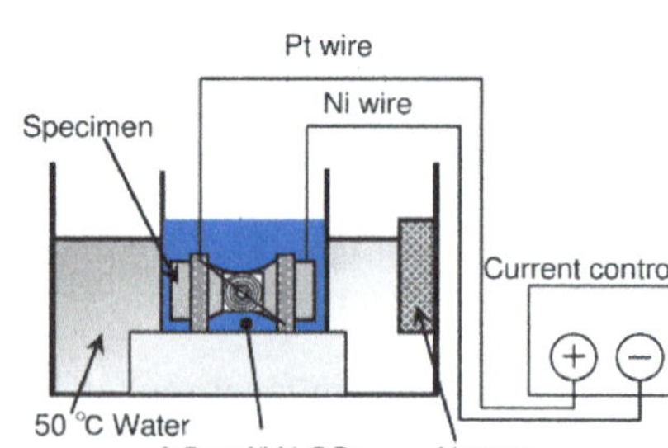

Fig. 4 Schematics of hydrogen charge

Tested material was stainless steel Japanese Industrial Standards JIS SUS316L. The tensile residual stress was introduced by an angle grinder. Figure 3 illustrates the schematic diagram of the test specimen. The notch which size was 5 mm in length, 0.5 mm in width and 0.25 mm in depth was introduced by a milling machine after cavitation peening to avoid the deformation during cavitation peening. The cathodic hydrogen charging as shown in Fig. 4 was conducted after introducing the notch. The residual stress was measured by a 2D X-ray diffraction method (21). The x and y were defined as Fig. 3. For example, the residual stress in longitudinal direction of the specimen was σ_x. The crack propagation was measured by using a displacement controlled plate bending fatigue test at stress ratio $R = -1$.

3 RESULTS

Figure 5 shows the aspect of the cavitating jet in air. The low speed water jet formed wavy pattern, and the cavitation cloud, which was white region, was shown in the low speed water jet as same as previous reports (18), (19). As the cavitation cloud around the high speed water jet was shedding periodically, the same frequency of the cloud shedding was appeared in the low speed water jet as wavy pattern (18).

In order to show the effect of cavitation peening CP on residual stress, Fig. 6 reveals the residual stress before and after cavitation peening. After the grinding, the tensile residual stress was introduced such as $\sigma_x = 276 \pm 25$ MPa and $\sigma_y = 680 \pm 25$ MPa. After cavitation peening at scanning speed of 6 mm/min, the compressive residual stress was introduced, and the residual stress σ_x and σ_y were -476 ± 25 MPa and $= -422 \pm 25$ MPa, respectively. Namely, cavitation peening using the cavitating jet in air improved the residual stress from tension to compression. Regarding previous report (19), in the case of thermal refining stainless steel, the cavitating jet in air introduced compressive residual stress about 350 μm in depth.

Figure 7 illustrates the crack growth with and without hydrogen charge for non-peened specimens and cavitation peened specimens in order to reveal effect of the cavitation peening on the crack propagation. Figure 7 (a) reveals the crack length $2a$ with the number of cycles, N, and Fig. 7 (b) shows the crack propagation rate da/dN with N. In the case of non-peened specimens, the crack of the specimen with hydrogen charge developed quicker than that without hydrogen charge. For example, the specimen without hydrogen charged was broken at 5.38×10^4, however, the hydrogen charged specimen was broken at 4.02×10^4. As shown in Fig. 8, many secondary cracks were

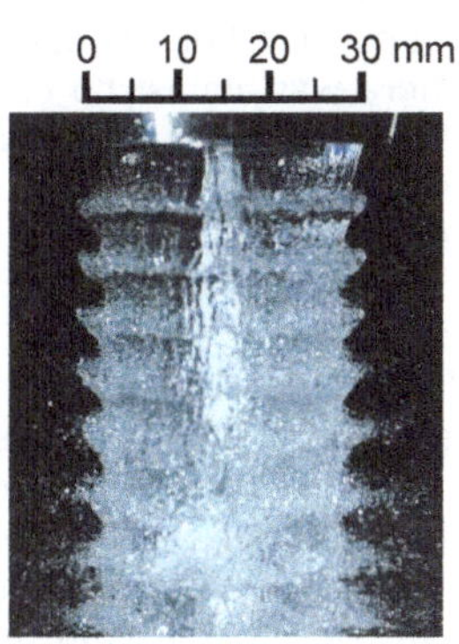

Fig. 5 Aspect of cavitating jet in air

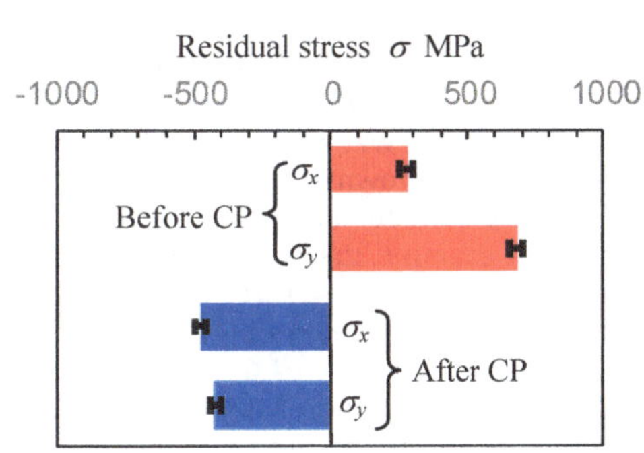

Fig. 6 Introduction of compressive residual stress by cavitation peening CP

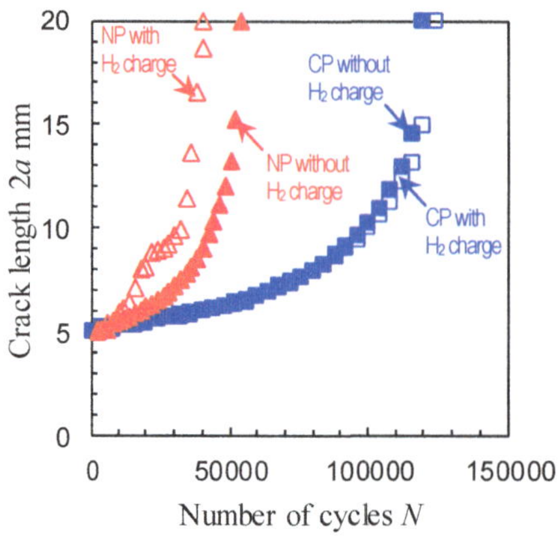

(a) Crack length with number of cycles

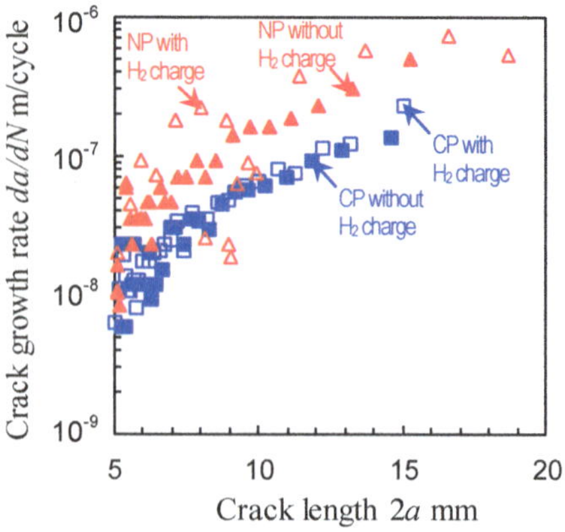

(b) Crack growth rate with number of cycles

Fig. 7 Reduction of crack growth of specimen with hydrogen charge by cavitation peening

(a) Without hydrogen charge (N = 18,000)

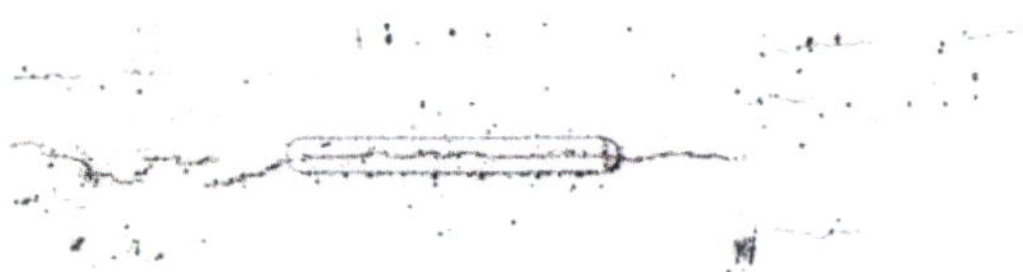

(b) With hydrogen charge (N = 18,000)

Fig. 8 Aspect of crack growth of non-peened specimen

(a) Without hydrogen charge (N = 18,000)

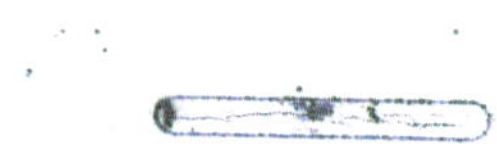

(b) With hydrogen charge (N = 18,000)

Fig. 9 Aspect of crack growth of cavitation peened specimen

generated on the surface of the non-peened specimen with hydrogen charge and combined together. On the other hand, the number of cycles to failure of cavitation peened specimen with and without hydrogen charge were 1.2×10^5 and 1.24×10^5. As shown in Fig. 9, the crack lengths of cavitation peened specimens were shorter than that of non-peened specimens, and the secondary crack was not observed on the surface of the cavitation peened specimen with hydrogen charge. There was almost no effect of hydrogen charge on the cavitation peened specimen. The life time of the hydrogen charged specimen was extend about three times by cavitation peening. In the case of da/dN with N as shown in Fig. 7 (b), the data of the non-peened specimen with hydrogen charge were plotted upper side of that of non-peened specimen without hydrogen charge. For example, the crack growth rate da/dN of the non-peened specimen without hydrogen charge at $2a$ = 12.1 mm was 2.3×10^{-7}, da/dN of the non-peened hydrogen charged specimen at $2a$ = 11.4 mm was 3.7×10^{-7}. Namely, da/dN of the non-peened specimen was accelerated 1.6 times by hydrogen charge. Note that da/dN of cavitation peened specimen without hydrogen charge was lower than that of the non-peened specimen. Even though the cavitation peened specimen exposed to hydrogen, da/dN was lower than that of the non-peened specimen without hydrogen charge. For example, da/dN of the cavitation peened specimen without hydrogen charge at $2a$ = 11.8 mm was

9.3×10^{-8} and that of hydrogen charged specimen at $2a$ = 12.2 mm was 1.2×10^{-7}. It was clearly observed that da/dN was suppressed by cavitation peening. It was concluded that cavitation peening reduced hydrogen embrittlement cracking.

4 CONCLUSIONS

In order to demonstrate the reduction of hydrogen embrittlement cracking improving the residual stress by cavitation peening, the tensile residual stress was introduced into stainless steel SUS316L by the angle grinder, and the specimen was treated by cavitation peening, then the hydrogen was put into the specimen by cathodic charge method. The crack propagation was evaluated by the plate bending fatigue test. It was concluded that the cavitation peening using the cavitating jet in air improved the residual stress from tension to compression and the hydrogen embrittlement cracking was reduced.

ACKNOWLEDGEMENTS

This work was partly supported by JSPS KAKENHI Grant Number 24360040. The authors thank Mr. M. Mikami, technician, Tohoku University for his help with the experiment.

REFERENCES

(1) Brass A.M. & Chene, J. (1991) *J. Mater. Sci.* **26**, 4517-4526.

(2) Soyama, H., Yamauchi, Y., Ikohagi, T., Oba, R., Sato, K., Shindo, T. & Oshima, R. (1996) *J. Jet Flow Eng.* **13**-1, 25-32.

(3) Soyama, H., Kusaka, T. & Saka, M. (2001) *J. Mater. Sci. Lett.* **20**, 1263-1265.

(4) Soyama, H., Saito, K. & Saka, M. (2002) *J. Eng. Mater. Technol.* **124**, 135-139.

(5) Odhiambo, D. & Soyama, H. (2003) *Inter. J. Fatigue* **25**, 1217-1222.

(6) Soyama, H., Sasaki, K., Odhiambo, D. & Saka, M. (2003) *JSME Inter. J.* **46**A, 398-402.

(7) Soyama, H., Macodiyo, D. O. & Mall, S. (2004) *Tribology Lett.* **17**, 501-504.

(8) Soyama, H. & Macodiyo, D. O. (2005) *Tribology Lett.* **18**, 181-184.

(9) Soyama, H., Shimizu, M., Hattori, Y. & Nagasawa, Y. (2008) *J. Mater. Sci.* **43**, 5028-5030.

(10) Soyama, H. & Yamada, N. (2008) *Mater. Lett.* **62**, 3564-3566.

(11) Sekine, Y. & Soyama, H. (2009) *Review of Automotive Eng.* **30**, 393-399.

(12) Takakuwa, O., Nishikawa, M. & Soyama, H. (2011) *Mater. Sci. Technol.* **27**, 1422-1426.

(13) Takakuwa, O., Nishikawa, M. & Soyama, H. (2011) *Key Eng. Mater.* **452-453**, 641-644.

(14) Seki, M., Soyama, H., Kobayashi, Y., Gowa, D. & Fujii, M. (2012) *J. Solid Mech. Mater. Eng.* **6**, 478-486.

(15) Soyama, H. (2014) *Materials Sciences and Applications*, **5**, 430-440.

(16) Takakuwa, O. & Soyama, H. (2012) *Inter. J. Hydrogen Energy*, **37**, 5268-5276.

(17) Soyama, H. (2004) *J. Eng. Mater. Technol.* **126**, 123-128.

(18) Soyama, H. (2005) *J. Fluids Eng.* **127**, 1095-1101.

(19) Soyama, H., Kikuchi, T., Nishikawa, M. & Takakuwa, O. (2011) *Surf. Coat. Technol.* **205**, 3167-3174.

(20) Soyama, H., Nishizawa, K. & Mikami, M. (2009) *Water Jetting Technology for LOHAS, International Academic Printing Co., Ltd.*, 133-137.

(21) Takakuwa, O. & Soyama, H. (2013) *Advances Mater. Phys. Chem.*, **3**, 8-18.

Evaluation for hydrogen embrittlement properties of ultra high-strength steel sheets by 4-Point Bending Technique

T. Hojo, F. Nishimura, H. Waki
Department of Mechanical Engineering, Iwate University, Japan

ABSTRACT

Hydrogen embrittlement properties of SCM435 and vanadium added steel sheets were investigated by using 4-Point Bending Technique (4PBT) to propose the standardization for the evaluation of hydrogen embrittlement properties. Maximum fracture strengths of steel sheets using 4PBT were decreased with increasing diffusible hydrogen concentration. vanadium added steels indicated high hydrogen embrittlement resistances compared with SCM435 steels. The tendencies between hydrogen concentration and maximum fracture strength obtained by 4PBT in SCM435 and vanadium added steels were similar to those using Conventional Strain Rate Technique (CSRT) although maximum fracture strengths using 4PBT were higher than those using CSRT in both steels.

1 INTRODUCTION

Recently, the requirements of applying ultra high-strength steel sheets for vehicles have risen to improve the car impact safety and to reduce the weight of the cars. In the present, the ultra high-strength steel sheets having tensile strength of more than 980 MPa grade have been applied for car impact safety parts. However, the occurrences of hydrogen embrittlement or delayed fracture were serious problems when tensile strength of ultra high-strength steel sheets exceeded more than 980 MPa as well as that of ultra high-strength structural steels.

In Japan, previous studies about hydrogen embrittlement were carried out for steel bars. It has been suggested that the estimation for occurrence of hydrogen embrittlement was carried out by making a comparison between critical absorbed hydrogen concentration (H_C) which was maximum hydrogen concentration without hydrogen embrittlement fracture of steels and environmentally absorbed hydrogen concentration (H_E) which was the hydrogen concentration entered from environment (1). In this method, Constant Load Technique (CLT) (1), Slow Strain Rate Technique (SSRT) (2) and Conventional Strain Rate Technique (CSRT) (3) were used to evaluate the H_C in the steel bars. In addition, it is possible to estimate the occurrence of hydrogen embrittlement by 4-Point Bending Technique (4PBT) (4) and U-Bend Method (5) in ultra high-strength steel sheets. However, it is expected that hydrogen embrittlement properties of steel sheets will be differ from those of steel bars because stress states in the steels between sheet and bar specimens are different. In this study, hydrogen embrittlement properties of SCM435 and vanadium (V) added steel sheets using 4PBT were investigated and compared with those using CSRT.

2 EXPERIMENTAL PROCEDURE

2.1 Steels

In this study, SCM435 and V added steel sheets of 1.6 mm in thickness with chemical compositions as listed in Table 1 were prepared. The SCM435 steels were annealed at 880°C for 60 min with quenching into oil and then tempered at 520°C for 60 min to make the tempered martensitic steels. On the other hand, the heat treatments of 920°C for 60 min with oil quenching, temporary tempering of 150°C for 30 min with air cooling and tempering at 600°C for 60 min were conducted to the V added steels. Tensile properties of these steels after heat treatment are shown in Table 2.

Table 1 Chemical compositions of steels used (mass%).

Steel	C	Si	Mn	P	S	Al	Cr	Mo	V	N
SCM435	0.35	0.24	0.79	0.023	0.016	0.036	1.09	0.15	-	0.005
V	0.41	0.20	0.70	0.005	0.005	0.035	1.19	0.65	0.30	0.004

2.2 Hydrogen analysis

Hydrogen charging was conducted by cathode charging method with hydrogen charging solution of 3 %-NaCl + 0-20 g/L-NH_4SCN solution at 25°C at current density of 0.1-10 A/m^2 for 48 hours. Diffusible hydrogen concentration (H_D) was controlled by change in the NH_4SCN content and the current density.

Table 2 Tensile properties of steels.

	TS	*YS*	*TEl*	*UEl*	*RA*
SCM435	1204	1056	5.9	2.8	6.5
V	1373	1239	11.5	7.5	13.0

TS (MPa): tensile strength, *YS* (MPa): yield strength or 0.2% proof stress, *TEl* (%): total elongation, *UEl* (%): uniform elongation, *RA* (%): reduction of area

The diffusible hydrogen concentration in the steels was quantified by TDS (Thermal Desorption Spectrometry Analysis) using gas chromatography. Specimens were heated from room temperature to 800°C at heating rate of 200°C/hour. The diffusible hydrogen was defined as hydrogen which evolved between room temperature and 300°C.

2.3 Hydrogen embrittlement tests by 4-Point Bending Technique

Rectangular samples with specimen size of 10 mm in width, 65 mm in length and 1.6 mm in thickness were used for 4-point bending specimens. Diffusible hydrogen was pre-charged to the samples by same conditions as hydrogen analysis without applying bending load for 48 hours to obtain the uniform hydrogen concentration in the steels. After hydrogen pre-charging, 4-point bending load was applied by 4-point bending jig shown in Fig. 1, and then 4-point bending cathode charging tests were conducted with hydrogen charging by the same condition as hydrogen pre-charging. In this case, applied stress was defined as a bending stress at outside of center of longitudinal direction of specimen, and calculated by following equation (1).

$$\sigma_b = 3P(L_2 - L_1) / bt^2 \quad (1)$$

where σ_b, P, L_1, L_2, b and t represent bending stress, applied load, distance of inner fulcrum (11 mm), distance of outer fulcrum (55 mm), specimen width and specimen thickness, respectively. Hydrogen embrittlement properties were evaluated by maximum fracture strength ($\sigma_{B\text{-}max}$) without hydrogen embrittlement fracture for 100 hours.

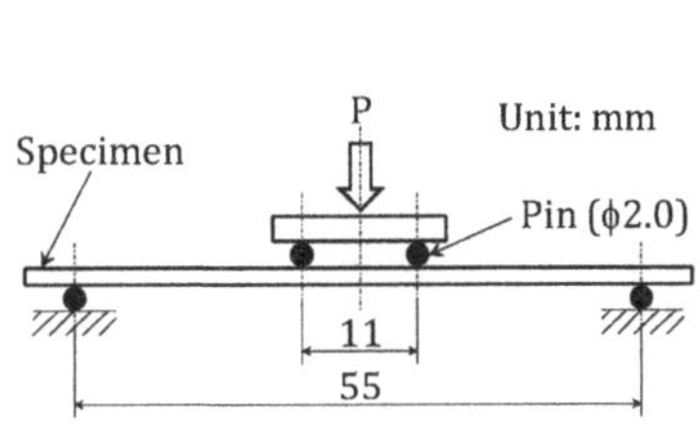

Figure 1 Illustration of 4-Point Bending testing jig.

3 RESULTS AND DISCUSSIONS

3.1 Hydrogen absorption properties

Hydrogen evolution curves of SCM435 steels varying NH_4SCN content and V added steels changing both NH_4SCN content and current density are shown in Fig. 2. In SCM435 steels, hydrogen was evolved at the temperature between room temperature and about 150°C. Diffusible hydrogen was increased with increasing in NH_4SCN content. On the other hand, hydrogen evolution was observed from room temperature to around 250°C in V added steels. Increase in the NH_4SCN content and the current density increased an amount of hydrogen desorption and shifted a peak temperature of the hydrogen evolution lower. Generally, it is known that the absorbed hydrogen in steels was trapped at grain boundaries (6), at lath boundaries, on dislocations (7) and at matrix/carbide interfaces (8) etc. Matrix of SCM435 steels exhibited fine and uniform martensite lath structure with high dislocation density. Thus it was considered that the hydrogen evolved between room temperature and 150°C corresponded to the hydrogen from grain boundaries, lath boundaries and dislocations. Moreover, Tsuchida et. al. (8) have reported that the diffusible hydrogen trapped at V carbide/matrix interfaces was evolved around 200°C, and this temperature was higher than those evolved from grain boundaries, lath boundaries and dislocations. Therefore it was suggested that V added steels indicated higher hydrogen evolution peak and amount of diffusible hydrogen than SCM435 steels due to a large amount of hydrogen trapping not only at grain boundaries, at lath boundaries and on dislocations but also at V carbide and at V carbide/martensite matrix interfaces.

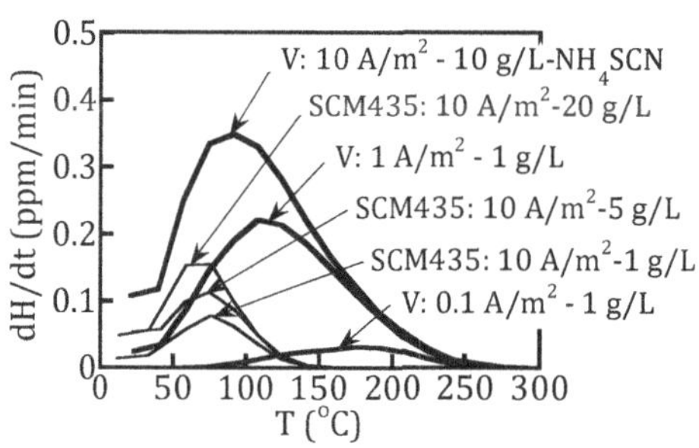

Figure 2 Hydrogen evolution curves of SCM435 and V added steels with different NH_4SCN content and current density.

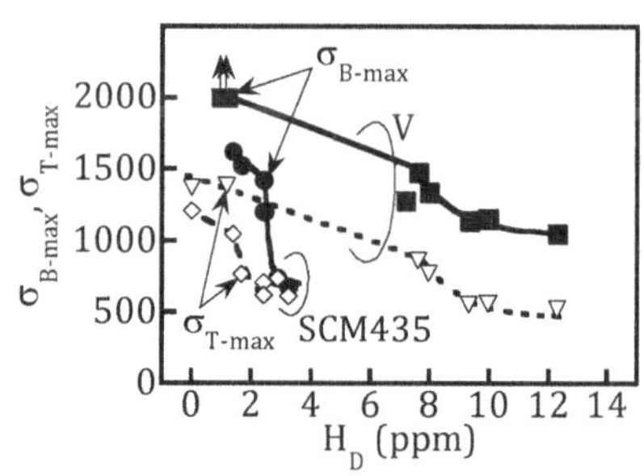

Figure 3 Variations in maximum fracture strength of 4PBT ($\sigma_{B\text{-}max}$) and CSRT ($\sigma_{T\text{-}max}$) as a function of diffusible hydrogen concentration (H_D) in SCM435 and V added steels.

3.2 Hydrogen embrittlement properties

Relationship between bending maximum fracture strength ($\sigma_{B\text{-}max}$) and diffusible hydrogen concentration (H_D) is shown in Fig. 3. Tensile maximum fracture strength ($\sigma_{T\text{-}max}$) obtained by Conventional Strain Rate Technique (CSRT) is also plotted in this figure. It was noted that there were two stages in $\sigma_{B\text{-}max}$ in SCM435 and V added steels, i.e. one stage was low H_D region which achieved small amount of deterioration of $\sigma_{B\text{-}max}$ and another was high H_D region which indicated significant decrease in the $\sigma_{B\text{-}max}$. A tendency between $\sigma_{B\text{-}max}$ and H_D of both steels using 4PBT was similar to that of CSRT even if $\sigma_{B\text{-}max}$ and H_D which dramatically deteriorated the $\sigma_{B\text{-}max}$ by 4PBT were higher than $\sigma_{T\text{-}max}$ and H_D which significantly deteriorated the $\sigma_{T\text{-}max}$ using CSRT in SCM435 steels. In the V added steels, it was confirmed that both $\sigma_{B\text{-}max}$ and $\sigma_{T\text{-}max}$ were deteriorated when H_D exceeded about 8.0 ppm, and $\sigma_{B\text{-}max}$ was higher than $\sigma_{T\text{-}max}$ as well as that in SCM435 steels. When hydrogen embrittlement resistance between SCM435 and V added steels was compared in 4PBT, it was noted that V added steels indicated more excellent $\sigma_{B\text{-}max}$ than SCM435 steels.

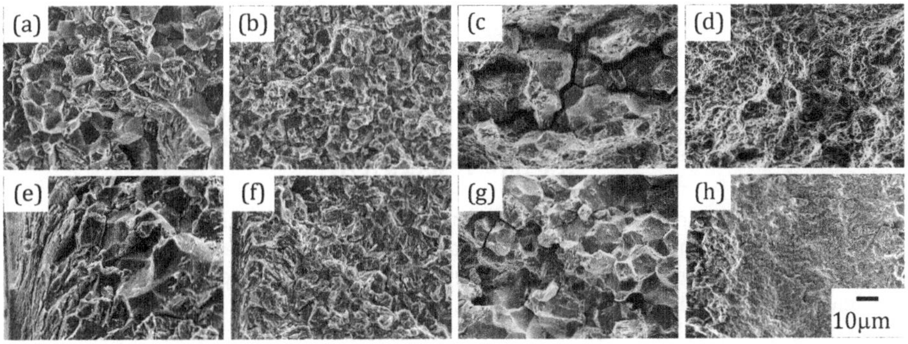

Figure 4 Scanning electron micrographs of (a-d) fracture surface and (e-h) fracture surface at edge conducted by (a, b, e and f) 4PBT and (c, d, g and h) CSRT of (a, c, e and g) SCM435 (H_D = 3.27 ppm) and (b, d, f and h) V added steels (H_D = 9.95 ppm).

Typical fracture surfaces of the steels after hydrogen embrittlement tests using 4PBT and CSRT are shown in Fig. 4. In the fracture surfaces using 4PBT, intergranular fracture mainly occurred in SCM435 and V added steels. Facet size of intergranular fracture of V added steels was small compared with that of SCM435 steels. The intergranular fracture with plastic deformation at initiation point of hydrogen embrittlement fracture was observed whereas the intergranular fracture without plastic deformation was appeared at inside from the initiation point when hydrogen embrittlement tests were conducted by 4PBT. On the other hand, if CSRT tests were carried out in the SCM435 steels, quasi cleavage and intergranular fractures were observed at high H_D region even if dimple fracture mainly occurred in low H_D region. In the V added steels, dimple fracture principally occurred at the several H_D. However, quasi cleavage and intergranular fractures were increased with increasing in the H_D as well as those of SCM435 steels. It was noted that intergranular fracture occurred at the side of the samples which was the initiation point of hydrogen embrittlement.

In this study, both SCM435 and V added steels exhibited fine and uniform martensite lath structure. In addition, V added steels possessed small size of prior austenite grain in comparison with that of SCM435 steels because of the precipitation of V carbides at the matrix during hot rolling. Moreover, it was also expected that fine vanadium carbides precipitated in the lath matrix while V added steels were tempered at 600°C. In general, it is known that hydrogen embrittlement properties depend on the grain size. Therefore, it was considered that high $\sigma_{B\text{-}max}$ of V added steels was attributed by the fine grain size compared with that of SCM435 steels. In addition, suppression of the increase in the hydrogen concentration at the grain boundaries due to the diffusible hydrogen trapping at the V carbides and carbide/matrix interfaces also contributed to the increase in the $\sigma_{B\text{-}max}$ in V added steels.

The H_D which dramatically deteriorated the $\sigma_{B\text{-}max}$ in 4PBT was higher than that in CSRT, and 4PBT indicted high $\sigma_{B\text{-}max}$ compared with $\sigma_{T\text{-}max}$ using CSRT at several H_D regions in SCM435 and V added steels. Fracture morphologies of 4PBT and CSRT were different as shown in Fig. 5.

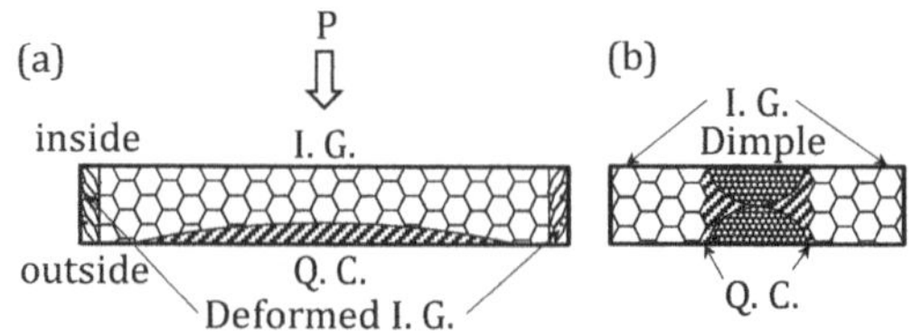

Figure 5 Illustration of fracture surface of samples obtained by (a) 4PBT and (b) CSRT, in which I. G. and Q. C. represent intergranular and quasi cleavage fracture, respectively.

If 4-point bending stress is applied to the specimens, tensile stress occurs at the outside of center of longitudinal direction whereas compressive stress is applied at the inside of specimen. According to the stress analysis by FEM at the center of the 4-point bending specimen, it was confirmed that shear stress acted at the side of specimen, and did not act at the internal of specimen although uniaxial tensile stress was applied to the specimen in the CSRT. This shear stress in the 4PBT likely caused intergranular fracture with the plastic deformation at the edge of specimen. Thus it was considered that $\sigma_{B\text{-}max}$ and the H_D which deteriorated the $\sigma_{B\text{-}max}$ of 4PBT were increased compared with those of CSRT because energy which supress the propagation of the hydrogen embrittlement cracks was increased.

4 CONCLUSIONS

To propose the standardization for the evaluation of hydrogen embrittlement, hydrogen embrittlement properties of SCM435 and V added steels were investigated by using 4-Point Bending Technique and Conventional Strain Rate Technique. The results are summarized as follows.

1) V added steels indicated the high hydrogen evolution peak at the high temperature region in comparison with SCM435 steels. This was caused by the hydrogen trapping not only at prior austenite grain boundaries, at martensite lath boundaries and on dislocations but also at vanadium carbides and V carbide/matrix interfaces in the V added steels.
2) V added steels possessed excellent hydrogen embrittlement resistance compared with SCM435 steels. This was achieved by the refinement of prior austenite grain size due to the precipitation of vanadium carbides in the matrix and the hydrogen trapping at the vanadium carbide and at V carbide/matrix interfaces.
3) It was confirmed that the tendencies between maximum fracture strength and diffusible hydrogen concentration using 4-Point Bending Technique and Conventional Strain Rate Technique were similar although the values of maximum fracture strength and diffusible hydrogen concentration which dramatically deteriorated the maximum fracture strength were different between 4-Point Bending Technique and Conventional Strain Rate Technique. This was attributed to the difference of hydrogen embrittlement fracture morphologies between 4-Point Bending Technique and Conventional Strain Rate Technique.

ACKNOWLEDGEMENT

The authors gratefully acknowledge financial support by Research Foundation for the Electrotechnology of Chubu.

REFERENCES

(1) N. Suzuki, N. Ishii, T. Miyagawa and H. Harada: Tetsu-to-Hagané, 79 (1993), 227.
(2) M. Wang, E. Akiyama and K. Tsuzaki: Mater. Sci. Eng. A, 398 (2005), 37.
(3) Y. Hagihara, T. Oba, N. Hisamori, H. Suzuki and K. Takai: Tetsu-to-Hagané, 97 (2011), 623.
(4) T. Hojo, K. Sugimoto, Y. Mukai and S. Ikeda: ISIJ Int., 48 (2008), 824.
(5) S. Takagi, Y. Toji, M. Yoshino and K. Hasegawa: ISIJ Int., 52 (2012), 316.
(6) A. Kimura and H. Kimura: J. Jpn. Inst. Met., 47 (1983), 807.
(7) T. Kushida: Advances in Delayed Fracture Solution, ISIJ, Tokyo, (1997), 40.
(8) T. Tsuchida, T. Hara and K. Tsuzaki: Tetsu-to-Hagané, 88 (2002), 771.

Hydrogen behavior in tensile-deformed Al-Zn-Mg alloy and Al-Mg alloy

T. Manaka[1], G. Itoh[2]
[1] Graduate School of Science and Engineering, Ibaraki University, Japan
[2] College of Engineering, Ibaraki University, Japan

ABSTRACT

Aluminum alloys with high content of Mg or with Zn and Mg are susceptible to stress corrosion cracking (SCC). Crack propagation in SCC is resulted from hydrogen embrittlement (HE) in these alloys. The mechanism of HE has not been elucidated yet. Therefore, it is necessary to analyse hydrogen behavior in these aluminium alloys. In this study, hydrogen behavior in tensile-deformed Al-Zn-Mg and Al-Mg alloys was investigated by means of hydrogen microprint technique. As a result, most grain boundaries with hydrogen emission were from 46 to 75 degree to tensile direction and nearly parallel to slip line in the two alloys.

1 INTRODUCTION

The Al-Zn-Mg alloys have been used as structural materials due to their high strength. The Al-Mg alloys with high Mg content have high strength and ductility. However, these alloys are susceptible to stress corrosion cracking (SCC). Crack propagation in SCC in these alloys is resulted from hydrogen embrittlement (HE) (1-3). The mechanism of HE has not been elucidated yet. Hydrogen in a metallic material is classified into two kinds: (a) impurity (internal) hydrogen introduced during the manufacturing process and (b) environmental (external) hydrogen introduced from the environmental in service. Environmental hydrogen plays major role in HE. However, the hydrogen behavior in materials has not been sufficiently revealed yet. Hence, it is necessary to investigate the hydrogen behavior in materials. Hydrogen microprint technique (HMPT) has been known as a method to visualize the microscopic location of hydrogen emission from specimen surface (4,5). It has been reported that hydrogen was emitted on slip line and some grain boundaries in Al-Mg alloy during tensile deformation by means of HMPT (6,7). Horikawa et al. observed that hydrogen emission on constitute particles, slip lines and grain boundaries in a tensile-deformed 7075 alloy (8). Hydrogen emission on slip lines is thought to be resulted from transportation by mobile dislocations (9). On the other hand, the mechanism of hydrogen emission on grain boundaries has not been realized yet. Preliminary results on Al-Mg alloy were reported elsewhere (10). In the present paper, hydrogen behavior of Al-Zn-Mg and Al-Mg alloys during tensile deformation was investigated by HMPT will be described.

2 PRINCIPLE OF HMPT AND CATHODIC ELECTROLYTIC CHRGING

2.1 HMPT

In HMPT, hydrogen atoms emitted from the specimen surface are visualized as silver particles using photographic emulsion layer containing silver bromide. The principle of HMPT is schematically shown in Fig.1. Following the oxidation-reduction reaction indicated in Eq. 1, the metallic silver atom will be produced by a strong reduction power of the hydrogen atom at the site where the hydrogen atom is emitted.

$$Ag^{+} + H \rightarrow Ag + H^{+} \quad (1)$$

The silver bromide particles that have not been reacted with hydrogen will dissolve out into the fixer, and finally only the metallic silver particles will remain on the specimen surface in gelatine film of the emulsion. Hence, it is possible to visualize the hydrogen emission site by observing the distribution of the remaining silver particles together with metallographic microstructure. When amount of hydrogen emission is small, soaking specimen into the developer before fixing is effective. The whole silver bromide particles will become metallic silver through this process.

2.2 Cathodic electrolytic charging

Cathodic electrolytic charging is an effective technique to enhance the invasion of hydrogen into materials. Although hydrogen atoms are generated in the cathode following Eq. 2, they usually recombine with each other to become hydrogen gas (Eq. 3), without invading the metal.

$$H^{+} + e^{-} \rightarrow H \quad (2)$$

$$2H \rightarrow H_2 \quad (3)$$

When ion of toxic element, such as arsenic, selenium and sulfur, are added into the electrolytic solution, recombination reaction is appreciably inhibited. As a result, the invasion of hydrogen is markedly promoted.

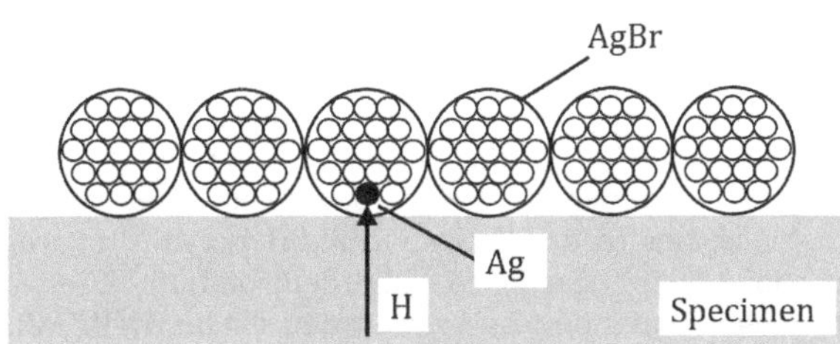

Fig.1 Principle of hydrogen microprint technique.

3 EXPERIMENTAL

The materials used in this study were an Al-Zn-Mg alloy supplied by Kobe steel Ltd. and Al-Mg alloy. The chemical composition of the specimens is shown in Table 1. In Al-Zn-Mg alloy, the ingot of 120 mm in thickness was homogenized at 450°C for 8h, hot-rolled at 400°C from 45 mm to 6mm and then cold-rolled to 1mm. The Al-9%Mg alloy was prepared from pure aluminum of 99.99% and pure magnesium of 99.9%. They were melted in air using $MgCl_2$ as a flux and cast in an iron mold. The ingot of 26.5mm in thickness was homogenized at 430°C for 18h, warm-rolled at 200°C from 13.0 mm to 1.1 mm and annealed at 400°C for 10min and then cold-rolled to 1.0mm. Tensile test pieces, with a gauge length of 12 mm along rolling direction and width of 12mm were cut from the sheets. The specimens were heat-treated as follows: Al-Zn-Mg alloy was solution-treated at 470°C for 1h, water-quenched and then aged at 120°C for 24h, the microstructure of which is presumed to consist of G.P. zones and /or η' ($MgZn_2$) precipitates beside the aluminium matrix (11). Al-Mg alloy was solution-treated at 500°C for 10min and then water-quenched, which will cause a microstructure composed only of the matrix with Mg in solution. The specimens were polished by abrasive paper up to #2000 in Al-Zn-Mg alloy, while #1500 and then electroltically polished in Al-Mg alloy. After that, hydrogen was introduced into Al-Zn-Mg alloy by cathodic electrolytic charging for 30min in a solution of pH 2.5 with 0.1mass% of NH_4SCN as hydrogen recombination inhibitor with current density of $10mA/cm^2$. A platinum wire was used as an anode. Hydrogen was charged only from gauge part in L-LT section and the other parts were insulated. The charged side of specimen was electrolytically polished after cathodic charging. Then, the electrolytically polished gauge portion of test pieces were coated with collodion film to improve the adhesion of emulsion to the specimen. Then, they were covered with photographic emulsion diluted by four times on the surface by the wire-loop method in a dark room. The emulsion-coated surface was shielded from light after it was dried. The test pieces were stretched at an initial strain rate of $6.94 \times 10^{-4}s^{-1}$ at room temperature. Plastic strain given was 4% in Al-Zn-Mg, and 9%, 12% and 21% in Al-Mg alloy. After stretching, specimens were immersed in formalin, soaked in developer, fixed with sodium thiosulfate aqueous solution, rinsed with tap water and then naturally dried. The specimens were observed with scanning electron microscope (SEM) equipped with an energy dispersive spectroscopy (EDXS) device. Also, surface relief of specimens in Al-Mg alloy was measured with a laser microscope.

Table 1 Chemical composition of the specimens in mass%.

Specimen	Si	Fe	Cu	Mg	Zn	Cr	Ti	Al
Al-Zn-Mg	0.008	0.005	0.002	2.391	5.791	0.000	0.014	Bal.
Al-Mg	<0.005	<0.001	<0.001	8.9	-	-	-	Bal.

4 RESULTS AND DISCUSSION

Figures 2 and 3 show the HMPT/SEM images. A number of white particles were observed at some grain boundaries of each specimen. These white particles were recognized as sliver particles from EDXS analysis. Hence, the hydrogen was emitted from some grain boundaries during tensile deformation. The number of grain boundaries with hydrogen emission was measured based on HMPT/SEM images of each specimen. The larger the tensile deformation amount, the greater the number of grain boundaries with hydrogen emission in Al-Mg alloys. Figure 4 shows the difference in the height between the two neighboring grains and the maximum gradient across the grain boundary with or without hydrogen emission. The amount of surface relief and maximum gradient of grain boundaries with hydrogen emission are relatively larger than those without hydrogen emission. It has been reported that the surface oxide film of matrix prevents hydrogen emission (12,13). The surface oxide film at grain boundaries was cracked by great surface relief and gradient during deformation, and hence the hydrogen was emitted from grain boundaries. Although the great surface relief and gradient were formed, grain boundaries without hydrogen emission were observed. This is probably associated with deformation mode in the grains. Figure 5 shows the angle of grain boundaries with hydrogen emission against slip line (θ_1) and against tensile direction (θ_2). Most grain boundaries with hydrogen emission are from 46 to 75 degree against tensile direction and nearly parallel to slip line in the two alloys. In Al-Zn-Mg alloy, θ_2 particularly concentrated from 46 to 60 degree. In a peak-aged Al-Zn-Mg alloy, precipitation free zones (PFZs) exist along grain boundaries. The PFZs deform more easily than the interior of the grains (11). It is suggested that hydrogen atoms are easily transported by dislocations in PFZ at the beginning of deformation. However, in this study, relationship between hydrogen behavior and PFZ was not revealed. Therefore, further study should be conducted to evaluate the effects of PFZ on hydrogen behavior.

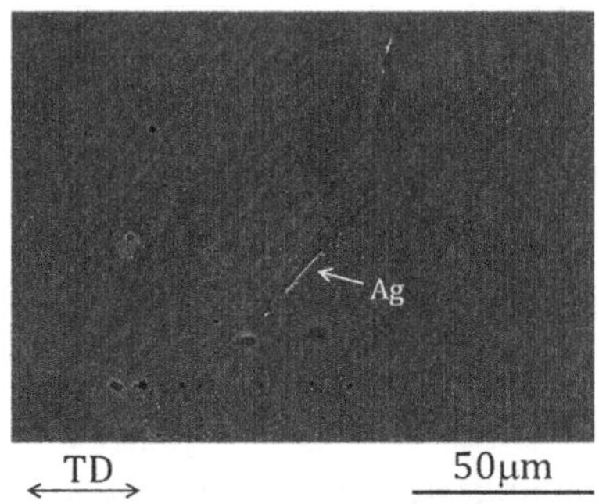

Fig.2 HMPT/SEM image of Al-Zn-Mg alloy stretched by 4%.

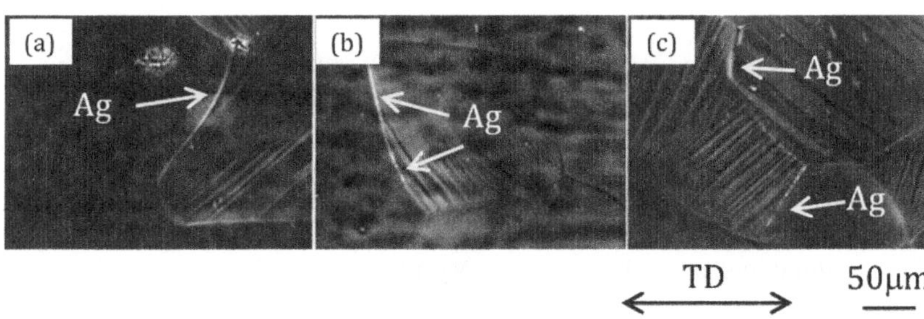

Fig.3 HMPT/SEM images of Al-Mg alloy stretched by 9%(a), 12%(b) and 21%(c), respectively.

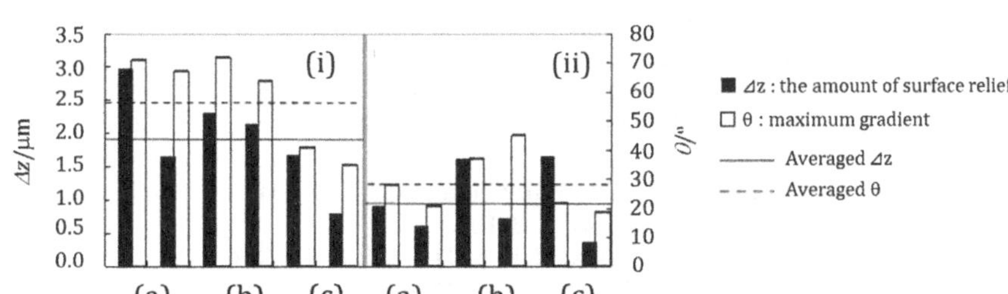

Fig.4 Difference in the height between the two neighboring grains, Δz, and maximum gradient angle in the surface profile, θ, for the grain boundaries with (i) and without (ii) hydrogen emission. Plastic strain is 9%(a), 12%(b) and 21%(c).

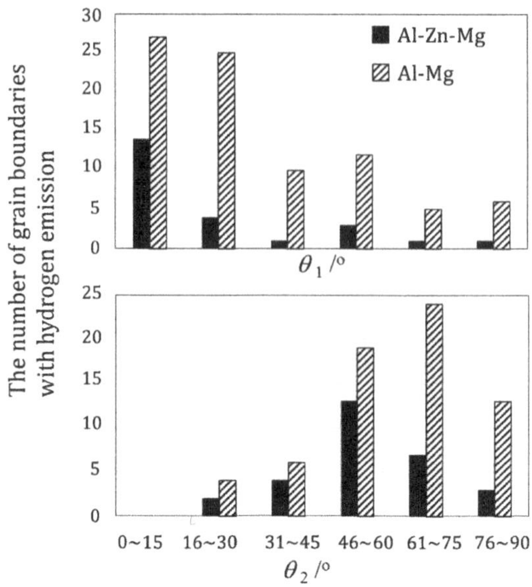

Fig.5 The angle of grain boundaries with hydrogen emission against slip line (θ_1) and against tensile direction (θ_2). The number in Al-Mg alloy is total of the three specimens.

5 CONCLUSION

Hydrogen behavior in tensile-deformed Al-Zn-Mg and Al-Mg alloys was investigated by HMPT. Most grain boundaries with hydrogen emission were from 46 to 75 degree to tensile direction and nearly parallel to slip lines. The amount of surface relief and maximum gradient of the grain boundary with hydrogen emission was relatively larger than those without hydrogen emission in Al-Mg alloy.

REFERENCES

[1] Murakami, Y. (1981) J. JILM **31**, 748-757.
[2] Ohnishi, T. & Higashi, K. (1984) J.JILM **34**, 675-681.
[3] Ohnishi, T. (1991) J.JILM **42**, 21-26.
[4] Nagao, A., Kuramoto, S. & Kanno, M. (1999) J. JILM **49**, 89-96.
[5] Kuramoto, S., Ichitani, K., Nagao, A. & Kanno, M. (2000) TETSU-TO-HAGANE **86**, 17-23.
[6] Horikawa, K. & Yoshida, K. (2004) Mater. Trans. **45**, 315-318.
[7] Izumi, T. & Itoh, G. (2006) JILM **56**, 478-482.
[8] Horilawa, K., Yamada, H. & Kobayashi, H. (2012) J.JILM **62**, 306-312.
[9] Itoh, G., Koyama, K. & Kanno, M. (1996) Scr. Mater **35**, 695-698.
[10] Koyama. R. & Itoh, G. (2014) Trans. Nonferrous Met. Soc. China **24**, to be published.
[11] Itoh, G., Eto, T., Miyagi, Y. & Kanno, M. (1988) J.JILM **38**, 818-839.
[12] Umeda, H., Itoh, G. & Kato, Y. (2006) J.JILM **56**, 203-209.
[13] Itoh, G., Izumi, T. & Tohyama, T. (2008) J.JILM **58**, 15-21.

Dynamic and quasi-static compressive properties of modified double-base propellant at low temperature

C. Sun[1], X. Chen[1], H. Li[2], Y. Zhen[1], J. Zhang[3]
[1] School of Mechanical Engineering, Nanjing University of Science and Technology, P.R. China
[2] Jinxi Industries Group Corporation, P.R. China
[3] Beijing Institute of Space Long March Vehicle, P.R. China

ABSTRACT

To obtain the dynamic and quasi-static mechanical properties of modified double-base propellant at low temperature, the uniaxial compressive experiments have been performed both on the universal testing machine and the Split Hopkinson Pressure Bar apparatus. True stress-true strain curves over a wide range of strain rate at -20℃ were presented, from 1.1×10^{-3} s^{-1} to 6×10^{3} s^{-1}. The results suggest that modified double-base propellant is a significant strain rate dependent material, and the initial modulus of elasticity and the yield strength increases along with the strain rate, which shows more strain rate sensitiveness at high strain compared at low strain rate.

1 INTRODUCTION

Propellant gain is the energy source of the solid rocket motor (SRM), and its structure integrity is the essential guarantee that the motor works well. Therefore, A successful grain design require any piece of solid propellant should hold its shape over an extended temperature range in a wide strain rate range (1). In general, the common approach to assess the safety and reliability of SRM is to analyze propellant grain structure integrity to conduct mechanical experiments and develop an accurate constitutive model. Many researchers have put great effort into this. Schapery studied nonlinear viscoelastic and viscoplastic behaviour of plastics and developed a thermodynamic formulation (2), (3). Hinterhoelzl developed a 3-D viscoelastic continuum damage model for solid propellant for wide use in Finite Element Model (4). Xu proposed a general 3-D nonlinear macroscopic constitutive law that models microstructure damage evolution upon straining through continuous void formation and growth (5). Matouš K was involved in modelling multiscale constitutive behaviour for solid propellant (6). Robert Neviere performed solid propellant quasi-static tensile test responses from -60℃ to +60℃ to investigate the HTPB non-linear viscoelastic behaviour (7). Baris Kalaycioglu simulated the extrusion process of double base solid propellant, whose model parameters were determined by quasi-static tensile test (8).

According to the aforementioned researches, these previous solid propellant works were focused on the quasi-static mechanical behaviour and the propellant response under high strain rate loading was not clear. Generally, high strain rate behaviour of materials is evaluated experimentally using Split Hopkinson Pressure Bar (SHPB) apparatus. After Kolsky (9) introduced the SHPB technique for dynamic testing, SHPB technique has become the widely used experimental technique to obtain stress-strain behaviour at high strain rate for materials, such as glass metals (10), ballistic gel (11), and polymeric materials (12) and so on. Based on the highlighted issues, the experiments were conducted here to obtain quantitative estimate of strain rates effect on static and dynamic compressive properties of Composite Modified Double Base(CMDB) propellant, at low temperature, -20℃. The compressive properties at room temperature, +20℃, has been researched by author, see more details in the reference (13).

2 EXPERIMENTAL METHODOLOGY

2.1 Materials and sample preparation

A typical Composite Modified Double Base (CMDB) propellant consists of double base binder, oxidizer (AP, RDX), fuel particle (powdered aluminium) and some other additives for improved bonding and burning, and usually the double base binder is nitrocellulose and nitroglycerine. The propellant studied here is RDX-CMDB, which the RDX content is approximately 45% by weight and Al particle content is approximately 10% by weight. Cylindrical specimens were machined from the tubular propellant grain. For static testing, the sizes of specimen are 10mm in diameter and 15mm in thickness, for SHPB testing, the sizes of specimen is 10mm in diameter and 4mm in thickness. After machined, both static and dynamic specimens were kept in thermal insulation chamber over 24 hours at 50℃ to get rid of the residual stress.

2.2 Static compression test

The CMDB specimens were tested under compression loading at -20℃, using universal testing machine. In order to minimize friction effect during the test, the slenderness ratio of specimen is 1.5. What's more, a thin film of lubricant was also applied to both specimens' interfaces in order to prevent any potential frictional constrain between the specimen's end and the anvils of the test machine. The temperature control box is set at -20℃ and all the specimens are kept in the temperature control box for at least 8 hours prior to starting. During the testing, four different strain rate deformations were applied on samples, with the velocity of the crosshead 1mm/min, 10mm/min, 100mm/min and 500mm/min. Three same measurements were conducted for each strain rate to make sure the data is well repeated.

2.3 Dynamic compression test

Basically, the SHPB apparatus consists of the striker bar, incident bar, transmitter bar and the energy absorber, as illustrated in Fig.1. Considering the CMDB propellant is kind of soft materials which wave impedance is low, all the bars are made in high strength aluminium alloy, LC4, to reduce the drastic impedance mismatch. The aluminium alloy's Young's modulus is 74GPa, yield stress is 490MPa, proportional limit is 370MPa, and density is 2.7×10^3kg/m^3. The diameter of the incident and transmitter bar is 14mm and the length is 1400mm, the striker bar's length is varied in the range of 200-500mm.

During the SHPB test, the specimen is located between incident bar and transmitter bar, when the striker bar impacts the incident bar, rectangular incident stress pulse ε_i is generated and travels along the incident bar until it hits the specimen. Part of the incident stress pulse reflects from the bar/specimen interface as reflected wave ε_r and part of it transmits through the specimen because of wave impedance mismatch. The transmitted pulse ε_t emitted from the specimen travels along the transmitter bar until it hits the end of the bar. It's assumed that the specimen is in force equilibrium and the specimen is deforming uniformly, it means this:

$$\varepsilon_{\mathrm{i}}(t)+\varepsilon_{\mathrm{r}}(t)=\varepsilon_{\mathrm{t}}(t) \tag{1}$$

Then the strain rate, strain and stress in the specimen can be obtained by the recorded strain history of incident and transmitter bars using the following equations：

$$\dot{\varepsilon}(t)=\frac{2C_0}{l_{\mathrm{s}}}\varepsilon_{\mathrm{r}}(t) \tag{2-1}$$

$$\varepsilon(t)=\frac{2C_0}{l_s}\int_0^t\varepsilon_{\mathrm{r}}(t)dt \tag{2-2}$$

$$\sigma(t)=\frac{E_0A_0}{A_{\mathrm{s}}}\varepsilon_{\mathrm{t}}(t) \tag{2-3}$$

Where C_0 is the elastic wave velocity in the bars, E_0 is Young's modulus of bars, l_s is specimen length, A_0 is cross-sectional area of the bars, A_s is cross-sectional area of specimen, and t is time duration.

The equation(2-2) and equation(2-3) are engineering strain and engineering stress, the true strain and true stress are related to engineering strain and engineering stress by:

$$\varepsilon_{\mathrm{T}}(t)=-\ln(1-\varepsilon(t)) \tag{3-1}$$

$$\sigma_{\mathrm{T}}(t)=(1-\varepsilon(t))\sigma(t) \tag{3-2}$$

3 EXPERIMENT RESULT AND DISCUSSION

3.1 Quasi-static compressive response

The compressive true stress–true strain curves of CMDB at low strain rates, $1.1\times10^{-3}s^{-1}$, $1.1\times10^{-2}s^{-1}$, $1.1\times10^{-1}s^{-1}$ and $5.5\times10^{-1}s^{-1}$, are shown in Fig.1. The data of these stress–strain curves were obtained by universal testing machine at -20℃, and each of the curves in the figures is repeated at least 3 times under identical loading conditions. In the Fig.1, all stress–strain curves of CMDB shared a common characteristic: a viscoelastic response in the initial stage, which is nearly linear, and then yields, coming into flow stage, at last, the drop at the end of the curve means failure behaviour of the materials.

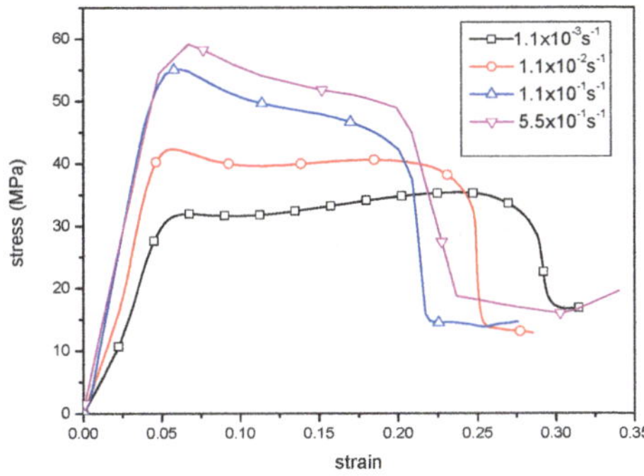

Fig.1 true stress-true strain curves of CMDB at quasi-static strain rates

Strain rate effect is apparent in the CMDB materials studied in this paper. It can be seen that the yield strength increases as the strain rate increases at quasi-static compressive loading. The yield stress is about 32MPa, at the lowest strain rate, $1.1\times10^{-3}s^{-1}$, and when the strain rate goes to $5.5\times10^{-1}s^{-1}$, the yield stress reach 59.1MPa, almost twice as much as the lowest strain rate. With regards to the yield strain, there is not very clear trend, it appears that yield strain increases slightly with increasing strain rate, which means the CMDB is ductile at quasi-static loading. Also, the initial linear region of the stress–strain diagram increases as the strain rate increases. At the lowest strain rate, $1.1\times10^{-3}s^{-1}$, elastic modulus is about 633MPa, and with increasing the strain rate to $5.5\times10^{-1}s^{-1}$, the elastic modulus raises to 1126MPa.

3.2 Dynamic compressive response

The true stress-true strain plot of CMDB at high strain rates is presented in Fig.2, including five different strain rates curves, $360s^{-1}$, $770s^{-1}$, $1000s^{-1}$, $2100s^{-1}$ and $6000s^{-1}$. All the dynamic specimens for SHPB experiment are kept at -20℃ for at least 8 hours prior to starting. In Fig.2, the stress–strain curves of CMDB at high strain rate exhibit same characteristic: a viscoelastic response in the initial stage, which is nearly linear, then yields, followed by a descending segment, which means failure in specimen. It could be explained that the material is very likely to be brittle at low temperature and high strain rate, so that the observed behaviour may have been compromised by fracture of the material.

It can be observed that the yield strength is enhanced significant at high strain rate loading compared with that at quasi-static loading. Further, it increases with increasing strain rate. At the $360s^{-1}$ strain rate, the deformation in specimen is smaller due to the low impact speed of striker, so the stress-strain curve is not complete. The yield stress is varied from 105MPa at 770 s^{-1} strain rate to as high as 165MPa at 6000 s^{-1} strain rate, which shows more strain rate sensitiveness at high strain compared at low strain rate. In opposition, the yield strain decreased significantly with and increasing strain rate. Also, the elastic modulus of CMDB is found to have the same trend as yield stress: the initial elastic modulus is increased with the strain rate, varied from about 4.5GPa at strain rate of $360s^{-1}$ to about 17.8GPa at stain rate of 6000 s^{-1}.

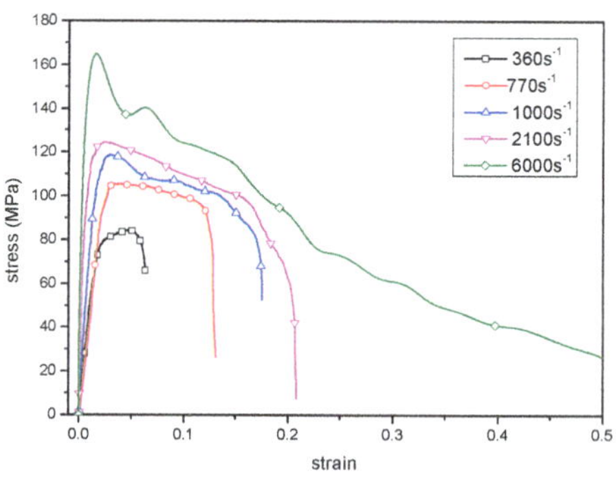

Fig.2 true stress-true strain curves of CMDB at dynamic strain rates

4 CONCLUSION

The quasi-static and dynamic compressive testing of CMDB materials at low temperature, -20 ℃, have been successfully performed at a wide strain rate range, from $1.1\times10^{-3}s^{-1}$ to $6\times10^{3}s^{-1}$, using the conventional universal testing machine and the Split

Hopkinson Pressure Bar apparatus, respectively. The results suggest that the mechanical properties of CMDB materials studied in this paper show significant strain rate sensitivity. The yield stress, compressive modulus increased as the strain rate increased, What's more, this increase is accentuated at high strain rate, compared to the increase at low strain rate. The stress–strain curves of CMDB at high strain rate exhibit same characteristic: a viscoelastic response in the initial stage, which is nearly linear, then yields, followed by a descending segment, which means the material is very likely to be brittle at low temperature and high strain rate.

REFERENCES

(1) Marimuthu R, Nageswara Rao B. Development of efficient finite elements for structural integrity analysis of solid rocket motor propellant grains[J]. International Journal of Pressure Vessels and Piping, 2013, 111:131-145.

(2) R A Schapery. Nonlinear Viscoelastic and Viscoplastic Constitutive Equations Based on Thermodynamics[J]. Mechanics of Time-Dependent Materials, 1997, 1: 209-240.

(3) R.A.Schapery. Nonlinear viscoelastic and viscoplastic constitutive equations with growing damage[J]. International Journal of Fracture, 1999, 97:33-66.

(4) R M Hinterhoelzl, R A Schapery. FEM Implementation of a Three-Dimensional Viscoelastic Constitutive Model for Particulate Composites with Damage Growth[J]. Mechanics of Time-Dependent Materials, 2004, 8:65-94.

(5) Xu F, Aravas N, at al. Constitutive modeling of solid propellant materials with evolving microstructural damage[J]. Journal of the Mechanics and Physics of Solids, 2008, 56(5): 2050-2073.

(6) Matouš K, Inglis H M, at al. Multiscale modeling of solid propellants: From particle packing to failure[J]. Composites Science and Technology. 2007, 67(7): 1694-1708.

(7) Nevière Robert. An extension of the time–temperature superposition principle to non-linear viscoelastic solids[J]. International Journal of Solids and Structures, 2006, 43(17): 5295-5306

(8) Baris Kalaycioglu, M Husnu Dirikolu. An elasto-viscoplastic analysis of direct extrusion of a double base solid propellant[J]. Advances in Engineering Software, 2010, 41(9):1110-1114.

(9) Kolsky H. An investigation of the mechanical properties of materials at very high rates of loading[J]. Proceedings of the Physical Society. 1949, 62(11)

(10) Wang Y S, Hao G J, at al. High strain rate compressive behavior of Ti-based metallic glass matrix composites[J]. Intermetallics, 2014, 52(0):138-143.

(11) Casem Daniel T, Dwivedi Ajmer K, at al. Compression response of a thermoplastic elastomer gel tissue surrogate over a range of strain-rates[J]. International Journal of Solids and Structures, 2014, 51(11-12): 2037-2046.

(12) Acharya S, Mukhopadhyay A K, at al. High strain rate compressive behavior of PMMA[J]. Polymer Bulletin, 2014, 71(1): 133-149.

(13) Sun Chaoxiang, Ju Yutao, at al. Experimental Research on the Compressive Properties of Modified Double Base Propellant at Low and High Strain Rates[J]. Acta Armamentarii, 2013, 34(6): 698-703.

Mechanical property of microstructure in die-cast magnesium alloy evaluated by indentation testing at elevated temperature

S. Fujisawa, A. Yonezu
Department of Precision Mechanics, Chuo University, Japan

ABSTRACT

$Mg_{17}Al_{12}$ is intermetallic compound in die-cast AZ91D alloy and will affect the macroscopic mechanical properties. Indentation tests on both eutectic β-$Mg_{17}Al_{12}$ phase and α-Mg matrix are carried out at room and elevated temperature. It is found that the yield stress and flow stress of $Mg_{17}Al_{12}$ phase are much larger than those of Mg matrix. Besides, the capabilities of plastic deformation of both phases are enhanced at elevated temperatures. The mechanical properties of both phases as a function of temperature are investigated. Based on the microstructural property, the temperature effect on macroscopic deformation behavior is also discussed.

1. INTRODUCTION

AZ91 alloy is the most popular material among cast magnesium alloy due to relatively high strength, excellent corrosion resistance, and good cast-ability. In order to improve the mechanical properties of die-casting AZ91 alloy, plastic forming is employed (1), and temperature plays a critical role in the forming process (2). Understanding the effect of temperature on the yield stress and flow stress will be helpful for the mechanical properties against the plastic forming. The AZ91 alloy contains a large fraction of intermetallic compound of β-$Mg_{17}Al_{12}$ phase, and usually distributes along the grain boundary of α-Mg phase (matrix). The β-phase increases the strength of the entire AZ91 alloy at room temperature (RT), and it decreases at elevated temperature. The β-phase is only several ten micrometers in size and the effects of its properties on the entire AZ91 alloys is still unclear. This study investigates the mechanical properties of microstructure in AZ91D by using an indentation testing and the properties of β-$Mg_{17}Al_{12}$ and α-Mg matrix are separately evaluated. In addition, FEM model of the AZ91D alloy containing both β-$Mg_{17}Al_{12}$ and α-Mg phase is built and compared with the experimental result. The thermal degradation of both phases is investigated, and then we discuss how these degradations affect the macroscopic deformation.

2. MATERIAL AND EXPERIMENTAL SETUP

A commercial die-cast magnesium alloy, AZ91D (Mg: 90%, Al: 9% and Zn: 1%) was used in this study. Two types of experiments were performed, microindentation test and conventional uniaxial compression test. In the microindentation test, the disk type specimen which is the size is 12 mm in diameter and 2 mm in thickness. After the mechanically polishing, the specimen surface was immersed in the etching solution (distillated water: 24 ml, 60% nitric acid: 1 ml, ethylene glycol: 75 ml). Figure1 shows the microstructure observed by scanning electron microscope (SEM). It is found that the β-$Mg_{17}Al_{12}$ phase precipitates along the grain boundary of α-Mg phase. We observed the large area of more than 1 mm^2 in the microstructure. It reveals that the β-$Mg_{17}Al_{12}$ distributes uniformly in the observed area, and the ratio of α-Mg and β-$Mg_{17}Al_{12}$ is 83 and 17. The grain size of α-Mg ranges from 15 to 30 μm, and is averaged to 22 μm. Because of non-plastic forming alloy (as-cast alloy), the grains of α-Mg are randomly orientated. Indentation test was carried out at RT and elevated temperature of 150°C, 200°C and 250°C. It is reported that an indentation curve is dependent on the loading rate. This study set the indentation strain rate of about 0.01 /s.

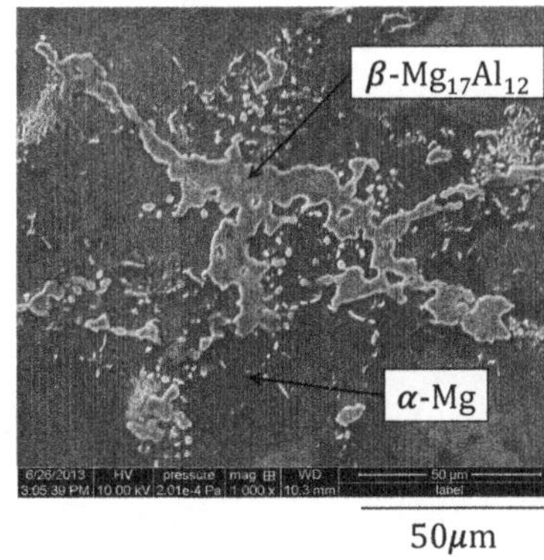

Fig.1 Microstructure of die-cast magnesium alloy (AZ91D).

In uniaxial compression test, a cylinder type specimen with diameter of 5 mm and height of 7.5 mm were prepared. The strain rate was set to 0.001 /s. This is different than the indentation strain rate (0.01 /s). However such a difference is reported to be minor in the macroscopic deformation behavior at RT (1). The compression test was carried out at room temperature. On the other hand, the stress-strain data at elevated temperature (150°C, 200°C, and 250°C) was obtained from the literatures (2) (3).

3. RESULTS AND DISCUSSIONS

3.1 α-Mg solid phase

A spherical indenter with 20 μm tip radius (R=20 μm) is used in this study. Since the spherical indenter is non self-similar geometry, it can mimic dual/plural indentation to enable the measuring elastoplastic properties (i.e. work-hardening law with non-linear deformation behavior), and we recently proposed simple framework which can apply to the Ludwick type hardening materials (4). The plastic constitutive equation of Ludwick law is

$$\sigma = \sigma_Y + K\varepsilon_p^n \qquad [1]$$

Where σ_Y is the yield stress, n is the work-hardening exponent, and K is the work-hardening strength. Note that ε_p is plastic strain. The alloy has random grain orientation of α-Mg. Several indentation tests (more than 20 tests) were, thus, conducted at different grains. The indenter impression size was about 40.6 μm which is larger than the average grain size (22 μm). The indentation curves are plotted in the inset figure of Fig.2. The maximum force is set to 1000 mN, such that the indentation depth h becomes deeper. In this figure, it is found that with the higher temperature, indentation curve is deeper. By using our reverse analysis (4), uniaxial stress - strain curve was estimated. For RT, the material constant of Eq.[1] are estimated to be σ_Y = 145 MPa, K = 443 MPa, and n=0.8. Figure2 includes the results at elevated temperature (150°C, 200°C and 250°C). It is found that with higher temperature, both yield stress and flow stress decrease. In particular, σ_Y=75.5 MPa at 250°C is smaller than that of RT(σ_Y=145 MPa), suggesting that the mechanical property of α-Mg phase is strongly dependent on temperature.

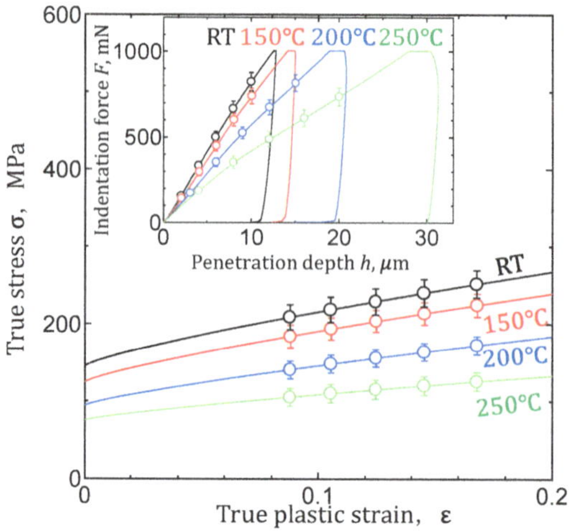

Fig.2 Indentation curves of α-Mg at RT, 150°C, 200°C, and 250°C, and the estimated stress-strain curves.

3.2 β-phase

This study next evaluates the properties of intermetallic compound (β-$Mg_{17}Al_{12}$ phase) at RT, 150°C, 200°C and 250°C. Since the β-phase has a small volume (small area) as shown in Fig.1, a sharp indenter with triangle shape was used. This is a self-similar indenter, leading that the strain distribution under indenter penetration is self-similar. It requires another shape to change the strain distribution when applying to the hardening materials. Therefore, this study employs two different shapes of triangle indenter whose the angle of the ridge lines are 115° and 100°. The indentation curves at RT and the indenter impression are shown in Fig.3.

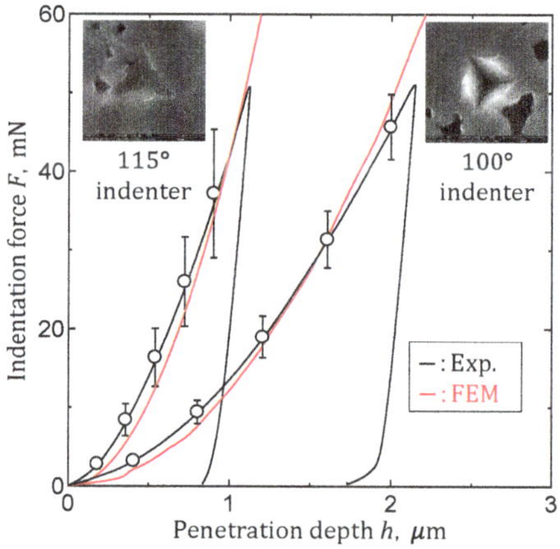

Fig.3 Indentation curves of experiment and FEM for β-phase at RT.

In order to identify the plastic properties of β-phase, FEM computations are carried out. This study creates two dimensional axisymmetric FEM model for forward analysis. The Young's modulus E is set to 71.9 GPa (from the literature (5)). Here we assumed the linear hardening law for the β-phase. The constitutive equation is in the Eq.[1], when the work-hardening exponent n is 1.0. Thus, this law involves two material constants; the yield stress σ_Y, and the hardening rate K. By changing of the yield stress σ_Y and the hardening rate K, parametric FEM study is conducted, such that the computed indentation curve agrees well with the experimental ones for both indenters. With the parametric study, we estimated material constants σ_Y=0.33 GPa and K=4.8 GPa for RT. The estimated plastic property is plotted in Fig.4. In order to verify this estimation, forward analysis was again carried out to compute the indentation curves for both indenters. The result is shown in Fig.3, indicating the good agreement between experiment and computation for both indenters.

Similarly, the indentation curves at elevated temperature are obtained. The indentation curve of 115° indenter is shown in the inset of Fig.4. At 150°C, the indentation curve agrees with that of RT. However, at higher than 200°C, the curves show plastically deformable. By using the above estimation process, we estimated plastic property (stress-strain curve) at elevated temperature, which is plotted in Fig.4. This reveals that an increasing of ambient temperature lets the β-phase deformable. In particular, at higher than 200°C, the intermetallic compound shows softening (i.e. plastically deformable).

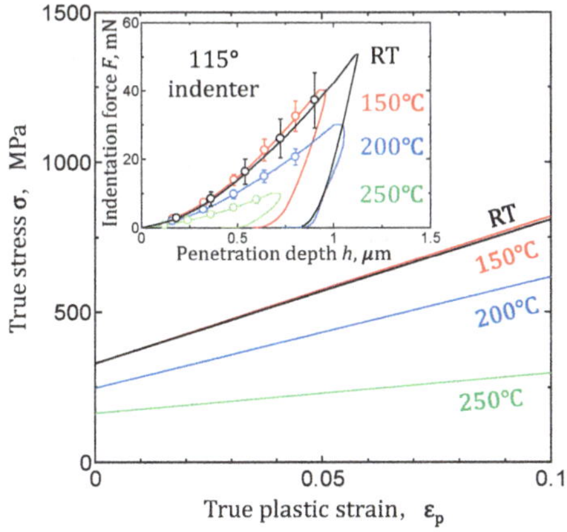

Fig.4 Indentation curves of β-$Mg_{17}Al_{12}$ phase obtain by 115° triangle indenter, and the estimated stress-strain curves β-$Mg_{17}Al_{12}$ at RT, 150°C, 200°C, and 250°C.

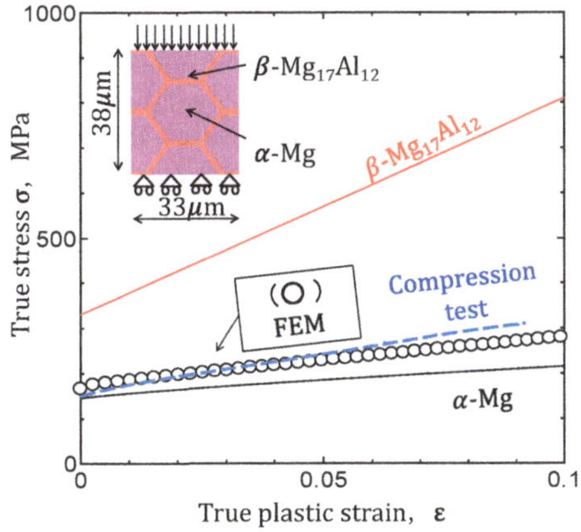

Fig.5 Stress-strain curve of experiment and FEM simulation. This also shows the curve of α-Mg and β- $Mg_{17}Al_{12}$ estimated by indentation method.

3.3 Macroscopic deformation

As mentioned above, this study evaluated the plastic property of each microstructure, i.e. α-phase and β-phase, through the indentation testing. Since the macroscopic deformation behavior is depended on the microstructural property, FEM is used to create the microstructural model for computing the macroscopic deformation. We assumed the two dimensional model under generalized plane strain. As mention above, intermetallic compound β-$Mg_{17}Al_{12}$ precipitated along the grain boundary of α-Mg matrix. The contained ratio for α- and β-phase is 83:17, and grain size is averaged 22 μm. Based on such structural information, the relatively simple model is created as shown in the inset of Fig.5. In order to employ a periodic boundary condition, both the right and left edge deforms uniformly along the horizontal direction, by adding the link option to each edge.

Figure5 shows the estimated macroscopic deformation. This figure also shows the stress-strain curves obtained by the compression test. It is found that the computation agreed well with the compression test. This indicates that the present FEM model is useful for the prediction of macroscopic deformation. Subsequently, we estimated macroscopic deformation at elevated temperature. Table1 shows the comparison of the yield

Table1 Comparison of yield stress obtain by experiment and FEM simulation at RT, 150°C, 200°C, and 250°C

Temperature	FEM[MPa]	EXP[MPa]
RT	167	150
150°C	151	139
200°C	115	132
250°C	90.3	93.3

stresses obtained by experiment and FEM computation. For each temperature, the computational result agrees with experimental one. It is revealed that the deformation resistance significantly decreases with the increasing of temperature. Such a softening (decreasing of deformation resistance) was observed in both phases as described in Section 3.1 and 3.2. These results suggest that the plastic deformation of microstructure affects the macroscopic deformation of entire AZ91D.

4. CONCLUSION

This study investigated the mechanical property of each phase (α- and β-phase) in die-casting magnesium alloy by using an indentation test. The microindentation was carried out at RT and elevated temperature (up to 250°C). For the indentation-based estimation, reverse analysis and forward analysis with finite element method (FEM) were carried out to estimate uniaxial stress and strain curve of both phases. It is found that the yield stress and flow stress of $Mg_{17}Al_{12}$ phase (β-phase) were much larger than Mg matrix (α-phase) at RT. In particular, the yield stress of β-phase achieves more than 300 MPa, which is twice larger than α-phase, suggesting that the β-phase may play a role of strength improvement of entire alloy.

Subsequently, elevated temperature indentation was carried out to estimate the property of both phases. At relatively low temperature (150°C), it is found that the α-phase softens, while the β-phase does not degrade, showing the same property at RT. At a elevated temperature (in particular, higher than 200°C), however, the plastic deformation of both phases is encouraged. In particular, at 250°C, the softening of β-phase is significant. This indicates that such temperature induces degradation of α-phase as well as β-phase. Since the β-phase locates along the grain boundary, such a thermal softening of β-phase is crucial for the strength and plastic forming of Mg alloy.

By using those properties of each phase, the macroscopic deformation was predicted via FEM model. The model was created with the each microstructure, in order to compute macroscopic compressive deformation. These responses (uniaxial stress and strain curve) agreed with the experimental ones in both RT and elevated temperature.

This work is supported by JSPS KAKENHI (Grant No. 26420025) and JGC-S Scholarship Foundation.

REFERENCES

(1) Ishikawa.K, Watanabe.H, Mukai.T. (2005) *Materials Letters*. 59, 1511-1515. doi: 10.1016/j.matlet.2005.01.012

(2) Cerri.E, Leo.P, De Marco.P.P. (2007) *Journal of Materials Processing Technology*. 189, 97-106. doi: 10.1016/j.jmatprotec.2007.01.010

(3) Mu.M, Zhi-min.Z, Bao-hong.Z, Jin.D. (2012) *Journal of Alloys and Compounds*. 513, 112-117. doi: 10.1016/j.jallcom.2011.09.102

(4) Yonezu.A, Akimoto.H, Fujisawa.S, Chen.X. (2013) *Materials and Design*. 52, 812-820. doi: 10.1016/j.matdes.2013.06.015

(5) Huang.W.Z, Zhao.H.Y, Hou.H, Han.D.P. (2012) *Physica B*. 407, 1075-1081. doi: 10.1016/j.physb.2011.12.132

Parametric study of the hydrogen diffusion in carbon steels under fatigue loading conditions using Green's function

B.H. Choi, W.H. Han, B.H. Kang
School of Mechanical Engineering, Korea University, Korea

ABSTRACT

Hydrogen diffusion is one of critical mechanisms of stress corrosion cracking (SCC) of engineering materials such as carbon steels. Especially, it is known that the degradation of pipe grade carbon steels under fatigue loading conditions is accelerated by hydrogen diffusion steels, but the classical diffusion equation cannot address this specific issue. So, in this paper, the conventional diffusion equation, i.e. Fick's second law, is modified by applying new boundary conditions considering fatigue loading conditions, and the solution is given analytically using Green's function. In addition, some parametric studies are provided by applying some key parameters of the solution, and the results are discussed to understand physical meanings of the final solution.

1 INTRODUCTION

Stress corrosion cracking (SCC) of pipes can be analyzed by understanding complex relationships between applied stresses, loading conditions, electrochemical environments, piping materials and so on. SCC is the universal long-term behaviour for most engineering materials. In Figure 1, a typical SCC observed in steel gas pipe and polyacetal pipe fitting. However, all of factors described above to control SCC are not easily related, and a synthetic approach for SCC is quite difficult. Most conventional models of SCC are developed to describe the formation of pitting mainly due to electrochemical approach and the development of a small crack at the early stage. However, the crack growth rate of SCC is very important due to a large portion of the crack propagation lifetime of the total lifetime of SCC. So, understanding of SCC growth is practically important to evaluate the residual lifetime of a pipe under SCC.

There have been a quite amount of researches on modeling of SCC growth, but most of them were constructed based on empirical observations of SCC or simple superposition methods. of most significant degradation behaviour during SCC formation is hydrogen embrittlement, which should be considered as an additional term to explain the reduction of fracture toughness around the crack tip. Choi et al. (1) simulated the SCC growth behaviour by comparing experimental data by developing a quite new model to explain the SCC using analytical model of hydrogen embrittlement based on the crack layer theory (2).

In this paper, the parametric study of hydrogen embrittlement model proposed by Choi et al. (1) is executed for various factors to controlling SCC. This model is formulated using Green's function by modification of Ficks' 2nd law considering fatigue loading conditions and new boundary conditions. Finally, the application of hydrogen embrittlement model to describing SCC growth behavior is also discussed.

2 MODELING OF HYDROGEN DIFFUSION UNDER FATIGUE LOADING

Choi et al. (1,3) proposed a SCC growth model based on the crack layer theory by using the concept of configurational mechanics of materials. By applying the crack layer theory, the separation of crack and process zone can be achieved. In case of the process zone, three regions can be identified as electrochemical zone, hydrogen-affected zone and plastic deformation zone. Among them, the second and the third regions can be approached together, but the first region should be separately considered. So, the crack growth equation considering the SCC driving forces can be proposed as,

$$\frac{da}{dt} = k_1[J_1^{mech} - 2\gamma(x, \rho(x,t))] + k_2(\varepsilon^P, T)J_1^{EC}, \tag{1}$$

where, k_1 and k_2 are kinetic coefficients for mechanical and electrochemical effects, and J_1^{mech} and J_1^{EC} are energy release rates related with mechanical and electrochemical terms. γ is the specific fracture energy, x is the distance from the crack tip and ρ represents the concentration of diffused hydrogen. In addition, ε^p is the electrochemical potential and T is temperature.

The effect of fatigue loading on the crack tip can be described by 'pumping effect' of hydrogen into the crack tip, and the flux of hydrogen ($q(x,t)_{|x=0}$) can be formulated by fatigue parameter (k) which is significantly related with the stress ratio (R) and frequency as,

$$-D\frac{\partial C}{\partial x} = q(x,t)_{|x=0} = -k(R,t)\cdot[\rho(x,t)_{|x=0} - \rho_*], \tag{2}$$

where, ρ_* is the saturation hydrogen density value, D is the hydrogen diffusion coefficient, and $\rho(x,t)_{|x=0}$ stands for hydrogen concentration at crack tip ($x=0$), which in its turn depends on the solution of the equation (2). In addition, k is the hydrogen influx coefficient that depends on the local environment, crack opening as well as on the rate of build-up and removal of the corrosion products. For simplicity, the diffusion is considered along the crack direction, and the boundary and initial conditions for the diffusion equation being as follows:

$$\rho(x,0) = 0 \text{ and } \rho(x,t)_{|x=0} = \rho_0(t), \tag{3}$$

Using the point source Green's function for the diffusion equation and boundary conditions (3), we obtain an integral equation for $C_0(t)$ as well as the hydrogen distribution $C(x,t)$ in front of the crack.

To evaluate hydrogen diffusion effect on SCC, we use the Green's function technique. Based on the standard solution of diffusion equation, the mass density of hydrogen $C(x,t)$ at a point "x" and a time instance "t" due to the unit mass concentrated at the point "ξ" at the instance "τ" $\leq t$, can be written as follows.

$$G(x,t;\xi,\tau)=\frac{1}{\sqrt{4\pi D(t-\tau)}}\exp\left[-\frac{(x-\xi)^2}{4D(t-\tau)}\right] \tag{4}$$

Hydrogen is diffused only from crack tip, $\xi=0$ with the initial density $\Delta M(t)$,

$$G(x,t;0,\tau)=\frac{\Delta M(\tau)}{\sqrt{4\pi D(t-\tau)}}\exp\left[-\frac{x^2}{4D(t-\tau)}\right], \tag{5}$$

where, $\Delta M(\tau)=\dot{\rho}_0\,\Delta\tau=k\{\rho_*-\rho(x,\tau)|_{x=0}\}\cdot\Delta\tau$.

Thus, the equation (5) may be rewritten as follows.

$$\rho(x,t)=\int_0^t\frac{k\{\rho_*-\rho(x,\tau)|_{x=0}\}}{\sqrt{4\pi D(t-\tau)}}\exp\left[-\frac{x^2}{4D(t-\tau)}\right]d\tau \tag{6}$$

Using the equation (3) and an expansion for $C_0(t)$, one obtains an integral equation for $\rho_0(t)$ from equation (6).

$$\rho_0(t)+\int_0^t\frac{k}{\sqrt{4\pi D(t-\tau)}}\rho_0(\tau)d\tau=\frac{k\rho_*}{\sqrt{4\pi D}}\cdot\sqrt{t} \tag{7}$$

Eq. (7) is the Volterra type equation of the second kind. It is conventionally solved by means of Laplace direct and inverse transforms. The solution of the integral equation depends on only one parameter, containing the ratio of the hydrogen influx and square root of hydrogen diffusion coefficients: $k\sqrt{t}/2\sqrt{D}$ term. In addition, the hydrogen influx coefficient k is proportional to the diffusion coefficient, $k=\beta D$, Therefore, the analytical solution of equation (7) can be written as follows,

$$\rho_0(t)=\rho_{\max}\left(1-\exp\left[\frac{\beta^2 Dt}{4}\right]\left\{1-erf\left[\frac{\beta\sqrt{Dt}}{2}\right]\right\}\right) \tag{8}$$

3 PARAMETRIC STUDY AND DISCUSSIONS

In Fig. 1, the distribution of mass density of hydrogen at the crack tip as a function of elapsed time is shown. It can be observed that mass density of hydrogen increases quickly once the fatigue parameter increases. So, it can be noticed that the fatigue loading can affect the accumulation of hydrogen which is the distribution of the mass density of hydrogen as a function of x (length from the crack tip). Hydrogen diffusion coefficient for carbon steels (4) selected as D=9.65x10^{-6} mm^2/s.

In Fig. 2, the increase of mass density of hydrogen at various locations away from the crack tip is shown. As x increases (away from the notch tip), the increase of mass density is delayed. So, it can be observed that hydrogen embrittlement effect may be localized, and the driving force of the SCC growth may be controlled by the accumulation of hydrogen around the crack tip.

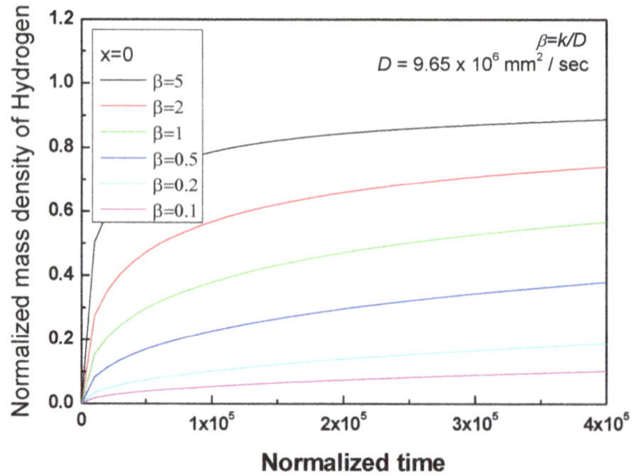

Fig. 1 Mass density of hydrogen at the crack tip for various fatigue conditions

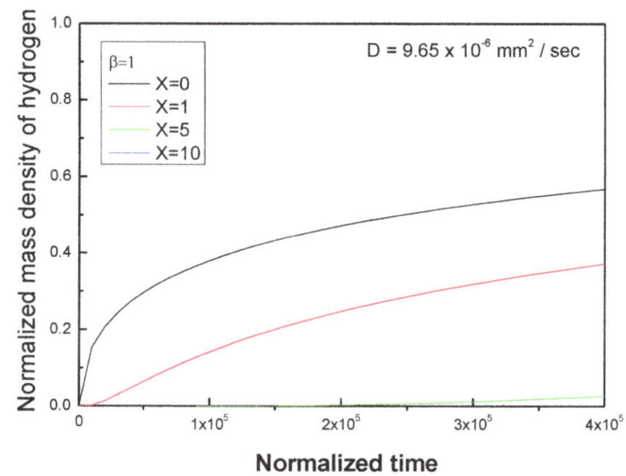

Fig. 2 Increase of mass density of hydrogen at various locations for k/D=1

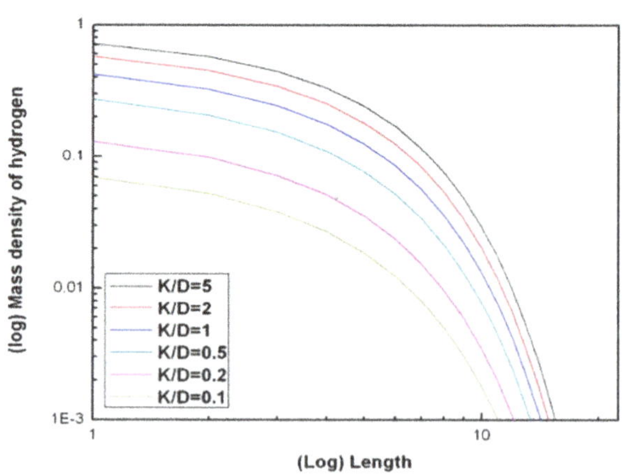

Fig. 3 Distribution of mass density of hydrogen at for various fatigue conditions

In Fig. 3, the distribution of mass density of hydrogen at 9 normalized time steps for various fatigue conditions. The distribution of mass density of hydrogen from the crack tip is mainly affected by the maximum hydrogen at the crack tip. It is known that many SCC phenomena are accelerated by the combination of fatigue loading (1,5), and it can be understood by applying the results in this study. Such behavior can be directly applied to model the SCC growth behavior expressed in equation (1).

4 CONCLUSION

In this study, the hydrogen diffusion equations using Green's function was reviewed and the effect of key parameters in the model was analyzed by parametric study. In the same location, k/D is the higher rate of diffusion, such as t (time) the higher the rate of diffusion from the k/D to find out there were enlarged. In other words, k/D the x (length from the crack tip) and an inverse relationship, t (normalized elapsed time) and the proportional relationship was observed. For SCC with fatigue loading conditions, k may affect the hydrogen diffusion behavior to the material, i.e. the hydrogen embrittlement behavior of carbon steels can be accelerated by fatigue loading at the crack tip. It will be investigated in detail as a further research.

ACKNOWLEDGEMENT

This work was financially supported by the Nuclear Power Core Technology Development Program (No.20131520000140) of the Korea Institute of Energy Technology Evaluation and Planning (KETEP) granted financial resource from the Ministry of Trade Industry and Energy.

REFERENCES

[1] Choi, B. H. & Chudnovsky, A. (2002) *Int. J. Fracture* **116**, L43-L48. doi: 10.1023/A:1022665831622

[2] Chudnovsky, A. The Crack Layer Theory. NASA Contract Report 174634 (1984).

[3] Choi, B.H. & Chudnovsky, A. (2011) *Metal. Mater. Trans. A* **42A**, 383-95. doi: 10.1007/s11661-010-0384-2

[4] Hirnyj S. The role of hydrogen in carbonate stress corrosion cracking of low carbon steel. Ph.D. Thesis, University Illinois at Chicago (1999).

[5] Zhang B. Understanding and modeling of stress corrosion cracking in high pressure pipeline steel. Ph.D. Thesis, University of Illinois at Chicago (2001).

Experimental investigation on mechanical property of an integrated thermal protection structure

X. Yang[+], Y. Sun[+], D. Shi[+], F. Cao[#], J. Feng[#]
+ School of Energy and Power Engineering, Beihang University, P.R. China
# College of Aerospace Science and Engineering, National University of Defense Technology, P.R. China

ABSTRACT

A new Integrated Thermal Protection Structure (ITPS) was prepared for flight vehicles to provide thermal protection as well as structural load bearing capabilities. Experiments were carried out to investigate mechanical properties of the ITPS, including flatwise tension, flatwise compression, edgewise compression and shear test. It is found that the stitched structure can overcome the insulation's shortcomings: low strength and modulus. Inner configurations of the ITPS were investigated by X-ray computed tomography system and Scanning Electron Microscope analysis. Mechanical modelling was conducted to investigate the contribution of panel, core and stitch to the load-bearing capability of the ITPS.

1 INTRODUCTION

The primary functions of the Thermal Protection System (TPS) are to protect vehicle and its contents from aerodynamic heating and maintain its structure integrity (1), and it is one of the most expensive and critical systems of the hypersonic vehicle (2). The current TPS technology for large surface area thermal protection consists of different concepts (3, 4, 5, 6).

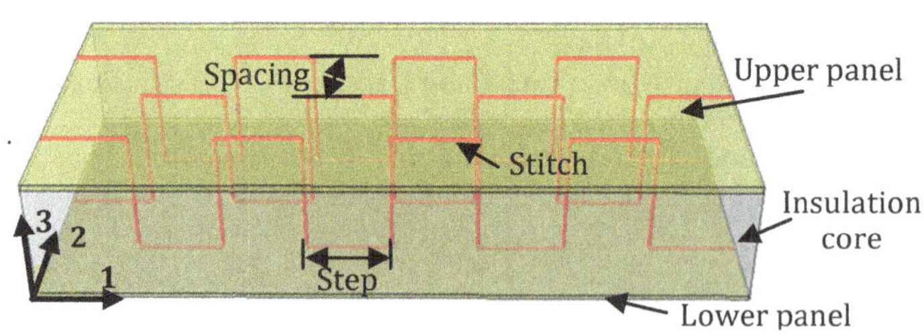

Fig 1. Schematic model of an ITPS with stitched sandwich structure.

Integrated TPS (ITPS) concept can significantly reduce the overall weight of space vehicles (7, 8), which will provide thermal protection as well as structural load bearing capabilities. As shown in Fig.1, a newly proposed ITPS concept is a stitched sandwich structure composed of an insulation core, two woven panels and the stitch fibers. It has been proved to be a potential candidate for primary load bearing and thermal protection integral structure (9).

Mechanical loads acted on the TPS include transverse aerodynamic pressure loads, in-plane inertial loads, acoustic and dynamic loads, and slight negative aerodynamic pressure loads (10). Hence, investigation on tension, compression and shear properties of the ITPS becomes important for its application.

In present research, experiments were carried out to study mechanical properties of the new ITPS. Mechanical modeling was also conducted to investigate the contribution of panel, core and stitch to the load-bearing capability of the ITPS.

2 MATERIAL AND TESTING METHODS

The test sample is a 20 mm thick stitched sandwich structure composed of an insulation core, two panels and some stitch fibers. The upper panel is made of ceramic-fiber-reinforced-resin-matrix composite. The lower panel is made of ceramic-fiber-reinforced-ceramic-matrix composite. The core is made of super insulation material: ceramic-fiber-reinforced SiO_2 aerogel (CFRSA) composite. The stitches are made of glass yarns.

Flatwise tension, flatwise compression, edgewise compression and shear tests were conducted according to ASTM standard (11, 12, 13, 14). Schematic graph of the above tests are shown in Fig.2.

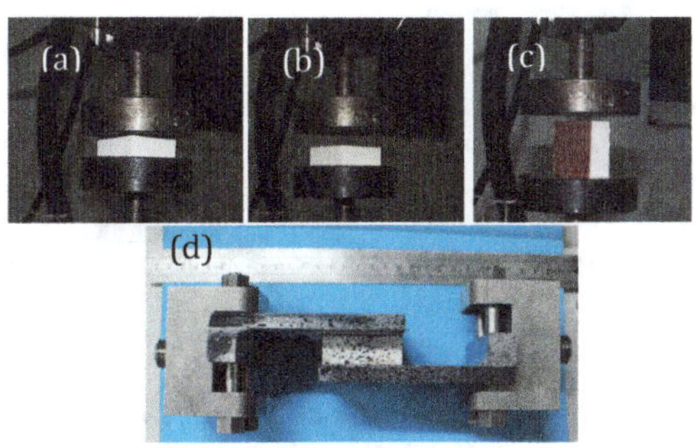

Fig 2. Fixtures of mechanical tests: (a) flatwise tension, (b) flatwise compression, (c) edgewise compression and (d) shear test.

3 RESULTS

Flatwise tension stress-strain curve (Fig.3a) showed initially a linear stage up to a point where a crack was found in the insulation core ($\sigma \approx 0.2$ MPa). The flatwise tension modulus and strength of the ITPS are nearly 8 times higher than that of the core material (15).

Flatwise compression stress-strain curve is obviously different from that of tension (Fig.3b). That may because of the tension-compression asymmetric of the core and stitch fiber. Flatwise compression modulus of the ITPS is nearly 2 times higher than that of the core material.

Edgewise compression curves (Fig.4) exhibited initially linear stage until buckling crack occurred on the sandwich panels ($\sigma \approx 1.2 \sim 1.5$ MPa). Edgewise compression modulus and strength of the ITPS are much higher than that of the core material.

Shear stress-strain curves (Fig.5) showed initially linear stage until cracks were found in the insulation cores. When the stresses increased to shear strengths (≈0.37 MPa), other cracks were found in the sandwich cores. Finally the sandwich failed at the interface between panel and core due to stitch fractures.

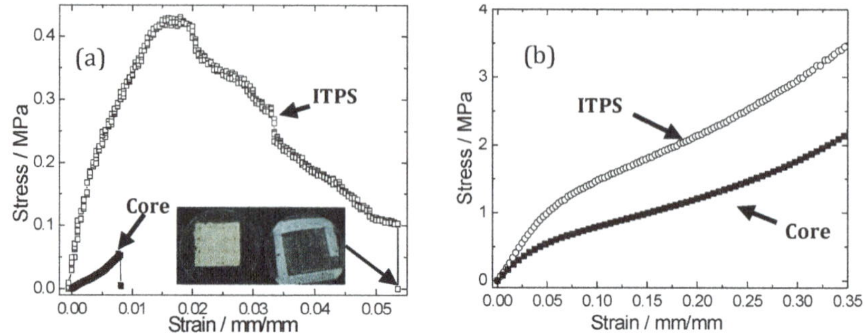

Fig 3. Flatwise stress-strain curve of the ITPS: (a) tension and (b) compression.

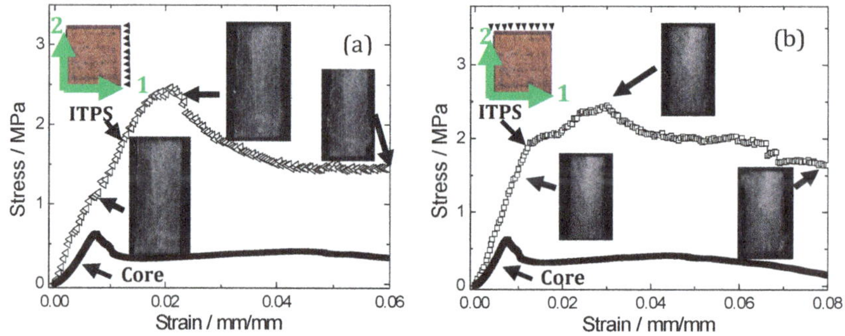

Fig 4. Edgewise compression stress-strain curves of the ITPS: (a) 1 direction and (b) 2 direction.

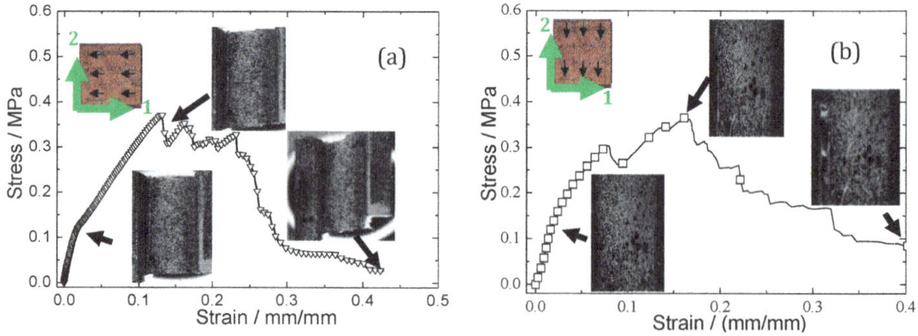

Fig 5. Shear stress-strain curves of the ITPS: (a) 1 direction and (b) 2 direction.

4 MODELING OF THE ELASTIC PROPERTIES

Mixed series-parallel spring models (MSPSMs) were constructed for Young's modulus predication based on analysis of a Representative Volume Element (RVE) of the ITPS. The RVE is a cube of 13.3×13.3×20 mm[3] composed of a 1 mm thick upper panel, a 0.7 mm thick lower panel, a 18.3 mm thick core, a stitch with radius 0.35 mm and the ceramic/resin around the stitch with external diameter 1 mm.

The MSPSMs are shown in Fig 6. Modelling results (Table.1) are numerically close to that from experiments, which indicates that the MSPSMs can be used to investigate the contribution of core, panel and stitch to the load-bearing capability of the ITPS.

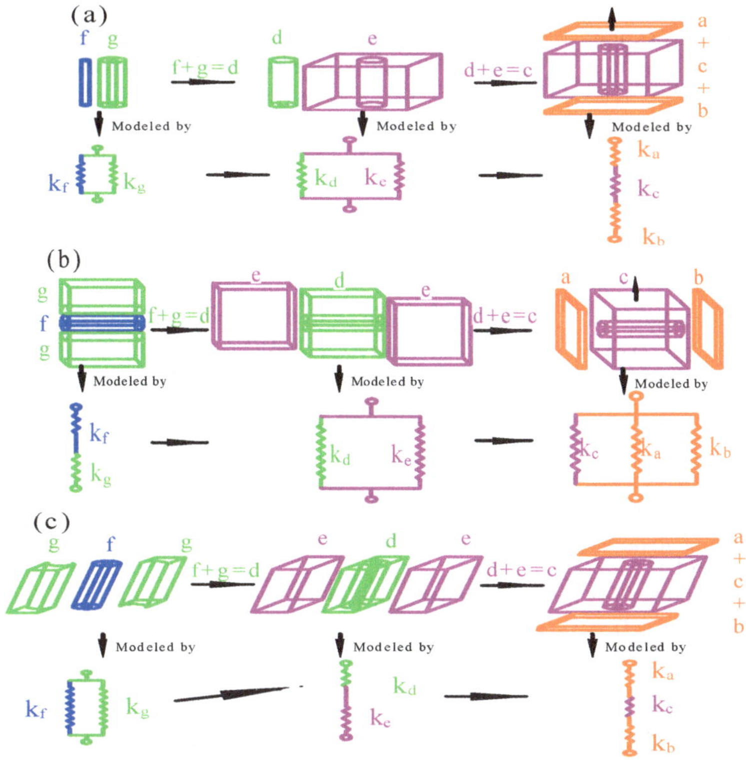

Fig 6. Schematic representations of the MSPSMs: (a) flatwise tension, (b) edgewise tension and (c) shear.

Table 1. Young's modulus obtained from experiments and MSPSMs.

Items	Experiments	MSPSMs
Flatwise tension modulus E_{t3} (MPa)	47	59.5
Flatwise compression modulus E_{c3} (MPa)	21.8	21.9
Edgewise compression modulus E_{c1} (MPa)	167	185
Edgewise compression modulus E_{c2} (MPa)	162	185
Shear modulus G_{13} (MPa)	7.16	8.1
Shear modulus G_{23} (MPa)	7.47	8.1

5 CONCLUSIONS

Experiments were carried out to investigate the mechanical properties of an ITPS composed of two composite panels, an insulation core and some stitches. It is found that the ITPS perform anisotropic, nonlinear and tension-compression asymmetric properties.

Mixed series-parallel spring models were constructed for Young's modulus predication of the ITPS, which proves to be feasible and valid.

6 ACKNOWLEDGEMENTS

The research was supported by National Natural Science Foundation of China (Grant NO. NSFC 51275023) and the Innovation Foundation of BUAA for PhD Graduates (Grant NO. YWF-12-RBYJ-007).

REFERENCES

[1] Blosser, M. L., Chen, R. R., Schmidt, I. H., Dorsey, J. T., Poteet, C. C. & Bird, R. K. (2002) AIAA Report 2002-0504.
[2] Behrens, B. & Muller, M. (2004) Acta. Astronaut. 55, 529-536.
[3] Cooper, P. A. & Holloway, P.F. (1981) Astronaut. Aeronaut. 19, 24-34.
[4] Kourtides, D. A., Tran, H. K. & Chiu, S. A. (1992) NASA Report TM-103836.
[5] Dorsey, J. T., Poteet, C. C., Wurster, K. E. & Chen, R. R. (2004) J. Spacecraft Rockets 41, 162-172.
[6] Blosser, M. L. (1996) NASA Report TM-110296.
[7] Martinez, O. A., Sankar, B. V., Haftka, R. T., Bapanapalli, S. K. & Blosser, M. L. (2007) AIAA J. 45, 2323-2336.
[8] Ravishankar, B., Sankar, B. V. & Haftka, R. T. (2011) 52nd AIAA/ ASME/ ASCE / AHS/ ASC Structures, Structural Dynamics and Materials Conference, Colorado, 4-7 April, 1-17.
[9] Yin, C. (2009) Ph.D. thesis, National University of Defense Technology, China.
[10] Bapanapalli, S. K. (2007) Ph.D. thesis, University of Florida, American.
[11] ASTM C297/C297-04M (2010) West Conshohocken: ASTM.
[12] ASTM C365/C365M-11a (2011) West Conshohocken: ASTM.
[13] ASTM C364/C364M-07 (2012) West Conshohocken: ASTM.
[14] ASTM C273/C273M-11 (2011) West Conshohocken: ASTM.
[15] Mi, C., Jiang, Y, Shi, D, Han, S, Sun, Y., Yang, X & Feng, J. (2013) Acta Mater. Comp. Sinica. http://www.cnki.net/kcms/detail/11.1801.TB.20131114.1019.003.html.

A novel positioning stage using piezoelectric actuator for antenna pointing

J. Chen[1], M.L. Xu[1], B. Feng[1], Q.G. Tian[2]
[1] State Key Laboratory for Strength and Vibration of Mechanical Structures, School of Aerospace, Xi'an Jiaotong University, P.R. China
[2] Xi'an Institute of Space Radio Technology, P.R. China

ABSTRACT

A novel positioning stage using piezoelectric actuator is presented for antenna pointing. The proposed stage works out-of-plane direction. It is driven by four piezoelectric actuators, which can amplify the motion of piezoelectric stack by a bridge type amplifier. The performance of actuator is verified by theoretical analysis and simulation. Modeling and analysis of antenna pointing mechanism are carried out. Prototype has been manufactured and tested. Experimental results indicate that the directional angle of antenna has range of ±0.08° while the pointing accuracy is 0.0005°.

1 INTRODUCTION

With the development of satellite technology, data transmission rate is getting higher and higher. The satellites must work at high-frequency band such as Ka-band to fulfill the requirement of communication. Accurate beam pointing is crucial for successful communication (1). That has, therefore, necessitated the development of a positioning stage with high accuracy.

To achieve high precision motion of positioning, linear motor and voice coil are commonly used. However, they have many shortcomings, such as backlash, friction, and stick-slip. By contrast, piezoelectric stack has great advantages, such as sub-nanometer level resolution, high precision, fast response and wide temperature range. However, it has low moving range. This shortcoming can be overcome using a displacement amplifier (2). The amplifier is integrated with piezoelectric stack as actuator to provide a large motion range.

There are a number of positioning stages using smart material. A 6-DOF magnetic levitation (maglev) positioner with air-core solenoids and permanent magnets was designed by Jung and Beak (3). Zhang developed 3-DOF out-of-plane stage which has 11.8 μm of z-axis working range and 22.7″ of tilt angle (4). Liu (5) presented a 3-DOF precision positioning stage actuated by PZT with a rotary range of 37.5 μm. However, the big challenge is to design a positioning mechanism which can achieve high precision as well as wide operating range.

In this paper, a novel positioning stage for antenna pointing is proposed. Piezoelectric stack is applied for the stage due to its incomparable advantage in high precision positioning. Additionally, a bridge type amplifier is designed for amplifying the motion of the piezoelectric stack. And the antenna mechanism can achieve long range, compact size, and high precision.

2 MECHANICAL SYSTEM DESIGN

2.1 The positioning stage configuration

In this section, the model of antenna pointing mechanism is purposed as shown in Fig. 1. Semi-parallel complaint mechanism is applied to the positioning stage, which consists of antenna, upper plate, four piezoelectric actuators and base plate. The upper plate that carries antenna is connected to the base plate via four actuators. And each actuator is located in each side of a square as illustrated in Fig. 2. Without the antenna, the size of stage is about 70 mm×120 mm×46 mm.

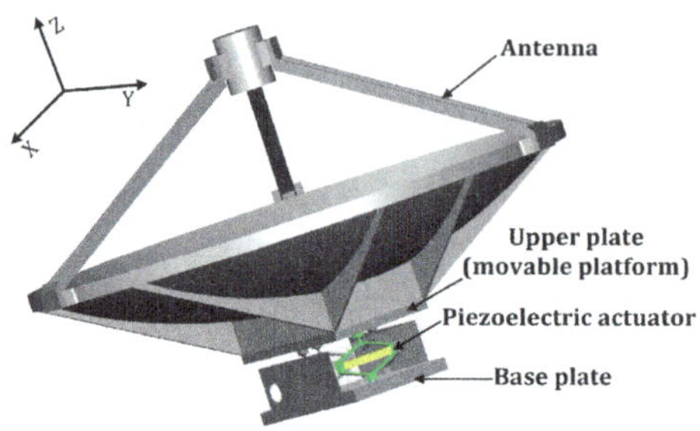

Fig. 1. Antenna pointing mechanism model

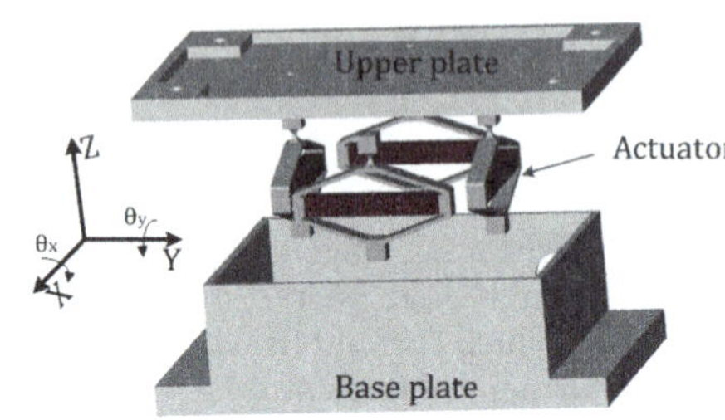

Fig. 2. Exploded diagram of stage

2.2 The piezoelectric actuator design

From the section above, the positioning stage is driven by four piezoelectric actuators, which consists of bridge type mechanism and piezoelectric stack as shown in Fig. 3. The piezoelectric stack is sandwiched between plane 1 and plane 2. When a voltage is applied to the piezoelectric stack, it would expand so that the bridge type amplification mechanism generates downward direction displacement. It can reduce the height of the stage and make the stage compact. Furthermore, as a compliant stage, it is expected that each actuator independently drives the stage in each direction without affecting each other. Hence, there is circular flexure hinge consisting of two orthogonal prismatic joints adopted since they possess the smallest center-shift when performing rotation.

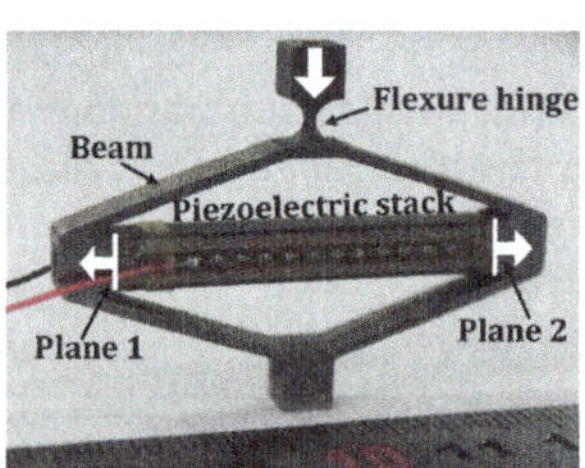

Fig. 3. Piezoelectric actuator

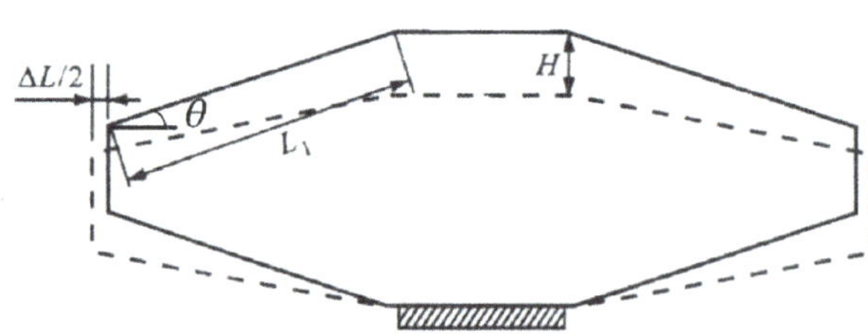

Fig. 4. Schematic diagram of the actuator

In consideration of that the piezoelectric stack is sensitive to shear forces or bending moment, the mechanism, hence can protect the piezoelectric stack as it can withstand pressure only.

Fig. 4 shows the working principles of piezoelectric actuator. The piezoelectric actuator fixed at bottom part. The relationship between the output displacement of amplifier and piezoelectric stack can be described as follows.

$$(L_1 \cos\theta + \Delta L/2)^2 + (L_1 \sin\theta - H/2)^2 = L_1^2 \quad (1)$$

Where L_1 is the length of beam of bridge structure, θ is the angle between beam and horizontal direction, ΔL is the motion of piezoelectric stack and H is the moving range of actuator. Under micro motion condition, the equation above could be written as:

$$H/\Delta L \approx \cot\theta \tag{2}$$

3 FINITE-ELEMENT ANALYSIS

3.1 Simulation of piezoelectric actuator

To check the performance of designed piezoelectric actuator, its finite element analysis is carried out using Pro/ENGINEER and Abaqus. The size of the piezoelectric stack (P-885.91, Physik Instrumente company) is 5 mm×5 mm×36 mm. And displacement range of unloaded stack is about 0 μm~+38 μm when the applied voltage range is 0 V~120 V. The size of the whole actuator is 5 mm×40 mm×20 mm. And the amplifier is made of an spring steel, 60Si2MnA. So it can provide a high yield stress for good repeatability. The amplifier fixed at bottom part and applied displacement at both sides. Fig. 5 shows simulation result of piezoelectric actuator. The simulation and theoretical results are compared in Fig. 6, which indicates that simulations are consistent with the theoretical theory results. The moving range of actuator is above 100 μm when the input displacement is 38 μm. Likewise, the actuator mechanism can linearly amplify the output displacement of stack.

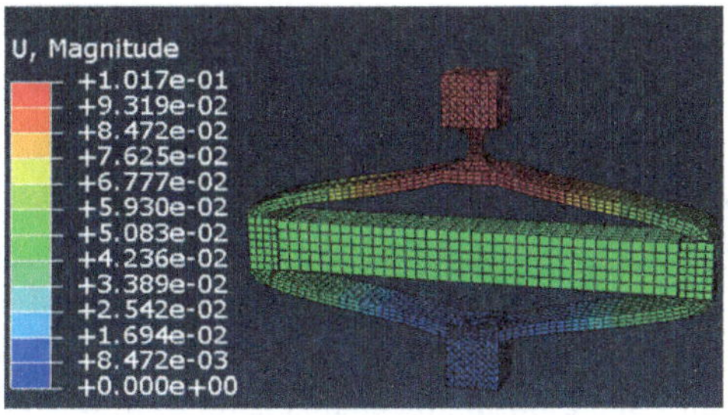

Fig. 5. FEA simulation

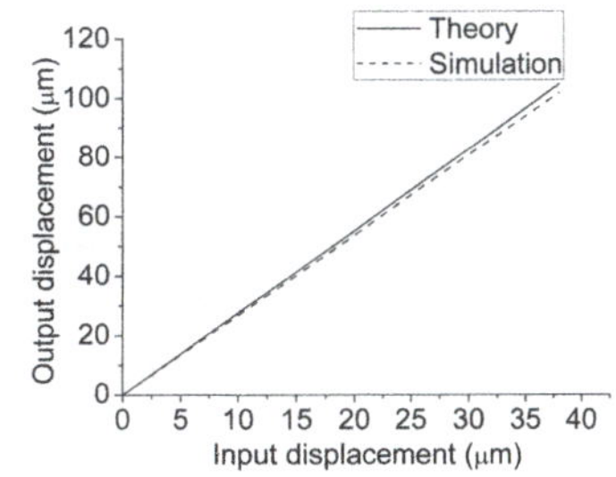

Fig. 6. Theory and simulation

Then the antenna pointing mechanism is simulated to evaluate the performance of the micro-motion stage. Firstly, the dynamic performance of the stage has been verified using Abaqus. The base of the antenna is constrained in all direction. Fig. 7 shows modals of the whole system. Mode 1 is rotation z-axis, Mode 2 is rotation x-axis, Mode 3 is rotation y-axis. Without the heavy antenna, the natural frequency of stage would increase to 213.70 Hz, 319.34 Hz and 319.99 Hz, respectively.

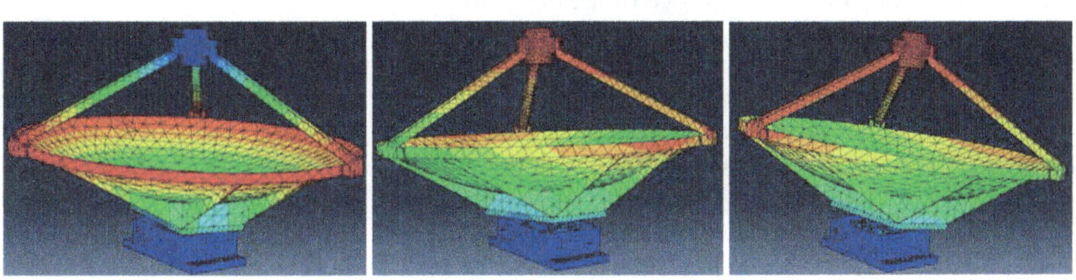

Mode 1(21.70 Hz) Mode 2(40.59 Hz) Mode 3(47.75 Hz)

Fig. 7. Modals of the whole system

In order to confirm the motion range of antenna pointing mechanism, each actuator is employed desired displacement, respectively. The constraints should be necessarily considered, such as maximum stress in the flexure hinge and distortion of piezoelectric actuator. The simulation results show that the greatest stress appears in the flexure hinge of actuator. When the antenna pointing mechanism reaches maximum motion range of ±0.09°, the maximum stress is 124.6 MPa, which is far less than the yield stress.

4 EXPERIMENTS

The experimental setup is shown in Fig. 8. National Instrument computer (PXI-1050) was applied as a real time controller. A high voltage piezo-amplifier was used to supply voltage above 120 V. And two laser displacement meters (LK-G80, Keyence) with a 0.1 μm resolution were used for measuring the displacement of the moving part of the stage. In order to improve the accuracy of measurement, the two laser displacement meters 1 and 2 employed differential measurement.

Firstly, the dynamic performance of antenna pointing mechanism was tested. The resonance frequency was obtained by impulse response of the system. Table 1 shows the first three resonance frequencies of the antenna pointing mechanism.

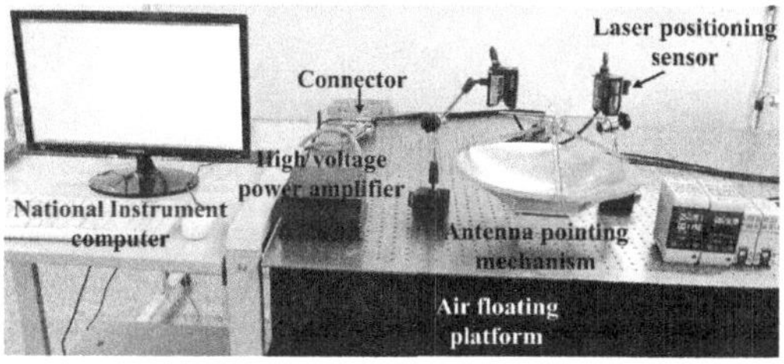

Fig. 8. Experimental setup

Table 1 Resonance frequency

Modal	Frequency(Hz)
Mode 1	17.65
Mode 2	37.14
Mode 3	43.25

To demonstrate the performance of the antenna pointing mechanism, its rotation about x-axis and y-axis were tested, respectively. From the section above, the actuator only has downward direction displacement. In order to make piezoelectric generate inverse displacement, a pre-load voltage should be applied to each actuator. When rotation about x-axis was tested, the two actuators at the y-axis were applied voltage signal as shown in Fig.9. However, the applied voltage of two actuators at x-axis remained unchanged. While the applied voltage of one actuator reached maximum voltage 120 V and the other was 0 V, the antenna mechanism achieved maximum range. For rotation about y-axis, it was reversed. Fig. 10 shows rotation range around x-axis and y-axis, respectively. They both can achieve range of ±0.08°.

The Preisach model was employed to predict the output displacement of each piezoelectric actuator. Hence, the actuators could apply specific voltages to obtain a desired rotation angle. Some of antenna pointing errors are listed in Table 2. The θ_x-directional and θ_y-directional precisions reach 0.0005°.

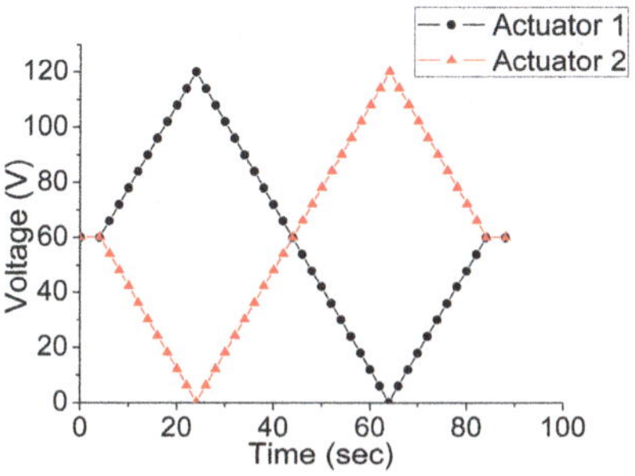

Fig. 9. Voltage signal

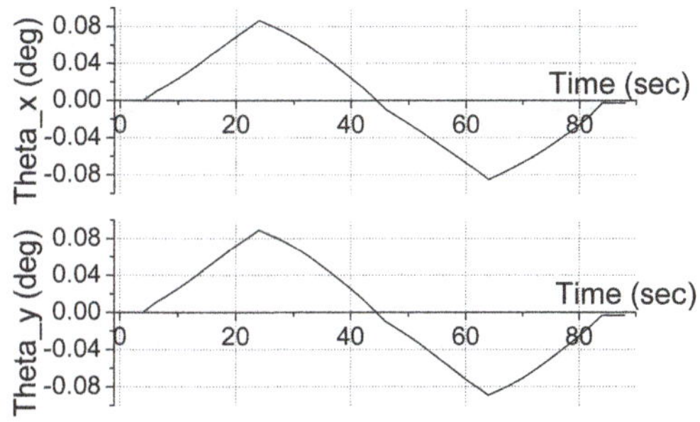

Fig. 10. Rotation range of the stage

Table 2 Measurement results of the pointing precision of stage

Rotation about x-axis			Rotation about y-axis		
θ_x Desired (deg)	θ_x Actual (deg)	Absolute error (deg)	θ_y Desired (deg)	θ_y Actual (deg)	Absolute error (deg)
0.005	5.227E-3	2.3E-4	0.005	5.178E-3	1.8E-4
0.020	1.979E-3	2.2E-4	0.020	2.033E-2	3.3E-4
0.035	3.468E-2	3.2E-4	0.035	3.457E-2	4.3E-4
0.050	5.032E-2	3.2E-4	0.050	5.047E-2	4.7E-4
0.065	6.457E-2	4.3E-4	0.065	6.472E-2	2.8E-4

5 CONCLUSION

In this paper, a novel stage driven by four piezoelectric actuators was proposed for antenna pointing. The piezoelectric actuator consists of piezoelectric stack and bridge type amplifier, which was developed to overcome the limited displacement of piezoelectric stack. The performance of actuator was researched by theory and simulation. The results have confirmed that actuator can linearly amplify motion of piezoelectric stack. Finally, the overall simulations and experiments have demonstrated that the antenna pointing mechanism can achieve good motion range of ±0.08° around x-axis and y-axis with accuracy of 0.0005°, respectively.

REFERENCES

(1) Nikulin, V. V., Bouzoubaa, M., Skormin, V. A., & Busch, T. E. (2001) *In SPIE Proceedings.* **(4272)** 72-82. doi:10.1117/12.430800

(2) Xu, W. & King, T. (1996) *Precision Engineering.* **19(1)**, 4-10. doi:10.1016/0141-6359(95)00056-9

(3) Jung, K. S. & Baek, Y. S. (2002) *Mechatronics, IEEE/ASME Transactions.* **7(1)**, 35-43. doi:10.1109/3516.990885

(4) Zhang, D., Chetwynd, D. G., Liu, X. & Tian, Y. (2006) *International Journal of Mechanical Sciences.* **48(12)**, 1401-1408. doi:10.1016/j.ijmecsci.2006.07.006

(5) Liu, Y. L. & Lee, G. Y. (2007) *Towards Synthesis of Micro-/Nano-systems.* 121-126. doi:10.1007/1-84628-559-3_20

Optimal design of the large stroke piezoelectric actuator using rhombic mechanism

S.B. Shao, M.L. Xu, J. Chen, B. Feng
State Key Laboratory for Strength and Vibration of Mechanical Structures, School of Aerospace, Xi'an Jiaotong University, P.R. China

ABSTRACT

A piezoelectric actuator using rhombic mechanism based on flexure hinges is developed to amplify the displacement of piezoelectric stack. The analysis of flexure hinge is performed then the maximum displacement amplification ratio of structure is obtained by optimization through analytic formulas, where the geometrical parameters of structure are selected as design variables. After that the output force of actuator is determined by stiffness design of flexure hinges. Therefore a mechanism with the specified magnification and output force is designed.

1 INTRODUCTION

Piezoelectric stack (PZT stack) has been widely used in many fields, such as active vibration control, precision positioning system and even MEMS, but the piezoelectric materiel has only very small induced displacement, which is about 1000μm/m. the length of most of commercial PZT stack is about 10cm to 30cm, so the induced displacement of PZT stack can hardly exceed 100μm. Recently, in order to amplify the deformation of PZT stack, a flexure hinge has been studied to design various kinds of amplification mechanism, because the flexure hinge has the advantage of compact structure, no backlash and no need of lubrication over conventional pivots. Several types of amplification mechanism were developed such as level-type, Moonie-type, rainbow-type and bride-type (1). In general, the amplification ratio of all these types can be estimated by level rule, but unfortunately, this is far from the real situation. The level rule will show large errors when one wants to get large amplification ratio in real systems. So a lot of work has been done to improve the accuracy of flexure hinge model (1-3). Paros and Weisbord (2) derived the stiffness of a circular hinge theoretically, Furukawa et al (3) modeled the bridge-type hinge as spring element but it still leads to large errors, Hong-Wen Ma et al (1) derived the real amplification ratio of bridge-type mechanism by analytic method. But few of them perform optimization design of amplification ratio. In this research, a rhombic mechanism is employed to amplify the displacement of piezoelectric stack, then the displacement amplification ratio is studied and optimized to get desired stroke, where the geometrical dimension of structure are selected as design variables. Finally, the output force of actuator can be determined by stiffness of flexure hinges which only depend on the geometrical size if the material is selected.

2 DESIGN OF ACTUATOR AND OPTIMIZATION OF AMPLIFICATION RATIO

2.1 Rhombic mechanism

Generally, a compact and light weight structure is always essential to the precision system. The Moonie-type, rainbow-type and bride-type mechanism are easier to design in a compact and light weight structure compare to level-type one. But among them only bride-type mechanism is capable of amplifying the displacement of piezoelectric stack. In this study, a rhombic mechanism using bridge-type flexure hinges is designed. Fig.1 shows the basic structure and its way of actuation. As shown in Fig.1. End D of actuator is clamped, when piezoelectric stack expands, the actuator outputs negative displacement along the y axis, and if piezoelectric stack shrinks the actuator outputs positive displacement.

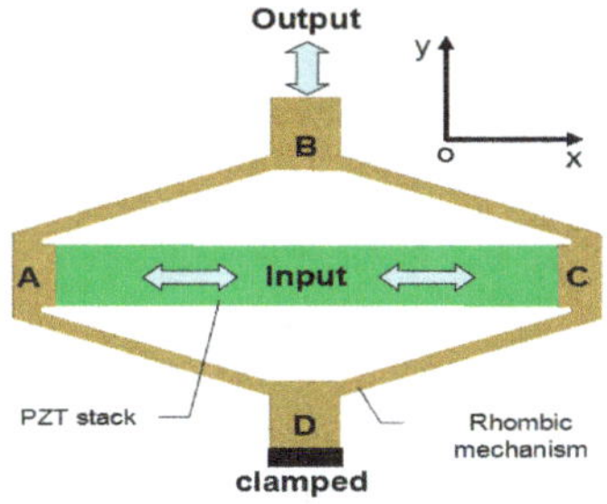

Fig.1. Rhombic actuator

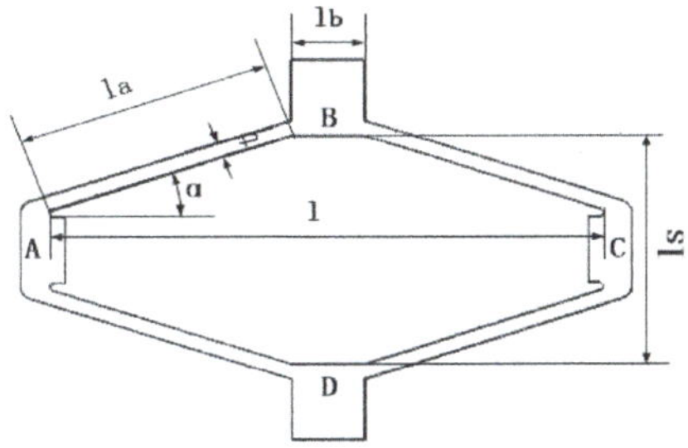

Fig.2. Geometrical dimensions

For a Rhombic mechanism, the input and output are only in the plan. Fig.2 shows the geometrical size of the actuator, and the ideal displacement amplification ratio can be derived using kinematic theory (1):

$$G_{ideal} = \frac{Output}{Input} = \cot\alpha \tag{1}$$

where α is the angle between the edge AB and the long diagonal AC, Eq.(1) illustrates that the ideal amplification ratio will increase with the decrease of the angle α, therefore it seems like that one can get very large amplification ratio if the angle α is set small enough, but in fact, flexure hinges posses bending and longitudinal stiffness as shown in Fig.4. These two stiffnesses can cause elastic energy stored in mechanism and complicate the relationship between amplification ratio and angle α.

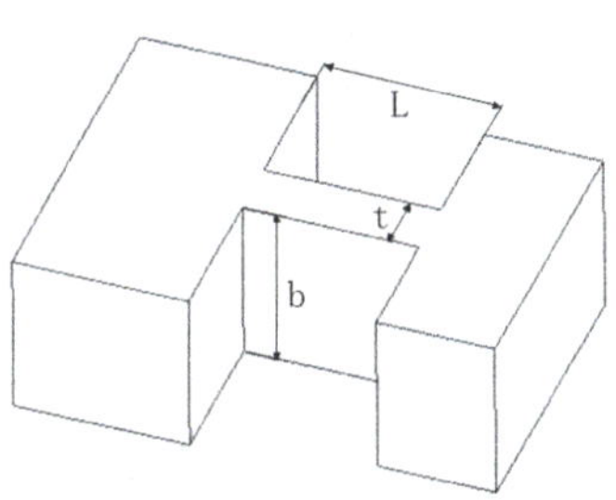

Fig.3. Flexure hinge

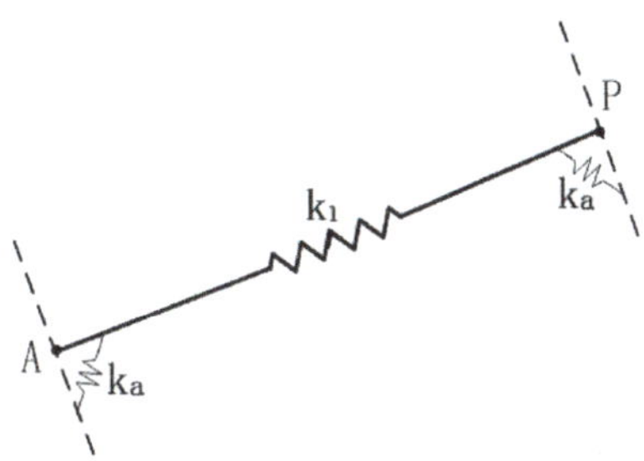

Fig.4. Mechanical model of flexure hinge

In Fig.4 the k_l and k_a represent longitudinal and bending stiffness respectively. They can be derived from beam theory as follows (4).

$$k_l = \frac{Ebt}{L} \quad (2)$$

$$k_a = \frac{Ebt^3}{12L} \quad (3)$$

where E is elastic module of material. Because of symmetric structure the model of rhombic mechanism considering stiffness of flexure hinges can be described as shown in Fig.5, where edge AB of rhombic mechanism is equivalent to two cascaded flexure hinges. Only longitudinal stiffness is considered at the point P which is the joint of two flexure hinges since the bending occurs in the middle region of edge AB can be equivalent to that occurs at the end of edge AB.

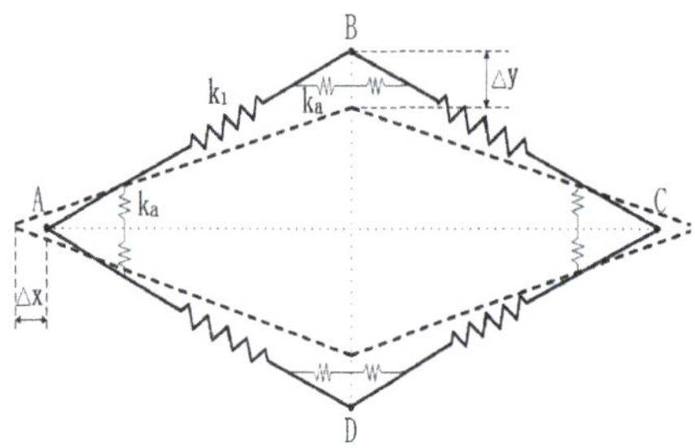

Fig.5. The model of rhombic mechanism

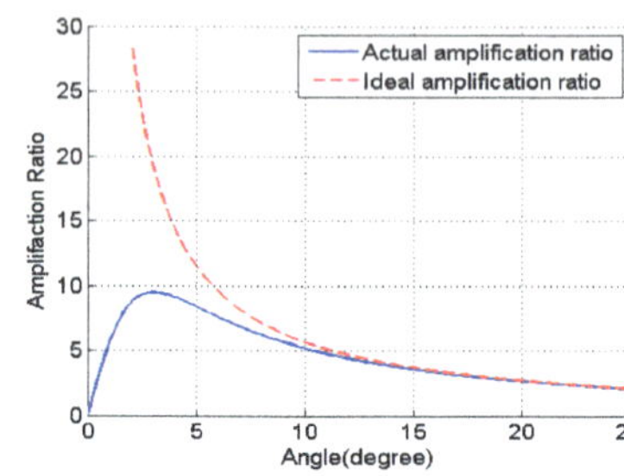

Fig.6. Amplification ration VS angle

If the stiffness of flexure hinges is considered, From Eq.(2) and Eq.(3), the actual displacement amplification ratio can be derived as (1):

$$G_{actual} = \frac{\Delta y}{\Delta x} = \frac{l_a \cos\alpha}{\dfrac{2k_{ae}\cos^2\alpha}{k_{le}l_a \sin\alpha} + l_a \sin\alpha} = \frac{l_a \cos\alpha}{\dfrac{t^2\cos^2\alpha}{1.5l_a \sin\alpha} + l_a \sin\alpha} \quad (4)$$

where $k_{le} = k_l/2$, $k_{ae} = 2k_a$ because of cascade of two flexure hinges. l_a and l are the length of the edge and long diagonal respectively, t is thickness of the edge AB. If fix t and l_a, the real and ideal relationship between amplification ratio and angle α are pictured in Fig.6.

2.2 Optimization of amplification ratio

Fig.6 shows that there is a optimum(which is maxim) point of actual amplification ratio when t and l_a are fixed. The optimum angle can be derived through derivation of α. But in fact, amplification ratio is associated with three variables which are thickness t, length l_a and angle α. It's reasonable to perform optimization with three variables simultaneously.

2.2.1 Design parameters and constraints

The geometrical parameters in Fig.2 are selected as design parameters. In this study, both length of output platform l_b and short diagonal l_s have dimensional constraints, that is $l_b = 5mm$ and $l_s < 10\,mm$. Furthermore the piezoelectric stack is selected beforehand and $l = 36mm$, then the constraint of angle α can be derived as Eq.(5) from geometrical relationship.

$$0 < \alpha < 15.52^\circ \quad (5)$$

According to the geometrical constraint, one can get Eq.(6) from Eq.(5).

$$l_a \cos\alpha = \frac{(l - l_b)}{2} = 15.5mm \tag{6}$$

Then the constraint of l_a is derived as Eq.(7) from Eq.(6).

$$14.6mm < l_a < 15.5mm \tag{7}$$

From Eq.(4), the hinge thickness t should be small to increase the amplification ratio, but to reduce the processing difficulty, the minimum thickness should be not less than 1mm.

$$t \geq 1mm \tag{8}$$

2.2.2 Stiffness of the flexure hinge

The maximum block force of actuator is the force needed at the output which made the output displacement zero when maximum input displacement is applied. It is usually employed to assess the ability of output force of actuator and can be described as:

$$F_{block} = \frac{4k_a \Delta\alpha}{l} \tag{9}$$

From Eq.(9), it's clear that F_{block} is proportional to k_a, so the k_a can be used to estimate the ability of output force too. On one hand, from Eq.(3) and Eq.(4) it's not hard to see that smaller thickness t leads to smaller stiffness k_a but lager amplification ratio G_{actual}, on the other hand the larger width of hinge b leads to larger stiffness k_a but has no effect on amplification ratio, so the thickness t along with the optimum G_{actual} can be derived through optimization of amplification ratio, then the width b can be determined from Eq(3) with certain k_a which is $k_a = 3881 N \cdot mm/rad$ in this research.

2.2.3 Object function and optimal results

In this research the displacement amplification ratio is selected as object function because it is the main concern of performance of actuator. The output force of actuator, as mentioned before, can be determined after the amplification ratio is optimized. One can substitute Eq.(6) into Eq.(4) to obtain another object function:

$$G_{actual} = \frac{15.5}{\dfrac{t^2 \cos^2\alpha}{23.25 \tan\alpha} + 15.5 \tan\alpha} \tag{10}$$

In Eq.(10) only two design variables are included which are thickness t and angle α, combine with their constraints, the sequential Quadratic Programming(SQP) method was used to get the optimum amplification ratio and corresponding design parameters. The optimal results are listed in table1 as follows. We can see that the maximum amplification ratio is get when the angle α equals 3.04°, that is one can not get arbitrarily large amplification ratio when angle α is small enough.

Table 1 Optimal results

Maximum G_{actual}	9.47
Optimum t	1mm
Optimum α	3.04°
Optimum b	5mm

3 SIMULATION AND EXPERIMENT

Fig.7 shows the static analysis of designed actuator by finite element analysis(FEA) simulation, where input $\Delta x = 5\mu m$ and the end D is clamped, so the amplification ratio $G = \Delta y / 2\Delta x \approx 6.2$. The difference between the theoretical result and simulation is mainly because the approximate equivalent model on point P in Fig.5.

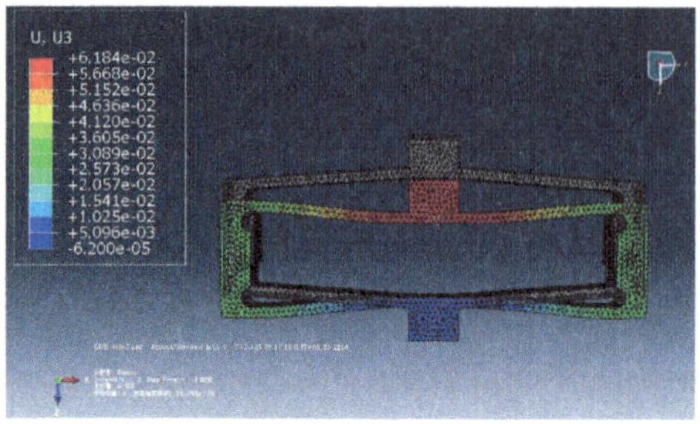

Fig.7. FEA simulation

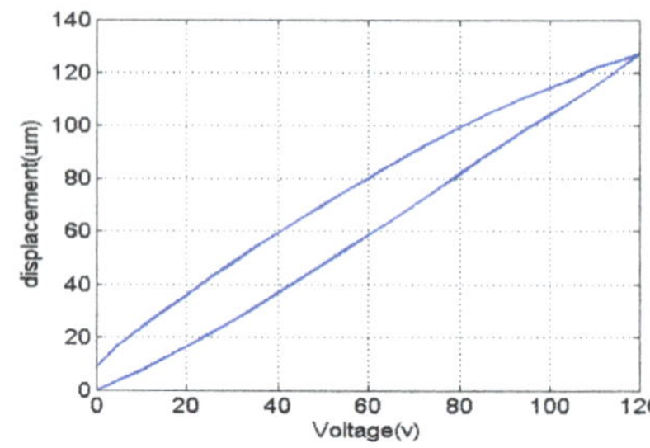

Fig.8. Experimental result

Fig.8 shows the experiment result of output displacement, the real output displacement of actuator presents hysteresis loop because of the hysteresis which is inherent in the piezoelectric material.

4 CONCLUSION

In this paper, a rhombic piezoelectric actuator is developed to get desired stroke and output force. The displacement amplification ratio of rhombic mechanism can be optimized by optimal method with analytic formulas and proper geometrical constrains. Then we get the maximum amplification ratio. After that the output force can be determined by given bending stiffness of flexure hinges. Finally, the FEA simulation and Experiment are made and verified the results of analytic method.

REFERENCES

(1) H.W. Ma, S.M. Yao, L.Q. Wang, Z. Zhong. (2006) *Sens. Actuators A.* **132**, p.730-736. doi: 10.1016/j.sna.2005.12.028

(2) J.M. Paros, L. Weisbord. (1989) *Machine Des.* **37**, p.151-156.

(3) E. Furukawa, M. Mizuno. (1990) *Int. J. Jpn. S. Prec. Eng.* **24**, p.285-190.

(4) J.H. Kim, S.H. Kim, Y.K. Kwak. (1990) *Sens. Actuators A.* **116**, p.530-538. doi:10.1016/2004.05.027

Active vibration control of cantilever beam using MFC sensor and actuator

W.K. Miao, M.L. Xu, C.S. Wu
State Key Laboratory for Strength and Vibration of Mechanical Structures,
School of Aerospace, Xi'an Jiaotong University, P.R. China

ABSTRACT

In this paper, three Macro Fiber Composite (MFC) patches are applied for vibration sensing and active control for a cantilever beam. Two of them are utilized as actuators: one provides the excitation force, the other creates the control force. The third one is used as a sensor. The sensitivity of MFC sensor is obtained by the test of sensing capability of MFC. Utilizing PD and fuzzy control algorithms, the active vibration control of the first two vibration modes is implemented. The effectiveness of the MFC sensor and actuator for sensing and controlling vibration is validated by experiment.

1 INTRODUCTION

Cantilever structures can be found in various important fields, for example, the solar panel of the satellite, the UAV with long wingspan, and the rotor wing of helicopter. The vibration will easily occur when they are subjected to the excitation from the external environment, which always threatens the safety of the structure. Thus, the vibration control becomes the key to ensure their structural safety.

Adopting piezoelectric actuators to control vibration has already been the subject of researches. However, traditional piezoelectric material, PZT, has extremely brittle characteristic which restricts its application (1). A kind of leading low-profile actuator and sensor, Macro Fiber Composite (MFC) invented by NASA, is flexible, durable, reliable, and easily surface-bonded. When subjected to voltage, it can deform materials, suppress or generate vibration. Conversely, it is a sensitive strain sensor with no voltage excitation (2). MFC have already been used as actuator for vibration control in aerospace devices by NASA's Langley Research Center (3). In recent years, MFC has been applied in various fields of vibration monitoring and control. Many relevant studies have been conducted by researchers. Salim Azzouz (4) modeled the MFC/AFC actuators and investigated the performance of MFC. In another study, Ro, Jeng-Jong (5) used MFC actuators to control the flexural vibration of the circular handlebars of a bicycle. In a work by Vadiraja (6), the experimental results revealed that MFC can be used to sense and control the vibration of rotating composite beams.

In this study, MFC sensor and actuator are utilized to sense and control the first two vibration modes of a cantilever beam using PD and fuzzy control algorithms.

2 EXPERIMENTAL OBJECT AND MODAL ANALYSIS

The experimental object, shown in fig. 1, is a cantilever beam whose cross section is rectangular. The length, width, and thickness of the cantilever beam are 370mm, 20mm, and 1.39mm respectively. The thickness of all MFC patches is 0.3mm. Two pieces of MFC patches (M2814-P1), whose length and width are 38mm and 20mm respectively, are surface-bonded on the cantilever beam as actuators. One MFC patch (M0714-P2), whose length and width are both 16mm, is boned on another side as a sensor. For convenience, the MFC actuator closer to the clamped end is defined No.1 to produce control force, and the other one is defined No.2 to provide excitation force. The MFC sensor is named No.3. In order to provide base for exciting the cantilever beam, the sine sweep frequency experiment is conducted. Here, we use laser displacement sensor to get the frequency response as shown in fig. 2, which indicates that the first three natural frequencies are respectively 8.27Hz, 51.14Hz, and 142.4Hz. A photograph of sweep frequency experimental setup is shown in fig. 3.

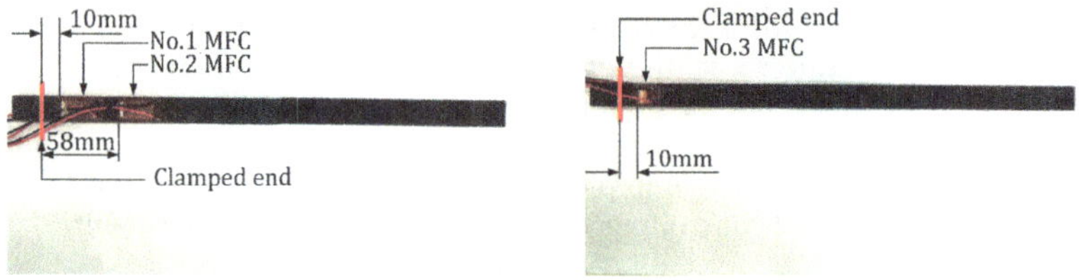

Fig. 1. Experimental object and MFC bonded locations in two sides of beam

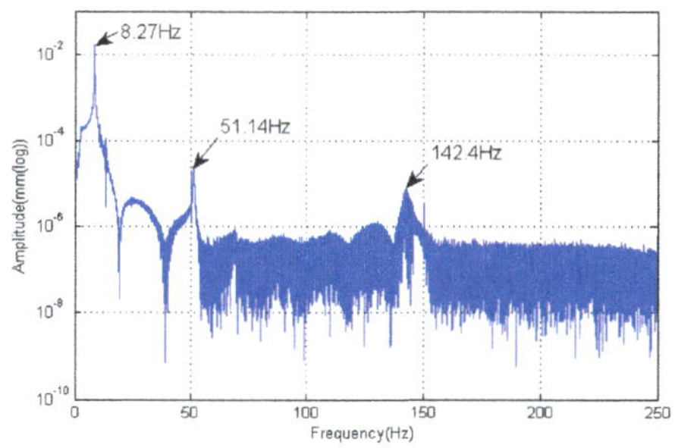

Fig. 2. Frequency response at free end

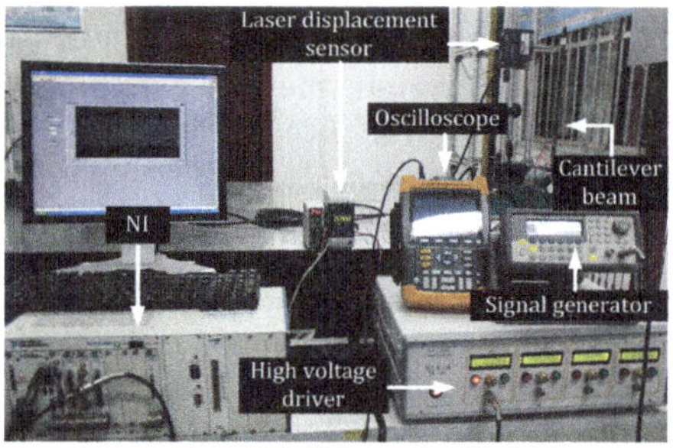

Fig. 3. Sweep frequency setup

3 SENSING CAPABILITY TEST OF MFC SENSOR

Given that MFC is a sensitive strain sensor, it is necessary to test the sensing capability of MFC to obtain its sensitivity. Fig. 4 presents the experimental setup of sensing test. The amplified sinusoidal excitation signal at 8.27Hz is applied to the No.2 MFC patch to make the cantilever beam vibrate as the first vibration mode. Meanwhile, the No.3 MFC patch is utilized to measure the strain of cantilever beam end. Use an oscilloscope to simultaneously observe the signal from the laser displacement sensor and charge-amplifier. The sensitivity of charge-amplifier is 1mV/pC. Thus, the charge generated by MFC sensor can be figured out. Corresponding strain can also be figured out according to amplitude measured by laser displacement sensor.

As can be seen in fig. 5, the output signal of the MFC sensor is well sinusoidal curve. And fig. 6 illustrates the relationship between output charge of MFC sensor and average strain of the cantilever beam end, which describes a well linear relationship. The result indicates that the sensitivity of MFC (M0714-P2) is $225.28 pC / \mu\varepsilon$. With the purpose of investigating the sensing capability of MFC much further, MFC sensor is utilized to obtain the frequency response of the cantilever beam. As shown in fig. 7, it is indicated that the data measured with MFC sensor is nearly consistent with the result acquired by using laser displacement sensor before. Actually, owing to the higher electromechanical coupling, the frequency response obtained with MFC sensor is more distinct. All above results reveal that MFC is a very excellent vibration sensor.

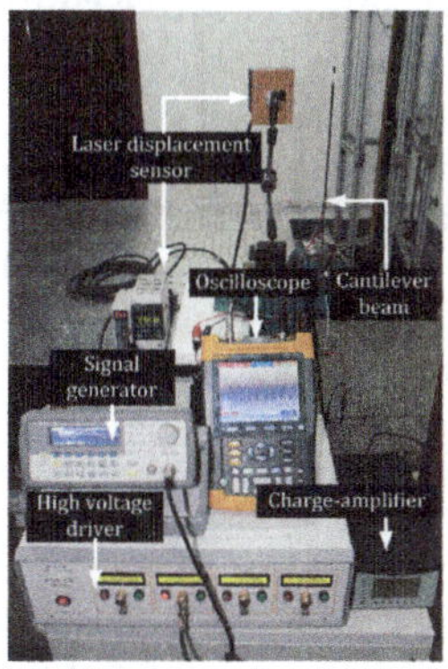

Fig. 4. Experimental setup of sensing test

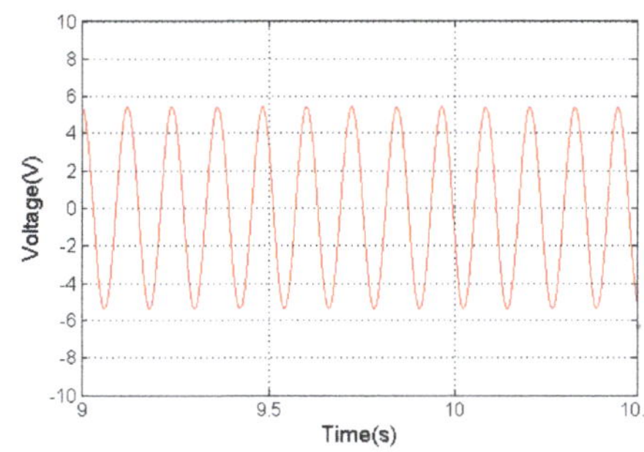

Fig. 5. Output signal of MFC sensor

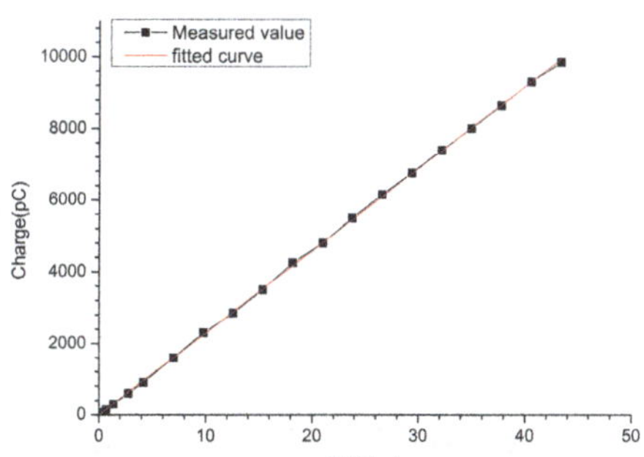

Fig. 6. Charge-strain relationship

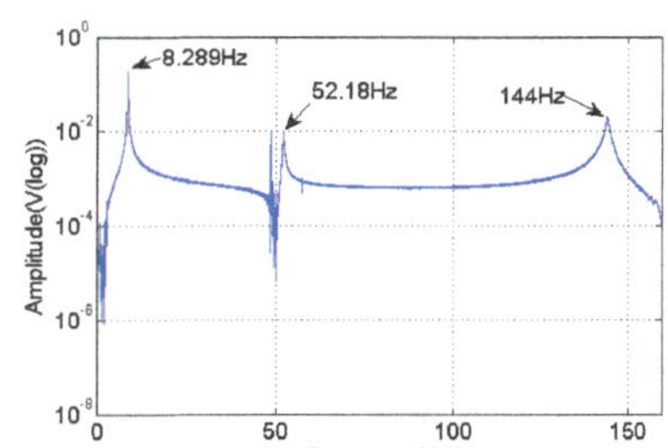

Fig. 7. Frequency response with MFC

4 ACTIVE VIBRATION CONTROL

As shown in fig. 8, utilizing PD and fuzzy control algorithms, a closed-loop control system which consists of a dSPACE system, three MFC patches, a signal generator, an oscilloscope, and a high voltage driver is built. The dSPACE system is utilized to implement the PD and fuzzy controller. Here, we experimentally investigate the suppression of first two vibration modes. The No.2 MFC patch is subjected to a sinusoidal signal at the first two natural frequencies of the cantilever beam, which leads to the first vibration mode and second vibration mode of the cantilever beam, respectively. Meanwhile, the No.3 MFC patch monitors the vibration. And the No.1 MFC patch is utilized to suppress the vibration.

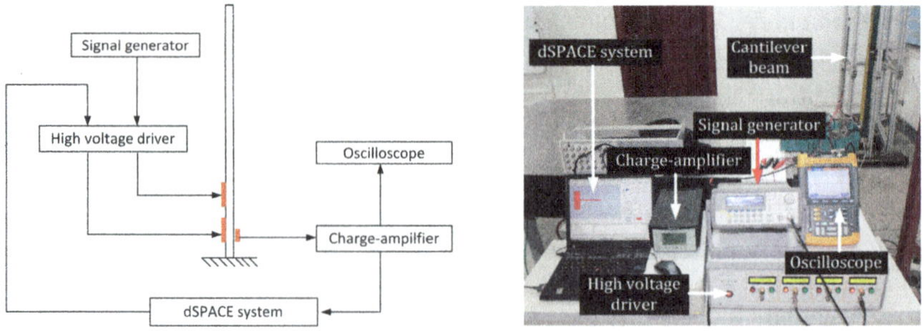

Fig. 8. Schematic diagram and experimental setup of active control

4.1 PD control results

As a common and mature algorithm, the PD control algorithm is selected to obtain the control voltage, which depends on two control parameters, namely k_p and k_d. While controlling, the appropriate control signal figured out by selecting suitable control parameters is applied to the No.1 MFC patch. The strain of the cantilever beam end can be figured out with the sensitivity of the MFC sensor and the charge-amplifier

Fig. 9 and fig. 10 respectively exhibit the time response and frequency response of the first two vibration modes before and after control. For the first vibration mode, the control parameters, k_p and k_d, equal 70 and 1 respectively while the corresponding parameters are -100 and -0.1 for the second vibration mode. Experimental results clearly indicate that the vibration attenuation of the first and second vibration mode is, respectively, 92.7% and 94.4%.

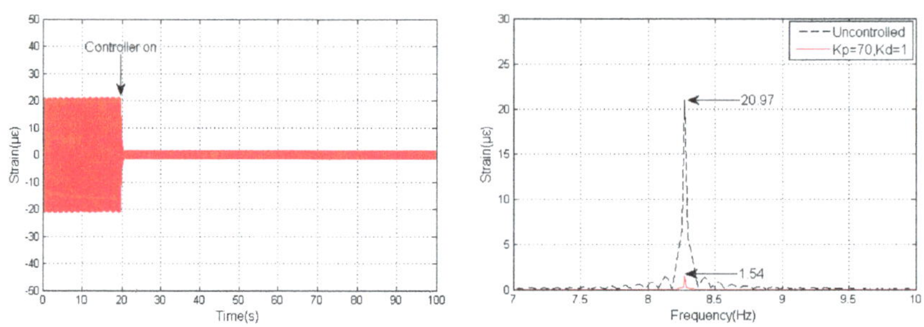

Fig. 9. Time response and frequency response of the first vibration mode before and after control by PD algorithm

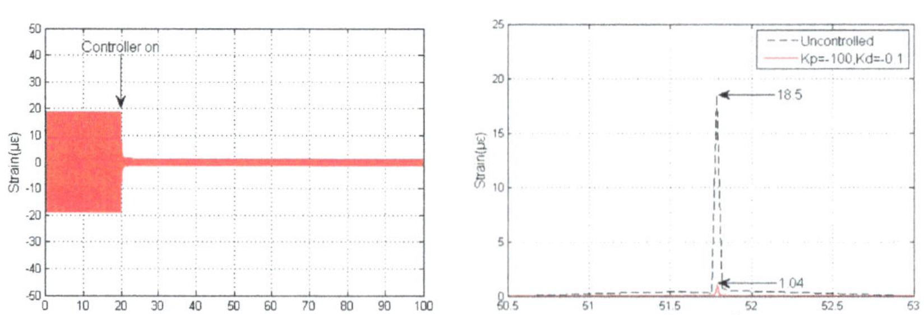

Fig. 10. Time response and frequency response of the second vibration mode before and after control by PD algorithm

4.2 Fuzzy control results

In this section, the vibration of the cantilever beam is controlled by fuzzy algorithm which has three control parameters, namely k_e, k_{ec}, and k_u. The control process is similar to the PD control. And the control parameters, k_e, k_{ec}, and k_u are -0.5, -0.5, and -9 for the first vibration mode and 10, 0.05, and-5.5 for the second vibration mode. The results are presented in fig. 11 and fig. 12, revealing that vibration attenuation of the first and second vibration mode is, respectively, 96.8% and 95.5%.

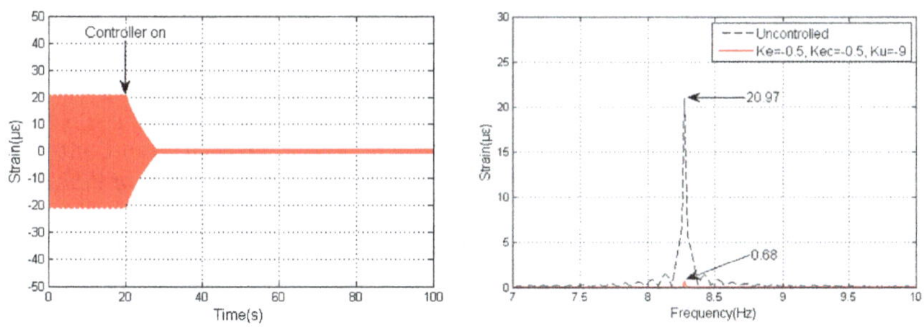

Fig. 11. Time response and frequency response of the first vibration mode before and after control by fuzzy algorithm

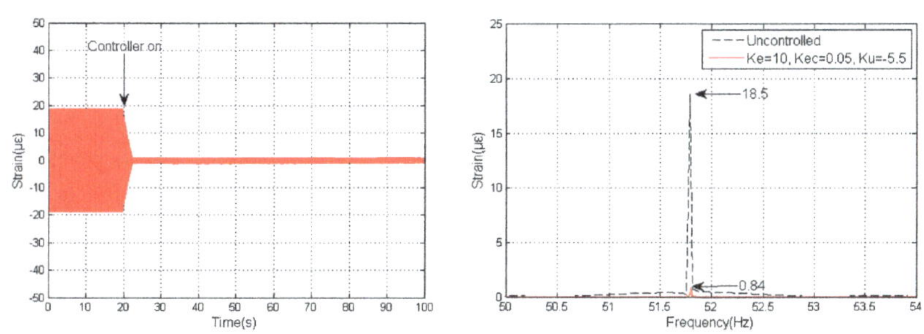

Fig. 12. Time response and frequency response of the second vibration mode before and after control by fuzzy algorithm

5 CONCLUSION

In this study, the Macro Fiber Composite (MFC) is utilized as actuator and sensor. Two MFC (M2814-P1) patches are utilized as actuators: one provides the excitation force, the other creates the control force. Meanwhile, one MFC (M0714-P2) patch is applied to sensing the vibration. Firstly, with a laser displacement sensor, the natural frequency is obtained by sweep frequency experiment. Then, the test of the sensing capability of MFC sensor is conducted to acquire the sensitivity, which demonstrates that MFC own the excellent sensing performance. Compared with the laser displacement sensor, more distinct frequency response is obtained by using MFC sensor, which further proves the excellent sensing property of MFC senor. Finally, the active vibration control for the cantilever beam is implemented, whose results clearly indicate that, using the PD algorithm and fuzzy algorithm, the first two vibration modes of the cantilever beam are obviously suppressed by the MFC sensor and actuator. This work provides an effective approach to control vibration of cantilever structures by utilizing MFC as actuator and sensor. The methods used in this study can be applied in many fields.

REFERENCES

(1) Sodano, H. A., Park, G. & Inman, D. J. (2004) An investigation into the performance of macro-fiber composites for sensing and structural vibration applications. *Mechanical Systems and Signal Processing*. **18(3)**, 683-697.

(2) http://www.smart-material.com.

(3) Wilkie, W.K., Bryant, R. G., High, J. W., Fox, R. L., Hellbaum, R. F., Jalink, A., Little, B. D., & Mirick, P. H. (2000) Low-cost piezocomposite actuator for structural control applications. *In SPIE Proc.*. 323-334.

(4) Salim Azzouz, M., Mei C., Bevan, J. S. & Ro, J. J. (2001) Finite element modeling of MFC/AFC actuators and performance of MFC. *Journal of Intelligent Material Systems and Structures*. **12(9)**, 601-12.

(5) Ro, J. J., Chien, C. C., Wei, T. Y. & Sun, S. J. (2007) Flexural Vibration Control of the Circular Handlebars of a Bicycle by Using MFC Actuators. *Journal of Vibration and Control*. **13(7)**, 969-987.

(6) Vadiraja, D. N. & Sahasrabudhe, A. D. (2009) Vibration analysis and optimal control of rotating pre-twisted thin-walled beams using MFC actuators and sensors. *Thin-Walled Structures*. **47(5)**, 555-567.

High temperature performance of a metal-packaged strain sensor based on a regenerated fiber Bragg grating in Boron–Germanium-codoped fiber

Y. Tu, S.-T. Tu, S.-P. Zhou
Key Laboratory of Pressure Systems and Safety (Ministry of Education),
School of Mechanical and Power Engineering,
East China University of Science and Technology, P.R. China

ABSTRACT

A metal-packaged strain sensor is successfully fabricated based on a regenerated fiber Bragg grating (RFBG) in boron–germanium-codoped photosensitive fiber by combining magnetron sputtering and electroplating. The strain characteristics of the sensor are evaluated by uniaxial tensile tests at different temperatures ranging from room temperature to 540 °C. The sensor exhibits higher strain sensitivity than that of the corresponding bare RFBG, as well as good linearity and repeatability of strain measurements. The strain sensor provides great promise for structural integrity monitoring of high-temperature components.

1 INTRODUCTION

Strain measurement has been referred as the most reliable method for structural integrity monitoring of high-temperature components since their principal mechanism of failure is creep damage that could be indicated by creep strain accumulation (1). However, it is difficult to guarantee reliability and durability of the sensors in high-temperature environments. Therefore, strain measurement at high temperature has been a long-standing challenge for many decades.

Optical fiber sensors are attractive candidates for structural integrity monitoring due to a number of advantages over their electrical counterparts, such as small size, light weight, electrically passive operation, high sensitivity, and resistance to electromagnetic interference and corrosion. In particular, wavelength encoded fiber Bragg grating (FBG) sensors have inherent self referencing and wavelength multiplexing capabilities that allow them to be easily multiplexed in a single fiber and spliced to telecommunication fibers for remote, distributed and multi-parameter sensing (2). However, conventional type-I FBGs exhibit poor thermal stability at high temperature (3). Many studies have been made to increase the thermal stability of these gratings (4-6). Recently, regenerated fiber Bragg gratings (RFBGs) with superior high temperature stability have been considered as an attractive potential for measurement at high temperature (7, 8).

It is hard to use bare RFBG as sensors directly without any protection since they become very brittle after high-temperature annealing. Accordingly, metallic packaging is proposed for protecting the bare RFBGs as well as easily attaching the RFBG-based strain sensors to metallic high-temperature components (9, 10). To our knowledge, metal-packaged RFBG sensors have mainly been fabricated based on RFBGs in standard telecommunication fibers that should be annealed at temperatures above 850 °C leading to a significant reduction in strength.

In this paper, we demonstrate a metal-packaged strain sensor fabricated based on a RFBG written in boron (B)-germanium (Ge)-codoped photosensitive fiber that can be annealed at relatively low temperatures. The sensor and its corresponding bare RFBG are evaluated by uniaxial tensile tests. The experimental and numerical results are presented, followed by a discussion about the strain characteristics of the sensor.

2 EXPERIMENTAL DETAILS

In this work, the seed FBG is written in a hydrogen-loaded B–Ge-codoped fiber (Fibercore Ltd., PS1250/1500) with a reflectivity of ~80%, a 3 dB reflection-bandwidth of less than 0.3 nm, and a grating length of 8 mm with about 10 mm acrylate coating stripped.

For fabrication of RFBG, the seed FBG was loosely placed in a miniature tube furnace to avoid any tension on the fiber and subjected to heat treatment. The furnace temperature was raised from room temperature (RT) to 500 °C within 50 min, and then held steady at 500 °C for ~120 min. As the temperature approached 500 °C, the reflection peak power of the FBG started to decay rapidly until it fell below the noise floor in a few min after the temperature reached 500 °C. During the 120 min isothermal annealing at 500 °C, a new grating regenerated at ~8 min followed by its reflection strength increased gradually to its maximum until stability. No variation was observed in the strength of the RFBG with the temperature dropped to RT over 90 min.

After that, a series of experiments was carried out for successfully packaging it into a P91 steel substrate to obtain a metal-packaged strain sensor based on the RFBG in B–Ge-codoped fiber by means of the same process of packaging used for the RFBG in standard telecommunication fiber described in our previous work (10).

In order to characterize the strain response, uniaxial tensile tests were performed on the metal-packaged strain sensor mounted onto a P91 steel specimen, which is similar to the previous tests we have done (10). The sensor was characterized in air at constant temperatures of RT (26.5 °C), 100 °C, 200 °C, 300 °C, 400 °C, 500 °C and 540 °C. The temperature was kept constant at each temperature for ~20 min before tensile testing. The load was held constant at every 1 kN or 0.5 kN for 2 min to obtain the average. For comparison, the uniaxial tensile tests were also conducted on a bare RFBG inscribed in B–Ge-codoped fiber by means of the same apparatus previously described (10).

3 RESULTS AND DISCUSSION

A prototype metal-packaged strain sensor fabricated based on the RFBG in B–Ge-codoped photosensitive fiber is shown in figure 1. Figures 2 (a)–(g) illustrate the wavelength shifts of the metal-packaged sensor as a linear function of strains determined from the applied forces, the cross-sectional area and Young's modulus of the P91 steel specimen at a constant temperature. The strain sensitivity derived from the slope of the straight lines in these graphs is 2.09, 2.14, 2.11, 2.17, 2.17, 2.11 and 2.10 pm $\mu\varepsilon^{-1}$ for the sensor under loading at constant temperatures of RT (26.5 °C), 100 °C, 200 °C, 300 °C 400 °C, 500 °C and 540 °C respectively, in contrast with that of 2.08, 2.10, 2.08,

2.15, 2.04, 2.07 and 2.06 pm $\mu\varepsilon^{-1}$ under unloading. The sensor preserves its linear behavior implying a good interfacial integrity between every two layers. The strain sensitivity under loading is little larger than that under unloading due to the elastic hysteresis that is caused by the relatively flexible structural substrate as the elastic sensitive element.

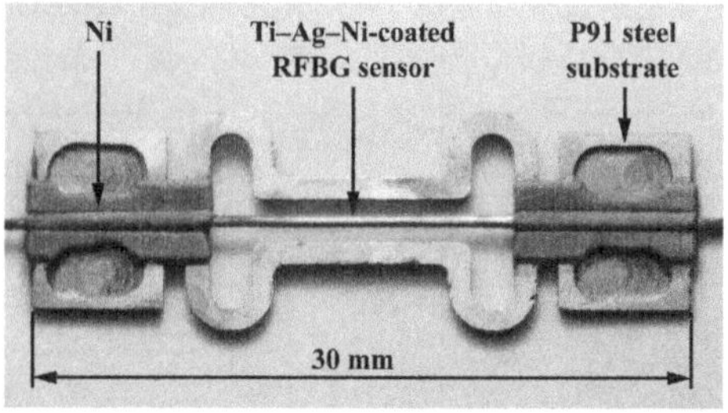

Figure 1. A prototype metal-packaged RFBG strain sensor based on the RFBG in B–Ge-codoped photosensitive fiber.

The results of the tensile tests performed on the bare RFBG in B–Ge-codoped fiber and a three-dimensional finite element (FE) model established in our previous work (10) are also plotted in figure 2 for comparison. We can obtain the strain sensitivity of 1.23, 1.23, 1.23, 1.25, 1.25 and 1.27 pm $\mu\varepsilon^{-1}$ for the bare RFBG sensor under loading at constant temperatures of RT (21 °C), 100 °C, 200 °C, 300 °C, 400 °C, and 500 °C respectively, in contrast to that of 1.23, 1.23, 1.24, 1.25, 1.26 and 1.30 pm $\mu\varepsilon^{-1}$ under unloading. The strain sensitivity of the metal-packaged strain sensor is larger than that of the corresponding bare RFBG sensor, which is primarily attributed to the flexible structure of the substrate. The numerical strain sensitivity can be calculated by substituting the strains of the optical fiber determined from the FE model into the following equation

$$\frac{\Delta\lambda_B}{\lambda_B} = \varepsilon_z - \frac{n_{eff}^2}{2}\left[p_{12}\varepsilon_z + \left(p_{11} + p_{12}\right)\varepsilon_r\right] \tag{1}$$

where λ_B is the Bragg wavelength, n_{eff} is the effective refractive index of the core, p_{11} and p_{12} are the components of the silica strain-optic tensor, and ε_z and ε_r are the axial and radial strain in the fiber respectively. For PS1250/1500 fibers, p_{11} = 0.113, p_{12} = 0.252, and n_{eff} = 1.482. Thus, the numerical strain sensitivity of 1.75 pm $\mu\varepsilon^{-1}$ is obtained according to equation (1). Comparisons of the experimental results and the numerical results show a satisfactory agreement with a relative error less than 19.4%.

Table 1. Strain sensitivity of the metal-packaged strain sensor based on the RFBG in B–Ge-codoped fiber obtained from mechanical load-cycling tests.

Temperature (°C)	Strain Sensitivity (pm $\mu\varepsilon^{-1}$)					
	Loading			Unloading		
	Run 1	Run 2	Run 3	Run 1	Run 2	Run 3
26.5	2.10	2.09	2.09	2.07	2.08	2.08
100	2.14	2.14	2.14	2.12	2.10	2.10
200	2.12	2.11	2.11	2.08	2.08	2.09
300	2.18	2.22	2.22	2.16	2.16	2.16
400	2.17	2.19	2.19	2.04	2.12	2.04
500	2.13	2.11	2.11	2.13	2.07	2.07
540	2.10	2.10	2.10	2.06	2.06	2.06

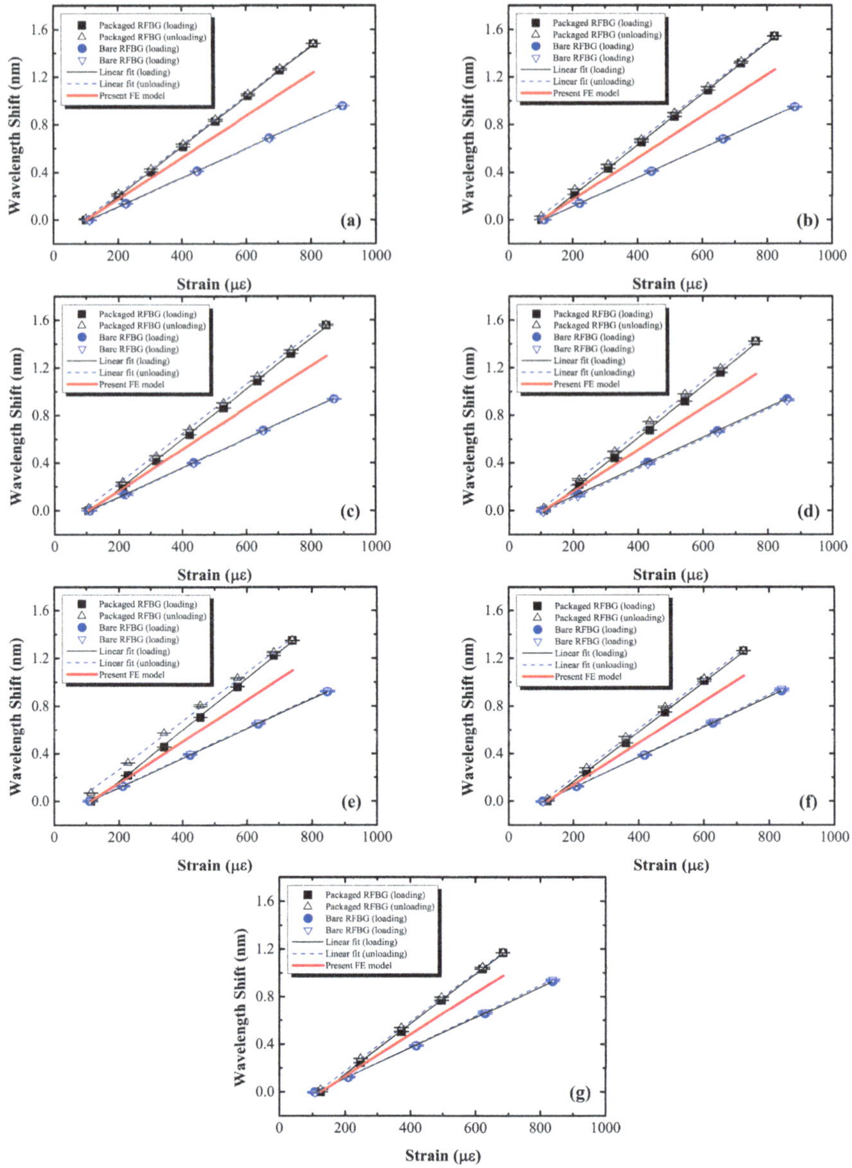

Figure 2. Shift in the Bragg wavelength as a function of strain obtained from tensile tests at constant temperatures of room temperature (a), 100 °C (b), 200 °C (c), 300 °C (d), 400 °C (e), 500 °C (f) and 540 °C (g).

Subsequently, the tensile tests were conducted on the same sensor sample at a constant temperature for three times to determine its repeatability. The tabulated strain sensitivity of the sensor obtained from the mechanical loading-cycling tests run agree well with one another, as given in table 1. The experimental results demonstrate that the metal-packaged strain sensor based on the RFBG in B–Ge-codoped fiber has good repeatability. It is also proved that the interfacial bonding between the optical fiber and the titanium layer is strong, as well as the nickel layer and the P91 steel substrate.

4 CONCLUSIONS

In this paper, the fabrication processes of the metal-packaged strain sensor based on the RFBG written in B–Ge-codoped fiber have been presented and the strain characteristics of the sensor have been evaluated by tensile testing. Some conclusions have been drawn from the investigation as follows:

(1) Metal-packaged strain sensors can be successfully fabricated by combining magnetron sputtering and electroplating based on RFBGs in B–Ge-codoped fibers, with strong bonded interfaces between the optical fiber and the titanium layer, as well as nickel layer and the P91 steel substrate.
(2) The metal-packaged strain sensor based on the RFBG in B–Ge-codoped fiber exhibits higher sensitivity than that of the corresponding bare RFBG due to the flexible structure of the substrate, as well as good linearity and repeatability of the strain measurements at constant temperatures ranging from room temperature to 540 °C.

The metal-packaged strain sensors provide great potential for strain measurement in high-temperature environments. In future work, more efforts will be needed to meet higher temperature challenges, and it is crucial to verify reliability and durability of metal-packaged strain sensors based on RFBGs.

ACKNOWLEDGMENTS

Y. Tu would like to thank Professor Lin Ye from Sydney University for his helpful suggestions. This work is supported by the Natural Science Foundation of China (Nos. 11172102 and 51175178) and the National High Technology Research and Development Program 863 (No. 2009AA044803).

REFERENCES

[1] Tu S.-T. (2009) *Proc. FM 2009 - Transferability and Applicability of Current Mechanics Approaches*. East China University of Science and Technology Press, 13-22.
[2] Majumder M., Gangopadhyay T. K., Chakraborty A. K., et al. (2008) *Sens. Actuator A-Phys.* **147**, 150-164. doi: 10.1016/j.sna.2008.04.008
[3] Erdogan T., Mizrahi V., Lemaire P. J., et al. (1994) *J. Appl. Phys.* **76**, 73-80. doi: 10.1063/1.357062
[4] Bartelt H., Schuster K., Unger S., et al. (2007) *Appl. Optics.* **46**, 3417-3424. doi: 10.1364/AO.46.003417
[5] Lowder T. L., Smith K. H., Ipson B. L., et al. (2005) *IEEE Photonics Technol. Lett.* **17**, 1926-1928. doi: 10.1109/lpt.2005.852646
[6] Butov O. V., Dianov E. M. & Golant K. M. (2006) *Meas. Sci. Technol.* **17**, 975-979. doi: 10.1088/0957-0233/17/5/s06
[7] Fokine M. (2002) *Opt. Lett.* **27**, 1016-1018. doi: 10.1364/OL.27.001016
[8] Bandyopadhyay S., Canning J., Stevenson M., et al. (2008) *Opt. Lett.* **33**, 1917-1919. doi: 10.1364/OL.33.001917
[9] Sandlin S., Kinnunen T., Ramo J., et al. (2006) *Surf. Coat. Technol.* **201**, 3061-3065. doi: 10.1016/j.surfcoat.2006.06.031
[10] Tu Y. & Tu S.-T. (2014) *Smart Mater. Struct.* **23**, 035001. doi: 10.1088/0964-1726/23/3/035001

Study on thermal physical properties of 304 stainless steel

F. Dong, J.M. Qie, H.H. Deng
School of Material and Metallurgy,
Inner Mongolia University of Science and Technology, P.R. China

ABSTRACT

The DIL402C thermal dilatometer and STA449C thermal analyzer were employed to measure the linear expansion and contraction coefficients, Cp and DSC curves of a 304 stainless steel. The result showed that the linear expansion and contraction coefficients were in the range of $20.9700\times10^{-6}\sim21.5712\times10^{-6}$ and $21.2528\times10^{-6}\sim21.9471\times10^{-6}$.The higher the coefficient of the slab the easier to form cracks. During 1450~1000 °C, the crystal phase transformation occurred, the curve of Cp fluctuated obviously and the curve of DSC was unsmooth. The thermal stability of slab was thus inferior during this temperature range.

1 INTRODUCTION

Thermal physical properties of materials were the important characteristic parameters of engineering materials. It will have significant scientific meaning and engineering values through studying the thermal physical properties of materials [1]. Some researchers conducted some studies about the thermal physical properties of casting blank. However, such studies were not systematic and comprehensive [2, 3]. The 304 stainless steel has enormous applications as a staple austenitic stainless steel. Currently, the studies on thermal physical properties to the 304 stainless steel were not enough, only limiting to the investigations of the expansion and contraction coefficient, but seldom on specific heat capacity and heat quantity with the temperature [4, 5]. Hence this paper presents a comprehensive result of the thermal physical properties of the 304 stainless steel in order to increase the understanding to the properties of the steel.

2 MATERIALS AND METHODS

The chemical composition of the 304 stainless steel sample is reported in Table 1.

Table 1. Chemical compositions of 304 stainless steel

Elements	C	Si	Mn	P	S	Cr	Ni
percentage composition (wt,%)	0.040	0.45	1.18	0.030	0.003	17.245	0.323

2.1 Thermal expansivity test

The DIL402C thermal dilatometer was used to measure the expansion and contraction coefficient from room temperature to 1000°C. The size of test sample was Φ4×25mm. The same speed of heating and cooling, 10°C/min, was selected. When the sample reaches 1000°C, heat equilibrium was kept for an hour, and then cooling was started with the same rate.

2.2 DSC (differential scanning calorimetry) test

The STA449F3 thermal analyzer was employed to measure the isobaric heat capacity (Cp) and heat quantity change with the temperature of the sample. The size of Cp sample was Φ3×1mm and the dimension of DSC sample was Φ5×1.5mm.

Heating rate of the Cp sample was 20°C/min from room temperature to 1400°C. Similarly, the DSC sample was heated to 1500°C at a heat velocity of 20°C/min. Argon was used as protective gas during the course of experiment.

3 RESULTS AND DISCUSSION

3.1 Thermal expansivity of 304 stainless steel

The curve of thermal expansivity and contractility with the experimental temperature was shown in Figure 1.

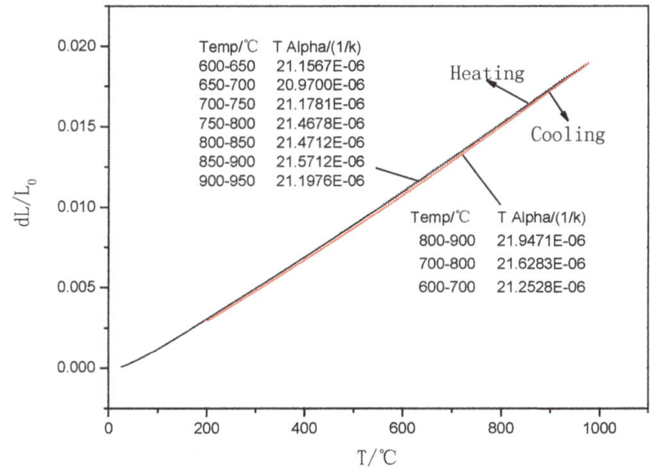

Fig.1. Corrosion rate of experimental steel in simulated industrial atmosphere

As is shown in Fig.1, The curves of thermal expansivity and contractility change linearly with the change in temperature during heating and cooling. The range of expansion coefficient was 20.9700×10^{-6}~21.5712×10^{-6} during heating and the contraction coefficient was 21.2528×10^{-6}~21.9471×10^{-6} during cooling. Compared with 2Cr13 martensitic stainless steel ($\alpha=12.08\times10^{-6}$) and 20CrMo structural alloy steel ($\alpha=14.05\times10^{-6}$), the expansion and contraction coefficient of 304 stainless steel was comparatively large [6, 7]. During the process of solidification, as the shrinkage of 304 stainless steel was relative larger on the surface than inside, the residue thermal stresses were non-uniform [8]. Hence the 304 stainless steel was easy to generate surface cracks and pits in process of production.

3.2 Isobaric heat capacity of the 304 stainless steel

Fig. 2 shows the curve of the isobaric heat capacity of 304 stainless steel sample in the heating process.

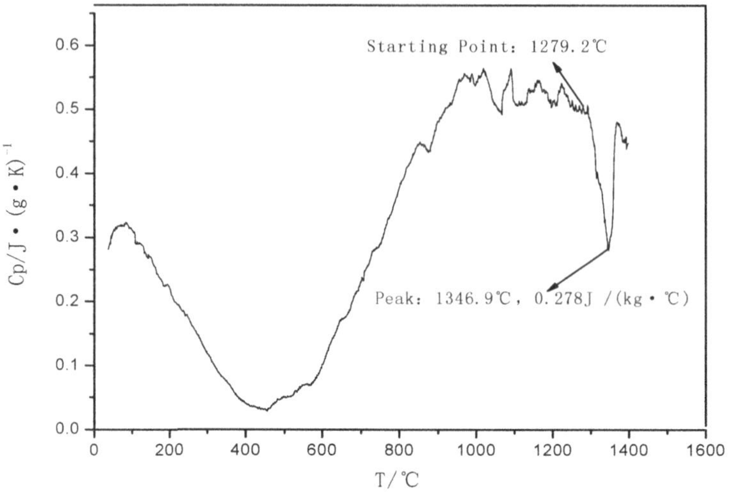

Fig.2 CP curve of 304 stainless steel with temperature increasing

As was shown in Fig. 2, the Cp curve descends gradually from room temperature to 500°C with temperature increasing, but the curve rises sharply in 600~900°C. The Cp curve fluctuates obviously in the temperature range of 1000~1400°C. It means that in this temperature range the crystal phase transformations and the phase transitions exist, so that the state of sample was unstable. When the curves pass through 1279.2°C where phase transition occurs at this temperature point, if the blank is located in this temperature range, like the upper of crystallizer, the brittleness of steel blank was comparatively large [9]. If the cooling of the crystallizer was quick and the heat transfer of slag film was non-uniform, the surface cracks and pits on the blank would generate more easily.

Temperature has an important impact on the isobaric heat capacity. At the low temperature stage, the ascendant tendency of Cp curve was comparatively slow due to the crystal shakes weakly. With temperature rising, the crystallization increases, the Cp increased evidently. At the same time, the tendency of Cp curve was related to the properties of the alloy phase, which is composing of the material. The change of thermal lattice vibration of the atoms may lead to the reduction of total energy and decreasing of the Cp [5]. The temperature range that the Cp curve generates fluctuation has an important meaning to the heat transfer and cooling of the casting blank in the process of being formed in crystallizer. It also has certain significance to the actual production process of casting blank.

3.3 DSC result of the 304 stainless steel

Fig. 3 showed the DSC test results of the sample upon heating. As was shown in the figure, an inflexion occurs at 1402.1°C and then the peak temperature reaches 1438.0°C. It is believed that the liquefaction temperature should be located in 1/3 from inflexion temperature to peak temperature according to the working principle of thermal analyser [10]. The liquefaction temperature was measured about 1414°C in this experiment.

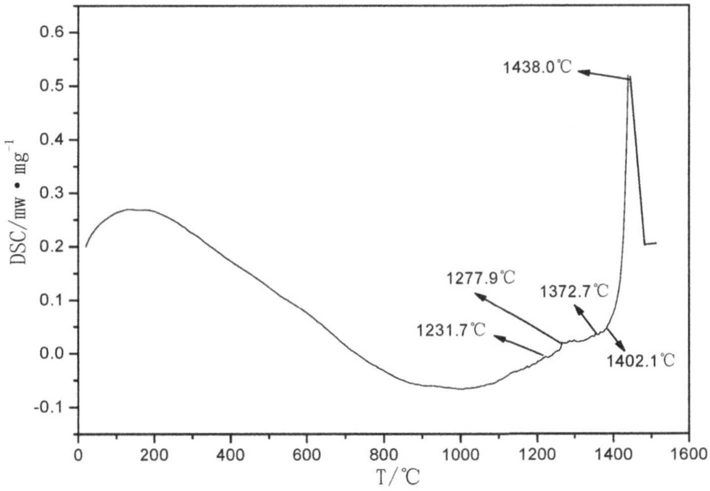

Fig.3 DSC curve of 304 stainless steel

The appearance of the peak temperature was related to the temperature that the velocity of the transformation of crystalline shape or phase transition reaches the maximum. When the crystal phase transformation or phase transition occur, sharp change in temperature and the quick heat change can cause the obvious endothermic or exothermic enthalpy, with the large peak shown on the DSC curve and the curve being unsmooth. At 1372.7°C, 1277.9°C, 1231.7°C, the crystal phase transformations would occur, so the emergence of the crystal phase transformation would lead to the variance of shell's volume. Therefore, the thickness of initial shell was non-uniform [11]. When the continuous casting slab was located in 1450~1100°C that contained these temperature points, like the upper of crystallizer, due to the existence of the crystal phase transformation, the occurrence of surface defects in casting slab were more easily under the influence of internal stress [12]. So when the initial shell formed in the middle and upper of the crystallizer, the control of slab withdrawing speed and crystallizer cooling intensity and uniformity would have become very important. It can avoid the occurrence of surface cracks in slab because of the greater crystal phase transformation.

4. CONCLUSIONS

The thermal physical properties of a 304 stainless steel have been studied. Thermal expansivity indicated that the expansion and contraction coefficient was greater and the 304 stainless steel belong to the crack sensitive steel. The range of expansion and contraction coefficient was 20.9700×10^{-6}-21.5712×10^{-6} and 21.2528×10^{-6}-21.9471×10^{-6} during heating and cooling. At the range of 1000~1400°C, due to the transformation of crystalline shapes, the Cp curve fluctuates obviously. At 1279.2°C, the sample generates phase transition. When the casting blank is located in the upper of crystallizer, the defects occurred easily due to the transformation of crystalline shape or phase transition. By analyzing of the results of DSC, it is confirmed that the 1414°C is the liquidising temperature of the 304 stainless steel. The thermal stability of slab was inferior and the DSC curve was unsmooth during 1450~1100°C, because of the existence of the crystal phase transformation. Moreover, in this temperature range, the thickness of initial shell was non-uniform and surface cracks in slab would occur easily.

REFERENCES

[1] Chen, G.S., Liao Y., & Zeng Y.G. etc: China Measurement Test, 2010, vol.36, pp.5-8.

[2] Collocott, S.J., & White, G.K.: Cryogenics, 1986, vol.26, pp.402-405.

[3] Wang, Q.M., Chen, D.F., & Gao, X. J. etc: Journal of the Chinese Science of Rare Earths, 2008, vol.26, pp.517-521.

[4] Daw, J.E., Rempe, J.L., Knudson, D.L.: Journal of Nuclear Materials, 2008, vol.376, pp.211-215.

[5] Ju, C.Y. The Study on Elevated Temperature Phase Transformation and Thermal Physical Properties of Two Stainless Steel[Dissertation]. Gan Su: Lanzhou University of Technology M.SC Thesis, 2009.

[6] Wang, W.X., Liu, C, L., & Wang, Y. etc: Iron Steel Vanadium Titanium, 2008, vol.29, pp.61-65.

[7] Pan, Y.H., Chen, D.F., & Dong, L.Y. etc: Journal of University ChongQing, 2006, vol.29, pp.68-71.

[8] Liu, D.J., Cai, Z.Z.: Journal of Iron and Steel Research, 1999, vol.11, pp.9-13.

[9] Deng, F.Y., Ma, Y.X., & Zou, A.Q.: Technical Development of Enterprise, 2007, vol.26, pp.19-21.

[10] Liu, P. Study on J4 Stainless Steel Continuous Casting Flux [Dissertation] Chongqing University M.SC Thesis, 2009.

[11] Wu, X. Research on high-temperature physical properties and mechanical properties of stainless steel [Dissertation]. Gan Su: Lanzhou University of Technology M.SC Thesis, 2010.

[12] Kang, L., Wang, Y., & Wang, E.G. etc: China Metallurgy, 2007, vol.1, pp.28-32.

Microstructure and properties of W-TiC/Cf composites prepared by spark plasma sintering

H. Chen[1], L. Luo[1,3], X. Tan[1], G. Luo[2], P. Li[1,3], X. Zan[1,3], X. Zhu[1,3], Y. Wu[1,3]
[1] School of Materials Science and Engineering, Hefei University of Technology, P.R. China
[2] Institute of Plasma Physics, Chinese Academy of Sciences, P.R. China
[3] Laboratories of Nonferrous Metal Material and Processing Engineering of Anhui Province, P.R. China

ABSTRACT

W-TiC/(0.5-2)wt.%Cf composites were successfully synthesized by spark plasma sintering. FE-SEM and HRTEM analysis, tensile test were used to characterize these samples. The grain size, relative density of the W-TiC/0.5wt.%Cf and W-TiC/2wt.%Cf samples were 4～5 μm and 98.3%, 97.62%, respectively. The Vickers micro-hardness of W-TiC/0.5wt.%Cf and W-TiC/2wt.%Cf samples are 688.36 and 1060.77Hv, respectively. Tensile strength values of W-TiC/2wt.%Cf samples were higher than others, reached 246.9 MPa.

1. INTRODUCTION

Tungsten is currently the most favored candidate for the plasma-facing material of nuclear fusion reactors due to its refractory nature, excellent surface erosion resistance and good thermal conductivity[1]. However, its inherent brittleness below the ductile-to-brittle transition temperature and embrittlement by neutron irradiation are the critical hurdles preventing structural application[2]. So it is necessary to perfect the properties of tungsten.

Fibers in the metal can effectively transmit and carry the force. When they are pulled out or debonded, the fibers are bond to consume energy and extend the crack propagation pathway, thereby improving the toughness of the material. In order to improve the tungsten toughness, we took a completely new approach, namely, carbon fibers (hereafter, all carbon fibers are Cf)-reinforced tungsten composites (W/Cf) whose toughening mechanism is based on that of fiber-reinforced ceramic matrix composites (CMC) [3-5]. In order to improve the relative density, we also doped 1wt% TiC into bulk tungsten[6]. The obtained powders were sintered by Spark plasma sintering (SPS) which can be used to synthesize refractory metal and its alloys, due to its fast heating rate and short holding time. After sintering, the composites were characterized.

2. EXPERIMENTAL PROCEDURES

2.1 Mixing and sintering

In the experiment, We dropped 0.5wt.% and 2wt.% Cf into tungsten powders respectively with 1wt.% TiC. Then the powders were mixed by ball-milling for 4h. After that the obtained powders were synthesized using SPS (FCT Group, SE-607, Germany). The sintering process was finished at 1800 °C under a pressure of 58 MPa and hold for 1.5 min in a flowing argon atmosphere. Furnace cooling to get a dense composite material, ultimately we gained samples of a desired shape by wire cutting.

2.2 Characterization

The density of the sintered specimens were measured by Archimedes method. And the Vickers micro-hardness of the specimens were measured by MH-3L micro-hardness tester along the cross-sectional surface of it, with a load of 300gf held for 15s. Field–emission scanning electron microscopy (FESEM, SU8020, Japan) and energy–dispersive X-ray spectroscopy (EDS) were performed to characterize the fracture morphologies of the sintered specimens. Fractured surface was obtained by tensile test at room temperature. And transmission electron microscopy (TEM, JEM–2100F, Japan) was performed to observe the microstructure of these specimens which were prepared via ion–thinning technology.

3. ANALYSIS DISCUSSION

Before the experiment, the tungsten powders and Cf were observed with FE-SEM equipped with EDS. The Cf has an average diameter of 7~8μm and the tungsten powders have an average grain size of approximately 1.2μm. The relative densities of all samples are higher than 97%. Especially the W-TiC/0.5wt.%Cf sample exhibits a relative density of 98.36%. Fig. 1(a), (b) show the image of W-TiC/0.5wt.%Cf and W-TiC/2wt.%Cf powders after balling mill mixing, respectively. During ball milling, the incremental surface energy and adsorption of powders make powders together to form agglomerated bundles easily. The after ball-milling grains become as large as 1~3 μm and look like spherical or lath shape. The Vickers micro-hardness of W-TiC/0.5wt.%Cf and W-TiC/2wt.%Cf samples are 688.36 and 1060.77Hv, respectively.

Fig. 1. (a) and (b) are SEM images of W-TiC/0.5wt.%Cf and W-TiC/2wt.%Cf, respectively

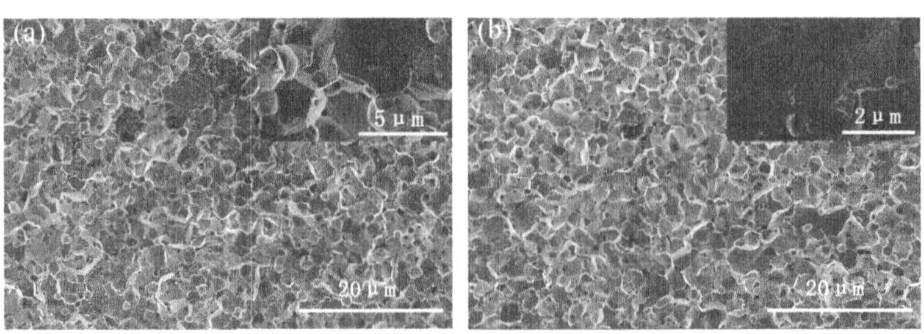

Fig. 2. SEM images of fractured morphology of W-TiC/(0.5-2)wt.%Cf

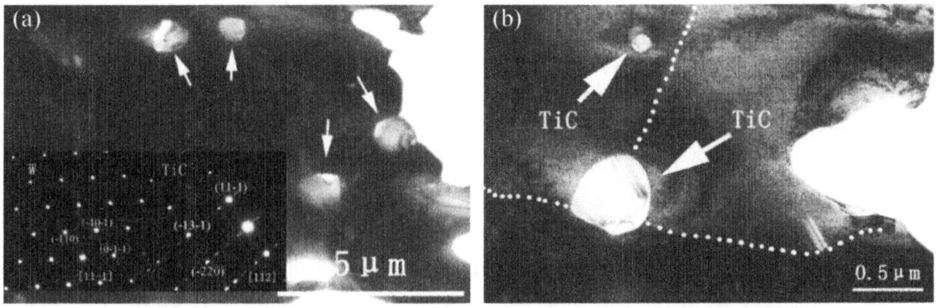

Fig. 3. Typical TEM images of W-TiC/2wt.%Cf. (a)TEM image of the distribution of TiC; (b)TEM bright field image of the distribution of TiC

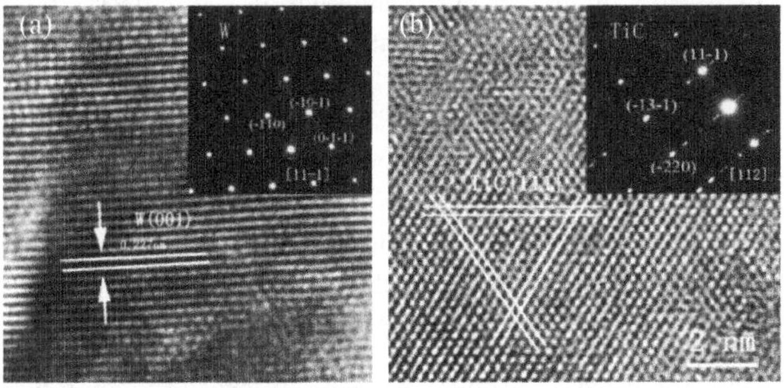

Fig. 4. High resolution TEM images and SAED patterns of W and TiC

The maximum tensile stress value of W-TiC/0.5wt.%Cf composite is 246.9MPa much larger than others. The addition of TiC makes the fracture mode a mix of intergranular and transgranular which improves the associativity of tungsten grains and the combination of tungsten grain boundary strength. The fracture micrograph of W-TiC/(0.5-2)wt.%Cf (Fig. 2) mainly consists of intergranular ruptures and some transgranular ruptures. Some micro-porous aggregation fractures can also be observed in the image of fractured morphology of W-TiC/0.5wt.%Cf (inset in Fig. 2a), which demonstrated the sample has a certain toughness. In the process of tensile test, dislocation movement lead to the concentration of stress or TiC and impurities particles in the matrix ruptured themselves, or got away from the interface of W matrix to nucleate. Micropores nucleated, grew up, aggregated and ruptured. However, this phenomenon is rare in W-TiC/2wt.%Cf composite. It implies that the content of Cf has an influence on the distribution of TiC and impurities in the alloys. These give reasonable explanation for the higher toughness of W-TiC/0.5wt.%Cf alloy than W-TiC/2wt.%Cf alloy.

The TEM bright field image of W-TiC/(0.5-2)wt.%Cf composites is showed in Fig. 3(a). W grains is represented as a dark or grey area and TiC particles appear as bright dots with a little variation of dimensions (arrows indicate the presence of TiC particles). X-EDS analysis performed on several of the dots showed very high TiC content. However, Cf could not be discovered in the bright field image because of its thick diameter. TiC particles distribute both in the grain boundaries and in the grains corresponding to the fractured morphology (Fig. 2). And the grain size of transgranular TiC particles seems more tiny than intercrystalline ones. Based on the selected area electron diffraction (SAED) pattern (inset in Fig. 4a), the W phase has a body-centered cubic structure and is along the zone axis of [11-1]. The SAED pattern along the [112] zone axis from the TiC (inset in Fig. 4b.) can be indexed to the face-centered cubic structure. The interplanar spacing of 0.227 nm, as shown by the vertical arrows in Fig.4(a), corresponds to the (001) crystalline plane of W. And in the same way, we can confirm that three different interplanar spacings of TiC belong to the same family of crystalline planes.

4. CONCLUSION

In this paper, microstructure and properties of W-TiC/Cf composites were investigated at different Cf content. The values of W-TiC/(0.5-2)wt.%Cf composite on density, micro-hardness and thermal conductivity were compared with each other. Results reveals that W-TiC/0.5wt.%Cf composite has higher relative density and toughness, whereas W-TiC/2wt.%Cf composite has higher micro-hardness, which is due to the content of Cf. The micro-porous aggregation fractures demonstrate the W-TiC/0.5wt.%Cf composite has higher toughness. TEM reveals that TiC particles both exist in grains and grain boundaries.

REFERENCES

(1) B.Y. Kim, J.H. You. Three-dimensional shakedown analysis of fiber-reinforced metal matrix composite (FRMMC) layered plasma-facing component for high heat flux loading. Fusion Engineering and Design. 84 (2009):1033-1037.

(2) J. Du, T. Höschen, M. Rasinski, J.-H. You. Shear debonding behavior of a carbon-coated interface in a tungsten fiber-reinforced tungsten matrix composite. Journal of Nuclear Materials. 417 (2011):472-476.

(3) J. Du, T. Höschen, M. Rasinski, J.H. You, Mater. Sci. Eng. A 527 (2010):1623.

(4) J. Du, T. Höschen, M. Rasinski, S. Wurster, W. Grosinger, J-H. You, Compos. Sci. Technol. 70 (2010):1482.

(5) J. Du, T. Höschen, M. Rasinski, J.-H. You. Interfacial fracture behavior of tungsten wire/tungsten matrix composites with copper-coated interfaces. Materials Science and Engineering A. 527 (2010):1623-1629.

(6) Tan Jun, Zhou Zhangjian, Qu Dandan, Ma Yao, Li Ming. Fabrication of Ultra Fine Grained W-TiC Composites by Spark Plasma Sintering. Rare Metal Materials and Engineering. 40 (2011):1990-1993.

Study on the microstructure and properties of different intermediate coatings to RuO_2-TiO_2-SnO_2/Ti Anodes

W. Li[1], W.Z. Chen[2], Q. Yan[1], Q.H. Guo[1], W.G. Wang[1]
[1] College of Materials Science & Engineering, Fujian University of Technology, P.R. China
[2] College of Materials Science & Engineering, Xiamen University of Technology, P.R. China

ABSTRACT

A normal RuO_2-TiO_2-SnO_2 /Ti coatings have been synthesized by a thermal decomposition method. The influence of the anodes with Ir, Ir-Co and Ir-Ta intermediate coatings is studied. The crystal phase structure and microstructure of the coatings are analyzed by XRD and SEM. The cyclic voltammetry curve and electrolytic life of anodes with different intermediate coatings are used to evaluate the electrochemical performance. The results show that cyclic voltammograms curve, chlorine evolution potential, the oxygen evolution potential and the electrolytic life change with different intermediate coatings. And the anode coating with Ir-Ta intermediate shows the best comprehensive performance.

Keywords: RuO_2-TiO_2-SnO_2 oxide; titanium anode; intermediate coating; Ir, Ir-Co, Ir-Ta

1 INTRODUCTION

Titanium anodes have been widely used in a variety of electrochemical processes like chlor-alkali industry, water purification, sea water decontamination, chlorate preparation, electrowinning of metals, metal foil production, cathodic protection, electroplating and so on(1) due to they have advantages in energy conservation, high efficiency, environmental protection and low cost(2). Electrocatalysts RuO_2-TiO_2-SnO_2 is considered to be one of the most commonly used anode material in electrochemical industry for their excellent performance such as low potential in chlorine evolution reaction(3). Because of the difference of thermal expansion coefficient of TiO_2 and RuO_2, electrolytic liquid will still be penetration and corrosion in the long time, which reduce the service life of the anodes(4).

Intermediate coating is an effective method to enhance the performance of the anodes. IrO_2 owns different expansion coefficient with the RuO_2. IrO_2 is more stable for strengthening the service life of anodes because it can form dense oxide film on the surface of Ti to obtain a good electrolytic durability(5). IrO_2 based coatings like IrO_2-CoO_2 can reduce the consumption of electrolysis process of Ir speed and effectively increase the coating adhesion to substrate(6). To clarify the effect of different Ir intermediate coatings on the electro-catalytic activity, a systematic study should be done. In this paper, the phase composition and microstructure of Ir, Ir-Co and Ir-Ta intermediate coatings based on the RuO_2-TiO_2-SnO_2/Ti oxide coatings are characterized by XRD and SEM, and the effect of different Ir intermediate coatings on the microstructure, electro-catalytic activity and service life of as-prepared anodes is discussed.

2. EXPERIMENTAL

The electrode is prepared using a 20×20×1 mm titanium plate as support. The plates are kept in the ethanol after sandblasted, etched in sulfuric acid and washed with distilled water. The solution of H_2IrCl_6, $Co(NO_3)_2 \cdot 6H_2O$ and $TaCl_5$, respectively are dissolved in the ethanol to prepare the corresponding sols. The sol is painted with a brush onto the support and the coating is dried by infrared light to improved adherence. After each step, the layer is calcined at 500℃ for 10 min to form the oxide. Ir and Co are added at a molar ration of 3:2 while Ir and Ta are added at a molar ration of 7:3. The solution of $RuCl_3$, $SnCl_4 \cdot 5H_2O$ and $C_{16}H_{36}O_4Ti$ are dissolved in the ethanol to prepare the corresponding sols. Ru, Ti and Sn are added at a molar ration of 3:5:2. The sol is painted with a brush onto the intermediate coating. Each coating is controlled coating liquid of 25 mL/m^2. After each step, the layer is calcined at 500℃ for 10 min to form the oxide. These procedures are repeated 6 times and finally the anodes are sintered at 500℃ for 1h.

X-ray diffraction (XRD) studies are carried out using a Bruker D8-advance instrument operated at 40 kV and 40 mA with Cu Kα radiation (λ=0.154 nm). The sample is analyzed by varying the 2θ angle from 20° to 80° with the scanning rate of 4°/min. The microstructure of coated anodes is evaluated by a Hitachi S3400 scanning electron microscopy. All electrochemical properties tests are carried out at Chinstruments model CHI660C. The potential of chlorine evolution is examined by the saturated NaCl solution. The scan rate cyclic voltammetry curves is 25 mv/s. The potential of oxygen evolution is examined by the H_2SO_4 solution, the test temperature is 25 ℃ and the reference electrodes are saturated calomel electrode. The anode service life and coating stability are evaluated by the accelerated life test which use the HDV-7c type transistor potentiostat.

3. RESULTS AND DISCUSSION

3.1 XRD analysis

Fig.1 shows the crystal phase structure of the coatings from different intermediate coatings RuO_2-TiO_2-SnO_2 /Ti anodes. The peaks at 2θ=28°, 36° and 54° are regarded as the rutile TiO_2 (JCPDS21-1276), which correspond the diffraction peaks of (110), (101) and (211), respectively. The spectra show the mixed oxides system formed solid solution of the RuO_2, TiO_2 and SnO_2. Since the ionic radius of Ru^{4+} (0.062 nm), Ti^{4+} (0.068 nm) and Sn^{4+} (0.069 nm) are very similar and the crystal RuO_2, TiO_2 and SnO_2 have the closely rutile-type crystal structure, it is easily for them to form a solid solution phase. Using the Scherrer equation: $D = k\lambda / \beta\cos\theta$, where D is the crystallize size, λ the x-ray wavelength, β the broading of the diffraction line, θ the Bragg angle and k a constant approximately equal to the unity, the sizes of the nano-crystals of coatings are evaluated. The mean sizes are about 5.0 nm, corresponding to the different intermediate coatings. The nano-crystals could improve the electrochemical properties due to their lager specific area and higher surface activity.

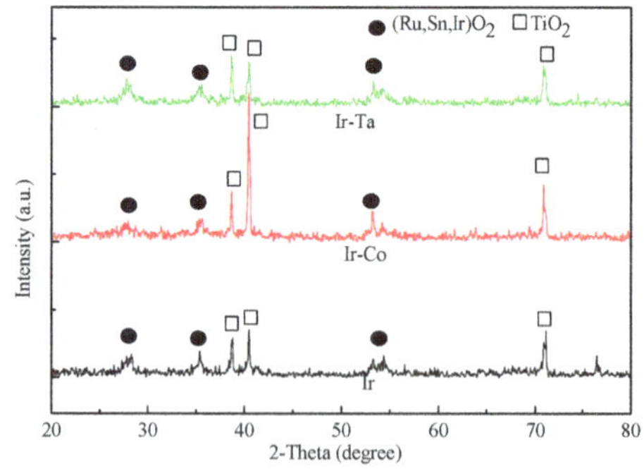

Fig.1 XRD patterns of coatings anodes with different intermediate coatings

3.2 Morphology and structure of the coatings

The surface morphology of anodes is significantly affected by the intermediate coatings. The morphology of the coating anode without intermediate coating, with Ir, Ir-Co and Ir-Ta intermediate coatings magnified 5000 times are shown in the Fig.2, respectively. It can be found that all the anodes consist of mud-cracks structure and flat area. There are crack and secondary crack on the coating surface. A crack is formed in the dry process of coating, and the secondary crack is formed in the heat treatment process. The crack can effectively improve the specific area of the anodes, leading to the increase of the electric catalytic activity of anodes. It also provides a convenient method for the penetration of the electrolyte. The oxygen precipitated from the anodes can quickly reach the substrate TiO_2 to form the passivation membrane, which increases the resistance between coating and substrate to the failure of the anodes. When the anodes own intermediate coatings, the surface coatings of the Ru-Ti-Sn are compact and uniform, with the 1μm crack width. The structure increases the surface activity of coatings, which also impends the oxygen and electrolyte penetration into the matrix, leading to the delay failure of the anodes. There are many needle-like crystallites along the inner side of the cracks. Which can be the rutile-type solid solution phase based on the RuO_2 and IrO_2.

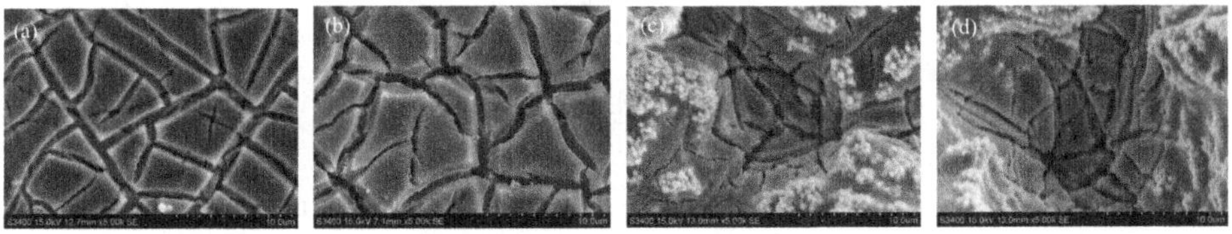

Fig.2 SEM images of coating anodes with none(a), Ir(b), Ir-Co(c) and Ir-Ta(d) intermediate coating respectively

3.3 Cyclic voltammetry analysis

Fig.3 shows the cyclic voltammetry curves of the coating anode with intermediate coatings under the scanning speed of 25mV/s. The area surrounded by the cyclic voltammetry curves of the anodes is proportional to the coating surface charge capacity, and reflects the catalytic activity of the anode. The area of the curves is proportional to the number of active points of the anode surface. When the anode coated with Ir-Ta intermediate, the most amount of activity spots exited. The area for the four samples are obvious different. In 0.35 ~ 0.65 V voltage range, anode oxygen evolution reaction, as the electrode current of electric double layer produced by the charge and discharge current, so the area surrounded by cyclic voltammograms curve can reflect the size of the electrode surface of the number of active points and electric catalytic activity. There is not oxygen evolution reaction in the anode when the voltage ranges from 0.35 to 0.65 V. So the current from the anodes is charged and discharged from the double layer. And the area surrounded by cyclic voltammetry curves can reflect the number of active points from the anode surface as well as the electric catalytic activity. The integral powers of the cyclic voltammetry for the coating anodes without intermediate coating, with Ir, Ir-Co and Ir-Ta intermediate coatings under the scanning speed of 25mV/s are 0.0236, 0.0568, 0.0550, and 0.0644, respectively. The integral power of the anode with the Ir-Ta intermediate coating is 2.73 times that of the anode without the coating. It can conclude that the anode with the Ir-Ta intermediate coating has the largest anode surface active points, which can increase the chemical reaction point and enhance the electric catalytic activity.

3.4 Values of chlorine and oxygen evolution potential

The oxygen evolution potential is an important measure of quality for the anode coatings. Fig 4 shows the oxygen evolution polarization curves of coating anode with different intermediate coatings. It is found that the three kinds of solvent coating oxygen evolution potential of anode rise sharply at the beginning of the reaction because the electrodes and the electrolyte are not stable. The oxygen evolution potential of the anode with Ir-Co, Ir-Ta and Ir intermediate coatings are 1.50, 1.52 and 1.54, respectively. The Co can reduce the oxygen evolution potential. The anode with Ir-Co intermediate coating showed the good electric catalytic activity than others. The higher oxygen evolution current efficiency gives them better performance and lower energy consumption.

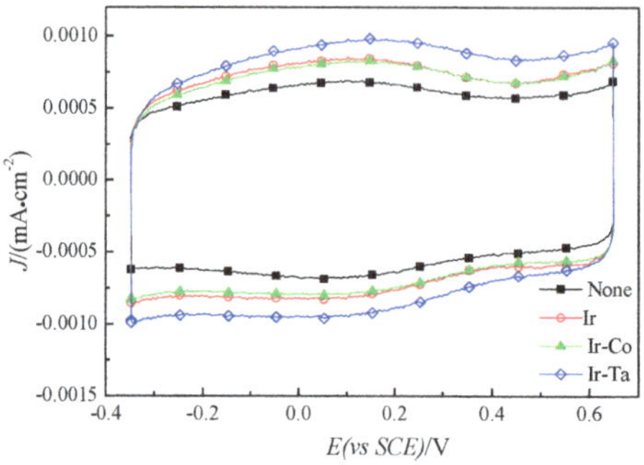

Fig.3 Cyclic voltammetry curves of the coating anode with different intermediate coatings

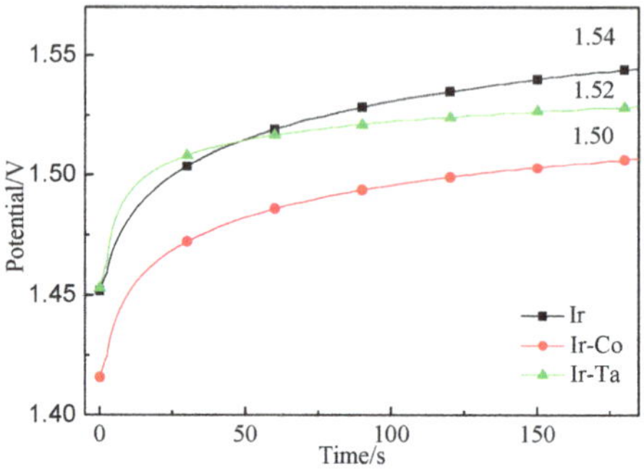

Fig.4 The oxygen evolution polarization curves of the coating anode with different intermediate coatings

Fig 5 shows the chlorine evolution polarization curves of the coating anode with different intermediate coatings. It is found that the chlorine evolution potential of the anode of the anode with Ir-Co, Ir-Ta and Ir intermediate coatings are 1.01, 1.05 and 1.11, respectively. The anode with Ir-Co intermediate coating has the roughness surface which can increase the effective area on the surface of the coating. The anode with Ir-Ta intermediate coating precipitate a great deal of IrO_2 can also increase the surface activation center numbers. The main reaction of the RuO_2-TiO_2-SnO_2/Ti oxide coating is the chlorine evolution reaction. So the lower chlorine forms, the less power consumes, suggesting that the anode with Ir-Co intermediate coating have the largest electric catalytic activity.

3.5 Accelerated corrosion test

The time dependence of the relative anode potential, obtains from the results of the accelerated corrosion test by increasing the density of the current. Fig. 6 is the accelerated corrosion test examined by the H_2SO_4 solution with 2 mol/L and the current density with 10A/cm^2. It is found that the accelerated life extends with the intermediate coatings. The anode with Ir-Ta intermediate coating has the longest accelerated life, which is 47% leaded for the anode without intermediated coating. Ir is generated TiO_2 between catalytic activity and the titanium substrate conductive which reduced the conductivity. The roughness surface Ir-Co coating can increase the specific surface area and reduce tank pressure during the electrolytic process. Co can reduce the consumption of electrolysis process of Ir speed. Ta can increase the stability of the coating. Therefore, long life of anode was achieved at anode with Ir-Ta intermediate coating.

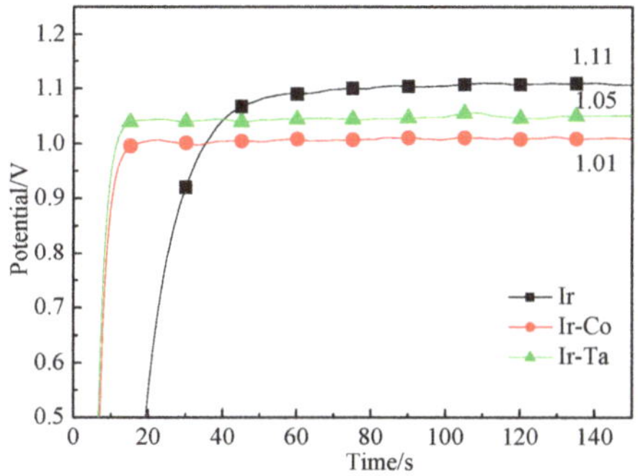

Fig.5 The chlorine evolution polarization curves of the coating anode with different intermediate coatings

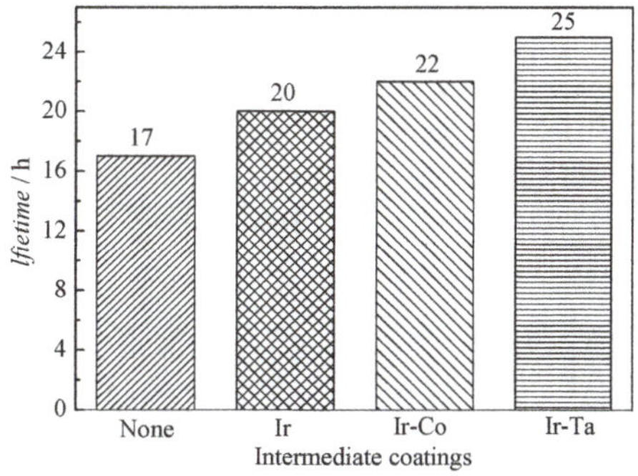

Fig.6 The accelerated life of the coating anode with different intermediate coatings

4. CONCLUSIONS

The different intermediate coating layers obtained by Ir, Ir-Co and Ir-Ta on RuO_2-TiO_2-SnO_2 /Ti oxide anodes have been prepared by a thermal decomposition method. The XRD shows the mixed oxides system formed solid solution of the RuO_2, TiO_2 and SnO_2. The SEM shows that all the anodes consist of mud-cracks structure and flat area, which can effectively increase the specific area. The electrochemical tests show that intermediate coatings can improve the stability of the electrode greatly, extend the life span as well as raise the electrode electric catalytic activity. The anode with Ir-Ta intermediate has the largest integral power of the cyclic voltammetry and the time dependence. The anode with Ir-Co intermediate owns the best oxygen evolution polarization curves and chlorine evolution polarization. The optimal comprehensive performance is the anode with Ir-Ta intermediate coating.

ACKNOWLEDGEMENT

This work is supported by the National Natural Science Foundation of China (No. 51271058).

REFERENCES

(1) Turro E, Giannis A, Cossu R, et al. Electrochemical oxidation of stabilized landfill leachate on DSA electrodes[J]. Journal of hazardous materials, 2011, 190(1): 460-465.

(2) Fan N, Li Z, Zhao L, et al. Electrochemical denitrification and kinetics study using Ti/IrO_2–TiO_2– RuO_2 as the anode and Cu/Zn as the cathode[J]. Chemical Engineering Journal, 2013, 214: 83-90.

(3) Zeng X, Zhang M, Wang X, et al. Effects of Sn content on Ti/RuO_2-SnO_2-TiO_2 anodes used in the generation of electrolyzed oxidizing water[J]. Journal of Electroanalytical Chemistry, 2012, 677: 133-138.

(4) Panić V V, Dekanski A, Milonjić S K, et al. RuO_2–TiO_2 coated titanium anodes obtained by the sol–gel procedure and their electrochemical behaviour in the chlorine evolution reaction[J]. Colloids and Surfaces A: Physicochemical and Engineering Aspects, 1999, 157(1): 269-274.

(5) Moradi F, Dehghanian C. Addition of IrO_2 to RuO_2+TiO_2 coated anodes and its effect on electrochemical performance of anodes in acid media.[J]. Progress in Natural Science: Materials International, 2014, 24(2): 134-141.
(6) Nunes F, Mendonça M H, da Silva Pereira M I, et al. Preparation and characterisation of spinel type cobalt and rhodium oxide coatings on titanium[J]. Materials chemistry and physics, 2005, 92(2): 526-533.

Influence of compound silicate gangue containing potassium in baiyunebo iron ore on solid-phase reactions during sintering process

G.P. Luo[1,2], X.L. Nie[1], S.L. Wu[2], Y.C. Wang[1,2]
[1] School of Material and Metallurgy,
University of Science and Technology Inner Mongolia, P.R. China
[2] School of Metallurgy and Ecology Engineering,
University of Science and Technology Beijing, P.R. China

ABSTRACT

Influences of the compound silicate gangue containing potassium on solid-phase reactions during sintering process were researched by means of differential thermal analysis and X-ray diffraction. The results show that Cuspidine are formed at 1077.5 °C in the Potash feldspar-CaO-CaF_2 system, and CaF_2 inhibits the formation of dicalcium silicate and gehlenite. Fe_2O_3 cannot promote the formation of complex calcium ferrite in the potash feldspar-CaO-Fe_2O_3 system, but it can reduce the forming temperature of dicalcium silicate and gehlenite. A small amount of calcium ferrite ($4CaO.7Fe_2O_3$) can be formed at 1140.3°C in the Potash feldspar-CaO-Fe_2O_3-CaF_2 system. When the temperature is more than 1200 °C potash feldspar can react with CaF_2 into gaseous SiF_4, which leads to larger pores in sintered samples.

The composition of Baiyunebo iron ore is complex, and contains high harmful elements F, K and Na. SiO_2 exists mainly in gangues containing K and Na, which will react with some minerals during sintering such as iron oxides, lime solvent and CaF_2, etc. These solid-phase reactions have an important impact on the agglomeration of sinter [1)]. Usually silicate minerals are the main existing form of gangues in iron ore [2)]. People is dedicated to the studying on solid-phase reactions of ordinary silicate gangues during sintering, but few people study special silicate gangues containing potassium and sodium[3-6)]. The research on solid-phase reactions of the compound silicate gangue containing potassium during sintering has important significance to make clear the special sintering behaviors of Baiyunebo iron ore.

1 EXPERIMENTAL MATERIALS AND METHODS

1.1 Experimental materials

Because the main mineral component of the compound silicate gangue containing potassium in Baiyunebo iron ore is potash feldspar, the commercial purity potash feldspar was selected for study, whose main chemical composition is $KAlSi_3O_8$. Raw materials used for the experiment are the commercial purity potash feldspar, chemical reagents CaO, Fe_2O_3 and CaF_2.

Properties of solid-phase reactions of the four systems were researched by means of differential thermal analysis and X-ray diffraction.

1.2 Experimental methods

Influences of potash feldspar on solid-phase reactions during sintering were researched by differential thermal analysis and X-ray diffraction. The sinter mixture was ground into powder smaller than 200 mesh, then pressed into small samples and heated from room temperature to 1350 °C at a rate of 10 °C /min in argon atmosphere, at the same time the DSC-TG curves were tested. If there were solid-phase reactions or crystal transitions in the heating process, some peaks would appear on DSC curves. The starting temperatures of solid-phase reactions could be determined by the peaks of DSC curve. To determine reaction products corresponding to the peak of DSC curve, the mixture sample with the same composition was heated to or above the peak temperature at 10 °C /min in nitrogen atmosphere and held for 4 hours. Then the sample was investigated by X-ray diffraction equipment to determine the reaction products and reveal influences of the compound silicate gangue containing potassium on solid-phase reactions of sintering.

2 EXPERIMENTAL RESULTS AND DISCUSSION

2.1 Potash feldspar—CaO—CaF_2 system

The DSC curve appears two endothermic peaks and an exothermic peak separately at 444.1 °C, 1309.3 °C and 1077.5 °C. The endothermic peak at 444.1 °C is due to the decomposition of $Ca(OH)_2$ into CaO and H_2O. The main mineral component is cuspidine ($3CaO.2SiO_2.CaF_2$) after sintered at 1080 °C. So the exothermic peak at 1077.5 °C indicates a solid phase reaction, which generates $3CaO.2SiO_2.CaF_2$. The sample sintered at 1310 °C had melted, so the endothermic peak at 1309.3 °C is due to the sample melting. The corresponding weightlessness indicates the reaction of SiO_2 in melting potash feldspar with CaF_2 into SiF_4 gas. The reaction is as follow: $SiO_2+2CaF_2=SiF_4+2CaO$.

By comparative analysis of the two systems referred above, CaF_2 in Baiyunebo iron ore can react with potash feldspar and CaO into $3CaO.2SiO_2.CaF_2$ at lower temperature, which consumes CaO and inhibits the formation of Ca_2SiO_4 and $Ca_2Al_2SiO_7$ at higher temperature.

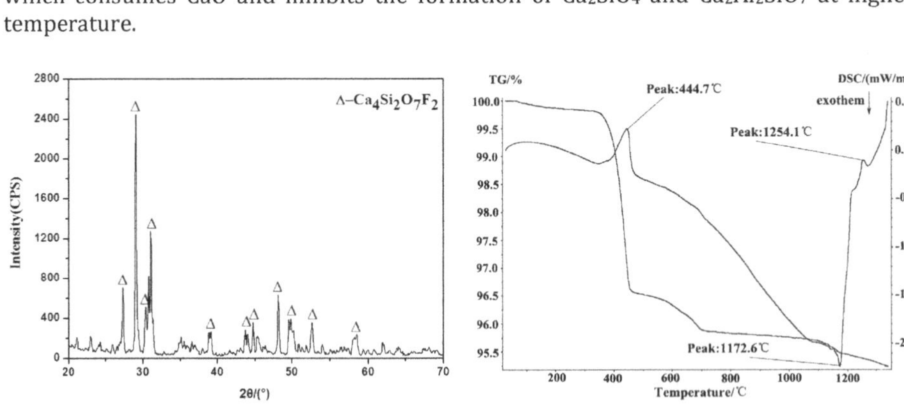

Fig.1. XRD patterns of potash feldspar-CaO-CaF_2 system sintered at 1080°C and DSC and TG curves of potash feldspar-CaO-Fe_2O_3 system

2.2 Potash feldspar—CaO—Fe_2O_3 system

The DSC curve appears two endothermic peaks at 444.7 °C and 1254.1 °C and an exothermic peak at 1172.6 °C. The endothermic peak at 444.7 °C and the corresponding weightlessness are results of decomposition of $Ca(OH)_2$ into CaO and H_2O.

The main mineral components are Ca_2SiO_4 and $Ca_2Al_2SiO_7$ after sintered at 1180 °C, and the corresponding solid phase reaction is $2KAlSi_3O_8+12CaO=5Ca_2SiO_4+Ca_2Al_2SiO_7+K_2O$. So the exothermic peak at 1172.6 °C is caused by this reaction. By comparative analysis of potash feldspar—CaO and potash feldspar—CaO—Fe_2O_3 systems, Fe_2O_3 promoted the solid phase reaction of potash feldspar and CaO into Ca_2SiO_4 and $Ca_2Al_2SiO_7$, and reduced the reaction temperature from 1227.8°C to 1172.6 °C. After sintered at 1260 °C the sample had melted, thus the endothermic peak at 1254.1 °C is due to the melting of sample.

2.3 Potash feldspar—CaO—Fe_2O_3—CaF_2 system

The DSC curve appears three endothermic peaks at 440.1°C, 1206.9°C and 1243.9°C and two exothermic peaks at 1100°C and 1140.3°C.

The sample sintered at 1122°Cconsists of more $3CaO.2SiO_2.CaF_2$ and less calcium ferrite ($Ca_2Fe_{22}O_{33}$) and Fe_2O_3. So the exothermic peak at 1100°C is caused by the formation of $3CaO.2SiO_2.CaF_2$ and $Ca_2Fe_{22}O_{33}$. CaF_2 in this system has promoting effect on formation of $3CaO.2SiO_2.CaF_2$ and $Ca_2Fe_{22}O_{33}$ at lower temperature, and inhibiting effect on the formation of Ca_2SiO_4 and $Ca_2Al_2SiO_7$ at higher temperature.

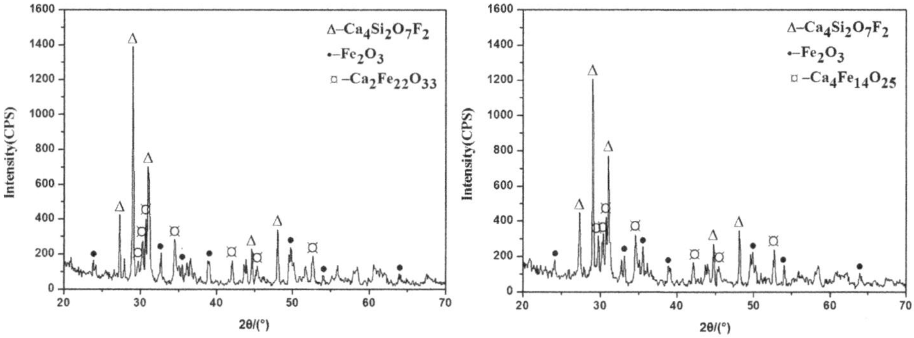

Fig.2. XRD patterns of potash feldspar-CaO-Fe_2O_3-CaF_2 system sintered at 1122° and XRD patterns of potash feldspar-CaO-Fe_2O_3-CaF_2 system sintered at 1140°

The sample sintered at 1140°C consist of more $3CaO.2SiO_2.CaF_2$ and less calcium ferrite ($4CaO.7Fe_2O_3$) and Fe_2O_3. By comparative analysis of Fig.5, the exothermic peak at 1140.3°C indicates that $Ca_2Fe_{22}O_{33}$ reacts with CaO into $4CaO \cdot 7Fe_2O_3$.

Some liquid appeared in the sample after sintered at 1207°C and the sample had melted after sintered at 1244°C. The peak at 1206.9°C is cause by the differentiation melting of $4CaO.7Fe_2O_3$, and the peak at 1244°C is cause by the sample melting. The continuous weightlessness appears on TG curve when temperature is more than 1200°C, the reason is that CaF_2 reacts with SiO_2 into SiF_4 gas, and the reaction is $SiO_2+2CaF_2=SiF_4+2CaO$.

3 CONCLUSIONS

(1) In potash feldspar-CaO-CaF_2 system, the exothermic peak at 1077.5°C indicates a solid phase reaction between $KAlSi_3O_8$, CaO and CaF_2 into $3CaO.2SiO_2.CaF_2$, which inhibits the formation of Ca_2SiO_4 and $Ca_2Al_2SiO_7$ at the higher temperature.

(2) In potash feldspar-CaO-Fe_2O_3-CaF_2 system, $3CaO.2SiO_2.CaF_2$ and a little amount of $Ca_2Fe_{22}O_{33}$ are formed in a wide temperature range of 1000-1122.3°C. The

exothermic peak at 1140.3°C indicates $Ca_2Fe_{22}O_{33}$ further reacts with CaO into $4CaO.7Fe_2O_3$.

(3) The compound silicate gangue containing potassium in Baiyunebo iron ore easily takes part in reactions into Ca_2SiO_4, $Ca_2Al_2SiO_7$, $Ca_4Si_2O_7F_2$ and silicate glass phase during sintering, which severely inhibits the formation of complex calcium ferrite and leads to reduced sinter strength as well as metallurgical properties.

ACKNOWLEDGEMENT

Foundation Item: Item sponsored by National Natural Science Foundation of China (51364030).

REFERENCES

[1] Zhou Chuandian. Technical manuals of blast furnace ironmaking, Metallurgical Industry Press, Beijing, (2003), 30.

[2] Liu Linrui, Wang Pizhen. Metallurgical Handbook (Second Edition), Metallurgical Industry Press, Beijing, (2001), 50.

[3] K. A. Zimmerma. Reasonably Controlling Chemical Composition to Increase Sinter Quality, Iron and Steel Society of AIVIE, Japan, (1997), 114.

[4] I. S. Nakazima, Y. Akizuki. Optimization sinter operation with high pellet feed fine ore ratio Process, technology Washington, 4(1986), 12.

[5] Matsuno F, Harard T. Changes of Mineral Phases during the Sintering of Iron Ore-Lime Stone System, transactions ISIJ, 21(1981), 318.

[6] Dawson P R. The Influence of Alumina on the Development of Complex Calcium Ferrites in Iron Ore Sinter, trans. Inst. Mining Metal. Section C, 94(1985), C71.

Dynamic mechanical behavior and microstructure evolution of an AM80 magnesium alloy

L.X. Li, P.C. Guo
State Key Laboratory of Advanced Design and Manufacturing for Vehicle Body, Hunan University, P.R. China

ABSTRACT

The dynamic mechanical behavior of a solution treated AM80 magnesium alloy was investigated at high strain rates and compared with the results at low strain rate (1×10^{-4} s^{-1}). The flow stress exhibits a visible positive strain rate sensitivity with strain rate ≤ 1850 s^{-1}, and it slightly decreases as strain increases to 2100 s^{-1} due to the local temperature rise promoting the slip of dislocations and finally decreasing the density of twins. A modified Johnson-Cook (J-C) model is employed in characterizing the mechanical response for the studied alloy and the calculated results are accordant well with the observed material responses.

1 INTRODUCTION

Magnesium and its alloys with high specific strength, good damping capacity and machinability have become attractive candidate materials for structural applications, especially, in aerospace and aeronautics, automotive and defence industries [1]. At present, most of the investigations have dealt with the mechanical behavior under quasi-static loading. However, components made by magnesium alloys will be potentially subjected to impact loads in some services, for example the automobile bumper beam. It is well known that the deformation behavior of materials under dynamic loading is very complicated and quite different from that under quasi-static loading due to the adiabatic effects, strong shockwave effects and localized deformation under dynamic loading [2]. Although dynamic deformation behavior and the corresponding microstructure evolution of magnesium alloys have been investigated by some authors [2-7], very limited research has been put on other magnesium alloys except for AZ31. Furthermore, temperature rise and microstructure evolution during dynamic deformation of magnesium alloys have not been clearly identified and a good understanding of dynamic deformation mechanism is much required.

The purpose of this work was to investigate the mechanical response and the evolved microstructure of a solution treated AM80 magnesium alloy under dynamic loading, with focus on the effect of strain rate on the mechanical response.

2 EXPERIMENTAL PROCEDURE

The investigated material was a cast AM80 magnesium alloy with chemical composition of Mg-8.0Al-0.1Mn (wt.%), which was solution treated at 723 K for 5h. The initial microstructure was formed of equiaxed grains, with average size of 200μm (Fig.1a). For both quasi-static and dynamic compressive testing, cylindrical samples with size of φ10 mm × 10 mm were cut from the cylindrical bar. The relative position between the cylindrical ingot and the impact samples is schematically shown in Fig. 1b.

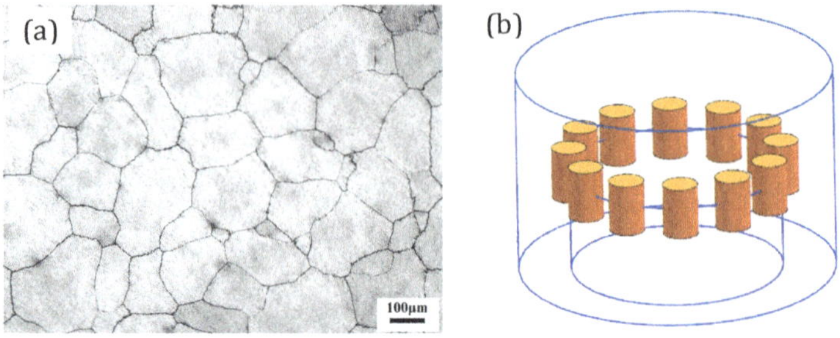

Fig. 1 (a) Optical microstructure and (b) schematic of samples for the as-received AM80 magnesium alloy

The quasi-static tests were performed using a conventional INSTRON-4206 testing machine with 1×10^{-4} s^{-1} strain rate, and the dynamic mechanic tests were carried out with the Split Hopkinson Bar (SHPB) at a strain rate range of 800-2100 s^{-1}. All the tests were performed at room temperature, and before the tests, the sample ends were lubricated with graphite to eliminate the friction effect. Samples for optical observations cut from the longitudinal section of cylinders were grinded with SiC papers up to 2000 grit and then mechanically polished before etching with acetic picral solution (4.2 g picric acid, 10 ml acetic acid, 10 ml water, and 70 ml ethanol).

3 RESULTS AND DISCUSSION

Fig. 2 illustrates the true stress-strain curves of the studied AM80 magnesium alloy at various strain rates. It can be clearly seen that, when the strain rate is less than 2100 s^{-1}, the flow stress increases as strain rate increases. This positive strain rate sensitivity indicates that the stress response of the studied material is strongly dependent on the applied strain rate by changing the dominated deformation model. Furthermore, the flow stress during quasi-static compression is distinctly lower than that under dynamic loading, showing a large increase in flow stress as train rate increases to 800 s^{-1} and then a slight rise as train rate further increases. However, it should be specially noted that, as strain rate increases from 1850 s^{-1} to 2100 s^{-1}, the flow stress doesn't increase any more, but showing a slight decrease, which is attributed to the effect of local temperature rise causing the formation of dynamic softening [8].

In order to make a good expression of the flow stress of the studied material under various strain rates, building a constitutive model is necessary for describing or predicting its dynamic mechanical response. The Johnson-Cook (J-C) constitutive model mainly considering the effects of strain, strain rate, and temperature, is widely used to simulate the mechanical response of metals subjected to large strains, high strain rates and high temperatures. In the present work, all the tests are conducted at room temperature. Therefore, the effect of temperature in the original J-C model can be neglected and the flow stress is expressed as follow:

$$\sigma = (A + B\varepsilon_p^{\ n})(1 + C\ln\dot{\varepsilon}^*) \qquad (1)$$

where σ is the flow stress, A is the yield stress at room temperature and reference strain rate, B and n are the strain hardening coefficient and exponent, respectively, ε_p is the plastic strain, C is the strain rate hardening coefficient, $\dot{\varepsilon}^* = \dot{\varepsilon} / \dot{\varepsilon}_0$ is the dimensionless strain rate with $\dot{\varepsilon}$ and $\dot{\varepsilon}_0$ being the strain rate and reference strain rate, respectively.

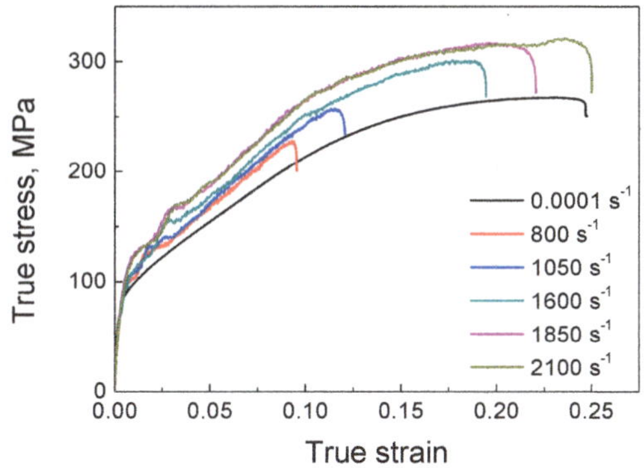

Fig. 2 Compressive true stress-strain curves of the studied AM80 magnesium alloy

According to the J-C model, strain and strain rate are two independent phenomena which can be isolated from each other, the total effects of strain and strain rate on the flow stress can be calculated by multiplying these two terms. In this paper, 1×10^{-4} s^{-1} is taken as reference strain rate for describing the flow stress of the cast AM80 magnesium alloy. At reference strain rate, Eq. (1) will reduce to: $\sigma = A + B_p^{\ n}$. The value of A is the yield stress calculated from the flow curve at 1×10^{-4} s^{-1}. Substituting the value of A in Eq. (2) and using the flow stress data at various strains for the same flow curve, ln (σ-A) vs. ln ε_p is plotted. Therefore, B and n are the intercept and slope of this plot, respectively. In order to make a good expression of the strain rate effect for the AM80 magnesium alloy, the parameter of C is defined as a function of $\dot{\varepsilon}$. Using the flow stress data beyond the yield strain at various strain rates, C is obtained by date fitting. So the Eq. (1) can be expressed as:

$$\sigma = (91.8 + 1300\varepsilon_p^{\ 0.90})[1 + (5.665\times10^{-6}\dot{\varepsilon} - 1.0659\times10^{-4})\ln\dot{\varepsilon}^*] \qquad (2)$$

Comparing between the experimental and the predicted data at various strain rates is shown in Fig. 3, it can be found that the calculated result is accordant well with the observed material response.

Fig. 4a shows the typical optical micrographs taken from the centre of fractured sample with strain of 0.246 at 1×10^{-4} s^{-1}. The observed microstructures clearly indicate that the grains are deformed seriously for the formation of abundant twins and some grain boundaries are hard to distinguish. Furthermore, it should be noted that the secondary twinning basically intersects with the initial twins, and all the microcrackes start from grain and twin boundaries and then develop along them. Figs. 4b-d shows the optical microstructures obtained from the centre of samples deformed under dynamic loading. Similar with quasi-static, a large number of mechanical twins including initial and secondary are generated. However, the density of twins generated at 1850 s^{-1} is obviously higher and the corresponding mean spacing between two adjacent twins is smaller, although the strain (ε=0.220) is lower than that under quasi-static loading (ε=0.246). These indicate that the mechanical twinning in the present AM80 magnesium alloy is sensitive to the applied strain rate, exhibiting obvious positive strain rate sensitivity. It is well known that twins acting as small angle grain boundaries increase the flow stress of materials during plastic deformation. It is precisely the positive strain rate sensitivity that increases the flow stress as strain rate increases. However, when the strain rate increases to 2100 s^{-1}, the density of mechanical twins decreases although the strain (ε=0.250) is larger than that at 1850 s^{-1}. It is generally accepted that the strain hardening rate is intimately related to the density of mechanical twins in magnesium alloys. Therefore, it can be concluded that the softening effect resulted from local temperature rise under dynamic loading is in a dominant position [9] in the comparison with strain rate hardening at 2100 s^{-1}. As a result, a slight decrease is observed at strain rate of 2100 s^{-1}.

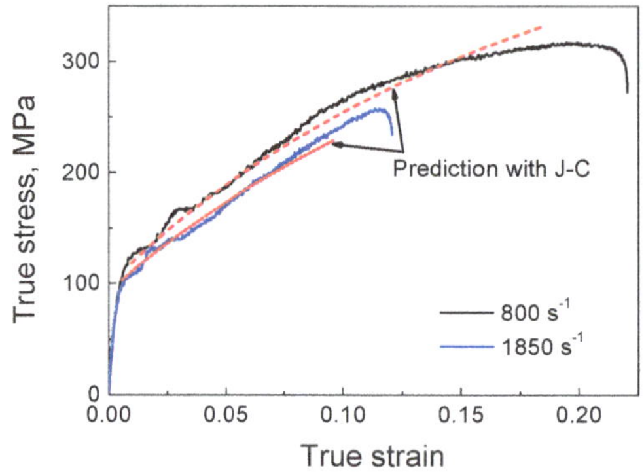

Fig. 3 Comparisons of stress-strain curves measured and calculated by J-C model

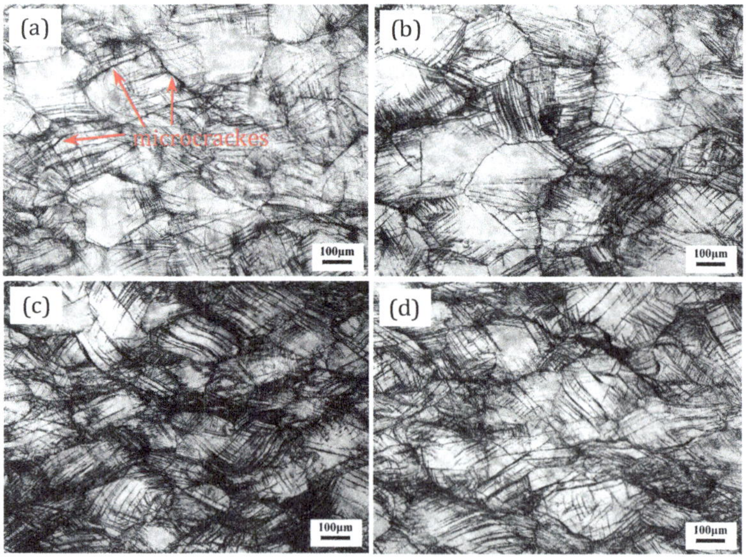

Fig. 4 OM images at different strain rates. (a) $\dot{\varepsilon}$ =1×10^{-4} s^{-1}, (b) $\dot{\varepsilon}$ =1050 s^{-1}, (c) $\dot{\varepsilon}$ =1850 s^{-1}, (d) $\dot{\varepsilon}$ =2100 s^{-1}.

4 CONCLUSIONS

The mechanical response and the evolved microstructure were investigated in a wide range of strain rates. The flow stress demonstrates a visible positive strain rate sensitivity with strain rate ≤ 1850 s^{-1} and then slightly decreases as strain rate increases to 2100 s^{-1}, which is attributed to the softening effect dominating in the comparison with strain rate hardening due to the local temperature rise at 2100 s^{-1}. Furthermore, a modified J-C model is employed in describing the flow stress behavior for the studied alloy and the calculated results are accordant well with the observed material responses.

REFERENCES

[1] Lou, X.Y., Li, M., Boger, R.K., Agnew, S.R. & Wagoner, R.H. (2007) *International Journal of plasticity*. **23**, 44-86. doi: 10.1016/j.ijplas.2006.03.005

[2] Ma, Q., El Kadiri, H., Oppedal, A.L., Baird, J.C., Li, B., Horstemeyer, M.F. & Vogel, S.C. (2012) *International Journal of Plasticity*. **29**, 60-76. doi: 10.1016/j.ijplas.2011.08.001

[3] Mao, P., Liu, Z. & Wang, C. (2012) *Materials Science and Engineering A*. **539**, 13-21. doi: 10.1016/j.msea.2011.12.068

[4] Feng, F., Huang, S., Meng, Z., Hu, J., Lei, Y., Zhou, M., Wu. D. & Yang, Z. (2014) *Materials & Design*. **57**, 10-20. doi: 10.1016/j.matdes.2013.12.031

[5] Yokoyama, T. (2003) *Strain*. **39**, 167-75. doi: 10.1046/j.1475-1305.2003.00086.x

[6] Wan, G., Wu, B.L., Zhang, Y.D., Sha, G.Y. & Esling, C. (2010) *Materials Science and Engineering A*. **527**, 2915-2924. doi: 10.1016/j.msea.2010.01.023

[7] Dudamell, N.V., Ulacia, I., Gálvez, F., Yi, S., Bohlen, J., Letzig, D., Hurtado, I. & Pérez-Prado, M.T. (2012) *Materials Science and Engineering A*. **532**, 528-535. doi: 10.1016/j.msea.2011.11.018

[8] Li, L.X., Zhou, J. & Duszczyk, J. (2006) *Journal of Materials Processing Technology*. **72**, 372-380. doi: 10.1016/j.jmatprotec.2005.09.021

[9] Ulacia, I., Salisbury, C.P., Hurtado, I. & Worswick, M.J. (2011) *Journal of Materials Processing Technology*. **211**, 830-839. doi: 10.1016/j.jmatprotec.2010.09.010

Effect of welding conditions on residual stress and stress intensity factor around remaining crack at seal welds

S. Okano, K. Torigata, M. Mochizuki
Department of Materials and Manufacturing Science, Osaka University, Japan

ABSTRACT

In this study, the effects of seal welding conditions, such as welding heat input, crack depth, and plate thickness, on welding-induced residual stress and stress intensity factor (SIF) around the remaining crack were numerically investigated. It was clarified that the residual stress and the SIF at the downside crack tip is governed largely by the ratio of weld heat input to square root of crack depth, which was derived from welding thermal conduction theory as a representative of the temperature distribution along the crack. Furthermore the residual stress at the downside crack tip becomes compressive and the SIF becomes zero when the ratio is smaller than approximately 1.5.

1 INTRODUCTION

Stress corrosion cracking (SCC) has been one of the issues in nuclear power plants (1). In particular, SCC has even occurred at weld heat affected zone in low-carbon austenitic stainless steel and weld metal in nickel-based alloy, which are SCC-resistant materials. That is why more reasonable repair welding technique has become increasingly important to ensure structural soundness and reliability of nuclear power plants. Seal welding, in which the crack remains within the structural members, has been developed as one of alternative repair techniques applicable when it is difficult to remove a crack completely or to weld it again with high integrity after removal of the crack. The residual stress is one of the driving forces for crack propagation and brittle fracture then it is important for ensuring structural soundness and the reliability at seal weld to evaluate and control the welding-induced residual stress around the remaining crack. In this paper, the systematic investigation using numerical analysis was performed to consider the effect of seal welding conditions on the welding-induced residual stress and the stress intensity factor (SIF) around the remaining crack. Based on these results, the effectiveness of appropriate seal welding conditions setting for the structural soundness and the reliability at weal welds was evaluated.

2 NUMERICAL PROCEDURE FOR ANALYZING RESIDUAL STRESS AND STRESS INTENSITY FACTOR AT SEAL WELDS

2.1 Problem description and its finite element model

The analytical object in this study is the shroud support, which is one of the internal core components in a boiling water reactor (BWR). The shroud support ring and cylinder constitute the shroud support, as shown in Fig. 1. The support is made of nickel-based alloy with a thickness of approximately 60 mm. Seal welding is regarded as one of the useful repair techniques against SCC recently observed at the shroud support because of the difficulty of gaining access to the shroud support. A two-dimensional plate model considering the plane strain conditions was used in the analysis, as shown in Fig. 1. This is because the effect of pipe geometry in the shroud support becomes negligible when the ratio of diameter to thickness becomes larger. The symmetrical model was used in numerical analysis because the seal welding parallel to the remaining crack was focused in this study. The material properties of nickel-based alloy used in the analysis were determined considering the temperature dependency and the work hardening with reference to data (2)(3). Only rigid-body motion was constrained in the mechanical model. Weld heat input was provided through the deposited weld metal. The relation between the cross-sectional area of the deposited weld metal and the weld heat input were estimated based on the measurements. The ratio of width to height of the deposited weld metal was set to 6:1 in any welding heat input conditions. After estimating the residual stress around the remaining crack, the stress intensity factor (SIF) was evaluated (4). An elastic analysis was performed using the estimated distribution of residual stress at the virtual crack propagation region of 0.05mm, as shown in Fig. 2. Then the SIF was estimated by the displacement method, as shown in Fig. 3. Although the relaxation of residual stress due to reduction of global stiffness for a large deep crack is not considered in this approach.

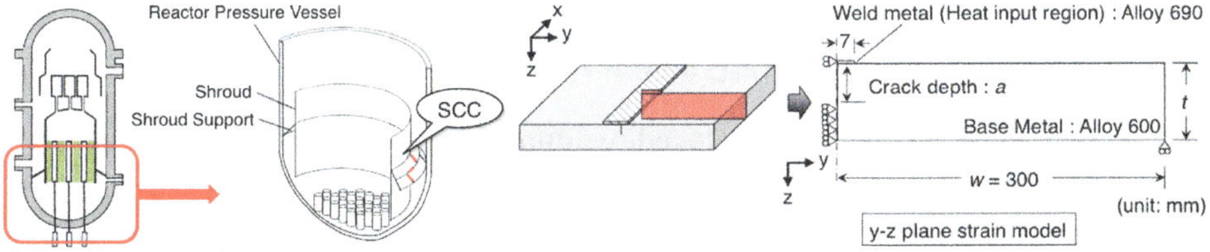

Figure 1 Schematic illustration of shroud support and its analytical model.

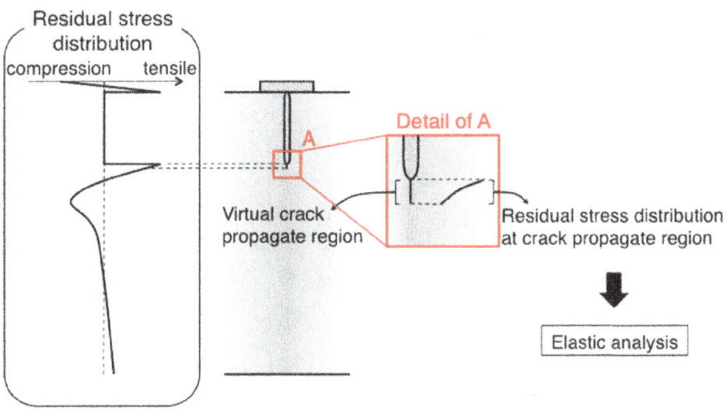

Figure 2 Scheme of superposition principle.

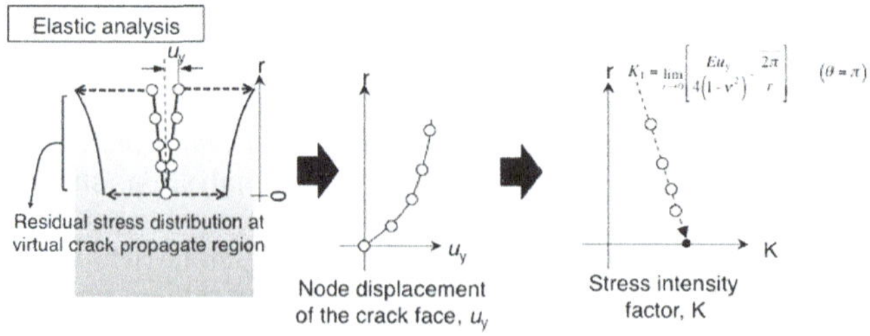

Figure 3 Scheme of displacement method.

2.2 Welding conditions in systematic investigation

The setup conditions of seal welding used in the systematic numerical investigation are shown in Table 1. The following conditions were investigated; three different plate thickness, five crack depths from 5 to 25mm and twenty weld heat inputs from 50 to 2000 J/mm were modelled.

Table 1 Welding conditions.

Plate thickness, h (mm)	40, 60, 120
Depth of Crack, a (mm)	5, 10, 15, 20, 25
Weld heat input, Q (J/mm)	50, 100, 150, 200, 250, 300, 350, 400, 450, 500, 600, 700, 800, 900, 1000, 1200, 1400, 1600, 1800, 2000

3 EFFECT OF WELDING CONDITIONS ON RESIDUAL STRESS AND STRESS INTENSITY FACTOR AROUND REMAINING CRACK AT SEAL WELD

3.1 Distribution of welding-induced residual stress

Representative analytical results of the distribution of the welding-induced residual stress along the plate thickness of 60 mm are shown in Figs. 4(a) and (b) for crack depths of 5 mm and 25 mm, respectively. In both cases of the crack depths, the effect of weld heat input on the distribution of residual stress along the thickness is obviously. The residual stress at the topside crack tip region is tensile stress under any conditions of weld heat input and crack depth. On the other hand, the residual stress at the downside crack tip region is possible to be compressive stress, depending on the combination of weld heat input and crack depth.

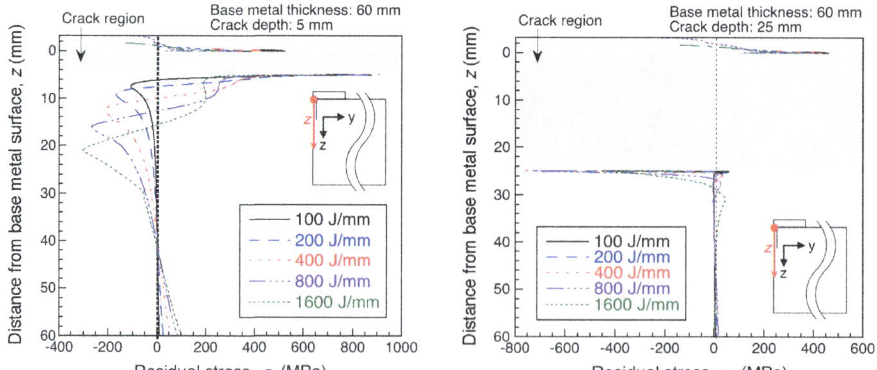

(a) Crack depth is 5 mm (shallow) **(b) Crack depth is 25 mm (deep)**

Figure 4 Distribution of welding-induced residual stress along crack depth.

The effect of weld heat input and crack depth on the residual stress at downside crack tip region is systematically investigated, as shown in Fig. 5. The location for evaluating the residual stress was at an element located 0.025mm from the downside crack tip. As shown in the figure, high tensile residual stress is generated in the range of large heat input for all crack depths, and compressive stress is generated in the range of lower heat input. The difference in weld heat input in the transition region seems to be dependent on the crack depth.

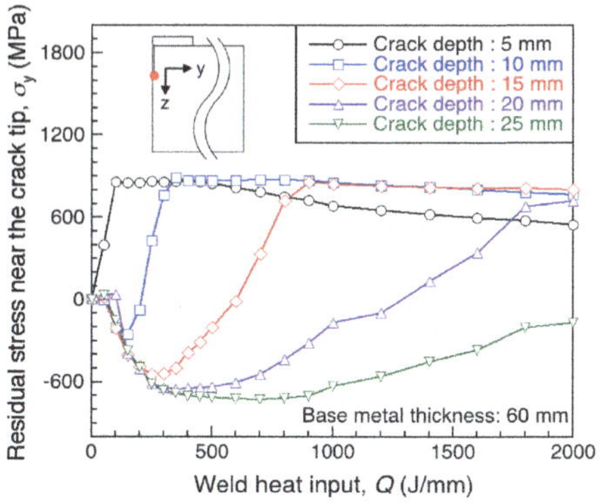

Figure 5 Effect of welding conditions on residual stress at downside crack tip region.

3.2 Factor dominating residual stress based on generation mechanism

In order to understand the mechanism of compressive stress generation, the stress and strain behaviors at the downside crack tip during seal welding was considered. The results in cases of heat input 250 J/mm and 900 J/m are shown in Figs. 6(a) and (b), respectively. In early stage, the tensile stress and the plastic strain are generated due to welding-induced tear force in both cases. In next stage, compressive stress is generated due to restraint of thermal expansion in both cases. However the difference in generation of plastic strain is shown between these cases because of difference in temperature rise. For this reason, the residual stress after welding is considered to be dependent on the temperature rise at the downside crack tip region. The influential factor dominating temperature rise at downside crack tip region was derived based on weld thermal conduction (5) theory as the following equation (I).

$$r/a = \sqrt{(1/\pi e)\cdot (1/c\rho T)}\cdot \sqrt{Q_{net}}\,/a \quad \cdots\cdots \text{(I)}$$

Here, r is the distance from weld center, α is crack depth, c is specific heat, ρ is density, T is temperature, and Q_{net} is net weld heat input per unit weld length. The above equation shows that the ratio of depth at a certain level of maximum temperature rise to crack depth is dependent on the ratio $\sqrt{Q_{net}}/\alpha$. The effects of weld heat input and crack depth on residual stress at downside crack tip region are arranged by a unified curve associated with the ratio $\sqrt{Q_{net}}/\alpha$, as shown in Fig. 7. Thus it is concluded that the residual stress at the downside crack tip region is dependent on the ratio $\sqrt{Q_{net}}/\alpha$. In addition, when the ratio $\sqrt{Q_{net}}/\alpha$ is smaller than approximately 1.5, the residual stress at the downside crack tip becomes compressive.

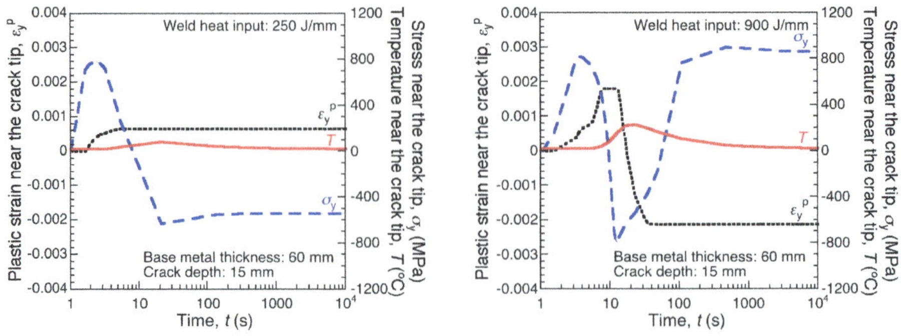

(a) Weld heat input is 250 J/mm **(b) Weld heat input is 900 J/mm**

Figure 6 Time histories of temperature, plastic strain and thermal stress.

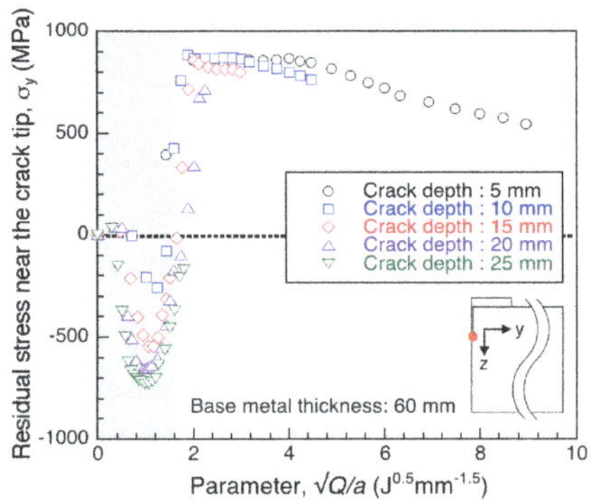

Figure 7 Arrangement of residual stress at downside crack tip by ratio $\sqrt{Q_{net}}/a$.

3.3 Stress intensity factor at downside crack tip region

The stress intensity factor (SIF) at downside crack tip region was estimated. As the result, the SIF was also collapsed into a single curve associated with the ratio $Q/\sqrt{a}$ in any conditions of plate thickness, as shown in Figure 8. It is also confirmed that the SIF becomes zero when the ratio $Q/\sqrt{a}$ is smaller than approximately 1.5. Appropriate seal condition setting in reference to the value of ratio $Q/\sqrt{a}$ is expected to become effective for ensuring and improving structural soundness and reliability at seal welds.

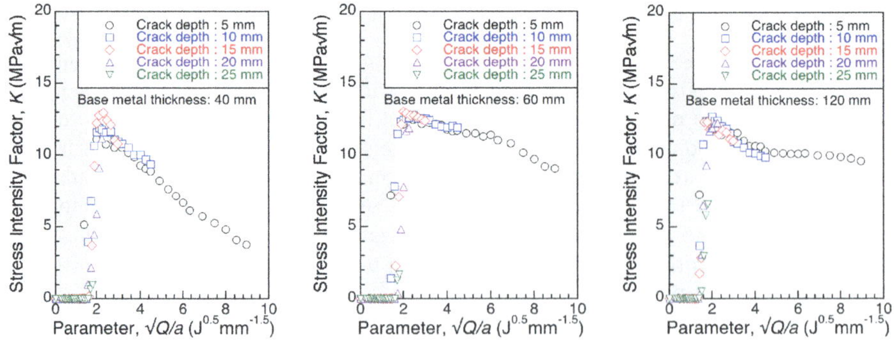

(a) Thickness is 40 mm **(b) Thickness is 60 mm** **(c) Thickness is 120 mm**

Figure 8 Arrangement of the SIF at downside crack tip by ratio $\sqrt{Q_{net}}/a$.

4 CONCLUSIONS

In this study, the effect of seal welding conditions on the welding-induced residual stress and the stress intensity factor (SIF) around the remaining crack was numerically investigated. It was clarified that the residual stress and the SIF at the downside crack tip is governed largely by the ratio $Q/\sqrt{a}$. The residual stress at the downside crack tip becomes compressive and the SIF becomes zero when the ratio $Q/\sqrt{a}$ is smaller than approximately 1.5.

5 REFFERENCES

(1) Okamura, Y. et al. (1976) *SMiRT 17* **WG01-1**.
(2) INCONEL ® Alloy 600 http://www.specialmetals.com/
(3) INCONEL ® Alloy 600 http://www.specialmetals.com/
(4) Shiratori, M. et al. (1984) *J. JSME* **51-467** 1828-1835. (in Japanese)
(5) Rosenthal, D. (1941) *W. J.* **20** 220-234.

Influence of thermal boundary conditions on melting behaviour of an Ag nanowire mesh induced by Joule heating

Y. Li, K. Tsuchiya, M. Saka
Department of Nanomechanics, Tohoku University, Japan

ABSTRACT

The influence of thermal boundary conditions was numerically clarified on the melting behaviour of an Ag nanowire mesh induced by Joule heating. The isotherm of the maximum temperature in the mesh, dependent on thermal boundary conditions, determines the melting behaviour. If such isotherm is a curve, global unstable melting happens where several nanowires melt simultaneously until the mesh is broken. The corresponding relationship of melting current and voltage becomes a data point. However, if the isotherm is a point, stable melting happens where only one nanowire melts at a time. The relationship of melting current and voltage becomes a curve.

1 INTRODUCTION

The combination of transparency, conductivity, and flexibility makes Ag nanowire mesh a promising alternative to indium tin oxide (ITO) as an ideal transparent conductive electrode (TCE) for next-generational optoelectronic devices [1, 2]. For example, the research group of Guo et al. [3] has contributed greatly to the fabrication of the Ag nanowire mesh by developing nanoimprint lithography. With such grid structure electrode, the organic solar cell has demonstrated the enhanced power conversion efficiency than devices made using ITO electrode. Groep et al. [4] has found that the best transparent Ag nanowire mesh can be obtained from thin wires and small pitch by identifying the physical mechanism determining the optical transmission of the mesh.

On the other hand, it has been experimentally [5] reported that the melting induced by Joule heating will cause the degradation of Ag nanowire mesh because of the reduced current flow area and therefore increased current density as compared to ITO. To clarify the melting behaviour of Ag nanowire mesh, a numerical algorithm [6] solving the electro-thermal problem was developed to obtain the relationship of melting current and melting voltage triggering the melting of the mesh segment. The observed zigzag pattern indicates that an Ag nanowire mesh at a specific working condition may experience repetitive unstable (where several wires melt simultaneously at a constant current/voltage) and stable (where melting progression needs the increase of current/voltage) melting until the mesh is open. Moreover, the consistent melting behaviour between Ag microwire and nanowire meshes was confirmed, where a dimensionless parameter, i.e., figure of merit for the mesh, was proposed to characterize the current carrying ability and to predict the melting behaviour of the mesh even with different wire [7]. However, it should be noted that the above report is only one example for an Ag nanowire mesh at a specified working condition. To obtain the general principal, it is important to investigate the relevant factors.

Therefore, the aim of the present work is to clarify the influence of thermal boundary conditions on melting behaviour of Ag nanowire mesh. Two cases of thermal boundary conditions were considered, which shows distinctive melting behaviours.

2 NUMERICAL ANALYSIS

As illustrated in Fig. 1, the Ag nanowire mesh with size of 10×10 employed here is composed of a rectangular wire network with 10 columns and 10 rows. The pitch size l is 200μm. The intersection of the (i+1)th column and the (j+1)th row is denoted as mesh node (i, j) (0≤i≤9, 0≤j≤9) The wire between two adjacent mesh nodes is denoted as mesh segment. Taking four segments centred with mesh node (i, j), the segments on the left, right, downside and upside are denoted as $S^{L}_{(i,j)}$, $S^{R}_{(i,j)}$, $S^{U}_{(i,j)}$ and $S^{D}_{(i,j)}$, respectively. Obviously, there are 100 mesh nodes and 180 mesh segments.

The geometrical and physical properties of Ag nanowire are listed in Table 1. Here, w and t are width and thickness, T_m, and λ are melting point and thermal conductivity, ρ_m is electrical resistivity at T_m, calculated from $\rho_0\{1+\alpha(T_m-T_0)\}$ with ρ_0 of the electrical resistivity at room temperature T_0 and α of temperature coefficient of resistivity.

The electrical boundary conditions are also given in Fig. 1. The current I is input from node (0, 0) and output from (9, 0), while there is no external input/output current for all the other nodes. Note that the electrical potential at node (9, 9) is zero.

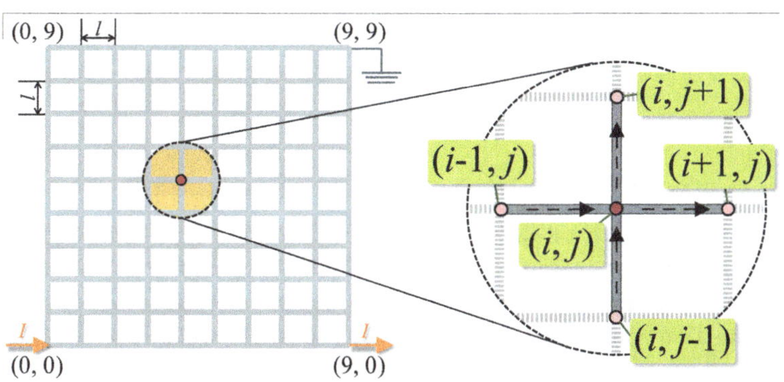

Fig. 1 Schematic Ag nanowire mesh 10×10

Table 1 Geometrical and physical properties of Ag nanowire

$w \times t$ [nm^2]	T_m [K]	λ [W/μm·K]	ρ_0 [Ω·μm]	α [/K]	ρ_m [Ω·μm]
100×100	873 [8]	3.346×10^{-4} [9]	0.119 [10]	3.8×10^{-3}	0.378

Two cases of thermal boundary conditions are discussed here. In **Case I**, only the nodes (0, 0) and (9, 0) are at room temperature, while no external input/output heat energy for all the other nodes. **Case II** is the same with those in Refs. [6, 7]. The peripheral nodes (i.e., (i, 0), (i, 9), (0, j), (9, j)) are at room temperature, while there is no external input/output heat energy for all the other nodes.

The theoretical analysis (6) is summarized as below. Firstly, taking a wire, i.e., mesh segment connected by nodes (i-1, j) and (i, j) as shown in Fig. 2a, the current I flows from node (i-1, j) to (i, j). Based on Ohm's law, the current density j in mesh segment is

$$j = -\frac{1}{\rho}\frac{d\phi}{dx}, \tag{1}$$

where ϕ is electrical potential, x is the axial coordinate. Note that for simplicity, ρ_m was used hereafter for ρ to approximate real conditions. By ignoring heat transfer of the mesh to the underlying substrate for simplicity, the heat conduction equation can be given as

$$\lambda\frac{d^2T}{dx^2} + \rho j^2 = 0, \tag{2}$$

where T is temperature. Note that the latent heat dissipation associated with phase transformation of Ag from solid to liquid is not considered here. Using Fourier's law, the heat flux q in $S^{L}_{(i,j)}$ is

$$q = -\lambda\frac{dT}{dx}. \tag{3}$$

In a similar way, the corresponding relationships can be obtained for all of the mesh segments. Secondly, taking a mesh node (i, j) with 4 adjacent nodes, the current and heat energy passing though are analysed as shown in Fig. 2b and 2c. According to the law of the conservation of current/heat energy, we have

$$I_{\text{external}} + I_{\text{internal}} = 0 \tag{4}$$

and

$$Q_{\text{external}} + Q_{\text{internal}} = 0 \tag{5}$$

by assuming the current/heat flows rightward/upward. Here, I_{external} and Q_{external} are the external current and heat energy at node (i, j), while I_{internal} and Q_{internal} are the sum of internal current and heat energy flowing through the node (i, j) from its adjacent nodes. By assuming the direction of flowing into the node is positive, we have

$$I_{\text{internal}} = \left\{ j_{S^{L}_{(i,j)}} - j_{S^{R}_{(i,j)}} + j_{S^{D}_{(i,j)}} - j_{S^{U}_{(i,j)}} \right\} A, \tag{6}$$

and

$$Q_{\text{internal}} = \left\{ q_{S^{L}_{(i,j)}} - q_{S^{R}_{(i,j)}} + q_{S^{D}_{(i,j)}} - q_{S^{U}_{(i,j)}} \right\} A, \tag{7}$$

where A is cross-sectional area of the wire, and the subscripts denote the mesh segment. By solving the systems of linear equations for all the nodes, the current density in any mesh segment and temperature at any mesh node can be obtained from Eqs. (1, 4, 6) and Eqs. (3, 5, 7), respectively. Finally, using the above results, the temperature in any mesh segment can be obtained by solving Eq. (3).

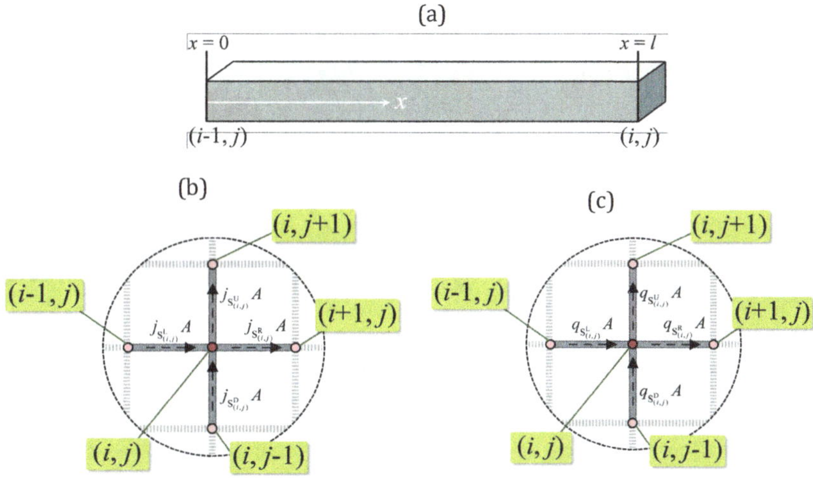

Fig. 2 Electro-thermal analysis (a) mesh segment, (b) current and (c) heat energy in mesh node

The computational algorithm [6] is briefly described as below. Firstly, it is to modulate the value of the input current I so as to make the maximum temperature T_{max} in the mesh reaching the T_m of the nanowire. At that time, the corresponding mesh segment melts and breaks with an arbitrary small force generated in actual operation. The current and voltage between input and output were recorded as melting current I_m and melting voltage V_m. The corresponding mesh resistance R_m can be calculated by dividing V_m with I_m. Subsequently, the cross-sectional area of the melted mesh segment was assigned an ultra-small value to approximate zero. The pathway of the current and heat in the mesh was therefore renewed. Similarly, the current triggering the melting of the next mesh segment can be determined. By repeating the above process until the mesh is open, the relationship of I_m and V_m as well as the variation of R_m with the number n_b of broken mesh segments can be obtained during the overall melting process of the mesh.

3 RESULTS AND DISCUSSION

The temperature distribution of the mesh is shown in Fig. 3, at which the maximum temperature T_{max} of the intact mesh reaches the melting point T_m. The obtained relationship between melting current I_m and melting voltage V_m, and the variation of R_m with the number n_b of broken mesh segments are given in Figs. 4a and 4b, respectively.

For the thermal boundary conditions of **Case I**, the centres at the middle segments of the intact mesh have the T_{max} as shown in Fig. 3a. The corresponding isotherm is a line. Once the T_{max} reaches the Tm, the central segments of the mesh will melt simultaneously as shown in Fig. 5a, making the mesh broken. Correspondingly, there is only one data point for the relationship of I_m and V_m in Fig. 4a, and the variation of R_m with n_b in Fig. 4b.

Therefore, it indicates that only global unstable melting (instantaneous melting occurs until the open mesh) will occur in the mesh under the thermal boundary conditions of Case I equipped with a current source. The corresponding melting current is 36.14μA in the present case.

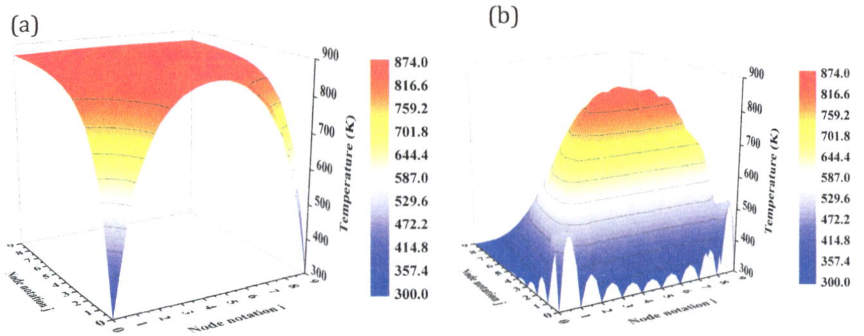

Fig. 3 Temperature distribution of the Ag nanowire mesh with T_{max} reaching T_m at different thermal boundary conditions (a) Case I, and (b) Case II

For the thermal boundary condition of **Case II**, the centre in one segment of the intact mesh has the T_{max} as shown in Fig. 3b. The corresponding isotherm is a point. When such T_{max} reaches the Tm, the segment will melt as shown in Fig. 5b. Accompanied with the melting of such segment, the current path in the mesh will be renewed. Afterwards, the melting of next segment will happen. During the overall melting process, the relationship of I_m and V_m is a curve with the repeated zigzag pattern as shown in Fig. 4a, where the Rm keeps increasing [6]. It implies that stable melting (further melting needs the increase of current/voltage), locale unstable melting (several mesh segments simultaneously melt at a constant current/voltage), and global unstable melting (mesh segments melt at the maximum current to make the mesh open) will occur for the mesh equipped with current source [6]. The current resulting in global unstable melting is 163.5 μA, which is much higher than that in Case I.

Therefore, it can be concluded that the thermal boundary conditions have great effect on the melting behaviour of the Ag nanowire mesh induced by Joule heating, as they changes the fundamental trend of the relationship between I_m and V_m (e.g., from one data point in Case I to the curve with repeated zigzag pattern in Case II).

4 CONCLUSIONS

By monitoring the temperature distribution, the melting behaviour of an Ag nanowire mesh was analysed with consideration of different thermal boundary conditions. It is found that the isotherm of the maximum temperature in the mesh relies on thermal boundary condition and determines the melting behaviour. If such isotherm is a curve, the relationship of melting current and voltage becomes a data point, where several wires melt simultaneously until the mesh is broken. On the other hand, if the isotherm is a point, the relationship of melting current and voltage becomes a curve with the repeated zigzag pattern, where local stable/unstable melting and global melting may occur.

ACKNOWLEDGEMENTS

The authors would like to thank Prof. H. Tohmyoh for his valuable discussion. This work was partly supported by JSPS KAKENHI Grant-in-Aid for Young Scientists (B) (No. 26820001) in Japan.

REFERENCES

[1] Ginley, D.S., Hosono, H., Paine, D.C. (2010) Handbook of Transparent Conductors. New York: Springer.

[2] Ellmer, K. (2012) Nat. Photonics. 6, 808-816.

[3] Kang, M-G., Park, H. J., Ahn, S. H., Xu, T., Guo, L.J. (2010) IEEE Journal of Selected Topics in Quantum Electronics. 16, 1807-1820.

[4] Groep, J.V., Spinelli, P., Polman, A. (2012) Nano Lett. 12:3138–3144.

[5] Khaligh, H.H., Goldthorpe, I.A. (2013) Nanoscale Res. Lett. 8, 235.

[6] Li, Y., Tsuchiya, K., Tohmyoh, H., Saka, M. (2013) Nanoscale Res. Lett. 8, 370.

[7] Tsuchiya, K., Li, Y., Saka, M. (2014) Nanoscale Res. Lett. 9, 239.

[8] Mayoral, A., Allard, L.F., Ferrer, D., Esparza, R., Jose-Yacaman, M. (2011) J. Mater. Chem. 21, 893–898.

[9] Xu, J., Munari, A., Dalton, E., Mathewson, A., Razeeb, K.M. (2009) J. Appl. Phys. 106, 124310.

[10] Liu, X.H., Zhu. J., Jin, C.H., Peng, L.M., Tang, D.M., Cheng, H.M. (2008) Nanotechnology. 19, 085711.

Study on structural lightweight design of oil tankers based on HCSR

W.T. Wang, H.L. Ren, G.Q. Feng, B.S. Zhang
College of Ship Building Engineering, Harbin Engineering University, P.R. China

ABSTRACT

The direct calculation for the yielding and buckling strength of the cargo hold of an oil tanker were carried out in terms of HCSR. Based on the analysis results, the influence of HCSR on the structural weight of oil tankers was analyzed. According to the requirements of HCSR, the mathematical model of the structural optimization of midship cross section was established, with the minimum section area as the object, and the hull girder strength, the local scantling and slenderness requirements as the constraint condition. The midship section optimization and design were performed by Multi-Island Genetic Algorithm, with the software Isight.

1. INTRODUCTION

With the release of Harmonized Common Structural Rules (HCSR)(1), undoubtedly the new requirements will be used on the structure design of the ships. The impact of HCSR on the structural weight attracts people's attention. On the other hand, there is a widespread concern on how to find the appropriate size of ship structure, meeting the conditions of the structural strength, stiffness and stability, to reduce the weight of the structure and improve the economics of the ship construction effectively. Consequently, the structural lightweight design of the oil tankers based on HCSR should be investigated.

So far many investigators have studied the optimization design of midship cross section structure of ship. Moe J(2) searched the optimal solution, using the Powell conjugate direction method. Jang C D(3) found the optimal results, using the method of multi-objective optimization. Optimal design theory is constantly researched by more and more investigators.

2. IMPACT OF OIL TANKER OF HCSRON THE STRUCTURAL WEIGHT

2.1 Basic information of ship model

An oil tanker of deadweight of 308000 tons was taken as an example, which complies to the Common Structural Rules (CSR), with main dimensions shown in Table 1. The ship is a double shell tanker with five cargo holds. Each cargo hold is divided into three parts of left, mid, right, by two oil-tight longitudinal bulkheads. The frame space of midship is 5.66m. Based on direct strength analysis, cargo hold finite element model of 2, 3, 4 cargo was built by Patran, as shown in Figure 1.

Table 1 The principle dimension of ship model

Item	Value	Unite
Overall length	330	m
Length between perpendiculars	320	m
Breadth	60.00	m
Depth	29.80	m
Clock coeffienct	0.825	-
Design draft	20.50	m

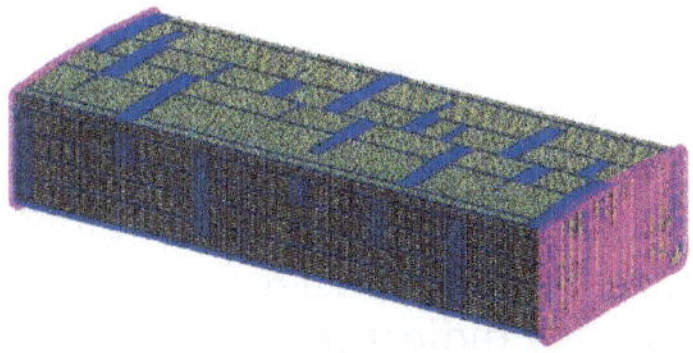

Fig.1 Cargo hold finite element model

2.2 Yield strength assessment and buckling strength assessment

Yield strength assessment is to comply with the following criteria:

$$\lambda \leq \lambda_{perm}$$

Where λ is yield utilization factor, $\lambda=\sigma_{vm}/\sigma_{perm}$. σ_{vm} is the von Mises stress. σ_{perm} is permissible yield utilization stress. λ_{perm} is permissible yield utilization factor defined in HCSR.

Buckling strength assessment is to comply with the following criteria:

$$\eta \leq \eta_{all}$$

Where: η is buckling utilization factor, $\eta=W_{act}/W_u$. W_{act} is applied equivalent stress due to the combined membrane stresses. W_u is equivalent buckling capacity. η_{all} is permissible buckling utilization factor defined in HCSR.

2.3 The result of yield and buckling assessment of ship model

Loading conditions include the static load conditions(S) and the static+dynamic load combinations(S+D). Table 2 shows the max yield and buckling factor in different component. Due to the limit of paper space, only list several components, which don't meet HCSR.

Table 2 Assessment result of ship in the cases of S and S+D

Mode	Yield				Buckling			
Load Case	S		S+D		S		S+D	
Name	λ	Y/N	λ	Y/N	η	Y/N	η	Y/N
Inner shell	0.730	YES	0.877	YES	0.868	NO	0.957	YES
Bottom shell plating	0.420	YES	0.758	YES	0.895	NO	1.060	NO
Longitudinal bulkhead	0.825	NO	0.964	YES	0.980	NO	1.090	NO
Transverse bulkhead	0.527	YES	0.792	YES	0.634	YES	1.100	NO
Horizontal girder	0.959	NO	1.225	NO	1.270	NO	1.550	NO
Longitudinal girder	0.635	YES	0.860	YES	1.290	NO	1.290	NO
Floor plating	0.844	NO	1.195	NO	1.312	NO	1.525	NO

As shown in Table 2, some components of the ship could not meet the requirement of HCSR. This situation illustrate that the requirements of HCSR are stricter than CSR. The structure design made based on HCSR will result in the increased weight of the ship.

3. OPTIMIZATION OF AMIDSHIPS OF SHIP MODEL

3.1 The design variables

In the overall design of the ship, the main dimensions and the cargo layout had been determined; they could be the constants in the cross-sectional structure parameter optimization design. The scantling of the longitudinal structural member was to be optimized. The design variables include three categories: thickness, longitudinal models and longitudinal spacing. In order to simplify the process of optimizing design, member was identified as a constant, which satisfy the Rules' requirements and satisfy a small contribution to the overall longitudinal strength, and the rest member was defined as design variables, which had a greater contribution to the longitudinal strength.

Firstly, a discrete set of 54 design variables was determined, as shown in Figure 2. The range of thicknesses variable were chosen from 8mm to 30mm, the interval was 0.5mm. The longitudinal of the ship is T profile. The discrete interval of height of web was 10mm, the discrete interval of width of face plate was 5mm, the discrete interval of thickness of web and face plate was 0.5mm. The longitudinal spacing was selected from 700mm to 900mm, its discrete intervals was 10mm. In order to reduce the number of design variables, the same longitudinal spacing of deck, inner bottom, and bottom shell plate was assumed. The same longitudinal spacing of longitudinal bulkheads, inner shell, and side shell was assumed.

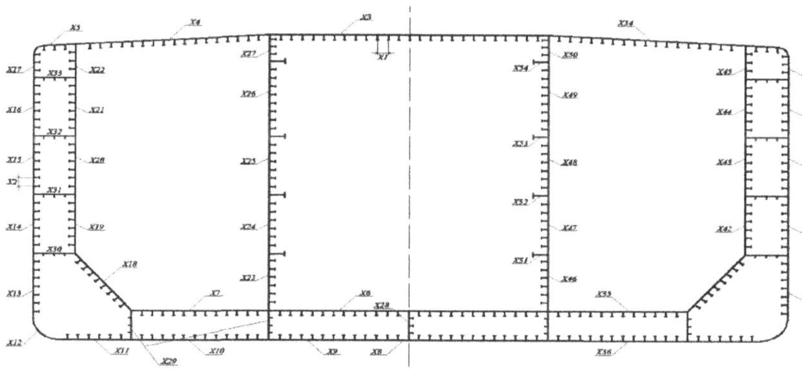

Fig.2 The design variables of cross-sectional

3.2 The objective function

The smallest area of the cross section was used as the objective in the optimization, the objective function was given by the following formula:

$$F(X)=\sum_{i}^{n}A_i(X)+\sum_{j}^{m}A_j(X)$$

Where n is number of design variables. m is number of constants. $A_i(X)$ is the cross section area of design variables. $A_j(X)$ is the cross section area of constants.

3.3 The constraints

(1)At the transverse section in the midship part, the net moment of inertia about the horizontal axis $I_{y\text{-}n50}$ is to be not less than I_{yR}, which is defined in HCSR. (2)At the transverse section in the midship part, the vertical hull girder net section modulus at deck and bottom $Z_{D\text{-}n50}$ and $Z_{B\text{-}n50}$ is to be not less than Z_R, which is defined in HCSR. (3)The normal stress σ_L at any point of the hull transverse section is to be not less than σ_{perm}, which is defined in HCSR. (4)The total vertical hull girder shear capacity Q_R is the minimum of the calculated values for all plates contributing to the hull girder shear of the considered transverse section. (5)The minimum net thickness for plating, stiffeners and primary supporting members of cargo should meet the HCSR requirement. (6)The net thicknesses of plates are not to be taken less than the greatest value calculated for all applicable design load sets. (7)The minimum net web thickness t_w of stiffeners is not to be taken less than the greatest value calculated for all applicable design load sets. (8)The minimum net section modulus Z of stiffeners is not to be taken less than the greatest value calculated for all applicable design load sets. (9)Slenderness of plates and stiffeners should meet the HCSR requirement.

3.4 Optimization------Multi-Island Genetic Algorithm

Multi-Island Genetic Algorithm(MIGA) is improved from Genetic Algorithm[4-6]. MIGA divides individual populations into several sub-groups, which may also be referred to as "islands". Selection, crossover and mutation are performed on each island, and at certain algebra, a certain number of individuals regularly migrate between the islands, to complete the exchange of individuals between populations, and then continue selection, crossover and mutation. The migration process can enrich types of individuals in subgroups, effectively inhibit premature phenomenon, and ensure the emergence of a global optimal solution. The flowchart of MIGA is shown in Figure 3.

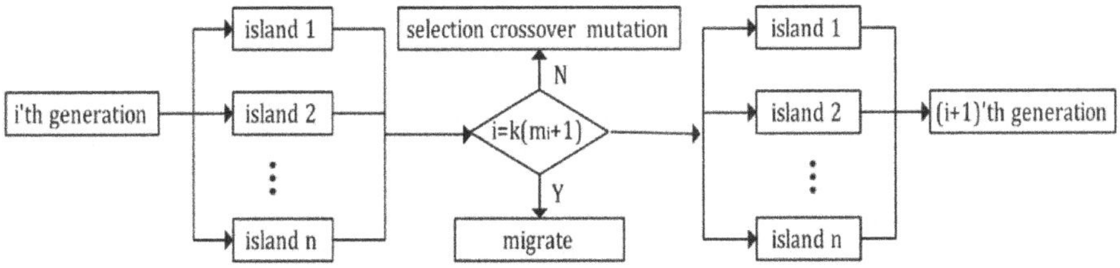

Fig.3 The flowchart of MIGA

3.5 The results

The midship sectional properties were written using excel form, in which include the design variables, objective function and constraints. Optimization software Isight[7] was used to integrate the form, using MIGA to optimize the longitudinal structure of the midship section. The excessive design variables would make it difficult to find the optimal solution in limited solution space. In response to this situation, hierarchical optimization was considered. Firstly, the value of the spacing and thickness were determined by optimization calculating, and then these values were taken as constants. Secondly, the variables of T bars were optimized. Due to the limit of paper space, only the thicknesses, which had changed after optimization, were listed, as shown in Table 3-4.

Table 3 Comparison of spacing and thicknesses mm

NO.	Name	Initial	MIGA	NO.	Name	Initial	MIGA
1	L.S1	850	880	21	Inner shell3	12.5	13.5
2	L.S2	850	860	22	Inner shell4	14.5	13.5
8	Keel	17.5	16.5	24	Longl.Bhd 2	16.0	16.5
9	Bottom shell1	15.0	14.5	26	Longl.Bhd 4	13.5	14
11	Bottom shell 3	15.0	14.5	27	Longl.Bhd 5	12.5	13.5
13	Side shell1	16.5	17.0	28	Centerline girder	15.5	14.5
16	Side shell 4	16.0	14.5	29	Side girder	12.0	11.5
17	Side shell 5	15.0	16.5	30	Horizontal girder 1	11.5	11.0
20	Inner shell2	14.0	15.0	33	Horizontal girder 4	12.0	10.5

Table 4 Comparison of T bars mm

No	Initial	MIGA	No	Initial	MIGA
34	420×9+160×21	420×8+160×20	41	420×9+160×22	430×8+160×25
35	600×10+175×29	600×10+195×26	42	580×9+160×22	550×9+150×20
36	600×10+175×29	640×10+195×24	43	500×9+160×22	480×8+150×20
37	700×10+185×30	660×9+160×20	44	480×9+160×22	480×8+150×21
38	580×9+160×22	550×8+150×20	45	500×9+165×22	420×9+160×20
39	500×9+160×22	480×8+150×21	46	650×11+175×33	620×9+150×20
40	480×9+160×22	480×8+150×20	47	580×10+165×23	550×8+150×21
51	1240×11+250×23	1000×8+200×20	48	480×9+160×23	480×8+155×20
52	1500×10+350×43	1000×8+200×20	49	480×9+160×23	480×8+155×20
53	1140×10+250×23	1000×8+200×20	50	450×9+160×12	430×9+160×12
54	1040×9+250×21	1090×10+220×25			

The initial area of the cross-sectional of midship was 50795.77mm^2, and was 49555.49mm^2 after optimization. The area was reduced by 2.44%.

4. CONCLUSION

The impact of oil tanker of HCSR on the structural weight was analyzed. It was shown that the structural strength requirements in HCSR were stricter than CSR, because of changes in the specific requirements, evaluation criteria. The updating of structural strength requirements would improve the structural strength of the ship and inevitably lead to the increased weight of the ship. How to reduce the construction cost due to the increased weight and achieve energy saving of operational process had become a hot topic. In order to improve the load capacity, reduce structural weight under HCSR, the mathematical model for the optimization and design of the oil tanker midship cross section structure was established. The structural lightweight design of oil tankers based on MIGA with the optimization software Isight was given. After the optimization, the midship cross section area was reduced by 2.44%. The weight of the hull was lower than the original weight.

REFERENCES

(1) IACS. Common structural rules for bulk carriers and oil tankers. International Association of Classification Society. February 2013.

(2) Moe J, Lund S. Cost and weight minimization of structures with special emphasis on longitudinal strength members of tankers. Trans RINA, 1968, 110: 43-70.

(3) Jang C D, Shin S H. A study on the optimum structural design for oil tankers using multi-objective optimization. IMDC, 1997: 217-230.

(4) Zhou G. A practical discrete optimization approach for ship structures based on genetic algorithms. Ship Technology Research, 1999, 46: 179-188.

(5) Hongwei Wang. Optimum design of midship section of large tanker. Harbin Engineering University, 2002.

(6) Lee K Y, Roth M I. An efficient genetic algorithm using gradient information for ship structural design optimization. Ship Technology Research, 2001, 48: 161-170.

(7) Yuyang Lai. Parametric theory and explain examples of Isight. Beijing Aerospace University Press. 2012.

Mechanical properties of PSZ-Ti composites fabricated by spark plasma sintering

K. Tohgo, T. Fujii, M. Harada, H. Isono, Y. Shimamura
Department of Mechanical Engineering, Shizuoka University, Japan

ABSTRACT

PSZ-Ti composites consisting of partially stabilized zirconia (PSZ) and pure titanium (Ti) were fabricated by spark plasma sintering (SPS), and their mechanical properties were evaluated. According to the X-ray diffraction (XRD) analysis, Ti_2O and Ti_2ZrO were detected on all composites and pure Ti was detected on the composites with Ti volume fraction more than 90%. The Vickers hardness and Young's modulus were higher than the predictions by rule of mixture. The bending strength and fracture toughness of the composites decreased up to 75%Ti and then increased with increasing Ti content. It was proposed that these results were caused by hardening and embrittlement due to the reaction phases such as Ti_2O and Ti_2ZrO.

1 INTRODUCTION

Partially stabilized zirconia (PSZ) and titanium (Ti) are used for implants in human body as kinds of biocompatible materials. The PSZ and Ti have both advantage and disadvantage as implant materials. To improve the biocompatibility and mechanical performances of implants, the concept of functionally graded materials (FGM), in which, for example, the material composition varies continuously from 100% of PSZ on the surface to 100% of Ti inside of implants, would be useful (1, 2). Fabrication of PSZ-Ti composites and their mechanical properties are key issues to develop the FGM consisting of PSZ and Ti.

In the previous investigation (3) on PSZ-Ti composites fabricated by hot pressing (HP), the Young's modulus and Vickers hardness were higher than the predictions by rule of mixture and bending strength decreased with increasing Ti volume fraction, and these results were attributed to the embrittlement due to reaction products such as Ti_2O and Ti_2ZrO which were formed during HP.

To derive the ductility of titanium in the PSZ-Ti composites, the reaction of PSZ and Ti, especially oxidization of Ti phase should be depressed during sintering. Thus, we focused our attention on spark plasma sintering (SPS), in which the sintering of powder is carried out by pulsed direct electric current and uniaxial press through the graphite die and plungers. Advantages of SPS are very short sintering time, relatively low temperature and easy control of the sintering condition. The short time and low temperature in sintering would depress the reaction of PSZ and Ti during sintering. The SPS has been applied to fabrication of metals, ceramics, ceramic-metal composites and FGM, ceramic-ceramic composites and FGM (4-7).

In this investigation, the PSZ-Ti composites with various volume ratios of PSZ and Ti were fabricated by the SPS, and their Vickers hardness, Young's modulus, bending strength and fracture toughness were evaluated and compared with those of the composites by the HP (3). The relationships between these mechanical properties and Ti volume fraction were discussed from a viewpoint of reaction products formed during sintering.

2 FABRICATION OF COMPOSITES BY SPS

Raw materials were commercially available partially stabilized zirconia (ZrO_2-3molY_2O_3, PSZ, manufactured by KCM Corporation Co. Ltd.) and pure titanium (Ti, manufactured by Toho Titanium Co. Ltd.) powders. The mean particle sizes of powders were 0.26μm for the PSZ and 24μm for the Ti, respectively. The PSZ and Ti powders were mixed in volume ratios of 100%PSZ, 75%PSZ-25%Ti, 50%PSZ-50%Ti, 25%PSZ-75%Ti, 10%PSZ-90%Ti, 5%PSZ-95%Ti and 100%Ti, respectively, and each mixture was milled for seven hours in vacuum by a vibrational ball mill (Nissin Giken Co. Ltd.). The composites fabricated by these powder mixtures are referred to as 100%PSZ, 25%Ti, 50%Ti, 75%Ti, 90%Ti, 95%Ti and 100%Ti, respectively. Spark plasma sintering (SPS) was carried out by using an SPS-211Lx apparatus (Fuji Electronics Industrial Co. Ltd.). The mixed powder is put into a graphite die with 10mm in inner diameter, 30mm in outer diameter and 40mm in height. Then the die and plungers with powder sample was set into SPS apparatus, and sintering of powder was carried out with a pulsed electric current and uniaxial press through the die and plungers. The present study is first step to fabricate a PSZ-Ti FGM. Because the FGM containing the full range of material composition would be fabricated by a certain sintering condition, in the present study the sintering condition was fixed for all composites with different material compositions. According to preliminary work, the sintering condition for the seven kinds of composites was determined as heating rate of 300°C/min up to 600°C, 200°C/min for 600°C to 1000°C and 100°C/min for 1000°C to 1200°C, sintering temperature of 1200°C, holding time of 5min and uniaxial pressure of 15MPa. The relative density of the composites was about 93% for pure PSZ and increased up to almost 100% with increasing Ti content.

Figure 1 shows microstructure of the composites by optical microscope. No large pores are observed in all composites. As the particle size of the PSZ powder was considerably smaller than that of the Ti powder, PSZ particles would have entered gaps between Ti particles on the fabrication. As a result, the PSZ phase is likely to construct the matrix in the composites. The Ti particles are dispersed in the PSZ matrix in the composites from 0% to 50% of Ti content, while the small PSZ particles are dispersed in the Ti matrix but distributed in the region between Ti powders in the Ti rich composites.

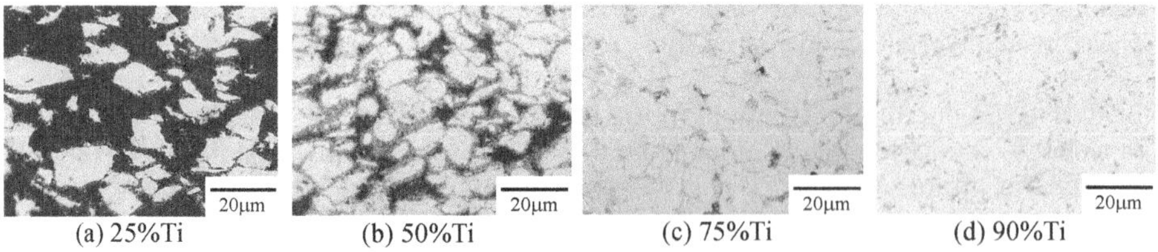

Fig.1 Microstructures of PSZ-Ti composites.

Phase constitution of all the sintered composites was analysed by X-ray diffraction (XRD) using Cu target at an accelerating voltage of 40kV, an accelerating current of 40mA and measurement area was 10 x 5 mm^2. The reaction products during SPS were identified by using the Powder Diffraction File. According to the XRD analysis, Ti_2O and Ti_2ZrO were detected on the composites with 25%Ti, 50%Ti and 75%Ti, and Ti_2O and pure Ti was detected on the composites with Ti volume fraction more than 90%. These products were produced in SPS process. Important thing in the results is that the pure Ti is recognized in the sintered composites with Ti content more than 90%. The same reaction products were created in the composites fabricated by HP as in reference [3].

3 EXPERIMENTAL PROCEDURE

Vickers hardness of the composites was measured by Vickers hardness tester (Matsusawaseiki Co. Ltd., S12-57). Test conditions were the applied load of 196N and holding time of 30sec to obtain the hardness from an indent large enough to contain PSZ and Ti phases. An average, maximum and minimum values of the Vickers hardness were obtained from ten trials. Three-point bending tests were conducted on a rectangular specimen as shown in Fig. 2 to obtain the Young's modulus and bending strength of the composites. The tests were carried out on two specimens for each composition at a cross head speed of 0.05mm/min and at room temperature. Tensile strain was measured by strain gauge pasted on the lower surface of the specimens. Bending stress-stain curves were obtained and elastic moduli which were used as the Young's modulus here were calculated from an initial slope up to the strain of 0.1% of the stress-strain curve. Fracture toughness of the composites was obtained by an indentation fracture (IF) method (Japanese Industrial Standard: JIS R 1607)(8).

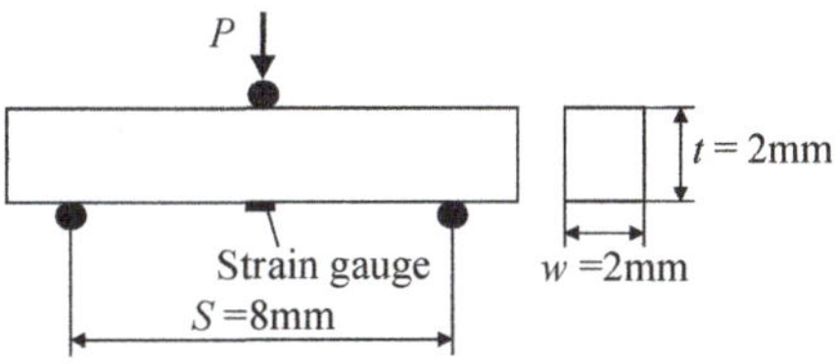

Fig.2 Three-point bending specimen.

The results for the composites fabricated by HP in the previous investigation (3) are used for the comparison with the present results.

4 EXPERIMENTAL RESULTS AND DISCUSSIONS

Figure 3 shows the Vickers hardness as a function of Ti content. For both composites fabricated by SPS and HP, the predictions by a rule of mixture are shown by dashed lines. As shown in Fig. 3, both composites exhibit the almost same characteristics on the Vickers hardness, and the hardness of the PSZ-Ti composites is higher than the prediction by rule of mixture due to creation of reaction phases such as Ti_2O and Ti_2ZrO.

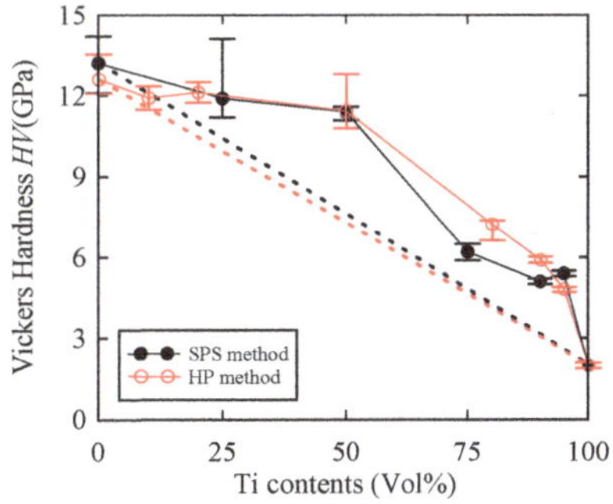

Fig.3 Vickers hardness as a function of Ti contents.

On all composites except for pure Ti, brittle fracture occurred after linear elastic deformation by the three-point bending tests. The Pure Ti exhibited large plastic deformation and fracture of the specimen did not occur in the present three-point bending test. From the stress-strain relations, the Young's modulus and bending strength were obtained for each composite. Figure 4 shows the relationship between the Young's modulus and Ti volume fraction. On both composites, as in Vickers hardness, the Young's modulus is higher than the prediction by rule of mixture. Furthermore, the Young's modulus of the composites by SPS is about 20 to 35% high as compared with that of the composites by HP. This might be attributed to the difference in test methods, namely three-point bending test for the composites by SPS and four-point bending tests for those by HP. Relationship between the bending strength and Ti content is shown in Fig. 5. On both composites fabricated by SPS and HP, the bending strength decreased up to 75%Ti and then increased with increasing Ti content. The bending strength of the composites by SPS is higher than that of the composites by HP in a full range of materials composition.

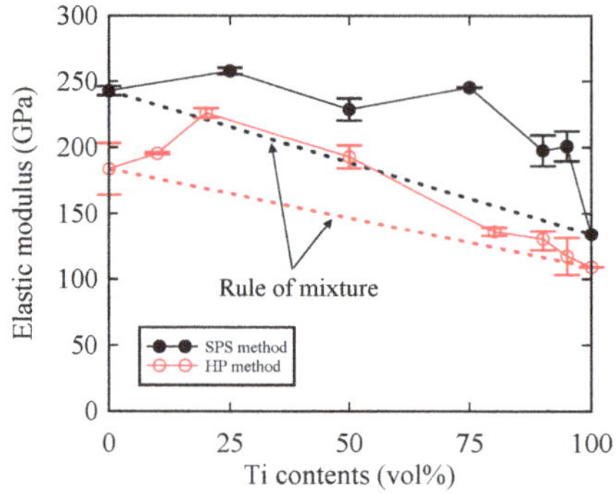

Fig.4 Elastic modulus as a function of Ti contents.

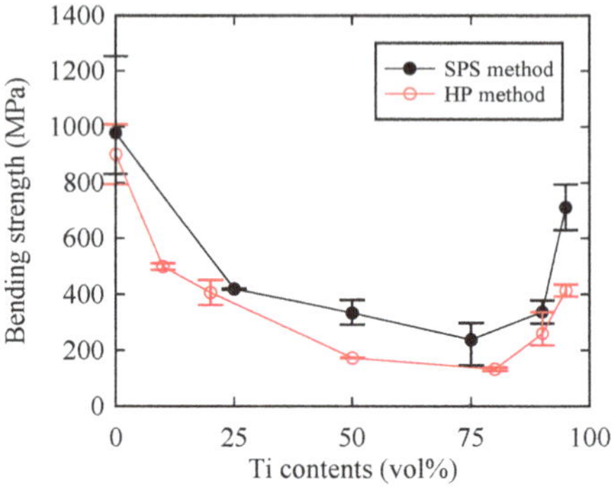

Fig.5 Bending strength as a function of Ti contents.

Figure 6 shows the fracture toughness obtained by the IF method for the composites by SPS and HP. Generally, for ductile materials, the fracture toughness cannot be obtained by the IF method because the cracking does not occur by Vickers indentation. However, for the present composites the fracture toughness was obtained by the IF method for the materials composition from 0% to 95%Ti. This shows that the most Ti phase in composites changed to Ti oxide in the sintering process and became brittle even in the 5%PSZ-95%Ti composite. The fracture toughness by the IF method for both composites decreases up to 75% Ti and the increases with increasing Ti content from 75% to 100%. It is noted that the fracture toughness of the composites by SPS is higher than that of the composites by HP.

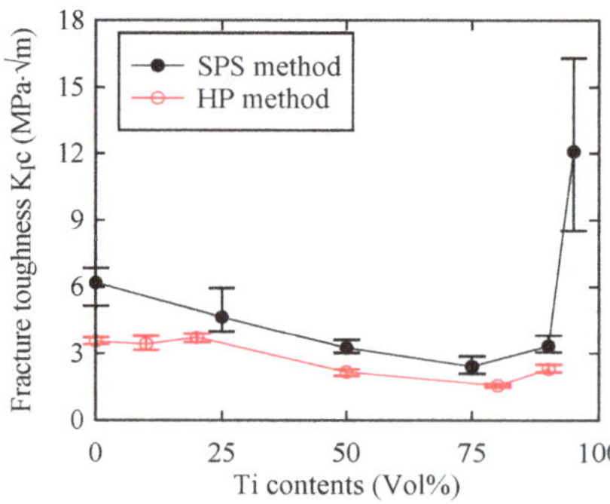

Fig.6 Fracture toughness as a function of Ti contents.

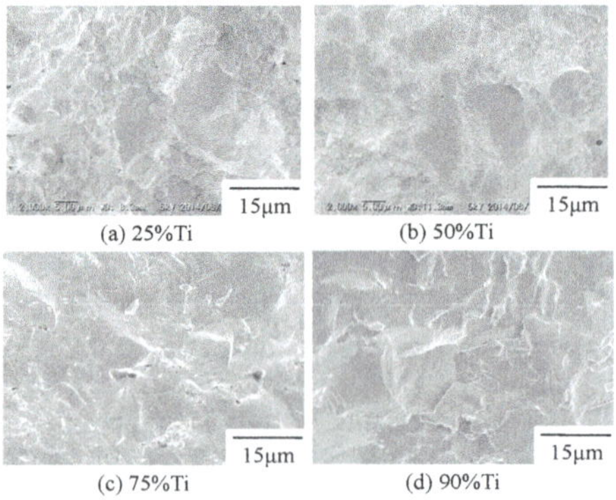

Fig.7 SEM micrographs of fracture surface of PSZ-Ti composites.

From these test results, it is concluded that the Vickers hardness and Young's modulus higher than the predictions by rule of mixture, the bending strength and fracture toughness decreasing up to 75%Ti are caused by creation of Ti_2O and Ti_2ZrO (6).

Figure 7 shows the micrographs of fracture surfaces of three point bending specimens observed by a scanning electron microscope. It is found from the fracture surfaces that all the composites except for pure Ti were fractured in brittle manner. In the composites with 25%Ti and 50%Ti, small pores are observed in both phases, the Ti phase is dispersed in the PSZ matrix and shows very flat fracture surface. In the composite with 75%Ti, relatively large pores are observed; this might be a reason of lowest fracture strength and fracture toughness. With increasing Ti content from 75% to 95%, roughness of fracture surface increases. The increase of fracture strength and fracture toughness with increasing Ti content from 75% to 95% might be caused by the crack path deflection inducing roughness of fracture surface in addition to the existence of pure Ti.

5 CONCLUSIONS

The obtained results in the present investigation are summarized as follows:

(1) The relative density of the composites is about 95% for pure PSZ and increases up to 100% for pure Ti with increasing Ti volume fraction even if 5 minutes sintering time for 1200°C.

(2) According to the X-ray diffraction (XRD) analysis, Ti_2O and Ti_2ZrO were detected on all composites and pure Ti was detected on the composites with Ti volume fraction more than 90%.

(3) The Vickers hardness and Young's modulus of the composites were higher than the predictions by rule of mixture. The bending strength and fracture toughness of the composites decreased up to 75%Ti and then increased with increasing Ti content from 75% to 100%. These results might be caused by hardening and embrittlement due to the reaction phases such as Ti_2O and Ti_2ZrO.

REFERENCES

(1) Miyamoto Y, Kaysser WA, Rabin BH, Kawasaki A, Ford RG. Functionally Graded Materials: Design, Processing and Applications, (1999), Kluwer Academic Publishers.

(2) Tohgo K, Suzuki T, Araki H. Eng. Frac. Mech. 2005;72:2359-2372.

(3) Fujii T, Tohgo K, Araki H, Wakazono K, Ishikura M, Shimamura Y, Journal of Solid Mechanics and Materials Engineering, 4 (2010) 1699-1710.

(4) Guo H, Khor KA, Boey YC, Miao X, Biomaterials 24 (2003) 667–675.

(5) Afzal M A F, Kesarwani P, Reddy K M, Kalmodia S, Basu B, Balani K, Materials Science and Engineering C, 32 (2012) 1164–1173.

(6) Fernandez-Garcia E, Gutierrez-Gonzalez CF, Fernandez A, Torrecillas R, Lopez-Esteban S, Ceramics International 39 (2013) 6931–6936.

(7) Gutierrez-Gonzalez C F, Fernandez-Garcia E, Fernandez A, Torrecillas R, Lopez-Esteban S, Journal of Materials Science, (2014) 49:3823–3830.

(8) Japanese Industrial Standards, Testing methods for fracture toughness of fine ceramics, Japanese Standard Association, JIS R1607, (1995).

A study on compressive behaviour of aluminium foam-filled tubes

Y. An[1], C. Yang[2,3], P.D. Hodgson[1]
[1] Institute for Frontier Materials, Deakin University, Australia
[2] School of Computing, Engineering and Mathematics, The University of Western Sydney, Australia
[3] Institute for Infrastructure Engineering, The University of Western Sydney, Australia

ABSTRACT

In this study, deformation behaviour and failure modes of aluminium tubes reinforced with aluminium foam fillers under compression are investigated. Both experimental tests and finite element modelling are employed to determine the influences of pertinent geometry parameters of hollow tube and foam filler, including wall thickness, and structure length on the compressive behaviour of the compound tubes. The obtained results show that the existence of foam fillers can significantly affect deformations and failure modes of these tubes. The information obtained is useful to facilitate the design of such foam-filled tubes as more efficient energy absorbing devices for automotive engineering applications.

1 INTRODUCTION

The effect of foam filler on the crush behaviours of thin wall tubes was first studied by Thornton in 1981 (1). After his research, more studies have been conducted on such structures. Foam-filled tubes offer lots of benefits for energy absorption since the foams filled in reinforce the tubes and thus the compounds have high energy absorption but low density (2, 3). A lot of investigations on the crush and energy absorption response of foam-filled tubes under uniaxial loadings have been published in last ten years. Foam-filled square and circular tubes are the most popular candidates as energy absorbers. Researchers (2-7) reported that the foam-fillers can not only stabilise the crush response of hollow tubes but also improve the energy absorption efficiency. Reid and other researchers predicted that an enhancement in the mean crushing loads of between 30 - 45% for foam-filled aluminium tubes (4, 5) and 40-60% for steel tubes (6). Overall, foam-filler is preferable in term of energy absorption capability since it can change the deformation behaviour and cause large load fluctuations during the impact process without increase the mass of the whole structures (7).

In this study, the foam-filled tubes using pure aluminium foam fillers are employed. The deformation behaviour and failure modes of these tubes under uniaxial compression are studied both experimentally and numerically. The effect of wall thickness and structure length of foam-filled tubes on the mechanical behaviour is further investigated.

2 EXPERIMENTAL PROCEDURE AND FINITE ELEMENT MODELLING OF FOAM-FILLED TUBE SAMPLES UNDER COMPRESSION

For sample preparation, the hollow aluminium tubes were chosen which are made of AA6063 and the pure aluminium foam-fillers have a porosity of 88.2% and an average cell size of 4.5 mm. Hollow tubes and foam-fillers were tightly fitted together to form the compound tubes. A series of cylindrical specimens were prepared to investigate the influences of main geometry parameters, i.e., structure length L, and wall thickness t. The four different structure lengths were chosen as 50, 100, 200 and 400 mm and three different thicknesses of tube walls were considered as 2, 4 and 5 mm. The compound tubes, aluminium foams, and hollow tubes were then investigated using uniaxial compression test individually for comparison purposes, which were performed in a standard MTS testing machine at a low strain rate of 0.02s-1 along the longitudinal direction of slender the samples.

The finite element models of foam fillers (8-10), hollow tubes and foam-filled tubes (11), which have the same dimensions of the specimens used in the experiments, were all developed using the commercial FEA software - Abaqus. The tube walls are modelled using the shell element S4R which is shown in Fig. 1(a), and the foam filler is modelled with the solid element C3D8R, which is shown in Fig. 1(b). The material models for the hollow tube and foam fillers were defined using the stress- strain data for AA6063 and the pure aluminium foams. The details of material model of metal foams can be found in author's previous publications (9, 10). The contact boundary conditions were considered in three ways: a) Tie-model, perfect connection between the tube and foam filler, which suits the scenarios where the foam filler is glued with the tube wall; b) Non-friction-model, free contact without considering friction between the tube and foam filler; and c) Friction-model, frictional contact considering the friction between the tube and foam filler with a friction coefficient f of 0.3 for aluminium and aluminium contact. To simulate the uniaxial compression, the top and bottom surfaces of the tubes foam-filled tubes were contacted with two rigid plates as shown in Fig. 1 (c). Moving down the upper rigid plate generates compression with a rate of 0.02s-1 while the lower rigid plate is kept stationary, which is the same as the experimental test. Thus, the clamped boundary conditions and velocity boundary conditions were applied on the lower and upper rigid plates, respectively.

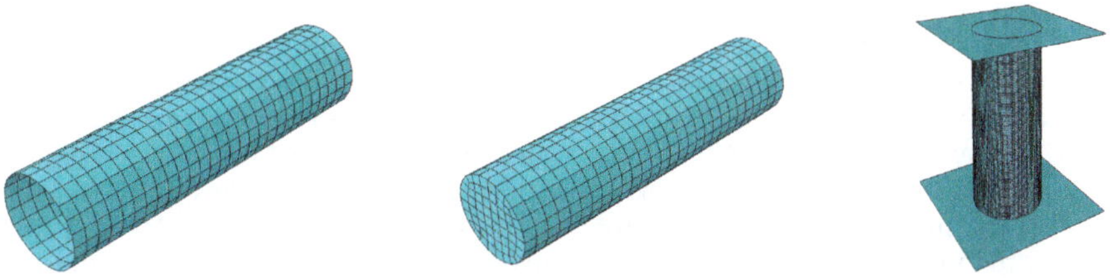

(a) Hollow tube part (b) Foam core part assembly (c) Compound tube

Figure 1 Finite element models of the tube sample under compression

Convergence testing and mesh sensitivity analysis were also conducted to obtain the reliable meshes. For instance, for the shell part, three different mesh sizes were employed in the tube models, which were designed as 5 mm, 2 mm and 0.5 mm, respectively, leading to 625, 3,487 and 31,400 shell elements used. The maximum von Mises stress of these models were found 83.33, 84.95 and 85.47 MPa and the crush stresses obtained from these models are 33.62, 34.42 and 34.68 MPa, respectively. The maximal difference in the crush stresses is 1.4% while the difference in maximum von

Mises stress is only 1.1%, which could be neglected in engineering applications. Hence, in terms of saving time and maintaining accuracy, the models were developed using the mesh size of 5 mm in the study.

3 RESULTS AND DISCUSSION

3.1 Experimental observation on failure modes and deformation behaviour of samples

Based on experimental observation, the crush mechanisms of these samples under uniaxial compression can be described by their deformation modes. The deformation mode of circular tubular structures depends on the geometry parameters, the ratio to their diameter and thickness (d/t) and the ratio to their length and diameter (L/D). In general, the following collapse modes were observed: (i) axisymmetric concertina bellowing (ring mode); (ii) non-symmetric buckling with a variable number of circumferential lobes or corners (diamond mode); (iii) mixed mode (combination of the ring and diamond modes); (iv) Euler or global buckling (Euler mode); and (v) other mode (3).

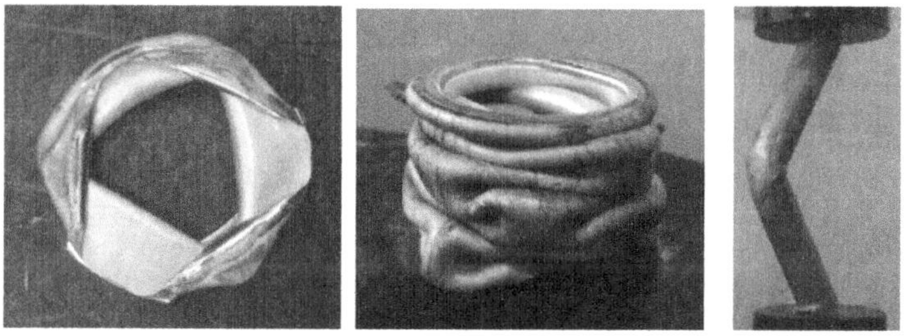

(a) (b) (c)

Figure 2 Deformation modes of hollow tubes subjected to uniaxial compression (a) Diamond mode (t = 2 mm, L = 50 mm); (b) Mixed mode (t = 2 mm, L = 100/200 mm); and (c) Euler mode (t = 2/4/5 mm, L = 400 mm)

Photographs of typical deformed hollow tubes obtained in this study are presented in Fig. 2. In this study, the ring mode, diamond mode, mixed mode and Euler mode were all observed. For the samples with the value of t = 2 mm and L = 50, 100, and 200 mm, the hollow tubes collapse in both the diamond mode as shown in Fig. 2(a) and mixed mode as shown in Fig. 2(b). The Euler mode shown in Fig. 2(c) can only be observed in the tubes with a larger length, L = 400 mm, or tubes with larger thickness, t = 4 mm and 5 mm. Compared to these three deformation modes of the hollow tubes, the foam-filled tubes deform much differently due to the contact between foam filler and tube. According to the experimental result, it is confirmed the main contribution of aluminium foam filler is to shift the deformation mode from diamond mode to ring mode (2).

As shown in Fig. 3 only ring and Euler modes have been observed in the compression tests for such samples. For the foam-filled tubes with a thickness of 2 mm and length of 50/100/200 mm, the ring mode was found. In the ring mode deformation, the lobes were found to form harder than the hollow tubes and the foam filler acts as an elastic-plastic foundation for the tube walls and thus affects their buckling behaviour. For the tubes with a larger length, L = 400 mm, Euler mode occurred. Meanwhile, for the samples with a larger thickness, i.e., 4 and 5 mm, only the Euler mode was observed.

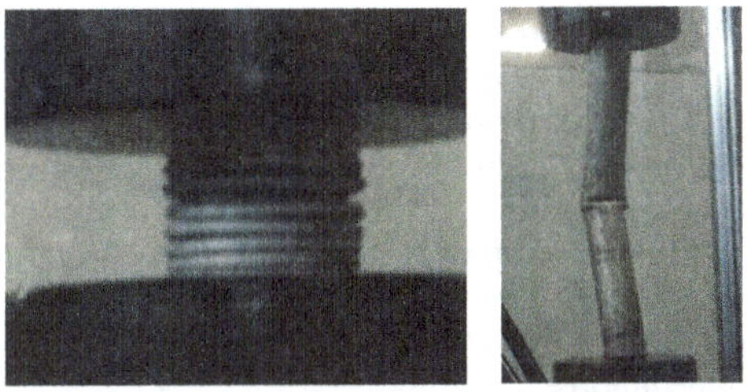

(a) (b)

Figure 3 Deformation modes of foam-filled tubes subjected to uniaxial compression: (a) Ring mode (t = 2 mm, L = 50/100/200 mm) and (b) Euler mode (t = 2/4/5 mm, L = 200/400 mm)

3.2 Numerical results on failure modes and deformation behaviour of samples

The developed finite element models of hollow and foam-filled tubes models were run to capture their deformation mechanisms which were termed as digital testing. Figs. 4 and 5 show the deformed hollow and foam-filled tubes with different wall thicknesses obtained from simulations. Diamond mode can be obtained from the collapsed hollow tubes with the value of t = 2 mm and L=50, 100, and 200 mm. Hollow tubes with L = 400 mm and t = 4 mm and 5 mm collapse in the Euler mode.

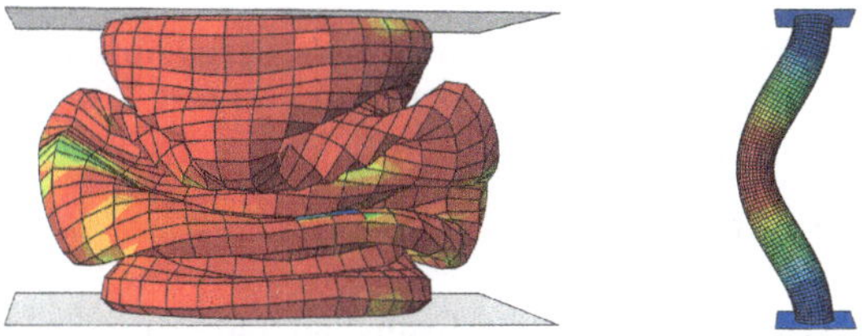

(a) Diamond mode (b) Euler mode

Figure 4 Deformed hollow tubes by digital testing

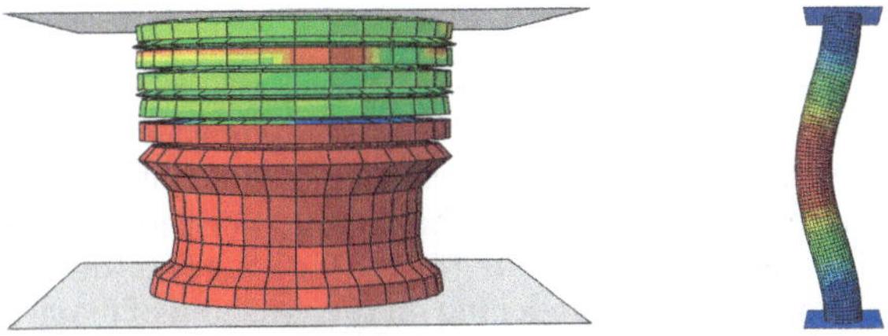

(a) Ring mode (b) Euler mode

Figure 5 Deformed foam-filled tubes obtained by finite element simulations (digital testing)

For the foam-filled tubes, both ring mode and Euler mode have been observed. The ring mode can be found in foam-filled tubes which have a thinner thickness, t = 2 mm and length of 50/100/200 mm. For such structures with a larger thickness, t = 4/5 mm, Euler mode occurred. Compared with the experiment results shown in Figs. 4 and 5, it can be found that these models can demonstrate the deformations of such structures precisely as well as the influence of foam fillers on the deformation modes.

3.3 Comparison on experimental data and numerical results

Fig. 6 shows the strain-stress curves obtained from both experimental and digital tests. The maximum discrepancies of 3.2% in main crush stress shown in Fig. 6(a) and 4.8% in absorbed energy presented in Fig. 6(b) have been found. Also, the elastic, plastic and collapse regions of hollow/foam-filled tubes can be identified clearly in the obtained stress-strain curves as shown in Fig. 6(a). Similar results were also obtained from the models with other geometric parameters. Due to flexibility, the numerical simulations can be further employed to perform more digital tests with a large range of geometric parameters and other parameters in future.

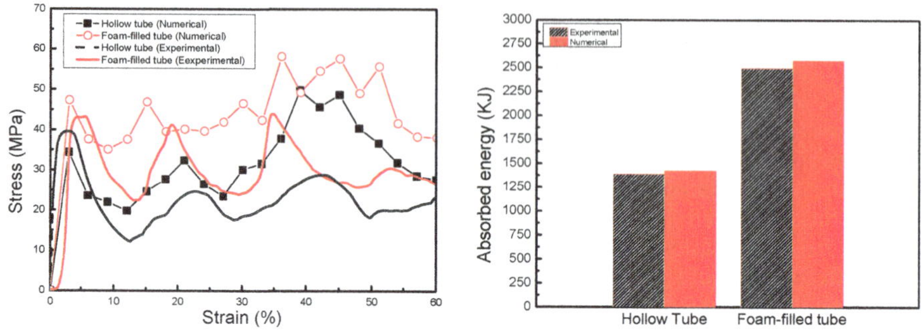

(a) (b)

Figure 6 Strain-stress curves and absorbed energy obtained from both numerical and experimental test

4 CONCLUDING REMARKS

The compressive behaviour of hollow tube, foam filler and foam-filled tubes subjected to uniaxial compressions have been studied experimentally and numerically in this research. Different deformation modes, such as ring mode, diamond mode, Euler mode and mixed mode have been identified for such structural members to show their deformation and failure mechanisms.

Based on the results obtained in the current study, the existence of the foam filler can significantly affect the deformation mode of the tubes. Among geometric parameters, the structural length L is the key parameter which has a significant influence on the compressive behaviour of foam-filled tubes. The deformation mode can be changed from diamond mode (hollow tube) or ring mode (foam-filled tube) to the Euler mode when the length increases. Finite element modelling on the crushing of foam-filled aluminium tubes can precisely capture the deformation mechanisms of such tubes and can give results with good agreement, compared with experimental results.

REFERENCES

[1] Thornton, P. H. (1981) Energy absorption of foam-filled structures. Society of Automotive Engineers, Inc.

[2] Mirfendereski, L., Salimi, M. & Ziaei-Rad, S. (2008) *International Journal of Mechanical Sciences*, **50** (6), 1042-1057.

[3] Gupta N. K. & Nagesh, K. (2006) *International Journal of Mechanical Sciences*, **48** (2), 210-223.

[4] Reid, S. R. (1993) *International Journal of Mechanical Sciences*, **35**, 1035-1052
[5] Reglero, J. A., Rodriguez-Perez, M. A., Solorzano, E. & de Saja, J. A. (2011) *Materials & Design*, **32** (2), 907-910
[6] Smith, B. H., Szyniszewski, S., Hajjar, J. F., Schafer, S. R. & Arwade, B. W. (2012) *Journal of Constructional Steel Research*, **71**, 1-10
[7] Ahmad, Z. & Thambiratnam, D. P. (2009) *Materials & Design*, **30** (7), 2393-2403.
[8] An, Y., Wen, C. E., Hodgson, P. D. & Yang, C. (2012) *Computational Materials Science*, **55**, 1-9.
[9] Yang, C., An, Y., Greenfield, R. & Hodgson, P. D. (2013) *Computer Methods in Materials Science*, **13** (4), 425-435
[10] Yang, C., An, Y., Tort, M. & Hodgson, P. D. (2014) *Computational Materials Science*, **81**, 89-97.
[11] An, Y., Wen, C. E, Hodgson, P. D. & Yang, C. (2012) *Applied Mechanics and Materials*, **152-154**, 436-439.
[12] Zhang, C., Feng, Y. & Zhang, X. (2010) *Transactions of Nonferrous Metals Society of China*, **20** (8), 1380-1386.

Trial fabrication of Al micro-materials by electromigration using buildup structure

Y. Kimura, M. Saka
Department of Nanomechanics, Tohoku University, Japan

ABSTRACT

The present research shows the trial fabrication technique for huge production of Al micro-materials by electromigration (EM) using a buildup structure. The objective of this study is provision of inspiring notable structure to overcome disadvantage that is low throughput production using EM technique. The fabrication of micro-materials using the presented buildup structure is partly demonstrated in this study. This study shows the dependence of fabrication of micro-materials on the conditions of holes introduced for discharging atoms, and potential for high throughput production.

1 INTRODUCTION

Electromigration (EM) is the physical phenomenon of atomic diffusion with dense electron flow in a metallic film [1], and is utilized as a fabrication technique of micro-materials which has been reported by our group [2-4]. In general, EM has been known as a negative phenomenon for electronic devices because EM deteriorates the metallic line of electronic devices by atomic accumulations and vacancies which are called hillocks and voids, respectively [5]. Conversely, our group applies erratic atomic diffusion caused by EM to fabricate Al micro-materials by intentionally generating concentration and discharge of atoms. We are considering and designing the sample structure for effective fabrication of micro-materials using atomic diffusion by EM.

So far, several sample structures have been developed. A part of reported sample structures are embedded structure as Al line is implanted into W line [6], and simple Al line like Blech structure which includes an artificial slit for damming up Al atoms and an artificial hole for discharging Al atoms intentionally [2]. The achievement using these sample structure is in the first development of fabrication technique by using EM and then improvement for effective fabrication. Furthermore, our group has investigated the capability of the fabrication technique by using EM and generated Al micro-material with high aspect ratio, single crystalline and arbitrary cross-section which is perpendicular to the discharge direction [2, 7-8]. However, single fabrication of Al micro-material is accomplished when we use the sample structure as described above.

Recently, the comb pattern with conductive passivation film was proposed by reference to [3, 8] for facilitating the process of sample preparation and for fabrication of multiple Al micro-materials [4]. In the previous study [4], formation of ten Al micro-materials was demonstrated by using the comb pattern and conductive passivation. This technique has the possibilities for multiple fabrication of Al micro-materials by EM. But there is still a room to improve the EM fabrication technique in terms of productivity.

We would like to modify the previous structure based on the use of conductive passivation. So far, the approach of improving sample structure is designed in the direction of planar like the comb pattern. One of the ideas for advancing the multiple fabrication is utilization of a buildup structure that is extended in the direction of thickness. We expect that the use of both buildup and comb pattern facilitates the fabrication of multiple Al micro-materials.

The primary objective of this study is to suggest the sample structure which is buildup structure for huge multiple fabrication. In this study, as unprecedented attempt, we carried out a trial fabrication using the buildup structure which has capability for generating two Al micro-materials through respective artificial holes at the same time. We verified the potential of the proposed structure in this study.

2 EXPERIMENT

2.1 Sample preparation

The process for preparation of sample is given herein. Figure 1 shows the presented buildup structure. By using RF sputtering, 300-nm-thick Ti for improving adhesion, 300-nm-thick TiN for serving as a bypass, and 600-nm-thick Al as the target for fabricating Al micro-materials were consecutively deposited onto SiO_2 composed by thermal oxidation. Photolithography was used to generate the sample pattern as the first objective layer of Al. After that, 300-nm-thick TiN and 600-nm-thick were serially deposited, and the Al pattern as the second objective layer was fabricated via photolithography. The conductive passivation that is 3000-nm-thick TiN was deposited as the topmost layer at the end. In this study, two square holes of 1.5 or 2.0 μm involved in discharging Al atoms simultaneously were introduced corresponding to sample A or B by a focused ion beam as shown in Fig. 1 (b)-(c).

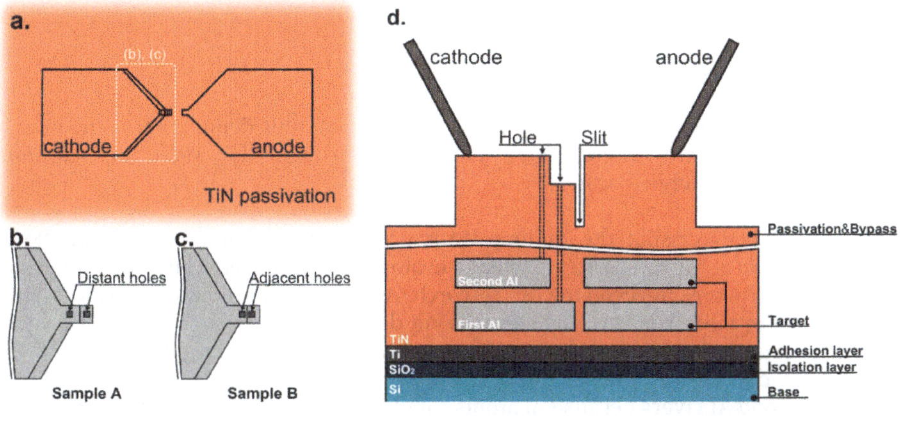

Fig. 1 Schematic of buildup structure involved in this study: (a) top view, (b) distant holes, (c) adjacent holes, and (d) cross-sectional view.

2.2 Fabrication principle

The general principle of fabricating Al micro-materials by EM has been reported in the previous work [2]. The Al atoms are conveyed with high-dense electron flow when the current is supplied. Electrons can be flowed in the first and second Al layers through the conductive passivation that is TiN. Whereas, the Al atoms are blocked up by a slit which discontinues Al line as shown in Fig. 1 (a). The accumulated Al atoms near the slit cause high compressive stress and this high compressive stress is released by discharging the Al atoms via artificial holes. The Al micro-materials are fabricated by the above procedures on a ceramic heater which gives off heat and facilitates the atomic diffusion.

The dominant factors for atomic diffusion and formation of Al micro-materials by EM are current density, substrate temperature and current stressing time according to Huntington–Grone equation [9] and preceding research [2, 10]. The Huntington–Grone equation is given as follows:

$$\mathbf{J} = \frac{ND_0}{kT}\exp(-\frac{Q}{kT})Z^* e\rho\,\mathbf{j} \tag{1}$$

where $\mathbf{J}$ is the atomic flux vector, N is the atomic density, D_0 is a prefactor, k is Boltzmann's constant, T is the absolute temperature, Q is the activation energy, Z^* is the effective valence, e is the electronic charge, ρ is the electrical resistivity, and $\mathbf{j}$ is the current density vector. Therefore, we need to describe the experimental conditions including current, substrate temperature and current stressing time for being profitable in the future. The current is expressed as one of the experimental conditions instead of current density because we cannot precisely calculate the current density due to the leakage of current through the conductive passivation. Table 1 shows the experimental conditions.

Table 1 Experimental conditions for fabricating Al micro-materials using buildup structure.

	Current [A]	Temperature [K]	Current stressing time [min]	Discharged hole
Sample A	1.3	763	200	Right
Sample B	1.3	763	210	Left

3 RESULTS AND DISCUSSION

Figures 2 and 3 show FE-SEM image of the Al micro-materials demonstrated by the trial fabrication. Al micro-material was fabricated through the right or left hole corresponding to sample A or B.

The final objective using the buildup structure is fabrication of huge multiple Al micro-materials in the future, so we need to evaluate the proposed trial fabrication using the buildup structure for evolving the research. As a result, the simultaneous formation of two Al micro-materials via two holes was not demonstrated in each sample. However, the formation of Al micro-material from respective holes, i.e. the right or left hole was demonstrated as shown in Figs. 2 and 3. These results indicate EM is happened in the first and second Al layers because Al atoms could be discharged through the right or left hole introduced into the first or second Al layer as shown in Fig. 1 (d), respectively. It is important that EM is happened in the second Al layer. This result denotes the potential of using the buildup structure for huge production of Al micro-materials by EM.

For advancement using the buildup structure, the sample structure is needed to modify. The dominant factor contributing to partly succeeded demonstration of forming Al micro-material is discharge force which depends on compressive stress caused by accumulation of Al atoms. The result of this study provides the presumption of that the compressive stress for discharging Al atoms around holes decreases when the Al atoms are discharged via another hole. Evidently, the Al atoms could not be discharged via both holes in this study. And yet, we have two options for discharging Al atoms via both holes, which are adjustment of the conditions including hole size and distance between two holes and increase of Al thickness.

The hole through which Al micro-material could be formed was changed when we used sample B which has different hole size and distance between two holes compared with sample A, as shown in Figs. 2 and 3. The different hole size and distance between two holes affects the ease of forming Al micro-material due to changing the area for releasing compressive stress. Therefore, if we optimize the hole conditions involved in size and distance, Al atoms can be discharged via both holes by optimal release of compressive stress from both holes. Additionally, other option is modification of Al thickness. That is increase of Al thickness for applying large current due to decrease in the resistance of Al line. Increase of Al thickness enlarges compressive stress due to large amount of atomic accumulation caused by flowing high-dense electron.

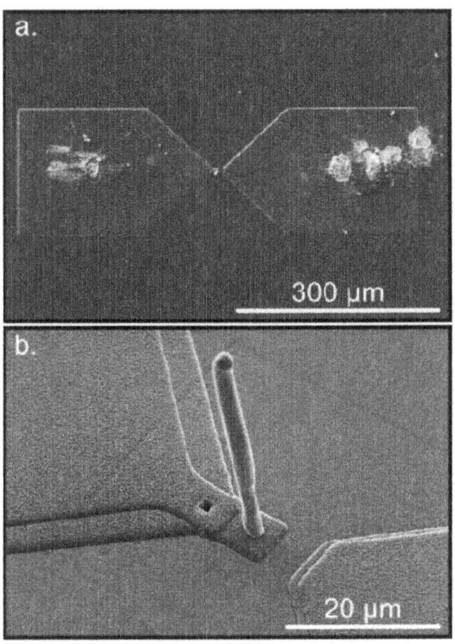

Fig. 2 FE-SEM images of Al micro-material using Sample A: (a) whole sample structure after applying current, and (b) a formed Al micro-material through right hole.

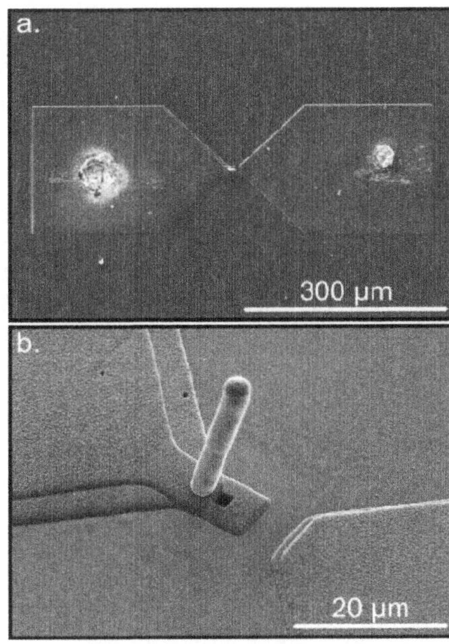

Fig. 3 FE-SEM images of Al micro-material using Sample B: (a) whole sample structure after applying current, and (b) a formed Al micro-material through left hole.

4 CONCLUSIONS

In this study, the new attempted sample structure which consists of the buildup structure including two Al layers was provided as a trial fabrication technique for huge production of Al micro-materials. We demonstrated the partial generation of Al micro-material which means that two Al micro-materials were not simultaneously formed through both holes. However, the important result in this study is in demonstration of formation of Al micro-material from the second layer using the buildup structure. The utilization of the second layer for fabricating huge multiple Al micro-materials has not been attempted, so far. We demonstrated the fabrication of Al micro-materials using the second layer and formation of Al micro-material from the second layer in the present study. This presented technique is one of methods for accomplishing huge production of Al micro-materials by EM.

ACKNOWLEDGEMENT

A part of this work was supported by JSPS KAKENHI Grant-in-Aid for Scientific Research (B) No. 26289001. Also a part was supported by Tohoku University Institute for International Advanced Research and Education.

REFERENCES

[1] H.B. Huntington, in: A.S. Nowick and J.J. Burton, editors. *Diffusion in solids*. New York: Academic Press Inc. 303 (1975).
[2] M. Saka and R. Nakanishi, *Mater. Lett.* **60**, 2129 (2006).
[3] R. Zanma and M. Saka, Int. *J. Mater. Struct. Integrity*, in press (2014).
[4] Y. Kimura and M. Saka, *Adv. Mater. Res.* **909**, 36 (2014).
[5] C.M. Tan and A. Roy, *Mater. Sci. Eng., R* **58**, 1 (2007).
[6] M. Saka and R. Ueda, *J. Mater. Res.* **20**, 2712 (2005).
[7] Y. Lu, H. Tohmyoh, and M. Saka, *J. Phys. D: Appl. Phys.* **44**, 045501 (2011).
[8] Y. Kimura and M. Saka, *Mater. Lett.* **116**, 278 (2014).
[9] H.B. Huntington and A.R. Grone, *J. Phys. Chem. Solids* **20**, 76 (1961).
[10] M. Saka, K. Kato, H. Tohmyoh, and Y. Sun, *J. Mater. Res.* **23**, 3122 (2008).

Evaluation of electromigration near a corner composed of dissimilar metals by analyzing atomic flux at the interface

Y. Kimura [1], X. Zhao [2], M. Saka [1]
[1] Department of Nanomechanics, Tohoku University, Japan
[2] Department of Mechanical Engineering, Akita University, Japan

ABSTRACT

This research evaluates the electromigration (EM)-induced damage at the interface near a right-angled corner, which is composed of dissimilar metals in a single-crystal line without passivation layer. The atomic flux at the interface is analyzed, and the effect of material combination on the accumulation or depletion of atoms is investigated. Ways for reducing the EM-induced damage are proposed, and are discussed in comparison with those reported in our previous work, where the countermeasures were deduced by analyzing the atomic flux divergence near the right-angled corner except for the interface.

1 INTRODUCTION

Electromigration (EM) refers to a phenomenon whereby metallic atoms are transported by electron flow at a high current density. It is significantly affected by current density and temperature. With accumulation and depletion of atoms, EM results in a dominant failure mechanism in microelectronics that leads to the appearance of hillocks and voids, which can cause short circuits and open circuits, respectively [1-6]. In particular, EM reliability issues in multilayer interconnection systems such as via structures, are more serious due to the current density concentration near the contact corner caused by the abrupt change of the geometry and mismatch in material properties [7, 8].

Recently, we have theoretically analyzed EM near a right-angled corner composed of dissimilar metals by investigating the effect of material combination on atomic flux divergence (AFD) [9], which essentially characterizes the EM-induced damage. The two-dimensional (2D) EM problem near the corner was analyzed by treating an angled line as a single-crystal line without a passivation layer and with a uniform width, and several countermeasures for reducing the EM-induced damage were proposed. Nevertheless, AFD at the interface was not treated in the reported work. At the interface, the atomic fluxes perpendicular to the interface are not continuous due to abrupt change of material properties, and hence differentiating the atomic flux to obtain AFD is unavailable. Unfortunately, the interface is usually a barrier that blocks the atomic flux into the other material, and the EM-induced damage is enhanced there.

This research evaluates the EM-induced damage at the interface near a right-angled corner, which is composed of dissimilar metals in a single-crystal line without passivation layer. The atomic flux at the interface is analyzed, and the effect of material combination on the accumulation or depletion of atoms is investigated. Ways for reducing the EM-induced damage are proposed and discussed.

2 EM-INDUCED DAMAGE AT INTERFACE

2.1 Problem representation

A 2D EM problem near a right-angled corner in the thin layered metal line shown in Fig. 1 is considered. A polar coordinate system (r, θ) is introduced in the vicinity of the corner. The line is assumed to be of uniform width b and thickness t, and is composed of homogeneous Materials 1 and 2 with respective electrical resistivity ρ of ρ_1 and ρ_2. The line is subjected to a steady direct current flow. The boundaries enclosing the line are assumed to be electrically insulated, except for both ends of the line where the uniform current density j_∞ is input and output far from the right-angled corner. The temperature T at the inner corner is denoted by T_C.

For simplicity in the theoretical analysis of atomic flux at the interface, the line shown in Fig. 1 is treated as a single-crystal line without a passivation layer. In this way, the EM-induced atomic flux is assumed to be dominated by lattice diffusion, and the effect of back flow on atomic flux [10] can be neglected. This consideration on the effect of back flow is also valid even for a metal line with a passivation layer when a short initial period of EM is examined in which there is no effective accumulation of atoms, and hence no active back flow is built up. In contrast, the atomic flux in polycrystalline structures dominated by grain boundary diffusion is more complicated, and the effect of grain boundary on the direction of atomic flux has to be considered, which is beyond the scope of this study.

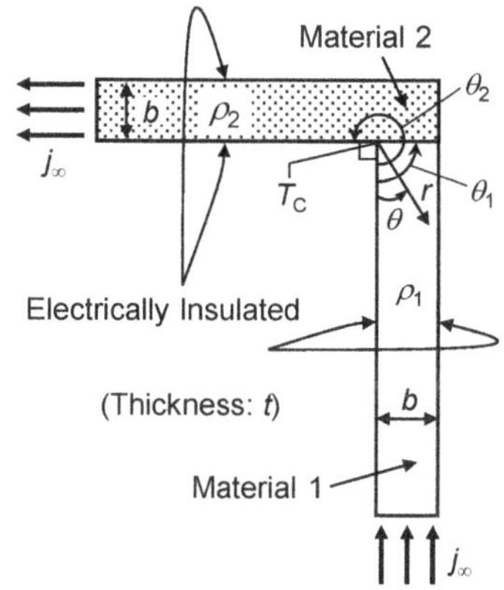

Fig.1 Right-angled line composed of dissimilar metals with uniform width and thickness, and $\theta_1 = \pi/2$ and $\theta_2 = 3\pi/2$.

2.2 Theoretical analysis

For a single-crystal line without a passivation layer, the EM-induced atomic flux is given by [11]

$$\mathbf{J}=\frac{ND_0}{kT}\exp\left(-\frac{Q}{kT}\right)Z^*e\rho\mathbf{j}, \tag{1}$$

where $\mathbf{J}$ represents the number of atoms migrating per unit time and unit area under the driving force $|Z^*e\rho\mathbf{j}|$, N is the atomic density, D_0 is a prefactor of diffusion coefficient, k is Boltzmann's constant, Q is the activation energy, Z^* is the effective valence, e is the electronic charge, and $\mathbf{j}$ is the current density vector. Equation (1) shows that $\mathbf{J}$ is significantly affected by current density and temperature. Note that Z^* is dimensionless and usually takes a negative value, which means that the atomic flux is opposite to the direction of current flow. In this study, the values of Z^* for Materials 1 and 2 are assumed to be negative.

At the interface, $\mathbf{J}$ is discontinuous due to abrupt change of material properties such as Z^*, Q, N, D_0, and ρ. Hence, the angular component of $\mathbf{j}$, j_θ, contributes to the occurrence of accumulation or depletion of atoms. The asymptotic solution for the distribution of j_θ at the interface near the corner can be obtained on the basis of our previous work [8] as

$$j_{\theta(1)} = j_{\theta(2)} = \frac{3\sqrt{3}}{4}\left(\frac{3\pi}{4}\right)^{-1/3} j_\infty \varepsilon \left(\frac{r}{b}\right)^{\varepsilon-1}, \quad (2)$$

where $\varepsilon\ [= \frac{2}{\pi}\sin^{-1}\sqrt{(2\rho_1/\rho_2 + 1)/2(\rho_1/\rho_2 + 1)}\ ]$ is a parameter affecting the singular field of current density, and the subscripts (1) and (2) represent Materials 1 and 2, respectively. By substituting Eq. (2) into Eq. (1), we get the general atomic flux contributing to the EM-induced damage at the interface, J_{int}, as

$$J_{int} = A j_\infty f_{(\varepsilon)}, \quad (3)$$

where

$$A = \frac{3\sqrt{3}}{4}\left(\frac{3\pi}{4}\right)^{-1/3} \frac{ND_0}{kT_C} \exp\left(-\frac{Q}{kT_C}\right) Z^* e\rho, \quad (4)$$

and

$$f_{(\varepsilon)} = \varepsilon\left(\frac{r}{b}\right)^{\varepsilon-1}. \quad (5)$$

Here, A is dependent on the material properties, and is obtained on the basis of the consideration that the temperature T in Eq. (1) approximately equals T_C near the corner. In contrast, $f_{(\varepsilon)}$ is dependent on ε, which takes a value in the range of $0.5 < \varepsilon < 1$ depending on material combination [8, 9].

For evaluating the EM-induced damage at the interface, the volume of atoms accumulated or depleted in unit time and unit length along the interface, V, is expressed as

$$V = |A| j_\infty t\Omega f_{(\varepsilon)}, \quad (6)$$

where Ω is the atomic volume.

3 DISCUSSION

Equation (6) is valid for both of Materials 1 and 2 by substituting the respective material properties into it. According to Eqs. (4) to (6), V monotonically decreases with decreasing ρ and $f_{(\varepsilon)}$ (> 0). Here, both ρ and $f_{(\varepsilon)}$ are dependent on the material selected. In particular, $f_{(\varepsilon)}$ containing ε is heavily dependent on the material combination.

Now let us consider how to reduce the EM-induced damage by decreasing the value of V. Taking Material 1 for example, depletion of atoms occurs at the interface. According to the discussion above, firstly, a material with a lower electrical resistivity is recommended for Material 1. In contrast, for decreasing $f_{(\varepsilon)}$, choosing a value of ε that yields a smaller $f_{(\varepsilon)}$ is considered. As shown in Fig. 2, the ranking of $f_{(\varepsilon)}$ varying with ε is different for the r/b range. However, note that the obtained V at the interface near the corner is only valid for $r/b \to 0$; hence only the corresponding $f_{(\varepsilon)}$ range should be considered. As a result, $f_{(\varepsilon)}$ monotonically decreases with increasing ε for a selected r/b. Therefore, "how to obtain a greater ε" becomes the focus for reducing V. According with our previous works [8, 9], ε increases with increasing ρ_1/ρ_2. When Material 1 with a lower ρ_1 has been selected as discussed above, the only way to increase ρ_1/ρ_2 is to select a material with a lower electrical resistivity ρ_2 from alternative materials as Material 2.

Similar to the discussion for Material 1 given above, to reduce V in Material 2, for which accumulation of atoms occurs at the interface, lower ρ_2 and higher ρ_1/ρ_2 are recommended. The above discussion on how to reduce the EM-induced damage is similar to that by treating AFD near the corner except for the interface in our previous work [9].

At the interface, the formation of hillocks in Material 2 and that of voids in Material 1 due to the respective accumulation and depletion of atoms, are determined by the direction of the electron flow, and are not affected by the temperature gradient. On the other hand, near the interface, AFD for Material 1 and that for Material 2 take the same sign being positive or negative [9]. That is to say, the formation of voids in the case of a positive value of AFD or hillocks for a negative value of AFD occurs in both of Materials 1 and 2 near the interface depending on the direction of electron flow and temperature gradient.

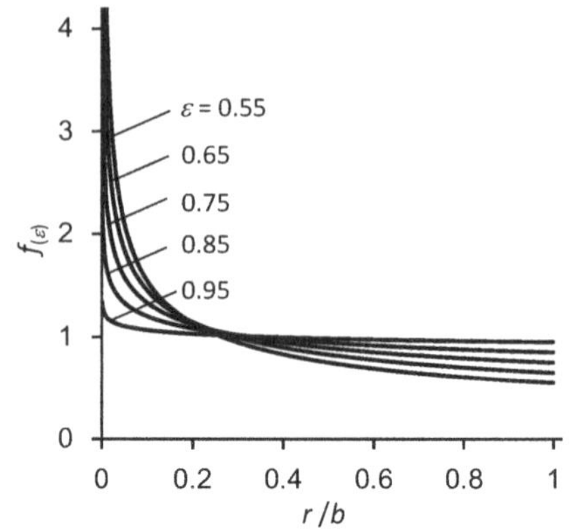

Fig.2 Graph of $f_{(\varepsilon)}$ varying with ε, versus r/b.

4 CONCLUSIONS

The EM-induced damage at the interface near a right-angled corner was evaluated. The effect of material combination on the accumulation or depletion of atoms was investigated. For reducing the EM-induced damage in Material 1, lower ρ_1 and higher ρ_1/ρ_2 were recommended; for that in Material 2, lower ρ_2 and higher ρ_1/ρ_2 were recommended.

ACKNOWLEDGEMENT

This work was supported by JSPS KAKENHI Grant-in-Aid for Scientific Research (B) No. 26289001.

REFERENCES

[1] J.R. Black, *Proc. IEEE* **57**, 1587 (1969).
[2] M.J. Attardo and R. Rosenberg, *J. Appl. Phys.* **41**, 2381 (1970).
[3] H.-U. Schreiber, *Solid-State Electron.* **28**, 1153 (1985).
[4] R.V. Goldstein, M.E. Sarychev, D.B. Shirabaikin, A.S. Vladimirov, and Yu.V. Zhitnikov, *Int. J. Fract.* **109**, 91 (2001).
[5] K. Zeng and K.N. Tu, *Mater. Sci. Eng. R* **38**, 55 (2002).
[6] H. Abé, K. Sasagawa, and M. Saka, *Int. J. Fract.* **138**, 219 (2006).
[7] E.C.C. Yeh and K.N. Tu, *J. Appl. Phys.* **88**, 5680 (2000).
[8] M. Saka and X. Zhao, *Int. J. Heat Mass Transfer* **55**, 6090 (2012).
[9] X. Zhao, Y. Kimura, and M. Saka, *Jpn. J. Appl. Phys.* **53**, 075504 (2014).
[10] I.A. Blech, *J. Appl. Phys.* **47**, 1203 (1976).
[11] H.B. Huntington and A.R. Grone, *J. Phys. Chem. Solids* **20**, 76 (1961).

Influence of substrate materials on deposition behaviour of cold spray emulated pure single Al particle

K. Ito, Y. Ichikawa, K. Ogawa
Fracture and Reliability Research Institute, Tohoku University, Japan

ABSTRACT

In a cold spray technique (CS), which used for making dense and thick metallic coatings, the deposition mechanism of fine solid particles is most important research target. In this study, in order to clarify the deposition mechanism, the influence of substrate materials on deposition behaviour of spherical pure Al particle with 1 mm diameter was investigated using the CS emulated single particle shot system. From the results, it was found that the conditions of natural oxides and deformability of each substrate material play important roles on the deposition behaviour of pure Al particle.

1 INTRODUCTION

Cold spray (CS) is well known as an effective coating technique to make dense and thick metallic coatings without high temperature oxidation and phase transformation (1). In this technique, fine solid particles are accelerated by high pressure gas flow up to supersonic velocities and subsequently impinge and deposit on a substrate material. Because the working gas temperature is normally below the melting point of particle materials, particle deformation and bonding are performed in a solid state during impact process. The property and performance of a CS coating significantly depend on the bonding state of particle-substrate and particle-particle interfaces. Therefore, the deposition mechanism of the CS particle has become a world-wide research target (2), (3). However, in spite of numerous experimental approaches, the deposition mechanism has not completely understood yet. It is attributed to the difficulties of accurate measurements, evaluations, and controls of particle velocity, particle size, particle shape, and particle surface condition. Therefore, we developed the CS emulated single particle shot system (SPSS). In this system, a spherical particle with 1 mm diameter can be accelerated up to 550 m/s, and the particle velocity can be accurately measured using lasers. The evaluations of the deposition behaviour of CS emulated particles are expected to make a huge contribution to understanding of the deposition mechanism of actual CS particles. In this study, using the SPSS, CS emulated pure Al particles with nominal diameter of 1 mm were impinged on various substrate materials. It was confirmed that the deposition behaviour of the pure Al particles is quite different for each substrate material. The difference is attributed to the adhesion energy, rebound energy, removal energy of natural oxides, and deformability. The rebound energy was calculated using finite element method (FEM), the removal energy of natural oxide was evaluated by X-ray photoelectron spectroscopy (XPS), and the deformability was assumed to correspond to the hardness. From the results, it was found that the conditions of natural oxides and deformability of each substrate material play important roles on the deposition behaviour of CS emulated pure Al particle.

2 EXPERIMENTAL

2.1 Used materials

High purity Al particles (>99.99%, Ohashi Steel Ball Corporation, Japan) with nominal diameter of 1 mm were used as the CS emulated particles. The Al particles were annealed at 500 °C for 1 hr under air atmosphere. Subsequently, the Al particles were chemical-polished by the acid solution ($H_2SO_4 : H_3PO_4 : HNO_3 = 5 : 14 : 1$, 90 °C, 1.5 min), because a lot of embedded alumina particle was observed on the as-received particle surface. The morphology of an Al particle after chemical-polishing is shown in Figure 1. The spherical shape was maintained, although several craters were formed on the surface due to removal of embedded alumina. Five kinds of substrate materials, pure Al (>99.7%), pure Cu (>99.96%), pure Ni (>99%), pure Ti (>99%), and austenite stainless steel 304 (SUS304), were employed. These substrates have cylindrical shapes with size of φ16 × 15 mm. These substrate materials except SUS304 were annealed under air atmosphere. After annealing, the surfaces of all substrates were mirror-polished using diamond paste and colloidal silica. The hardness of each material after annealing measured with micro-Vickers hardness testing equipment (HMV-1, Shimadzu Corporation, Japan) was summarized in Table 1. At this time, the vertical load and holding time were fixed at 0.98 N and 10 s, respectively.

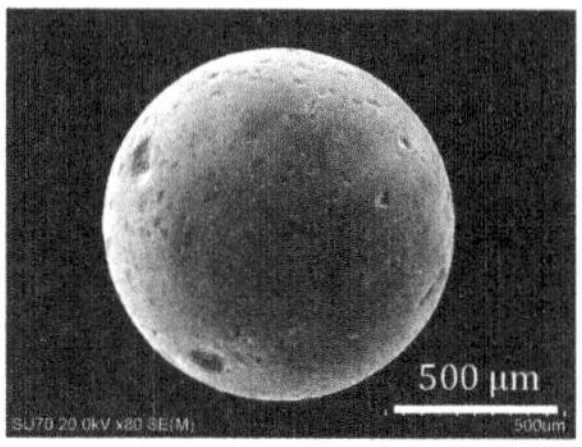

Figure 1 Morphology of an Al particle after chemical polishing

Table 1 Hardness of materials used

	Particle	Substrate				
	Al	Al	Cu	Ni	Ti	SUS304
Hardness ($HV_{0.1}$)	19.0 ± 0.4	25.1 ± 0.7	60.1 ± 1.8	126.7 ± 3.4	151.7 ± 13.1	175.4 ± 7.7

2.2 Single particle impingement tests

The schematic illustration of the SPSS is shown in Figure 2. Helium (He) was used as acceleration gas. The particle velocity was controlled by the He gas pressure and calculated from the time gap between two continuous wave lasers placed at 60 mm interval obstructed by the particle. The pressure in the vacuum chamber was reduced to approx. 0.5 kPa before impingement. The deposition behaviours of the Al particles impinged on each substrate at various velocities were investigated. The substrate surfaces after impingement were observed by scanning electron microscopy (SEM).

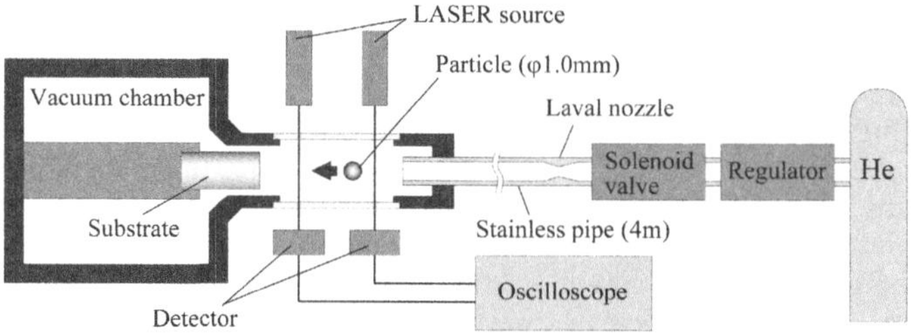

Figure 2 Schematic illustration of the SPSS

2.3 XPS analysis

XPS analysis was selected to evaluate the removal energy of natural oxide on each substrate material. Scanning ESCA Microprobe Quantum 2000 (ULVAC-PHI, Inc., Japan) was used as the XPS equipment. In the XPS analysis, Ar sputtering was carried out at suitable time intervals to reduce the natural oxides gradually. The X-ray spot size was 100 μm, and the accelerating voltage for Ar ions and the sputtering area were fixed at 2 kV and square 2 mm on a side, respectively.

2.4 FEM model

FEM was used for the estimation of rebound energy. The three dimensional model that represents the impingement of a spherical Al particle with 1 mm diameter on each substrate material was constructed using commercial FEM software (ABAQUS/CAE 6.12-3, Dassault Systemes, France). The plastic deformation of metallic material was assumed to comply with the Johnson–Cook plasticity model. A linear Mie–Gruneisen equation of state (EOS) was employed as an elasticity model. And also, it was assumed that 90% of the plastic strain energy dissipates into heat. The material parameters required for these calculations were taken from the literatures (4)-(7). In order to evaluate the true rebound energy of Al particle, the interaction between a particle and a substrate, such as friction or adhesion, was not taken into account.

3 RESULTS

Typical SEM images of each substrate surfaces after impingement of Al particle with 1 mm diameter are shown in Figure 3. In the case of Al substrate, the Al particle was deposited on the substrate at 504 m/s, and large plastic deformation of both the particle and the substrate was observed. And also, the Al particles were deposited on the Cu and Ni substrates at 462 m/s and 447 m/s, respectively. However, although the deposited particles had larger plastic deformation compared with Al substrate, the Cu and Ni substrates were hardly deformed. On the other hand, in the cases of Ti and SUS304 substrate, the Al particles were not deposited at 478 m/s and 477 m/s, respectively, although small craters were formed at the central region of contact area. The relationship between deposition behaviour of Al particle impinged on each substrate material and particle velocity were summarized in Figure 4. From Figure 4, it was confirmed that the deposition behaviour of the Al particle is quite different for each substrate material. The Al particle primarily began to deposit on the Ni substrate at around 300 m/s, in other words, the critical velocity was estimated as 300 m/s under this condition. Secondly, the Al particle was easier to deposit on the Cu substrate, and the critical velocity was approx. 350m/s. And also, the critical velocity of the Al particle impinged on the Al substrate was around 440 m/s. On the other hand, in the cases of the Ti and SUS304 substrates, the depositions of the Al particles were not observed in examined ranges of particle velocity, although only one particle deposited on the SUS304 substrate was separated by ultrasound (US) cleaning. It is supposed that the difference for deposition behaviour between each substrate material is attributed to the adhesion energy, rebound energy, removal energy of natural oxide, and deformability.

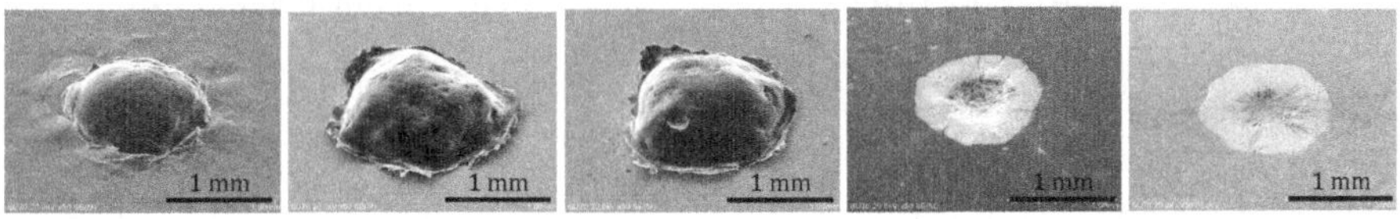

(a) Al on Al (b) Al on Cu (c) Al on Ni (d) Al on Ti (e) Al on SUS304

Figure 3 Typical SEM images of Al particle deposited on each substrate material

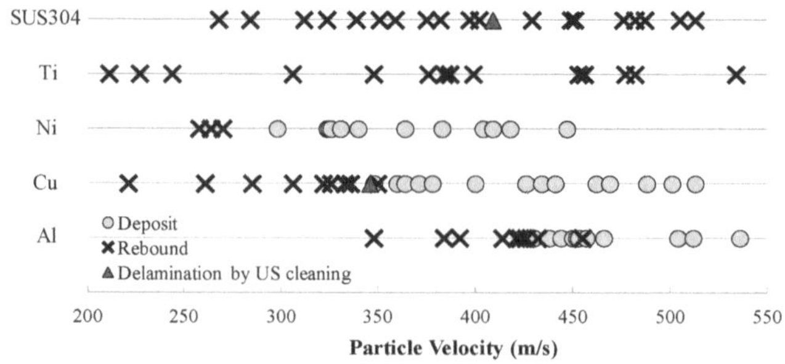

Figure 4 Relationship between deposition behaviour of Al particle impinged on each substrate material and particle velocity

The relationship between oxygen contents on each substrate material and Ar sputtering time obtained by XPS analysis is shown in Figure 5. The oxygen contents were normalized by the primary values. The removal energy of natural oxide is proportional to the Ar sputtering time, because the condition of Ar sputtering was fixed. As seen in this figure, the oxygen on the surface of Ni substrate was almost instantaneously removed by the Ar sputtering. And also, the oxygen on the surface of Cu substrate was rapidly decreased. On the other hand, in the cases of Al and Ti as passive metals, these natural oxides were strong, and a lot of energy was required to remove the natural oxide. In the case of SUS304 substrate, although SUS304 is also known as one of passive metal, the natural oxide was removed in a shorter time compared with the Al and Ti substrates. As shown in Figure 6, the rebound energies of Al particles impinged on each substrate material at 300 m/s were calculated using FEM. Each rebound energy were normalized by the rebound energy on Al substrate. The initial particle velocity was selected so as to correspond to the critical velocity of Al particle impinged on Ni substrate. The rebound energies were increased in the order corresponding to Ni, SUS304, Cu, Ti, and Al substrate. This result indicates that large bonding energy is required to deposit on Al substrate, on the other hand, Al particle can be deposited on Ni substrate at small bonding energy. In addition, these rebound energies tended to be inversely proportional to the Young's modulus of substrates.

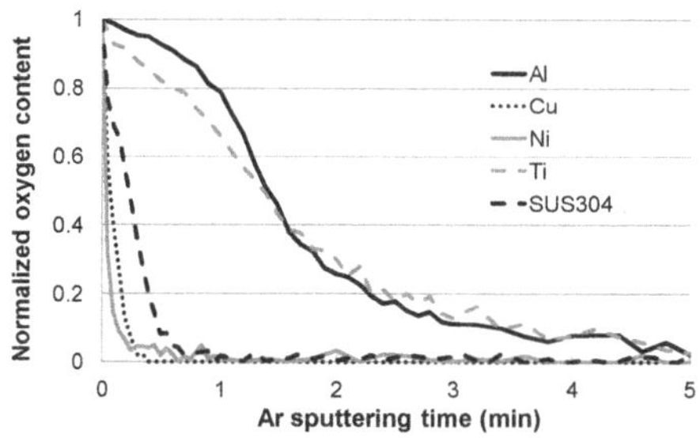

Figure 5 Relationship between oxygen contents on each substrate material and Ar sputtering time

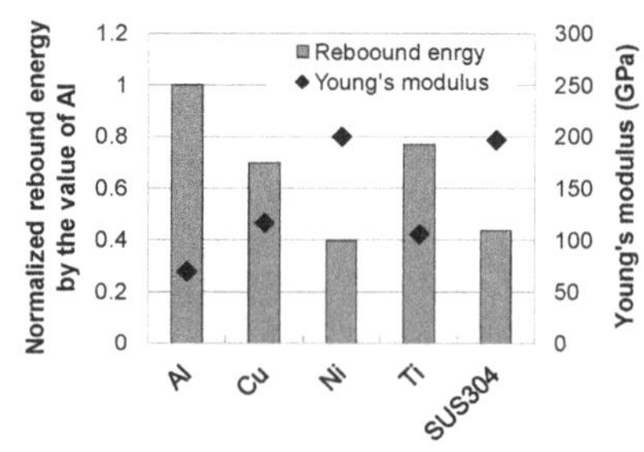

Figure 6 Rebound energies of Al particles impinged on each substrate material at 300 m/s

4 DISCUSSIONS

As mentioned above, the Al particle could be deposited on the Ni, Cu and Al substrates. The critical velocities for Ni, Cu, and Al substrates were approx. 300, 350, and 440 m/s, respectively. In other words, the Al particle was easy to deposit on the Ni and Cu substrates compared with the Al substrate. On the other hand, the Al particle was difficult to deposit on the Ti and SUS304 substrates. Firstly, we focus on the deposition behaviour of Al particle on Al substrate. From the result of XPS analysis, a lot of energy was required to remove the natural oxide on Al. This means that nascent surfaces of Al particle and substrate, which contributes to a formation of metallic bonding, are generated exclusively by severe plastic deformation. In addition, the rebound energy on of Al particle impinged on Al substrate was relatively large. However, because not only the Al particle, but also the Al substrate are deformed owing to their comparable hardness, the Al particle could be deposited at high particle velocity. Secondly, in the cases of Ni and Cu substrates, the Al particles impinged on these substrates were severely deformed, in contrast, these substrates were hardly deformed due to the higher hardness of substrate materials compared with Al particle. However, the natural oxides on both Ni substrate and Cu substrate were amazingly easy to remove. It is supposed that the nascent surfaces on these substrates were formed enough not by plastic deformation, but by relative slip between particle and substrate. In addition, the rebound energies on Ni and Cu substrates were lower than that on Al substrates. Consequently, the Al particle could be easily deposited on the Ni and Cu substrates at low particle velocity. Thirdly, in the case of Ti substrate, it is thought that the nascent surface on Ti substrate was hardly formed due to the tough natural oxide and the lack of plastic deformation. Therefore, the Al particle could not be deposited on the Ti substrate. Finally, in the case of SUS304, the Al particle was hardly deposited. However, the results of XPS and FEM analysis indicate that there is a possibility of deposition of Al particle. In fact, the deposition of only one Al particle was observed, although it was separated by ultrasound (US) cleaning. This means that the balance between the bonding energy and rebound energy was exceedingly sensitive. And also, it is supported that the deposition behaviour was significantly affected by the slightly larger removal energy of natural oxide on SUS304 compared with Ni and Cu, or by other factors such as adhesion energy.

5 CONCLUSION

The influence of substrate materials on deposition behaviour of a spherical pure Al particle with 1 mm diameter was investigated using the CS emulated single particle shot system. From the results, it was found that the conditions of natural oxides and deformability of each substrate material play important roles on the deposition behaviour of CS emulated pure Al particle.

REFERENCES

(1) A.P. Alkhimov, A.N. Papyrin, V.F. Kosarev, N.I. Nesterovich, M.M. Shushpanov (1994) *U.S. Patent*, No. 5,302,414.

(2) H. Assadi, F. Gärtner, T. Stoltenhoff, H. Kreye (2003) *Acta Materialia*, 51, 4379-4394. doi:10.1016/S1359-6454(03)00274-X

(3) K. Ogawa, K. Ito, K. Ichimura, Y. Ichikawa, S. Ohno, N. Onda (2008) *Journal of Thermal Spray Technology*, 17(5-6), 728-735. doi:10.1007/s11666-008-9254-5

(4) S.A.A. Akbari Mousavi, Sh. Ranjbar Bahadori, A.R. Shahab (2010) *Materials Science and Engineering A*, 527, 3967–3974. doi:10.1016/j.msea.2010.02.077

(5) G. Bae, Y. Xiong, S. Kumar, K. Kang, C. Lee (2008) *Acta Materialia*, 56, 4858-4868. doi:10.1016/j.actamat.2008.06.003

(6) S. Yina, X. Wanga, W.Y. Li, H. Jie (2011) *Applied Surface Science*, 257, 7560–7565. doi:10.1016/j.apsusc.2011.03.126

(7) C.Y. Ni, Y.C. Li, F.X. Xin, F. Jin, T.J. Lu (2013) *Composites: Part A*, 46, 69–79. doi:10.1016/j.compositesa.2012.07.019

Stress-induced anisotropic diffusion of component elements in stacked thin-film multi-layer structures

K. Suzuki, M. Ochi, H. Miura
Fracture and Reliability Research Institute, Tohoku University, Japan

ABSTRACT

The mechanism of the directional coarsening (rafting) of γ' phases of Ni-base super-alloy under uniaxial strain at high temperatures was investigated by molecular dynamics (MD) analysis of stacked thin-film multi-layer structures. The strain-induced anisotropic diffusion of constituent atoms perpendicular to the interface was observed clearly in Ni(001)/Ni_3Al(001) and Ni(001)/Cu(001) interfaces. In contrast, no strain-induced anisotropic diffusion occurred in the interface structure consisted of the face-centered cubic (FCC) metals with (111) crystallographic orientation. The estimated strain-induced anisotropic diffusion was validated by experiment using the Ni/Cu stacked thin films structures.

1 INTRODUCTION

Directionally solidified or single crystal Ni-base superalloys have been widely applied to turbine blades or vanes in combustion power plants and jet engines. High temperature mechanical properties of the Ni-base superalloy are strengthened by the fine cuboidal γ' (Ni_3Al) precipitates orderly-dispersed in the γ matrix (Ni-rich matrix) because the dispersed texture in a grain inhibits dislocation motion. However, directional coarsening of the γ' phase perpendicular to a principal stress, which is called "rafting," occurs when an uniaxial external stress is applied to the superalloy at high temperatures. The γ' precipitates start to grow perpendicularly to the direction of the externally applied load, and the initial finely-dispersed cuboidal texture changes to large layered texture. Once the layered texture appears, the strength of the alloy at high temperatures decreases and cracks starts to propagate along the layered interface between newly grown γ' and γ phases. Similar micro texture change was observed in this alloy near crack tips after fatigue loading (1). This means that the rafting occurs not only during creep loading but also during cyclic loading. Therefore, it is very important to understand mechanisms of the rafting for improving both creep and fatigue resistance of the Ni-base superalloy at high temperatures. Although not a few research activities have been continued actively (2, 3), the mechanism of the micro texture change has not been clarified fully yet.

Since the rafting tends to occur when the Ni-base superalloy are subjected to a small tensile stress at elevated temperatures, the rafting process is basically considered as high-temperature diffusion-controlled process. Atomic diffusion in solids is expressed by the chemical potential gradient. The combination of the externally applied stress and the lattice mismatch between γ' and γ phases produces the anisotropic strain field in the Ni-base superalloy. Since strain can alter chemical potentials of constituent atoms, atoms and vacancies in the Ni-base superalloy anisotropically diffuse depending on the local strain state. As a result, distribution of alloying elements is gradually changed with the development of the rafting. We have considered, therefore, that the strain-induced anisotropic diffusion is one of the most important factors that dominate the rafting phenomenon in the Ni-base superalloy. In this study, in order to make clear the mechanism of the rafting, molecular dynamics (MD) method was applied to investigate the strain dependence of diffusion properties of constituent atoms around the Ni/Ni_3Al coherent interface. In particular, the anisotropic diffusion of Al atoms around the interface, which is presumed to be dominant process for the rafting evolution, under uniaxial strain was investigated. In addition to the atomic scale analyses, the estimated strain-induced anisotropic diffusion was validated by experiments using the stacked thin film structure. The change of the diffusion profile before and after the annealing of the stacked structure under tensile strain applied parallel to the stacked interface was measured by x-ray photoelectron spectroscopy (XPS).

2 MOLECULAR DYNAMICS ANALYSIS OF STRAIN-INDUCED ANISOTROPIC DIFFUSION IN STACKED THIN-FILM MULTI-LAYER STRUCTURES

2.1 Analytical method and model

In this study, simple crystallographic matched interface structures between single-crystalline Ni as γ phase and Ni_3Al as γ' phase were used for the analysis of diffusion properties around the γ/γ' interface under tensile strain. Figure 1 shows the supercell models of the Ni/Ni_3Al interface. Two interface structures of Ni/Ni_3Al having different interfacial crystallographic planes, (001) and (111) respectively, were modelled. The coherent (001) interface is mainly observed in the actual γ/γ' interface. To examine the effects of alloying (dopant) elements on diffusion properties in the Ni/Ni_3Al interface, some Ni atoms in the Ni layer were replaced by dopant atoms randomly as shown in Fig. 1(c). In addition to the Ni/Ni_3Al interface, a bimetal structure model of Ni(001)/Cu(001) was also used to validate that the strain-induced anisotropic diffusion occurs in stacked thin-film multi-layer structures consisting of face-centered cubic (FCC) metals. The MD simulations were carried out using the Large-scale Atomic/Molecular Massively Parallel Simulator (LAMMPS) code (4). The generalized embedded atom method (GEAM) interatomic potential (5) was used and temperature was fixed at 1573 K. In order to analyze the diffusion behavior of constituent atoms under uniaxial tensile strain, tensile strain of 2% was applied parallel to the interface and MD simulations were carried out for 500 ps under the fixed volume condition (no fluctuation of lattice constants during the simulation). The uniaxial tensile strain parallel to the interface was applied by increasing the length of the lattice parameters along the x-axis from the equilibrium value. The diffusion constants were evaluated from the slope of the mean square displacement (MSD).

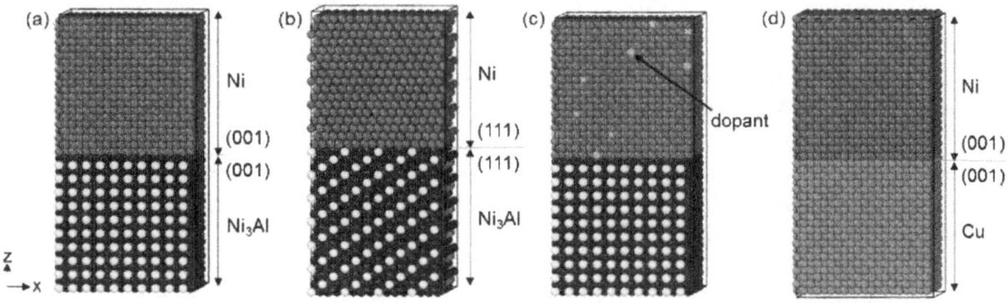

Fig. 1 MD simulation models of the interface structure; (a) Ni(001)/Ni_3Al(001), (b) Ni(111)/Ni_3Al(111), (c) doped-Ni(001)/Ni_3Al(001) and (d) Ni(001)/Cu(001) interfaces.

2.2 Analytical results and discussion

Figure 2 shows the MSD values of Al atoms in the Ni(001)/Ni_3Al(001) interface under 0% strain (no strain) and 2% tensile strain along x-direction. The MSD value of Al atoms perpendicular to the interface (z-direction) was clearly increased when the tensile strain of 2% was applied parallel to the interface. The diffusion coefficient of Al atoms perpendicular to the interface was increased by about 20 times. This anisotropic diffusion of Al atoms in Ni(001)/Ni_3Al(001) interface under tensile strain corresponds well to the fact that the finely-dispersed γ' phase starts to grow and form thin layered structures perpendicular to the direction of the applied uniaxial stress. The strain dependence of the atomic diffusion in the Ni(111)/Ni_3Al(111) structure was also analyzed. Figure 3 shows the atomic configuration of the Ni(001)/Ni_3Al(001) and Ni(111)/Ni_3Al(111) interface structures under 2 % strain. In the Ni(001)/Ni_3Al(001) interface, both Ni and Al atoms around the interface diffused significantly, and the mixing of Al and Ni atoms around the interface occurred. On the other hand, the slip deformation occurred in the Ni(111)/Ni_3Al(111) interface under strain, as shown in Fig. 3(b). Hence, since the strain applied parallel to the interface of Ni(111)/Ni_3Al(111) caused the slip deformation, no stress-induced anisotropic diffusion of both Al and Ni occurred around the interface. When strain was applied to the crystal structure, the structure deforms to relieve the applied stress or the high strain energy. However, the slip generation was strictly prohibited in the Ni(100)/Ni_3Al(100) interface structure because the resolved shear stress within the slip system was not large enough to activate the slip generation. Thus, the anisotropic diffusion perpendicular to the interface occurred to relieve the elastic strain energy.

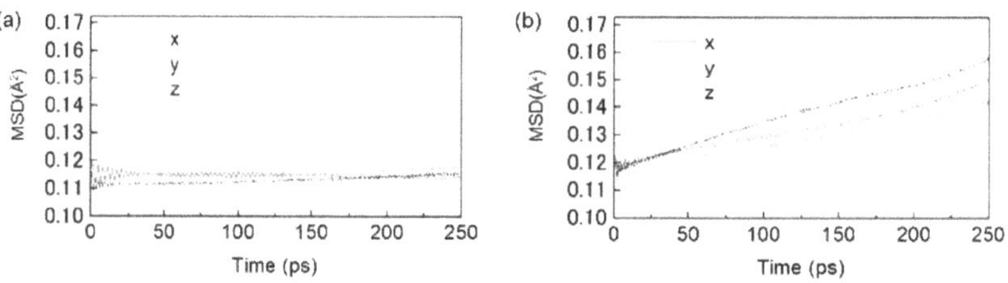

Fig. 2 MSD of Al atoms in Ni(001)/Ni_3Al(001) interface with (a) no strain and (b) 2% tensile strain parallel to the interface.

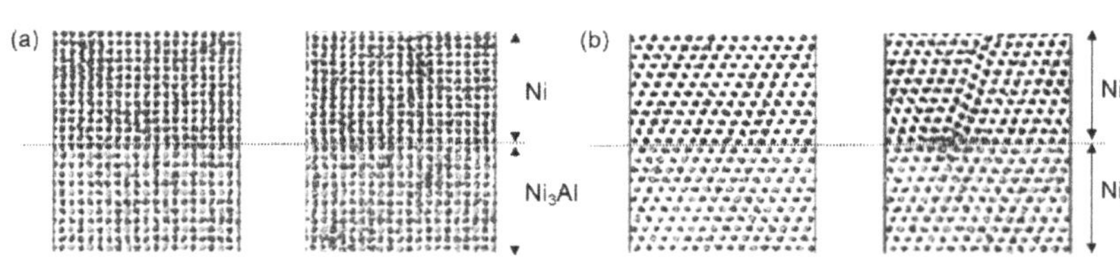

Fig. 3 Snapshots of Ni/Ni3Al interface structure under 2% strain parallel to the interface; (a) Ni(001)/Ni_3Al (001) and (b) Ni(111)/Ni_3Al(111) interfaces.

Figure 4 summarizes the effect of the dopant element on the diffusion coefficient of Al atoms perpendicular to the interface under 2% strain. The broken line in this figure shows the diffusion constant of Al atoms in no dopant elements. The diffusion constant of Al atoms was changed drastically by the addition of the dopant elements. Since the anisotropic diffusion of Al atoms plays a key role in the rafting phenomenon of γ' phase, the reduction of the diffusion of Al atoms perpendicular to the γ/γ' interface should decrease the evolution of the rafting. From this analysis, Pd is one of the most effective elements that restrain Al atoms from the diffusion around the interface.

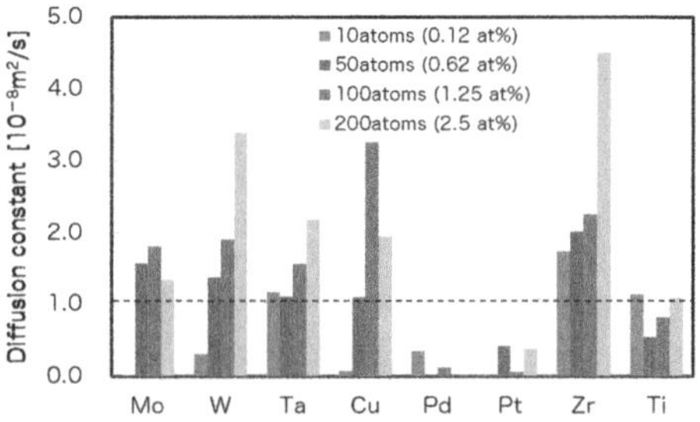

Fig. 4 Effect of dopant elements on the strain-induced anisotropic diffusion of Al atoms.

3 EXPERIMENTAL VALIDATION OF STRAIN-INDUCED ANISOTROPIC DIFFUSION AROUND FCC(001)/FCC(001) INTERFACE

In order to discuss that the strain-induced anisotropic diffusion occurs only in the Ni and Ni_3Al system or not, the diffusion behavior in the similar face-centered cubic structure consisting of Ni(001)/Cu(001) interface was analyzed. Figure 5 shows the MSD values of Ni atoms in the interface under 0, 1 and 2% tensile strain parallel to the interface at 1200 K. The MSD values perpendicular to the interface (z-direction) were clearly increased when the tensile strain was applied parallel to the interface. Thus, the strain-induced anisotropic diffusion would occur when the interface consisting of FCC(001)/FCC(001) and tensile strain is applied parallel to the interface.

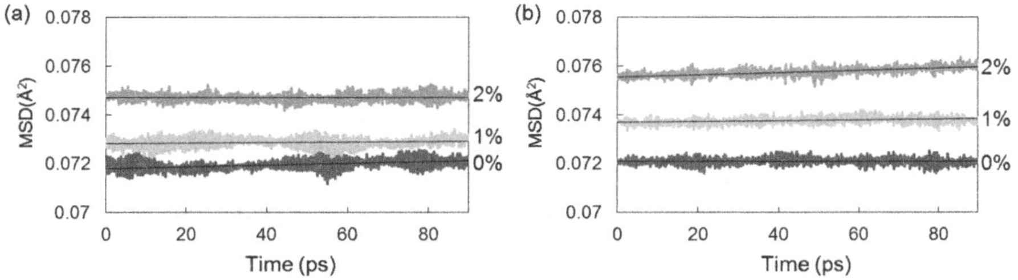

Fig. 5 MSD of Ni atoms in Ni(001)/Cu(001) interface under 0, 1 and 2 % strain parallel to the interface; (a) MSD in x-direction and (b) in z-direction.

In order to validate the estimated strain-induced anisotropic diffusion phenomenon, thin film stacked structures which consisted of FCC metals interfaces were made by using an electron beam deposition method. A heavily rolled copper foil with strongly (001) orientation was used for the substrate. The Ni layer about 500 nm was deposited on the substrate to make Ni(001)/Cu(001) and Ni(111)/Cu(001) thin-film stacked structures. These stacked structures were annealed 200°C for 192 hours under tensile strain applied parallel to the Ni/Cu interface. A four-point bending method was used for the loading. After the annealing, the change of the depth profile of the atomic composition of the stacked structure was measured by x-ray photoelectron spectroscopy. Figure 6(a) shows an example of the measured change of the diffusion depth of Ni and Cu atoms around the Ni(001)/Cu(001) interface after the four-point loading. When tensile strain of 0.2% was applied to the stacked Ni(001)/Cu(001) interface, the diffusion depth of Ni atoms into the Cu substrate increased significantly. This result clearly indicates that the diffusion of Ni atoms perpendicular to the interface was enhanced drastically by the application of tensile strain parallel to the interface.

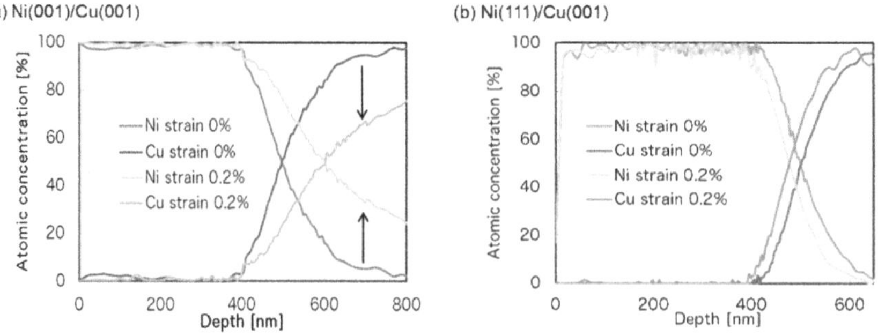

Fig. 6 Change of the depth profile of the atomic composition around the interface measured by XPS.

Figure 6(b) shows an example of measured depth profile of Ni and Cu atoms in the Ni(111)/Cu(100) interface. Comparing with the result of the Ni(001)/Cu(001) interface, the slopes of the depth profile of Ni and Cu did not change around the Ni(111)/Cu(001) interface even after the application of 0.2 % tensile strain. This result suggests that no strain-induced anisotropic diffusion around the Ni(111)/Cu(001) interface under the tensile strain applied parallel to the interface. From these results, the estimated strain-induced anisotropic diffusion was validated and it can be concluded that this strain-induced anisotropic diffusion occurs significantly only when the interface consists of FCC(001)/FCC (001) and tensile strain is applied parallel to the interface.

4 CONCLUSION

In this study, MD simulation was applied to explicate the mechanism of the directional coarsening (rafting) of γ' phases in Ni-base superalloys during creep loading. The strain-induced anisotropic diffusion of Al atoms in the Ni/Ni_3Al interface structures was found to occur under the tensile strain applied parallel to the interface. The diffusion constant of Al atoms was changed drastically by the dopant elements and their atomic concentration. This estimated result was confirmed by measuring the change of the depth profile of the atomic composition of the stacked FCC(001)/FCC(001) thin-film structure with and without tensile strain applied parallel to the interface. Since the strain-induced change of the micro texture causes abrupt fracture of the alloy, it is very important to develop the countermeasure which minimizes the strain-induced anisotropic diffusion of Al atoms in order to assure the long life reliability of the alloy in actual operation.

REFERENCES

[1] Miura, H., Akahoshi, K., and Suzuki, K. (2009) J. of NDT and E International, **42**, pp. 188-192.

[2] Ratel, N., Calderon, H. A., Mori, T., and Withers, P. J. (2010) PHILOSOPHICAL MAGAZINE, **90**, pp. 585-597.

[3] Zhou, N., Shen, C., Mills, M. J., and Wang, Y. (2009) ACTA MATERIALIA, **56**, pp. 6156-6173.

[4] LAMMPS Molecular Dynamics Simulator, http://lammps.sandia.gov

[5] Zhou, X. W., Wadley, H. N. G., Johnson, R. A., Larson, D. J., Tabat, N., Cerezo, A., Petford-Long, A. K., Smith, G. D. W., Clifton, P. H., Martens, R. L., and Kelly, T. F. (2001) ACTA MATERIALIA, **49**, pp. 4005-4015.

Thermal shock resistances of AlO-CrO/ NiCoCrAlYTa and AlO/NiCoCrAlYTa coatings deposited by atmosphere plasma spraying

C. Tao [1)2)], L. Wang [1)], S.D. Huang [2)], H.F. Hu [1)2)], Y. Liu [1)], X. Song [1)]
[1)] Key Laboratory for Anisotropy and Texture of Materials (Ministry of Education), Northeastern University, P.R. China
[2)] Baosteel Group Shanghai Meishan Co. Ltd, P.R. China

ABSTRACT

AlO-CrO/NiCoCrAlYTa and AlO/NiCoCrAlYTa coatings were deposited onto 316L stainless steel substrate by atmosphere plasma spraying (APS), in order to improve the thermal shock resistance. The influences of CrO content on thermal shock resistance of the coatings were investigated at 700°C, 900°C and 1100°C, respectively. The results indicate that AlO-CrO/NiCoCrAlYTa coating has better thermal shock resistance, attributing to its lower oxide rate and better thermal expansion matches, as well as better wettability than that of AlO/NiCoCrAlYTa coating.

1 INTRODUCTION

Powder material can be heated by atmosphere plasma spraying(APS) to a molten or semi-molten state, then deposited onto the substrate to form a dense uniform coating due to the power heat source itself and the additional high velocity gas flow,. For example, MCrAlY is one kind of widely used powder materials, and the MCrAlY-based coatings obtained by APS can be used to increase the resistance to high temperature oxidation and corrosion (1-4). More and more experts have paid close attentions to the mechanisms of high-temperature oxidation and corrosion of MCrAlY-based coatings. Compared with the MCrAlY coating, the AlO/NiCoCrAlTaY coating added with oxide ceramics has a better high temperature wear performance and resistance to high temperature oxidation (5-7). In this research, a AlO-CrO/NiCoCrAlYTa coating has been developed from the basis of traditional binary AlO/NiCoCrAlYTa coating, in order to improve the thermal shock resistance. And the influences of the CrO content on thermal shock resistance of the coatings were investigated at 700°C, 900°C and 1100°C, respectively.

2 EXPERIMENTAL PROCEDURES

2.1 Materials and APS thermal spraying process

Three kinds of powders were selected for the present research, named AlO, CrO and NiCoCrAlYTa, respectively. The size of powders is in a range of 15–45μm. Stainless steel 316L was used as the substrate. Before the coating, 316L substrate was degreased and grit blasted until its roughness was about 15μm. Then the coatings were obtained with above mentioned powders using an APS machine (Prax-atmosphere-3710, USA). Spraying processes were executed with a distance of 80 mm. The APS processes were performed at a voltage of 40V and a current of 500A. The powder feed rate was fixed at the rate of 5 rpm with the aid of a computerized powder feeder station. The coatings were deposited to a thickness of 180 ± 20μm after 6 passes.

Table 1 Samples of coating (wt. %)

Samples	AlO	CrO	NiCoCrAlYTa
M1	18	12	70
M2	30	0	70

2.2 Thermal shock test

Based on the aviation industry standard HB7269-96, the thermal shock resistance tests were carried out on the coatings. The size of thermal shock tests specimens was 20mm×20mm×20mm. The specimens were heated to 700°C, 900°C and 1100°C, respectively, keeping 15 minutes, then quenched into the water. The specimens were dried in a drying box, and then examined by a stereomicroscope. The process were repeated and thermal cycling tests didn't stopped until macroscopic cracks or more than 10% broken area had been found on the surface of the specimens, The number of thermal cycles was recorded.

2.3 Microstructural characterization

In order to identify the phase composition, X-ray diffraction (XRD) (Bruker D8-Advanced) was performed on as-received powders and as-sprayed coatings with Cu-K$_\alpha$ radiation. The Vickers micro-hardness of coatings was evaluated (HXD 1000TC). The porosity of coatings was measured using image analyzer under a magnification of 500. An optical microscope (OM) (OLYMPUS BX51M) was used to examine the microstructure of the coatings. In addition, the microstructure of the coating was examined by a scanning electron microscope (SEM) (Hitachi S-3400N) before and after the thermal cycling tests respectively.

3 RESULTS AND DISCUSSION

3.1 Phase composition

The XRD patterns of the feedstock powder and as-sprayed coating are shown in Fig 1. It is clear that $Cr_2O_{2.4}$, Ni_3Al and Al_2O_3 are detected in the feedstock powder. However, after APS, a new phase of chromiumoxide can be found. It is obvious that there is a distinct diffuse diffraction peak around 2θ≈45°, indicating the formation of an amorphous/nanocrystalline phase in the coating during the APS process due to higher flame temperature and fast cooling velocity. The similar phases have also been mentioned by earlier reports in other APS coatings (8-10).

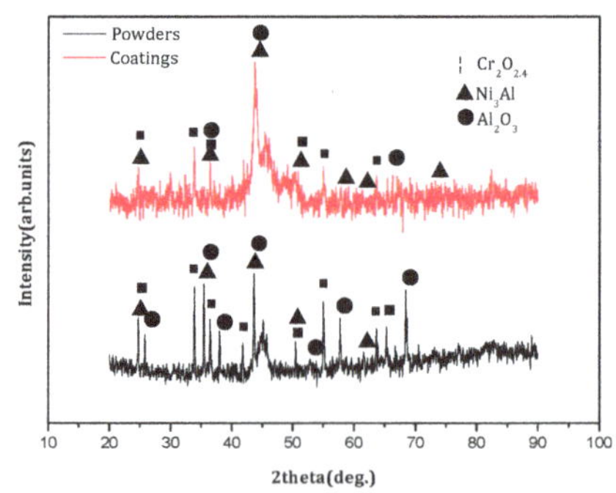

Fig 1 XRD patterns of the feedstock powder and as-sprayed coating

3.2 Characterization of the coatings

The microstructures of as-sprayed coatings on a transverse surface are shown in Fig 2. It reveals that the thickness of coatings is about 0.18±0.02mm which shows a fine and close structure and a good adhesion to the substrate. In Fig 2, it is found that the coatings present a lamella structure almost paralleling to the interface between coating and substrate. Besides, some melted or semi-melted particles and micro-cracks are observed, which is caused by high velocity spray and high melting point of the powders (11-12). In Table 2, the average porosity of the coatings is determined to be 15.06% and

17.88%, respectively. And the average micro-hardness is 413$HV_{0.1}$ and 381$HV_{0.1}$, respectively, which is approximately as 2 times as that of the substrate (200 $HV_{0.1}$).

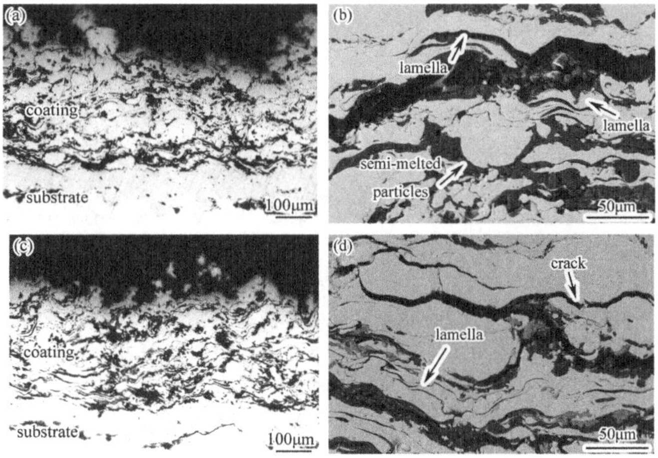

Fig 2 Images of transverse section of as-sprayed coatings (a), (b) OM and SEM of AlO-CrO/NiCoCrAlYTa coating; (c), (d) OM and SEM of AlO/NiCoCrAlYTa coating

Table 2 Powder compositions and characteristics of testing coatings

Coating	Feedstock composition	Substrate(HV0.1)	Coating hardness ($HV_{0.1}$)	Porosity (%)
M1	AlO-CrO/NiCoCrAlYTa	198	413	15.06
M2	AlO /NiCoCrAlYTa	201	381	17.88

3.3 Thermal shock characteristics

Fig 3 shows the thermal cyclic test results of as-sprayed coating. Both coatings have the largest number of cycles at 700°C. The numbers of thermal shock cycles to failure of M1 at 700°C, 900°C, 1100°C, are 153, 47, 3 times, respectively, while they are 79, 39, 2 times for M2, respectively.

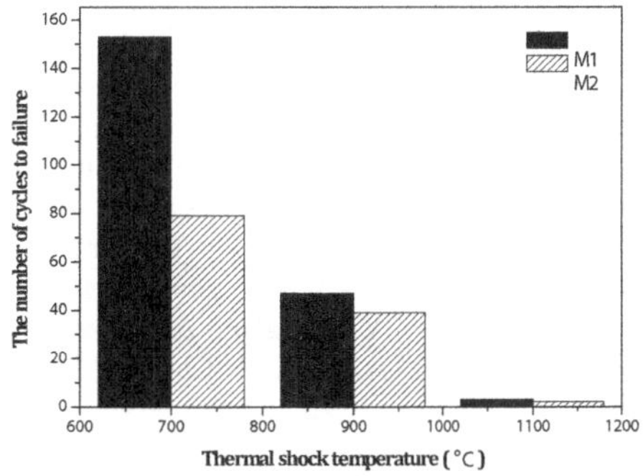

Fig 3 Thermal cyclic test results of as-sprayed coating

Fig 4 (a) and (d) show the damage morphologies of the as-sprayed coatings thermal shocked at 700°C. There are a lot of chipping and spalling with small pieces on the surface of coatings. With the increasing of thermal shocking cycles, the number of spalled pieces increases and the area of spallation is enlarged.

Fig 4 (b) and (e) show the damage morphologies of two coatings thermal shocked at 900°C. As a fact, the failure mode of the as-sprayed coatings is so much different from that at 700°C. Only after several thermal cycles, some cracks appeare on the surface of the coating. Then the delaminated area of coating becomes obviously larger and larger. Finally the coatings are interfacial delaminated along the micro-cracks.

At 1100°C, the critical cracks can be observed at first thermal shock cycle (Fig 4 c and f). Therefore, the coating is not capability enough for the thermal shock at 1100°C.

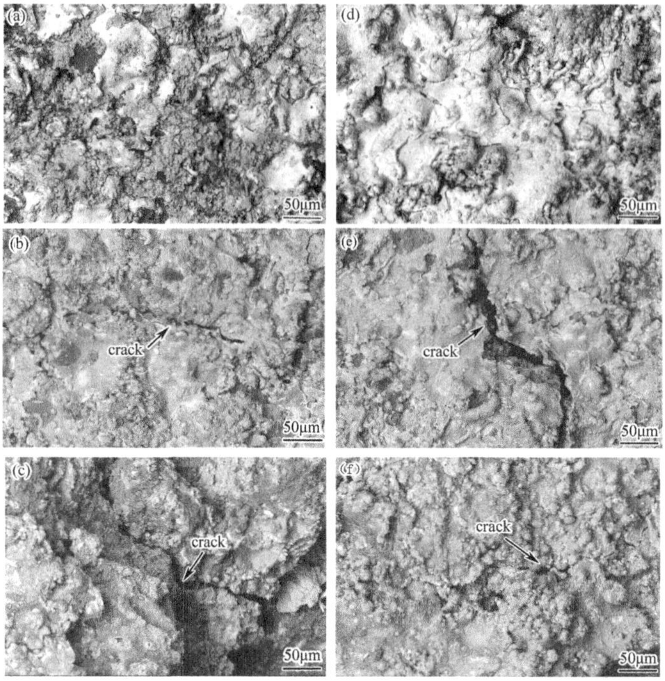

Fig 4 SEM image of M1 coating surface after thermal cycles at 700°C(a), 900°C(b), 1100°C(c) and M2 at 700°C(d), 900°C(e), 1100°C(f)

According to analysis of the microstructures and crack propagation modes, two type coatings have different thermal shock resistances during thermal cycles. Both coatings show so many pores in them, but the quantity and the morphology of these pores are obviously different from each other. Because of adding of CrO powders, the pores in AlO-CrO/NiCoCrAlYTa coating are sphericaler and more uniformly distribute than that in CrO/NiCoCrAlYTa coating. The uniformly distributed spherical moderate pores could relax tensile stress and restrain the crack propagation s, so that the thermal shock resistance can improved (11). The cracks always initials from the pores which are in strip shape (13). The lower porosity and better homogeneous morphology of pores in AlO-CrO/NiCoCrAlYTa coating playe more important roles on improving thermal shock resistance than that of CrO/NiCoCrAlYTa coating.

4 CONCLUSIONS

AlO-CrO/NiCoCrAlYTa and CrO/NiCoCrAlYTa coatings were deposited by APS. The microstructure, phase constitution, mechanical properties and thermal shock behavior of the coatings were studied. The following conclusions can be drawn:

(1) The amorphous / nanocrystalline phases are found in the coatings due to higher flame temperature and fast cooling velocity of APS process.

(2) Both mechanical properties and porosity of AlO-CrO/NiCoCrAlYTa coating can be improved compared with that of CrO/NiCoCrAlYTa coating.

(3) The AlO-CrO/NiCoCrAlYTa coating shows better thermal shock resistance than that of the CrO/NiCoCrAlYTa coating. It can be attributed to lower porosity and finer pores with more uniform distributions in the AlO-CrO/NiCoCrAlYTa coating.

REFERENCES

(1) W. P. Xu, B. S. Xu, W. Zhang, Y. X. Wu, Erosion-corrosion behaviors of high velocity arc sprayed Fe–Al/Cr_3C_2 coating, Journal of Wuhan University of Technology-Mater. Sci. Ed., 21 (2006) 29-31.

(2) M. Uusitalo, P. Vuoristo, T. Mäntylä, Elevated temperature erosion–corrosion of thermal sprayed coatings in chlorine containing environments, Wear, 252 (2002) 586-594.

(3) H. Koivuluoto, J. Lagerbom, P. Vuoristo, Microstructural studies of cold sprayed copper, nickel, and nickel-30% copper coatings, Journal of Thermal Spray Technology, 16 (2007) 488-497.

(4) S. Chinnadurai, S. Bahadur, High-temperature erosion of Haynes and Waspaloy: effect of temperature and erosion mechanisms, Wear, 186 (1995) 299-305.

(5) H. Wang, D. Zuo, J. Yan, M. Huang, X. Li, Effects of nanometer Al_2O_3 particles on oxidation behaviors of laser cladding low Al NiCoCrAlY coatings, Oxidation of metals, 74 (2010) 49-60.

(6) H. Li, Q. Wang, S. Jiang, J. Ma, J. Gong, C. Sun, Oxidation and interfacial fracture behaviour of NiCrAlY/Al_2O_3 coatings on an orthorhombic-Ti_2AlNb alloy, Corrosion Science, 53 (2011) 1097-1106.

(7) K. Khor, Y. Gu, Z. Dong, Plasma spraying of functionally graded yttria stabilized zirconia/NiCoCrAlY coating system using composite powders, Journal of thermal spray technology, 9 (2000) 245-249.

(8) K. Yang, J. W. Feng, X. M. Zhou, S. Y. Tao, Microstructural Characterization and Strengthening-Toughening Mechanism of Plasma-Sprayed Al_2O_3-Cr_2O_3 Composite Coatings, Journal of Thermal Spray Technology, 21 (2012) 1011-1024.

(9) V. P. Singh, A. Sil, R. Jayaganthan, Wear of Plasma Sprayed Conventional and Nanostructured Al_2O_3 and Cr_2O_3, Based Coatings, Transactions of the Indian Institute of Metals, 65 (2012) 1-12.

(10) Y. Murakami, T. Hirata, Y. Tsuru, Metastable phase relationship in the Al_2O_3-Cr_2O_3 system and phase equilibria of 1 : 1 composition at 1200 degrees C, J. Ceram. Soc. Jpn., 110 (2002) 541-543.

(11) B. Liang, C. Ding, Thermal shock resistances of nanostructured and conventional zirconia coatings deposited by atmospheric plasma spraying, Surface and Coatings Technology, 197 (2005) 185-192.

(12) Y. Wang, W. Tian, Y. Yang, Thermal shock behavior of nanostructured and conventional Al_2O_3/13 wt% TiO_2 coatings fabricated by plasma spraying, Surface and Coatings Technology, 201 (2007) 7746-7754.

(13) W. Tian, Y. Wang, Y. Yang, Three body abrasive wear characteristics of plasma sprayed conventional and nanostructured Al_2O_3-13% TiO_2 coatings, Tribology International, 43 (2010) 876-881.

Vibration control of sandwich structure by integration of shear thickening fluid (STF)

M.T. Hasib, L. Ye, L. Chang
Center for Advanced Materials Technology (CAMT), School of Aeronautics, Mechanical & Mechatronic Engineering, The University of Sydney, Australia

ABSTRACT

The experimental findings are focused on assessing the mechanical property of shear thickening fluid (STF) afterward its transition to solid phase. On account of determine the shear modulus of STF, bending test was performed on simply supported sandwich beam fabricated with double carbon epoxy beams with STF layer. The stiffness of sandwich structure with STF was evaluated at different midpoint displacements varying from 3mm to 8mm with fixed speed 50 mm/s and also at different speed of crosshead 10 mm/s to 50 mm/s with constant displacement of 8 mm. Further investigation of STF is carried on accompanying with two perspex cantilever beams for the sole purpose of gaining the dynamic stiffness and damping capacity under force vibration state.

1. INTRODUCTION

A group of materials that have recently gained a lot of attention in research world are shear thickening fluids (STF). A shear thickening fluid (STF) is a material whose viscosity increases significantly when the shear strain rate is above a critical value [1]. The integration of shear thickening fluid into composite structures has received considerable amount of interest in recent past years in order to investigate stiffness and damping capacity [2]. Most shear thickening fluids are formed by using a colloidal suspension of solid particles in a liquid matrix. This allows solidification of the fluid by congregation of the particles under stress [3].

There have been numbers of research and development already carried out last decade on the integration of shear thickening fluid into composite structure [2, 4-6]. Jolly et al [7] tested shear thickening fluid which provided variable damping. When subjected to a predetermined shear rate, the shear thickening composition would undergo a dramatic and substantial increase in viscosity and shear stress. Seshimo and Kiyoshi [8] invented a viscoelastic damper filled with STF fluid in order to control the vibration of a structural member, a tank, a pipe etc. Fischer et al [9] demonstrated that STF integrated structure under dynamic loading condition performed better damping and stiffness properties. On the other hand Zhang et al [4] showed that integrating an STF into a damper can lead simultaneously to changes in stiffness and damping under dynamic loading as the frequency is varied.

In this study, focus is highlighted to investigate the feasibility of integrating STF into sandwich cantilever structure with the aim of evaluate the damping capacity and stiffness. Vibration beam testing is being proposed to conduct in order to find the response of the sandwich cantilever with STF incorporating between the beams interfaces. The research is also focused on determine the shear modulus of the STF (styrene/acrylate particles in ethylene glycol) integrating with two carbon fiber-epoxy (CF/EP). STF is placed between two (CF/EP) beams interface and three point bending test is executed. Tests are conducted several times with varies deflections as 3mm, 4mm, 5mm, 6mm and also different load speed rates as (20-30-40-50) mm/sec.

2. DETERMINING SHEAR THICKENING FLUID (STF) PROPERTY

2.1 Material and experimental method

The STF used in this experiment is supplied by BASF AG (Germany). The STF is prepared with 58 vol. % dispersion of styrene/acrylate particles in ethylene glycol. Two carbon fiber epoxy beams are used with size dimension 100 mm long, 20.84 mm width and 1.82 mm thickness. The STF is placed between the gaps of 0.01 mm of the two beam interfaces. Before start the sandwich beam experiment, firstly conducting the bending test to measure only the elastic modulus of single carbon epoxy fiber rectangular beam, which is 27.47 GPa. ASTM D 790 standard is followed to perform the bending test. The 0.01 mm thickness of STF (as shown in figure1) and later oil is placed between carbon epoxy fiber beams and performed bending test.

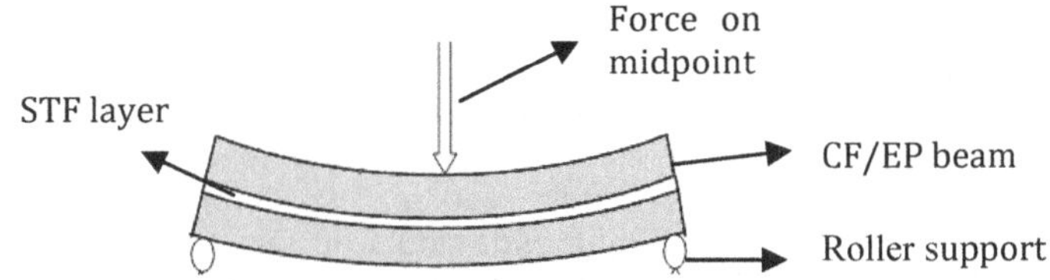

Figure 1: Schematic diagram of three point bending test of STF sandwich structure

Experiment is performed under two setup conditions. In order to calculate the modulus of the STF which acted as the adhesive interface between the two beams, the laminate beam theory is evaluated. The deflection of the sandwich beam with STF layer under simply supported condition is calculated by following equation

$$\delta = \frac{WL^3}{48D} + \frac{WL}{4AG_c} \quad (1)$$

Where, W= load, D = equivalent rigidity of structure, L= span length, G_c = Shear modulus of Core/STF material.

2.2 Determine shear modulus of STF

Results obtained from the tests are categorised two cases. Case one is for the deferent deflections with fixed speed rate 50mm/sec and case two is the fixed displacement 8 mm with various speed rates. Data Acquisition rate is setup at 500Hz for each case. The stiffness of sandwich structure with STF displays higher value than oil structure. The total deflection is the sum of bending and shearing stress which expressed in equation 1. The value of the load is obtained from machine and deflection δ was also measured during experiment. The shear modulus of STF is obtained 0.167 MPa. Figure 2(a) shows the almost constant shear modulus values with 0.0019 standard deviation of STF in respect of various midplane displacements. For figure 2(a) the speed rate was fixed at 50 mm/s. However from figure 2(b), the slowly increasing of G values with respect of different speed rates of crosshead are observed. As progressively higher speed rare, the transition of STF status from liquid to solid occurred swiftly which displays the higher shear modulus in figure 2(b).

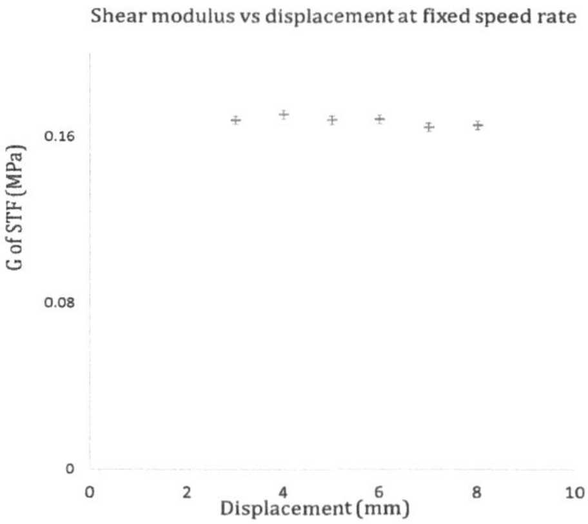

Figure 2(a): Shear modulus of STF vs. displacement

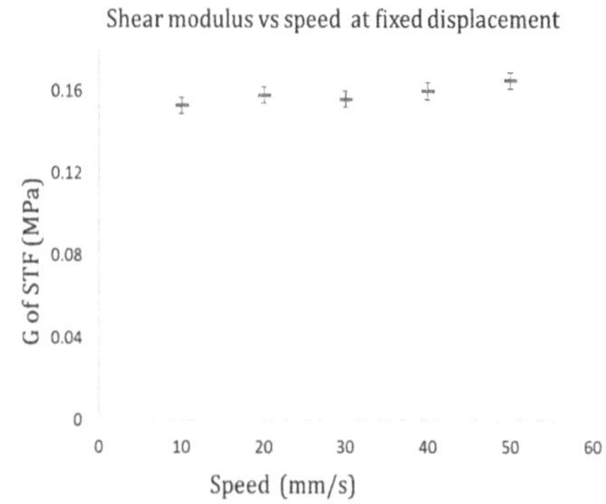

Figure 2(b): G vs. crosshead speed (20-30-40-50 mm/s)

3. VIBRATION REDUCTION WITH SHEAR THICKENING FLUID (STF)

3.1 Materials and method

The same STF is applied for this test but the gap of the beam interface was 400 μm. Mono ethylene glycol is used for carrier fluid in order to compare the rheological response of STF corresponding oscillatory loading. Two Perspex beams were chosen with dimension of 350mm long, 30mm wide and 2.98 mm thickness. The literature values assumed for E is 3100 MPa and density ρ is 1200 kg/m^3. To hold STF and carrier fluid between the interfaces of the beams, the silicon glue was applied to seal the gap. The one end of the sandwich beam was fixed by clamp holder. The sandwich structure is experienced dynamic sinusoidal force which showing flexural deformation with a point to point displacement at specific chosen point as shown in figure 3.

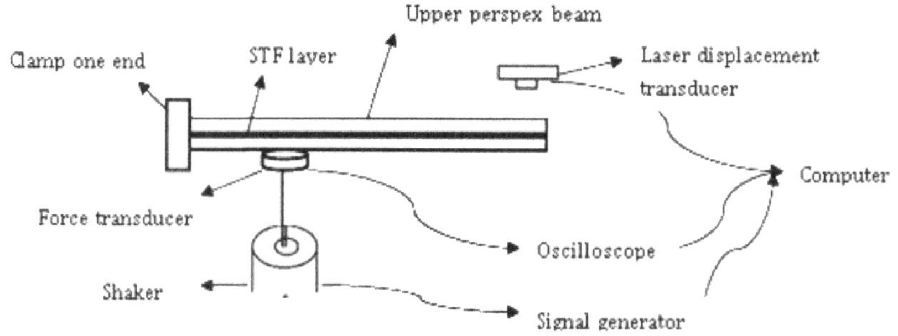

Figure 3: Schematic diagram of experiment setup of force vibration for STF with sandwich cantilever beam

Frequency response function (FRF) is applied to isolate the inherent dynamic properties of mechanical structure. The input is measured by forced transducer and output is obtained from displacement transducer. Dynamic stiffness is calculated by the compliance where compliance is the ratio of displacement to force. The force vibration of cantilever structure was conducted according to the guide line of ASTM E756. The displacement of the sample tip was controlled using laser displacement transducer and the amplitude of the excitation continuously adjusted so as to maintain the displacement amplitude constant. For both STF and carrier fluid the tip displacement recorded for 1-1.5-2 mm. The Bernoulli–Euler equation leads to an analytical solution for the resonant frequency f_n associated with the nth mode of a cantilever beam

$$f_n = \frac{\Omega_n^2}{2\pi}\left(\frac{EI}{\rho AL^4}\right)^{1/2} \quad (2)$$

Where Ω_n is constant characteristic of the vibration mode, ρ is the beam density and A is the cross sectional area of the beam.

3.2 Result and discussion

In figure 4 shows the clear idea of the influence of the STF on the dynamic properties of the beam. At 1.5 mm tip displacement, the flexural compliance increases for carrier fluid, where STF shows higher stiffness with lower value of flexural compliance.

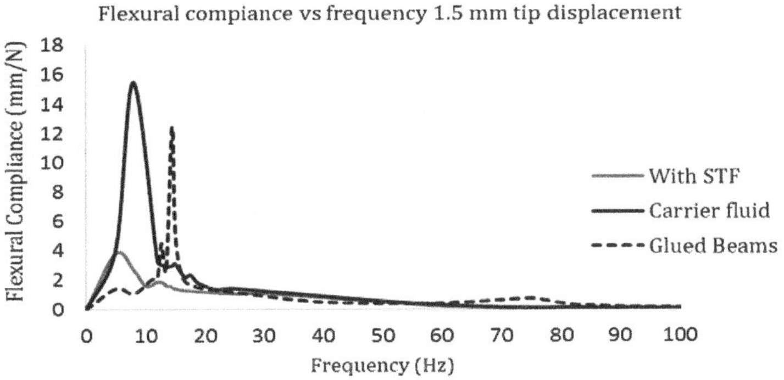

Figure 4: Frequency sweeps for sandwich beam with STF, carrier fluid and glued beam with resect 1.5 mm tip displacement

STF with beam structure, the first resonance occurs between 5Hz to 10Hz where carrier fluid it appears between 17Hz to 20Hz. For case of second mode of resonance the STF shows much higher dynamic stiffness than carrier fluid and glued beams structure. Results also show greater increase of the structures dynamic stiffness for 1.5mm and 2mm tip displacement.

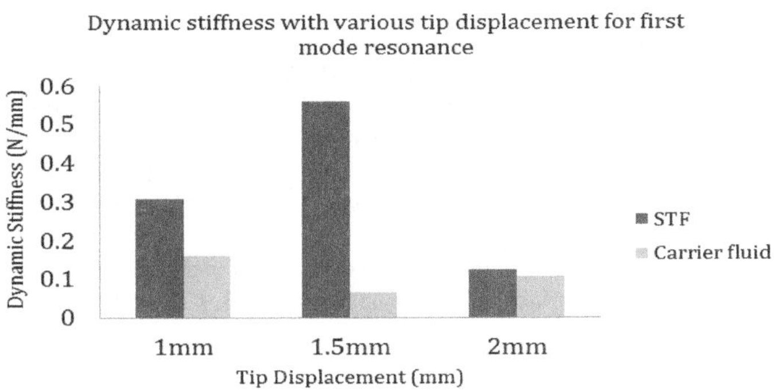

Figure 5: Dynamic stiffness with tip displacement 1-1.5-2 mm at first mode resonance

From the figure 5 for the first mode resonance the values of the dynamic stiffness of STF structure are higher than the stiffness of carrier fluid structure at first mode. Compare to carrier fluid, STF (1mm displacement) proves better performance to reduce vibration as the structure damping capacity increases 66% against carrier fluid. In case of 1.5mm displacement the damping capacity increases 45% compared to carrier fluid structure.

4. CONCLUSION

In this study, the integration of sandwich beam structure with Shear thickening fluid (STF) was developed to find the property of STF after the transformation of the solid state under loading condition. Numerical calculation was derived from laminar beam theory to determine the shear modulus of STF which is 0.167 MPa (average value). As mentioned earlier two experimental conditions were taken under consideration, the shear modulus values obtained for STF were close in range with standard deviation of 0.0019 and 0.004 respectively.

Forced vibration on cantilever sandwich structure with STF was studied. With different tip displacements the structure with STF performed better against carrier fluid structure on account of damping capacity. Results showed good enhancement for STF structure on dynamic stiffness. This work has presented that integrating STF with strong rheological characteristic into sandwich structure can play vital role in stiffness and damping under dynamic flexural loading. In future study it is intended to investigate damping response of the STF sandwich structure with various thicknesses of STF layers and establish the correlation of shear strain at beam- beam interface with the critical strain of the STF determined from rheological measurements.

5. REFERENCES

[1] Subranian, S., *Non-Newtonian Flows.* 2003.

[2] Fischer, C., et al., *Structural damping of model sandwich structures using tailored shear thickening fluid compositions.* Smart Materials & Structures, 2010. **19**(3).

[3] Bossis, G., & Brady, J. F, *The rheology of Brownian suspensions.* The Journal of chemical physics, 1989. **91**(3): p. 1866-1874.

[4] Zhang, X.Z., W.H. Li, and X.L. Gong, *The rheology of shear thickening fluid (STF) and the dynamic performance of an STF-filled damper.* Smart Materials & Structures, 2008. **17**(3).

[5] M. Parlinska, J.A.B., V. Michaud, J.-E. Bidaux, J.A.Manson and R. Gotthardt *Vibration response of adaptive composites.* J.Phys, 2001. **IV**(11).

[6] Fischer, C., et al., *Shear thickening fluids as a tunable damping element: experimental results and modeling - art. no. 652513.* Active and Passive Smart Structures and Integrated Systems 2007, 2007. **6525**: p. 52513-52513.

[7] Jolly, M.R., Bender, Jonathan W, *Field responsive shear thickening fluid.* 2006.

[8] Seshimo, K., *Viscoelastic damper* 1998.

[9] Fischer, C., et al., *Dynamic properties of sandwich structures with integrated shear-thickening fluids.* Smart Materials & Structures, 2006. **15**(5): p. 1467-1475.

Simulation of material behaviour of engineered cementitious composites under uniaxial tension

T. Huang, Y.X. Zhang
School of Engineering and Information Technology,
The University of New South Wales, Australian Defence Force Academy, Australia

ABSTRACT

In this paper, a numerical approach with the cohesive zone model (CZM) implemented through extended finite element method (XFEM) is employed to model the overall tensile stress-strain curve of Engineered Cementitious Composites (ECC). A crack bridging relation-based hardening cohesive law is developed to better represent the characteristics of ECC materials such as tensile strain-hardening behaviour and ultra-high ductility. The overall tensile behaviour of a typical polyvinyl alcohol (PVA) fibre-based ECC is modelled using the proposed numerical approach. The agreement of the stress-strain curve obtained from the numerical modelling with that obtained from the experiment demonstrates the effectiveness of the proposed numerical approach.

1 INTRODUCTION

Engineered Cementitious Composites (ECC) are a special class of High Performance Fibre Reinforced Cementitious Composites (HPFRCC) which are distinct for the tensile strain-hardening behaviour and tensile ductility contrasting to the quasi-brittle nature of ordinary concrete and fibre reinforced concrete (FRC). ECCs have been developed to achieve superior tensile ductility using only moderate quantity of fibre reinforcement through material design on the basis of fracture principles and micromechanics (1, 2). Under tensile loading, an important characteristic of the ECC is the development of multiple stable tiny cracks bridged by fibres, before one of them finally leads to fracture.

It was observed experimentally that the tensile ductility of the ECC strongly correlated with the multiple cracking behaviour (3). In order to derive the overall tensile stress-strain curve of the ECC, it is necessary that the multiple cracking behaviour is properly interpreted. Some theoretical models have been developed to describe the tensile behaviour of the ECC (4-6). The dependence of the tensile behaviour of ECC on the crack bridging law, in relation to the development of the cracks, and on the heterogeneity in the material, in relation to the initiation of the cracks, were clarified in these models. However, very few numerical studies have been conducted to analyse the multiple cracking behaviour of the ECC and its effect on the strain-hardening response.

The consideration of the cohesive zone model (CZM) was pioneered for concrete by Hillerborg et al. under the name of fictitious crack model (7) to predict concrete fracture using the finite element method. Since then, the CZM has been widely used in finite element analysis for crack propagation problems. In most of the cases, the CZMs are based on intrinsic formulations which require a pre-defined crack path and penalty stiffness prior to the softening behaviour. This approach introduces an artificial softening into the structure. The development of the extended finite element (XFEM) in the last decade not only brings about new possibilities in the application of the extrinsic CZMs, but also makes the implementation of extrinsic CZMs independent of mesh configuration (8, 9).

It is essential that the evolution of the failure of the cohesive zone is governed by the traction-separation law (TSL). So far, different TSLs such as the linear, bilinear, trilinear and exponential softening curves have been developed to predict the fracture behaviour of concrete and FRCs. However, these softening cohesive laws insufficiently represent the crack bridging behaviour in ECCs.

In the present work, a micromechanics-based crack bridging law is adopted, instead of the softening cohesive laws, as the TSL. The CZM is implemented into elements in the form of strong discontinuity through the XFEM. After a brief review of implementation of the CZM into XFEM, the specific TSL is discussed aiming at clarifying the necessity of developing a new TSL. Comparison between the computational and the experimental tensile stress-strain curves validates the numerical approach.

2 XFEM FOR CZM

The key of the XFEM lies in the enrichment of the FE model by introducing additional degrees of freedom (DOFs) to the nodes of the elements intersected by the crack. In this manner, the discontinuity can be directly embedded in the numerical model. The displacement field is approximated as

$$\mathbf{U}=\sum_{i=1}^{N} N_i(x)\left\{\mathbf{a}_i+\left[H(x)-H(x_i)\right]\mathbf{b}_i\right\}=\left[\mathbf{N}^{\mathrm{std}}\quad \mathbf{N}^{\mathrm{enr}}\right]\left[\mathbf{a}\quad \mathbf{b}\right]^{\mathrm{T}}=\mathbf{N}\mathbf{u} \tag{1}$$

where $\mathbf{a}_i$ and $\mathbf{b}_i$ represent the nodal standard and enriched DOFs, respectively. The discontinuous function across the crack is given by $H(x)=1$ on or above the crack, and $H(x)=-1$ otherwise. The relative displacement across the crack can be obtained as

$$\Delta\mathbf{U}=2\left[\mathbf{0}\quad \mathbf{N}^{\mathrm{std}}\right]\left[\mathbf{a}\quad \mathbf{b}\right]^{\mathrm{T}} \tag{2}$$

The normal and tangential separation of the crack is then determined by transforming the global relative displacement to the crack normal and tangential directions as

$$\boldsymbol{\delta}=\mathbf{R}\Delta\mathbf{U}=\mathbf{B}_{\mathrm{coh}}\mathbf{u} \tag{3}$$

where $\mathbf{R}$ is the coordinate transformation matrix, and $\mathbf{B}_{\mathrm{coh}}$ is separation-displacement matrix.

The finite element equation after implementing the CZM into XFEM is given by (8-10)

$$\left(\mathbf{K}_{\mathrm{bulk}}+\mathbf{K}_{\mathrm{coh}}\right)\Delta\mathbf{u}=\mathbf{f}_{\mathrm{ext}}-\mathbf{f}_{\mathrm{int}} \tag{4}$$

where $\mathbf{K}_{\mathrm{bulk}}=\int_{\Omega}\mathbf{B}^{\mathrm{T}}\mathbf{D}\mathbf{B}\,\mathrm{d}\Omega$, $\mathbf{K}_{\mathrm{coh}}=\int_{\Omega}\mathbf{B}_{\mathrm{coh}}{}^{\mathrm{T}}\mathbf{T}\mathbf{B}_{\mathrm{coh}}\,\mathrm{d}\Omega$, $\mathbf{f}_{\mathrm{ext}}=\int_{\Gamma_t}\mathbf{N}^{\mathrm{T}}\bar{\mathbf{t}}\,\mathrm{d}\Gamma_t$ and $\mathbf{f}_{\mathrm{int}}=\int_{\Omega}\mathbf{B}^{\mathrm{T}}\boldsymbol{\sigma}\,\mathrm{d}\Omega+\int_{\Gamma_c}\mathbf{B}_{\mathrm{coh}}{}^{\mathrm{T}}\mathbf{t}\,\mathrm{d}\Gamma_c$. $\mathbf{B}=\nabla^{s}\mathbf{N}$ is the strain-displacement matrix, and $\bar{\mathbf{t}}$ is the prescribed traction imposed at the boundary Γ_t. Here, $\mathbf{D}$ is the tangential stiffness of the bulk material, and $\mathbf{T}$ is the tangential cohesive stiffness due to the nonlinear behaviour in the cohesive zone. The traction acting on the discontinuity Γ_c is related to the crack separation which can be written in the incremental form as

$$\dot{\mathbf{t}} = \mathbf{T}\dot{\boldsymbol{\delta}} \tag{5}$$

3 CRACK BRIDGING LAW-BASED CZM

The fibre bridging behaviour of short fibre reinforced cementitious composites has been extensively investigated based on different interface properties and fibre failure modes, and relevant crack bridging laws have been established. A brief review of these crack bridging models can be found in (6). These crack bridging laws assume that the fibres are randomly and uniformly distributed in the matrix, and relate the statistic averaged normal traction to the normal crack separation, as well as micromechanical properties. Here, the normal crack bridging law is generically written as

$$t_{\mathrm{n}} = t_{\mathrm{n}}\left(p_1, p_2, \cdots, p_i, \cdots, \delta_{\mathrm{n}}\right) \tag{6}$$

where p_i denotes one of the micromechanical properties, such as the fibre geometries and mechanical properties, interfacial bond strength and interaction properties. For specific expressions, one may refer to the literatures (6).

It should be noted that the normal crack bridging law is derived in the context that the crack pre-exists. Consequently, the cohesive law in Eqn. (6) starts from zero separation as well as zero traction as shown in Fig. 1, which may not agree with real situation. By setting a threshold value equal to the matrix cracking strength f_{mc} for the normal cohesive law, the cohesive zone acts from the status where it was formed (see Fig. 1). In addition, it worth noting that the maximum fibre bridging stress f_{t} should be greater than the matrix cracking strength in ECC. In such manner, the material can further bear load even after the first cracking, and more cracks can thus be initiated. Conventional softening cohesive laws fail to embody this significant characteristic. Therefore, a hardening cohesive law is proposed here, which means the use of the CZM is not limited to the softening regime as in most of the previous applications. As the maximum fibre bridging stress and corresponding crack separation δ_{np} are most crucial, among the other characteristics of the crack bridging behaviour, the hardening portion of the normal cohesive law is assumed to be linear for simplicity. After these modifications, the normal cohesive law takes the form as given by Eqn. (7). A plot of Eqn. (7) can be found in Fig. 1.

$$t_{\mathrm{n}} = \begin{cases} f_{\mathrm{mc}} + \left(f_{\mathrm{t}} - f_{\mathrm{mc}}\right)\delta_{\mathrm{n}}/\delta_{\mathrm{np}} & \text{if } \delta_{\mathrm{n}} \le \delta_{\mathrm{np}} \\ t\left(p_1, p_2, \cdots, p_i, \cdots, \delta_{\mathrm{n}}\right) & \text{otherwise} \end{cases} \tag{7}$$

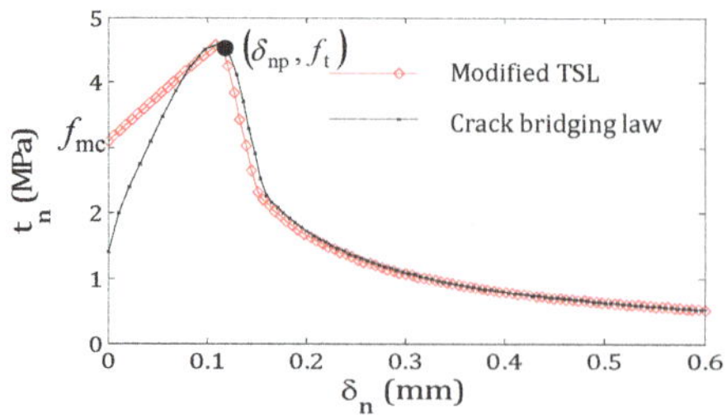

Figure 1. Crack bridging law-based normal TSL

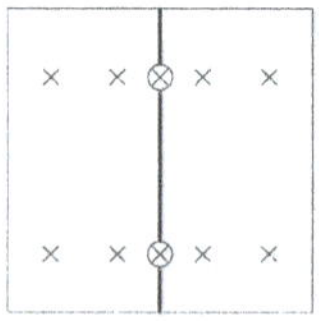

Figure 2. Integration scheme for an enriched 4-node quadrilateral element

The tangential traction is assumed to be provided by the pulled-out part of the fibre. Consider the fibre as a Timoshenko beam spanning over a crack of width δ_n, the shear stress due to crack tangential sliding δ_t can be obtained by solving Timoshenko beam problem (11). The tangential cohesive law is generically written as

$$t_t = t_t(p_1, p_2, \cdots, p_i, \cdots, \delta_n, \delta_t) \tag{8}$$

The tangential cohesive stiffness matrix can now be derived as

$$\mathbf{T} = \begin{bmatrix} \partial t_n / \partial \delta_n & 0 \\ \partial t_t / \partial \delta_n & \partial t_t / \partial \delta_t \end{bmatrix} \tag{9}$$

4 COMPUTATIONAL IMPLEMENTATION AND EXAMPLES

An extended finite element-based CZM with the crack bridging law-based cohesive law is implemented as a UEL subroutine in ABAQUS. Plane stress conditions are assumed and four-node quadrilateral element is used. Some assumptions are adopted in this work. During a calculation, a cohesive zone is introduced as a straight segment through the entire element, if the maximum principal stress at any of the integration points in the element exceeds f_{mc}. And each element can be intersected by no more than one crack. Furthermore, the cohesive zone is embedded within the element perpendicular to the loading direction and right on the center line, given that the crack direction under uniaxial tensile loading is fairly predictable and that the fact of the occurrence of the cohesive zone outweighs its location in this case. The material outside the cohesive zone is assumed to have a linear-elastic behaviour. Elements which are not crossed by a discontinuity are integrated by standard four-point Gauss quadrature. When an element is intersected by a discontinuity, the element is divided into two sub-quadrilaterals, each on either side of the discontinuity. Within each quadrilateral sub-domain, four-point Gauss quadrature is applied. In addition, two integrations points are positioned on the cohesive zone in order to compute the cohesive stiffness and traction (see Fig. 2).

A polyvinyl alcohol fibre-based ECC (PVA-ECC) with the PVA fibre having a 1.2% by weight oil coating is studied herein using the proposed numerical approach. A list of the micromechanical properties of the material can be found in literature (3). Rectangular coupon specimens measured 185 mm × 76 mm (L × w) were tested under uniaxial tensile loading with displacement control. After testing, the average crack spacing x_d is determined as around 3 mm. The specimen is treated as a series of representative volume elements (RVE) where each element potentially contains a cohesive zone (Fig.

3(a)). An arbitrary element is analysed in this work for initial validation. Due to symmetry, only half of the representative volume element is considered (Fig. 3 (b)). The deformed model shape is shown in Fig. 3 (c), and the computational overall tensile stress-strain curve is plotted along with the experimental data (3) in Fig. 4. As can be seen, the numerical result agrees very well with that from the experiment. This demonstrates the feasibility and reliability of the numerical approach.

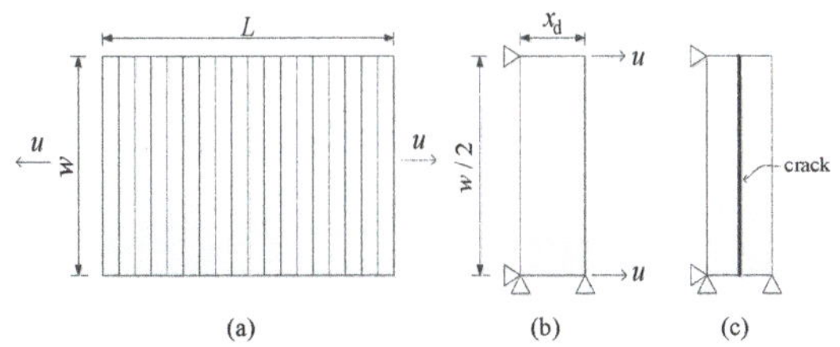

Figure 3. (a) Finite element model of the tensile specimen; (b) RVE model with boundary condition; (c) Cracked RVE

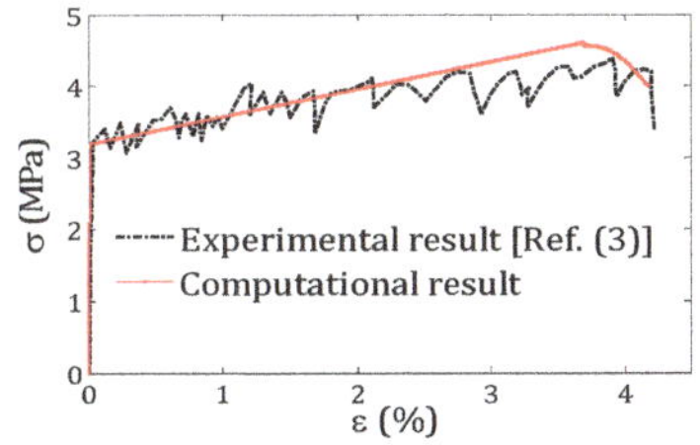

Figure 4. Comparison of computational and experimental overall tensile stress-strain curve

5 CONCLUSION

In the present work, an extrinsic CZM is implemented through the XFEM to predict the overall tensile stress-strain curve of ECC. The constitutive law of the CZM is defined as a hardening traction-separation relation, which is based on the micromechanical crack bridging law. Simulation of the tensile behaviour of a PVA-ECC demonstrates the effectiveness of the proposed numerical approach.

REFERENCES

[1] Li, V. C., Wang, S. & Wu, C. (2001) *ACI Materials Journal.* 98(6), 483-492.

[2] Li, V. C., Mishra, D. K. & Wu, H.-C. (1995) *Materials and Structures.* 28(10), 586-595.

[3] Li, V. C., Wu, C., Wang, S., Ogawa, A. & Saito, T. (2002) *ACI Materials Journal.* 99(5), 463-472.

[4] Kanda, T., Lin, Z. & Li, V. C. (2000) *Journal of Materials in Civil Engineering.* 12(2), 147-156.

[5] Kabele, P. (2007) *Engineering fracture mechanics.* 74(1), 194-209.

[6] Cai, X. R., Xu, S. L. & Fu, B. Q. (2011) *Engineering fracture mechanics.* 78(6), 1091-1100.

[7] Hillerborg, A., Modéer, M. & Petersson, P. E. (1976) *Cement and concrete research.* 6(6), 773-781.

[8] Wells, G. & Sluys, L. (2001) *International journal for numerical methods in engineering.* 50(12), 2667-2682.

[9] Moës, N. & Belytschko, T. (2002) *Engineering fracture mechanics.* 69(7), 813-833.

[10] Xu, Y. & Yuan, H. (2011) *Engineering fracture mechanics.* 78(3), 544-558.

[11] Kabele, P. (2002) *Engineering Mechanics (Association for Engineering Mechanics, Czech Republic).* 9(1/2), 75-90.

Evaluation of martensite transformation process in austenitic stainless steel due to transient resistivity measurement

H. Date
Dept. of Mechanical Engineering and Intelligent Systems,
Tohoku Gakuin University, Japan

ABSTRACT

The volume fractions of martensite induced in austenitic stainless steel during compressive deformation at temperatures of 77K, 187K and 293 K and the transformation process were examined using the resistivity measured with the four-point probe method. It was discovered that martensite transformation processes induced by compressive deformation are apparently different from those induced by tensile deformation. In addition, since the volume fraction of martensite was expressed as a linear function of the resistivity ratio, the resistivity was also available as an index for estimating the volume fraction of martensite by compressive test.

1 INTRODUCTION

The resistivity method was used to estimate the martensite volume of a compressive deforming specimen. However, it is not easy to estimate deformation-induced martensite in a deforming specimens using the resistivity measurement because resistivity is influenced by not only martensite formation but also temperature, impurities, defects and so on[1].

The tensile test was previously performed using a specimen made of stainless steel [2], and the following results were obtained:

(1) The negative resistivity verifying ε phase formation which is a precursor of martensite was observed on a strain- resistivity curve.
(2) The inherent characteristics of martensite decrease resistivity, but resistivity increases with the increase of martensite due to multiple defects generated by the formation of martensite.
(3) The resistivity ratio is available as an index for estimating the volume fraction of martensite of a deforming specimen.

The effects of the martensite formation on resistivity were examined from the aspects of strain, temperature and stability of the austenite. Additionally, a temperature changing test during deformation was also performed to clarify the effect of ε phase on the martensite transformation process. Finally, an approximation method for estimating the volume fraction of martensite was proposed using the direct relation between the resistivity and martensite contents.

2 MATERIALS AND EXPERIMENTAL METHODS

Two types of austenitic stainless steel termed 304ss and 316ss, which indicate the different stabilities of the austenite phase, were used in the experiment. A dumbbell-type specimen with the diameter and gauge length of 5 mm and 5 mm respectively was used in the compressive test [3]. The specimen was annealed at a temperature of 1323 K for 3.6ks. To measure the electrical resistance of the gauge length during deformation, four-point probes were installed on both sides of the gauge length. The specimens were loaded at a strain rate of 10^{-3} s^{-1}. The temperatures of 77K, 187K and 293K were accepted as the temperatures of the specimen. The volume fraction of martensite induced by deformation was measured by the electromagnetic induction method (Feritescope, MP30, Fisher) [4]. The resistivity ratio of the specimen was estimated using the following equation:

$$\Delta\rho/\rho_0 = (\rho - \rho_0)/\rho_0 = (R/R_0)\left(1/(1+\varepsilon)^2\right) - 1 \quad (1)$$

where ρ is electrical resistivity, R is electrical resistance, ε is strain and subscript *0* is the initial value.

3 RESULTS AND DISCUSSION

3.1 Estimation of Martensite by Feritescope

The volume fractions of martensite plotted against the plastic strain are shown in Fig. 1. The martensite volume induced by compression almost linearly increases from zero strain. Although the martensite in the tensile test increases according to the quadratic form from a specific strain depending on temperature because the strain region of tensile test is larger than that of compression, the linearity seen in Fig.1 is due to the small strain region in the compressive test.

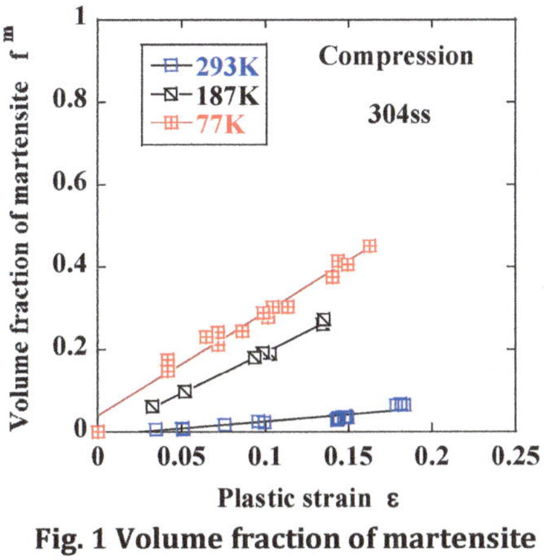

Fig. 1 Volume fraction of martensite

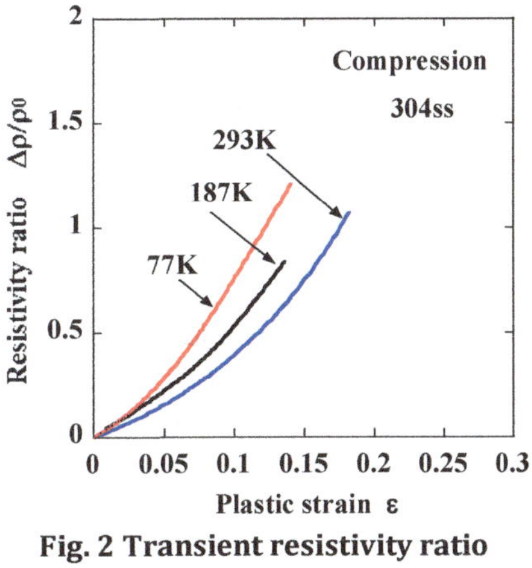

Fig. 2 Transient resistivity ratio

3.2 Effects of temperature on the resistivity curve

The resistivity curves of 304ss measured during the compressive test are shown in Fig. 2. The resistivity curves regularly increase according to the quadratic form, although the resistivity curves generated by the tensile deformation increase with the increase of strain after passing through a negative minimum value depending on temperature. The cause of the decrease in the resistivity during deformation is the absorption of vacancies by mobile dislocation [5]. On the other hand, it is reported that the ε phase, which is formed by a stacking fault and the precursor of the martensite, also decreases resistivity [6]. Accordingly, it is deduced from Fig. 4 that ε phase was not formed by compressive deformation because the resistivity does not indicate negative value. The appearance of positive resistivity is related to martensite formation, as described later.

Resistivity is decreased by martensite formation and increased by the growth of defects [5]. However, the resistivity increases regularly with martensite formation, as shown in Fig. 2. It should be noted that since the absolute value of the increment of resistivity due to the defects derived from the martensite formation is much larger than that of the decrement of resistivity due exclusively to martensite formation, the observed total resistivity increases with the increase of martensite.

3.3 Transformation process from austenite phase to martensite phase

In order to clarify the transformation process of compressive deformation [7], a compressive test on the 316ss type which does not induce martensite at a temperature of 293K was carried out under abrupt temperature change. The temperature change adopted in the experiment was from 293K to 77K at any strain. The resistivity curves and the volume fraction of the martensite plotted against plastic strain are shown in Fig. 3 (a) and (b), respectively. As shown in Fig. 3 (a), three resistivity curves indicate very similar processes except the drastic change corresponding to temperature change. Additionally, it is clear from Fig. 3 (b) that the pre-strain has no influence on the volume fraction of martensite and is dependent on only the post-strain. This suggests that pre-deformation at 293K hardly induces ε phase which is a precursor of martensite in the 316ss matrix.

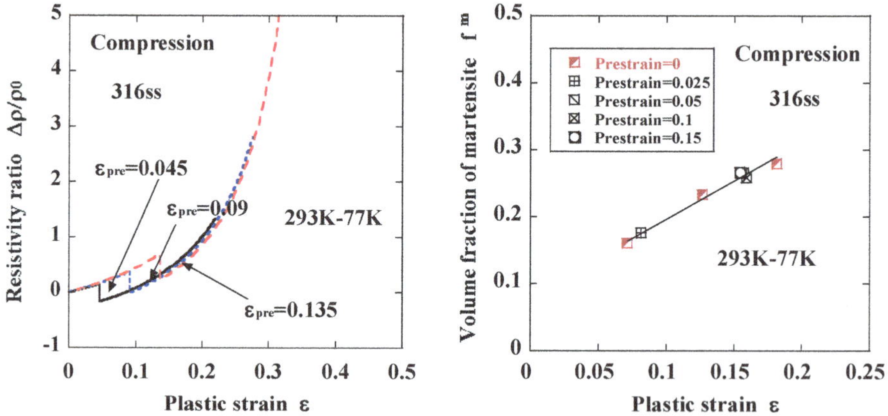

(a) Effect on resistivity **(b) Effect on martensite volume**

Fig. 3 Effects of temperature change on martensite transformation

As described in the introduction, in tensile deformation under abrupt temperature change, the pre-deformation of 316ss specimen at 293K induced ε phase which is a precursor of martensite and has an hcp structure produced by a stacking fault. Since the ε phase is transformed to martensite by tensile deformation at 77K, the resistivity and volume fraction of martensite attained by post-deformation are strongly dependent on pre-deformation at 293K. The supplementary 310ss test showed that although it is deduced that the ε phase is formed in 310ss because the resistivity shows a negative value, the ε phase was not transformed to martensite in the 310ss type even at 77K.

Accordingly, the stability of the austenite phase may be related to the stability of the ε phase. Those results suggest that most the austenite phase is transformed directly to the martensite phase by compressive deformation.

3.4 Estimation of martensite by resistivity

The relation between the volume fraction of martensite and plastic strain is almost linear as shown in Fig.1 because the strain region is small. Deducing from the results of tensile deformation, a quadratic form is suitable as a regression curve between the volume fraction of martensite and plastic strain. On the other hand, since the relation between resistivity and plastic strain can also be closely approximated by quadratic form, as seen in Fig. 2, the relation of volume fraction of martensite and resistivity ratio is deduced to be a linear one. The relation is shown in Fig. 4. Although the gradient is dependent on the specimen temperature, it is clear that the relation of volume fraction of martensite and resistivity ratio is closely approximated by the linear regression curve. It is therefore concluded that the resistivity ratio is available as an index for estimating the volume fraction of martensite. Finally, although the volume fraction of martensite against plastic strain can be deduced using a linear relationship, the gradient increases with the decrease of the specimen temperature. On the other hand, since the gradient measured by tensile deformation increases with the increase in specimen temperature, both transformation processes are also confirmed to be different due to the gradient of the relation between the volume fraction of martensite and the ratio of resistivity.

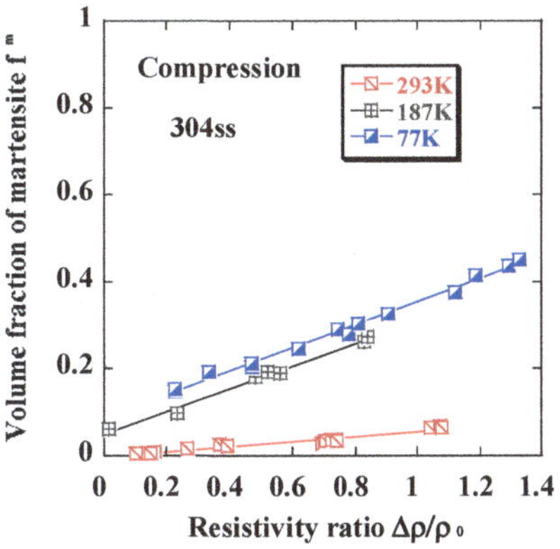

Fig. 4 Linear relation between martensite volume and resistivity

4 CONCLUSIONS

The effects of compressive deformation on the resistivity curve used for estimating the martensite induced in austenitic stainless steel were examined at 77K, 187K and 293K.

1. The resistivity curve and volume fraction of martensite are strongly dependent on specimen temperature.
2. Since martensite induced by compressive deformation generates many matrix defects, and the defects increase resistivity, the resistivity is strongly affected by defects in compressive deformation.
3. The volume fraction of martensite is expressed as a linear function of the resistivity ratio: therefore the resistivity ratio is available as an index for estimating the volume fraction of martensite.

ACKNOWLEDGMENT

The author would like to thank Mr. Toshiyuki Kudo. The research was accomplished because of his devotion.

REFERENCES

[1] A. Takimoto, R. Nishihara and S. Soda: Relationship between the Electrical Resistivity and the Volume Fraction of Martensite Induced Quenching and Deformation, J Japan Inst. Metal, Vol. 49 (1985), p.313.

[2] H. Date: Estimation of Volume Fraction of Martensite Induced in Austenitic Stainless Steel during the Tensile Deformation by the Resistivity and Specimen Configuration Dependency, J. of the Society of Materials Science Japan, Vol. 60, (2011) pp.721~727.

[3] H. Date: Effects of Strain Rate on Strain-induced Martensite and Resistivity of a Deforming Stainless Steel, Abstracts Book of the 8th International Symposium on Impact Engineering, (2013), p.136.

[4] T. Ogata, T. Yuri and Y. Ono: Load-displacement Curves, Specimen Heating and Strain-induced Martensitic Transformation of Austenitic Stainless Steel at Cryogenic Temperatures, J. Cryo. Soc. Japan, Vol. 42 (2007) pp.10~17.

[5] S. Komatsy, T. Hata, T. Sugumoto, K. Akamatsu and K. Kamei: Resistivity change of Age-Hardening Aluminum Alloys by Low Temperature Deformation, J Japan Inst. Metal, Vol. 42 (1978), pp.763~769.

[6] Y. Shugo, K. Sakazume, S. Kato and T. Honma: Studies on the Martensitic Transformation in Fe-Mn and Fe-Mn-C Alloys, Bulletin of the Research Institute of Mineral, Dressing and Metallurgy, Tohoku University, Vol. 32 (1976), p.109.

[7] H. Date, Effects of Deformation Mode on a Resistivity Curve used for Estimating Martensite induced Stainless Steel, Material Science Forum, Vols. 638-642 (2010), p.2992.

Image-guided morphological measurement for the rabbit aortic arch in the non-loaded state

X. Zhang [1, 2], Y. Tang [2], X. Li [1]
[1] School of Mechanical Engineering and Applied Electronics Technology, Beijing University of Technology, P.R. China
[2] School of Computer Science, Engineering and Mathematics, Flinders University, Australia

ABSTRACT

In order to quantify residual strains in aortic arch, elliptic curve fitness which combines computerized measurements of images, are presented. And the cross-sectional configuration parameters of rabbit aortic arch in the non-loaded state, such as semi-major/minor axis, thickness of arterial wall, were measured. The axial distribution of configuration parameters indicates that aortic arch is elliptic cross-section in its curved section but is circular cross-section in its straight section. Furthermore, the whole aortic arch is cone-shaped in axial direction. Simultaneously, the distribution of averaged thickness is nearly uniform in circumferential direction, but not in axial direction.

1 INTRODUCTION

It has been known since 1960 [1] that when a ring segment is cut from an aortic and a radial cut is made in the ring, it uncoils like a watch spring, but this phenomenon wasn't explained with residual strains and stresses in blood vessel until 1983 [2]. They suggested that residual stress developed in growing arteries so as to reduce stress gradients across the wall thickness. Another illustration, which termed the uniform strain hypothesis, was adopted by Takamizawa and Hayashi [3] who assumed that, under physiological conditions, the circumferential strains are constant across the thickness of a blood vessel wall. Studies on the vascular residual strains have become one of the most attractive subjects of biomechanics in last 20 years [4].

In order to quantify the residual strains, we must know the deformation fields when all outer acting loads are removed, as those that exist in a solid. The popular technique is computerized measurement of the vessel images, through which required geometrical parameters in the non-loaded state, subsequently, in the zero-stress state, can be measured. And then, residual strains can be calculated. Matsumoto [5] divided the vessel profile into 32 segments and residual strains were derived from the local curvature of each segment. Li and Hayashi [6] have proposed an alternative approach based on the measurement of the edge angle. Han [7] and Zhao [8] measured the transverse residual strains by optically tracking the surface displacement.

However, researches on residual strains and stresses of blood vessel nearly concentrate on the straight segments of aorta, and studies on curved and bifurcate vessel are limited in theoretic modeling. For example, Li [6] presented a theoretical model to analyze the non-linear elastic properties of the aortic arch wall. Fortunately, computerized measurements of images have developed to carry out experimental study on residual strains in curved aorta recently.

2 METHOD

In order to calculate the circumferential residual strains, cross-sectional configuration parameters of aortic arch, such as semi-major/semi-minor axis, mean wall thickness of the arterial wall in the non-loaded state, were measured.

2.1 Elliptic curve fitness of the cross-sectional boundary

In the non-loaded state, the cross-sectional boundary line of aortic ring was fitted with elliptic curve, and the corresponding elliptic equation is as following.

$$x^2 + bxy + cy^2 + dx + ey + f = 0 \tag{1}$$

In order to guarantee the higher accuracy of the fitted elliptic equation, the linear parameters equations are deduced as square matrix.

$$\begin{bmatrix} \sum_i x_i^2 y_i^2 & \sum_i x_i y_i^3 & \sum_i x_i^2 y_i & \sum_i x_i y_i^2 & \sum_i x_i y_i \\ \sum_i x_i y_i^3 & \sum_i y_i^4 & \sum_i x_i y_i^2 & \sum_i y_i^3 & \sum_i y_i^2 \\ \sum_i x_i^2 y_i & \sum_i x_i y_i^2 & \sum_i x_i^2 & \sum_i x_i y_i & \sum_i x_i \\ \sum_i x_i y_i^2 & \sum_i y_i^3 & \sum_i x_i y_i & \sum_i y_i^2 & \sum_i y_i \\ \sum_i x_i y_i & \sum_i y_i^2 & \sum_i x_i & \sum_i y_i & n \end{bmatrix} \begin{bmatrix} b \\ c \\ d \\ e \\ f \end{bmatrix} = \begin{bmatrix} -\sum_i x_i^3 y_i \\ -\sum_i x_i^2 y_i^2 \\ -\sum_i x_i^3 \\ -\sum_i x_i^2 y_i \\ -\sum_i x_i^2 \end{bmatrix} \tag{2}$$

According to the Gaussian Method of Elimination algorithm, the parameters, such as *b, c, d, e* and *f*, could be solved in the equation (2).

Thus, the coordinates of central point of fitted ellipse is:

$$x_0 = \frac{-2cd + be}{4c - b^2}, \quad y_0 = \frac{-2e + bd}{4c - b^2} \tag{3}$$

The directional angle of fitted ellipse is:

$$\theta = 0.5\tan^{-1}(b / (1 - c)) \tag{4}$$

And corresponding semi-major and semi-minor axis of the fitted ellipse are:

$$A = \sqrt{\frac{ff(cos^2\theta - sin^2\theta)}{c\,sin^2\theta - cos^2\theta}}, \quad B = \sqrt{\frac{ff(cos^2\theta - sin^2\theta)}{sin^2\theta - c\,cos^2\theta}} \tag{5}$$

where $ff = f + 0.5(dx_0 + ey_0)$.

2.2 Thickness of the aortic segments

During the experimental measurements, positional thickness of aortic wall was defined by the length of perpendicular line between the inner and outer fitness curves. For the convenience of computation, the three-point-parabola fitness of the inner boundary of the cross-sectional segments was applied. So the general parabola fitness equation is:

$$F(x) = ax^2 + bx + c \tag{6}$$

The equations of vertical lines which depart from the point $P_i(x_i, y_i)(i = 1,2,3)$ should be:

$$Q(x) = -\frac{1}{2ax_i + b}x + \frac{x_i}{2ax_i + b} + y_i \tag{7}$$

Thus, the thickness of the aortic segments could be calculated according to the two interaction points between the perpendicular line and two boundary lines.

2.3 The cross-sectional ellipticity of the aortic segments

The cross-sectional configuration of aortic ring in the non-loaded state should be circle ideally. However, the aortic ring configuration was found to be elliptic in the experiment. Thus, the cross-sectional ellipticity was defined by

$$e = B / A \tag{8}$$

3 MATERIALS AND MEASUREMENTS

Four 4-5 months old healthy purebred rabbits are chosen in the experiment. And the experiment was conducted within the animal welfare regulations and guidelines of China. Surgery was performed under intravenous celiac pentobarbital anesthesia (30mg/kg), and aortic arch in the non-loaded state was dissected from each rabbit, which was recorded (Fig.1a) in the computer via a CCD camera (WV-CL700, Panasonic, Osaka, Japan). As seen in Fig.1b, in order to investigate different local circumferential positions, four representative directions were marked, which were labeled as Inner, Outer, Front and Back.

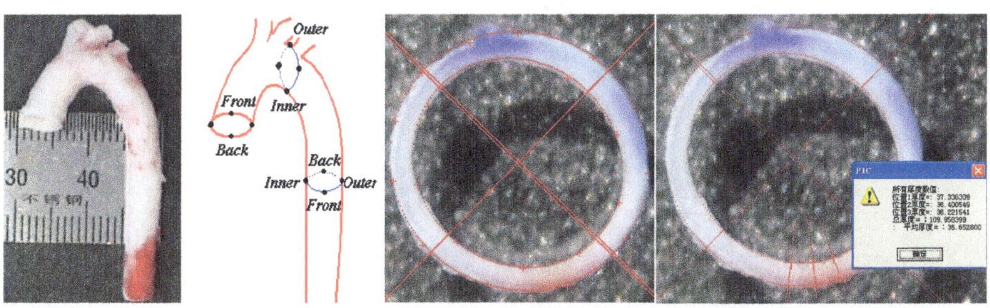

(a) Rabbit aortic arch (b) Four directions (c) Boundary fitness (d) Thickness measurement

Figure 1. The schematic diagram of experimental measurement

About 1mm-long ring specimens in the non-loaded state were obtained after axial cutting along each rabbit aortic arch, independently, and were recorded in the computer via CCD camera after immersed in saline solution at room temperature.

In the non-loaded state, as seen in the Fig.1c, the cross-sectional boundary line of aortic ring was fitted with elliptic curve, and corresponding measurements of thickness were investigated (Fig.1d). And statistic results of the semi-major and semi-minor axis of the fitted ellipse are given in Table 1.

Table 1 Measurement results in the non-loaded state

	12108 N=29		12115 N=38		12123 N=32		12129 N=47	
	Mean	*S.D.*	*Mean*	*S.D.*	*Mean*	*S.D.*	*Mean*	*S.D.*
a_i	2.158	0.178	2.286	0.303	2.845	0.257	3.631	0.193
b_i	1.711	0.185	1.812	0.232	2.255	0.259	2.887	0.202
a_e	2.507	0.199	2.689	0.328	3.057	0.209	2.986	0.322
b_e	2.097	0.192	2.249	0.265	2.574	0.323	2.502	0.251
e_i	0.796	0.092	0.792	0.091	0.793	0.089	0.795	0.079
e_e	0.839	0.078	0.836	0.078	0.842	0.082	0.838	0.088

where, a_i, b_i-semi-major and semi-minor axis of inner fitness ellipse respectively (*mm*)
a_e, b_e-semi-major and semi-minor axis of outer fitness ellipse respectively (*mm*)
e_i, e_e-ellipticity of inner wall and outer wall separately

4 RESULTS AND DISCUSSION

4.1 Distribution of cross-sectional ellipticity

And the ellipticity, which can be seen in Table 1, was calculated according to measured semi-major axis and semi-minor axis. Furthermore, the ellipticity distribution of inner and outer wall along aortic arch is presented in Fig.2, in which the left vertical line marked location of carotid corresponding to the curved section of aortic arch, whereas right vertical line marked the location of intercostals artery corresponding to the straight section of aortic arch (Same followed).

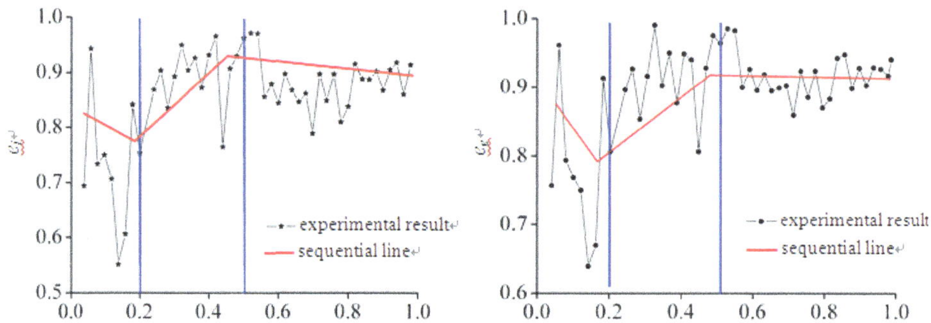

(a) Inner boundary (b) Outer boundary

Figure 2. The axial distributions of averaged ellipticity of aortic arch in non-loaded state

In brief, only results about the rabbit specimen No.12108 were given in the paper, which are consistent well with results of other rabbit specimens. It is concluded that the cross-sectional configuration is elliptic in curved aortic section but is circle nearly in straight section.

4.2 Distribution of aortic wall thickness

Circumferential distribution of the aortic wall thickness was investigated. Comparing the wall thicknesses in four representative positions in Table 2, there is nearly no difference in statistic. Therefore, the wall thickness of aortic arch in the non-loaded state is uniform in circumferential direction.

Table 2 Thicknesses in the particular positions

	12108 *N*=29	12115 *N*=38	12123 *N*=32	12129 *N*=47
	Mean±S.D.	*Mean±S.D.*	*Mean±S.D.*	*Mean±S.D.*
T_o	0.349±0.060	0.384±0.070	0.494±0.058	0.486±0.068
T_i	0.352±0.065	0.407±0.053	0.513±0.045	0.527±0.061
T_f	0.385±0.060	0.433±0.075	0.535±0.049	0.533±0.065
T_b	0.388±0.061	0.408±0.069	0.483±0.057	0.526±0.059

where, T_o, T_i, T_f, T_b—thickness in *outer, inner, Front* and *Back* directions separately (*mm*)

Furthermore, axial distribution of averaged aortic thickness is presented in Fig.3. The results show that the amplitude of variation in aortic wall thickness is big in the curved section but is small in the straight section separately.

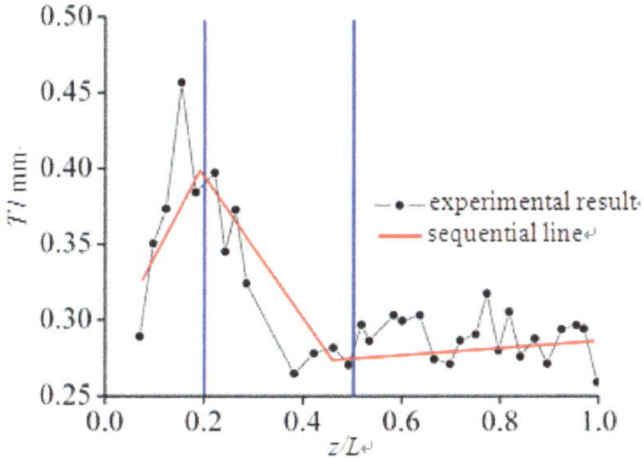

Figure 3. The axial distributions of averaged thickness of aortic arch in non-loaded state

5. CONCLUSIONS

White rabbits were taken as an example in the experiment, and elliptic curve fitness, which combines computerized measurements of images, is applied to measure the cross-sectional configuration parameters of aortic arch in the non-loaded state. The results of the experiment and measurement indicate that aortic arch in the non-loaded state is elliptic cross-section in its curved section but is circular cross-section in its straight section. Furthermore, the whole aortic arch is cone-shaped in axial direction, which agrees well with the physiologic practice. Simultaneously, circumferential distribution of averaged thickness is nearly uniform in the non-loaded state. However, axial distribution of averaged thickness is not uniform any longer.

ACKNOWLEDGEMENT

This work was financed by the research fund of National Natural Scientific Foundation of China (Grant No: 10372010).

REFERENCES

[1] Bergel, D.H. (1960) The viscoelastic properties of the arterial wall, *Ph.D. thesis*, University of London.

[2] Chuong, C.J. and Fung, Y.C. (1983) Three-dimensional stress distribution in arteries, *J. Biomech. Eng.* **105**, 268-274.

[3] Takamizawa, K. and Hayashi, K. (1987) Strain energy density function and uniform strain hypothesis for arterial mechanics, *J. Biomech.* **20**, 7-17.

[4] Rachev, A. and Greenwald, S.E. (2003) Residual strains in conduit arteries, *J. Biomech.* **36**, 661-670.

[5] Matsumoto, T., Hayashi, K. and Ide, K. (1995) Residual strain and local strain distributions in the rabbit atherosclerotic aorta, *J. Biomech.* **28**, 1207-1217.

[6] Li, X. and Hayashi, K. (1996) Alternative method for the analysis of residual strain in the arterial wall, *Biorheology* **33**, 439-449.

[7] Han, H.C. and Fung, Y.C. (1996) Direct measurement of transverse residual strains in aorta, *Am. J. Physiol.* **270**, 750-759.

[8] Zhao, J., Lu, X., Zhuang, F. and Gregersen, H. (2000) Biomechanical and morphometric properties of the arterial wall referenced to the zero-stress state in experimental diabetes, *Biorheology* **37**, 385-400.

Estimation method of residual stress and plastic strain in austenitic stainless steel by single indentation test

R. Kusano, T. Hiyoshi, A. Yonezu
Department of Precision Mechanics, Chuo University, Japan

ABSTRACT

This study proposed a method to evaluate the residual stress σ_{res} and plastic strain (pre-strain) ε_{pre} of austenitic stainless steel SUS316L using Berkovich indentation test. This method deals with SUS316L that obeys the Ludwik work hardening law and involves in-plane equi-biaxial residual stress. Numerical experiment with parametric finite element method (FEM) study by changing of σ_{res} and ε_{pre} was conducted in order to deduce relationship between indentation curve and materials parameter (ε_{pre} and σ_{res}). This relationship can be simply expressed by dimensionless functions. Thus, this method can estimate both ε_{pre} and σ_{res} simultaneously, when a single indentation test is applied to SUS316L.

1. INTRODUCTION

Tensile residual stress often induces environmentally assisted cracking and fatigue crack initiation, resulting in crucial damage (1). Thus, the evaluation of residual stress is very important for industrial steels and structures. For the typical measurement of residual stress, X-ray diffraction is widely used. However, the method cannot evaluate plastic strain (called pre-strain) which may contribute to generate residual stress. Therefore, the present study proposed a simple method to evaluate both residual stress and plastic strain (plastic properties) using an indentation test.

2. MATERIALS

The material considered in this study is austenitic stainless steel (SUS316L). The true stress and true strain curve in plastic deformation region can be approximated by Ludwick law. The constitutive equation in elastic and plastic region is described as following.

$$\sigma = E\varepsilon \quad \text{for } \sigma \leq \sigma_Y, \quad \text{and} \quad \sigma = \sigma_Y + K\varepsilon_p^n \quad \text{for } \sigma \geq \sigma_Y \qquad \text{Eq. (1)}$$

Where E is the Young's modulus, σ_Y is the yield stress, n is the work-hardening exponent, and K is the work-hardening strength. Note that ε_p is plastic strain. The yield stress of pre-strained steel becomes higher than that of as-received steel due to work-hardening, and then the increased yield stress can be plotted on the stress-strain curve of as-received steel when it shifts to the value of pre-strain. Since this study addresses only

SUS316L, the yield stress of pre-strained steel can be obtained by substituting the pre-strain into the Eq. (1). Therefore, the plastic properties can be expressed by only the parameter of pre-strain ε_{pre}, and then we describe the yield stress as $\sigma_{Y(\varepsilon pre)}$ hereafter.

3. NUMERICAL ANALYSIS

The conical indenter whose half apex angle is 70.3° was employed for the two dimensional model in order to approximate Berkovich indenter. The semi-infinite model was created using more than 20,000 four-nodes wherein fine meshes were created around the contact region and a mesh convergence study was carried out. The rigid indenter was created and the contact option was employed. The Coulomb's law of friction was assumed with a friction coefficient of 0.15. The employed pre-strain ε_{pre} was 0, 5, 10, 20, 30, 40, 60 and 100% (i.e. the total is 8 cases) in this study. In addition, we introduced the various residual stress σ_{res} in the material before indenter penetration. Note that σ_{res} normalized by $\sigma_{Y(\varepsilon pre)}$ ($\sigma_{res}/\sigma_{Y(\varepsilon pre)}$) is used for the assigned parameter in the parametric FEM study. The value of $\sigma_{res}/\sigma_{Y(\varepsilon pre)}$ was changed -0.75, -0.25, 0, +0.25 and + 0.75 (5 cases as the total). Therefore, the computations of 40 cases (=8 x 5) were carried out in this study. In addition, the Young's modulus E and the Poisson's ratio were set to 195 GPa and 0.3 for all computation. We simulated the representative material case in order to investigate the effect of ε_{pre} and σ_{res} on the indentation curve. The result of no residual stress shows that the indentation force increases when the ε_{pre} increases. It is also found that the indentation force increases with the larger compressive residual stress, while the indentation force decreases with the larger tensile residual stress. Thus, it is concluded that both pre-strain ε_{pre} and residual stress σ_{res} affect the indentation curve.

4. ESTIMATION METHOD

There are many previous studies on the estimation of elastoplastic properties based on dimensionless function (2), (3). Such a simple function may be useful for reverse analysis. Similarly, this study conducted dimensionless analysis to develop the estimation method of both ε_{pre} and σ_{res}. In other words, the dimensionless function which correlates the indentation curve with material parameters to be identified (ε_{pre} and σ_{res}) was established through the parametric FEM study. A schematic of an indentation curve is shown in Fig. 1. In this figure, the area of indentation curve corresponds to a total work volume W due to indenter penetration. Contrary to this, the area of unloading part (as shown by the solid diagonal line area, W_u) indicates the work of elastic recovery (unloading work). Here, the recovery depth h_e can be obtained by $h_{max} - h_r$. Note that h_r is the depth of permanent impression. This study focuses on the total work W and the elastic work W_u, since these two parameters are independent. Dimensionless analysis with Π theory dictates the following relationships for W and W_u.

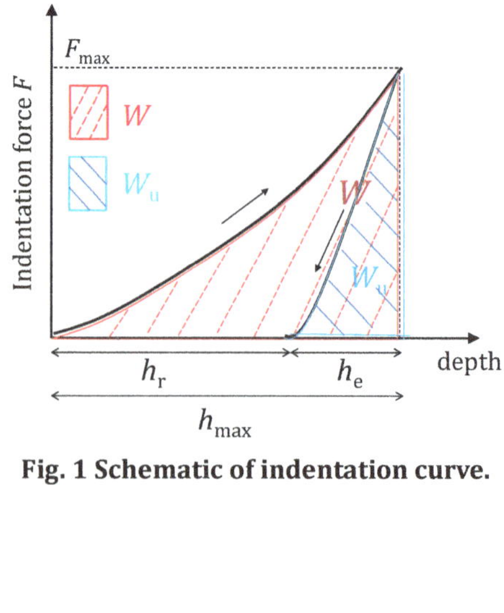

Fig. 1 Schematic of indentation curve.

$$\frac{W}{\sigma_{Y(\varepsilon_{pre})}{h_{max}}^3} = \Pi(\frac{E^*}{\sigma_{Y(\varepsilon_{pre})}}, \frac{\sigma_{res}}{\sigma_{Y(\varepsilon_{pre})}}) \quad \text{Eq. (2)}$$

$$\frac{W_u}{\sigma_{Y(\varepsilon_{pre})}{h_e}^3} = \Pi(\frac{E^*}{\sigma_{Y(\varepsilon_{pre})}}, \frac{\sigma_{res}}{\sigma_{Y(\varepsilon_{pre})}}) \quad \text{Eq. (3)}$$

As mentioned above, the yield stress $\sigma_{Y(\varepsilon pre)}$ is related to pre-strain ε_{pre} with the Eq. (1).

In both Eqs. (2) and (3), E^* is the reduced Young's modulus, which can be expressed by Eq. (4).

$$\frac{1}{E^*} = (\frac{1-{\nu_s}^2}{E_s} + \frac{1-{\nu_i}^2}{E_i}) \quad \text{Eq. (4)}$$

Note that for the subscript of E and ν (Poisson's ratio) "s" is that of specimen and "i" is that of indenter. The reults of parametric FEM studies on 40 cases spanning of the material parameter space, were introduced into the Eqs. (2) and (3). Their results are shown in Figs. 2 and 3.

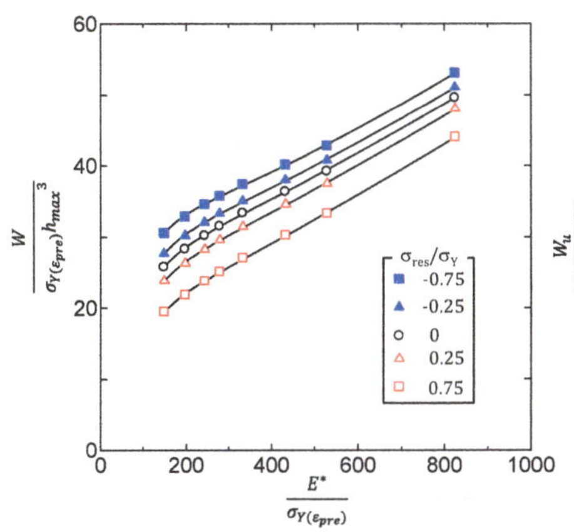

Fig. 2 Relationship between $\frac{W}{\sigma_{Y(\varepsilon_{pre})}{h_{max}}^3}$ and $\frac{E^*}{\sigma_{Y(\varepsilon_{pre})}}$ at each $\frac{\sigma_{res}}{\sigma_{Y(\varepsilon_{pre})}}$.

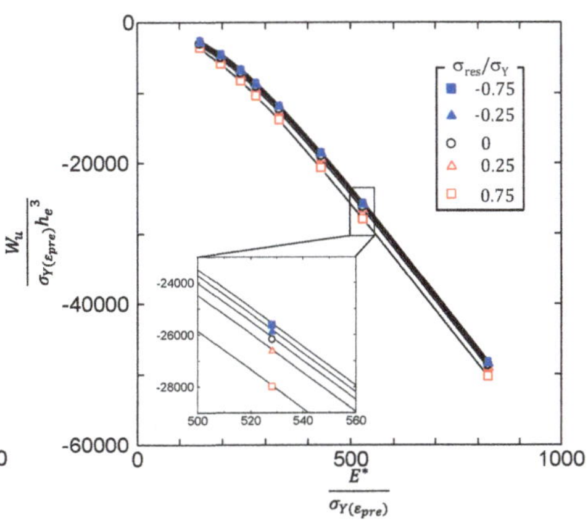

Fig. 3 Relationship between $\frac{W_u}{\sigma_{Y(\varepsilon_{pre})}{h_e}^3}$ and $\frac{E^*}{\sigma_{Y(\varepsilon_{pre})}}$ at each $\frac{\sigma_{res}}{\sigma_{Y(\varepsilon_{pre})}}$.

These figures show that the Eqs. (2) and (3) functions are dependent on $\frac{E^*}{\sigma_{Y(\varepsilon_{pre})}}$ and $\frac{\sigma_{res}}{\sigma_{Y(\varepsilon_{pre})}}$. The values of $\frac{W}{\sigma_{Y(\varepsilon_{pre})}{h_{max}}^3}$ increases in $\frac{E^*}{\sigma_{Y(\varepsilon_{pre})}}$ monotonically (Fig. 2), on the other hand, the values of $\frac{W_u}{\sigma_{Y(\varepsilon_{pre})}{h_e}^3}$ decreases in $\frac{E^*}{\sigma_{Y(\varepsilon_{pre})}}$ monotonically (Fig. 3). As shown by the solid lines, the data of $\frac{W}{\sigma_{Y(\varepsilon_{pre})}{h_{max}}^3}$ and $\frac{W_u}{\sigma_{Y(\varepsilon_{pre})}{h_e}^3}$ were separately approximated by polynomial function for each $\frac{\sigma_{res}}{\sigma_{Y(\varepsilon_{pre})}}$. Since these polynomial functions can be interpolated as a function of $\frac{\sigma_{res}}{\sigma_{Y(\varepsilon_{pre})}}$, any function at a given $\frac{\sigma_{res}}{\sigma_{Y(\varepsilon_{pre})}}$ can be deduced. Thus, we obtained the explicit form of Eqs. (2) and (3). By introducing the values of W, h_{max}, W_u, and h_e (from a result of an indentation test) into the Eqs. (2) and (3), the relationship between $\frac{E^*}{\sigma_{Y(\varepsilon_{pre})}}$ and $\frac{\sigma_{res}}{\sigma_{Y(\varepsilon_{pre})}}$ can be deduced uniquely. In other words, two independent relationships of between $\frac{E^*}{\sigma_{Y(\varepsilon_{pre})}}$ and $\frac{\sigma_{res}}{\sigma_{Y(\varepsilon_{pre})}}$ can be obtained from Eqs. (2) and (3). The intersection point of these curves is the solution of $\frac{E^*}{\sigma_{Y(\varepsilon_{pre})}}$ and $\frac{\sigma_{res}}{\sigma_{Y(\varepsilon_{pre})}}$. Here, the reduced modulus E^* is known prior. Thus, this two independent $\frac{E^*}{\sigma_{Y(\varepsilon_{pre})}}$ and $\frac{\sigma_{res}}{\sigma_{Y(\varepsilon_{pre})}}$ relationships can provide the yield stress $\sigma_{Y(\varepsilon pre)}$ (which further leads to the pre-strain) and residual stress σ_{res}.

5. VALIDATION OF THE METHOD

The present method is now applied to the numerical experiment in order to investigate the estimation accuracy of our method. As mentioned above, parametric FEM study was carried out to establish the dimensionless function. Those data was employed to estimate the residual stress and pre-strain. Furthermore, two additional sets, (ε_{pre}=0.15, $\frac{\sigma_{res}}{\sigma_{Y(\varepsilon pre)}}$ =0.5) and (ε_{pre}=0.5, $\frac{\sigma_{res}}{\sigma_{Y(\varepsilon pre)}}$ = - 0.4), were also investigated to verify our method. Note that these two material cases are not included for the FEM parametric study (to establish the dimensionless function of Eqs. (2) and (3)).

For all cases, we extracted loading data (W, h_{max}), and unloading data (W_u, h_e) from the computational indentation curve, and then estimated the materials properties. The estimated properties (ε_{pre}, $\frac{\sigma_{res}}{\sigma_{Y(\varepsilon_{pre})}}$) by our method are shown in Fig. 4. The square symbol indicates the input value (solution) and the symbol of "X" shows our estimation. It is found that the estimations of σ_{res} and ε_{pre} are good agreement. All calculated results agree with the input data, with an error of less than about 20%.

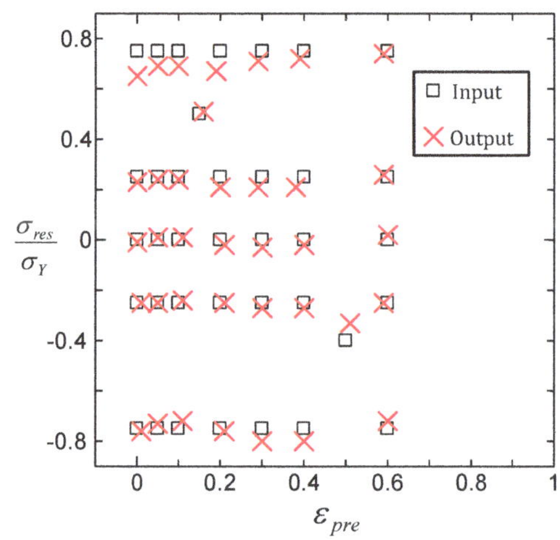

Fig. 4 Comparison of the estimated result by reverse analysis with input value.

Table 1 Representative materials for sensitivity analysis.

	Case 1	Case 2	Case 3	Case 4
$\sigma_{res}/\sigma_{Y(\varepsilon pre)}$	-0.4	0.25	-0.25	0.75
ε_{pre}	0.5	0.4	0.1	0.2

In addition, we investigate the robustness of the method. In this study, we conducted sensitivity analysis for four representative materials (in Table 1). For the indentation curve (i.e. W and W_u), the perturbations are set to 2 %, since the previous studies have used this value for sensitive analysis (4). These perturbations are introduced to each case separately. The result is shown in Fig. 5. In Fig. 5(a), the deviations (errors in the estimated values compared to the input values) about ε_{pre} was within 15% in all cases. On the contrary in Fig. 5(b), the σ_{res} error shows large within about 90% for the perturbation cases. Several previous studies on sensitivity analysis reported that their methods induced 50 % error at a maximum (4). This indicates that the present method may be relatively underperforming for perturbations due to potential experimental errors, but these error cases may represent extreme example.

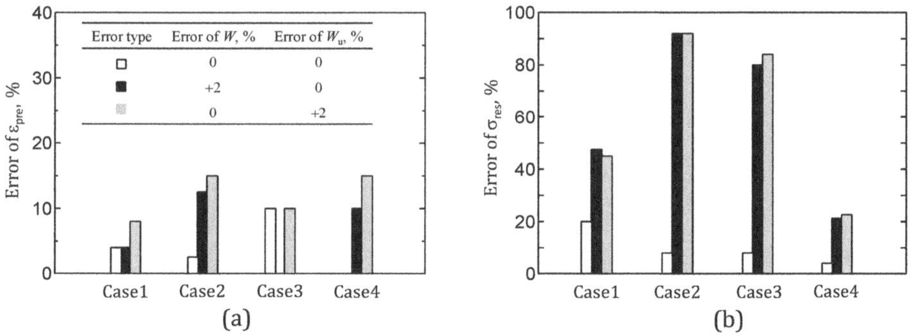

Fig. 5 Sensitivity study of representative materials; (a) pre-strain ε_{pre}, and (b) residual stress σ_{res}.

6. CONCLUSION

This study proposed a new indentation method to simultaneously evaluate the residual stress and plastic strain of an austenitic stainless steel. This method focuses on an austenitic stainless steel, SUS316L that obeys the Ludwick work hardening law and involves in-plane equi-biaxial residual stress (yet the framework is applicable to any material with little modification). First, a parametric FEM study by changing of both residual stress σ_{res} and pre-strain ε_{pre} was conducted in order to deduce the relationship between the indentation curve and the materials parameters of ε_{pre} and σ_{res} (which was employed for the FEM study). The relationship could be expressed by two dimensionless functions with simple formulae. Thus, the present method can estimate both ε_{pre} and σ_{res}, when a single indentation test is applied to SUS316L. Finally, the estimation accuracy and the robustness of our method were investigated using a numerical experiment.

This work is supported by JSPS KAKENHI (Grant No. 26420025) and JGC-S Scholarship Foundation.

REFERENCES

(1) Lo KH, Shek CH, Lai JKL. (2009) *Materials Science and Engineering*. R 65:39-104. DOI: 10.1016/j.mser.2009.03.001
(2) Dao M, Chollacoop N, VanVliet KJ et al. (2001) *Acta Materialia*. 49:3899-3918. DOI: 10.1016/S1359-6454(01)00295-6
(3) Cheng YT, Cheng CM. (2004) *Materials Science and Engineering*. R 44:91-149. DOI: 10.1016/j.mser.2004.05.001
(4) Zhao M, Chen X, Yan J, Karlsson AM. (2006) *Acta Materialia*. 54:2823-2832. DOI: 10.1016/j.actamat.2006.02.026

Measurement of the strength of a grain boundary by using the combination of focused ion beam and electron back-scatter diffraction methods

T. Nakanishi, K. Suzuki, H. Miura
Fracture and Reliability Research Institute,
Graduate School of Engineering, Tohoku University, Japan

ABSTRACT

A novel micro-tensile test method that can measure the interface strength of grain boundaries has been developed by applying scanning electron microscopy and a focused ion beam technology. A micro-scale cantilever structure was fabricated from a single-crystalline silicon substrate and a fine nano-scale sample cut from a thin film using a focused ion beam was attached at one end to the cantilever by deposition of tungsten vapour in the microscope. The other end was attached to a micro probe similarly. Finally, the micro probe was activated to pull the sample and the deformation of the sample was observed by a scanning electron microscope. The fracture strength was measured by detecting the deflection of the cantilever at the breakage of the sample.

1 INTRODUCTION

Recently, mechanical properties of polycrystalline thin films have been found to vary drastically depending on their micro texture. In particular, the electroplated copper thin films used for advanced semiconductor devices vary widely from brittle characteristics to ductile ones (1)-(3). The main reason for the variation can be attributed to their micro texture which consists of fine columnar grains with porous grain boundaries (4). The crystallinity of the grain boundaries was found to dominate both their mechanical and electrical properties and the long-term reliability. Therefore, it is very important to evaluate the characteristics of the grain boundaries quantitatively. The authors have already developed a new method for evaluating the crystallinity of grain boundaries quantitatively by applying an electron back-scatter diffraction (EBSD) method. By analysing the image quality of Kikuchi Line obtained from the analysis, it is possible to measure the order of the atomic alignment in the area observed by an electron beam with 50-nm in diameter (5). In this paper, a novel micro-tensile test method that can measure the interface strength of grain boundaries quantitatively has been developed by applying scanning electron microscopy and a focused ion beam technology, in order to evaluate the effect of the crystallinity of grain boundaries on their strength quantitatively. The strength of the grain boundaries of which the position was determined by the EBSD analysis was successfully measured quantitatively. In was confirmed that the strength of the grain boundaries with low crystallinity was much lower than that with high crystallinity. Thus, both the variation and degradation mechanisms of the crystallinity of grain boundaries can be investigated quantitatively by this method.

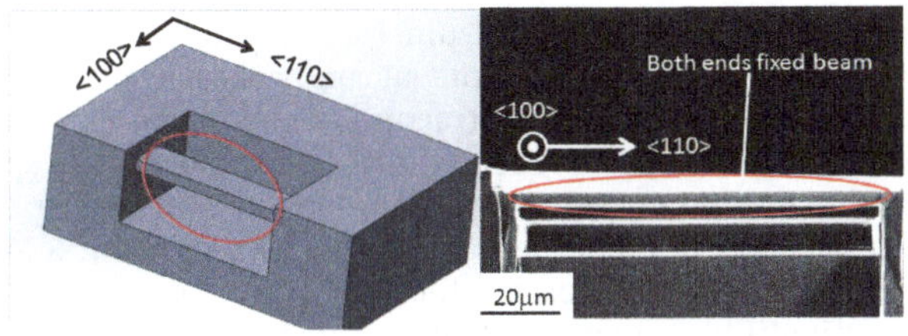

(a) Formation of a micro cantilever using a single-crystalline silicon substrate for a three-point bending test

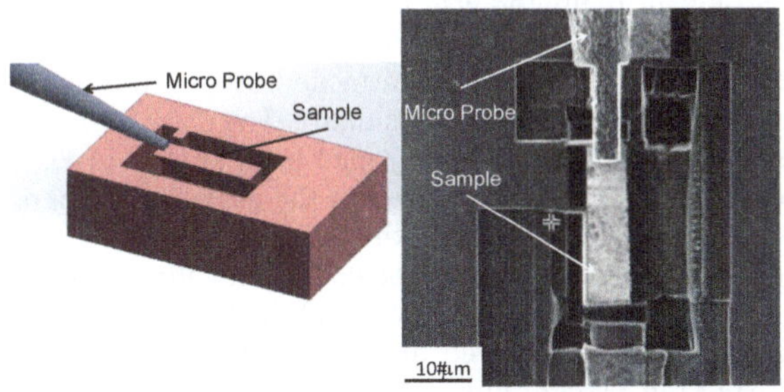

(b) Preparation of a sample using a focused ion beam

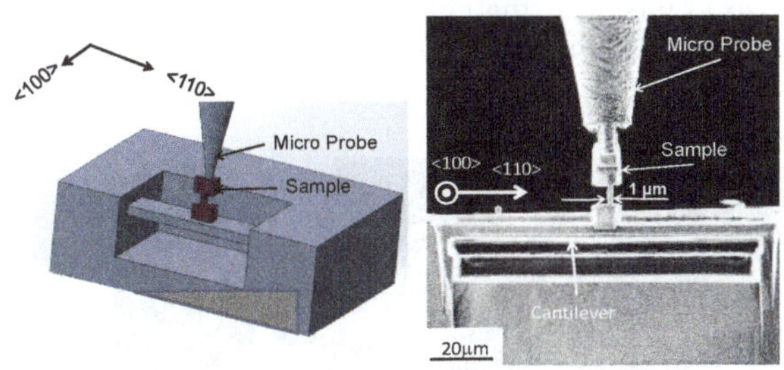

(c) Outlook of a micro tensile test system

Fig. 1 Preparation method for a micro tensile test of a grain boundary

2 MEASUREMENT METHOD

The micro tensile test system consists of a three-point bending loading tool which is cut from a singe-crystalline silicon substrate and a test sample which consists of one grain or two grains with a grain boundary. The loading tool was manufactured by using a focused ion beam as shown in Fig. 1(a). It was cut from (100) single-crystalline silicon wafer along <110> crystallographic direction. The typical length of the beam was 80 μm, and its width and thickness were 10 μm and 3 μm, respectively. The spring constant of this both ends supported beam is 360 N/m when a load is applied to the center of the beam. A test sample was similarly cut from a thin film material as shown I Fig. 1(b). The sample consisted of two columnar grains in series, and thus, there was a grain boundary between them. The position of the grain boundary and the crystallographic direction of the grains were determined by EBSD observation explained in detail later. One end of the sample was attached firmly to a edge of a micro probe which also consisted of single crystalline silicon with the axial direction of <110> by deposition of tungsten, then the other end was attached to the center position of the silicon beam as shown in Fig. 1(c).

The average diameter of the sample was 1 μm. Next, the micro probe was pulled up by uniaxial actuator and the deformation of the silicon beam was observed by a scanning electron microscope until the sample was fractured as shown in Fig. 2. The load applied to the sample was calculated by the observed deformation using the spring constant of the beam. Since the maximum error in the observed deformation was 72 nm, the maximum error in the estimation of the fractured stress was about 5 MPa in this study. The position of a grain boundary for the tensile test was determined by an electron back-scatter diffraction (EBSD) method as shown in Fig. 3 (5). By analyzing the Kikuchi pattern obtained from the EBSD analysis, the position of a grain boundary was determined by the confidence index (CI) map shown in Fig. 3(b). When a focused electron beam of 50-nm in diameter was scanned on the surface of a polycrystalline thin-film, this index goes down to zero at a grain boundary where two grains with different crystallographic orientation are attached. Thus, the distribution of grain boundaries can be visualized as shown in this figure. The quality of each grain boundary is evaluated by using the image quality (IQ) map as is shown in Fig. 3(a). This image quality value indicates the order of the period of the atomic alignment in the observed area. Since the Kikuchi pattern consists of the Kikuchi lines which are formed by the interaction between the irradiated electron beam and the atomic plane in the irradiated area based on Bragg's law, the average sharpness of the Kikuchi lines corresponds to the crystallographic quality of the observed area. The IQ value was calculated by using Hough transform (6) of the Kikuchi line in the observed area. Thus, the quality of each grain boundary can be evaluated quantitatively by using these two parameters. It was possible to cut two grains with the characterized grain boundary from the thin-film material by using a focused ion beam. Finally, the strength of various grain boundaries can be measured by using this method.

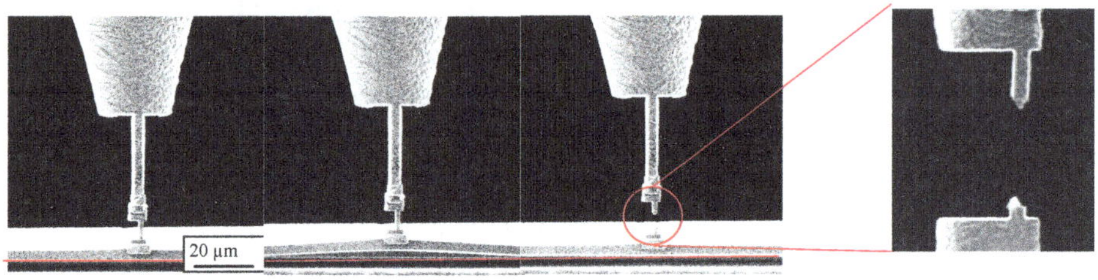

(a) Before test (b) During loading (c) After fracture (d) Fractured sample

Fig. 2 Example deformation of the silicon beam during the tensile test

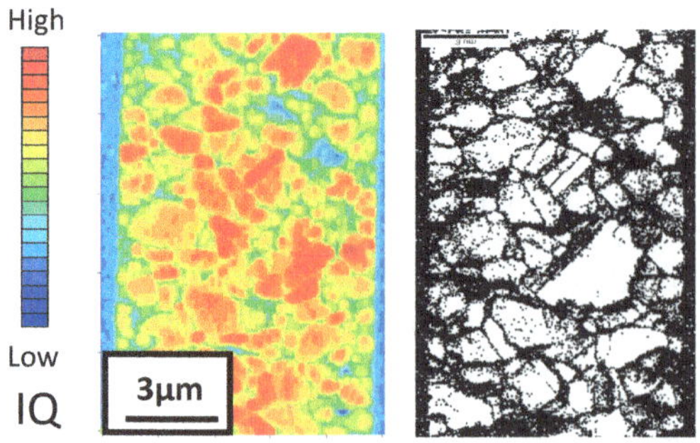

(a) Image quality map (b) Confidence index map

Fig. 3 Determination of the position of a grain boundary and its quality by using EBSD analysis

3 MEASUREMENT RESULT

Electroplated copper thin films were used for the tensile test. The films were electroplated on a silicon wafer on which a thin tantalum and copper thin films were sputtered as a seed layer for electroplating. The plating bath consisted of 80 g of CuO powder and 186 g of H_2SO_4 into 1000 ml of purified water. The current density during the electroplating was fixed at 10, 30, 50 mA/cm^2. By changing the current density, both the average size and quality of grain boundaries were varied drastically. In order to increase the average grain size, some films were annealed in an argon gas at temperatures from 100°C to 400°C for 30 min., respectively. Figure 4 shows an example of the observed change of the crystallographic quality of grain boundaries in the annealed electroplated copper thin films. Various grain boundaries with different quality were cut from the films and the tensile strength of each grain boundary was measured by the developed micro tensile test system. Figure 5 shows an example of a fractured sample with a low quality grain boundary. The brittle fracture occurred at the grain boundary with low quality as was expected. By measuring the area of the fracture surface, strength of the fractured grain boundary was simply estimated by dividing the fracture load by the area of the cross-section. The fracture in a single crystal copper occurred on a (111) slip plane as was expected. The critical resolved shear stress was also calculated by considering the area of the slip plane. Thus, this micro tensile test system is effective for evaluating the strength of a small sample quantitatively.

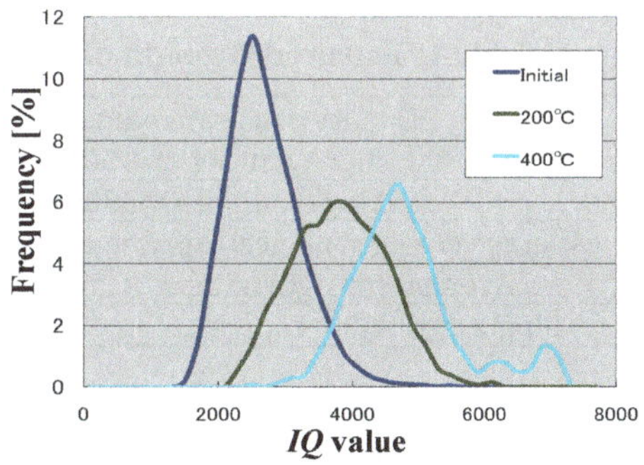

Fig. 4 Change of the distribution of the quality of grain boundaries in annealed electroplated copper thin films

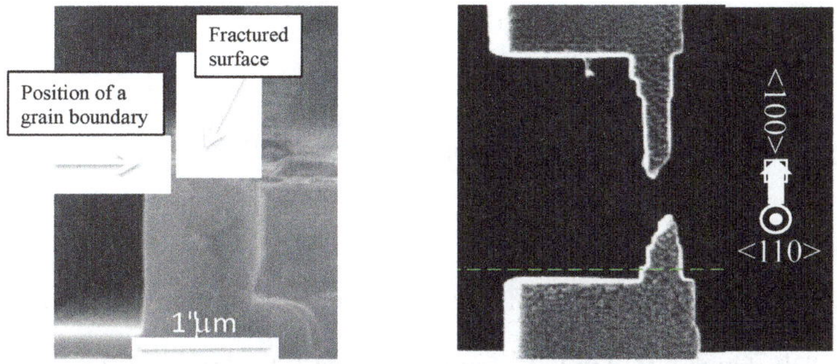

(a) Low quality grain boundary **(b) Single-crystalline copper**

Fig. 5 Outlook of an example of a fractured sample

The measured strength of various samples is summarized in Table 1. The total number of the samples was 5 in each category. The strength of a grain boundary was found to vary drastically depending on its crystallographic quality. In this study, it was confirmed that the electroplated copper thin films with average IQ value lower than 3500 showed brittle fracture at grain boundaries. It is clear that the average strength of grain boundaries with high quality was much higher than that of average strength of electroplated copper thin films, while it was much lower than that when the quality of a grain boundary was lower than 3500. This result clearly indicated that the brittle fracture of the electroplated copper thin films was dominated by the low strength of grain boundaries with low quality.

Table 1 Measured strength of grain boundaries with different quality

Quality of a grain boundary	Strength (MPa)
High quality (IQ>3500)	500 - 800
Low quality (IQ<3500)	170 - 190
Average strength of electroplated copper thin films	350
Average strength of bulk copper (C1020)	220
Single crystalline copper along <100> direction	80 - 120

4 CONCLUSIONS

A novel micro-tensile test method that can measure the interface strength of grain boundaries has been developed by applying scanning electron microscopy and a focused ion beam technology. A micro sample was attached to the center of a both ends supported beam cut from a single crystalline silicon wafer. The spring constant of the beam was about 360 N/m. The deformation of the beam was observed by a scanning electron microscope during the tensile test until the fracture of the sample. The fracture load of the sample was estimated by the critical deformation of the beam using the spring constant. The resolution of the fracture strength measurement was about 5 MPa. The effect of the crystallographic quality of a grain boundary in electroplated copper thin films was evaluated by using this system. It was confirmed that the strength of a grain boundary with high quality, which was measured by an EBSD method, was much higher than the average strength of the films, while that with low quality was much lower than the average strength. Brittle fracture of the films was found to be dominated by these grain boundaries with low quality. Finally, this micro tensile test system is effective for evaluating the strength of a micro sample quantitatively.

ACKNOWLEDGEMENTS

This research activity has been supported partially by Japanese special coordination funds for promoting science and technology, and Tohoku University.

REFERENCES

(1) Lin C. T. and Lin K. L., Journal of materials science, Materials in electronics, Vol. 15, No. 11, pp. 757-762, (2004).
(2) Chang S. H., Shieh J. M., Lin K. C., Dai B. T., Wang T. C., Chen C. F., Feng M. S., Li Y. H., and Lu C. P., Journal of vacuum science and technology B, Microelectronics and nanometer structures, Vol. 19, No. 3, pp. 767-773, (2001).
(3) Muto, H., Sakai, M., "The Large-scale Deformation of Polycrystalline Aggregates: Cooperative Grain-boundary Sliding," Acta Materialia, No. 48, pp. 4161-4167, (2000).
(4) Tamakawa, K., Sakutani, K., and Miura H., Journal of The Society of Materials Science Japan, 56, pp. 907-912, (2007) [in Japanese].
(5) Murata N., Suzuki K., and Miura H., Proc. of the ASME IMECE2012, No. IMECE2012 87426, pp. 1-6, (2012), (CDROM).
(6) Shapiro L. and Stockman G., "Computer Vision", Prentice-Hall, Inc. (2001).

Tensile characteristics of Cu/Sn IMCs estimated by using miniature composite solder specimen

K. Ohguchi
Department of Materials Science and Engineering, Akita University, Japan
K. Kurosawa
Graduate School of Engineering and Resource Science, Akita University, Japan

ABSTRACT

An experimental method employing a miniature composite type solder specimen is proposed to estimate deformation characteristics of the Cu/Sn intermetallic compounds (IMCs) of Cu_3Sn and Cu_6Sn_5 by tensile tests. The cross-section of the specimen has concentric layers of the Cu/Sn IMCs and Cu. Considering the structure of the cross-section, a data processing method for the tensile tests results is also proposed based on the rule of mixture (ROM) to estimate the stress-strain relation of the Cu/Sn IMCs.

1 INTRODUCTION

The strength reliability of the solder joints in electronic equipment is evaluated by conducting a finite element analysis (FEA). The solder joints have the Cu/Sn intermetallic compounds (IMCs) layer, which consists of Cu_3Sn and Cu_6Sn_5, at the interface between solder and copper wiring. Since the IMCs are considered to be brittle, fatigue cracking is likely to occur at the IMCs layer when the solder joints are subjected to cyclic deformation. Then, the FEA of the solder joints should be conducted with consideration for the deformation characteristics of the IMCs. To conduct such a FEA, the deformation characteristics of the IMCs must be clarified, and therefore some studies to investigate them have been conducted (1-4). Deng et al. measured Young's modulus of Cu_3Sn and Cu_6Sn_5 in Cu/solder joints by nanoindentation (1). Jiang et al. made micropillar specimens of Cu_3Sn and Cu_6Sn_5 by using FIB and the specimens were used in compression tests employing a nanoindenter with a flat tip (2). Although the FEA needs the material parameters for the Cu/Sn IMCs estimated by tensile tests, these studies employ compression loadings to evaluate the deformation characteristics of the IMCs. This is because there is no easier method to make the Cu/Sn IMCs specimen such as dog-bone type which is necessary to conduct tensile test.

In the present study, we propose an experimental method to evaluate the deformation characteristics of the IMCs of Cu_3Sn and Cu_6Sn_5 by tensile tests. The method employs a miniature dog-bone type specimen of Sn-3.0Ag-0.5Cu (SAC) lead-free solder the cross-section of which has concentric layers of the Cu/Sn IMCs and the Cu electrodeposit. The specimen can be made by using widely used equipment. Three types of the specimen each of which has different thicknesses of the concentric layers were prepared, and the tensile tests using them were conducted. Tensile characteristics of the Cu/Sn IMCs were estimated by employing a data processing method for the tests results.

2 EXPERIMENTAL PROCEDURES

2.1 Specimen

The specimen used in this study was made by three processes shown in Fig. 1. Firstly, a miniature specimen which has a gauge length of 2 mm and a gauge diameter of 0.5 mm was made by casting Sn-3.0Ag-0.5Cu (SAC) lead-free solder with a mold made of carbon tool steel. The appearance of the specimen is shown in Fig. 1 (a). Secondly, electrodeposition of Cu was performed on the miniature specimen in a mixture of 1 M $CuSO_4$ and 1 M H_2SO_4 at a constant current density of 50 mA/cm^2. Figure 1 (b) shows the schematic of the electrodeposition process. In the process, a copper plate was employed as the anode while the solder specimen was employed as the cathode. The thickness of deposits was aimed at 3μm. To equalize the Cu plating thickness, the copper plate used for the anode was bent in a circular shape and then it was set around the solder specimen. Finally, the electroplated specimen shown in Fig. 1 (b) was vacuum-encapsulated as shown in small photo in Fig. 1 (c), and then it was put in an electric muffle furnace the inside of which was kept at 453 K for a definite period of time. Three kinds of time period of 36, 48 and 60 hours were employed for the heat treatment to make three kinds of specimens which have different thickness of IMCs layer.

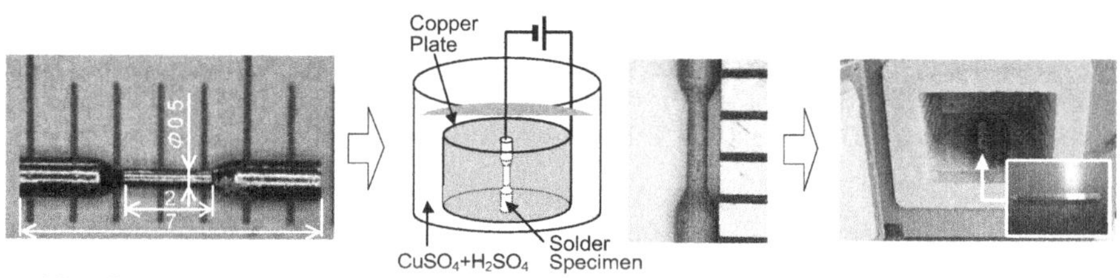

(a) Molten miniature solder specimen (b) Electrodeposition process (c) Heat treatment process

Fig. 1 Processes for making miniature solder specimen with Cu/Sn IMCs layer.

2.2 Tensile testing machine and loading condition

A small size tensile testing machine EZ-S (Shimadzu Co., Ltd., Japan) was employed as the base of the apparatus. However, the displacement resolution of the machine was not enough to precisely conduct the tensile test of the miniature specimen used in this study. Therefore, an electronic actuator (ORIENTAL MOTOR Co., Ltd., Japan) which has the displacement resolution of 20 nm was installed to the machine as a loading equipment.

The distance between the upper and lower grip of the specimen was measured by using high-speed and high-accuracy digital sizer LS-7500 (KEYENCE CORPORATION, Japan) during the tensile test, and the measured value was used for calculating the strain. Also, the load was measured by a load cell incorporated in the testing machine, which has the load capacity of 500 N. The stress was calculated by dividing the measured load value by the cross-section area of the specimen. All the tensile tests were conducted at room temperature under a constant tension rate of 20 μm/s.

3 RESULTS AND DISCUSSIONS

3.1 Structure of the specimen

Figure 2 shows the longitudinal section of the specimen made by the heat treatment of 48 hours. Figures 2 (a) and (b) respectively show an optical micrograph and an elemental map of Cu. The IMCs layer can be observed between the solder and the electrodeposited Cu layer in Fig. 2 (a), and it consists of two layers each of which has different colors. In Fig. 2 (b), there are two layers which have different elemental concentration of Cu between the solder and the Cu. Elemental analyses by EPMA were conducted in the two layers, and the results showed that the layer near the solder side was Cu_6Sn_5 while the layer near the Cu side was Cu_3Sn as shown in Fig. 2 (b). Namely, the structure of the cross-section of the specimen has concentric layers of the Cu/Sn IMCs and the Cu electrodeposit. Then, the specimen can be regarded as a composite material in which the SAC solder is reinforced by the Cu_6Sn_5, Cu_3Sn and Cu. Therefore, we termed the specimen miniature composite solder (MCS) specimen.

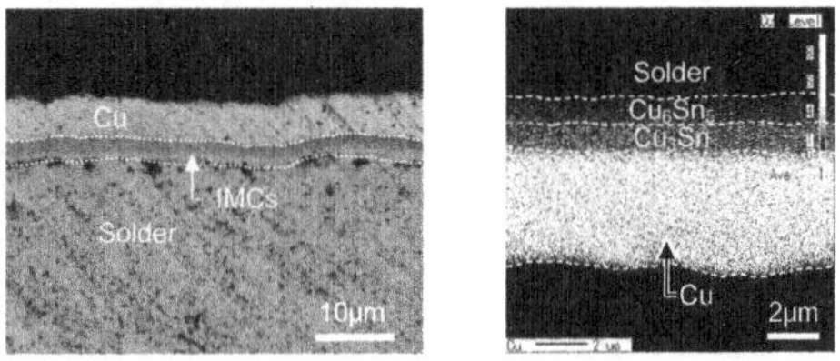

(a) Optical micrograph (b) Elemental map of Cu

Fig. 2 Longitudinal section of specimen.

Meanwhile, the difference in color observed in IMCs layer in Fig. 2 (a) is due to the difference of chemical composition as shown in Fig. 2(b). This means that the layer of Cu_6Sn_5 and that of Cu_3Sn can be distinguished by the optical microscopic observation. Then, the thicknesses of each layer of the Cu_6Sn_5, Cu_3Sn and Cu were measured using the micrographs of the MCS specimens made by the heat treatment time t_{ht} of 36, 48 and 60 hours. The average thickness values of the Cu_6Sn_5, Cu_3Sn and Cu for t_{ht} of 36 h were respectively 1.4, 1.4 and 2.8 μm. Those for t_{ht} of 48 h were 1.2, 1.9 and 2.6 μm, while those for t_{ht} of 60 h were 0.91, 2.2 and 2.8 μm. Using these values with measured diameter of the MCS specimen, the variation of the area ratio of each layer in the cross-section of the MCS specimen, such as Cu_6Sn_5, Cu_3Sn and Cu, with the time period for the heat treatment was investigated. As a result, the area ratios of the Cu_6Sn_5, Cu_3Sn and Cu were respectively expressed as a function of the heat treatment time as follows:

$$\left(r_a\right)_{Cu_6Sn_5} = -2.47\times10^{-1} t_{ht}^{\ 1/2} + 2.62\ (\%), \tag{1}$$

$$\left(r_a\right)_{Cu_3Sn} = 3.16\times10^{-1} t_{ht}^{\ 1/2} - 7.26\times10^{-1}\ (\%), \tag{2}$$

$$\left(r_a\right)_{Cu} = -3.40\times10^{-2} t_{ht}^{\ 1/2} + 2.38\ (\%), \tag{3}$$

where the r_a denotes the area ratio and its subscript corresponds to each material. Since the variation of the area ratio is due to the elemental diffusion, Eqs. (1) - (3) are expressed as a function of the square root of heat treatment time t_{ht}.

Equations (1) and (2) suggest that the area ratio of Cu_3Sn increases while that of Cu_6Sn_5 decreases, with increase in the heat-treatment time. Namely, the MCS specimens have different area ratio of IMCs in the cross-section for different heat-treatment condition.

3.2 Tensile characteristics of the MCS specimen

Figure 3 shows the representative stress-strain relations obtained from tensile tests. The symbols of the open circles, the open squares and the open triangles show the stress-strain relations of the MCS specimen made by the heat-treatment time of 36, 48 and 60 hours, respectively. The lines show the stress-strain relations of the miniature SAC solder specimens which were treated with heat under the same conditions as that for the MCS specimens. The solid line, the dotted line and the dashed-dotted line respectively correspond to the stress-strain relations of the SAC solders prepared by the heat-treatment time of 36, 48 and 60 hours. These stress-strain relations can be regarded as the stress-strain relations of the SAC solder inside the MCS specimens. In Fig. 3, the stress levels in the stress-strain relations of the MCS specimens are higher than those of the heat-treated solder. This means that the MCS specimen is the composite material in which the SAC solder is reinforced by a layer which consists of the IMCs and the Cu. Then, we termed the layer reinforced material (RM) layer.

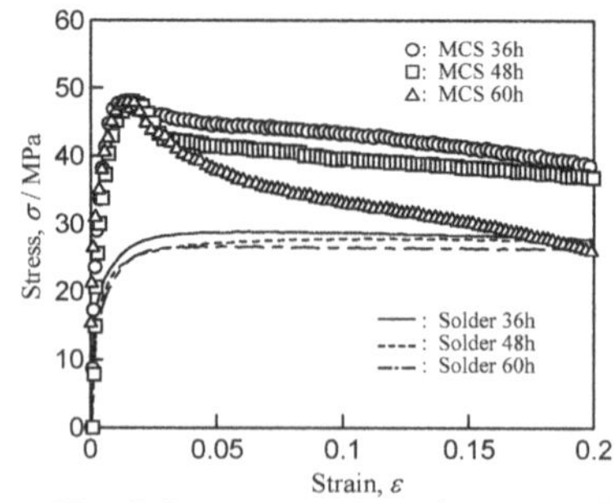

Fig. 3 Stress-strain relations of the MCS specimen and the solder.

On the other hand, the ultimate tensile strength of the MCS specimen is observed at the strain of slightly over 1 % irrespective of the heat-treatment time. After the ultimate tensile strength has been observed, the stress-strain relations of the MCS specimens approach the stress-strain relations of the solder. The tensile tests employed 5 MCS specimens for each heat-treatment time, and the above-mentioned phenomena were observed in all the tests. Observing the MCS specimen after use for a tensile test, it was found that the stress drop was due to the failure of the IMCs layer. Therefore, the uniform elongation of the IMCs layer was estimated at approximately 1 %.

3.3 Estimation of the stress-strain relation for the Cu/Sn IMCs

Considering the structure of the MCS specimen, the stress generated in the RM layer during the tensile test using the MCS specimen can be estimated by the following equation based on the rule of mixture (ROM):

$$\sigma_{\mathrm{RM}} = \left[\sigma_{\mathrm{MCS}} - \left\{1-\left(r_{\mathrm{a}}\right)_{\mathrm{RM}}\right\}\sigma_{\mathrm{SAC}}\right] / \left(r_{\mathrm{a}}\right)_{\mathrm{RM}}, \tag{4}$$

where σ_{RM}, σ_{MCS} and σ_{SAC} are the stresses generated in the RM layer, the MCS specimen and the SAC solder respectively at the same strain. $\left(r_{\mathrm{a}}\right)_{\mathrm{RM}}$ is the area ratio of the RM layer in the cross-section of the MCS specimen which is obtained as a sum of Eqs. (1) - (3).

The symbols in Fig. 4 show the estimated stress-strain relations for the RM layers the heat-treatment time of which are different. For the estimation, the average stress-strain relation of 5 specimens in the strain range from 0 to 1% of the MSC specimens and the heat-treated solder were applied to Eq. (4) with Eqs. (1) - (3). The solid line, the dotted line and the dashed-dotted line in Fig. 4 are the approximate curves for the stress-strain relations. The following Ramberg-Osgood law was employed for the approximation:

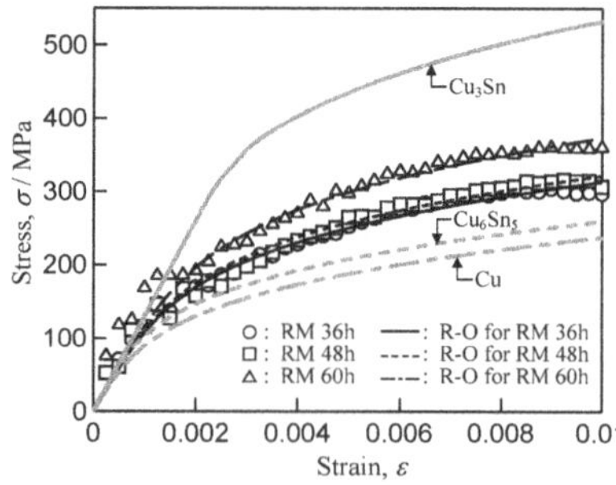

Fig. 4 Estimated stress-strain relations for the RM, Cu and IMCs.

$$\varepsilon = \frac{\sigma}{E} + 2.0\times10^{-3}\left(\frac{\sigma}{\sigma_{0.2}}\right)^{m}, \tag{5}$$

where E is the Young's modulus, $\sigma_{0.2}$ is the 0.2 % proof stress and m is the hardening exponent. As a result of the approximation, it was found that $\sigma_{0.2}$ and m were expressed as a function of the heat-treatment time as follows:

$$\sigma_{0.2} = \exp(8.93 \times 10^{-3} t_{ht} + 5.09)(\text{MPa}), \quad (6)$$

$$m = \exp(3.79 \times 10^{-3} t_{ht} + 1.29). \quad (7)$$

Using Eqs. (6) and (7) with literature data for E (134 GPa for Cu_3Sn (1), 112 GPa for Cu_6Sn_5 (1) and 110 GPa for Cu (5)), the stress-strain relation of the IMCs were estimated by the following procedure.

(i) Since the RM layer by the heat-treatment time t_{ht} of zero consists only of Cu, the $\sigma_{0.2}$ and the m for Cu were calculated by plugging in zero for t_{ht} in Eqs. (6) and (7). Applying the calculated parameters to Eq. (5), the stress-strain relation of the Cu (SS-Cu) in the RM layer was estimated.

(ii) The $\sigma_{0.2}$ and the m for the RM layer by the heat-treatment time t_{ht} of 5.28 h (RM5.28h) were calculated to estimate the stress-strain relation of the RM5.28h (SS-RM5.28h) by Eq. (5). Plugging in 5.28 h for t_{ht} in Eq. (2) makes the $(r_a)_{Cu_3Sn}$ become zero. This means that the RM5.28h consists of the Cu_6Sn_5 and the Cu. Then, we estimated the stress-strain relation of the Cu_6Sn_5 (SS-IMC1) by applying the SS-RM5.28h and the SS-Cu to the ROM with Eqs. (1) and (3).

(iii) The stress-strain relation of the Cu_3Sn (SS-IMC2) was estimated by applying the SS-IMC1, the SS-Cu and the stress-strain relations of the RM layer shown in Fig. 6 to the ROM with Eqs. (1) - (3).

The grey bold lines in Fig. 4 show the estimated stress-strain relations for the Cu (dashed-dotted line), the Cu_6Sn_5 (dotted line) and the Cu_3Sn (solid line). Both IMCs of the Cu_6Sn_5 and the Cu_3Sn show the plastic deformation behavior. In addition, the stress-strain relation of the Cu_3Sn has larger stress than that of the Cu_6Sn_5.

4 CONCLUSIONS

This paper proposed an experimental method which employs the MCS specimen to estimate tensile characteristics of the IMCs of the Cu_3Sn and the Cu_6Sn_5. A data processing method was also proposed to estimate the stress-strain relations of the Cu/Sn IMCs from the tensile tests results of the MCS specimen. As a result, the following conclusions were obtained:

(1) The uniform elongation of the Cu/Sn IMCs is approximately 1 %.
(2) Both the IMCs of the Cu_6Sn_5 and the Cu_3Sn show the plastic deformation behavior.
(3) The stress-strain relation of the Cu_3Sn has larger stress than that of the Cu_6Sn_5.

REFERENCES

(1) Deng, X., Koopman, M., Chawla, N., and Chawla, K.K. (2004) *Mater. Sci. Eng. A* **364**, 240-243.
(2) Jiang, L. and Chawla, N. (2010) *Scr. Mater.* **63**, 480-483.
(3) Deng, X., Chawla, N., Chawla, K.K., and Koopman, M. (2004) *Acta Mater.* **52**, 4291-4303.
(4) BALAKRISNAN, B., CHUM, C.C., LI, M., CHEN, Z. and CAHYADI, T. (2003) *J. Electron. Mater.* **32**, 166-171.
(5) W. D. Callister, Jr. (2007) *Materials Science and Engineering: An Introduction*, 137.

In situ micro-mechanical testing of grain boundaries combined with environmental TEM

Y. Takahashi, H. Kondo, R. Asano, M. Takuma, K. Saitoh
Department of Mechanical Engineering, Kansai University, Japan
S. Arai, S. Muto, N. Tanaka
EcoTopia Science Institute, Nagoya University, Japan

ABSTRACT

In order to evaluate the effect of gaseous environment on grain boundary fracture, a novel micro-mechanical experimental method was proposed. A micro-cantilever containing a single grain boundary was fabricated from a bulk poly-crystalline intermetallic compound (Ni_3Al). The micro-cantilever was then loaded by using a nano-indenting specimen holder operated in a special high-voltage transmission electron microscope (HVTEM) equipped with an environmental cell (EC). It was shown that, among two specimen types (V-notched and unnotched), grain boundary cracking was only observed in a V-notched specimen loaded in a gas environment containing hydrogen (H_2).

1 INTRODUCTION

The strength of a material is dependent on various microstructures. Grain boundary (GB) is one of such microstructures. Since a GB acts as a barrier against slip transmission and, moreover, a potential site of cracking, direct strength measurement of each GB is important for clarifying the mechanism of macroscopic deformation and/or failure. Recent studies have demonstrated that the mechanical testing of a single GB is possible by utilizing a usual nano-indenter tool (1,2). These studies, however, are limited to the evaluation conducted under an atmospheric environment. Very few reports have been made on the strength property of an environmentally affected single GB (3). The aim of our study is to develop a completely novel approach that enables the mechanical testing of a single GB under various gaseous environments combined with an in-situ microscopy. Here, we report on our recent achievements regarding the issue of brittle GB fracture triggered by hydrogen gas.

2 MATERIALS AND METHODS

The wafer of a stoichiometric Ni_3Al was electropolished to remove a damaged surface layer. The crystallographic orientation of the wafer surface was analysed by electron backscattered diffraction (EBSD) measurement. Figure 1 shows an example of the inverse pole figure map of the wafer. The average grain size was measured as approximately 110 μm. A random-type GB having a relatively large orientation difference (> 20°) was selected. A micro-mechanical specimen was fabricated by focused ion beam (FIB) technique, see e.g. (4). A block containing a GB was firstly cut out and picked out from the wafer. The block was then transferred and adhered onto the flattened end of a wire. The block was further milled and a cantilever shape was formed. In order to evaluate the effect of gas environment on GB fracture behaviour, several specimens were sampled from the same GB.

Figure 1 Inverse pole figure color map of a Ni_3Al poly-crystal used for study.

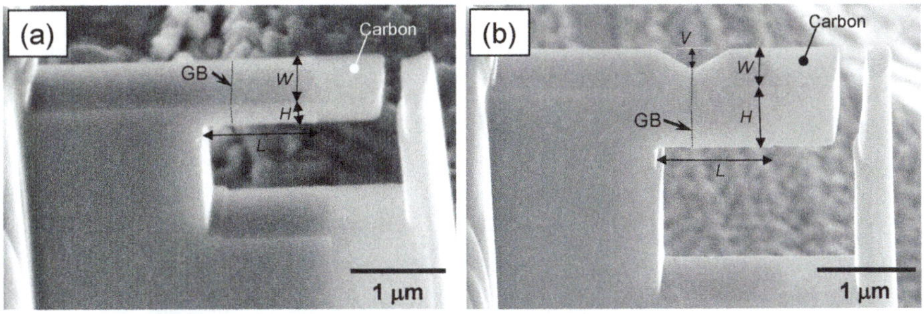

Figure 2 SEM micrographs of cantilever specimens containing a grain boundary (tilted view): (a) unnotched and (b) V-notched.

Figure 2 shows the scanning electron microscope (SEM) images of cantilever specimens. Two types of specimens were prepared: one with a smooth lever shape and the other with a V-notch. The former contains a GB located close to the fixed end of the lever. The shape of the GB (i.e. interface) free-edge for this type is 90°/90°. The latter contains a GB located at the tip position of the V-notch. The shape of the upper GB free-edge for this type is 135°/135°. The approximate size of the cantilevers are shown in Table 1.

Table 1 Specimen dimensions (in nm)

Specimen type	Length, L	Height, H	Width, W	Notch depth, V
Smooth(unnotched)	1000	400-500	400-500	-
V-notched	1000	1000	400-500	300-400

The micro-mechanical experiment was conducted inside a high-voltage transmission electron microscope (HVTEM) equipped with an open-type environmental cell (EC) (5). A nano-indenting TEM specimen holder was used for the experiment. The detail of these special tools will be reported elsewhere. The position between the micro-specimen and diamond indenter tip was aligned under TEM observation by the fine movement of a piezo-actuator. Hydrogen gas diluted with nitrogen gas (N_2:H_2 = 4:1) was admitted to the EC until the total gas pressure reached about 5000 Pa. The specimen was loaded by piezo-displacement control at 0.5 nm/s. Note that the load was applied at an extra carbon layer of the cantilever which was hard enough for preventing indent mark formation. The applied load and the in-situ TEM image were simultaneously recorded during the experiments. A high electron acceleration voltage (1 MeV) was used for the observation because electron scattering by the gas layer was significant particularly under a relatively high gas pressure (6).

3 RESULTS AND DISCUSSION

Figure 3 shows test results of unnotched specimens. Since the displacement value shown in the charts is the operational value of a piezo-actuator, it does not directly represent the deflection of the specimens. However, it is clearly seen that significant amount of plastic deformation follows after elastic deformation. No eminent hardening occurs during plastic deformation which is accompanied by extensive slip around the fixed end of the cantilever. The slip is apparently blocked by the GB (orientation difference: 25.8°) and concentrated to one side of the grain. The loading was continued even after the contact position between indenter tip and specimen shifted from the initial position due to the severe bending deformation. No cracking along the GB was, however, observed both in vacuum and H_2-containing gas environments.

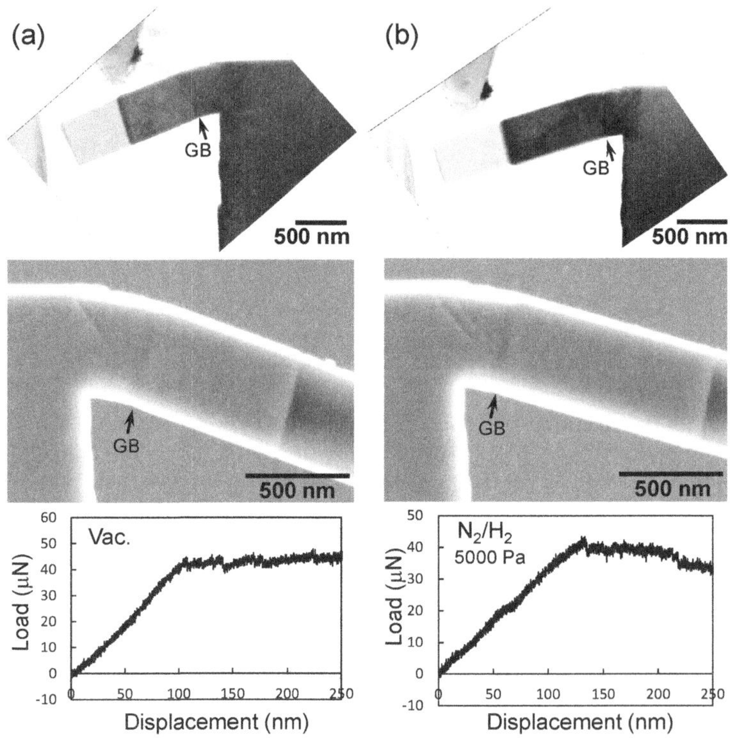

Figure 3 Test results of unnotched specimens: (a) in vacuum and (b) in N_2/H_2 gas. Top: TEM image after deformation, middle: SEM image after test, bottom: load-displacement relation.

Figure 4 shows test results of V-notched specimens. The orientation difference for this GB is 53.7°. Similar to unnotched specimens, significant amount of plastic deformation with no eminent hardening continues after elastic deformation. In situ TEM and the subsequent high-magnification SEM revealed that slip appeared in both grains, particularly around the notch tip region. The deformation behaviour in both environments was generally similar. The marked difference was the emergence of a sudden cracking along the GB in H_2-containing gas environment. This phenomenon was clearly observed both in TEM image and load-displacement chart. The sudden drop of load indicates that the crack initiation was a rather brittle event. The crack, however, stopped propagation soon after initiation. This was found to be the result of crack arrest at a slip plane intersecting the GB (e.g. see the lowermost of Fig. 4(b)). In vacuum, the specimen showed no GB cracking even at a severer deformation level.

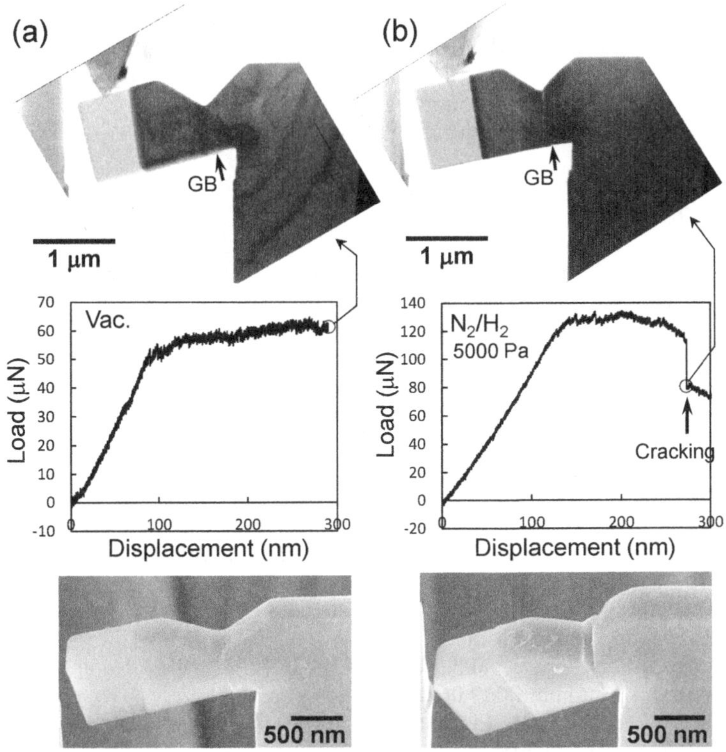

Figure 4 Test results of V-notched specimens: (a) in vacuum and (b) in N_2/H_2 gas. Top: TEM image during test, middle: load-displacement relation, bottom: SEM image after test.

The important points are summarised as follows: (i) the influence of H_2 gas environment (i.e. GB fracture) was only observed in V-notched specimen, (ii) GB fracture occurs after severe plastic deformation. The former point may suggest that hydrogen embrittlement (HE) of the current GB is facilitated by the magnitude of local stress concentration. A rough calculation of stress state around the upper GB free-edge by an elastic analysis showed that the stress singularity around a V-notch tip was much higher than 90°/90° free-edge. The high level of local hydrostatic stress might effectively collect solute hydrogen around the notch tip (7). Such an effect alone, however, seems not enough for causing GB fracture as it was not attained during elastic loading. The later point may then suggest that the accumulation of plastic deformation around a GB assists GB embrittlement; transport of solute hydrogen to a GB by dislocation (8) may increase hydrogen concentration around the GB and promote its weakening. These interesting phenomena provide important information as to the fundamental mechanisms of HE that have not been directly obtained by conventional macroscopic tests.

4 SUMMARY

In this study, a novel testing method for evaluating the strength of a single grain boundary (GB) subjected to gaseous environments was proposed. The method combined a high-voltage TEM with environmental cell and a nano-indenting specimen holder. The method was applied to a micro-cantilever specimen having a GB sampled from a bulk poly-crystalline Ni_3Al. It was shown that, among two specimen types (V-notched and unnotched), grain boundary cracking was only observed in a V-notched specimen loaded in a H_2-containing gas environment.

ACKNOWLEDGMENTS

The authors are grateful to financial support by: JSPS KAKENHI Nos. 23234567 (Research Activity Start-up), 25709003 (Grant-in-Aid for Young Scientists A), 25000012 (Specially Promoted Research); the Kansai University Grant-in-Aid for progress of research in graduate course (2012). The present study was also performed under the auspices of Reaction science/Nanomaterials science Research support base of Microstructure Analysis Platform in 'Nanotechnology Platform Project' by MEXT, Japan.

REFERENCES

(1) Armstrong, D.E.J., Wilkinson, A.J., Roberts, S.G. (2011) *Philos. Mag. Lett.* **91**, 394-400. doi: 10.1080/09500839.2011.573813
(2) Kupka, D., Huber, N., Lilleodden, E.T., (2014) *J. Mech. Phys. Solids* **64**, 455-467. doi:10.1016/j.jmps.2013.12.004
(3) Armstrong, D.E.J., Rogers, M.E., Roberts, S.G. (2009) *Scr. Mater.* **61**, 741-743. doi:10.1016/j.scriptamat.2009.06.017
(4) Takahashi, Y., Hirakata, H., Kitamura, T., (2008) *Thin Solid Films* **516**, 1925–1930. doi:10.1016/j.tsf.2007.08.062
(5) Tanaka, N., Usukura, J., Kusunoki, M., Saito, Y., Sasaki, K., Tanji, T., Muto, S. & Arai, S. (2013) *Microscopy*. **62**, 205-215. doi: 10.1093/jmicro/dfs095
(6) Butler, E.P. & Hale, K.F., Dynamic experiments in the electron microscope, in: Glauert, A.M. (Ed.), Practical Methods in Electron Microscopy, vol.9, North Holland, Amsterdam 1981.
(7) Troiano, A.R., (1960) *Trans. ASM* **52**, 54-80.
(8) Tien, J.K., Thompson, A.W., Bernstein, I.M., Richards, R.J., (1976) *Metall. Trans.* **7A**, 821-829. doi: 10.1007/BF02644079

Stress estimation of polyamide material by X-ray diffraction

T. Doi
Kobe City College of Technology, Mechanical System Course, Japan
M. Nishida
Kobe City College of Technology, Mechanical Engineering, Japan

ABSTRACT

In this study, residual stresses in polyamide (PA) materials were measured by the x-ray stress measurement technique. There are two problems for measuring residual stresses in polymer. Firstly, the diffraction peak from the polymer appears at the low 2θ angle region. Thus the measurement accuracy for strains reduces. Secondly, the low 2θ angle region is very difficult to use the $\sin^2\psi$ method. In this study, we employed the transmission diffraction in Ω-diffractometer method to resolve these problems. X-ray elastic constant (XEC) of PA was estimated as the material characteristics.

1 INTRODUCTION

The fiber reinforced plastics always keeps residual stresses from each the fiber and the matrix. Because there is large different thermal expantion coefficient between the fibre and the matrix (1,2). For example, few reports is published that in x-ray stress measurement of the polymer (3). However, in many cases, residual stresses in polymer materials were calculated indirectly from the stress value of the metallic particle or fiber phase in the composite (4). It is better for it to be able to directly measure the stress in polymer materials by the x-ray diffraction.

In this study, residual stresses in polymer constituting the matrix phase in the composite were measured by x-ray stress measurement. This measurement is widely used to measure residual stresses, and so-called $\sin^2\psi$ method is common non-destructive technique for evaluation residual stresses in near surface layer. On the other hand, an x-ray beam penetrates into polymeric materials because they are the low absorption coefficient. Thus, we used transmission method of x-ray diffraction instead of the traditional reflection method (5, 6).

2 X-RAY STRESS MEASUREMENT

Fig.1 shows the coordinate system of the x-ray stress measurement. In the theory of x-ray stress measurement, the ψ angle is defined between the normal direction of the diffraction plane and the normal axis of the sample surface. Then the equation of d-$\sin^2\psi$ method under the plane stress condition can be expressed as

$$\sigma_x = \frac{E}{1+\nu} \cdot \frac{1}{d_0} \cdot \frac{\partial d_\psi}{\partial \sin^2 \psi} \quad , \qquad (1)$$

where E is Young's modulus, ν is Poisson's ratio, d_0 is stress-free lattice spacing, d_ψ is lattice spacing in tilt angle ψ and direction of σ_X is coincided with the principle direction in the plane stress. The term $\partial d_\psi / \partial(\sin^2\psi)$ in Eq.(1) is calculated from the gradient of the regression line in the d-$\sin^2\psi$ diagram. Thus σ_X can be calculated from Eq.(1) with the d-$\sin^2\psi$ diagram. $E/(1+\nu)$ is called the x-ray elastic constant (XEC) that can also be expressed by

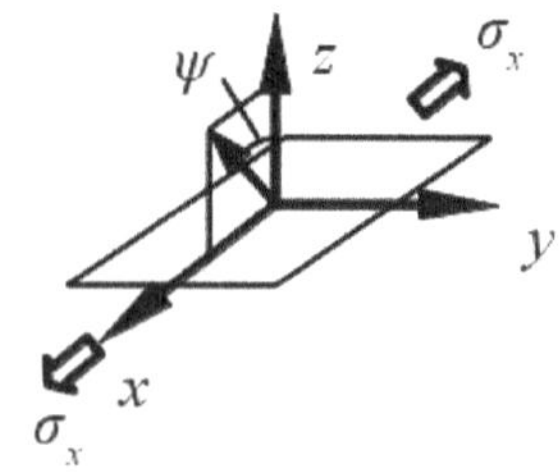

Fig.1 Difinition of coordinate system and ψ angle.

$$\frac{1}{2}s_2 = \frac{1+\nu}{E} = \frac{1}{d_0} \cdot \frac{\partial M}{\partial \sigma} \quad , \qquad (2)$$

where M is the gradient of the regression line in the d-$\sin^2\psi$ diagram. The XEC connects macroscopic stresses and the lattice strain calculated from the lattice spacing d_ψ. In this study, it was not only the calculation of the XEC, but also including parameters of the elastic constant: E and the Poisson's ratio: ν were separated from the XEC by following equations,

$$E = \frac{d_0}{\dfrac{\partial M}{\partial \sigma} + \dfrac{\partial d_{\psi=0}}{\partial \sigma}} \quad , \qquad (3) \qquad\qquad \nu = \frac{-\dfrac{\partial d}{\partial \sigma}}{\dfrac{\partial M}{\partial \sigma} + \dfrac{\partial d_{\psi=0}}{\partial \sigma}} \quad , \qquad (4)$$

where $\partial d_{\psi=0}/\partial\sigma$ is the gradient of regression line in $d_{\psi=0}$-σ diagram, and $d_{\psi=0}$ is lattice spacing when the ψ=0° angle.

In this study, the diffraction profile of the PA samples was measured by the Ω-diffractometer method combining the transmission method and the parallel beam slits (5-7). Here parallel beam slits were essential for measurement to the x-ray transmission diffractions, since the x-ray beam will diverge if it's not used. On the other hand, in this study, this diffraction method was employed in the transmission configurations ($0.4 \leqq \sin^2\psi \leqq 1.0$). It should be noted that the d-$\sin^2\psi$ method is based on the assumption of the plane stress condition because of a shallow penetration of x-rays. On the other hand, though x-rays penetrate deeply into polymeric materials, we use the d-$\sin^2\psi$ method in this study. Because these samples were very thin (0.068mm), therefore we assumed that there was no stress component normal to surface. Furthermore, PA samples used in this study consist of two phases: an amorphous phase and a crystalline phase. Residual stresses are only able to be measured from the crystalline phase of PA by x-ray diffraction.

3 MEASUREMENT METHOD

In this study, measurement samples were cut out from a commercial PA sheet. The thickness of samples is 0.068mm. These samples were used for the tensile testing and the x-ray stress measurement. Fig.2 shows results of tensile testing of the PA samples and sample dimensions are 80mm×10mm. Three kinds of tensile speeds are used. From

0.5mm/min to 3mm/min and these tests were repeated three times, respectively. We decided to use the result of the slowest tensile speed 0.5mm/min in this report.

In this study, three types of sample were prepared in each characteristic deformation. In Fig.2, first is the position A in the elastic deformation and near the yield point area. Second is the position B in the softening area and third is the position C in the oriented hardening area. Stresses in PA samples were measured by x-ray stress measurement under several loads. Fig.3 shows a photograph of one example of transmission method for longitudinal direction with a parallel beam system.

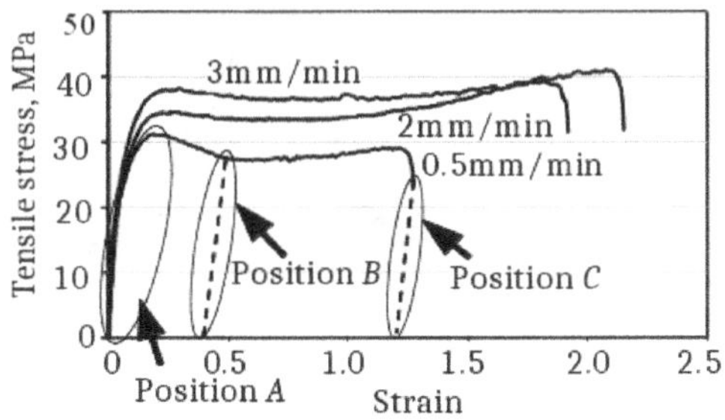

Fig.2 Stress-strain diagram in PA and sample picking places for the x-ray stress measurement.

Fig.3 Photograph of small tensile system mounted on a Ω-diffractometer and maximum plate weights loaded.

In Fig.3, loads were applied by the plate weights. The weight of one plate is 4.9kN and a maximum load, four plates total, is 19.6kN. In elastic deformation of position A, the maximum load could be employed. However in positions B and C the maximum load could not be used, because the deformation levels were over the elastic deformation. Therefore, positions B and C used until tensile loads of 9.8kN, respectively. Table 1 shows the conditions of the x-ray stress measurement.

Table 1. Conditions of stress measurement

Characteristic x-rays	CrKα
X-ray optics	Parallel beam slits
Tube voltage	30kV
Tube current	20mA
$\sin^2\psi$	0.4~1.0 0.1step
hkl plane	Unknown
Diffraction angle	2θ= 32.55
Filter	Vanadium
irradiated area	5mm×2mm
Peak deciding method	HWHM method with absorption correction

4 RESULTS AND DISCUSSION

Fig.4 shows the peak profile from PA and the results of a peak separation. In this figure, three peaks are included in the measured peak profile. There are a halo from the amorphous phase and two diffraction peaks from the crystalline phase. X-ray stress measurement is not applicable for a multi peak in usual cases (8). However, it was impossible for other PA peaks to use the x-ray stress measurement because of the weak intensity. Therefore, in this study we tried to measure this multi peak. Fig.5 shows the diffraction peak shift by applied loads. The right side line in the peak profile is shifted to the low angle with increasing applied loads. However, the left side line in the peak profile is not shifted. In the x-ray stress measurement method, the alternation of lattice spacing is estimated from the shift of the peak profile and stresses are calculated. From the results in Fig.5, though the peak shift of the peak profile is only in the right side of the peak profile, it seems that there is a possibility of the x-ray stress measurement for PA material.

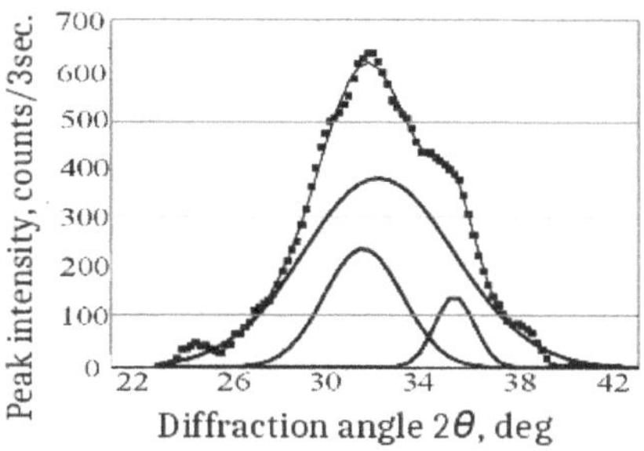

Fig.4 Diffraction peak in PA. This peak consists of three peaks.

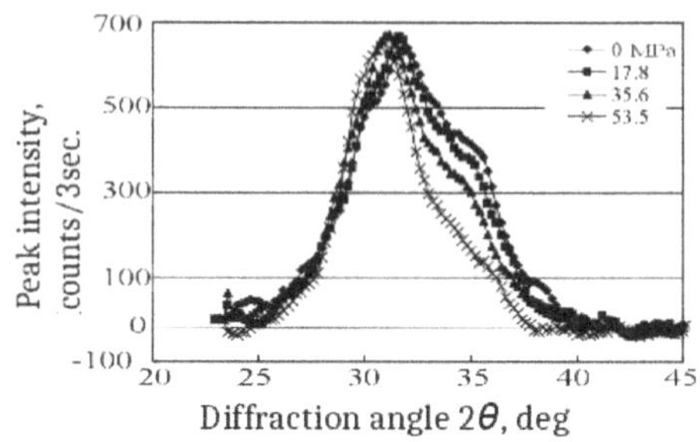

Fig.5 Example of the peak shift by several loads. Right side of diffraction peak shift to the low angle region with increased applied loads.

Fig.6 shows one example of the d-$\sin^2\psi$ diagram under several applied loads. This measurement employed the Ω-diffractometer method with the transmission x-ray diffraction. From this result, regression lines of the d-$\sin^2\psi$ diagram show the good linearity in each line. Therefore, these results can expect to have highly accurate calculations for residual stresses. The gradient of the regression line increases with the load rising. These phenomena indicate that the crystalline phase responded to the external forces. Fig.7 shows the relationship between the gradients of d-$\sin^2\psi$ diagram and the applied stresses for comparing of the results in position A, B and C. From this results, the gradient in position A is increased with the applied load rising. On the other hand, position B and C don't respond the applied loads. Therefore, it is assumed that the residual stress in the elastic region can be measured by x-ray stress measurement with d-$\sin^2\psi$ method. However, when the sample includes the x-ray residual stress measurement is not available. The cause of this phenomenon is considered as follows. In the x-ray

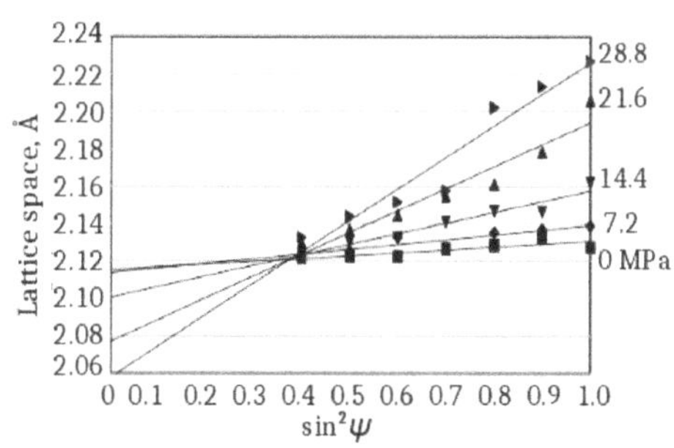

Fig.6 d-$\sin^2\psi$ diagram of Ω-diffractometer method in position A.

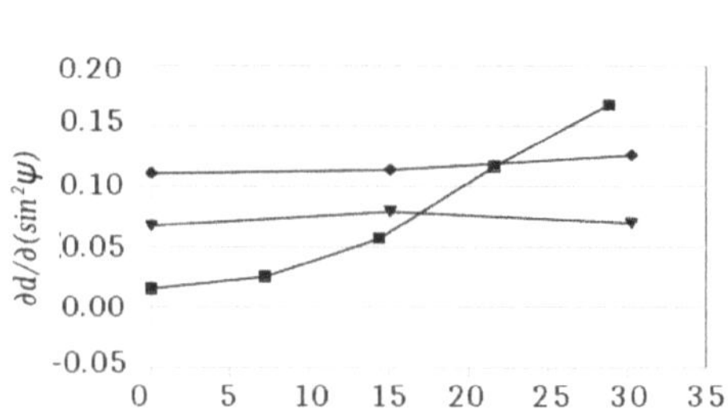

Fig.7 Relationship between $\partial d_\psi/\partial(\sin^2\psi)$ and applied stresses for comparing position A, B and C.

stress measurement, stresses are calculated by the diffraction peak from a specific diffraction plane. In the initial condition, it is assumed that the orientation of the crystalline phase is a random condition. Therefore, the x-ray residual stress is available in the elastic region in position A. However, when the plastic deformation were generated in the PA sample, it is considered that the orientation which arranges the crystalline phase unidirectionally was also generated in PA. In the result, the crystalline phase contributed the x-ray diffraction changes and decreases. Therefore, the x-ray stress measurement is not available in the PA sample including plastic deformation in positions B and C. Furthermore, though the crystalline phase contributing the diffraction peak decreased, the whole peak profile did not disappear, because the multi peak is used for the x-ray stress measurement in this study.

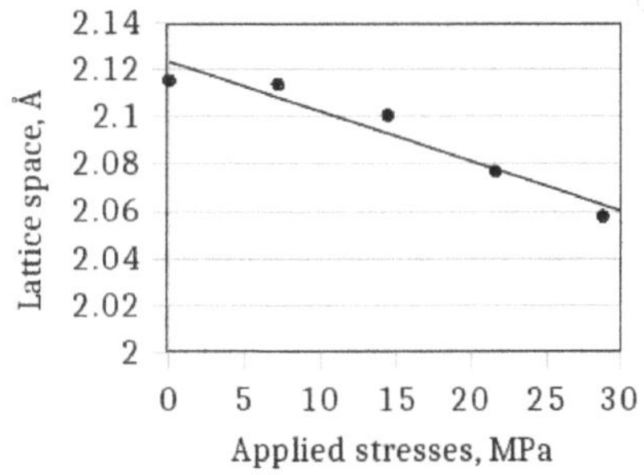

Fig.8 The relationship between $d_{\psi=0}$ with applied stresses in position A.

Fig.8 shows the relationship between $d_{\psi=0}$ with applied stresses in position A. The $d_{\psi=0}$ is the intercept of regression line in d-$\sin^2\psi$ diagram. This diagram is used to calculate x-ray elastic constant: E and Poisson's ratio: ν in Eq.(2) and (3). In this study, calculated results of XEC, E and ν are 0.384GPa, 0.622GPa and 0.6, respectively.

5 CONCLUSIONS

1. Though the PA peak profile in this study includes multi peaks, the right side line of this peak profile shifted to 2θ low angle side with increasing applied loads. Therefore, using the ordinary d-$\sin^2\psi$ method, this peak is able to measure the residual stresses of the PA.
2. In the elastic region, d-$\sin^2\psi$ method is available to measure in PA. However, the samples included plastic deformation, this method is not available.

ACKNOWLEDGMENT

This work was supported by JSPS KAKENHI Grant Number 25420037.

REFERENCES

[1] Y. Ikeuchi, T. Matsue and T. Hanabusa: 5th Int. Conf. on Residual Stresses (1997), 958-963.
[2] Y. Ikeuchi, K. Iga and T. Hanabusa: J. Soc. Mater. Sci., Japan (1994), 792-798.
[3] V. Hauk, A. Troost and D.Ley: Advanced in Polymer Technology (1987), **7**, 389-396.
[4] R. Meske and E. Schnack: Mater. Sci. Forum (2000), **347-349**. 217-222.
[5] Yoshitaka Watanabe, Masayuki Nishida, Takao Hanabusa: Advances in X-ray Analysis (2009), **52**.271-278.
[6] M. Kitamura, M. Nishida and T. Hanabusa: Mater. Sci. Forum (2011), **681**. 381-386.
[7] M. Nishida, T. Hanabusa and Y. Ikeuchi: Key Engineering Mater (2007). **353-358**, 2423-2426.
[8] The Society of Materials Science Japan ed. *Standard Method for X-Ray Stress Measurement* (2005), ISBN 4-901381-27-X, The Society of Materials Science Japan.

Prediction of the resistance to machining-induced cracking in zirconia by nanoindentation

A.-R. Alao, L. Yin
College of Science, Technology and Engineering, James Cook University, Australia

ABSTRACT

Zirconia is a unique material widely used in engineering, medicine and dentistry as load-bearing structures. Zirconia structures can be shaped either in pre-sintered or sintered states by abrasive machining. However, abrasive machining inevitably induces surface and sub-surface cracks in both pre-sintered and sintered zirconia materials, resulting in poor surface quality and shortened lifespans of zirconia products. This research was undertaken to predict the resistance to machining-induced cracking for pre-sintered and sintered zirconia in nanoindentation using the Sakai model at a peak load of 10 mN and loading rates of 0.1–2 mN/s. The results show that pre-sintered zirconia yielded higher ductility indices and better resistances to machining-induced cracking than sintered zirconia. Both materials revealed the loading rate-independent resistances to machining-induced cracking. This work suggests that pre-sintered zirconia can sustain more mechanical damage and absorb more energy and therefore can be more easily machined than sintered zirconia.

1 INTRODUCTION

Zirconia ceramics exhibit high strength and fracture toughness due to their phase transformation toughening. These properties make them attractive for load-bearing applications in mechanical bearings. Zirconia ceramics also display good biocompatibility and chemical inertness which make them a wide usage in medicine and dentistry as orthopedic bones and joints, dental implants, crowns and bridges [1]. Currently, zirconia products are manufactured by either machining of pre-sintered blocks followed by post-machining sintering or direct machining of fully sintered blocks [2]. In both processing routes, diamond abrasive tools make numerous cuts on pre-sintered and sintered zirconia blocks, inducing surface/subsurface damage in both zirconia materials and resulting in poor surface quality [3]. Further, these machining-induced flaws in zirconia are stress concentration sites for crack nucleation and propagation, reducing the fatigue strength and lifetimes of zirconia products. To minimize machining-induced cracking for pre-sintered and sintered zirconia during manufacturing, it is important to predict their resistances to fracture. The resistance to machining-induced cracking of pre-sintered and sintered zirconia can be simulated in nanoindentation due to similar micromechanical involvements in indentation and abrasive machining [4]. In indentation, the Sakai model [5], which incorporates fracture energy, R_c, and ductility index, D, can be used to determine the resistance to machining-induced cracking, M. The model was based on the premise that the extent of the plastic process zone ahead of a crack tip controls the fracture resistance. Therefore, the aim of this study was to predict the resistances to machining-induced cracking, M, for pre-sintered and sintered zirconia by nanoindentation using the Sakai model.

2 MATERIALS AND METHODS

Pre-sintered zirconia (IPS e.max ZirCAD, Ivoclar Vivadent) blocks commonly used in dental CAD/CAM machining process to make crowns and bridges was selected. The material consisted of 87–95 wt% ZrO_2, 4–6 wt% Y_2O_3, 1–5 wt% HfO_2 and 0.1–1 wt% Al_2O_3 [6]. The density, porosity, strength and grain size of this material were 3.09–3.21 g/cm^3, 47.3–49.3%, 50–90 MPa and 0.3 μm, respectively [6]. The fracture toughness was 3.17 MPa.m$^{0.5}$ [7]. Indentation samples were made with a thickness of 2 mm. They were metallographically ground and polished using successively finer diamond paste to obtain mirror surfaces [8]. Some polished pre-sintered zirconia samples were sintered in a furnace (MTI GSL1500X) to 1300°C for 2 hours at 10°C/min heating rate and then naturally cooled to room temperature. The sintered samples became rougher due to zirconia crystal growth and were metallographically re-polished to obtain mirror surfaces. The density and strength of the sintered zirconia were approximately 6.0 g/cm^3 and 900 MPa, respectively; the porosity was less than 0.5%; the grain size was 0.5 μm and the fracture toughness was 5.5 MPa.m$^{0.5}$ [6].

Nanoindentation was conducted on the pre-sintered and sintered zirconia samples using a nanoindentation system (Triboscope, Hysitron, USA) equipped with a diamond Berkovich indenter at a peak load of 10 mN and 0.1–2 mN/s loading rate. The system has load and displacement resolutions of 1 nN and 0.0002 nm, respectively. Other nanoindentation details are reported in [8]. At each loading rate, six indents were made on each specimen to generate force (p) and displacement (h) data for the p-h curves. The p-h curves were used to extract the ductility index, D, which is defined as [5]:

$$D = U_r / U_t \tag{1}$$

where U_r and U_t are the indentation absorbed energy and total energy, respectively. The U_t is the total area under the p-h curve, which can be expressed as:

$$U_t = \int_0^{h_{\max}} P dh = \int_0^{h_{\max}} k_1 h^n dh = k_1 h_{\max}^{n+1} / (n+1) \tag{2}$$

where k_1 and n are the loading curve constant and exponent, respectively; h_{max} is the maximum indentation depth. The elastic strain energy, U_e, is the area of the unloading portion of the p-h curve, which is expressed by:

$$U_e = \int_{h_f}^{h_{\max}} P dh = \int_{h_f}^{h_{\max}} k_2 \left(h - h_f\right)^m dh = k_2 \left(h_{\max} - h_f\right)^{m+1} / (m+1) \tag{3}$$

where k_2 and m are the unloading curve constant and exponent, respectively; h_f is the final depth. The U_r is, therefore, the difference between U_t and U_e. The potential strain energy for machining-induced cracking is the U_e while the total energy consumed during surface cracking is the fracture energy, R_c, and the resultant fracture surface area for the mechanically induced i-th crack, $\sum_{i=1}^{n} A_i$. The R_c is expressed as:

$$R_c = K_{IC}^2 / E' \tag{4}$$

where K_{IC} is the fracture toughness and E' is the plane strain modulus given by [9]:

$$E' = E / \left(1 - \nu^2\right) = \left(\left(1/E_r\right) - \left(\left(1 - \nu_i^2\right)/E_i\right)\right)^{-1} \tag{5}$$

where E_r is the reduced modulus, E_i and ν_i are the Young's modulus and Poisson's ratio of the indenter respectively, E and ν are the Young's modulus and Poisson's ratio for the material. The E_i and ν_i for the diamond indenter are 1141 GPa and 0.07, respectively [9]. Assuming that all the potential strain energy is converted to surface energy, it is obtained:

$$U_t - U_r = R_c \sum_{i=1}^{n} A_i \tag{6}$$

Substituting Eq. (1) into Eq. (6), and defining M is expressed as the inverse degree of damage for a unit applied work (4):

$$M = 1\Big/\left[\sum_{i=1}^{n} A_i\right]\Big/U_t = U_t\Big/\sum_{i=1}^{n} A_i = R_c \sum_{i=1}^{n} A_i \Big/ (1-D)\sum_{i=1}^{n} A_i = K_{Ic}^2 / E'(1-D) \tag{7}$$

Thus, if $D = 0$, M reduces naturally to the fracture energy; if $D = 1$, M becomes infinite.

Mean values and standard deviations of ductility indices, D, and resistances to machining-induced cracking, M, were evaluated from six indentations at each loading rate. Single factor ANOVA was performed at 5% confidence interval to examine the loading rate effect in D and M.

3 RESULTS AND DISCUSSION

Figure 1 shows the Force-displacement (*p-h*) curves for pre-sintered and sintered zirconia at 2 mN/s loading rate, in which pre-sintered zirconia has higher U_t and U_r than sintered zirconia.

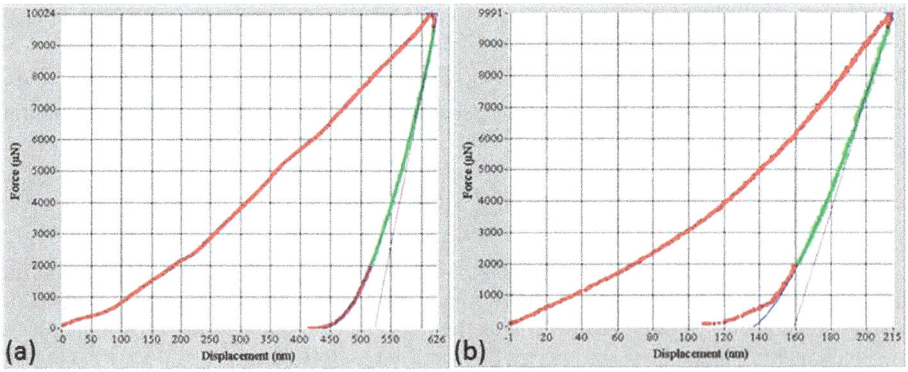

Figure 1. Force-displacement (*p-h*) curves for (a) pre-sintered zirconia (b) sintered zirconia at 2 mN/s loading rate

Figure 2(a) shows ductility indices, D, for pre-sintered and sintered zirconia versus loading rate. For pre-sintered zirconia, the ductility indices remained relatively constant regardless of the loading rate. For sintered zirconia, their ductility indices increased as the loading rate increased from 0.1–0.5 mN/s and then remained constant from 0.5–2 mN/s loading rates. ANOVA analyses show that there was no significant influence of loading rate on ductility indices ($p > 0.05$) for both materials. The ductility index of pre-sintered zirconia at any loading rate was higher than those of sintered zirconia.

The loading rate-independent ductility indices for pre-sintered and sintered zirconia in Figure 2(a) indicate that ductility index, *D*, is a material characteristic parameter. Therefore, the average values of the indices from 24 indentations for pre-sintered and sintered zirconia of 0.75 ± 0.08 and 0.58 ± 0.16, respectively, can be used to rank the material ductility. Pre-sintered zirconia with a higher average index consumed or dissipated more energy in plastic deformation than sintered zirconia during nanoindentation, displaying higher quasi-plasticity than sintered zirconia. This is in agreement with the literature where porous hydroxyapatite was found to be more ductile than sintered hydroxyapatite [10]. The average ductility index of pre-sintered is slightly lower than copper's index (≅ 0.9), which is a well-known ductile material [4].

Figure 2(b) shows resistances to machining-induced cracking, *M*, for pre-sintered and sintered zirconia versus loading rate. For pre-sintered zirconia, their resistances to machining-induced cracking increased with the loading rate from 0.1–0.5 mN/s, and then decreased when the loading rate was increased to 1 mN/s. From 1–2 mN/s, the resistance remained constant. For sintered zirconia, their resistances to machining-induced cracking remained constant regardless of the loading rate. In both cases, ANOVA analyses show insignificant loading effect on resistance values ($p > 0.05$). At any loading rate, the average resistances of pre-sintered zirconia were higher than those of sintered zirconia.

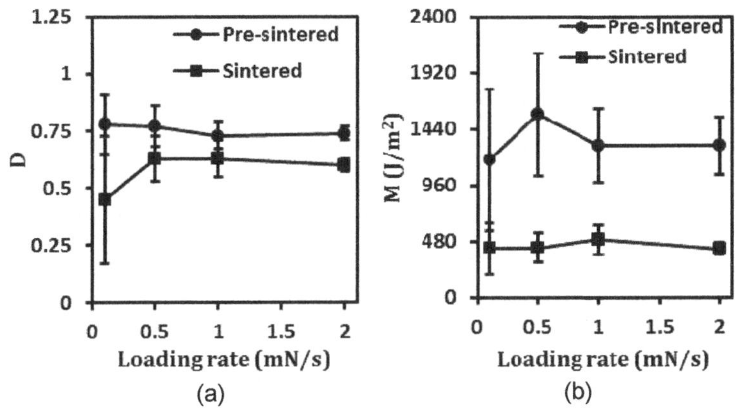

Figure 2. (a) Ductility index, *D*, versus loading rate, (b) Resistance to machining-induced cracking, *M*, versus loading rate

Because resistances to machining-induced cracking for pre-sintered and sintered zirconias were not affected by the loading rate, they can represent a material parameter. The large standard deviations for pre-sintered zirconia can be attributed to the high porosity in the material [8]. Thus, the average resistance values from 24 indentations for pre-sintered and sintered zirconia are 1333.88 ± 440.76 J/m^2 and 438.78 ± 135.58 J/m^2, respectively. These values are comparable to magnesia-partially stabilized zirconia's resistance (a very tough ceramic) for pre-sintered zirconia and silicon nitride's resistance (a very brittle ceramic) for sintered zirconia [4], respectively. This indicates that pre-sintered zirconia can sustain more mechanical damage than sintered zirconia in indentation. This observation also agrees with the previous study where highly porous alumina exhibited better resistance to Hertzian contact damage than sintered alumina

[11]. Further, sintered zirconia with higher fracture toughness became less damage-resistant than pre-sintered zirconia. This indicates that it is the responses ahead of crack tips, such as plastic deformation and micro-void coalescence that govern the ceramic fracture resistance behaviour.

4 CONCLUSIONS

This work applied nanoindentation and the Sakai model to predict resistances to machining-induced cracking for pre-sintered and sintered zirconia. The results show that both loading rate-independent ductility indices and resistances to machining-induced cracking represent the deformation and fracture resistance behaviors of pre-sintered and sintered zirconia. The higher ductility indices for pre-sintered zirconia predict its higher quasi-plasticity. The higher resistances to machining-induced cracking for pre-sintered indicate its better machinability.

ACKNOWLEDGEMENTS

The authors thank Dr. Shane Askew of the Advanced Center at James Cook University (JCU) and Mr. Phillip McGuire of Northern Petrographics for their experimental assistance. A.-R.A acknowledges the JCU PhD (JCU IPRS) scholarship. This research was supported by the JCU Collaboration Grants Scheme awarded to L.Y.

REFERENCES

1. Piconi, C. & Maccauro, G. (1999) *Biomaterials* **20**, 1–25. doi:10.1016/S0142-9612(98)00010-6
2. Denry, I. & Kelly, J. R. (2008) *Dent. Mater.* **24**, 299–307. doi:10.1016/j.dental.2007.05.007
3. Rekow, D. & Thompson, V. P. (2007) *J. Mater. Sci. Mater. Med.* **18**, 47–56. doi:10.1007/s10856-006-0661-1
4. Sakai, M. & Nowak, R. (1992) Fracture toughness and brittleness of ceramic materials. In: Bannister M. J. (Ed), Ceramics-Adding the Value, Australian Ceramic Society (Austceram 92), 2, 922–931
5. Sakai, M. (1999). *J. Mater. Res.* **14**, 3630–3639. doi:10.1557/JMR.1999.0490
6. Ritzberger, C., Apel, E., Höland, W., Peschke, A., & Rheinberger, V. M. (2010) *Materials* **3**, 3700–3713. doi:10.3390/ma30633700
7. Mitov, G., Gessner, J., Lohbauer, U., Woll, K., Muecklich, F., & Pospiech, P. (2011) *Dent. Mater.* **27**, 684–691. doi:10.1016/j.dental.2011.03.010
8. Alao, A. -R. & Yin, L. (2014) *J. Mech. Behav. Biomed. Mater.* **36**, 21–31. doi:10.1016/j.jmbbm.2014.03.019
9. Oliver, W. C. & Pharr, G. M. (1992) *J. Mater. Res.* **7**, 1564–1583. doi:10.1557/JMR.1992.1564
10. He, L. -H., Standard, O. C., Huang, T. T. Y., Latella, B. A., & Swain, M. V. (2008) *Acta Biomater.* **4**, 577–586. doi:10.1016/j.actabio.2007.11.002
11. Latella, B. A., O'Connor, B. H., Padture, N. P., & Lawn, B. R. (1997) *J. Am. Ceram. Soc.* **80**, 1027–1031. doi:10.1111/j.1151-2916.1997.tb02940.x

Lamb wave characterisation and damage imaging for isotropic plate-like structures using 3D laser vibrometry

P. Aryan[1], A. Kotousov[1], C.T. Ng[2], B.S. Cazzolato[1]

[1] School of Mechanical Engineering, The University of Adelaide, Australia

[2] School of Civil, Environmental & Mining Engineering, The University of Adelaide, Australia

ABSTRACT

The scattering of Lamb waves has been adopted for effective detection of different types of mechanical damage in both metal and composite structures. The scattering field from a defect is normally obtained through various types of sensors, which can quantify this field in only one direction. This paper presents experimental investigations adopting 3D laser vibrometry and the root mean square (RMS) of the velocity fields in isotropic plate-like structures. Experimental case studies with defects are considered. The results clearly illustrate the benefits of the method and utilisation of the full 3D characterisations of Lamb wave propagation and scattering.

1 INTRODUCTION

Lamb waves signify a type of Guided Wave (GW) generated in plates and shell components. There are two fundamental modes of Lamb waves: symmetric and anti-symmetric. Damage prognostic systems and failure prediction techniques have recently been the subject of demanding research and development in a broad range of engineering applications. The potential benefits of such systems and techniques are significant. These include a considerable increase in reliability and cost reduction of maintenance procedures for critical structures by changing the maintenance strategies from schedule-driven to condition-based.

Lamb wave based testing in particular has been confirmed to be a capable tool among various structural health monitoring (SHM) techniques. Lamb waves have been proven to be reliable tools to identify most types of mechanical damage amongst a range of non-destructive methods. Lamb wave-based damage detection methods have been adopted for numerous research studies recently for both composite [1, 2] and metallic structures [3, 4, 5].

In terms of sensing the Lamb wave propagation, piezo and optical sensors are widely utilised for measurements of Lamb waves due to their high sensitivity and low response time [3, 4]. However, the conventional sensors (including 1D laser vibrometry) can only quantify the scattering field in one direction. Consequently the interpretation of the information obtained with conventional sensors produces significant uncertainties and difficulties. In addition, to achieve a satisfactory resolution for a large inspection area, a dense array of sensors is normally required. These challenges can impact on both the cost and reliability of the measurements.

The application of Laser vibrometry for Lamb wave propagation and scattering characterisation measurements and damage detection purposes has been growing significantly over the past twenty years [6, 7, 8, 9]. The main reason for utilising laser vibrometry is its ability to perform accurate and non-contact surface velocity measurements with high resolution images of wave propagation [10, 11]. The perception of Lamb waves propagating in various structures and their interaction with structural variations can be advanced by utilising 3D laser vibrometry technology.

Considering all the aforementioned limitations and challenges, this paper presents a Lamb wave based characterisation and damage imaging method, adopting advanced 3D laser vibrometry. The advanced, high sensitivity and resolution technology combines a high sensitivity resolution and processing time.

2 EXPERIMENTAL SETUP

The test sample represents large aluminium plates 3 mm in thickness with in-plane dimensions of 400 mm by 800 mm. A crack of 1.5 mm depth and 10 mm length were milled in the centre of the plate, as it is shown in Figure 1. To generate Lamb waves, a piezoelectric transducer of disk shape was glued 70 mm away from the centre of the plates.

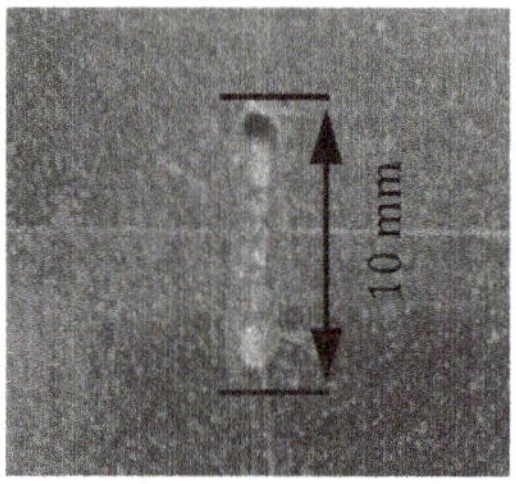

Figure 1: Picture of 10mm length crack at the centre of the plate.

The experimental equipment included a signal amplifier, 3D laser vibrometer and a built-in signal generator as illustrated in Figure 2. 5.5 cycle Hanning windowed tone burst signals at frequencies between 100 to 300 kHz were generated using the signal generator. Then, the signal was amplified up to ±50 V using a power amplifier and applied to the transducer mounted on the surface of the plate. The transducer transformed the amplified electrical signal to the surface displacements, which generate the incident Lamb wave burst. The out-of-plane and in-plane velocities were measured with a Polytec PSV-400 3D scanning laser vibrometer and recorded in a file.

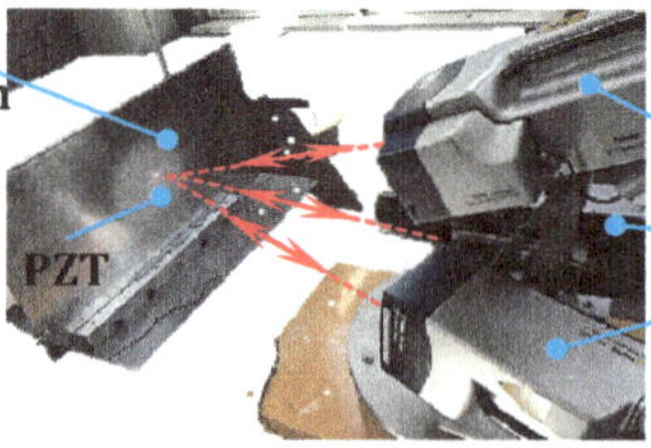

Figure 2: Photograph of the Polytec PSV-3D laser vibrometer focused at an aluminium specimen with a 10mm crack at the centre.

The velocity components were measured at different grid points of 100mm by 100mm, covering the defect and a sufficient area near the defect. In order to build a high resolution wave field image, a fairly small measurement grid size considering the wavelengths of the Lamb waves and close to the transducer was set (giving approximately more than 8 points per wavelength). A time averaging of 100 was used for the time response at each measurement point to improve the signal-to-noise ratio (SNR), also a band pass filter with lower and higher cut off frequencies based on the signal envelope energy of each centre frequency was applied to reduce the noise outside the frequency band. A sensitivity of 10 mm/s/V and a sampling rate of 1.5 MHz were employed. Each laser head can be autofocused for each scan point but this will seriously increase the scanning time. Instead, interpolated focusing was utilised and then all the other scan points were interpolated. According to the literature [10], this approach is suitable for planar surfaces.

3 EXPERIMENTAL RESULTS

A plate with a crack was considered in this method. The two fundamental modes of Lamb wave (A_o and S_o) were measured. Four snapshots of the Lamb wave propagation can be seen in Figure 3. The frames of the snapshots represent the out-of-plane velocity field across the surface of the plate. The incident waves, the propagation and scattering due to the presence of the crack, can be clearly seen. Further to this, the crack location, representing a defect, is circled in Figures 3 and 4.

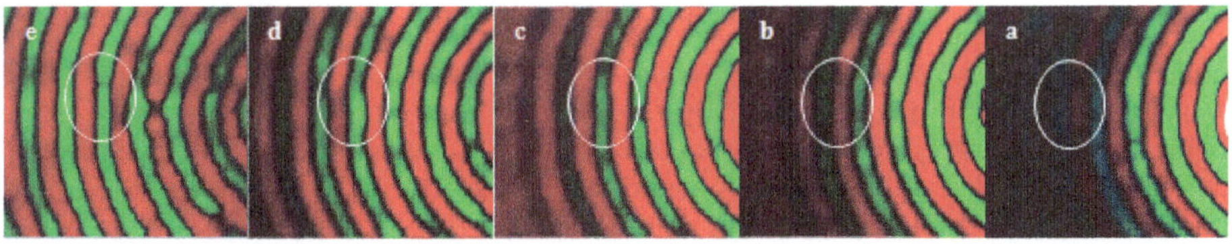

Figure 3: Snap shots of z-component velocity field (out-of-plane) at (a) 30.08µs (b) 32.20 µs (c) 35.40 µs (d) 40.69µs and (e) 46.20µs Location of the crack is marked by a circle.

The Root Mean Square (RMS) value of a set of values is the square root of the arithmetic mean (average) of the squares of the original values. In the case of a set of n values of $\{a_1, a_2, a_3, \ldots, a_n\}$ the RMS value is given by this formula:

$$a_{RMS} = \sqrt{\frac{1}{n}(a_1^2 + a_2^2 + \cdots + a_n^2)} \quad (1)$$

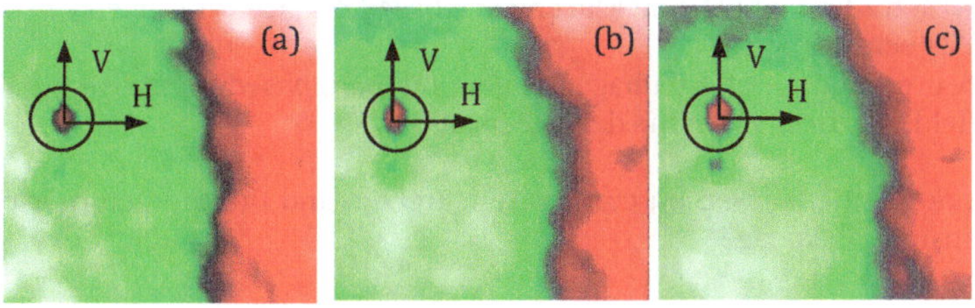

Figure 4: RMS velocity fields for 10 mm length crack in (a) x, (b) y and (c) z components.

To investigate more about Lamb wave characterisation, particularly in damage location, the signal ratio was defined. The signal ratio here means the ratio between the maximum RMS velocity value and the average RMS velocity value. The ratio was calculated along both horizental (H) and vertical (V) axis, crossing the damage location. The ratio can provide a good perspective for comparison of all three velocity components, along with their sensitivity and itraction to the damage.

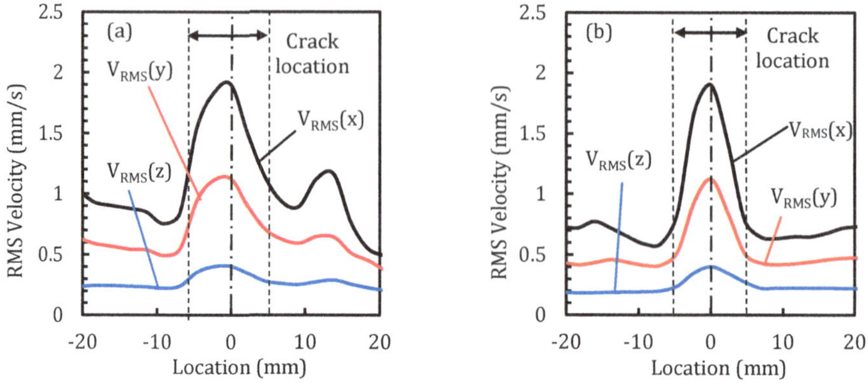

Figure 5: Section views of RMS velocity fields along the (a) H and (b) V axis through the 10 mm crack for the specimen.

Considering Figure 5 and Table 1, it can be concluded that damage detection is more sensitive to in-plane modes, especially the x-component, rather than the out-of-plane mode. This illustrates that 3D characterisation of Lamb waves provides improved understanding of Lamb wave propagation and scattering for damage detection purposes.

4 CONCLUSIONS

In the current paper, imaging and localization of various mechanical types of damage for isotropic plate-like structures were investigated. The investigation considered whether the application of 3D laser vibrometry can seriously advance the identification and localization of different types of mechanical damage, based on the scattering of Lamb waves. The main outcomes of this work are:

(1) The merits of adopting the 3D scattering field characterisation in comparison with traditional (1D) measurements, based on laser vibrometry or an array of PZT sensors and
(2) A presentation of a damage imaging method, based on utilising RMS velocity fields, particularly at the damage location.

Future work will be focused on composite materials and the characterisation of more common structural damage.

ACKNOWLEDGEMENT

This work was supported by the Australian Research Council under Grant Number DE130100261.

REFERENCES

1. H. Sohn, D. Dutta, J. Y. Yang, H. J. Park, M. DeSimio, S. Olson and E. Swenson, 2011, "Delamination detection in composites through guided wave field image processing", Journal of Composites Science and Technology, vol. 71, no. 9, pp. 1250-1256.
2. C. M. Yeum, H. Sohn, H. J. Lim and J. B. Ihn, 2013, "Reference-free delamination detection using Lamb waves", Journal of Structural Control and Health Monitoring, vol. 21, no. 5, pp. 675-684.
3. L. Ge, X. Wang and C. Jin, 2014, "Numerical modeling of pzt-induced Lamb wave-based crack detection in plate-like structures", Journal of Wave Motion, vol. 51, no. 6, pp. 867-885.
4. R. A. Osegueda, V. Kreinovich, S. Nazarian and E. Roldan, 2003, "Detection of cracks at rivet holes in thin plates using Lamb-wave scanning", In NDE for Health Monitoring and Diagnostics, International Society for Optics and Photonics, pp. 55-66.
5. A. S. Purekar, D. J. Pines, S. Sundararaman and D. E. Adams, 2004, "Directional piezoelectric phased array filters for detecting damage in isotropic plates", Journal of Smart Materials and Structures. vol. 13, no. 4, , pp. 838-850.
6. H. Sohn, E. D. Swenson, S. E. Olson, M. P. DeSimio and D. Dutta, 2012, "Delamination detection in composite structures using laser vibrometer measurement of Lamb waves", International Society for Optics and Photonics , pp. 76500-76500.
7. P. Malinowski, T. Wandowski, P. Kudela and W. Ostachowicz, 2010, "Laser vibrometry for guided wave propagation phenomena visualisation and damage detection" vol. 1253, no.1 , pp. 140-149.
8. D. Adams, D. Koester and S. Underwood, 2014 "Damage detection using laser vibrometry", U.S. Patent No. 8,656,779.
9. W. J. Staszewski, B. C. Lee, L. Mallet and F. Scarpa, 2004, "Structural health monitoring using scanning laser vibrometry: I. Lamb wave sensing", Journal of Smart Materials and Structures, vol. 13, no. 2, pp. 251-260.
10. E. D. Swenson, H. Sohn, S. E. Olson and M. P. Desimio, 2010, "A comparison of 1D and 3D laser vibrometry measurements of Lamb waves", International Society for Optics and Photonics, pp. 765003-765011.
11. H. Weisbecker, B. Cazzolato, S. Wildy, S. Marburg, J. Codrington and A. Kotousov, 2012, "Surface strain measurements using a 3d scanning laser vibrometer", Journal of Experimental mechanics, vol. 52, no. 7, pp. 805-815.

Testing localized microstructural architectures with miniaturised cantilever beams

W. Costin, A. Kotousov
School of Mechanical Engineering, The University of Adelaide, Australia

ABSTRACT

The fracture resistance of many alloys is largely determined by their microstructure which may consist of diverse constituents with characteristic architectures and fracture properties. However, the fracture properties of individual microscopic constituents cannot be directly measured with conventional mechanical tests. Micro-fracture testing is a powerful tool to overcome such experimental limitations and can be used to develop an understanding of the relationship between bulk materials properties and the localized mechanical properties of individual constituents. This paper focuses on the application of miniaturized fracture tests to determine the fracture properties of selected microstructural constituents with distinctive architectures. Site specific Focused Ion Beam (FIB) micro-machining was utilized to fabricate a sharply notched micro-cantilever, which was loaded by utilizing a nano-indenter with a large spherical tip. The fracture characteristics were assessed with a simplified plastic hinge model assisted with FE simulations.

1 INTRODUCTION

The ongoing trend to miniaturize mechanical, optical and electronic components requires an understanding of the mechanical behavior of materials at small scales. Additionally, recent advances in material science focus on the development of design concepts for materials based on specific localized architectures and mechanical properties. These reasons are the main driving forces behind the development of experimental techniques for the evaluation of mechanical properties at micro- and nano-scales.

Over the past decade advances in the miniaturisation of mechanical test specimens have been largely associated with the development of Focused Ion Beam (FIB) micro-fabrication techniques. The increasing efficiency and accuracy of modern FIB workstations allow nowadays a reproducible, site-specific fabrication of micro and nano- scaled mechanical test specimens with diverse complex shapes. Mechanical loading of such small samples requires the application of nano-indentation techniques or micro-manipulators.

In the current paper we attempted to probe the fracture properties of a selected microstructural constituent (upper bainite) with a unique architecture. Micro-scaled sharply notched cantilever beams were manufactured using FIB techniques. The samples were loaded with a nano-indenter utilising a spherical tip of a sufficiently large diameter to avoid local plastic deformations in the vicinity of the contact area. The load displacement curve was recorded and analysed with a simplified plastic hinge model, as the tested microstructure demonstrated an essentially ductile behaviour. The simplified model provided values of the critical plastic crack tip opening displacement and critical tip opening angle. Further analysis can include a detailed finite element modelling to evaluate other fracture controlling parameters. The developed technique can be applied to investigate fracture properties of other constituencies and microstructural architectures to identify main fracture mechanisms and the role of individual microstructural architectures in fracture resistance of bulk materials (1).

2 MICROSTRUCTURAL ARCHITECTURE

Ferritic weld metal is due to the large variety of its intricate constituents with inherent mechanical properties a particularly interesting material to study the relationship between microstructural architecture and fracture properties (2, 3). A common constituent in weld metal is upper bainite. This characteristic microstructural region is typically an aggregate of adjacent clusters of ferrite plates with elongated cementite particles between the boundaries of the plates. This microstructure has been selected for the present investigations. In the next section we outline the fabrication procedure.

3 FABRICATION

FIB milling is a unique, versatile technique, which allows the fabrication of relatively complex structures into various materials by using an accelerated, focused beam of Ga+ ions to selectively remove material from the surface of a sample. The scale of these structures can range from a couple of nanometres to a few hundred micrometres. Since the fabrication time strongly depends on the total volume of the removed material, this technique becomes largely impractical for applications at larger scales.

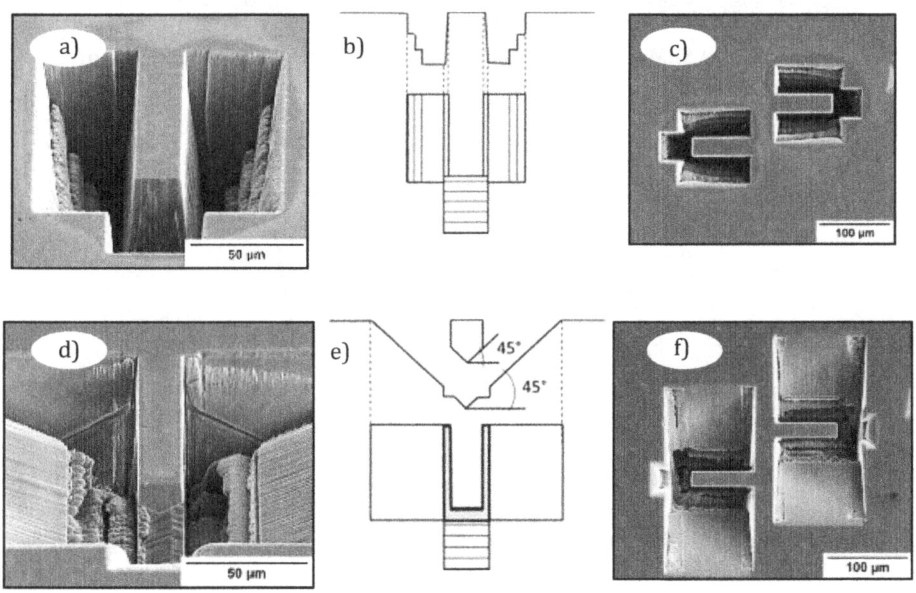

Figure 1: (a-f) SEM images illustrate the fabrication process of the micro beams.

The selected test specimen shape was a micro-beam with a pentagonal profile. If the specimens are fabricated in the vicinity of free edge, it is also possible to machine micro-beams with a rectangular profile as it was demonstrated by (4). This could approach the shape of the micro-specimen to the standard (i.e. rectangular cross-section) shapes used for fracture testing, such as in SENB or CB fracture specimens. However, for the objective of the current work the fabrication of the specimens had to be very site specific to target a specific microstructural architecture; and therefore such an approach was not feasible

4 TESTING RESULTS

A nano-indenter (IBIS Authority) with a closed loop motorized sample stage with an accuracy of 0.25 μm was used as a load cell to apply the bending load to the free end of the fabricated notched cantilever samples, see Fig. 2(a-b) as well as to measure the magnitude of the load and displacement shown in Fig. 3.

Figure 2: (a) SEM image of specimen before testing; (b) SEM image of specimen after testing.

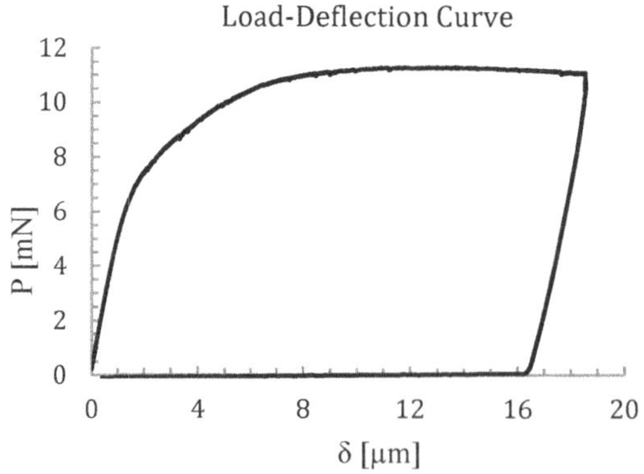

Figure 3: Load-deflection curve, i.e. the applied load, P, versus deflection δ.

The loading type was displacement-control with a loading rate of 0.02μm/sec and a maximum displacement of 18 μm. A large spherical tip with a diameter of 40μm was utilised for the loading of the samples to avoid penetration and reduce the deformations at the contact area. The measured displacements were therefore dominated by the beam deflection rather than the contact deformation process. Load-deflection curve, see Fig. 3, is perceptibly linear at low loads; this confirms that the measured deflection is dominated by beam deflection. The maximum force and the corresponding deflection are identified from the load-deflection curve, which are utilised in fracture resistance characterisation.

The testing of the notched micro-cantilever reveals largely ductile fracture behaviour of this particular microstructural architecture. Therefore, the methods of LEFM are considered to be inapplicable to characterise the fracture resistance and a simplified plastic hinge model was instead utilised as described below. It is interesting to note that upper bainite is considered to be brittle at macro-scale.

5 DISCUSSION

The presence of a large yielding region ahead of the crack tip requires non-linear approaches to the evaluation of fracture resistance. The most popular non-linear criteria, which are widely utilised for this purpose at macro-scale, are based on a the J-integral, crack tip opening displacement (CTOD) and crack tip opening angle (CTOA). These criteria assume that there are some critical values of these parameters, which to some extend are geometry and loading independent.

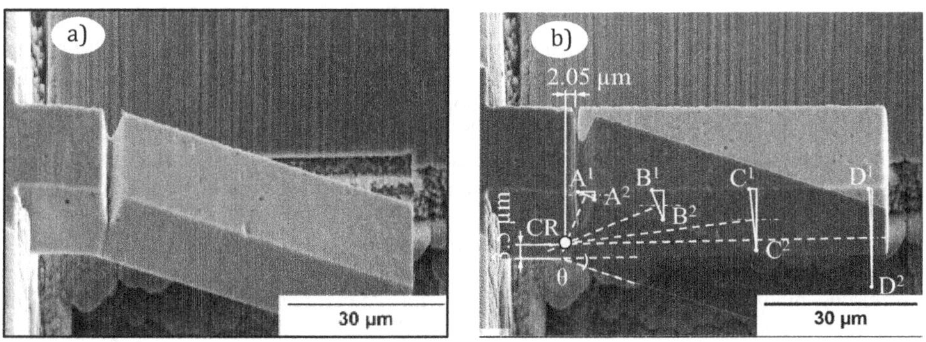

Figure 4: Determination of the rotation centre in plastic hinge model from the deformed sample.

The CTOD and CTOA are particular appealing because of its readily-grasped physical significance and the opportunity that it offers for direct measurement. The latter is very important for the purpose of the current study as the material properties at the micron-scale are very different to the bulk properties, and there are no possibilities for instrumentation and other direct measurements except using SEM or similar imaging tools. It also should be recognized that there are two different definitions of CTOA: a crack tip angle that reflects the actual slope. It can be linked directly to the fracture process but the measurement of this parameter represents a formidable measurement task. Another definition for CTOD is an averaged value of the slope, which can be measured but it has less physical justification (5).

Plastic CTOD and CTOA can be relatively easy evaluated from a simplistic plastic hinge model as illustrated in Fig. 4(b). The critical values of CTOD and CTOA can be found by substituting into (1) and (2) the displacement, δ_c, which corresponds to the maximum load, P_c. The values were found to be 7.40 and 1.5 μm respectively. The first value (CTOA) is in the range of the corresponding values for structural materials (5). As a benchmark example, for a tungsten single crystal, a material which is too tough and too soft at room temperature, CTODC were reported from 230 nm to 570 nm, which resulted into fracture toughness of 9.2 $MPa\sqrt{m}$ to 14.4 $MPa\sqrt{m}$ (6). The critical plastic CTOD can also be linked to J-integral or specific surface energy and fracture toughness (7), however, this evaluation requires knowledge of material properties, which are obviously different to the bulk properties at this scale. Moreover, it is expected that the tested microstructural constituent (upper bainite) possesses a significant anisotropy, which makes the evaluation of other fracture characteristics quite problematic.

6 CONCLUSION

In the current paper we attempted to probe the fracture properties of upper bainite, which is a common constituent in weld metal. It has a very distinctive architecture, which represents an aggregate of adjacent clusters of ferrite plates with elongated cementite particles between the boundaries of the plates. Microscopic sharply notched cantilever beams were manufactured using FIB techniques. The samples were loaded with a nano-indenter thereby utilising a spherical tip of a sufficiently large diameter to avoid local plastic deformations in the vicinity of the contact area. The load displacement curve was successfully recorded and analysed.

The respond of the sample to the applied loading had an initial linear region, which indicated that the deformations are largely dominated by beam deflections rather than the localised contact. The linear region was followed by almost a flat loading part (the force stayed constant with the continuous increase of beam deflections). Such mechanical behaviour normally indicates the development of large plastic deformations prior failure. The simplified plastic hinge model was therefore utilised to analyse the fracture behaviour, and non-linear critical parameters of crack tip field were calculated.

The developed technique can be applied to investigate fracture properties of other constituents and microstructural architectures within the same material, to identify, for example, main fracture mechanisms and the role of individual microstructural architectures in the overall fracture resistance. Ultimately, the described experimental technique can provide a unique insight on how microstructural design can be implemented more efficiently to develop new materials with superior fracture properties.

REFERENCES

[1] Dodds Jr., R. H., Shih, C. F. and Anderson, T. L. (1993) *International Journal of Fracture* **64**, 101-133.

[2] Díaz-Fuentes, M., Iza-Mendia, A., and Gutiérrez, I. (2003) *Metallurgical and Materials Transactions* A, **34**: 2505-2516.

[3] Thewlis, G. (2004) *Materials Science and Technology* 20: 143-160.

[4] Motz, C., Schöberl, T., and Pippan, R. (2005) *Acta Materialia* **53**: 4269-4279.

[5] Wang, J.Q. and Shuai, J. (2012) *Engineering Fracture Mechanics* **79**: 36-49.

[6] Wurster, S., Motz, C., and Pippan, R. (2012) *Philosophical Magazine* **92**:1803-1825.

[7] Tagawa, T., Kayamori, Y., Ohata, M., Handa, T., Kawabata, T., Yamashita, Y., Tsutsumi, K., Yoshinari, H., Aihara, S. and Hagihara, Y. (2010) *Engineering Fracture Mechanics* **77**: 327-336.

AUTHOR INDEX

Lightning Source UK Ltd.
Milton Keynes UK
UKOW07f2032260515

252292UK00004BA/24/P